Diagnostics for
Experimental Thermonuclear
Fusion Reactors

Diagnostics for Experimental Thermonuclear Fusion Reactors

Edited by

Peter E. Stott

JET Joint Undertaking
Abingdon, Oxfordshire, England

Giuseppe Gorini

University of Milan
Milan, Italy

and

Elio Sindoni

University of Milan
Milan, Italy

Plenum Press • New York and London

Library of Congress Cataloging-in-Publication Data

On file

Proceedings of the International Workshop on Diagnostics for ITER,
held August 28 – September 1, 1995, in Varenna, Italy

ISBN-13:978-1-4613-8020-7   e-ISBN-13:978-1-4613-0369-5
DOI: 10.1007/978-1-4613-0369-5

© 1996 Plenum Press, New York
Softcover reprint of the hardcover 1st edition 1996
A Division of Plenum Publishing Corporation
233 Spring Street, New York, N. Y. 10013

PREFACE

This book of proceedings collects the papers presented at the Workshop on Diagnostics for ITER, held at Villa Monastero, Varenna (Italy), from August 28 to September 1, 1995. The Workshop was organised by the International School of Plasma Physics "Piero Caldirola." Established in 1971, the ISPP has organised over fifty advanced courses and workshops on topics mainly related to plasma physics. In particular, courses and workshops on plasma diagnostics (previously held in 1975, 1978, 1982, 1986, and 1991) can be considered milestones in the history of this institution. Looking back at the proceedings of the previous meetings in Varenna, one can appreciate the rapid progress in the field of plasma diagnostics over the past 20 years. The 1995 workshop was co-organised by the Istituto di Fisica del Plasma of the National Research Council (CNR).

In contrast to previous Varenna meetings on diagnostics, which have covered diagnostics in present-day tokamaks and which have had a substantial tutorial component, the 1995 workshop concentrated specifically on the problems and challenges of ITER diagnostics. ITER (the International Thermonuclear Experimental Reactor, a joint venture of Europe, Japan, Russia, and the United States, presently under design) will need to measure a wide range of plasma parameters in order to reach and sustain high levels of fusion power. A list of the measurement requirements together with the parameter ranges, target measurement resolutions, and accuracies provides the starting point for selecting a list of candidate diagnostic systems. Several factors make the implementation of diagnostics much more difficult on ITER than on present experiments. It is also anticipated that diagnostics will be used much more extensively as input to control systems on ITER than on present fusion devices and this will require increased reliability and long-term stability.

The workshop opened with a series of papers on general issues concerning ITER diagnostics. Other oral sessions dealt with the application to ITER of magnetic diagnostics, reflectometry and ECE, interferometry and polarimetry, Thomson scattering, spectroscopy, fusion products and divertor diagnostics. The many excellent papers that were presented showed the strong interest in the ITER project by diagnosticians throughout the world fusion community. The workshop was attended by 82 participants from the ITER parties with strong participation from the United States, Japan, and Russia, as well as from Europe. There were 45 invited papers presented orally and 34 contributed papers presented in 2 poster sessions. A particular and very productive feature of the workshop was the extensive time allowed for discussion during the oral sessions.

In the interest of speed of publication, these proceedings have been reproduced directly from the camera-ready copy provided by the authors without refereeing or editing.

v

Thanks are due to the institutions that have contributed financially to the organisation of this workshop, in particular the Istituto di Fisica del Plasma of CNR, the Commission of the European Communities, the Department of Physics of the University of Milan, the Province of Como, the ENEA, the Camera di Commercio of Lecco, and the Gilardoni S.p.A.

P E Stott, G Gorini, E Sindoni

CONTENTS

THE ITER PROJECT

MAGNETIC DIAGNOSTICS

REFLECTOMETRY AND ECE

INTERFEROMETRY, POLARIMETRY AND THOMSON SCATTERING

SPECTROSCOPY

FUSION PRODUCTS

DIVERTOR DIAGNOSTICS

DIAGNOSTICS OF OTHER FUSION EXPERIMENTS

Diagnostics for Experimental Thermonuclear Fusion Reactors

THE ITER DEVICE

ITER Joint Central Team and Home Teams
R. R. Parker

ITER Garching Joint Work Site
Max-Planck-Institute für Plasmaphysik
Boltzmannstrasse 2
85748 Garching-bei-München, Germany

INTRODUCTION

The ITER Engineering Design Activities (EDA) phase is now at the mid-point in its six year duration and substantial progress has been made in both the design of the tokamak device itself and its supporting facilities. The overall objective of ITER as stated in the ITER Agreement is to demonstrate the scientific and technological feasibility of fusion energy by "demonstrating controlled ignition and extended burn of deuterium-tritium plasmas, with steady-state as an ultimate goal, by demonstrating technologies essential to a reactor in an integrated system, and by performing integrated testing of the high-heat-flux and nuclear components required to utilize fusion energy for practical purposes." The Special Working Group 1 (SWG 1) has provided more specificity to this general objective, for example, by requiring ITER to produce ignition and extended burn for pulses with flat top duration in the range of 1000 s and to produce a wall loading of about 1 MW/m^2 for component testing. As the design has progressed from the Conceptual Design (CD) final report[1] (1991) to the Outline Design (OD) report[2] (1994) to the present design submitted to the Parties in the Interim Design Report[3] (IDR), these objectives have remained invariant. While changes to the basic plasma configuration and parameters have occurred during the evolution from the OD to the IDR design, they have been modest and lead to nearly identical plasma performance when modifications in confinement projections, based on expansion of the confinement data base, are taken into account. The main result of this progression has been rather to simplify and improve the design of key components and systems and, as would naturally be the case, to provide more design detail and documentation. A top-level constraint in this process has been the control of projected construction cost. Further modifications aimed at more cost-effective design solutions can be expected.

The purpose of this paper is to present an overview of the ITER design as described in the IDR. Emphasis is on the engineering design of the tokamak device. Those areas where significant modifications from the OD to the IDR design have occurred are highlighted. An attempt is made to identify global issues which impact the design of diagnostics and other ancillary systems. The paper concludes with a brief discussion of the physics performance and effects of disruptions on the design in-vessel components.

TOKAMAK DESIGN DESCRIPTION

A cross-sectional view of the ITER device is shown in Figure 1, and the main parameters and characteristics of the device are given in Table 1. The configuration is that of a

Diagnostics for Experimental Thermonuclear Fusion Reactors
Edited by P. E. Stott *et al.*, Plenum Press, New York, 1996

conventional single-null diverted tokamak with moderate elongation κ = 1.6 and an aspect ratio (2.9) typical of modern tokamaks. The Nb₃Sn toroidal field (TF) coils produce 5.7 T on axis, limited by the field strength of 12.3 T at the coil. These values together with the requirement of producing sustained ignition with an average neutron wall load of 1 MW/m² yields a minimum major radius of about 8 m when scaling laws for favorable modes of confinement are used.

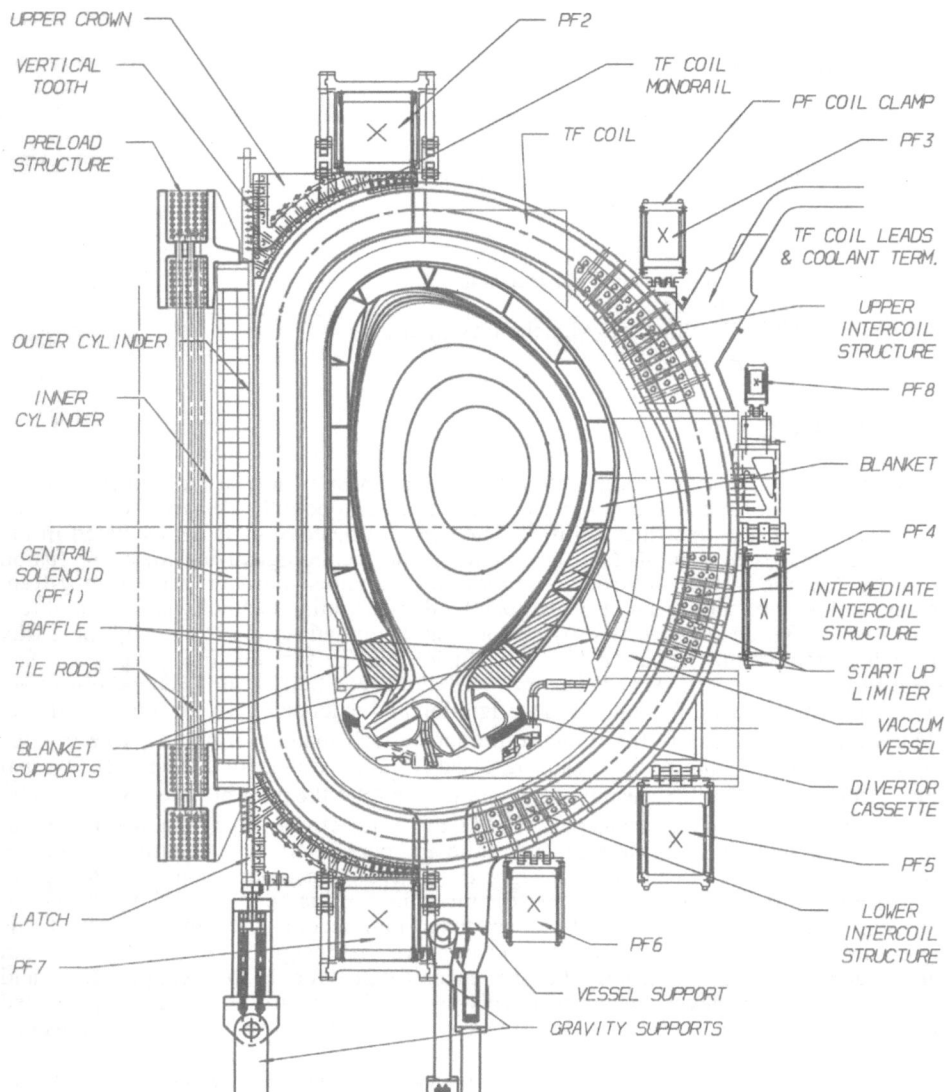

Figure 1. Cross-sectional view of the ITER Interim Design

Magnets

As was the case for the Outline Design, the TF magnets buck against the central solenoid (CS) which is encased in inner and outer bucking cylinders. The bucking cylinders are connected at top and bottom and partially react the out of plane loads on the TF coils by

Table 1. Major parameters of the ITER Interim Design

Parameter	Symbol	Value
Major radius	R	*8.14 m*
Minor radius	a	*2.80 m*
Plasma configuration	—	*Single-null*
Plasma elongation	κ_{95}, κ_X	*~1.6, ~ 1.75*
Plasma triangularity	δ_{95}	*~ 0.24*
Nominal plasma current	I	*21 MA*
Toroidal field	B	*5.68 T (at R = 8.14 m)*
MHD safety factor	q_{95}	*3.05*
Fusion power (nominal)	P_{fus}	*1.5 GW*
Average wall loading	Γ_n	*~1 MW/m^2*
PF flux swing	$\Delta\phi_{PF}$	*530 Wb*
Flux swing for burn	$\Delta\phi_{burn}$	*84 Wb*

means of crown structures which engage the TF coils near the top and bottom of the CS. Steel plates in the TF coils react the forces on the TF conductors, thus preventing a build-up of compressive stress that would occur in a wound coil in which the turns are allowed to directly bear on each other. This was an important feature in the Outline Design, but in the present case the plates are radial rather than toroidal. As a result, the coils have far greater in-plane stiffness which, together with the crown system, has made two substantial improvements possible: elimination of the complex system of keys, which were a main concern relating to the fabrication of the TF system in the earlier design, and simplification of the TF support structure, which in the Outline Design completely enveloped the TF coils. Each TF coil weighs 670 tonnes and has outer dimensions of 12.2 x 18 m. When energised, the 20 coils store a magnetic energy of 101 GJ.

The CS is layer wound, with each of the 14 layers wound "four-in-hand". Layer to layer connections are made about 1.5 m above and below the ends of the windings in order to make the connections in a relatively low field region. While this configuration is simple and reliable to construct, a uniformly wound central solenoid limits plasma shaping and control possibilities. In particular, a vertically segmented CS would enable higher triangularity and this could be important for increasing the beta limit, enhancing confinement and accessing "high-performance" regimes. A segmented solenoid design is under consideration. The present solenoid is 12 m high, has inner and outer radii of 1.9 and 2.7 m, and weighs with the bucking cylinders 1350 tonnes. When energized, the peak field at the conductor is 13 T and the stored energy is 12.3 GJ.

Cryostat and Tokamak Building

The tokamak is located in a cylindrical cryostat (height ≈ OD = 36.5 m) which is nested within a cylindrical biological shield with OD = 40 m. The bioshield is in turn housed inside a 71 m OD cylindrical pit which extends 53 m below grade. The cryostat is a double-wall stainless steel structure fabricated from 20 mm thick plate. The total thickness is typically 240 mm. The bioshield and pit are concrete (composition to be determined) and are respectively 1.2 m and 2 m thick in the equatorial plane.

A north-south section through the center of the pit is shown in Figure 2. Of interest here is the 70 x 70 m^2 space located under the assembly hall, which will be used for a neutral beam test stand and auxiliaries, and a staging area for diagnostics. The latter will be located in the machine, e.g., magnetic and divertor diagnostics, in plugs inserted into the vacuum vessel ports and/or in the annular area between the bioshield and the pit wall. Viewing the plasma from top ports is made difficult by the dense array of first wall and blanket cooling pipes which penetrate the machine through top ports and are routed to heat exchangers located in vaults in the tokamak hall. Requirements for top port diagnostic access need to be specified in order to determine compatibility with the routing of the cooling pipes and other penetrations required to service the tokamak.

Access to the pit can be gained via lifts which service the equatorial, divertor and basemat floors. The use of these lifts will require removal of shield plugs in the floor of the tokamak hall and the top of the lift. This can easily accomplished with the aid of an overhead crane. Alternatively, the pit can be accessed through lifts and tunnels from either the laydown hall or from the hot cell to the divertor floor.

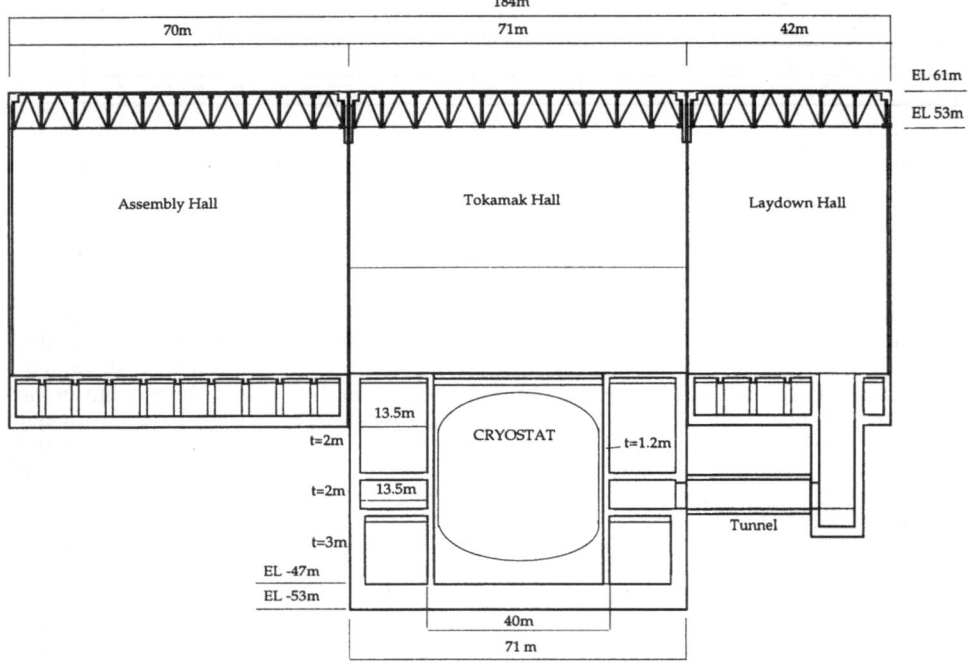

Figure 2. North-South elevation of the tokamak building. Grade level is at the floor of the Assembly, Tokamak and Laydown Halls.

Many diagnostics will see sufficient radiation to preclude hands-on maintenance. Tritium contamination either by direct exposure inside the vessel or by accident must also be considered. Hot cell repair and servicing operations for all diagnostics need to be specified in order to define the RH tools, transport systems and decontamination procedures that will be needed for their maintenance.

The vacuum vessel and cryostat form, respectively, the first and second confinement barriers to the release of volatile substances contained within the vessel. The vessel is designed to maintain confinement to a maximum internal pressure of 0.5 MPa and the cryostat to a maximum internal pressure of 0.2 MPa. These two barriers must be preserved by every penetration of the vessel and/or cryostat. Providing two such *independent* barriers is one of the most difficult challenges confronting the design of the ITER diagnostics.

Vacuum Vessel

The vacuum vessel provides a number of important functions, among them: it forms the primary vacuum boundary and must, together with the pumps, establish the high vacuum conditions required for operation of the tokamak; as just discussed, it forms the first safety barrier against the accidental release of volatile radioactive substances; it provides, together with the blanket, sufficient shielding to limit heating and radiation damage in the magnets to acceptable levels; it must be sufficiently robust to support both itself and all in-vessel components against the effects of electromagnetic loads arising from disruptions; and it must provide the access ports necessary for heat transport, auxiliary heating, blanket test modules and diagnostics.

The vacuum vessel is made 316 LN stainless steel and has a double wall structure. The two toroidal shells are 40 mm thick and are joined together by poloidal ribs. The total thickness varies with poloidal location and is in the range 0.45 - 0.83 m. Additional plates are inserted between the shells, and segmented and attached in such a way that the effect on the net toroidal resistance is minimal. The combined toroidal resistance of the vessel and other in-vessel components is $\sim 4\ \mu\Omega$. This is the minimum permissible resistance, since the resistive loss of volt-sec during start-up could limit the pulse length to less than 1000 s if the resistance falls below this value.

The vessel has 20 upper, mid-plane and divertor ports which are used for utility feedthroughs, component installation and maintenance, auxiliary heating systems, diagnostics, test modules, and vacuum pumping. The top ports are trapezoidal in cross-section ($0.34 \rightarrow 1.356 \times 3.34\ m^2$) and are used for blanket and first wall cooling penetrations, a viewing system and diagnostics; the horizontal ports are rectangular ($1.8 \times 3.0\ m^2$) and used for auxiliary heating, test blanket modules, diagnostics and remote handling systems; and the divertor ports (trapezoidal, $0.965 \rightarrow 1.647 \times 2.545\ m^2$) are used for divertor cooling penetrations, vacuum pumping and remote handling systems used for the introduction and removal of divertor cassettes. The allocation of the mid-plane ports depends on the type of heating used since the number required is different for each heating method. It is likely that at least two methods will be selected to meet the 100 MW heating and current drive power requirement. The present allocation of mid-plane ports for each combination of two heating technologies (assuming 50 MW each) is given in Table 2.

Table 2. Present allocation of mid-plane ports for each combination of heating methods

Port Number	NBI/ICRF Layout	NBI/ECH Layout	ICRF/ECH Layout
1	ICRF	Diagnostics	Diagnostics
2	Test module	Test module	Test module
3	Test module	Test module	Test module
4	Remote handling	Remote handling	Remote handling
5	Test module	Test module	Test module
6	Pellet injector / Torus rough pump	Pellet injector / Torus rough pump	Pellet injector / Torus rough pump
7	Diagnostics	Diagnostics	Test module
8	Test module	Test module	Diagnostics
9	Remote handling	Remote handling	Remote handling
10	Diagnostics	Unallocated	Diagnostics
11	NBI	NBI	Diagnostics
12	NBI	NBI	ICRF
13	NBI	NBI	ICRF
14	Remote handling	Remote handling	Remote handling
15	ICRF	Diagnostics	ECH start-up
16	ICRF	ECH	ECH
17	ICRF	ECH start-up	ICRF
18	Diagnostics	Diagnostics	ICRF
19	Remote handling	Remote handling	Remote handling
20	Diagnostics	Diagnostics	Diagnostics

Blanket and First Wall

The blanket and first wall system is comprised of approximately 700 modules which are attached to a toroidal backplate. The latter is made from 80 -100 mm thick stainless steel plate, has a horseshoe shape in the poloidal plane, and is attached to the vessel at 20 toroidal locations near the inside and outside open end of the horseshoe (See Figure 1). Most of the electromagnetic loads due to disruptions are either reacted by hoop stress in the backplate, as would be mainly the case for a centered disruption, or transferred through the blanket supports to the vessel, as would be the case for a vertical displacement event (VDE).

The modules are divided further according the quantity and type of heat flux expected to fall on their surface. The majority of modules, the so-called primary modules, are expected to see only radiation from the plasma; assuming as an upper limit that 80% of the alpha power is radiated, one estimates the surface heat flux to be 0.2 MW/m^2 if the radiation is uniform. Allowing for a peaking factor of 2.5, which might be on the optimistic side if the radiation is localized in a MARFE, gives the design value of 0.5 MW/m^2. In addition to the primary modules are limiter and baffle modules, which are designed to handle significantly higher heat flux. The limiter modules are located in two toroidal belts under the outside midplane. Their function is to serve as start-up and shutdown limiters during the period when the plasma is not diverted (typically when I < 16 MA). They are designed to dissipate a steady-state power density of 5 MW/m^2, a value which should give a comfortable margin for ohmic regimes but which may, depending on the scrape-off thickness and alignment, limit the amount of auxiliary power that can be applied in non-diverted regimes to significantly less than 100 MW. The baffle modules are located at each end of the backplate, just above the divertor. Their function is to prevent the backflow of neutrals from the divertor to the main plasma. The peak power conducted to the baffle surfaces is expected to be lower than to the limiters since they are located along the flux surface passing 5 cm outside the separatrix on the outside mid-plane and are then not in contact with the hot plasma in the SOL. However, power loss by charge-exchange is expected to be significant due to charge-exchange of divertor neutrals with the SOL plasma in the vicinity of the X-point. Further, charge-exchange erosion could be severe in this region, depending on the selection of baffle wall material. For this reason tungsten is being considered as the baffle first wall material, particularly along the surface just above the divertor. The reference surface material for all other modules is beryllium.

Construction of a typical module is shown in Figure 3. The modules are typically 1x 2 m^2 in width and height, and are 0.4 m thick, and are composed of a 20 - 25 mm thick first wall bonded to a stainless steel shield block The first wall consists of a poloidal array of 10 mm diameter steel pipes of thickness 1 mm. The pipes are embedded in a copper mat which is attached, for example by brazing or hot isostatic pressing (HIP), to the shield block. The plasma facing side of the mat is lined with 5-10 mm tiles of beryllium. Coolant water is supplied to the first wall pipes via manifolds running toroidally inside the shield block . These manifolds are in turn fed through branch connections to the main blanket cooling manifolds which run poloidally along the back plate. In the figure, the modules are shown to be mechanically attached to the backplate. The very strong torque applied to the module by disruption loads is reacted by a system of shear keys as shown. An alternative scheme for attachment involves welding rather than bolting. However, a mechanical attachment is preferred since it inherently maintains better tolerances and results in less debris during the attachment and removal process.

Replacement of a module begins by inserting a toroidal rail into the plasma chamber through two dedicated remote handling ports. Up to four vehicles equipped with appropriate manipulator arms, end effectors and tools travel along the rail to the position of the affected module. Meanwhile, pipe crawlers have been inserted into the poloidal manifolds at a position above the vertical port, and have travelled through the manifold to a position where the two branch pipe connections are cut. The module is then gripped and supported by the vehicle arm and the attachment bolts are removed. The vehicle then transports the module to another dedicated RH port where it is exchanged for a new module (see Figure 4), and the process is reversed. A complete changeout of the blanket, for example from one which only shields to one which also breeds, has been estimated to be possible within two years.

Divertor

The power to be exhausted to the divertor is up to 80 % of the alpha power which, with a 20 % excursion in an ignited plasma, totals 300 MW. Assuming a 10 mm power scrape-off width at the midplane and requiring that the angle of the field lines relative to the normal of the divertor plates is not less than 1.5 ° implies a normal power density on the plates of ~ 30 MW/m^2, far above that power density which can be reliably removed by even the most advanced cooling technology. Thus adoption of the conventional high-recycling divertor fails for ITER and a different approach must be developed. Fortunately, modern divertor experiments have indicated a solution through the discovery of detached regimes[4,5,6,7,8] in which both power and particle fluxes to the divertor plates fall well below attached levels. Power is removed by charge exchange and radiation, with radiation in the divertor channel or near the X-point the dominant loss process.

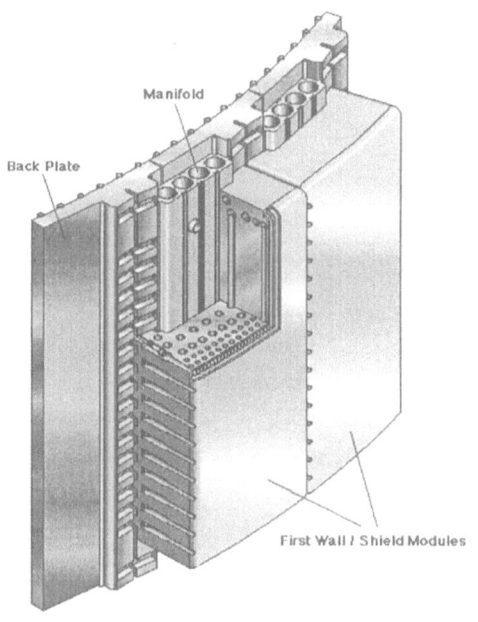

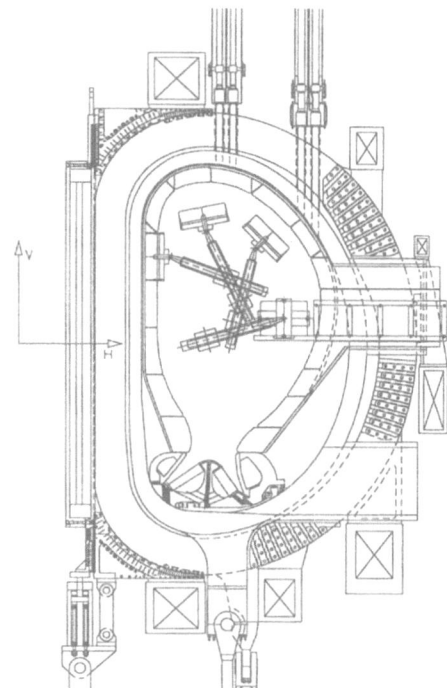

Figure 3. Cutaway of a blanket module.

Figure 4. Removal of a blanket module. (Toroidal rail and vehicle supporting arm not shown)

Radiating the bulk of the plasma power flowing into the SOL while maintaining plasma pressure balance along the open field lines requires interaction with neutrals. For moderate neutral densities, e. g. $n_0 \sim 10^{20}$ m^{-3} and low plasma temperatures, $T < 5$ eV, the coupling is strong and plasma pressure balance is easily maintained by transfer of plasma momentum to the neutrals. However, for ITER, radiating the entire alpha and auxiliary power by impurities which are in coronal equilibrium in the SOL requires impurity concentrations which are somewhat higher than the maximum level that can be tolerated in the core, if ignition is to be maintained. In this case, one must invoke either non-coronal processes in the SOL, thereby enhancing the effectiveness of impurity radiation, or effective trapping of SOL impurities, which would allow somewhat higher concentrations of impurities in the SOL than in the core plasma. Fortunately, only about half of the total heating power needs to be dissipated in the SOL and divertor channel since about 100 MW is radiated by bremsstrahlung and synchrotron radiation and additional power will be radiated inside the separatrix. Therefore, only small enhancement factors are required.

The development of a quantitative, experimentally validated model that can be used to predict and optimize divertor performance in ITER is a high priority within the worldwide effort providing the physics basis for ITER. Excellent progress has been made during recent years thanks to the coordinated efforts of the community of modelers, theorists and experimental teams which work on diverted tokamak experiments. Extrapolating to the end of the EDA, and into the construction phase, one can have full confidence that successful divertor operation in ITER will be achieved. A key aspect of the present design of the ITER divertor is that it is sufficiently flexible to incorporate the results of future R&D efforts, such as effects of geometry, materials, operational window, ELM's, and impurity transport.

The divertor assembly consists of 60 cassettes (3 per sector). A detail of the reference "vertical target" design is illustrated in Figure 5. This concept, similar to that adopted in Alcator C-MOD, has the advantage of providing good confinement of the neutrals and a target whose shape and inclination to the field lines can be optimized to minimize power flux to the plates during attached operation. In the private flux region, supported by the dome, are wings (or vanes) which act as a semi-transparent wall; radiation and charge-exchanged neutrals emitted from the divertor region are intercepted with high efficiency, by this structure but gas is nevertheless able to freely circulate or to be injected into the divertor channel.

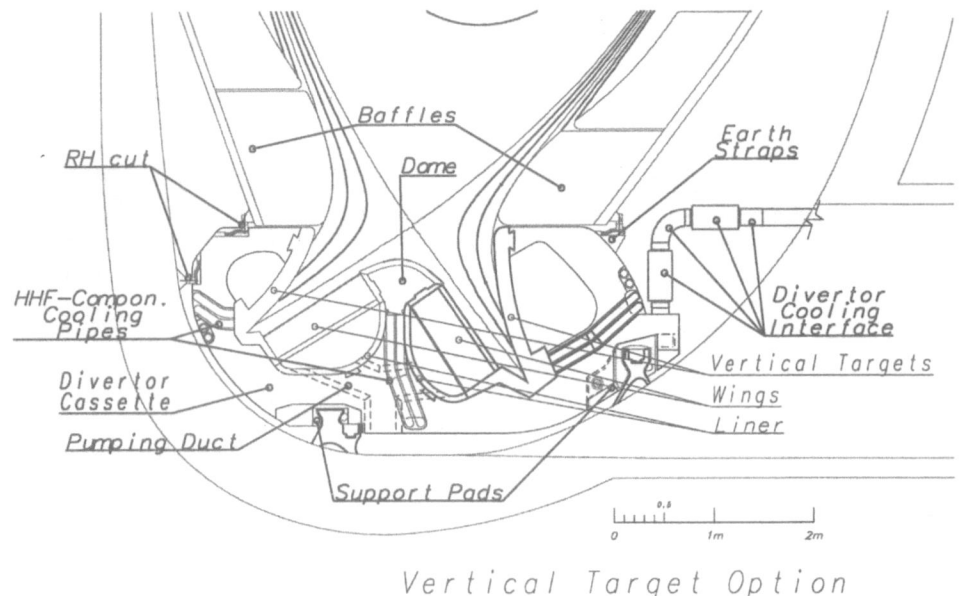

Figure 5. Cross-section of the divertor cassette.

The main structure of the cassette is made from 316 LN stainless steel. All high heat flux components are designed to be expediently removed from the cassette by hot cell operations. This approach provides flexibility and permits adjustment to new developments in divertor physics. The cassettes are serviced by straight radial pipes penetrating through each divertor port, thus allowing the use of CO_2 laser welding and cutting techniques. Two remote handling machines, the so-called toroidal and radial movers, have been conceptually designed to manipulate each cassette into one of four dedicated remote handling ports where it can be transported to the hot cell for maintenance. Full exchange of all divertor cassettes is possible in less than 6 months.

All high heat flux components are designed to accommodate a power density of 5 MW/m^2. The vertical target consists of a steel strongback and a HHF structure consisting of Cu alloy monoblocks brazed to Cu alloy tubes containing swirl tapes. The monoblocks are clad with either Be, C, or W (thickness 10 to 40 mm). The choice of material(s) depends on results from the R&D program and a selection will be made only at the end of the EDA. Although the nominal design heat flux is 5 MW/m^2, the CHF is about 20 MW/m^2 so heat fluxes up to 15 MW/m^2 are thermohydraulically acceptable with adequate margin albeit with rapid erosion due to melting and vaporization of the cladding surface.

PHYSICS REGIMES AND MARGINS

The main physics objective set out for ITER by the SWG 1 is to produce an ignited, steady burn for a duration of 1000 s and a neutron wall loading of 1 MW/m^2. Meeting this objective requires that ITER should be able to operate in regimes which are permitted by established principles of tokamak operation, for example those governing confinement, β limits, density limits, disruption avoidance, impurity control, particle transport, etc. A detailed treatment of any of these topics lies well outside the scope of this paper. However, a discussion which will be useful in identifying regimes in which ITER is likely to operatewill be given below. Deeper treatments of these and related topics can be found elsewhere[3,9].

Energetically accessible regimes can be determined through a simple power balance using representative profiles and a global confinement model. Using the empirically derived expression for ELMy H-Mode,

$$\tau_{E,th}^{ELMy} = H_H * 0.85\,\tau_{E,th}^{ELM-free} \ ,$$

where

$$\tau_{E,th}^{ELM-free} = 0.053 I_p^{1.06} B_T^{0.32} P_L^{-0.67} M_i^{0.41} R^{1.79} \overline{n}_e^{0.17} \varepsilon^{-0.11} \kappa^{0.66} \ ,$$

and where H_H is a scale factor equal used to assess the confinement margin relative to H-mode ($H_H=1$), the curves shown in Figure 6 can be produced[10]. The curves passing through the open circles indicate ignited contours possible with H-mode ($H_H=1$) confinement, and with various degradation factors ($H_H=0.95$, 0.9, $\cdots$). In obtaining these curves, the power appearing in the confinement expression has been set equal to the net heating power, i.e., the alpha power diminished by the power radiated by bremsstrahlung. Also, the helium particle confinement time has been set to $\tau^*_{He}/\tau_E = 10$.

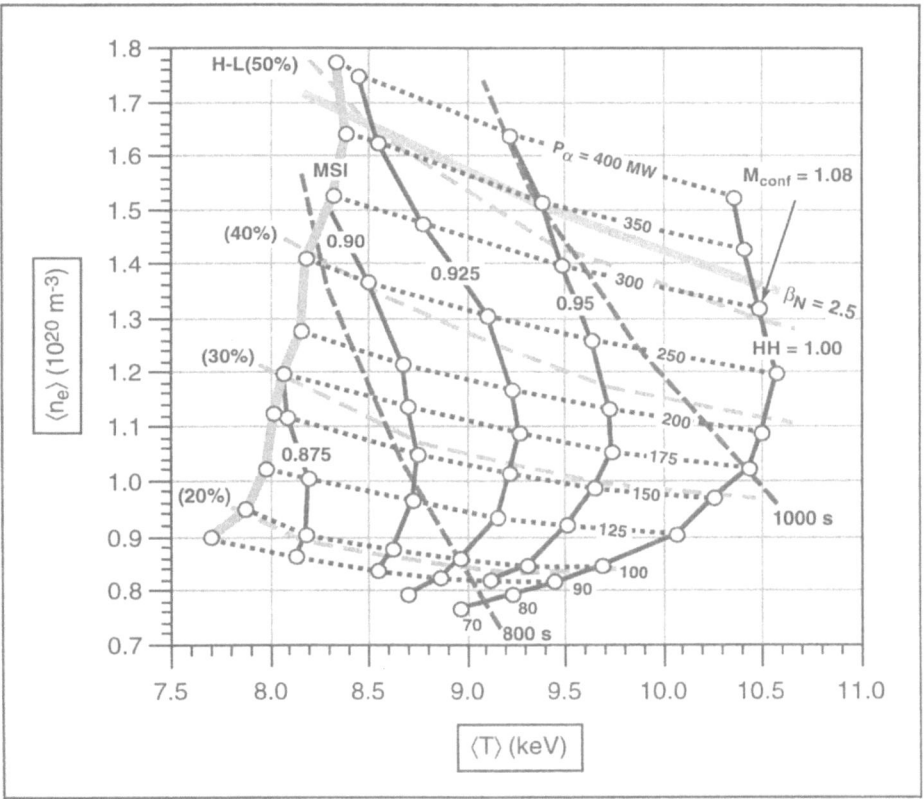

Figure 6. Ignited operating contours (curves passing through open points) energetically permitted by H-mode confinement (HH=1) and effect of degradation (HH=0.95, 0.9 $\cdots$). The solid curve labelled MSI corresponds to the bremsstrahlung limit. The curve labelled β_N = 2.5 corresponds to a normalized β of 2.5; the curves labelled 900 and 1000 s are contours along which the pulse length is constant at the value indicated; the curves labeled P_α = 400 MW, 350 MW,.$\cdots$ correspond to contours of constant alpha power, and the curves labelled H-L= 50 %, 40 % $\cdots$ show contours where the power crossing the separatrix equals 0.5, 0.4,$\cdots$ of the L → H transition threshold.

Inspection of Figure 6 shows that the margins for achieving the SWG 1 objectives are small, and that any reduction in machine performance for example by reducing the size or field could jeopardize the success of the ITER mission as strictly defined by SWG 1. However, further analysis considering the *driven* performance using 100 MW of auxiliary power shows that high Q regimes are possible with considerably more confinement margin, for example one finds Q = 15 with H_H = 0.75. Such regimes must be considered as fall back modes of operation in the (unlikely) event that no further progress is made in confinement optimization. In any event, ITER is an experimental machine and optimization of its performance will be possible only with a complement of diagnostics which is at least as extensive in scope as is found on any of today's tokamaks.

DISRUPTIONS

Disruptions of ITER plasmas containing ~1 GJ of both thermal and magnetic energy produce extremely large forces and stresses on all in-vessel components and are the determining factor in their structural design. All components such as diagnostics, heating systems, test modules, etc., which are to be inserted into ports will likewise be subject to disruptive loads and the design of such systems must take full account of their consequences.

Disruptions can be grouped according to three classifications: i) those which rapidly (10's msec) release both the thermal and magnetic energy without vertical motion; ii) those which release the thermal energy and then undergo a Vertical Displacement Event; and iii) those undergo a Vertical Displacement Event without appreciable loss of thermal energy. In class i) disruptions, the toroidal current transfers to the blanket and vessel systems and its evolution and distribution, and therefore the EM forces, are reasonably straightforward to calculate. The forces on the wall can be accounted for by a magnetic pressure which peaks at about 1.5 MPa on the inner wall. The energy deposition to the wall is small (~1 MJ/m^2) since the plasma energy is more or less uniformly radiated. The effects of class ii) and iii) disruptions are much less predictable, since plasma motion and contact with the first wall occurs before the final disruption. In particular halo currents up to 1/3 of the plasma current are possible and these give rise to net vertical forces of ~150 MN. Further, it is known that halo currents can be non-axisymmetric with a 100 % modulation by an n = 1 component. The associated forces are transmitted through the blanket and vessel supports, and produce high stresses in the vessel. Some rearrangement of the vessel supports and stiffening of the cross-section will be required to bring such stresses within allowables in the present design. Also, better characterization and modelling of halo currents is needed and this is an area of active physics R&D.

In class iii) disruptions, substantial energy (up to 60 MJ/m^2) can be delivered to the first wall during plasma contact. As this can occur over a period of a few seconds, substantial wall damage can occur as well as possible structural damage due to coolant burnout. The latter possibility is mitigated by design measures which are intended to keep the maximum heat flux below the CHF. Wall damage may be limited by vapor shielding, but the effectiveness of this effect has not yet been established. In view of the serious consequences of class ii) and iii) disruptions, it is intended to attempt to mitigate such events with pre-emptive intervention by means of an impurity pellet. The aim would be to turn these disruptions into class i) events. Some limited success in this regard has been demonstrated, but further development and understanding are required.

SUMMARY

The status of the ITER tokamak design at the mid-point of the EDA has been described. Relative to the Outline Design, major improvements have been incorporated in the design of the TF magnets, the cryostat and the blanket-first wall system. In each case the modifications have simplified the design and facilitated fabrication, assembly and maintenance. The divertor concept remains similar to that of the OD, except a vertical target has been adopted as the first option. The cassette approach retains flexibility and permits rapid changeout of the divertor, which will allow ITER to take full advantage of future progress in divertor physics. The projected plasma performance of the IDR design is essentially unchanged from that of the OD, and allows realization of the ITER objectives based on extrapolation of present confinement behavior. Disruptions produce substantial loads on all in-vessel components and their effects must be carefully taken into account in design of systems which extend into the vessel ports.

ACKNOWLEDGEMENT

This report has been prepared as an account of work performed under the Agreement among the European Atomic Energy Community, the Government of Japan, the

Government of the Russian Federation, and the Government of the United States of America on Cooperation in the Engineering Design Activities for the International Thermonuclear Experimental Reactor ("ITER EDA Agreement") under the auspices of the International Atomic Energy Agency (IAEA).

REFERENCES

1. *ITER Conceptual Design Report*, ITER Documentation Series, IAEA, Vienna (1992)
2. P-H. Rebut et. al., *ITER Outline Design Report* and related documentation (unpublished)
3. R. A. Aymar et.al., *ITER Interim Design Report* and related documentation (unpublished)
4. T. W. Petrie, D. Buchenauer, D. N. Hill, et. al., *Divertor heat flux reduction by D_2 injection in DIII-D*, J. Nucl. Mater., 196-198:848 (1992)
5. G. Janeschitz, s. Clement, N. Gottardi, et. al., *High power operation with a radiative divertor in JET*, Proc. 19th EPS Conference on Controlled Fusion and Plasma Physics, Innsbruck, 1992, 16C (Part 2):727
6. V. Mertens, K. Büchl, W. Junker, et. al., *Experimental investigation and interpretation of marfes and density limit in Asdex Upgrade*, Proc. 20th EPS Conference on Controlled Fusion and Plasma Physics, Lisbon 1993, 17C (Part 1):267
7. N. Nagami, JT-60 Team, *Latest results from JT-60U divertor research* J. Nucl. Mater., 220-222:3 (1995)
8. B. Lipschultz, J. Goetz, B. LaBombard, et. al., *Dissipative divertor operation in the Alcator C-MOD tokamak*, J. Nucl. Mater., 220-222:3 (1995)
9. G. Janeschitz, ITER Joint Central Team, ITER Home Teams, *Status of ITER*, Proc. 22nd EPS Conference on Controlled Fusion and Plasma Physics, Bournemouth 1995 (to be published)
10. J. Wesley, Private communication

THE ITER DIAGNOSTIC PROGRAMME

K.M. Young,[1] A.E. Costley,[2] T. Matoba,[3]
D. Orlinski,[4] and P.E. Stott[5]

[1]Princeton Plasma Physics Laboratory,
P.O. Box 451, Princeton, NJ 08543 USA.

[2]ITER Joint Central Team, San Diego Joint Work Site, La Jolla, CA 92037
USA.
[3]Japan Atomic Energy Research Institute, Naka Fusion Research
Establishment, Naka-machi, Naka gun, Ibaraki-ken 311-01, Japan.
[4]Kurchatov Institute of Atomic Energy, Ploshchad'Akademika Kurchatova
46, Moscow 123182, Russia.
[5]JET Joint Undertaking, Abingdon, Oxfordshire, OX14 3EA United
Kingdom

INTRODUCTION

The diagnostics for ITER have to enable the device to meet its missions of achieving ignited plasmas lasting for over 1000 seconds while producing an average neutron wall loading of 1 MW/m^2. The diagnostics must provide the information for interpreting the plasma performance, but must clearly also provide signals to permit device protection and to allow real-time control of many plasma parameters using many more signals than is currently customary. ITER will be the first device in which the behavior of ignited plasmas can be studied[1]. For example, because the impact of major disruptions on the divertor first-wall and the potential for large halo currents in the first wall structure is so severe, fast diagnostic information to provide a preemptive response must be provided. In its first ten years of operation, the Basic Performance Phase, physics understanding and operational experience must be developed for moving forward to the Enhanced Performance Phase, in which high-fluence neutron testing is the main goal. In this latter phase, the number and sophistication of the diagnostic systems will probably be reduced to accommodate more technological modules. However, since the program calls for further improvement in the plasma parformance, many of the sophisticated profile-measuring diagnostics will have to beretained.

This paper will describe the organization that has been set up to develop the designs of the diagnostics during the Engineering Design Activity (EDA), a project scheduled to last another three years. While complete designs for every diagnostic will not be possible, the quality of designs must be such that the overall performance, safety and reliability of

Diagnostics for Experimental Thermonuclear Fusion Reactors
Edited by P. E. Stott *et al.*, Plenum Press, New York, 1996

13

the tokamak as a whole can be credibly demonstrated at the end of the EDA period. Since the resources are very limited, the design process must be as efficient as possible, within the constraints of organizing a large international project. The primary deliverable products are clearly the designs of diagnostic systems and the documentation validating the proposed performance.

There are secondary benefits of the program which are of significant importance. These are the involvement of the world-wide community in providing the system designs and critiquing them. The best possible physics input from operational experience at many fusion devices will be available. Many expert diagnostic physicists will be involved in this first advance into the concepts and technology necessary for the measurement of the plasma parameters in the environment of an ignition device. Also many experienced diagnostic physicists will be considering the impact of long-pulse operation on control integration and on reliability and maintainability that has to be factored into the equipment design. Real-time control integration is a rapidly evolving aspect of the plasma studies in the fusion program, and ITER should benefit as much as possible from this experience.

To clarify the scope of the work required within the EDA, table 1 shows the plasma parameters whose measurement is expected to be required to control the tokamak plasma, to evaluate and optimize the plasma performance, and for understanding the underlying physics[2]. This table is the result of extensive discussion and review by the ITER Diagnostics Expert Group, and has been circulated widely for comment to the tokamak operational community, especially members of other ITER Expert Groups and members of the ITER Physics Committee. Over 50 different measurements are indicated: some may use the same diagnostic techniques, but more probably more than one diagnostic technique may have to be used to provide the full range of necessary information expected for one plasma measurement, e.g. full spatial coverage, sufficient temporal resolution, specific parametric dependence of a technique or maintenance of the calibration over the required dynamic range. Thus there is a huge amount of work involved in arriving at the optimum diagnostic techniques, developing interfacing designs for them with the other tokamak components and with the other diagnostics to preserve the integrity of the shielding, vacuum vessel and cryostat.

The first category in table 1 includes diagnostics necessary to assure protection of the tokamak. The divertor and first wall are liable to significant erosion; runaway electrons could cause serious damage and only a few full-power disruptions can be accommodated. Hence diagnostic measurements are required to provide some precursor indication of these events. During the physics phase the appropriate responses to such a precursor message by heating, fuelling or positioning systems will be developed in a fully integrated control system. Similarly, a capability to maintain steady operation for long times at high powers will have to be created. The instrumentation to allow an extended physics study of the interactions of the different sensors and the controlling responders must be available early in the device lifetime to be able to move forward to long-pulse operation. The second category of measurements, some of which could easily become control-participating diagnostics as more is learned about the key operational parameters in determining the long-time behavior (identified by the symbol †), includes those considered essential for developing an underlying understanding of the performance, so allowing the operational physicists to improve that performance and optimize the plasma for high neutron flux, and hence high fluence, operation.

The third category is included to account for the measurement needs for understanding the physics of this first device where the heat source, and the ignition, is provided by the confined fusion-product alpha-particles. The measurements listed are similar to those used on operating tokamaks where physicists are trying to understand the potential impact of instabilities in limiting confinement. The presence of fast particles, with velocities greater than the Alfven velocity of the background plasma, in toroidal devices is known to be a potential driver of classes of MHD instabilities so that the large population of high energy alpha-particles could affect the confinement.[3] The detailed requirements on the accuracies and resolutions for these measurements, together with the techniques presently being considered to carry them out are given in the next paper.[4] Theserequirements have evolved from those developed for the Conceptual Design Activity (CDA).[5] Reference 5 remains however a concise summary of the issues involved in ITER diagnostics.

This list of measurements is almost certainly too extensive for the space that will be available in ports on the tokamak for diagnostic equipment, taking account of the additional space probably required for optical periscopes and remote-maintenance features required for many diagnostics. Present budget planning also does not consider completion of the full set of diagnostics in the construction phase. Thus a key activity during the EDA will be honing down the number of measurements in this table, and of the techniques in the associated table of potential candidate diagnostics, as the design and R&D programs progress.

Table 1: Plasma Measurement Requirements
Category (i): Measurements for Machine Protection and Plasma Control

#	Plasma Parameter to be measured	Purpose
1	Plasma Current	Total current required for control
2	Plasma Position and Shape	Poloidal field feedback
3	Loop Voltage	l_i, plasma startup
4	Beta	Disruption avoidance
5	Total Radiated Power	Disruption avoidance
6	Plasma Line-averaged Density	Plasma operation, disruption avoidance
7	Total Neutron Flux and Emission Profile	Burn control, fuelling control
8	Locked-modes	Disruption avoidance
9	m=2 MHD Modes, Sawteeth	Disruption avoidance
10	Plasma Rotation	Disruption avoidance, confinement optimization
11	nT/nD in Plasma Core	Fuelling control
12	Impurity Species Monitor	Impurity control, wall protection
13	Zeff (line-averaged)	Impurity control, He accumulation
14	ELMs, L/H Mode Indicators	H-mode realization, optimization control, divertor prot.ection
15	Runaway Electrons	Runaway avoidance
16	Key Divertor Parameters	
	Divertor Plate Temperature*	Protection of divertor structure
	Divertor Radiated Power	Divertor operation
	Divertor Plate Ablation	Divertor plate erosion
	Ionization Front Position	Divertor optimization
	Divertor Gas Pressure*	Divertor operation
17	Edge Light & First Wall Temperature*	Startup, position and hot spot monitor (360° Coverage?)
18	Base Pressure*	Tokamak readiness
19	Gas Pressure in Duct*	Divertor functionality
20	In-vessel Inspection*	Inspect for internal damage
21	'Halo' Currents*	Monitor forces on in-vessel components
22	Toroidal Magnetic Field*	Cyclotron resonance position, q-value
	Indicates a machine diagnostic	

Category (ii): Measurements for Performance Evaluation and Optimization

#	Plasma Parameter to be measured	Purpose
23	Electron Temperature Profile†	Profile control by aux. heating; transport
24	Electron Density Profile†	Transport, fueling optimization
25	q(r) profile†	Plasma stability, transport
26	Zeff profile†	Impurity transport
27	Fishbones, TAE Modes†	Burn optimization, beta limit indication
28	Ion Temperature Profile	Burn optimization and transport
29	Helium Density in Plasma Core	Burn optimization and transport
30	Confined Alpha-Particles (r)	Alpha particle transport, heating profile
31	Escaping Alphas	Performance evaluation
32	Impurity Density	Condition, high-Z use in radiative loss
33	Edge n_T/n_D, n_H/n_D	Fueling optimization, plasma dilution
34	Long-term Neutron Fluence	Calibration of neutron dets., blanket data
35	Impurity and D,T Influx in Divertor	Divertor optimization
36	I_{Sat} in Divertor†, n_e, T_e at target	Divertor optimization
37	Rad. Profiles (Core, X-pt./MARFE, Div.)	Divertor optimization
38	Heat Loading Profile in Divertor	Divertor optimization
39	Divertor Helium Density	Divertor optimization, He pumping physics
40	n_T/n_D, n_H/n_D in Divertor	Fueling optimization, plasma dilution
41	n_e and T_e in Divertor	Divertor optimization
42	Ion Temperature in Divertor	Divertor optimization
43	Divertor Plasma Flow	Divertor optimization
44	n_H/n_D in Plasma Core	Plasma dilution
45	ICRF Anenna Coupling	Heating optimization
	† *Potential Category (i) measurement*	

Category (iii): Additional Measurements for Physics Understanding

#	Plasma Parameter to be measured	Purpose
46	Pellet Penetration	Fuelling optimization
47	Plasma Facing Material Erosion	Erosion physics
48	Density Fluctuations	Instabilities (e.g., TAE mode)
49	Edge Turbulence	Understanding of edge transport
50	MHD Activity in Plasma Core	Instability studies
51	Te Fluctuations	Transport understanding
52	E Field and E Fluctuations	Transport understanding
53	ICRF Physics Measurements	Wave-plasma interaction

THE ORGANIZATION FOR DESIGNING DIAGNOSTICS

A tight-knit organization has been put together for the ITER Project and, within that, a similar organization has been set up to provide the diagnostic design and supporting R&D. This organization is shown in a simplified organization chart in Fig. 1.

ITER DIAGNOSTICS ORGANIZATION (EDA)

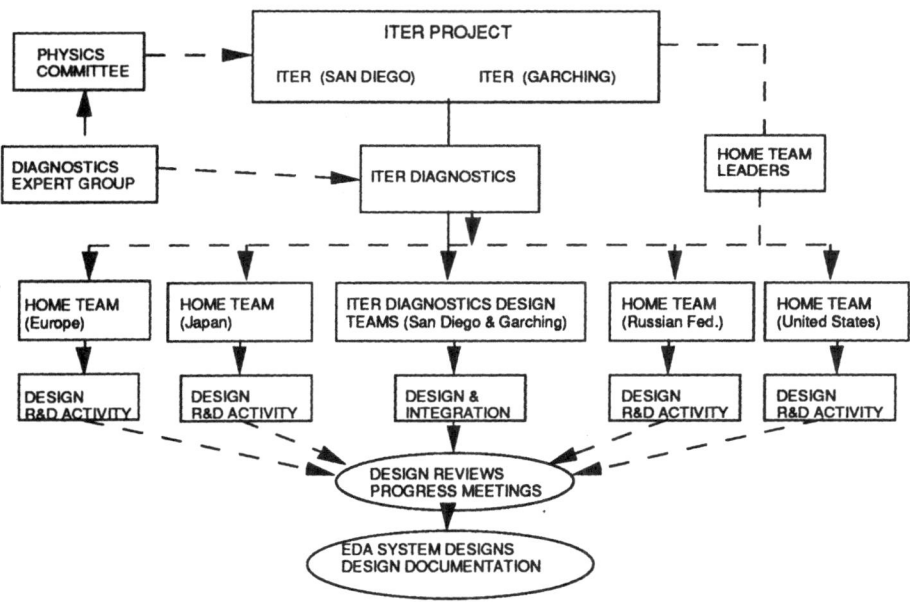

Figure 1. Simplified organization chart for the design of diagnostics.

There are two main components of the organization for designing ITER. There is a core Joint Central Team (JCT) and a supporting team made up of members of Home Teams for the four Parties which set up the ITER Project. The diagnostics design will follow the same pattern. The organizational scheme is shown in fig. 1. The solid arrows show the communication links within the JCT and, at the foot of the figure, lead to the necessary activities which the JCT oversees. Similarly the dashed lines show the linking from the JCT to the Home Teams and the activities which the Home Team organizations control and contribute to within the diagnostics area. There are a variety of communication paths between the JCT Group and the Home Teams; one of the most important is shown explicitly and that is the attendance at and participation in Progress Meetings and Design Reviews which are called for it the Task Agreements betweek the JCT and the Home Teams. On the left of the figure, an advisory organization, the Diagnostics Expert Group and the Physics Committee, supports the ITER Diagnostics and ITER Physics management, is also shown. The roles of these organizational elements will be discussed in the following sections of the paper.

The anticipated deliverable products at the conclusion of the EDA will be a set of diagnostic system designs and documentation supporting these designs, including the results of the validating R&D program. There will also be significant design documentation demonstrating that the interfaces with the other ITER systems satisfy all the criteria by which this design will be judged.

THE ROLE OF THE JOINT CENTRAL TEAM

It is the Director and his Joint Central Team who are responsible for providing the design of the tokamak and its supporting systems. The Director works with an oversight Council and two Advisory Committees, one Technical and one Administrative, to assure that the design meets technical objectives and is meeting with the expectations of the Parties participating in the Project. The JCT is separated into groups at three design sites. For diagnostics, design responsibility is split between the San Diego and Garching sites, and is led from San Diego. Because much of the diagnostics system has very complex interfaces with the vacuum vessel and components inside it, such as the blanket modules and the divertor modules, a design group has been provided at Garching. The relationship with the activity at Naka is not so immediate, so there is no diagnostic presence there but constant communication on control issues and interfacing requirements of the diagnostics with the blanket and facilities systems is needed.

The JCT Diagnostic Group manages the diagnostic design activity and plans and integrates the diagnostic effort. It also integrates the diagnostic needs with the other prinicpal tokamak systems. Members of this Group determine the necessary activity and develop the manpower and funding plans necessary to fulfil these goals in agreement with their management. They carry out some specific design work, where there are particularly urgent or important design integration concerns. For example, the magnetic diagnostics, all of which are inside the vacuum vessel, are being designed by the JCT, but with considerable consultation with experts within the Home Teams.

The JCT Group prepares Task Agreements for the performance of design within the Home Teams. The Director has agreed with the Home Team leaders that the total design effort for the remainder of the EDA period (1995 -1998) will be 16.5 years of professional design in the JCT and 25 in the Home Teams. The JCT professional staff will be supported by CAD designers. While the JCT does not pass funds over to the Home Teams, satisfactory performance of these Task Agreements leads to the Home Team being creditted with the allocated amount of professional years.

To provide guidance and control over the design of the diagnostics with so many organizations involved the ITER Diagnostics Group has prepared a General Design Requirements Document (GDRD) and individual Diagnostic Design Description Documents (DDDs). The former document relates the diagnostic designs to the interfaces with other systems and will be updated as these interfaces become better defined or there are significant changes in the tokamak configuration. The DDDs will evolve as the design of each individual diagnostic progresses and will be modified by the design group involved. Expert CAD groups exist in both San Diego and Garching to handle all the drawings and to help the ITER diagnostic team in its integration task. A mailbox of drawings of in-vessel tokamak components, selected for their likely interaction with diagnostics penetrating in some way into this region, is available at Garching for people engaged in diagnostic design.

There is also an R&D scope within diagnostics, also agreed in scale between the Director and the Home Team Leaders. 7300 IUA (1 IUA = $1000 on January 1, 1989) has been set aside for diagnostics for the years 1995 -1998. About two-thirds of this total is devoted to studies of radiation effects on diagnostic components and the remaining third will be allocated to specific components development determined to be necessary in the design process. Again, the detailed tasks have to be agreed between the JCT Group and the Home Teams. In this case, however, all the R&D studies are carried out within the Parties.

THE ROLE OF THE HOME TEAMS

The Home Teams perform a large fraction of the design scope and all of the R&D. Within the Home Teams there is a similar organization to that within the JCT. Task Area Leaders (TAL) for Diagnostics, the managerial contacts for the JCT Diagnostics Head, report to the Home Team Leader, in most cases through a Physics Manager as in the United States. This TAL is responsible for ensuring that the required work, expressed in the Task agreements, is properly managed and carried out within the Party and that it is suitably documented through design drawings and descriptive paperwork. He is also responsible for managing the R&D program.

For the three remaining years of the EDA, a total of 25 Professional Person Years (PPY) of effort has been allocated to diagnostic design inside the Home Teams. Specific numbers have been agreed between the Director and the Home Team Leaders for all areas of design for ITER, such that the aggregate commitments by each Party are met. The effort committed to diagnostics is relatively high compared to other systems, but the number of individual designs and the variety of specialist expertise required means that the effort is spread rather thin.

The agreed allocations (in PPY) are given below:

EU	5
JA	2
RF	12
US	6
Total	25

With these thinly spread resources it is extremely important that there should be effective collaboration between groups in different parties working on similar diagnostic systems, leading to efficient sharing of concepts and critiquing of each others' designs. The TALs are responsible for organizing such relationships, though the implementation will have to be done by the working designers.

It is clear that very close coordination is called for between all the groups involved. Thus, wherever possible, the JCT personnel involve the TALs in setting up the work and priorities. This is done through meetings contiguous to other scheduled meetings and exchange of documents for review prior to formal promulgation. The co-authors of this paper are the four Home Team TALs and the Head of the ITER Diagnostics Group.

THE EXPERT GROUP ON DIAGNOSTICS

The ITER Director and the Head of the ITER Physics Group have set up advisory groups in physics areas to help integrate physics requirements into the ITER Design process. These Expert Groups are made up of two people from each of the Parties and two from the JCT (because of the split of the JCT Diagnostic team into two sites, there are four JCT members in the Diagnostics Expert Group); the experts relate to the physics programs in the four Parties. These Expert Groups report both to the Physics management within the JCT but also to a Physics Committee for ITER whose membership includes prominent physicists in the Parties' fusion programs, typically the heads of tokamak experiments. The Expert Groups are asked to provide the JCT with coherent assessments of the available information in their area and to arrange for experiments if that information is inadequate. A particularly important aspect of their work is organizing data-bases out of which the extrapolations of performance out to the ITER scale can be made with confidence.

There are seven Expert Groups, one in Diagnostics. The others are a) Confinement and Transport, b) Confinement Modelling and Database, c) Disruptions, Plasma Control and MHD, d) Divertor, e) Divetor Modelling and Database and f) Energetic Particles, Heating and Current Drive. Since diagnostics is an area of physics interest, but is also a very key element in the engineering design, this Group's mission is somewhat different from the other Groups. In its first two meetings, the Group has worked hard with the JCT members to develop the tables in defining the measurements (table 1 above), specifying measurement requirements and in selecting diagnostic techniques. Members of the Group have participated in the Progress Meetings in support of the JCT. But now the Group's mission should change. Important elements of the Group's task will be to help to identify research needs, to propose research programs and facilitate testing of systems within the Parties' programs. One of the Group's future roles will be in assuring that developments in diagnostics throughout the world fusion program will be considered as possible improvements to the chosen ITER set and in reviewing and helping the JCT in achieving its design goals. It will also try to take an unbiassed oversight position as far as trying to anticipate future diagnostic requirements and innovations and work with the JCT Diagnostics Group to see that the ITER design does not preclude these possibilities. Because of the cross-national make-up of the committee, it will be possible to nourish collaborations between groups to bring the best possible diagnostic design capability to the EDA.

The Expert Group structure is designed to ease communication within expert branches of the fusion community to get the best possible advice for the ITER design. Everybody is encouraged to bring their interests and concerns to the members of the Expert Group in their Party.

RESEARCH AND DEVELOPMENT PROGRAM

A major component of the ITER budgetting is devoted to a Research and Development (R&D) program. The most costly elements in this program will be demonstration prototypes of key elements in the tokamak such as a model toroidal field coil and a mock-up of the divertor module. These prototypes will add credibility to the design since they show not only operational capability, but also manufacturing and handling feasibility. Similarly funding has been provided for two principal areas of diagnostic activity to provide similar demonstrations of feasibility of specific diagnostics.

The first is a program for investigation of the sensitivity of diagnostic components to the intensely unpleasant neutron and gamma radiation environment at various locations around ITER. To compound the problem for components close to the plasma, there will also be big changes in temperature and the need for the components to maintain their reliability and calibration through the 1000 second pulses. This program has been under way for about two years and, in the remaining three years (1995 -1998) has been allocated about 7300 IUA (1 IUA = $1000 at January 1, 1989, the time of the CDA cost estimates). All four Parties are participating in different aspects of this program. Table 2 shows the main elements of the program at present; more detailed information on the active studies can be obtained from the Diagnostics Group in Garching.

The second area, for which 2500 IUA has been allocated, will be applied after the design stage of the diagnostics is more advanced, starting with the series of Progress Meetings planned for early in 1996. It is realized that there are many aspects of the diagnostics which may need validation through testing of new concepts or new components, development of new sources or detectors to match the particular requirements set by the ITER plasma properties, or altogether new techniques of measurement. It is highly desirable that all the new initiatives in the diagnostics should have had some degree of testing in a less hostile operational environment. The funding will be allocated once the need has been demonstrated to the ITER JCT Task Officer.

Table 2. Main Elements of the ITER Diagnostics R&D Program in Radiation Effects on Diagnostic Components (Task T246)

Diagnostic Component	Examples of Radiation Testing in Progress or Necessary
Detector sensitivities to neutrons and gammas	Full system test of a Mirnov loop in a fission source; Neutron heating of a bolometer.
Ceramic insulators	Radiation-induced conductivity in Al_2O_3 and silica glass; Radiation-induced electrical degradation in Al_2O_3 and silica glass; Development and testing of new mineral-insulated cables; Mechanical degradation of structural elements.
Windows and mirrors	Absorption and luminescence in windows as functions of flux and fluence; materials to cover wide range of wavelengths; Study of "hard" vacuum seals to withstand mechanical changes under irradiation; Irradiation effects on anti-reflection coatings; Front-end mirrors probably metal; erosion and sputtering issues; Effects on dielectric mirrors, LSMs.
Fiberoptics	Influence of temperature on absorption and luminescence; Development of fluorine-doped fibers; "Hardening" of fibers by pre-irradiation in gammas.

SUMMARY

The intent of this paper has been to lay out the main organizational elements, the participants and the communication techniques that have been set up to enable diagnostic design to be done effectively and efficiently with the worldwide distribution of resources. The Joint Central Team has clear responsibility for assuring that the diagnostics being designed give the full capability for machine protection and control of the plasmas and for evaluating the plasma performance and optimizing it during the Basic Performance Phase of operation. The JCT must also ensure that a) all these diagnostics can be integrated into the tokamak facility, with sightline access through the cryostat, the vacuum vessel wall and shield blankets, b) those diagnostics that need to be integrated with the divertor modules are, c) appropriate services are provided for the diagnostics, including possible water-cooling connections inside the vacuum vessel and d) that space is available for the equipment and for laboratory testing. The ITER tokamak environment is unique and the communication of the unique aspects such as the high neutron fluxes, very high power loads and remote-maintenance integration to the designers in the Home Teams will be a constant requirement and will need careful documentation.

The design phase has started with initial assignments to the Home Teams, and internal to the JCT, for diagnostic design having been made. There is very limited manpower available for the task so careful prioritization in establishing credible designs for the end of the EDA will be required, with the understanding that the final designs for all except a few diagnostics will be completed during the construction phase. An associated R&D program has been started to tackle the new problems for diagnostics, associated with operation of diagnostics in a radiation environment. Initial work in identifying issues and in the use of shielding to limit background noise levels has been done on TFTR[6], but there the source strength is lower and the pulse length is only ~1 second. An intensive set of studies using radiation sources is necessary to define potentially serious issues for ceramic insulators inside the vacuum vessel and for mirrors which will almost certainly have to be used as the first optical element close to the plasma.

We must not lose sight of the fact that ITER will not operate until more than ten years from now. The advances in new diagnostics and creative uses of old techniques that have arisen in the last ten years make it clear that we must retain some level of flexibility in our overall design. There may be a need for new techniques to meet all the requirements for a specific measurement and this need should be fulfilled through the R&D funding which ITER has available for diagnostics. Other than that, it is incumbent on all of us in the diagnostics community to work to bring the best possible information about new developments to the JCT and to try to come up with better methods and to test them on currently operating tokamaks. The ITER Expert Group has especial responsibility in this area.

The work of the first author is supported by USDOE Contract No. DE-AC02-76-CHO-3073.

REFERENCES

1. R. Parker, "The ITER Device", these Proceedings. See also Presentations to the Technical Advisory Committee (TAC-8) by the ITER Director and Joint Central Team Presenters, ITER Document # TAC-95-19 (July 1995).

2. Minutes of the Second Meeting of the ITER Diagnostics Expert Group. ITER Document # S CX MI 3 95-03-23 F 1 (March 1995).

3. S.J. Zweben et al., "Alpha Physics and Measurement Requirements for ITER ", these Proceedings.

4. A.E. Costley et al., "Requirements for ITER Diagnostics", these Proceedings.

5. V. Mukhovatov, H. Hopman, S. Yamamoto et al, "ITER Diagnostics", ITER Documentation Series #33 (IAEA, Vienna, 1991).

6. A.T. Ramsey, Rev. Sci. Instrum., 66, 871 (1995).

REQUIREMENTS FOR ITER DIAGNOSTICS

A E Costley[1], R Bartiromo[2], L deKock[3], E Marmar[4], T Matoba[5],
V Mukhovatov[1], K Muraoka[6], A Nagashima[5], D Orlinski[7],
M Petrov[8], P E Stott[9], V Strelkov[7], S Yamamoto[3] and K M Young[10].

1 ITER Joint Central Team, San Diego, USA
2 ENEA, Frascati, Italy
3 ITER Joint Central Garching, Germany
4 Plasma Fusion Center, MIT, USA
5 JAERI, Naka, Japan
6 Kyushu University, Fukuoka, Japan
7 Kurchatov Institute, Moscow, Russia
8 Ioffe Institute, St Petersburg, Russia
9 JET, Abingdon, UK
10 PPPL, Princeton, USA

ABSTRACT

The requirements for diagnostics on ITER are determined from a consideration of the role plasma measurements will play in the operation and evaluation of the tokamak. The proposed list of measurements and the detailed measurement specifications - parameter ranges, target measurement resolutions and accuracies - are presented. The principal practical problems that will have to be overcome in implementing the diagnostics are outlined and areas where particular difficulties arise are identified.

INTRODUCTION

In order to operate and control ITER at high levels of fusion power (~ 1.5 GW) it will be necessary to measure accurately and reliably a wide range of plasma parameters. Measurements of key first wall and divertor parameters will also be required. The measurements will have to be made in the presence of high levels of radiation and with limited access to the plasma. Together these requirements pose a demanding measurement task.

The measurements will be made with an extensive diagnostic system. The

Diagnostics for Experimental Thermonuclear Fusion Reactors
Edited by P. E. Stott *et al.*, Plenum Press, New York, 1996

23

system will play a key role in the operation of ITER, and it will represent a substantial investment both in terms of hardware and facilities and valuable space in the ports and around the tokamak. It is essential that the design of the system is optimised.

The first step in designing the system is to determine the measurement requirements which must be met. These are obtained from a careful consideration of the role the plasma measurements will play in the operation of ITER. The measurement requirements serve as a target for the design of the individual diagnostic instruments which together make up the diagnostic system.

In this paper we consider three principal topics. First we consider the origin of the measurement requirements. We discuss briefly the planned operational scenarios for ITER, and we consider the measurements required for machine protection and plasma control, for evaluation and optimisation and for understanding key physical phenomena which may limit ITER performance. Second, we consider the detailed measurement requirements. Tables which summarise the parameter ranges in which the measurements are required, and the target measurement resolutions and accuracies are presented. Third, we consider the principal practical difficulties and problem areas which will have to be overcome in the implementation of the diagnostic system.

ORIGIN OF THE MEASUREMENT REQUIREMENTS

The measurement requirements have their origin in the ITER mission objectives. ITER is intended to demonstrate the scientific and technological feasibility of fusion energy for peaceful purposes. In order to do this it is necessary to

(i) demonstrate controlled ignition and extended burn at reactor relevant power levels ($\sim$ 1.5 GW);

(ii) perform an operational programme and provide facilities aimed at understanding and documenting the operational characteristics of an ignited plasma; and

(iii) provide sufficient flexibility to explore steady state operation as an ultimate goal.

It is intended that ITER will provide the physics and technology data bases on which a demonstration fusion reactor (DEMO) can be designed. In brief, the diagnostic requirements for ITER are those necessary to achieve these objectives.

OPERATIONAL SCENARIOS

The reference operational scenario has six main phases: (i) plasma breakdown and current initiation are achieved via a Townsend avalanche with electron cyclotron (EC) ionization and heating assist; (ii) current rampup and plasma cross-section expansion in a limiter configuration followed by divertor configuration formation and attainment of constant current; (iii) plasma heating to ignition or driven-burn conditions; (iv) sustained burn with controlled fusion power level; (v) burn termination and fusion power shutdown; and (vi) plasma current shutdown in a limiter configuration (figure 1). The nominal operating conditions are an inductive flat-top under ignited conditions of 1000s, a fusion power of 1.5 GW, maximum fusion power excursion of ± 20%, plasma current of 21 MA, and pulse repetition time of 2200s. Full control of the plasma involving plasma measurements and real-time signal processing and analysis is required in all phases of the operational scenario.

The design has sufficient flexibility for variants of the reference scenario to support plasma operation in sub-ignited, extended-duration and/or steady-state modes with a lower plasma current as well as non-fusion operation with H, D and He plasmas. 'Advanced' plasma operation modes based on active control of plasma profiles by current drive or other non-inductive means are foreseen in extended ITER operation. Preliminary experiments will be carried out in the basic performance phase of ITER operation ; that is the first 10 years of ITER operation.

REQUIRED MEASUREMENTS

Plasma measurements are required for (i) machine protection and plasma control, (ii) for evaluating and optimising plasma performance, and (iii) for attempting to understand the physical phenomena which limit ITER performance. We consider separately the measurements required in these categories.

Measurements for Machine Protection and Plasma Control (Category (i))

The plasma control requirements are, to a first approximation, the same as those employed successfully in presently operating single-null divertor tokamaks. As in these tokamaks, control is required of the sequencing of the machine operation and plasma scenario, the plasma magnetic equilibrium, and the kinectics of the plasma and the divertor. In addition, since ITER will operate at high levels of fusion power a facility for fast plasma shutdown is required to prevent damage to key machine components in the event of the plasma approaching off-normal conditions [1]. Each aspect of control leads to different measurement requirements.

Machine Operation and Plasma Scenario This involves the PF coil premagnetisation, plasma initiation, current ramp up, divertor formation, auxillary heating, current ramp down and termination. It provides the operational framework within which the subordinate plasma equilibrium, plasma kinetics and fast shutdown systems effect their control actions. It leads to requirements for measurements on the PF system, plasma initiation, plasma current and the auxillary heating systems. Mainly engineering parameters are involved; few plasma measurements are required.

Plasma Equilibrium and Control This encompasses the control and stabilisation of the plasma magnetic configuration - mainly the plasma shape and position including the control of selected plasma-to-first-wall clearance gaps, and the nominal divertor magnetic configuration. The essential requirement is to maintain the plasma suitably positioned so that the kinetic control can be effected. In addition, this control must provide active stabilisation of vertical displacement and disturbances of the plasma which may lead to disruption. It leads to

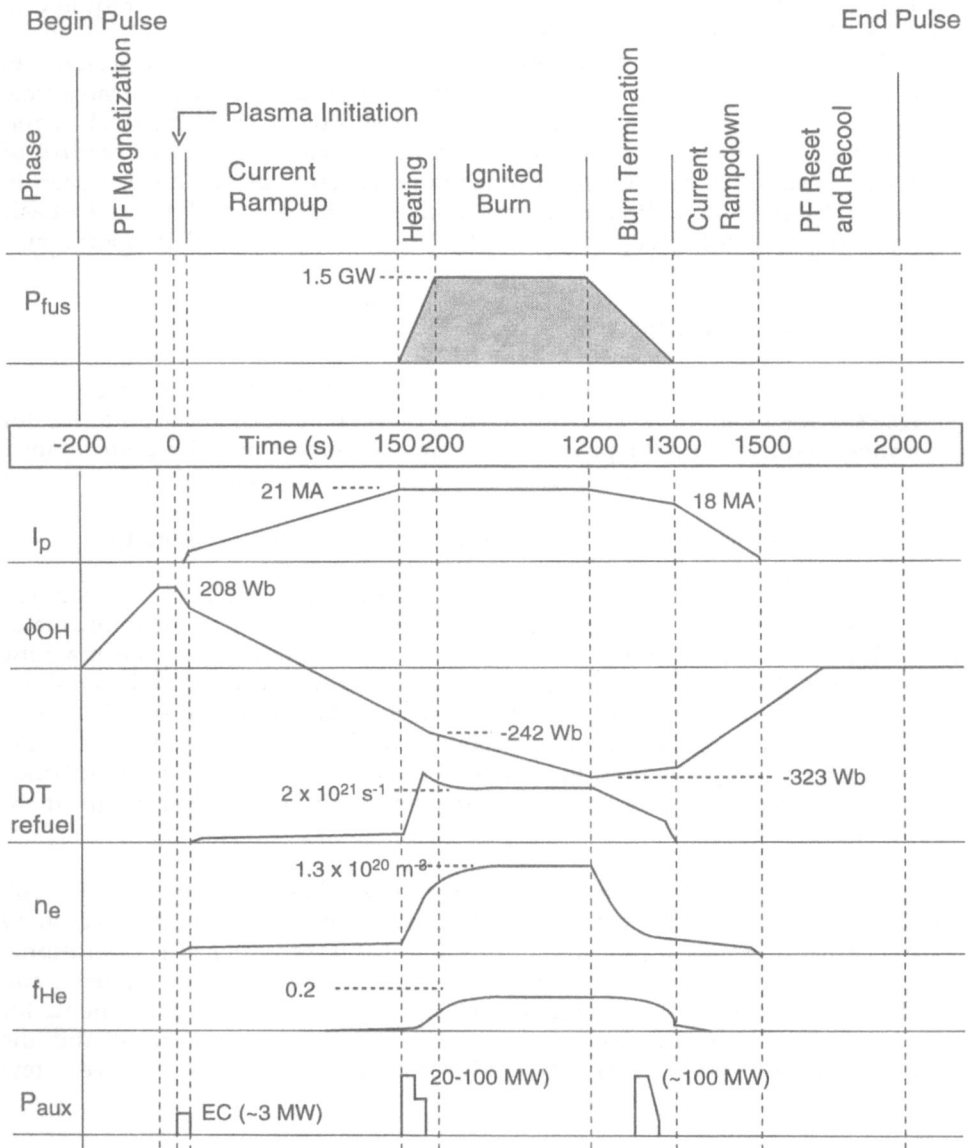

Figure 1. Nominal ITER plasma pulse operation scenario and wave forms.

requirements for measurements of the *plasma current,* and the *plasma shape and position* in both the main chamber and the divertor.

Plasma Kinetics and Divertor Control The kinetics control addresses the problem of controlling the core plasma bulk parameters - especially the plasma density and/or the fusion power - while simultaneously satisfying the various physics constraints and operational limits that effect energy confinement, MHD stability, and flow of power from the core to the plasma boundary and divertor. It leads to requirements for a wide range of plasma measurements including *plasma density, fusion power, impurity content, radiated power, locked modes, MHD (m=2 and sawteeth), ELMs, plasma rotation, fuelling (nT/nD in the plasma core), radiated power in the divertor, and position of the ionisation front in the divertor.* Advanced operational scenarios require core plasma profile control and hence require measurements of the *current profile, and the electron density and temperature profiles.*

Fast Plasma Shutdown This will prevent damage to key machine components which could arise from positive power excursions in situations where the normal burn control is inadequate. The proposed method is to trigger a radiative collapse of the plasma by the injection of an impurity pellet. It leads to requirements for very reliable measurements of the *fusion power,* and key engineering parameters such as the *divertor plate temperature.*

All the plasma measurements and some of the key engineering parameters presently foreseen for machine protection and plasma control during the basic performance phase of ITER operation are listed in the first column of Table 1.

Measurements for Evaluation and Optimisation of Plasma Performance (Category (ii))

During the operational phases of ITER, there will be extensive programmes to evaluate and optimise the plasma performance. Many different topics will be studied and a wide range of plasma measurements will be required. Four principal topics will be plasma energy and particle confinement, optimisation of the fusion burn, alpha particle effects, and optimisation and evaluation of the divertor operation.

Plasma Energy and Particle Confinement. As on present tokamaks, the determination of the particle and energy confinement will be an important undertaking. There will be operation in the different confinement regimes (L-mode, quiet H-mode, ELMY H-mode, Supershot, etc) and studies will be made of energy confinement scaling, H-mode power threshold, particle and momentum confinement, ELMs and sawteeth. In order to carry out these studies measurements of *the electron temperature and density profiles, the ion temperature profile, the radiated power, fuelling rate, speed of plasma rotation, and the impurity density* will be necessary.

Optimisation of the Fusion Burn. Optimisation of the fusion burn will be a new topic addressed for the first time by ITER. It will involve precise control of the plasma kinetics (above) achieved by a combination of measurements of key parameters, processing of the data in real time, and action via actuators such as the

Table 1 Summary of Required Measurements

Category (i): Measurements for Machine Protection and Plasma Control	Category (ii): Measurements for Performance Evaluation and Optimization	Category (iii): Additional Measurements for Physics Understanding
Plasma Current	Electron Temperature Profile†	Pellet Penetration
Plasma Position and Shape	Electron Density Profile†	PFC Material Erosion
Loop Voltage	$q(r)$†	Density Fluctuations
Beta	Zeff Profile†	Edge Turbulence
Radiation	Fishbones, TAE Modes†	MHD Activity in Plasma Core
Plasma Line-Averaged Density	Ion Temperature Profile	T_e Fluctuations
Total Neutron Flux and Emission Profile	He Density in Plasma Core	E field and E fluctuations
Locked-modes		
m=2 MHD Modes, Sawteeth	Escaping Alphas	
Plasma Rotation	n_H/n_D in Plasma Core	
nT/nD	Confined Alpha-Particles	
	Escaping Alphas	
Impurity Species Monitor	Impurity Density	
Zeff (Line-Averaged)	Edge n_T/n_D	
ELMs, L/H Mode Indicators	Neutron Fluence	
Hard X-ray Emission	Impurity & D/T Influx in Divertor with Spatial Resolution	

Key Divertor Parameters:
  Plate Temperature*
  Radiated Power
  Plate Ablation
  Ionization Front Position
  Gas Pressure*
	Ion Saturation Current in Divertor;
	n_e, T_e at Divertor Target	

Edge Light & First Wall Temperature*	Radiated Power (Edge and Divertor Region)	
Base Pressure*	Radiation Profile in Divertor	
Gas Pressure in Duct*	Heat Loading Profile in Divertor	
In-vessel viewing*	Divertor Helium Density	
Halo' Currents*	$n_T/n_D, n_H/n_D$ in Divertor	
	n_e, T_e and T_i of Divertor Plasma	
Toroidal Magnetic Field*	Divertor Plasma Flow	
* Important Engineering Parameters	† Potential Category (i) Measurements	

DT fuelling and the additional heating systems. The principal parameters that have to be measured for the optimisation are *the neutron emission profile, the ion temperature profile, the density of the helium ash, and the fuelling ratio n_T/n_D in the core and edge regions.*

Alpha Particle Effects. In order for high levels of fusion power to be achieved it is essential that a high fraction of the alpha particles is confined. On the other hand, it is possible that confined fast alpha particles can drive kinetic instabilities such as TAE modes which may lead to enhanced alpha particle loss. Optimisation and studies of the confinement of the alpha particles, and any alpha particle driven instabilities, will be an important part of the experimental programme. In order to carry out these studies measurements will be required of *the lost and confined alpha particles (energy spectrum and density), the wall loading, and kinetic instabilities such as fishbones and TAE modes.*

Divertor Optimisation. Optimisation of the operation of the divertor will be one of the most important topics undertaken on ITER. It will involve the control of the magnetic configuration, the kinetics of the divertor plasma (fuelling, exhaust, impurities) and the interaction with the main plasma. It leads to requirements for many measurements in the divertor region including *the plasma shape and position, heat fluxes on the divertor plates, impurity and D,T influxes, radiation profiles, helium density, n_T/n_D ratio, electron density and temperature, ion temperature, and plasma flow in the divertor region.*

The plasma measurements and key engineering parameters required for evaluation and operation are listed in the second column of Table 1. Under advanced operation of ITER, some of these parameters may be used as control parameters. These are marked with a dagger (†).

Measurements for Physics Understanding (Category (iii)).

As in the case of present tokamaks there will be limits to ITER performance. For a given plasma current and magnetic field, there will be a density limit and a beta limit. Under high levels of fusion heating, instabilities may occur which may lead to a loss of particles and energy. In order to try to gain an understanding of the underlying physical phenomena, detailed plasma measurements are required. Measurements are required of *density and temperature fluctuations, the MHD activity in the plasma core, TAE modes, the electric field and fluctuations in the electric field, egde turbulence, pellet penetration etc.* These measurements constitute the third category of measurements and are listed in the third column of Table 1.

All measurements in categories (i) and (ii) are regarded as essentail for plasma operation. As many as possible of category (iii) measurments will be made but the available port space, budget, and the associated complication of the machine will limit the number of these measurements that can be made.

REQUIRED PARAMETER RANGES, TARGET MEASUREMENT RESOLUTIONS AND ACCURACIES

For each measurement parameter there is a range within which the measurements are required. Similarly the intended use of the measurement will determine the required temporal and spatial resolutions and accuracy.

As an example, the plasma current is required to be measured in the range 0.1 - 28 MA. Measurements during the flat top and ramp-up/ramp-down require a rate of change in the range 4×10^1 to 5×10^5 As^{-1}. The study of transients during the flat top, for example just before a disruption, leads to a requirement for a time resolution of 1 ms. The measurement will be used in the determination of the plasma beta and this gives rise to a requirement on the accuracy of 1%. For measurement during disruptions, a rate of change of $< 5 \times 10^9$ As^{-1} and a time resolution of 0.1 ms are required. The measurement would be used for model validation and in this case an accuracy of 30% would probably be sufficient.

As a second example, the measurements of the plasma position and shape must give an accuracy in the determination of the selected gaps of 1 cm. In order to be able to cope with the expected transients the time resolution during the flat-top must be 1 ms. The rate of change (drift in position) is expected to be of order 1 cms^{-1}. During disruptions, the requirement on the rate of change rises to 1 ms^{-1}, the required time resolution is 1 ms and the accuracy is 2 cm.

A systematic study based on the expected use of all the measurements has been carried. The determined parameter ranges, target resolutions and accuracies are shown in Table 2 (begins page 32, following references).

THE PRINCIPAL PRACTICAL DIFFICULTIES

Several factors will make implementation of diagnostics more difficult on ITER than on present machines. The large size of the tokamak - approximately 2.5 times bigger in linear dimension and 20 times bigger in volume than JET - will mean that there are potential problems with thermal expansion, vibration and with supporting the large diagnostic equipment. It will necessitate the use of long transmission paths and time consuming calibration and alignment procedures. The high levels of neutron flux and annual neutron fluence at the first wall - about 10 and 100 times higher respectively than on JET - could lead to damage of the in-vessel diagnostic components. The effects of the relatively high stray magnetic fields (~ 150 mT at the mid-plane of the pit) will have to be mitigated. The requirement for extensive shielding in the ports, and the use of the ports for additional heating systems, other machine systems and blanket testing modules means that the access to the plasma will be very restricted. All diagnostic equipment inside the biological shield and inside the vessel will have to be capable of being maintained by remote handling equipment. Very long pulse (near steady state) operation will mean that in some cases novel measurement approaches will be required, and the requirement for very high reliability will drive the use of more than one technique for the measurement of some of the key parameters and will necessitate some duplication in measurement instrumentation.

KEY PROBLEM AREAS

The combination of the demanding measurement requirements, the harsh environment and the limited access means that it will be difficult to make some of the category (i) and (ii) measurements. Of particular concern are the measurements of the helium ash in the plasma core, the energy and density of the confined alpha particles, the n_T/n_D ratio in the plasma core, and the q(r) profile. In these cases substantial practical difficulties are foreseen in the implementation of the relevant diagnostic techniques. Viewing the high heat flux components during the pulse, and obtaining two dimensional measurements in the divertor region, also appear to be very difficult. The resolution of these difficulties is an important part of the work on diagnostics being undertaken during the EDA.

CONCLUSIONS

In order to meet the operational requirements for control and evaluation of ITER, it will be necessary to measure a substantial number of plasma parameters and key engineering parameters (~ 50). Many of the measurements will be used for machine protection and plasma control and so will have to be made at a very high level of reliability and availability. The detailed measurement requirements - determined from a careful consideration of the role each measurement will play - represent a demanding measurement task. The measurements will be made with an extensive diagnostic system. Implementation of the diagnostic system requires the solution of some difficult practical problems arising mainly from the harsh environment and restricted access.

We wish to encourage the widest possible involvement in the development of the design of ITER diagnostics. If any reader has specific comments or suggestions please forward them to any of the authors of this paper.

REFERENCES

[1] ITER Interim Design Report, Chap. 5, published by ITER -EDA, July 1995. See also *'Plasma Control Requirements and Concepts for ITER'* by J Wesley et al to be published in Fusion Technology.

Table 2. Parameter Ranges, Target Measurement Resolutions and Accuracies.
a) Measurements for Machine Protection and Plasma Control

1) Plasma Current

Parameter	Parameter range	Time res.	Rate of change	Accuracy
Plasma current	0.1-28 MA	1ms	$40\text{-}5\text{x}10^5$ As^{-1}	1% (for $I_p > 1$MA)
		0.1ms	$<5 \times 10^9$ As^{-1} †	~30%

† *At current quench at a disruption*

2) Plasma Position and Shape

Parameter	Time res.	Rate of change	Accuracy*
$\Delta_{in}, \Delta_{out}, \Delta_{top}$,	1 ms	1 cm s^{-1}	1 cm
	1 ms	1 m s^{-1}†	2 cm
Divertor channel position (r-direction)	1 ms	1 cm s^{-1}	1 cm
	1 ms	1 m s^{-1}†	2 cm

Δ is a width of gap between separatrix (or layer with n_e=TBD) and first wall
  † *At current quench at a disruption*
  * *For full bore plasma*

3) Loop Voltage

Voltage range	Numb. of Loops	Time res.	Rate of change	Accuracy
0.01 - 30 V	TBD	1 ms	5 V s^{-1}	0.005 V
< 500 V †	TBD	1 ms	5 kV s^{-1}	10%

† *At current quench at a disruption*

4) Plasma Beta

Beta range	Time resolution	Rate of change	Accuracy
$.01 < \beta_p < 3$	1 ms	0.05 s^{-1}	5% at β_p~1
	1 ms	$< 1 \times 10^5$ s^{-1} †	~30%

† *At thermal quench of a disruption.*

5) Total Radiated Power

	Power Range	Time res.	Accuracy
Main plasma	≤ 0.6 GW	10 ms	10%
	<100 GW †	3 ms	20%
X-point/MARFE	≤ 0.6 GW	10 ms	10%
Divertor	≤ 0.6 GW	10 ms	10%

† *At current quench at a disruption.*

6) Line-Averaged Electron Density

Density range	Time res.	Accuracy
$10^{18}\text{-}2\text{x}10^{20}$m^{-3}	1 ms	1%

7) Total Neutron Flux and Emission Profile

Parameter	Parameter range	Spatial res	Time res.	Accuracy
Total neutron flux	10^{14}-10^{21} n s^{-1}	integral	1ms	10%
Neutr./α source	10^{14}-4×10^{18}s^{-1}m^{-3}	30 cm	1ms	10%
Fusion power	$\leq$2 GW	integral	1ms	10%
Fusion power density	$\leq$10 MWm^{-3}	30 cm	1ms	10%

8) Locked Modes

Parameter	Parameter range	Mode Number	Time res.	Accuracy
$\tilde{B}_r / B_p$	10^{-4}-10^{-2}	m/n=2/1	1ms	30%

9) m=2 MHD Mode, Sawteeth

	Parameter	Frequency	Spatial res.	Accuracy
m=2 Mode	$\tilde{B}_r / B_p$	0-3kHz	-	10%
Sawteeth	$\tilde{T}_e / T_e$	0-3kHz	10 cm	10%

10) Plasma Rotation

Parameter	Parameter range	Spatial res.	Time res.	Accuracy
$v_{toroidal}$	1-50 km/s	50 cm	10ms	30%

11) n_T/n_D Ratio in Plasma Core

Parameter	Parameter range	Spatial res.	Time res.	Accuracy
n_T/n_D	0.1-3	30 cm	100 ms	20%

12) Impurity Species Monitor

Parameter	Parameter range	Spatial res.	Time res.	Accuracy
Be influx	TBD	Several points	10 ms	10%(rel.)
C influx	TBD	TBD	10 ms	10%(rel.)
Cu influx	TBD	Several points	10 ms	10%(rel.)
Ne influx	TBD	TBD	10 ms	10%(rel.)

13) Zeff (Line-Averaged)

Zeff range	Time res.		Accuracy
1 - 3	10ms		20%

14) ELMs

Parameter	Time res.
D_α bursts	0.1 ms
n_e fluctuations	0.1 ms

15) Runaway Electrons

Parameter	Parameter range	Time res.	Accuracy
Max. energy	1-500 MeV	10 ms	20%
Runaway current	TBD	10 ms	30% rel.

16) Key Divertor Parameters

Parameter	Parameter range	Spatial res.	Time res.	Accuracy
Max. surface plate temp.	200-2500°C	1 cm	1 ms	10%
Rad. power	≤ 0.4 GW	10 cm	1 ms	10%
	$\leq 10^3$ GW †	10 cm	0.1 ms	30%
Plate Ablation	To be defined			
Gas pressure	≤ 0.1 Torr	Several points	5 ms	10%
'Ionization front' position	0-2 m	~10 cm	1 ms	10%

† At thermal quench at a disruption.

17) Edge Light & First Wall Temperature

Parameter	Parameter range	Spatial res.	Time res.	Accuracy
Edge light	-	10 cm	100 ms	-
Surf. temper.	200-1500°C	-	10 ms	20°C

18) Base Pressure and Residual Gas Analysis

Parameter	Parameter range	Time res.	Accuracy
Base pressure	10^{-3} - 10^{-9} Torr	1 s	20%
Gas composition	A=1-100; ΔA$\leq$0.5	10 s	50%

19) Gas Pressure and RGA in Duct

Parameter range	Time res.	Accuracy
$\leq$5 x 10^{-3} Torr	100 ms	20%

20) In-Vessel Inspection
(will be designed by the Remote Handling Unit)

	Inspected surface	Spatial resolution
First wall	100%	1 mm
Divertor	100%	1 mm

21) Halo Currents

	Current range	Time res.	Accuracy
Poloidal current (many (TBD) points in tor. and poloidal direction)	≤ 0.5 x I_p	1 ms	20%

22) Toroidal Magnetic Field

Parameter	Parameter range	Time res.	Rate of change	Accuracy
B_T	3-6T	1s	steady state	0.1%

Table 2b: Measurements for Performance Evaluation and Optimisation

23) Electron Temperature Profile

	Temp. range	Spatial res.	Time res.	Accuracy
Core	0.5-30 keV	10 cm	10 ms	10%
Edge	0.05-10keV	0.5 cm	10 ms	10%

24) Electron Density Profile

	Density range	Spatial res.	Time res.	Accuracy
Core	$(0.3\text{-}3)10^{20}\text{m}^{-3}$	30 cm	10 ms	5%
Edge	$(0.05\text{-}3)10^{20}\text{m}^{-3}$	0.5 cm	10 ms	5%

25) q(r)

Range of q	Spatial resolution	Time resolution	Accuracy
0.5-5	30 cm	10 ms	10%

26) Z_{eff} Profile

	Z_{eff} range	Spatial res.	Time res.	Accuracy
Steady state	1 - 3	30 cm	100 ms	10%
Transients		30 cm	0.1 ms	20%

27) Fishbones, TAE Modes

Type of perturbation	Frequency range	Mode numbers
Fishbones ($\tilde{B}_p/B_p, \tilde{T}_e/T_e, \tilde{n}_e/n_e$)	3 - 30 kHz	m/n=1/1
TAE modes ($\tilde{B}_p/B_p, \tilde{T}_e/T_e, \tilde{n}_e/n_e$)	30 - 500 kHz	n=1-20

28) Ion Temperature Profile

Temp. range	Spatial resolution	Time resolution	Accuracy
0.5-50 keV	30 cm	100 ms	10%

29) Helium Density in Plasma Core

n_{He}/n_e range	Spatial resolution	Time resolution	Accuracy
1-20%	30 cm	100 ms	10%

30) Confined Alphas

Parameter	Parameter range	Spatial res.	Time res.	Accuracy
Energy spectrum	0.1-3.5 MeV	30 cm	100 ms	20%
Density profile	$(0.1-2)10^{17}$ m^{-3}	30 cm	100 ms	20%

31) Escaping Alphas

	Flux on first wall	Spatial res.	Time res.	Accuracy
Steady state	≤ 2 MW/m^2	30 cm	100 ms	10%
Transients	≤ 20 MW/m^2	-	10 ms	30%

32) Impurity Density Profile

Impurity	Content range	Spatial res.	Time res.	Accuracy
Z$\leq$8	0.5-20%	30 cm	100 ms	20%
Z>8	0.01-0.3%	30 cm	100 ms	20%

33) Edge n_T/n_D, n_H/n_D

Parameter	Parameter range	Spatial res.	Time res.	Accuracy
n_T/n_D	0.1-10	-	100 ms	20%
n_H/n_D	0.01-0.1	-	100 ms	20%

34) Long-Term Neutron Fluence

	Fluence range	Spatial res.	Time res.	Accuracy
First wall	0 - 3 MWy/m^2	~10 locations	10s	10%

35) Impurity and D,T Influx in Divertor

Parameter	Parameter range	Spatial res.	Time res.	Accuracy
Γ_{Be}, Γ_W	10^{17} - 10^{22} at/sec	3mm	1ms	30%
Γ_D, Γ_T	10^{19} - 10^{25} at/sec	3mm	1ms	30%

36) Ion Saturation Current, n_e and T_e at Divertor Target

Parameter	Parameter range	Spatial res.	Time res.	Accuracy
Γ_D	10^{19} - 10^{25} ion/sec	3mm	1ms	30%
n_e	10^{19} - 10^{22} m^{-3}	3mm	1ms	30%
T_e	1 eV - 1 keV	3mm	1ms	30%

37) Radiation Profile

Region	Rad. power range	Spatial res.	Time res.	Accuracy
Main plasma	0.01-1 MWm^{-3}	20cm	10ms	20%
X-point/MARFE	$\leq$300 MWm^{-3}	20cm	10ms	20%
Divertor	$\leq$100 MWm^{-3}	5cm	10ms	30%

38) Heat Loading Profile in Divertor

Parameter	Parameter range	Spatial res.	Time res.	Accuracy
Surface temp.	200 - 2500 $^\circ$C	3mm	2ms	10%
Power loading (calc. from temp.)	$\leq$20 MWm^{-2}	3mm	2ms	10%

39) Divertor Helium Density

Parameter	Parameter range	Spatial res.	Time res.	Accuracy
n_{He}	10^{17}-10^{20} m^{-3}	-	1ms	20%

40) n_T/n_D, n_H/n_D in Divertor

Parameter	Parameter range	Spatial res.	Time res.	Accuracy
n_T/n_D	0.1-10	-	100 ms	20%
n_H/n_D	0.01-0.1	-	100 ms	20%

41) Electron Density and Temperature in Divertor

Parameter	Parameter range	Spatial res.	Time res.	Accuracy
n_e	10^{19}-10^{22}m^{-3}	10 cm (along legs) 0.3 cm (across legs)	1 ms	20%
T_e	1-200 eV	10 cm (along legs) 0.3 cm (across legs)	1 ms	20%

42) Ion Temperature in Divertor

Parameter	Parameter range	Spatial res.	Time res.	Accuracy
T_i	1-200 eV	10 cm (along legs) 0.3 cm (across legs)	1 ms	20%

43) Divertor Plasma Flow

Parameter	Parameter range	Spatial res.	Time res.	Accuracy
v_p	TBD	10 cm (along legs) 0.3 cm (across legs)	1 ms	20%

44) n_H/n_D Ratio in Plasma Core

Parameter	Parameter range	Spatial res.	Time res.	Accuracy
n_H/n_D	0.01-0.1	30 cm	100 ms	20%

ITER PLASMA DIAGNOSTICS GENERIC ACCESS

C.I. Walker, L. de Kock

ITER JCT, Garching, Germany

INTRODUCTION

The effort on generic access for diagnostics to date has addressed the port requirements and types of solutions feasible and required to be incorporated in the conceptual designs of diagnostics. These must be further refined and detailed in each diagnostic area and more importantly incorporated in the design of the other, non-diagnostic ITER systems. The approach shall still be the same, to look at the design of specific diagnostic systems or of the component solution for specific systems and then to broaden the design to include other similar systems. The ITER diagnostic design problems originate from radiation shielding and damage; removal of nuclear heating; increased reliability and safety related to the tritium confinement barriers.

In the first part of this paper we have looked in more detail at the types of access, optical, microwave, and electrical, and proceeded to assembly and vacuum extensions access considerations. In the second part particular ports and sites for diagnostic equipment are considered.

1. DIAGNOSTIC ACCESS

1.1 Optical Access Routes

When considering the optical generic access; an early subdivision is into the categories i) Narrow beams, ii) Imaging systems (up to 10^0 divergence) and iii) Light Collecting systems. Some or all of the following elements may be required in each type. Optical elements are described in the optical progression from plasma to Cryostat window as shown in Figure 1.

Shutter
Mechanical actuation is required, near the very front of a port plug. The shutter will be designed as a modular unit within port plug with simple plug-in electrical and mechanical actuation connectors. Hydraulic and pneumatic connection require rather more special connections for which a suitable generic solution must

Diagnostics for Experimental Thermonuclear Fusion Reactors
Edited by P. E. Stott *et al.*, Plenum Press, New York, 1996

39

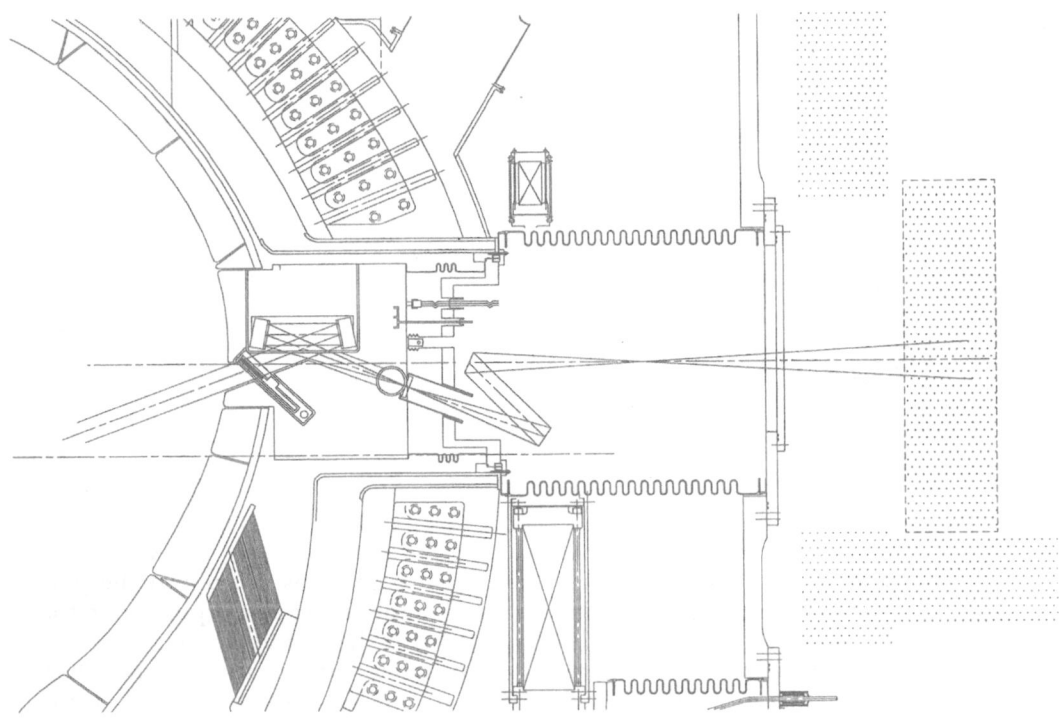

Figure 1. Typical Arrangement of Optical Access (imaging) in Mid-Port Plug

still be proposed. Standard solutions for vacuum bearings or flexipivots can be used but tests under irradiation will be required. For the prime mover standard solutions are available but reliability is paramount, failure must be open. For Linear actuation avoid long strokes through the vacuum boundary. For Rotary motion feed through, existing standard solutions will require high level of qualification and proving. It should be possible to make use of the perpetual presence of B_T with electromagnetic actuation requiring only an electrical current (a few Amp) through the vacuum boundary. Rods and wires can be used if necessary, for which vacuum lubrication must be assured.

Status information (open/shut) is required as a fail safe closed contact.

A shutter consumes twice the optical opening area to plasma and intrudes into the plug cooling design. It is desirable to minimise its volume and thus nuclear heating, although normally open it will not see the plasma.

The first detail design for a system (say LIDAR) will expose and solve the generic access problems for others.

For an <u>Aperture,</u> where actuation is required (e.g. for neutron cameras), these can be treated as a shutter with the possible extra requirement of positional instrumentation. They are permanently in line to plasma and must, therefore, be reflective on plasma facing surface and with enhanced radiative heat transfer to cooled plug on non plasma facing surfaces.

First Mirror

This must be fluid cooled, to remove nuclear heating and equilibrate temperature for optical stability. Use water at 6 bar if vented during bake-out, or Helium. This will be of a metallic substrate, with copper as a preliminary reference choice.

(High power industrial laser developments using OFHC Cu or W with internal labyrinth cooling are directly applicable). The choice of material for a reflective surface layer and tests of longevity are required. Beryllium (particularly foamed) should be investigated.

Mirror Mount

This is required to carry the first and second mirrors in one frame of reference and will be the basis of a "sub-plug". It will be fluid cooled with first and second mirrors for optical stability. Thermal equilibration will be more important than heat removal. The mirrors will be quasi-kinematically attached to the Mount. The decision whether the mirror alignment adjusters are fixed or motorised will have important consequences on the early design.

Second Mirror and other Reflective Optics

Radiation may be 10^{-3} of that on the first mirror, but cooling with the first mirror and mount will guarantee optical stability. Mirror adjustment will be as for the first mirror.

Refractive and Fibre Optics

The radial position from which refractive optics can be used is to be defined and materials chosen. Optical fibre routes require definition with the typical bundle size and typical bend radius. The need for fibre annealing or cooling must be considered as this affects size and radius. A typical, generic, feed through design is required. The acceptability of Epoxy seals in metal feed through flange has to be decided. A connection philosophy is required (e.g. Lemo, automatic or handled) and the design of an automatic connector.

Vacuum Vessel Window Assembly

The principle of a <u>double boundary</u> will be universally applied to 'glass' windows. The innermost window element will normally be between the vacuum vessel vacuum and one of the cryostat volumes. This vacuum vessel window will be considered to be the primary tritium boundary. Safety aspects for 5 bar fault over pressure have to be defined. There will be a second window element between the Cryostat volume and the Pit air. Where possible this double boundary shall be by two distinct window assembly elements either vacuum sealed together or individually sealed to the vacuum vessel boundary and the cryostat boundary.

The <u>Window Material</u> for optically transparent windows must be specified first for transmission properties. Acceptable materials will probably include: Fused Silica, Crystal Quartz, Sapphire, Zinc Selenide, (others candidates should be identified now). Stainless Steel 316 for neutrons, Beryllium and Mylar for X-rays should be dealt with separately along with vacuum extensions, they raise different safety issues. An acceptable Aspect Ratio (diameter/thickness) should be redefined given ITER safety requirements, for the present it will be reasonable to assume that the conventional 10:1 is still sufficiently conservative. For large size windows (>~175mm diameter) a source of material and bond must be established and qualified.

For the <u>Ferrule & Bond</u> pure metal diffusion bonds are likely to be acceptable with the correct specification and proof testing. Alloy materials of solder (e.g. AgPbSn) will need to be assessed, but out at the radius of the PF Coils transmutation problems should not prevail. Ferrule geometry at the bond region should be defined for the bonding process only, once established this will remain constant for all window applications. A flexible tubular region must be incorporated as near as possible to the ferrule to isolate the window and bond from vertical displacement event (VDE) forces (an estimate of their magnitude at main port positions is required) and asymmetric thermal loading.

Attachment to structural component, window tube on the port seal plate for example, will be by one of the set of techniques standard to ITER (RH edge weld, bolted 'O' ring seal, etc.) dependant on reliability and required replacement frequency of the particular window. At the window tube joint the window assembly must be made of stainless steel 316L(N?) and the window assembly will, have some permanent handling features (to be defined in conjunction with RH Group).

A <u>Window Development Programme</u> must be established as soon as relevant window designs are proposed and normal and fault conditions are defined. There is no opportunity to develop windows in service. Full life testing and failure mode testing of full quality windows is required before incorporation on the Torus. Testing will be independent of the manufacturer.

Alignment Optics, between Vacuum Vessel and Cryostat

Radial and vertical relative movement between the vacuum vessel flange and the cryostat flanges is of the order of 60mm between operating temperature and temperature during assembly. During operation movements are not expected to exceed ±5mm. Alignment optics must accommodate these movements and mechanical support must be provided for this. Typical optical pointing accuracy required is ~ <1mrad over ~100mrad of movement. Intrinsically tolerant designs should be considered, where alignment achieved during build can be relied on in the operating state and alignment is independent of vibration of the mechanical support. Vacuum compatible bearings or pivots are required. Intrinsic, passive damping must be considered. Relative movement between the cryostat flange and the pit, earth, diagnostic hall equipment etc., is also required to be accommodated.

Cryostat Window

Applying the same constraints and designs to the cryostat windows as to the vacuum vessel windows, will give greater safety factors here.

1.2 Microwave guide Routes Windows

Two main categories are considered: i) Oversized wave guide, rectangular or corrugated circular and ii) Gaussian beam transmission systems. Each access route is required to be broadband to allow multiple use except where only a specific application is foreseen.

Receiving and transmitting antennae will be mainly determined by the available space. For applications with high demands on antenna pattern a trade off with space has to be made.

All microwave access routes can almost naturally incorporate a labyrinth for radiation shielding; no radiation damage problems are foreseen although cooling of nuclear heat will be required.

Calibration facilities are required for both the reflectometer and ECE systems in the form of shutters, return paths, in- and ex-vessel sources.

Both in the port plug and in the cryostat volume there is the need for a remote joint in wave guide. This can be considered to be a development of the electrical connection without the spring contact requirement but with enhanced assembly accuracy.

Windows are required in the vacuum vessel and cryostat tritium confinement barriers. In-wave guide windows will have potential advantages of compactness and reliability but may be narrow band.

Multiple wave guide bundles from cryostat, labyrinthed through the biological shield, to diagnostic hall need to be defined. Figure 2 shows the environment for the routing of signals, micro wave or others, from the torus to diagnostic hall.

Application of ECRH may lead to unwanted pick up and will need blocking provisions.

The main access routes are in the mid plane port, in the divertor cassette and in the blanket structure.

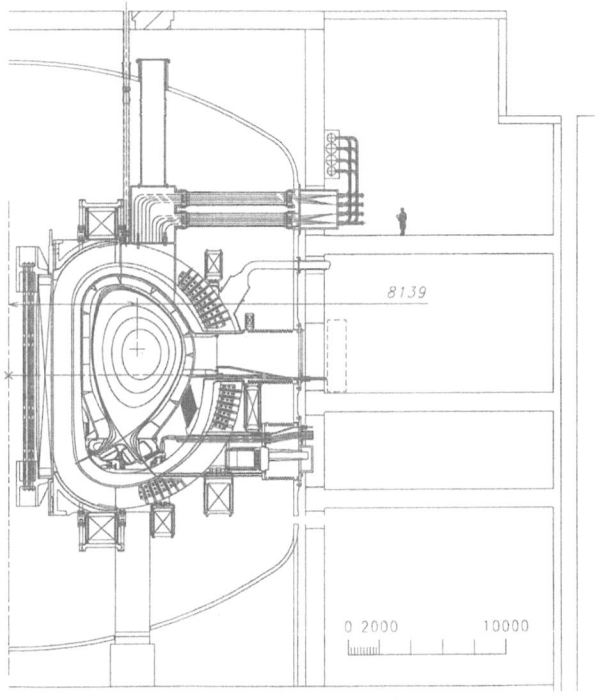

Figure 2. Vertical Cross Section of ITER showing the torus, the cryostat, the biological shield and the pit.

1.3 Neutron Diagnostic Access

Multi-channel 2D cameras require extensive coverage of the plasma cross-section resulting in channels with extreme elevations and some of the neutron spectrometers require a tangential access. All viewing chords are straight channels crossing the vacuum vessel and cryostat boundaries through thin sections in the wall < 5mm.

All cameras and spectrometers require well shielded collimators and detectors leading to massive shielding blocks occupying space in the port, in- and outside the cryostat and the biological shield.

1.4   Electrical Signals Access

Diagnostic Signals

Generally twisted pairs, aiming at a pitch of 10x diameter, correctly aligned to the poloidal field at untwisted sections such as in the connector. There will be a limited set of in-vessel wire types, including twisted pair (Copper), twisted pair (Thermocouple) and coaxially screened (screen not earthed). Requirements must be specified now for conductivity, inductance, capacitance, core diameter, material, screening, twisting, microphony etc. so the list of wire types can be drafted. Control Signals will be similar to diagnostic signals of the simplest, twisted pair or single wire form. Signals should be closed contact where possible. Power Signals, with high current requirements, such as those for pressure gauges & calibration sources, must take into account the magnetic load on wires and the fact that ohmic losses are difficult to dissipate in vacuo.

Mineral insulated wires only will be considered. These must be vacuum sealed, even where they do not cross vacuum boundaries, to minimise the outgassing and to guarantee the integrity while in store prior to installation. The wire sealing component will be identifiably different from the joining component, whether in a permanent or a removable connection.

Wiring will be terminated in multiway connectors and loomed to enable simple installation within conduit channels in sub-plugs and modules.

Permanent and removable connections must include an adequate amount of elastic pretension, Permanent joints will be burnt-in (fritted) and spot welded where necessary and must be proven at low signal levels before acceptance. Care must be over the design at metal transitions.

Multiple way Removable Connectors are required in standard, modular formats. Self welding of contacts must be sufficient for no signal interruption, not enough to prevent RH disassembly. Pins must be self aligning in plugs and plugs must be self aligning in larger assemblies even when wired up. Plugs will be provided with a standard set of plug condition connections (plug inserted fully, plug disconnected fully, reference temperature, continuity check, calibration flux measurement). The components must be damage-proof when disconnected and open and débris tolerant with a wiping contact action on insertion. There will be standard sized modular units within the variously sized plugs and sockets. Each unit will have the equivalent area of approximately 50 signal cores, or for other electrical, fibre optic (requiring development), microwaveguide (higher precision). A vacuum tight fluid service connector is unlikely to fit this simple insertion action requirement, but would be of significant use.

There are of two generic types of Removable Connector:

i) Those required to be connected in-vessel. To minimise installation access time these must connect automatically where possible as a by-product of installing the main component (Automatic Connector). For these a Torus compatible edge connector can be investigated. The connectors are mounted flexibly within the main component, picking up their own guide during installation.

ii) Those required to be connected within some external module, e.g. in port plug. Drive-in / drive-out capability within connector, i.e. no special RH end effectors. These must still be able to be handled remotely (Handled Connector).

Electrical Feedthroughs will be high density arrays of mineral insulated cables individually vacuum sealed at either end, all vacuum brazed into metal ferrules welded into the port plug sealing plate and cryostat sealing plate.

1.5 Assembly & Installation Access

The access for *built-in* diagnostic equipment, such as wiring and connectors for blanket modules is a problem of spatial interface within the other permanent machine features. The integrity and available redundancy of these features will be high as there is no planned scope for replacement

For removable equipment, one of the most fundamental of the generic access aspects is the <u>installation and removal route.</u> In figure 3 the space around the torus outside the cryostat is shown. Removal from this area is through a toroidal transport of the equipment to the lift that bring it to the level of the hot cell. The design of this access is dependant on whether the diagnostic, or the larger module (e.g. Divertor Cassette) in which it is assembled, is required to be regularly maintained, replaced with new components, salvaged or abandoned after disfunction. As a starting point the diagnostic equipment should be specified to have a lower access requirement than the component in which it is installed. Thus equipment in the blanket modules will be replaced with the new breeding blanket module, equipment on the Divertor Cassette will be regularly removed for maintenance in the remote handling Hot Cell, but duplicate replacements will be already installed in spare cassettes. There will be standard solutions to installation routes to each access point. Again as a starting point for design the diagnostic equipment will be required to be transferred by these routes in no way unusually from other Torus components.

Diagnostic modules are of two types:

i) Those required to be installed directly in-vessel. These are almost invariably going to be of the in-built type, not requiring in-vessel maintenance time or equipment. Equipment that does require maintenance access must be designed to minimise installation access time and the design must include the necessary installation tooling.

ii) Those required to be connected within some external module, e.g. in a port plug. It is possible for these to involve more remote handling operations, with simple, less dedicated tools used within the hot cells.

1.6 Vacuum Extensions

The diagnostics that require vacuum from plasma to detector without intervening windows, or with windows of inadequate strength for the operating or fault pressure conditions will require vacuum extensions, typical examples are the NPA and VUV Spectrometers. These have to be enumerated.

A systematic approach to the requirements of the vacuum systems will be adopted, pioneered by the design of the first typical system. The systematic nature of the solution indicates the advisability of a diagnostic pumping crown, provided the diagnostic equipment inputs are properly controlled. The torus vacuum pumping and tritium handling equipment has then only one source of diagnostic exhaust irrespective of the appendant diagnostics. The actual position of the vacuum extensions may be rather dispersed, so the trade off will be the accommodation of extensive pipework in non diagnostic areas.

The philosophy of isolation valves, fast shutters, bakeout, pressure gauging, valve control, roughing, venting and purging must all be addressed with the respective systems of ITER, and the safety case made.

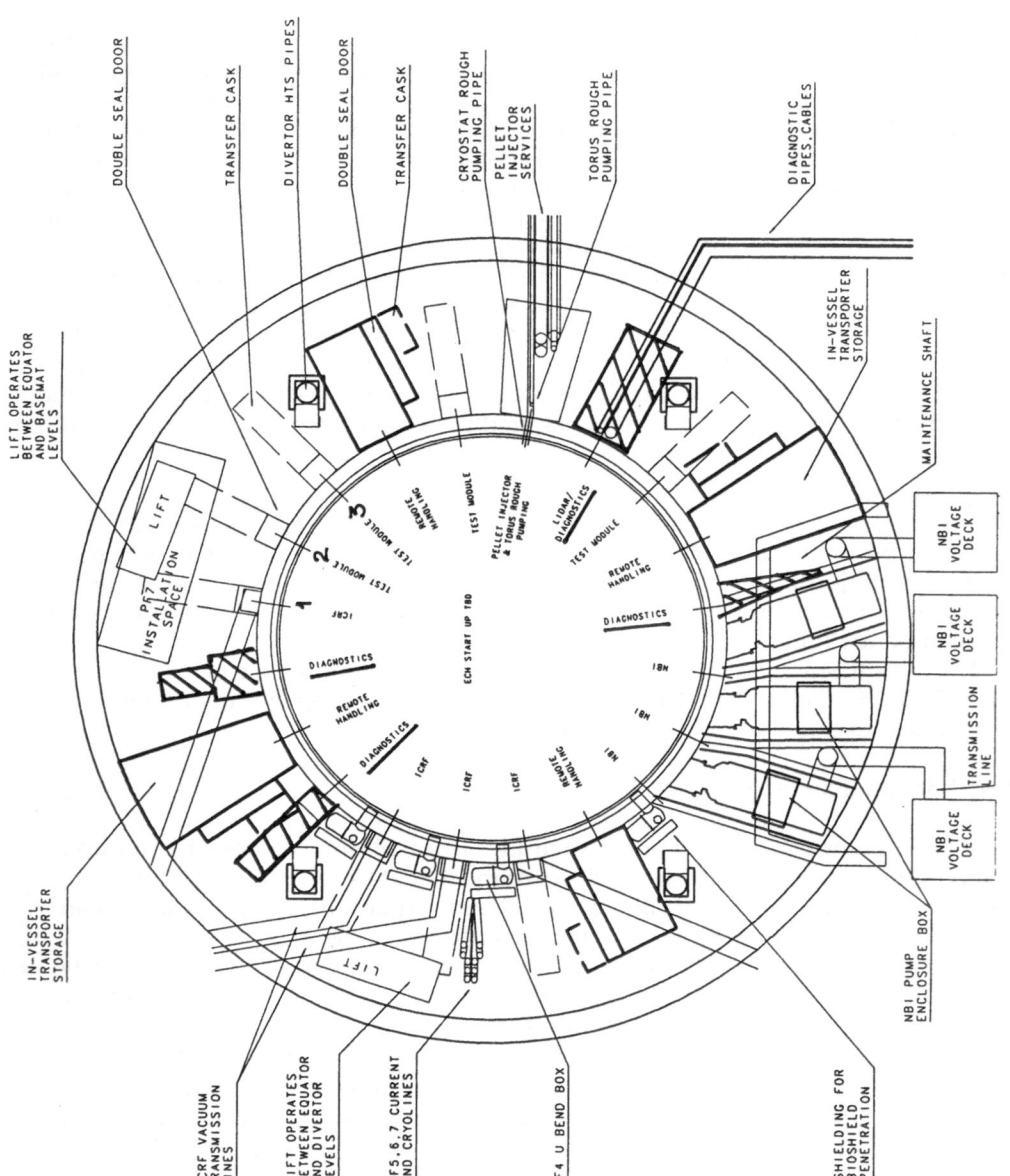

Figure 3. Plan view of Equipment Distributed in the Pit

2. ITER PORTS

2.1. Mid-Plane Port

Any Mid-Plane Port Plug will be handled as one whole non-special unit. Those Plugs with diagnostic equipment in them will be prepared in advance invisible to the installation and removal operations. The plug will be provided with a method of translating radially ~7m and then locking to a precise position defined on and by the Blanket Back Plate. The vacuum seal will be made with an ITER standard lip weld and mechanical clamp at the vacuum vessel flange 2.7m out from the Back Plate. Thermal flexibility is thus required between the front and the back of the Port Plug.

Depending on the choice of the combination of addional heating methods 4 or 5 mid-plane ports will be available for diagnostics (see Figure 3).

Sub-Plug modules will contain the diagnostic equipment, pre-assembled, wired, commissioned etc. These modules are inserted in a simple manner in a routine way. The design will aim to limit the number of routes: vertically downwards (normally preferred), horizontally to the right. Vertically up will not be used as this will conflict with the provision of a railway system for handling. The design of Port Plug components will be modified to suit the requirements of the diagnostic equipment, while preserving their original function.

Plug Shielding is normally water cooled and the cooling requirements of sub-plug components must be assessed on an individual basis. The first non-shield plug component will be >0.5m from the plasma facing surface The physical design solutions to the cooling requirements will be from a standard set. Where sub plugs are sufficiently small (possibly <50mm dia) they may rely on passive cooling to the Plug shielding. If somewhat larger (say <150mm diameter) then surface cooling of the sub plug will be used. Larger components will require internal cooling provision. Cooling will be with either water at 6 bar (which will require venting and imply no diagnostic cooling during bake-out of the vessel at 350°C) or Helium which could be left on to cool diagnostic equipment during bake-out.

The Port Plug Seal Plate will have some common fixed features such as the mechanical arrangement for attaching it to a railway carriage, and independent vertical jacks for the shield plug end and the seal plate. These will be designed identically on each port plug. There will also be a number of common elements that will be arranged on the seal plate to suit the space available, remote handling requirements etc. These include Feedthroughs for electrics, hydraulics, pneumatics, mechanical actuation, optical sight lines (windows), microwave guides, etc..

Near its outer surface the port plug will incorporate a number of Service Ducts into which the looms of wires and pipes with their end connectors are laid (without topological links or knots), fastened and protected. Internally the port plug will carry Mechanical Support and Service Connections for the various components, sub-plugs etc.

2.2. Vertical Port

The requirements for vertical access originate from the neutron profile camera, optical observations and neutral particle analysers. The neutron camera requires a slot over the full radial width of the port to be partly filled with a pre-shield of minimum width of 100mm with a collimator reaching to outside the biological shield via a 5mm thick SS "window". The neutral particle analyser requires a number of unobstructed straight extensions of the vacuum of 200mm diameter to outside the biological shield. Optical access can be obtained from a number of labyrinth channels of 100mm diameter and by telescope structures requiring up to 400mm diameter. All have classical windows.

The space requirements and the interference with cooling manifolds, the blanket modules, etc. are under study.

2.3. Divertor Port

The concept for diagnostic access to the divertor has been developed as part of the design of the divertor cassette. The remote maintenance is through 4 toroidally equidistant ports and the last/first cassette to be moved is the diagnostic cassette. This cassette has special channels and provisions for optical and microwave diagnostics. The two neighbouring cassettes are used for electrical diagnostics. The space under the cassette will be used for the optical signal transmission originating from the target (IR) or the divertor plasma (visible). Microwave access uses wave guide embedded in the cassette body. The signals of the diagnostic cassette are routed to a large shielding block in the port where due to the low radiation levels transitions to optical fibre or even detection will be possible reducing the number of vacuum windows. The block may also contain a VUV spectrometer.

3. BLANKET

Most Blanket Modules are installed by in-vessel handling. The module are bolted to the back plate. The water cooling pipes have to be welded (internally). Diagnostic equipment embedded in a blanket module will preferably be fitted with an automatic connector which will engage a female connector behind the module. Slots for neutron cameras required in the middle of the vertical post modules must be considered equipment in that they require derogation of a standard blanket module albeit with no additional materiel. The back plate will carry the services in conduits and the requirements for this and any perforation of the plate must be specified soon.

A Handled Connector can be envisaged perhaps on the module edge provided the handling required is equivalent to one of the module installation bolting activities. Diagnostic equipment (e.g. bolometer pinholes) can be considered in the vertical and horizontal gaps between the blanket modules. For the latter case there is no competition with the installation bolts and handling space requirements

There is a desirable candidate space for diagnostics using the filler gap between blanket modules. These spaces run toroidally round the machine and will be provided with electrical sockets. Some magnetic coils and bolometers have been identified already for this space connected by an Automatic electrical Connector. These systems will have no active cooling system an so must be reflective towards the plasma and emissive towards the cooled blanket module. Conduited wires are required on and through the Blanket Back Plate.

Blanket Module 8 is distinct in that, being the module that sit directly under the vertical port, it is the one type of module that is installed pre-connected to its cooling as a port plug. This might lend itself to the design of particular diagnostic equipment where connections are to be avoided for any reason or if a connection further out is advantageous.

4. BIOLOGICAL SHIELD EXTENSION

The method of support must be defined and the means of removal, retraction or opening. Design solutions proposed for more advanced systems will be plagiarised. Location must be provided for the remote handling contamination containment flask. Use will be made of the common transporting equipment (railway) round the pit and to the RH cells.

5. SUMMARY

Starting from diagnostic and machine requirements a number of feasible solutions to the diagnostic access problem in ITER have been proposed. In Figure 1 an example is given for optical measurements.

The conceptual stage is under active study for all diagnostic systems, in particular addressing the extent that diagnostic access affects the design of the machine components.

The problems of the detailed design are indicated in this paper as far as can be foreseen to date. The detailed design of the diagnostic systems is the challenging task for the remainder of the ITER Engineering Design Activity.

RADIATION HARDENING OF DIAGNOSTIC COMPONENTS

D.V. Orlinski

NFI RRC Kurchatov Institute, Moscow, RF

1. INTRODUCTION

The peculiarity of the diagnostic system in fusion reactor in particular consist in rather hard radiation environment for some diagnostic components located near the plasma. Intense fluxes of neutron and gamma radiation can result in electric property degradation of conductors and insulating materials, in the change of optical properties of lenses, windows and mirrors, mechanical properties of soldered joints etc. Diagnostic elements most of all subjected to the nuclear and gamma radiation are listed in the Table 1.

Table 1. The expected neutron flux for some diagnostic components in ITER.
Absorbed doses of neutron and γ radiations are approximately equal.

Diagnostic components	neutron flux n/cm^{2}s	neutron fluence,n/cm^2
Magnetic field sensors, microwave antennas, corner reflectors, mirrors in the straight plasma visibility	3×10^{14}	3×10^{22}
Lenses, fiberoptics and mirrors in diagnostic channels	$(1-6) \times 10^{12}$	$(1-6) \times 10^{20}$
Bolometers, foils, X-ray crystals and different components behind the straight channels	$(1-5) \times 10^{10}$	$(1-5) \times 10^{18}$
Ceramic to metal soldered joints, bellows, gates and so on behind the straight diagnostic channels	$(5-10) \times 10^{9}$	$(5-10) \times 10^{17}$
Windows and fiberoptics behind the periscope (dog leg) diagnostic channels	$(1-5) \times 10^{8}$	$(1-5) \times 10^{16}$
Various detectors, scintillators etc. behind the shield	$(1-10) \times 10^{5}$	$(1-10) \times 10^{13}$

Diagnostics for Experimental Thermonuclear Fusion Reactors
Edited by P. E. Stott *et al.*, Plenum Press, New York, 1996

51

A lot of data on the radiation resistance of different materials were collected for nuclear reactor and military system needs and some of them were published[1,2]. But as it has turned out these data are not enough for the fusion reactor diagnostic system:
-   the composition of the nuclear radiation in fusion reactor differs from that one of the fission-type reactor,
-   the set of tested earlier materials is not adequate to materials usually used in the plasma diagnostic systems,
-   the neutron fluence for the all lifetime of ITER will be much larger than it may be at the nuclear reactor.

Several years ago the radiation resistance study for some of listed in the Table 1 materials was started in different laboratories. To define advisable test conditions (neutron flux and fluence and γ dose rate) it was necessary to have a picture of the expected radiarion environment in supposed positions of diagnostic components.

2. RADIATION ENVIRONMENT

In addition to data given in the Table 1 some remarks have to be done.

(I) The gamma radiation absorbed dose in the ITER is approximately equal to the absorbed dose of the neutron radiation (1 rad in Si corresponds roughly to 10^9-10^{10} n/cm^2 depending on the neutron energy).

(2). The necessity of the plasma control and plasma behavior information lead to the large number of diagnostic systems. The limited number of ports and their small sizes result in the large number of channels in the radiation shield. To protect surrounding space from neutron and gamma fluxes the local additional shield will be placed behind the port. Depending on the diagnostic channel number and sizes the background of the neutron and gamma radiation between the port and the shield may reach from 0.1 to 10 % of fluxes at the straight channel axes. For example, estimations (A.N.Svetchcopal, 1994, 3-D MCNP code[3]) for the channel length of 6 m shows that for 5 channels with diameter of d=10 cm neutron flux at the axis is $F_n = 2 \times 10^9$ $n/cm^2 s$ and background $F_b = 6 \times 10^6 n/cm^2 s$. For one channel of d=50 cm, one of d=35 cm and two of d=30cm behind the channel of d=35 cm $F_n = 6.8 \times 10^{10}$ $n/cm^2 s$ and $F_b = 4 \times 10^9$ $n/cm^2 s$. These figures, which were received for the primary neutron first wall loading of 1 MW/m^2, means that similar calculation have to be done for each diagnostic port and results must be accounted for in the design of diagnostic systems.

(3) Some diagnostic components must be located near the first wall and in the shield channels. Magnetic probes will be placed at the first wall level, metal mirrors - at the end of the short channel, but in the direct vision of the plasma, mirrors with coating and optical objectives - inside the channels beyond the first metal mirror. Neutron and secondary gamma radiation fluxes and spectral distributions are different in different places. The neutron flux in the straight channel may be roughly estimated by the simple expression: $F \sim F_0 \, r^2 / L^2$, where F_0 is the neutron flux at the first wall (about $2 \times 10^{14} n/cm^2 s$ for loading of 1 MW/m^2), r – is the radius of the cylindrical channel and L – the distance from the first wall to the point of the estimation. At the Fig.1 relative neutron energy spectra at the first wall, at the end of the straight channel axis and in the background are shown. Neutron flux at the end of the channel almost totally consist in unscattered primary neutron with the energy of 14.1 MeV. At the Fig.2 one version of the channel for the optical radiation of the plasma and first wall observation is shown. Neutron flux and gamma dose rate distribution along the channel with

diameter of 20 cm for the neutron load of 1 MW/m² are indicated in the Table 2 and their energy group spectra - at the Fig.3.

Table 2. Total neutron flux and γ-dose rate distribution along the channel.

Points at Fig.2	Distance from the previous point, cm	neutron flux, n/cm²s	gamma dose rate, Gy/s
1	0	1.3×10^{14}	320
2	20	5.4×10^{13}	150
4	30	1.3×10^{13}	45
5	27	2.5×10^{12}	9.5
6	27	4.2×10^{11}	2.0
6a	40	7.0×10^{10}	0.35
7	90	4.0×10^{9}	0.023
8	100	8.0×10^{8}	0.005
9	110	3.0×10^{8}	0.0025

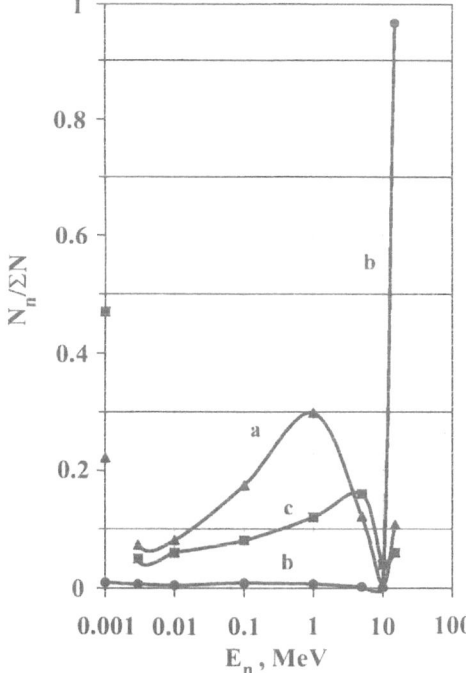

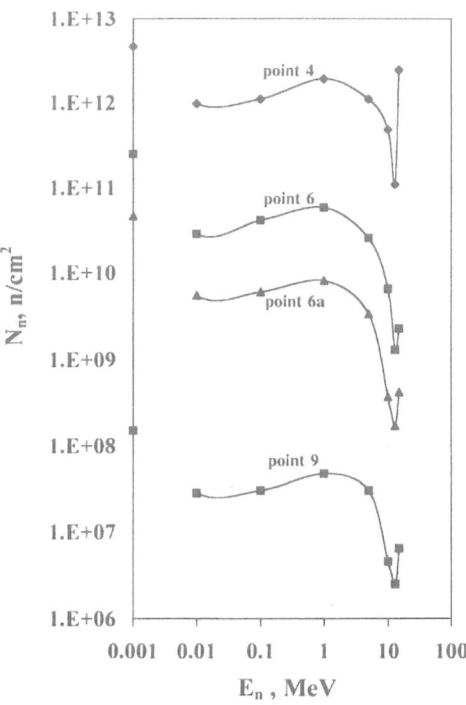

Figure 1. Expected neutron spectra (relative number of neutrons N_n in an energy interval as a function of E_{max} in each interval): near the first wall (a), at the axis of the straight channel outlet (b) and in the background between the diagnostic port flange and the aadditional local radiation shield (c).

Figure 3. Expected neutron spectra in different places of the image transmission optical channel, shown in the Fig.2 (next page) ,with diameter of 20 cm.

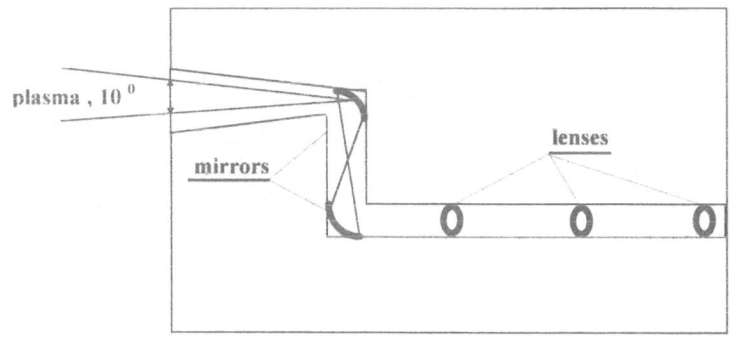

Figure 2. A version of the optical image transmission channel in the Fe-H$_2$O shield.

3. RADIATION TESTING FEATURES

Investigations of radiation effect on materials for the fusion reactor are bound up with some difficulties.

1). Radiation tests of material specimens must be done in the same radiation and other condition at which a real diagnostic components manufactured from these material will be in the reactor. Unfortunately at the moment there are no equipments which can correspond to this requirement in full measure. Only large tokamaks TFTR and JET working with the DT mixture can give fusion neutrons (neutron flux up to 5×10^{12} n/cm^{2}s and fluence up to 2×10^{15} n/cm^2)[4,5], but with an energy spectrum, which differs from the expected at ITER owing to different structure and volume of the shield and surrounding materials.

2). Maximum neutron fluence will be accumulated in materials during 10-15 years of the ITER operation with many breaks. It is impossible to reproduce such conditions in radiation tests, which have to be done in a shoter space of time but at much larger neutron fluxes than that one in ITER.

3). All diagnostic components have to preserve their properties not only after, but in the time of irradiation. It means that tested material characteristics must be measured during irradiation, *in situ.*, what in many cases can not be realized.

4). Neutron and γ radiation is not only factor which exert influence on material properties. Some components, especially located near the first wall, are subjected to a nuclear heating, visible and X-ray irradiation, neutral particle bombardment, electric and magnetic fields. Some of these factors may improve material characteristics, but some make them worse. It is very difficult to reproduce these factors in simulating testing experiments and some of them even impossible.

All indicated difficulties necessitate a detail investigation of radiation characteristics by measuring dependencies of material specimen properties on the neutron fluence an flux and the gamma dose and dose rate at different environment conditions - a temperature, radiation and particle fluxes etc. It is necessary to note that such experiments are very expensive and therefore needs in careful planning and preliminary choice of materials on the basis of known data.

In the next sections are given some examples of data on recent investigations of magnetic sensors, optical materials for windows and lenses, fiberoptic and X-ray crystals and multilayer mirrors. Ceramics and reflectors will be the subjects of other lectures.

4. MAGNETIC SENSORS

Magnetic sensors must be placed as near to the plasma as possible. Almost definitely they will be made in a traditional form of induction magnetic probes, which are the most radiation resistant in comparison with other types of magnetic sensors. They include a ceramic insulation and metal conductor as which a cable with a mineral insulation (MI cable) may be used and MI cable to conduct the signal out of the vacuum vessel. The most dangerous phenomena for ceramics are the radiation induced degradation of the conductivity and the breakdown voltage and for the probe as a whole - heating and a radiation induced electromotive force (RIEMF), which will be discussed separately.

Magnetic probe is a multiturn coil made of MI cable and wounded on the alumina bobbin[7,8,9] or without any bobbin placed into a box with a ceramic powder fill in[9]. Some examples of magnetic probes are shown in Fig.4 a and b. The results of magnetic probe testing in nuclear reactor JMTR (Japan) are shown in Fig.5. It is seen that the most critical is the probe inductance dependence on the temperature.

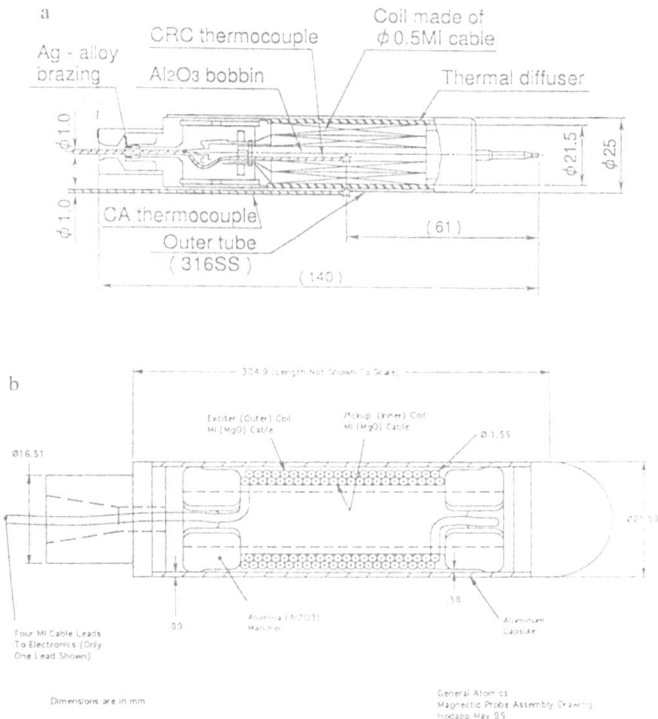

Figure 4. Design examples of magnetic probes for radiation testing: specimen with a temperature sensor[7] (a) and specimen with exciter and pickup coils[8] (b).

5. OPTICAL MATERIALS — WINDOWS AND FIBEROPTICS

5.1. Windows

For windows were proposed a few materials and best of them are quartz, sapphire, cerium glass and spinel ($MgAl_2O_4$). The first criterion for the choice is the neutron and gamma

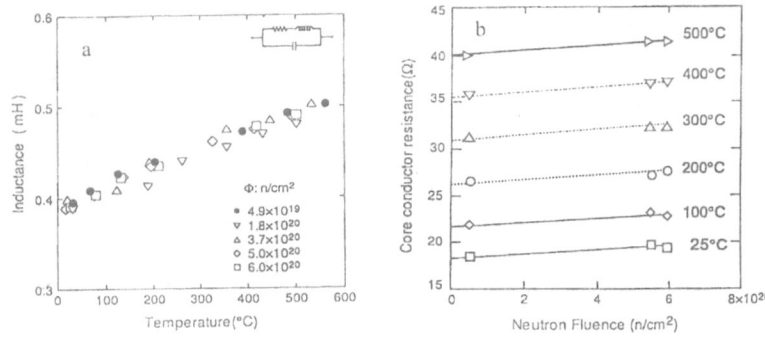

Figure 5. Magnetic probe (Fig.4a) inductance (a) and cable core conductor resistance (b) dependencies on the temperature and the neutron fluence[7].

radiation influence on the material transparency in the visible region of the spectrum (as is generally known the irradiation influence on the transparency in the IR region is much weaker). Transparency measurements of these materials[10] have shown that quartz and sapphire have lost this one in the UV region almost totally already at small dose an fluence ($\sim$0.1 Mgy or $\sim 10^{15}$ n/cm^2). The transparency region of the unirradiated spinel and cerium glass starts only at $\lambda > 400$ nm, cerium glass is activated with a very long decay time and spinel is very expensive in a manufacture. At Fig.6 transparency spectral distributions for the quartz and sapphire are shown. It is obvious that quartz is better. Besides that sapphire luminescence under the neutron and gamma irradiation is much more intensive than that one of quartz (according to[11] the ratio of radiated energy in the region of 200-800 nm to the absorbed energy of the γ radiation for sapphire is of the order of 0.01; in contrast to sapphire the quartz luminescence was not observable in equal conditions). Therefore the quartz is the most suitable material for windows and lenses in ITER.

To transfer the vessel wall image through the channel shown at Fig.2 two type of transparent optic may be used: either set of composite objectives (one quartz lens and the second from a material with other refraction index) or fiberoptics. The problems are: for the first version - the choice of the second radiation resistant material and for the second version - radioluminescence of the fiberoptic.

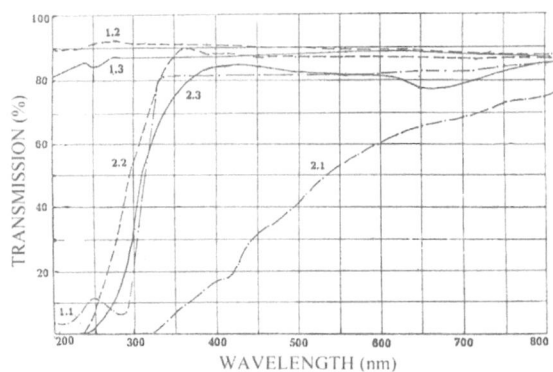

Figure 6. Transparency spectral distribution for sapphire (curves 1.1 and 2.1), quartz KU (OH content 10^{-3}ppm, curves 1.2 and 2.2) and KUVI (OH $\sim 10^{-6}$ppm, curves 3.1, 3.2). Curves 1.2, 1.2, 1.3 are received before and 2.1, 2.2, 2.3 - after irradiation in nuclear reactor up tu the fluence of 10^{17} n/cm^2.

5.2. Fiberoptics

During several years the fiberglass radiation resistance was investigated in Japan[12] (at the neutron generator FNS and the reactor JMTR), United States[8] (under spallation neutrons and gamma rays and at TFTR) and at JET[13]. The light absorption and luminescence intensity measurements are easier for fiber glass than for optical material small specimens owing to the large length and to the relatively small part of the scattered light. The fiberglass made of the quartz core and the quartz clad with different admixtures, which are required to have different refractive indexes of core and clad, was tested in different radiation sources. It was shown[12] that Ge doped SiO_2 is less resistant than the pure SiO_2. Therefore now in all experiments is used a F-doped quartz for the clad (it was shown[13], that admixture of F to the core reduces its resistance) and pure SiO_2 for the core. As to the OH amount influence, measurements at the gamma source[14] indicate on radiation resistance increasing with OH content, Fig.7 (similar results were received at the irradiation of window specimens in the nuclear reactor[10]).

The main results of the fiberoptic radiation resistance study are the next:

- irradiation effect is rather weak in the near IR region ($\lambda > 700$ nm),
- pure silica core fibers are permanently hardened by high dose of the γ-ray preirradiation[8] (Fig.8,);
- the hardening is speeded by the presence of light (photobleaching), from external source or owing to radio luminescence, but at high OH content this effect is negligible[8];
- noticeable part of the luminescence is a Cherenkov radiation[12] (Fig.9).
- data given in papers[13] allows to estimate the ratio of the luminescence energy to the absorbed energy of JET plasma nuclear radiation ($\sim 10^{-5}$).

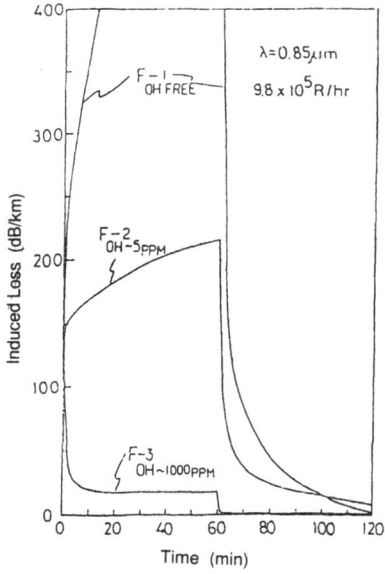

Figure 7. Effect of the OH content on an absorption loss for pure silica core optical fiber under gamma irradiation.

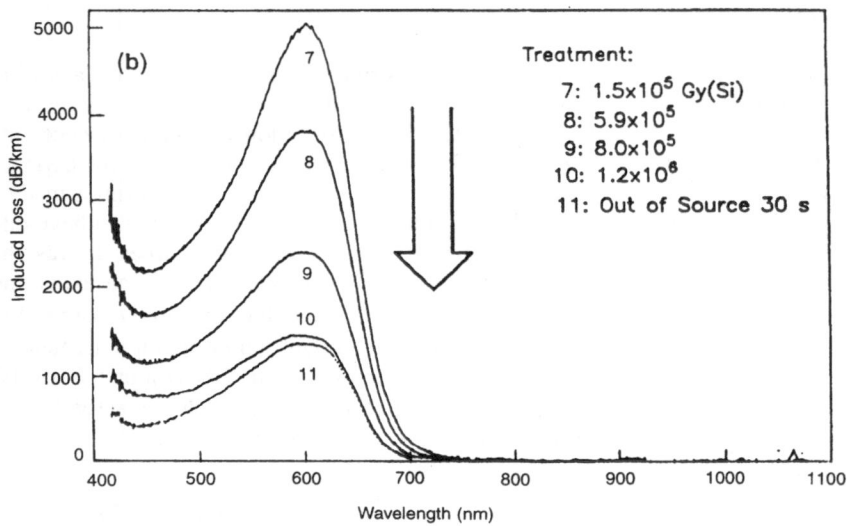

Figure 8. Preirradiation influence on the induced
absorption loss in quartz optical fiber.

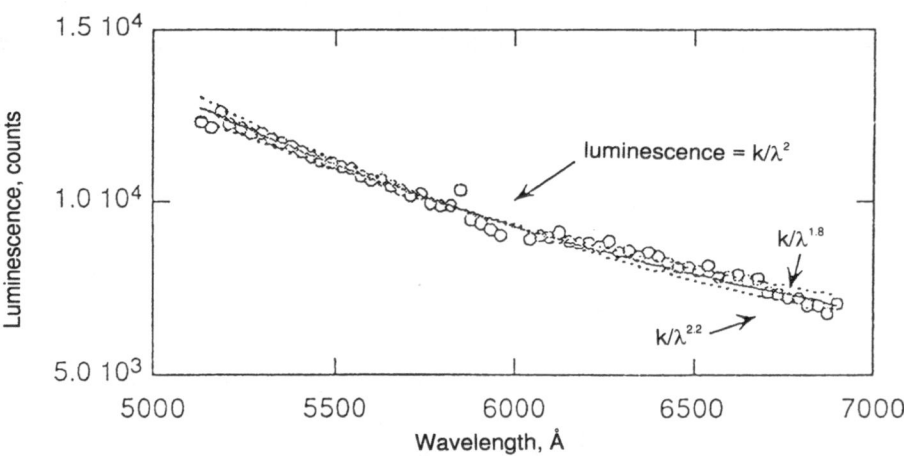

Figure 9. Radio luminescence mechanisms: relative spectral distribution of the quartz
optical fiber luminescence under irradiation in the nuclear reactor (Ref. 17).
(The spectrum is consistent with Cerenkov radiation.)

6. MIRRORS AND REFLECTORS

No one transparent optical material can withstand the nuclear radiation near the first wall.
Therefore only the pure metal mirror can be used as a first optical element receiving the
straight plasma nuclear and neutron radiations, as is shown in Figure 2. All other optical
components, including mirrors with a surface layer coating, must be placed behind the more
or less thick radiation shield. Some data on the experimental investigations of metal mirrors
and mirrors with another material covering are given in the lecture of V.S. Voitsenja[15].

7. X-RAY CRYSTALS AND MULTILAYER MIRRORS

Radiation hardening of different crystal structures was studied already for a long time[1]. It is known that lattice constant, for instance, of the quartz is changed by 0.3 - 2.2 % along different axis at the neutron fluence of 10^{20} n/cm^2. But next to nothing is known about crystal reflectivity .Therefore several crystals (Ge, Si, three cuts of SiO_2, graphite and some organic crystals and

after that the mica and LiF) were irradiated in the nuclear reactor BOR–60 (neutrons with $E_n > 0.1$ MeV flux of 3×10^{11} $n/cm^2 s$) up to fluence of 2×10^{19} n/cm^2 [16]. All organic crystals and LiF were destroyed at fluence of $\sim 10^{17}$ n/cm^2. Reflectivity of SiO_2 was changed nonmonotonically increasing at 10^{19} and decreasing at 2×10^{19} n/cm^2. Other specimens were not changed.

Multilayer mirrors (Cr/C, N=41; W/Si, N=60; Fe/C, N=40; $MoSi_2$/Si, N= 40; Mo/Si, N=25) testing has shown that the more is the layer number the less is the departure from the initial characteristics.

8. CONCLUSION

The next irradiation effects have to be investigated in the nearest future:
- the charge accumulation in the magnetic sensor and cable insulation;
- radioluminescence of optical materials and photobleaching effect; to choose the second material for compound objectives;
- the mutual influens of the nuclar irradiation and sputtering for metal mirrors,
- the irradiation influence on the X–ray crystals spectral resolution in a range of

$$\lambda / \Delta\lambda \sim 10^4.$$

REFERENCES

1. J.F.Kircher and R.E.Bowman, *Effects of Radiation on Materials and Components*. Reinhold Publ.Corp., 1964.
2. J.F.Baur,B.A.Engholm,M.P.Hacker et al., *Radiation Hardening of Diagnostics for Fusion Reactors*, Rep. GA-A16614, Dec. 1981.
3. J.Blesmeister (Editor), "MCNP-A General Monte Carlo Code for neutron and photon Transport, Version 3A", LA-7396-m, Rev.2, Sept.1986.
4. ITER DIAGNOSTIC NEEDS as reported by the ITER Physics Expert Group on Diagnostic, Febr.8-10, 1995, JAERI, NAA, Japan.
5. A.T.Ramsey, *Rev.Sci.Instrum*.66, 871, (1995).
6. S.F.Paul, J.L.Goldstein, R.D.Durst, R.J.Fonck, *Rev.Sci.Instrum*. 66:1252,(1995).
7. H.Sagawa, S.Ikeno, T.Sekine, E.Ishitsuka, H.Kawamura, T.Matoba, M.Saito, J.Nucl. Materials 212-215; 431-434(1994). "94 Task (T28) Results in the Japanese Home Team" presented by T.Nishitani at the Meeting on Irradiation Testing of diagnostic Components, Garching, 22-24 May 1995.
8. "*Ceramics and Optics for Diagnostic Systems*", presented by K.Young, E.Farnum et al.at the Meeting on Irradiation Testing of Diagnostic Components,Garching, 22-24 May, 1995.
9. "*Magnetic Measurements and magnetic sensors for ITER*", presented by S,Bender at the Technical Meeting on Radiation Effects on In-Vessel Components. Garching, Nov.15–19 1993.
Yu.K.Kuznetsov, I.K.Tarasov et al. *Radiation induced EMF on MI probes under intense gamma ray flux. DIAGNOSTIC FOR ITER International Workshop, Varenna, Aug.1995.*

10. A.B.Berlizov, O.V.Kachalov, A.R.Matevosov, D.V.Pavlov et al. *"Different quartz type, sapphire and cerium glass behavior under the nuclear reactor irradiation"*, The 3 Int. Conf.on Radiation Effects on Fusion Reactor Materials, St.Petersburg,Sept.23-25,1994.

11. "EU 94 results", presentation by E.Hodgson at the Meeting on Irradiation Testing of Diagnostic Components, Garching, 22-24 May, 1995.

12. T.Shikama,M.Narui,T.Kakuta et al. *Sci.Rep.RITU* A40 (1994) pp 147-152; *J.Nucl.Material*. 212-215: (1994) 421-425. T.Iida, Sh.Ie, K Sumita et al.,*J.Nucl.Sci.and Techn*. 24, (1987) No.12, p.1073.

13. P.D.Morgan, *Irradiation of optical fibers at JET through 14 MeV neutron production*, JET-P (92) 72, p.167.

14. Y.Morita and W.Kawakami, *IEEE Trans.NS*, 36, (1989) 584.

15. V.S.Voitsenja *Effects on the inner elements of spectroscopic and submillimeter diagnostics* Lecture at the Workshope on Diagnostics for ITER, Varenna (Italy), 1995.

16. V.M.Kosenkov,V.A.Neverov,Yu.L.Revyakin,D.V.Orlinski, *J.Nucl.Mater* 212-215:1056(1994).

17. *"The TFTR-JET Collaboration on Optical Fibers: Radiation Effects and Remediation"*, Presented by K.M.Young, ITER/US/D4/PH-07-26

IMITATION OF FUSION REACTOR ENVIRONMENT EFFECTS ON THE INNER ELEMENTS OF SPECTROSCOPICAL, MM AND SUB-MM DIAGNOSTICS

V.S.Voitsenya,[1] A.F.Bardamid,[2] V.L.Berezhnyj,[1] Yu.N.Borisenko,[3] V.I.Gritsyna,[1] V.T.Gritsyna,[3] V.G.Konovalov,[1] V.L.Ocheretenko,[1] D.V.Orlinski,[4] R.O.Pavlichenko,[1] L.V.Poperenko,[2] V.V.Ruzhitskij,[1] V.F.Rybalko,[1] A.N.Shapoval,[1] A.I.Skibenko,[1] N.V.Vinnichenko,[2] and K.I.Yakimov[2]

[1]Kharkov Institute of Physics and Technology, 310108, Kharkov
[2]Kiev University
[3]Kharkov State University
[4]RSC Kurchatov Institute, Russia

This paper consists of two parts and concerns the influence of fusion reactor conditions on inner elements of some diagnostic systems. The first part is devoted to results of imitation experiments on the long-term operation of plasma-viewing mirrors, and in the second part there are discussed the results on a possibility to use carbon-graphite materials for fabricating the active elements of mm and sub-mm diagnostics (radiating and receiving antennae, reflectors). It is evident that results of the second part can be useful only in the case that carbon-based materials but not a beryllium will be chosen as the first wall protection.

1. Metal-made plasma-viewing mirrors (PVM)

1.1.Introduction

According to ITER-90 project (e.g. [1]) the plasma-viewing mirrors (PVM) have to be placed rather close to a hot plasma. The main aim of PVM use is to scan the large part of plasma volume and to direct the chord integrated plasma radiation of chosen spectral ranges into the periscopic diagnostic channels. Due to their locations, PVM will be simultaneously subjected to influence of several factors which can lead to the degradation of mirror optical properties for much shorter time than the whole planned period of reactor operation. Potentially the most dangerous for optical quality safety of metallic PVM are the following components of plasma radiations: (1) flux of neutrons with a high portion ($\leq 20\%$) of 14-MeV energy neutrons and (2) charge exchange atoms (CEA) of a quite broad energy spectrum with a mean energy of several hundred eV.

It is evident that details of environment effects on PVM characteristics can be obtained only directly in a fusion reactor, but some estimations of these effects on PVM optical properties can be obtained taking account results of proper model experiments. For example, it was found in imitation experiments with copper, stainless steel and beryllium mirrors [4-6],

Diagnostics for Experimental Thermonuclear Fusion Reactors
Edited by P. E. Stott *et al.*, Plenum Press, New York, 1996

61

that neutron irradiation itself up to dose typical for the ITER first wall (~10dpa) will not lead to a significant degradation of metal - made PVM reflectance (R). At the same time, due to small ranges of CEA in solids the effects of these particles on PVM optical properties can be very strong as a result of different processes in a mirror presurface layer: sputtering, accumulation of gases, creation of defects.

In this connection we continue our imitation experiments [5] in the direction of revealing effects of long-term mirror surface bombardment with CEA. For imitation of CEA flux (which will consist of deuterium and tritium atoms, mainly) we use the ion component of a reflex discharge plasma in hydrogen or deuterium as well as He ions of gas- discharge ion source. In all cases the ion flux bombarding the mirror surface contained some portion of impurity ions with their fractions much less that the light ion fractions.

1.2. Copper mirrors

Fig. 1 shows spectral dependencies ($450 < \lambda < 600$ nm) of a copper mirror reflectance for three cases: before any irradiation (1), after irradiation with 3 MeV Cu^{2+} ions up to dose ~25 dpa, as an imitation of neutron irradiation (2), and after additional bombard-ment with ions of a hydrogen plasma of about 1.2 keV mean energy (3). From the total ion fluence the thickness of sputtered layer for curve (3) was estimated as 0.7 μm, i.e., it was less than calculated maximum range of 3 MeV Cu ions in a copper (~1.2 μm) [6].

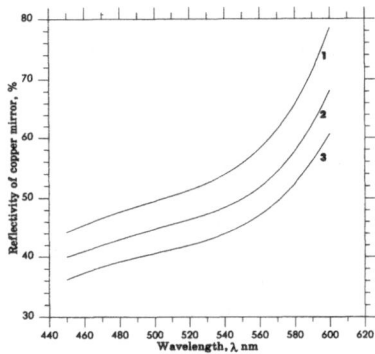

Figure 1. Spectral dependence of reflectance of copper mirror. 1 - as fabricated, 2 - after Cu^{2+} ions bombard-ment, 3 - after Cu^{2+} and plasma ions bombardment.

It is seen some small decrease of R after high energy ions bombardment with further decreasing R after sputtering of the presurface layer.

These results were qualitatively confirmed by those ones obtained after sputtering with ~ 4 keV He ions of similar thickness layer (~0.7μm) by measuring ellipsometric parameters Δ and ψ versus angle of incidence at $\lambda = 632.8$ nm. From ellipsometric data the refractive index (n) and the extinction coefficient (k) were calculated giving, in turn, the normal incidence reflectance values R: 71% for the part of mirror nonirradiated with Cu ions and 63% - for preliminary irradiated. Thus, the defects accumulated due to high energy ion irradiation result in faster mirror quality degradation when mirror surface is subjected to the He ions bombardment.

The SEM pictures of Cu mirror after He ion bombardment also show the difference in surface micromorphology for both parts of the same copper sample.

1.3. Mirrors made of stainless steel

The R dependencies on the thickness of sputtered layer, δ, for mirrors made of stainless steel (OCr16Ni15Mo3B type) after step by step sputtering with D^+ ions (mean energy ~ 0.65 keV) are shown in Fig.2 for the wavelength λ = 0.5 μm. Qualitatively similar R dependencies are observed for the whole range of spectrum where such measurements were carried out (0.2 < λ < 0.8 μm).

SEM pictures of this sample show that after sputtering the layer of 1.6 μm in thickness, on the preirradiated surface there appeared carters with 2-3 μm in diameter with the whole surface fraction of several per cent. Besides, there observed much higher density of etch pits of lesser size (~ 1 μm).

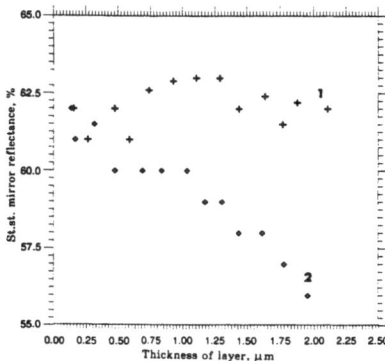

Figure 2. Dependence of st. st. mirror reflectance on thickness of layer sputtered with D^+ ions. 1 - nonirradiated sample, 2 - sample preirra-diated with Cr^{2+} ions (up to dose ~ 15 dpa).

Thus the data for st. st. mirrors also demonstrate a little less resistance of mirror optical properties under low energy deuterium ions bombardment for that part of the mirror which was irradiated with high energy ions as compare to the mirror area that have not been irradiated preliminary.

According to measurements of sample mass loss during every step, the maximum thickness of a sputtered layer here is substantially higher than the calculated maximum range of 3 MeV Cr^{2+} ions (~ 1.2 μm [6]) in a stainless steel. And as it is seen, still the optical properties of samples are not too much worse than at the beginning of sputtering procedure. Using data on the CEA flux to the first wall from [1] (3×10^{16} -10^{17} at/cm²·s) we can estimate the time of reactor operation when such thickness (~2 μm) of PVM presurface layer will be sputtered. Taking account the sputtering coefficient of a stainless steel due to CEA flux (0.015-0.02 at/at) and the CEA flux density to PVM surface ,10^{15} at/cm²·s (after attenuation in 1.5-2 orders in magnitude as compare to the flux estimated for the first wall) we obtain ~ 1000 pulses.

Thus, our test is equivalent to only small portion (~1%) of the planned time of ITER Technological phase.

1.4. Beryllium mirrors

Similar experiments with mirrors made of pressed beryllium showed much faster decrease of reflectance under bombardment with D^+ ions. This effect is especially substantial for the shorter wavelength range of spectrum, Fig.3.

The pictures of surface structure obtained with TEM and SEM show that the degradation of optical properties of beryllium mirrors is connected with significant change of a microrelief. After ~ 0.65 keV D^+ ions bombardment there appear many etch pits with diameters of the order of hundreds nanometers already after the low dose ($>10^{17}$ ions/cm^2). Very often the pits are localized in the grooves of mechanical polishing. The number and size of etch pits initially increase with increasing exposure time of Be samples to plasma. At much higher fluence of D^+ ions ($> 10^{19}$ cm^{-2}) the number of pits decreases but pinholes are observed with a maximum diameter ~ 1μm.

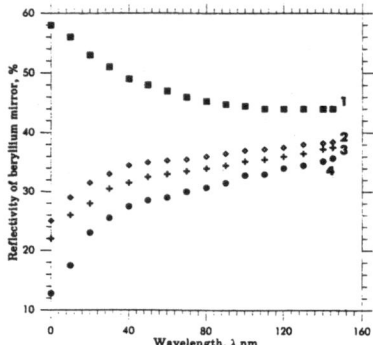

Figure 3. Spectral dependence of reflectance of beryllium mirror. 1 - as fabricated and after bombardment with D^+ ions to different fluence: 2 -10^{17}, 3 - 5×10^{17}, 4 - 10^{19} ions/cm^2.

1.5. Plasma ion influence on mirror with antireflection coating

In the frame of imitation of fusion reactor conditions on characteristics of optical diagnostics there was investigated also an influence of low energy ions on Be mirror with an antireflection coating. The coating (the combination of Al and Al_2O_3 films) improves essentially the reflectance of Be mirror in the wavelength range λ = 0.25 -1.0 μm as compare to the mirror without any coating (Fig. 4). The mirror was exposed to ions of hydrogen ECR discharge plasma. The ion energy (25 - 50eV) was near the threshold energy for sputtering with H^+ ions [7]: E_{thr} (Al) ~ 35eV. After exposure to ion irradiation the coating was strongly destroyed and R of this mirror became much lower than its initial value. At the same time after similar procedure for Be mirror without coating there was not found measurable change of optical properties.

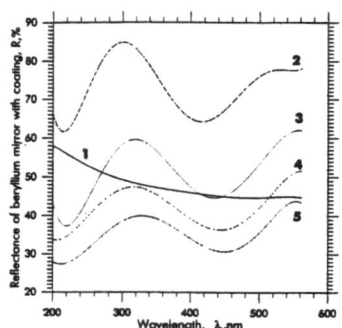

Figure 4. Effect of low energy ions bombardment on reflectance of beryllium mirror with antireflection coating. 2 -as fabricated, 3 -after 4 hours of exposure to plasma, 4 - after 7 hours, 5 - after 10 hours. 1 - data for beryllium mirror without coating.

The estimations show that during this test the total ion flux to mirrors at the end of an exposure time was approximately the same as the charge exchange atom flux for about 1000 pulses in a fusion reactor to the mirror placed deeply in the channel with attenuation of the flux in about 100 times as compare to the flux to the first wall. So, it follows from these results that mirrors with similar coating cannot be used in a straight view of a confined plasma even being placed inside the channel.

1.6. Discussion and conclusion

1.Results of our imitation experiments on CEA influence on metallic PVM indicate a very strong dependence of mirror rate degradation under low energy ion bombardment upon the mirror material and its structure. Between tested mirrors the stainless steel one is much more stable and the pressed beryllium has very low resistance. One of the reason of this difference can be connected with difference in sputtering coefficients values: $Y(Be) \cong Y(Cu) \cong 2.5 \times Y(st. st.)$ [7]. But the ratio of Y(st.st.) to Y(Be) is not so high comparing to the difference in R rate degradation between Be and st.st. samples, especially in UV. In this case very important is, probably, the rate of microrelief change under identical conditions of mirror irradiation for these metals that have different structures because of different technology of manufacturing.

The very important corollary from these results is as follows: for the proper choice of PVM material it is necessary to take into account not only its optical properties but the possible mirror surface erosion rate which is defined, in a high degree, by the value of sputtering coefficient (if there are no big differences in structure). As a first approximation in attempt to solve the problem of searching for PVM materials we propose to use the distinctive "figure of merit" which is simply the ratio of $R(\lambda)$ to Y. In Fig.5 there are presented the R/Y ratios for several metals calculated using values of R(l) measured by different authors [8-11] and Y values for 0.3 keV D^+ ions from [7] (such values of CEA mean energy were typically measured in experiments on PLT and ASDEX tokamaks [12,13]). It is seen that according to R/Y criterion, mirrors made of such widely used optical materials as Al, Ag and Au are much behind the mirrors made of refractory metals. The accounting of the fact that CEA flux will consist of the mixture of D, T and small portion of He atoms does not change significantly the relation of R/Y for different metals. Such a comparison gives a chance to have a better choice of such a mirror material that demonstrates the greatest resistance to the erosion due to CEA bombardment and simultaneously has appropriate optical properties.

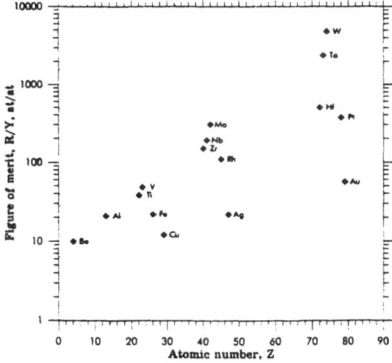

Figure 5. Values of R/Y criterion for different metals (λ = 500 nm).

Such an approximation (the use of R/Y criterion) means that the equal thickness of eroded layers (due to sputtering with CEA) leads to the equal degree of mirror reflectance degradation for different materials. In other words, in this case there is ignored the above mentioned remark about different consequences for differently structured metals in identical irradiation conditions. But this difference could be taken into account during further accumulation of experimental data.

2. It follows from results of many calculations and imitation experiments (e.g., the review paper [14]) that defect structures created in a nearsurface layer of a mirror under MeV energy range ions bombardment are in much degree similar to ones appeared due to neutron irradiation. And some of our results reveal faster degradation of mirror quality during long-term bombardment with keV range energy light ions (D and He) if such a defect structure exists. In application to PVM it means, that simultaneous influence of neutron and CEA fluxes will result in faster degradation of mirror reflectance as compare to such imitation experiments when only low energy ions are used for long-term sputtering. In turn from comparison of results for Cu and st.st. samples one can conclude that the role of defects produced due to irradiation with high energy ions is less important for mirror made of a material which is more resistant when is being subjected to keV ions bombardment.

3. Conditions of the tests carried out are equivalent to rather small portion of the whole reactor operation time even if the CEA flux to the mirror surface will be attenuated 1.5-2 orders in magnitude as compare to the flux to the first wall. Thus in addition to R/Y criterion use, there will be probably a need to invent such a construction of mirror assembly that would result in much stronger attenuation of the mean CEA flux to the mirror surface (i.e., using special blend).

2. Graphite-made elements of microwave and sub-mm diagnostics

2.1. Introduction

In the case of carbon-graphite materials will be used for the ITER first wall protection, some of protection elements can be also used as reflectors in μ-wave and sub-mm regions. Previous results [15-17] indicate that reflectance (R) of different graphites sharply decreases only near the infrared region ($\lambda < 0.1$ mm) but in microwave and sub-mm regions R values are close to ones measured for metallic reflectors. In addition, it was shown that characteristics of funnel-shaped metallic and graphite-made antennae are not too much different [15], and the polarization conservation as well as transmissivity of graphite waveguide in sub-mm range is close to 100%, like for a quartz waveguide [17], being significantly better than for metallic waveguide of the same shape. In the present paper we discuss results of imitation experiments of environment effects on graphite-made active elements of μ-wave and sub-mm diagnostics.

2.2. Imitation of fusion reactor conditions on reflectors

Results of our measurements of the reflectance (R) of different carbon-based materials are presented in Table 1.
Two columns for every frequency in this Table are results of R measurements for plates of indicated graphite types and averaged microrelief values. The first column for every frequency presents R values (relatively to R of a metallic plate) for graphite plates as they were fabricated, and the second one - after exposure to high power pulsed plasma streams with energy density ~ 1 kJ/cm^2 and time duration ~ 100 μs, as imitation of probable conditions during the disruption event. Totally the number of pulses was several tens. Accuracy of R measurements is estimated was not better than ± 0.05.

Table 1.

Type of graphite	h, μm	f=37.5 Ghz λ=8 mm		f=75 Ghz λ=4 mm		f=105 Ghz λ=2.86 mm		f=890 Ghz λ=0.337 mm		f=2521 Ghz λ=0.119 mm	
RG-Ti-91	7	0.99	0.95	0.88	0.99	0.92	1.04	0.93	0.99	0.70	0.73
RG-Ti-91	12	1.00	1.00	0.99	1.00	0.93		0.92		0.72	0.65
RG-Ti-91	19	0.99	1.05	0.92	1.00	0.92	0.99	0.86	0.90	0.57	0.56
RG-Ti-91+0.8%B	13	0.97	0.95	0.95		0.93	1.02	0.90		0.82	0.60
RG-Ti-91+0.8%B	17	0.92	1.00	0.91	1.00	0.93	0.99	0.82		0.64	0.43
RG-Ti-91+0.25%B	12	0.94	0.99	0.94	0.99	0.93	1.09	0.91		0.74	0.67
RG-Ti-91+0.25%B	13	0.95	1.02	0.94	1.01	0.90	1.02	0.90		0.59	0.64
UAM-92-3Gr	66	0.98	0.98	0.84	0.94	0.89	0.90	0.55	0.40	0.33	0.17
UAM-92-3Gr	69	0.98	0.99	0.82	0.94	0.85	0.91	0.55	0.48	0.36	0.27
UAM-92-5d	37	0.93	0.98	0.77	0.95	0.90		0.49	0.38	0.32	0.25
UAM-92-5d+0.1%B	26	0.95	0.93	0.90	0.91	0.93	0.92	0.52	0.88	0.45	0.50
MPG	9	0.96	0.99	0.95	1.01	0.95	0.99	0.78	0.79	0.72	
GSP	22	0.97	0.99	0.88	0.92	0.87	1.02	0.90		0.68	
ARV	53	0.96	1.02	0.93	1.02	0.91	1.00	0.68	0.69	0.12	
Reaktor graphite	40	0.96	0.96	0.96	0.97	0.91	1.00	0.70		0.14	
ATJ	13							0.95		0.50	
POCO	3	0.94	0.97	0.99	1.01	0.96	0.97	0.97		0.85	
Carbon-graphite-composite	30							0.33		0.35	

It is evident from these data that the material of a plate does not play a significant role in the determination of R. Much more important is the microrelief mean quantity (h) for a given sample, and the comparison of data for different frequencies indicate that R dependence on h values becomes more pronounced with increasing frequency. In Fig.6 measured R values of several graphites (for the wavelength 2.8 mm) are shown as function of microrelief mean quantity.

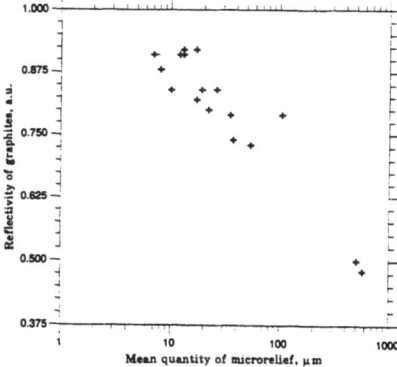

Figure 6. Dependence of reflectivity of graphites on the microrelief mean quantity.

At the same time, comparing data of the first and second columns of the same frequency, we can see that after exposure to plasma streams, as a rule, there observed some increasing of R in such a degree that to within the error bars results obtained are identical to ones measured for metallic plate. One of the reason of R increase after plasma influence is connected with decreasing h, as have been measured by means of the optical microscope.

In connection with data concerning the strong influence of neutron irradiation on some properties of graphites, and especially on their heat conductivity [18,19], we measured the reflectance of several graphite samples at the incidence angle 45° after they have been exposed in a fission reactor. The irradiation doses (neutrons per sq.cm) for cylindrical samples (diameter 5.0mm) of three graphites types (GR-280, PGT-91, UAM-91) together with d.c. electrical (σ, Ohm^{-1}·cm^{-1}) and heat (κ, W/m·K, for room temperature) conductance, and reflectance (R) at $\lambda = 0.337$mm are presented in Table 2.

Table 2

Type of graphite	As received			After irradiation								
				T=690°C D=3.5×10^{21}			T=400°C D=2×10^{22}			T=700°C D=2×10^{22}		
	κ	σ	R	κ	σ	R	κ	σ	R	κ	σ	R
GR-280	~100	1.0×10^3	0.76	<10	6.0×10^2	0.82	---	---	---	---	---	---
PGT-91	~500	7.0×10^3	0.90	~100	---	---	~100	8.0×10^2	0.82	~150	1.2×10^3	0.80
UAM-91	220	1.8×10^3	0.77	---	---	---	84	4.5×10^2	0.69	28	3.8×10^3	0.66

It is seen that in spite of strong change of d.c. and heat conductance (almost one order in value) the relative change of reflectance is rather small and does not exceed 15%.

2.2. Imitation of fusion reactor conditions on antennae

The test of 4 funnel-shaped antennae made of two sorts of graphite and stainless steel has been conducted using the same source of power plasma streams (70 shots). After test the characteristics of st. st. antenna decreased because of appearing many tracks of melting on its working (inner) surfaces and small change of a shape due to local overheating. At the same time working characteristics of so called "thin wall" antenna made of MPG type of graphite significantly improved. The experimental results for these two antennae (the directivity diagram) are shown in Fig.7 in relative units.

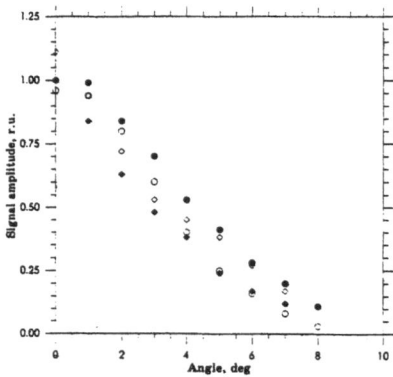

Figure 7. Directivity diagrams for metallic (oo) and graphite (qq) tunnel-shaped antennae before (xx) and after exposure to power plasma streams.

It is seen that the shape of directivity diagram does not change appreciably after exposure to plasma streams for both antennae. But the absolute amplitude of a signal near the diagram axis for graphite antenna became several per cent higher after the test and for st. st one - several per cent lower.

The effect of heating on the radiating properties was measured only for the conical antenna made of MPG-7 graphite. It was found that within experimental error bars the hot antenna (heated up to ~600 C) has the same directive diagram as the cold one.

2.3. Conclusion

It follows from presented results that carbon-based materials can be successfully used as inner components of mm and sub-mm plasma diagnostic instead of reflectors and antennae (both radiating and receiving ones) made of metals. The use of graphites for fabricating these active elements looks logical in the case that similar materials will be used as the first wall protection.

High resistance of graphite reflectors to the neutron irradiation found in our measurements gives the foundation to hope that antennae made of carbon-based materials will also be highly resistant in saving their working characteristics during quite long period of ITER operation..

REFERENCES

1. V.S.Mukhovatov, H.Hopman, S.Yamamoto et al., ITER Diagnostics, ITER documentation series, No 33, Vienna 1990.
2. In Physics Design Description Document for the ITER Plasma, June 1995.
3. W.W.Heidbrink and Sadler, The behaviour of fast ions in tokamak experiments, Nucl. Fusion. 34: 535 (1994).
4. J.S.Hartman, Optical effects of energetic copper-ion irradiation on copper mirrors, Appl. Optics. 20: 4062 (1981).
5. V.S.Voitsenya, Yu.N.Borisenko, V.V.Bryk et al., Simulation of radiation effects on reflectors using heavy ion beams, J.Nucl. Mater. 212-215: 1640 (1994).
6. Yu.N.Borisenko, V.V.Bryk, V.V.Gann et al., Effect of MeV-range heavy ion irradiation on the properties of metallic mirrors, Plasma Devices and Operations. 3: 157 (1994).
7. Y.Yamamura and H.Tawara, Energy dependence of ion-induce sputtering yields from monoatomic solids at normal incidence, Research Report NIFS-DATA-23, March 1995.
8. R.Blickensderfer, D.K.Deardorff and R.L.Linkoln, Normal total emittance at 400-850 K and normal spectral reflectance at room temperature of Be, Hf, Nb, Ta, Ti, V and Zr, Journal of the Less-Common Metals. 51: 13 (1977).
9. C.W.Allen. *Astrophysical Quantities.* Publisher, University of London, The Athlone Press (1955).
10. D.W.Juenker, L.J.Le Blanc and C.R.Martin, Optical properties of some transition metals, JOSA. 58: 164 (1968).
11. J.K.Coulter, G.Hass, and J.B.Ramsey, Jr., Optical constants and reflectance and transmittance of evaporated rhodium films in the visible, JOSA. 63: 1149 (1973).
12. S.Cohen, D.Ruzic, D.E.Voss et al., Measurements of low energy neutral hydrogen efflux during ICRF heating. - Preprint PPPL- 2133, Princeton, September 1984.
13. H.Verbeek and the ASDEX Team, Low energy neutral particle fluxes to the walls of ASDEX during He and D discharges, J.Nucl Mater. 145-147: 523 (1987).
14. D.J.Mazey. Fundamental aspects of high-energy ion-beam simulation techniques and their relevance to fusion material studies, J.Nucl.Mater.174:196 (1990).

15. V.V.Chebotarev, I.P.Fomin, R.O.Pavlichenko et al., The prospects of using carbon-graphite materials as construction elements of the microwave plasma diagnostic in a fusion reactor, J.Nucl.Mater.212-215:1157(1994).

16. V.L.Berezhnyj, V.S.Voitsenya and V.L.Ocheretenko, On possibility of graphite elements usage in submillimeter plasma diagnostics on fusion devices. - Preprint KFTI 93-31, Kharkov, 1993.

17. M.Nagatsu, N.Takada, T.Tsukishima and M.Shimada, Reflectivity measurements of graphite in the infrared and submillimeter wave regions, J.Nucl. Mater. 209: 204 (1994).

18. T.Maruyama and M.Harayama, Neutron irradiation effect on thermal conductivity and dimensial change of graphite materials, J.Nucl.Mater. 195: 44 (1992).

19. P.A.Platonov, V.J.Karpukhin, A.A.Mitrofanskii et al., Properties of neutron irradiated carbon-based materials for fusion reactor application, Plasma Devices and Operations. 3: 79 (1994).

OVERVIEW OF MAGNETIC DIAGNOSTICS PLANNED FOR ITER

L. de Kock[1], S. Ali-Arshad[2], V. Belyakov[3], S. Bender[3], A. Costley[1],
O. Gruber[4], A. Kellman[5], Yu. Kutznetsov[6], J. Leuer[5], T. Matoba[7],
P. McCarthy[8], V. Mukhovatov[1], D. Orlinski[9], P. Stott[2], R. Snider[5],
T. Todd[10], C. Windsor[11], I. Yasin[6], K. Young[12]

[1] ITER Joint Central Team
[2] JET Joint Undertaking, UK
[3] Efremov Institute, St. Petersburg, RF
[4] IPP Garching, Germany
[5] GA, San Diego, USA
[6] IPP, Kharkov, Ukraine
[7] JAERI, Naka, Japan
[8] Univ. of Cork, Ireland
[9] Kurchatov Institute, Moscow, RF
[10] AEA Culham Lab., UK
[11] AEA Harwell Lab., UK
[12] PPPL, Princeton, USA

INTRODUCTION

The magnetic diagnostic for ITER has to meet the classical requirements of the magnetic diagnostic on the present day large divertor tokamaks and has in addition the requirement to work during the long pulse in excess of 1000s and in a hostile radiation environment.

Inductive measurements need a front end analogue integrator to obtain the flux or field and to correctly process unexpected fast transients. The drift caused by small thermo-electric voltages and the operational amplifier must be compensated on the time scale of more than 1000s. Several promising schemes have been proposed and are under test. The main design is therefore based on classical inductive measuremnts of fluxes and fields by means of loops and coils respectively.

The steady state conditions during the flat top, however, suggest a direct measurement of the magnetic field or a measurement of the location of the boundary by non-magnetic means. An example of the first category is the Hall probe, well shielded from radiation and of the second category the reflectometer.

Diagnostics for Experimental Thermonuclear Fusion Reactors
Edited by P. E. Stott *et al.*, Plenum Press, New York, 1996

In the latter case only a full bore plasma can be measured, smaller bore plasmas as occur during start up can only be measured by magnetic means. In ITER some of these techniques will be used to gain experience and to have the possibility to detect drifts of the inductive system. This paper will be restricted to the inductive system only.

High radiation levels will be experienced in locations close to the plasma where coils need to be placed for the measurement of fast fluctuations and for the compensation of eddy currents in the structures. High radiation levels lead to the need for active cooling and possibly to the need to replace the coils during the life time of the machine. These are important complications in the design.

The measurement of the location of the plasma boundary is only obtained from numerical calculations. Several methods are available and all of them are well developed. In ideal conditions the reconstruction can be as accurate as 1 cm, which is the target for ITER. It results from the need to diagnose the plasma edge by independent other diagnostics for which the magnetic information provides the reference.

A subset of the output of the numerical codes is used for the plasma current and shape control. Five gaps between the plasma boundary and the critical first wall components will be used as feedback parameters. It is the intention to use the fluctuation measurement system as monitor of dangerous MHD activity.

REQUIREMENTS

The physics and technical requirements originate from the role of the magnetic measurements in establishing and controlling the plasma configuration. This requires both accuracy and reliability. In Table 1 the parameters, purpose and possible diagnostic techniques are listed and additionally the parameter range, the required accuracy and spatial and time resolution are given(from [1]).

The most important application of the diagnostic is its input to the current and shape control. The target accuracy of 1 cm for the diagnostic use exceeds the required accuracy for control. Also the time resolution is an order of magnitude better. There are several candidate reconstruction codes to provide the input parameters, which are basically a subset of the overall equilibrium. A very important aspect of the control application is that it must be reliable and robust. For the purpose of recovery from unwanted situations or to start protective or evasive action this diagnostic must remain operational far outside the required scenarios.

Table 1: Specification of the parameters to be measured by the magnetic diagnostic

#	Plasma Parameter to be measured	Purpose	Candidate Diagnostics
1	Plasma Current	Total current required for control	Rogowski Coil for ≤1000s; study for longer times

Parameter	Parameter range	Time res.	Rate of change	Accuracy
Plasma current	0.1-28 MA	1ms	$40-5 \times 10^5$ As^{-1}	1% (for $I_p > 1$MA)
		0.1ms	$<5 \times 10^9$ As^{-1} [†]	~30%
	[†] At current quench at a disruption			

Table 1 (Continued)

2	Plasma Position and Shape		Poloidal field feedback		Position Loops for ≤1000s, study for longer times; Reflectometry
Parameter		Time res.	Rate of change		Accuracy*
D_{in}, D_{out}, D_{top},		1 ms	1 cm s^{-1}		1 cm
		1 ms	1 m s^{-1}†		2 cm
Divertor channel position (r-direction)		1 ms	1 cm s^{-1}		1 cm
		1 ms	1 m s^{-1}†		2 cm

D is a width of gap between separatrix (or layer with n_e=TBD) and first wall
   † *At current quench at a disruption*
   * *For full bore plasma*

3	Loop Voltage		l_i, plasma startup			Flux Loops	
Voltage range		Numb. of Loops	Time res.		Rate of change		Accuracy
0.01 - 30 V		TBD	1 ms		5 V s^{-1}		0.005 V
< 500 V †		TBD	1 ms		5 kV s^{-1}		10%

   † *At current quench at a disruption*

4	Beta		Disruption avoidance		Diamagnetic Loops; Flux Contour Reconstruction
Voltage range		Numb. of Loops	Time res.	Rate of change	Accuracy
Beta range		Time resolution	Rate of change		Accuracy
$0.01 < \beta_p < 3$		1ms	0.05 s^{-1}		5% at β_p~1
		1ms	$<1 \times 10^5$ s^{-1} †		~30%

   † *At current quench at a disruption*

8	Locked-modes		Disruption avoidance		Short Flux Loops, ECE, Reflectometry
Parameter		Parameter range	Mode Number	Time res.	Accuracy
$\tilde{B}_r / B_p$		10^{-4}-10^{-2}	m/n=2/1	1ms	30%

(continued)

Table 1 (Continued)

9	m=2 MHD Modes, Sawteeth	Disruption avoidance	Mirnov Loops, ECE, Reflectometry, Neutron Flux Array, Soft X-ray Array
	Parameter	Frequency	Spatial res.	Accuracy
m=2 Mode	$\tilde{B}_p / B_p$	0-3kHz	-	10%
Sawteeth	$\tilde{T}_e / T_e$	0-3kHz	10 cm	10%

In addition the magnetic diagnostic should have the capability to measure the static error field generated by asymmetries in the TF and PF coil sets.

For following the development of small bore plasmas and movements during disruptive events it is useful to have the opportunity to construct the average properties of the plasma from the Shafranov integrals. These are also required for the derivation of the energy content from the measurement of the diamagnetism.

For fluctuation measurements the frequency range has to be extended to 100-200kHz, the range of TAE modes. For equilibrium measurements the plasma should be followed on a <100ms time scale, smaller than the L/R times of the conducting shells, requiring compensation of eddy currents.

From the technical side the sensitivity of the coils and loops must be determined in relation to the input circuits and the bit resolution of the data acquisition system. Further requirements originate from the ITER environment: vacuum, temperature, radiation, remote replacement and reliability.

PROPOSED CONFIGURATION

The proposed configuration is shown in figures 1 and 2. It consists of discrete measurements of differential flux and poloidal field along the contours of the vessel (24 positions) and of the back plate (15 positions). Both sets have a substantial number of voltage loops (typically 10). The set on the back plate has tangential poloidal field coils for equilibrium and for the measurement of high frequency oscillations. These coils are mounted in the poloidal gaps between the modules. The set on the inner wall of the vessel has a mixture of tangential and normal field coils

Both sets have a 5 or 10-fold toroidal distribution for analysis of the toroidal mode number and for the necessary redundancy.

The set on the inside wall of the vessel is in a low radiation field and can survive for the life time of the machine; the set on the back plate (deep in the blanket structure) may need replacement.

In the divertor cassette special sets are embedded. They are housed in the standard cassettes next to the 4 toroidally equally spaced diagnostic cassettes, 8 in total (but not equally spaced).

In addition 4 Rogowski loops and 4 diamagnetic loops, external to the vessel and mounted on the TF casing, are planned.

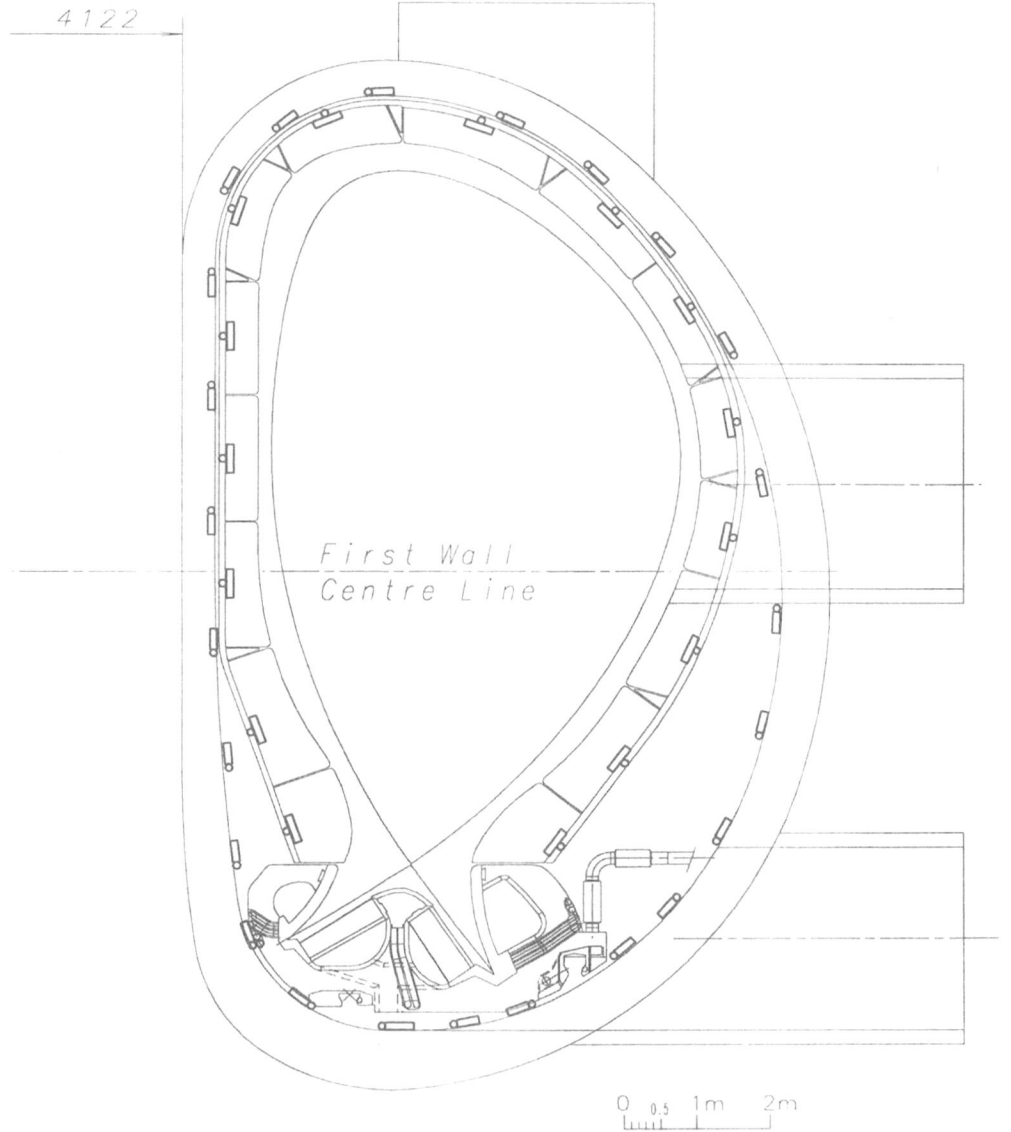

Figure 1. The locations of the flux and field measurements on the inner wall of the vessel and on the plasma side of the back plate are indicated by open circles and small boxes respectively. The flux measurements are a mixture of voltage loops (full flux measurements) and saddle loops (differential flux measurements). The poloidal field measurements on the vessel wall are a mixture of tangential and normal field measurements; on the back plate there are tangential equilibrium and h.f. coils at all locations. To be optimised.

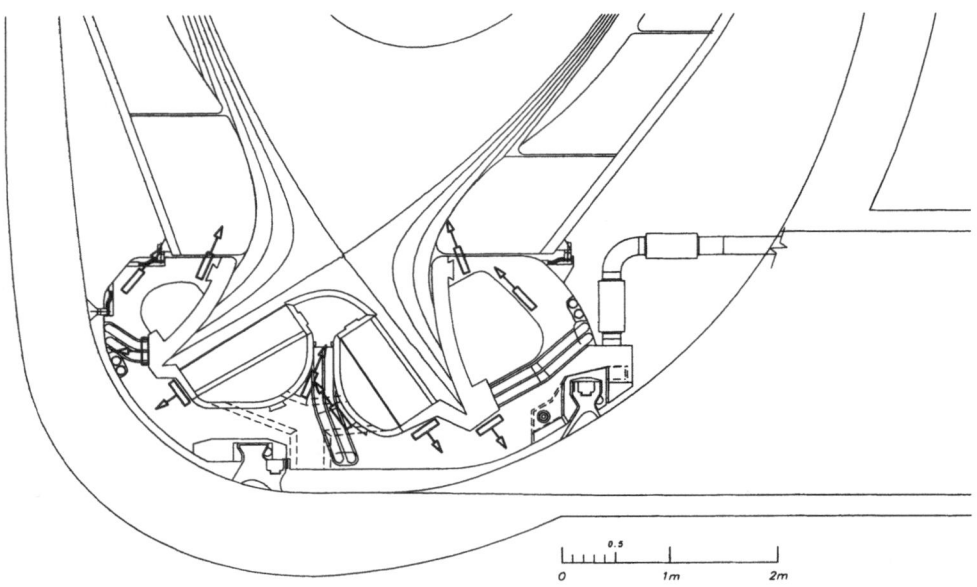

Figure 2. The locations of the poloidal field coils in the divertor. A somewhat arbitrary mixture of normal and tangential coils is used in various locations in the divertor. The choice depends on technical constraints. To be optimised.

SENSITIVITY AND S/N RATIO.

The factors that affect the lowest detectable signal are: the effective area of the coils and loops; the gain of the front end integrator or amplifier related to the bit resolution of the analogue to digital conversion; the expected noise and uncompensated drift (off-set) of the input circuit; and finally the interference.
Both in the poloidal field measurement and in the flux measurements by means of saddle loops a small component of the TF will be present. In ITER where the TF coils are super conducting the field will be switched on continuously with only a small variation of 1% allowed during the plasma pulse. Even under large misalignments this will give a negligible contribution to the expected signal and we will ignore it for this discussion.

The poloidal field measurement

For the estimate of the S/N ratio of the poloidal field measurement the following assumptions are made:
+ 16 bit ADC with maximum range of 10 V and one bit for the polarity (2^{15}=32768 or 1 bit= 300μV)
+ Effective area of the pick-up coil NA=0.3 m^2
+ Noise ≈ 1 mV pp   estimates for the integrator
+ Offset ≈ 5 mV
The measured signal is Φ_p = 0.3 Wb assuming B_p = 1 T at full plasma current I_p = 21 MA.
Assuming that 10 V of the ADC signal represents 0.5 Wb with a fixed gain of the integrator, the signal is 6 V +5mV (offset) + 1 mV (noise).

The error in this measurement is 0.1% at full plasma current.

A signal of 120 mV has therefore 5% maximum believable error corresponding with I_p=0.5 MA, similarly 0.6 V gives 1% at I_p = 2.5 MA.

The errors do not depend strongly on the poloidal position except near the X-point, where the relative errors become larger. This is not relevant for the start-up. The bit noise is so small that it is negligible. Calibration errors can be kept below 1%.

The flux measurements

The differential fluxes on the poloidal contour of the vessel or any other toroidally symmetric structure are measured by means of saddle loops. For the estimate of the voltage measured by a typical saddle loop the important parameter is the value of $d\Psi/dr$ giving the separation between flux surfaces, where r is the minor radius. The flux difference between two surfaces Δr apart is:

$$\Delta\Psi \cong 2\pi R\Delta r B_p$$

for a separation Δr = 5 cm near the outer mid plane R=12 m and B_p =1 T (at full current) we find $\Delta\Psi$= 3.76 Wb. This is the poloidal flux difference that typical neighbouring full flux loops would measure and represents a normally occurring mismatch between the poloidal contour of the vessel and the plasma boundary. On the top of the plasma and near the divertor the differential flux, as it is measured by saddle loops, is much larger due to the strong vertical component of the poloidal field.

Assuming now that the saddle loops cover a 1/20 toroidal sector the measured flux is only 1/20 of the flux calculated earlier. We obtain $\Delta\Psi_m$= 0.19 Wb. Again we neglect the erroneous TF pick-up.

Let us now assume that 10 V on the ADC corresponds to 0.2 Wb and that the values for noise and offsets are the same as for the poloidal field measurements. For the full plasma current the signal is:

9.5 V + 5 mV (offset) + 1 mV (noise)

with a 0.06% relative error.

At I_p= 1.5 MA there would be a 1% relative error in the flux measurement at the outer mid plane. This error of 1% represents an error in the reconstruction of 0.3mm using the expression for $\Delta\Psi$. Note that this expression cannot be used for accurate calculations to locate the boundary since the approximation is to crude.

The true signal can vary enormously following a series of differential flux measurements along the poloidal contour of the vessel. On the top and bottom one expects a large signal due to the contribution of the vertical field. Near the outer and inner mid plane the saddle loops may lay on a flux surface bringing the true signal close to zero. This does not affect the accuracy of the reconstruction since the possible error does not represent a large distance between the flux surfaces, although the relative error may be larger than 100%.

RECONSTRUCTION OF THE EQUILIBRIUM

Various techniques are available to calculate the magnetic equilibrium configuration from magnetic measurements. Several investigators have used the following methods to investigate the accuracy of the reconstruction from the magnetic measurements[2]:

a) Variable and Fixed Current Filament models (VCF and FCF)
b) Local Field Expansions (LFE)
c) Full MHD reconstruction (Grad-Shafranov) in an infinite domain: EFIT [3]
Under study are reconstructions based on statistical methods: Function Parametrisation (FP) and Neural Network Analysis (NNA).
Case a) and b) are fast methods for finding the boundary only and suitable for control; Case c), EFIT, provides the complete MHD equilibrium which includes internal parameters. This method is presently too slow for on-line control, but attempts are underway to improve the speed and, with improved computational hardware, some variation of this technique should be capable of real time control. The CF method is used in a mixed mode of VCF and FCF for efficiency.
The results of the sensitivity studies are shown in Table 2. The SOB (start of burn) plasma of TAC 4 [4] was used in the analysis. All reconstructions examined the effects of statistical errors on the "measurement" to obtain variations in the plasma parameters.

Table 2: Comparison of results from reconstruction codes on the plasma shape from random measurement errors

	EFIT	CF	LFE
inner boundary R (cm)	2.37	2.68	4.78
outer boundary R (cm)	1.67	2.54	3.56
X-point Z (cm)	3.72	0.87	1.52
R (cm)	0.53	1.74	1.46
imposed rms errors	3%	3%	3%

The table shows the results for the nominal equilibrium and 3% error in the signals; actual errors are expected to be much smaller. The full set of probes has been used in the above table but without the flux measurements on the back plate and using the old TAC 4 configuration.

EDDY CURRENTS

The response time of the magnetic diagnostic is one of the factors that have to be taken into consideration for the analysis of the plasma shape feedback control system. For diagnostic purposes we wish to find the equilibrium on time scales short relative to the L/R time of the axisymmetric structures surrounding the plasma.
Firstly there are fixed delays due to the sampling rate and the cycle time of the algorithm that calculates the equilibrium. The maximum specified sampling rate is 1 msec. Typical delay times due to the processing time of the algorithm are around 1 msec. Data transfer times have to be minimised.
Secondly there are phase delays occurring when magnetic fields penetrate conducting structures. The eddy current penetrates the structures with the skin time constant $\tau_{sk} = \mu_0 d^2/2\eta$
where d = thickness of the shell

η = resistivity in Ωm
We find for the first wall (5 mm copper) 1 msec; for the back plate (10 cm SS) 8 msec. During this time the description of the fields is inaccurate because the

currents flow in a variable location for the skin time quoted.

We propose to treat these deviations as error fields, which are relatively small. After this the eddy current is uniform and decays with a L/R-time given by:

$$\tau_{pen} = \mu_0 rd/2\eta$$

where r = radius of the shell

We find for each of the shells in ITER approximately 0.5 sec. With a known τ_{pen} the compensation can be done by adding a correction $+ \tau_{pen} dB/dt$ to the integrated signal. Alternatively, a circuit model can be included to simulate the eddy currents as an integral part of the reconstruction process.

When the eddy currents are measured on-line, the currents can be directly introduced in the analysis code.

Non-axisymmetric eddy currents which occur around major ports and in shield modules can be treated in a similar manner. Shield modules are not connected to the neighbouring modules to form a closed shell. These effects can be modelled as soon as the design is completed. The expected L/R times are around 10 msec .

TECHNICAL SOLUTIONS

All technical solutions are in the proposal stage. This chapter is therefore a design status report.

Equilibrium coils

The standard poloidal equilibrium coils are wound from 2 mm diameter Mineral Insulated (MI) coaxial cable. For the tangential field coil 4 layers on a body 20 cm long and a external diameter of 5 cm give an effective area of 0.3 m^2. For the normal field the coil is flat with a diameter of 12.4 cm and a length of 3 cm, also giving an effective area of 0.3 m^2. The equilibrium coil in the blanket might need active cooling.

High frequency coils

The high frequency coil consists of a ceramic body with grooves to hold the (single) layer of turns, again of 2 mm diameter MI cable. The typical dimensions are 20 cm long and diameter 2 to 5 cm depending on available space. If active cooling is required the cooling tube is wound between the active turns.

The response of the coil is determined mainly by two effects: attenuation due to the metal shell around the windings, and a damped resonance involving the inductance of the coil and the capacitance of the cable. The resonance produces a large phase shift near the resonance frequency but careful matching of the coils produces a useful range up to 200 kHz.

Flux loops

Full flux loops on the inside wall of the vacuum vessel are made of sectors attached to the vacuum vessel units. They are supported from alumina supports. Depending on the assembly sequence they are connected by crimping or an automatic connector. On one of the vessel units a connection to a twisted pair is made to take the signal out. On the plasma side of the back plate the full flux loops are mounted after completion of the back plate and are made of 5mm diameter MI cable. There are typically 10 full flux loops in both locations.

All saddle loops are 2mm diameter MI cable directly mounted on the vessel wall or back plate. On the vessel wall they are extending over 1/20 of the toroidal circumference; on the back plate it can be as much as 1/10. Special saddle loops can be installed for specific purposes such as detection of locked modes.

Integrators

Two schemes have been proposed to compensate the long term drift of the integrator. Both have a measurement time in which the signal plus drift is recorded and a similar time in which only the drift is measured. A software step is required before the actual signal is obtained.

In the first scheme I, proposed in ref. [5], the first step is a passive integration to take out large transients and smooth the signal. This signal is fed into an integrator and digitised. The input of the integrator is alternately switched to a dummy load to record the drift level. The signals are stored at alternate places in the digital memory (see Fig. 3a).

A small auxiliary periodic signal is added before digitisation to increase the resolution between bits. This scheme has passed the first encouraging tests.

In the other scheme II, proposed in ref. [6], two, not necessarily identical, integrators are used: integrator A will measure the signal+drift and integrator B is connected to a dummy to measure the drift of channel B. After a number of data points the role is switched. The transition is such that there are a few data points of signal + drift of both integrators overlapping so that a continuous signal can be obtained (see Fig. 3b).

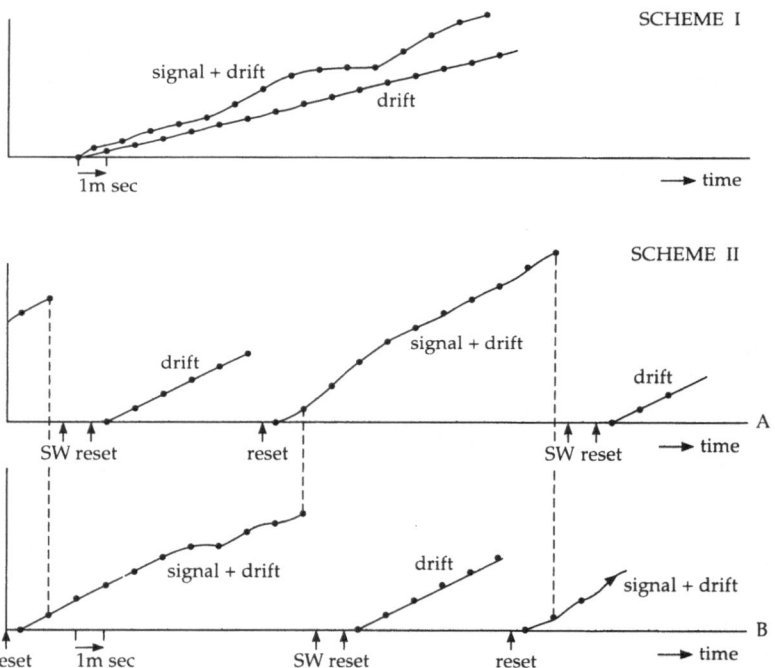

Figure 3. Schematic of the data taking for the two different schemes for integrator drift compensation. For clarity only a small number of data points is shown. The data sampling rate is chosen to be 1 kHz. In scheme I **(a)** there is a switching action between all data points; in scheme II **(b)**SW means cross over switch between coil and dummy.

Although both schemes need a software step to calculate the corrected signal the sampling rates required are modest and can easily be met.

Since both schemes operate continuously there is in principle no time limit, but there is a fundamental limit posed by the detectability of the input signal.

CONCLUSION

Due to the availability of drift compensation schemes for pulses longer than 1000s a classical inductive magnetic diagnostic can be used. It will be enhanced by a limited number of non-inductive measurements to be able to detect drifts and to gain experience for true steady state conditions.

A substantial number of flux measurements will be done by voltage loops to avoid toroidal asymmetries. Mode number analysis will be possible using the proposed 5 or 10 fold toroidal periodicity of the sets of sensors.

The trial reconstructions done so far indicate that the required accuracy of the diagnostic can be met under actual conditions.

Eddy current effects can be avoided by placing the magnetic sensors in front of the back plate and leave slots in the first wall at the locations of the poloidal field coils. Also high frequency pick up coils need this facility and the frequency response can be in the range of 100-200 kHz.

If used in combination with a fast boundary reconstruction code the proposed system will fulfil the plasma control requirements.

REFERENCES

1. Minutes of the 2nd meeting of the ITER Experts Group on Diagnostics, Naka Joint Work Site, Japan, February 1995
2. Report on the IAEA Technical Committee Meeting on Magnetic Measurements for Large Fusion Devices, Kharkov, Ukraine, 5-7October 1994. Nucl. Fusion to be published.
3. L.L. Lao, et.al. "Reconstruction of Current Profile Parameters & Plasma Shapes in Tokamaks", Nuclear Fusion, **25**, (1985), 1611.
4. TAC 4, ITER Documentation.
5. D. Baker, L.L. Lao, E.J. Strait. "Magnetic Diagnostics for ITER", Report GA-C21299, June 1993.
6. S. Ali-Arshad, L. de Kock, Rev. of Sci. Instrum. **64** (1994) 2679-82.

MAGNETIC EQUILIBRIUM RECONSTRUCTION TECHNIQUES
FOR TOKAMAK REACTORS

Enzo Lazzaro

Istituto di Fisica del Plasma
CNR-ENEA-Euratom Association
via Bassini, 15 - 20133 Milano, Italy

INTRODUCTION

The successful operation and performance of both present-day experimental tokamaks and future tokamak reactors, such as ITER, relies heavily upon an accurate determination of plasma position, shape and energy content, and on other global and local equilibrium quantities. Part of this information is required on a real-time basis to allow control of the discharge and possibly, in a reactor, on-line improvements of performances. For in-depth scientific analysis of the results and comparison with theory, however, it is also necessary to perform, both as snapshots and as historic time-evolution, off-line equilibrium reconstructions from the available measurements. As a rule, this is required with high space- and time-resolution and high accuracy.

Over the last decade, the reconstruction of the equilibrium configuration of a tokamak has become a multi-diagnostic problem of treatment of input data of diverse nature. A simpler mathematical problem involving fewer variables is embedded in this problem. Strict agreement must result among the measurable and the calculated quantities involved in the mathematical formulation. Further, no serious contradiction should arise between the reconstruction and measured quantities which are not explicitly present in the equations considered but are nevertheless expected to be involved in a broader definition of the equilibrium.

Purposes of this paper are: i) to review critically which quantities can be reliably reconstructed on the "prompt" and long processing time-scales; ii) to review critically which accuracy and resolution can be reasonably expected from treating a *finite* number of input data; iii) to discuss the propagation of errors effectively limiting the amount of information useful to obtain results of required accuracy.

In a long-pulse reactor environment, standard magnetic measurements are expected to suffer deterioration of the signal-to-noise ratio and lose reliability because of intense neutron fluxes and γ radiation damage[1]. Here it is discussed how the information provided

Diagnostics for Experimental Thermonuclear Fusion Reactors
Edited by P. E. Stott *et al.*, Plenum Press, New York, 1996

83

by non-destructive internal measurements, such as electron cylotron emission (ECE) spectra, polarimetry and reflectometry, motional Stark effects, etc., can be used to obtain a consistent reconstruction within a classical model of MHD equilibrium. The role of other data, like CCD camera images, X ray tomography, etc., and the perspectives of model-free analysis based on maximum entropy methods and neural networks, are also briefly discussed.

RECONSTRUCTION PROBLEM AND DATA SELECTION

The basic formulation of the problem relates to the ideal MHD volume-force balance condition,

$$\underline{J} \times \underline{B} = \nabla \cdot \underline{\Pi}, \tag{1}$$

involving the vector fields $\underline{J}$ and $\underline{B}$, and the pressure tensor field $\underline{\Pi}$. In the cylindrical coordinates (R, ϕ, Z) and for an ideal axisymmetric tokamak, the equilibrium condition given by eq. (1) can be expressed in terms of the single scalar magnetic stream function $\Psi(R, Z)$, with a magnetic field of the form

$$\underline{B} = RB_\phi \nabla\phi + \nabla\phi \times \nabla\Psi \tag{2}$$

and an isotropic scalar pressure,

$$P(R, Z) = n_e(R, Z)T_e(R, Z) + n_i(R, Z)T_i(R, Z), \tag{3}$$

constructed with the charged species densities and temperatures. The field function $\Psi(R, Z)$ is given by the famous Grad-Shafranov (G-S) elliptic equation:

$$R^2 \nabla \cdot \left[R^{-2} \nabla \Psi \right] = -\mu_0 R \left[P' + \frac{FF'}{\mu_0 R^2} \right], \tag{4}$$

featuring two functions $P(\Psi)$ and $F(\Psi) = RB_\phi$ which can be assigned freely.

For any arbitrary choice of the source functions and given boundary conditions of the Dirichlet type, $\psi|_\Gamma = f(s)$ on a contour Γ, a direct problem (DP) is formulated for the field $\Psi(R, Z)$. The experimental measurements here are a finite number, N, of scalar values $g_n = F_n\{\Psi(J)\}$ which are (or are supposed to be) functionals of Ψ. A typical inverse problem[2,3,4] (IP) for eq. (4) aims at reconstructing the field $\Psi(R, Z)$, and more specifically the source current-density profile $J_\phi(R, \Psi)$, by fitting, with prescribed tolerance, N functionals $F_n\{\Psi\}$ to N data g_n, subject to the condition of satisfying eq. (4) with boundary conditions $\psi|_\Gamma = f(s)$.

The workhorse of equilibrium measurements in contemporary tokamaks is provided by a finite number of magnetic flux measurements on Γ and a finite number of tangential magnetic field measurements along the same contour[1], possibly interpolated to provide the values of both $\psi|_\Gamma = f(s)$ and its normal derivative, $\underline{n} \cdot \nabla\psi|_\Gamma = b(s)$. In these terms the problem consists in solving an elliptic equation subject to Cauchy boundary conditions. This problem however is *ill posed*[5-7]. Small perturbations in the boundary data will in fact propagate as large changes in the solution a short distance away from the measuring contour.

In principle, in absence of measurements internal to the plasma it is impossible to determine the spatial structure of the current density J. The Green theorem applied to the G-S equation, eq. (4), actually shows[2,8,9] that a suitable surface-current can produce the same Ψ external to that surface as any distributed current inside it. However, it has been shown that the plasma boundary and a set of formal moments of the toroidal current distribution can be obtained exactly, and without any need of information from the interior, by appropriately weighted contour integrals[10] of the tangential and normal components of the magnetic field, B_τ and B_v, measured by pick-up coils placed around the meridian contour Γ of the tokamak vessel. One has:

$$Y_m \equiv \frac{1}{I_p} \int J_\phi f_m dS_\phi = \frac{1}{\mu_0 I_p} \int [B_\tau f_m + RB_v g_m] dl_\theta, \tag{5}$$

where $\nabla \cdot R^{-2} \nabla f_m = 0$. The moments of lowest order give the coordinates of the centroid of the current distribution, thereby also providing information about the dominant scale-length of the regular solution in the domain. Accordingly, magnetic data can be said to have a basic resolution in space of the same order of minor radius. Fine structures with shorter scale lengths may still exist but will be invisible to magnetics.

The ill-posed problem of full equilibrium reconstruction can be reduced to a well-posed problem, within the given domain, by exploiting the freedom in the choice of the source functions to express them in a parameterized form[3,4,9,11] of the type:

$$J_\phi(R, \Psi; a_r, b_r) = R \left[P'(\Psi; a_r) + \frac{FF'(\Psi; b_r)}{\mu_0 R^2} \right], \tag{6}$$

with M parameters, and by using suitable regularisation constraints which restrict the class of admissible solutions. The regularisation technique is a form of the well-known 'constrained maximum likelihood' estimation[12] and is obtained by requiring the best fit of various supplementary data through minimization of the function

$$\chi^2 = \sum_{k=1}^{N} \left(\frac{F_k(\Psi, \underline{a}, \underline{b}) - g_k}{\sigma_k} \right)^2 + \kappa\Omega(\Psi) \xrightarrow[\Delta^*\psi = -\mu_0 RJ_\phi]{} min, \tag{7}$$

subject to the constraint of satisfying eq. (4) with given total current[2-4]. The smoothness of the solution is controlled by the regularizing (penalty) term $\kappa\Omega(\Psi)$. This in turn can take either a "tensioning" form[3,9],

$$\kappa\Omega(\Psi) = \kappa_1 \int_0^1 \left(\frac{\partial^2 J}{\partial \psi^2} \right)^2 d\rho + \kappa_2 \int_0^1 \left(\frac{\partial^2 n_e}{\partial \psi^2} \right)^2 d\rho, \tag{8}$$

or a "maximum entropy" form[13], $\kappa\Omega(\Psi) = \kappa \int \Psi \ln(\Psi) d\tau$. If the number of free parameters (typically the nodal values of an interpolating tensioned spline[3,9]) is M< N then, only provided that χ^2 is convex[7], the resulting (overdetermined) nonlinear least-square problem will have a unique solution. Since for $v \equiv (N - M) \to \infty$ the function χ^2 becomes normally-distributed[12] with mean v and standard deviation $\sigma_{\chi^2}^2 = 2v$, its minimum should be ~ N - M. If M>N, instead, the solution of the underdetermined problem is no more unique. Nevertheless, among the infinity of degenerate solutions there will still always be a principal solution of minimal norm (obtainable by 'singular-value decomposition' techniques) which is unique[5-7]. In particular, if $\kappa = 0$ and M>N a condition of meaningless overfit[3,12], $\chi^2 \to 0$, can occur while, for N data, the expected value should be $\chi^2 \approx N$. It is thus seen that the number of free parameters which can be determined meaningfully is limited in all cases[3,4,9,11].

In modern tokamaks, the N = I + J + K + M input data for the inverse problem could be the measurements

$$
g_{N=I+J+K+M} = \left\{ B_{\tau_i}\big|_{\Gamma} ; \alpha_{Stark\,j} ; \alpha_{Far\,k} ; P_{Kin}(R_m) \right\}, \tag{9}
$$

which include I values of the magnetic field tangential to the contour Γ, J values of the motional-Stark-effect pitch angle, K values of the Faraday-rotation angle along chords crossing the plasma and M values of the kinetic pressure, each of which with its statistical standard deviation. The success of the procedure will depend on the convexity[2,7] of the function χ^2. Adding measurements which are not strictly related to the governing equations may destroy the convexity!

The information achievable from non-magnetic data will be valuable in two cases: i) if it contributes to reduce the resolution scale-length without deteriorating the accuracy, i.e. the response to errors in the data; ii) if it provides additional constraints on the solution strictly compatible with the equilibrium equation. In the case of model-free reconstruction techniques, the basic estimate of resolution is provided by the Nyquist sampling criterion for discrete data[12]. In this case the resolution scale-length is twice the sampling interval. In the case of polarimetry, for instance, the basic scale-length which can be trusted for any fine structure found in the reconstructed $n_e B_z$ distribution[14-17] is the distance between the measuring chords. The "true" solution might still have shorter scale-lengths: simply, they are not contained in the available data and as such are inaccessible for a reconstruction. The reconstruction from finite data is always a finite-dimensional image of the "true" solution.

The problem always being that of a projection on a finite-dimensional space of an ill-posed infinite-dimensional problem which lacks continuous dependance on the data, by increasing the number of free parameters and of additional data to improve the resolution, also the errors in the data will be amplified. As a consequence, the accuracy may decrease[6] to the point that no more true information about the solution is actually gained. Instead, if a model of the object to be identified (e.g., the G-S equation) is used (parametric identification) then the Nyquist criterion is no longer so relevant. The resolution in this case does no more depend strictly, or solely, on the sampling interval.

As Tab. 1 clearly shows, actually no measurements in fusion devices have uncertainties of less than 1%. On the other hand, no "model-free" reconstruction, based for instance on multivariate regression, can produce results whose accuracy is better than the uncertainty in the signals used. The addition of other input data, therefore, will be useful only provided that the associated statistical error is smaller. Furthermore, not all the new data will "belong" to the equations of the problem (to be useful, they must usually be combined with other data) and as a result the model-free reconstruction could be biased by the choice of these data itself.

Table 1. List of relevant measurements and associated uncertainties .

Magnetic flux	$\psi_i = \Psi(R_i, Z_i)$	$\varepsilon_{\psi_i} \approx 1 - 3\%$
Tangential magnetic field	$b_i = B_\tau(R_i, Z_i)$	$\varepsilon_{b_i} \approx 1 - 3\%$
Line-integrated density	$N_k = \int_{L_k} n_e(R_k, Z) dZ$	$\varepsilon_{N_k} \geq 5\%$
LIDAR electron pressure profiles	$P_e(R_{min} < R < R_{Max})$	$\varepsilon_{P_{ef}} \geq 10\%$
ECE electron temperature profiles	$T_e^{ECE}(B(R > R^*))$	$\varepsilon_{T_{ece}} \approx 10\%$
Soft X-ray emission profiles	$I_{SX}(T, n_e, Z_{eff})$	$\varepsilon_{I_{SX}} \approx 5 - 10\%$
Charge-exchange ion temperature profiles	$T_i^{CX}(R_1 < R < R_2)$	$\varepsilon_{T_{CX}} \geq 10\%$
Z_{eff} profiles	$Z_{eff}(R, Z)$	$\varepsilon_{Z_{eff}} \geq 20\%$
Charge-exchange ion toroidal rotation profiles	$\omega_{CX}(R_{0CX} < R)$	$\varepsilon_{\omega_{CX}} \approx 20\%$
Faraday-rotation angle along L_F chords	$\alpha_{Far} = c_3 \int_{L_F} n_e(z, R) B_z dz$	$\varepsilon_{\alpha_{Far}} \leq 5\%$
Motional Stark-effect pitch angle	$\alpha_{Stark} = \arctan(B_z / B_\phi)$	$\varepsilon_{\alpha_{Stark}} \approx 1\%$

ACCURACY AND RESOLUTION

The shortcomings of ill-posedness also affect the simpler homogeneous problem of boundary identification[2,8,18,19]. Accuracy and space resolution are competing requirements and it is important to recognise that there is an optimum number of data[18] (say, Fourier harmonics) allowing reconstruction of the most accurate ψ. The existence of an optimum is the outcome of the competition between a poor resolution, as in the case of too few Fourier harmonics, and an accuracy which, because of the ill-posedness of the Cauchy problem, degrades with increasing the number of harmonics.

As an illustration, it may be useful to extend to the toroidal geometry a classical example by Hadamard[5]. Let us consider the solution of the homogeneous G-S equation inside a domain D with associated Cauchy boundary conditions on the surrounding contour Γ,

$$R = \frac{R_0 Sh\xi}{Ch\xi - \cos\eta}; \quad Z = \frac{R_0 \sin\eta}{Ch\xi - \cos\eta}; \quad \phi = \phi. \tag{10}$$

Writing the Beltrami operator in these coordinates, the Grad-Shafranov equation becomes:

$$\left[\frac{\partial^2}{\partial\eta^2} + \frac{\partial^2}{\partial\xi^2}\right]\Psi + \frac{\sin\eta}{Ch\xi - \cos\eta}\frac{\partial\Psi}{\partial\eta} - \frac{1 - Ch\xi\cos\eta}{Ch\xi - \cos\eta}\frac{\partial\Psi}{\partial\xi} = 0, \tag{11}$$

which can be trasformed in a separable form by the substitution:

$$\psi = U(\xi,\eta)/\sqrt{Ch\xi - \cos\eta} \tag{12}$$

So one obtains:

$$U_{\xi\xi} + U_{\eta\eta} - Coth\,\xi U_\xi + \frac{1}{4}U = 0. \tag{13}$$

Setting $U(\xi,\eta) = F(\eta)\,G(\xi)$, the "angular" equation for F gives $F(\eta) = M_m \cos(m\eta) + N_m \sin(m\eta)$. The "radial" equation for $G(\xi)$ in turn is a hypergeometric equation whose solution is:

$$G_m(\xi) = \frac{\Gamma(m - 1/2)}{\Gamma(m + 1/2)}Sh\xi\left[iA_m P^1_{m-1/2}(Ch\xi) - B_m Q^1_{m-1/2}(Ch\xi)\right]. \tag{14}$$

On a surface $\xi_s > 0$ and in the range $-\pi < \eta < \pi$ we assume the following Cauchy boundary conditions:

$$U(\xi_s,\eta) = 0; \quad \frac{\partial U(\xi_s,\eta)}{\partial\eta} = \frac{\sin(m\eta)}{m}. \tag{15}$$

Now, since for m growing to infinity both boundary conditions go to zero, the solution of the homogeneous equation goes to $U = 0$. In order to have continuous dependence on the data at least in a neighborhood of $\xi = \xi_s$ then one should have

$$\lim_{m\to\infty} U_m(\xi_s,\eta) = 0. \tag{16}$$

However, this does not occur. From eqs. 14 and 15 one in fact gets

$$U_m(\xi,\eta) = c_m\left[\frac{Q^1_{m-1/2}(Ch\xi_s)}{P^1_{m-1/2}(Ch\xi_s)}P^1_{m-1/2}(Ch\xi_s) - Q^1_{m-1/2}(Ch\xi_s)\right](Sh\xi)\sin(m\eta), \tag{17}$$

the constants c_m being determined by the equation

$$\left.\frac{\partial U_m(\xi,\eta)}{\partial \xi}\right|_{\xi_s} \equiv c_m \left[\frac{Q^1_{m-1/2}(Ch\xi_s)}{P^1_{m-1/2}(Ch\xi_s)} \left(P^1_{m-1/2}\right)' - \left(Q^1_{m-1/2}\right)' \right] (Sh\xi)^2 \sin(m\eta) = \frac{\sin(m\eta)}{m}. \quad (18)$$

Therefore, taking into account the asymptotic expansions of the associated Legendre functions for large index and fixed argument (e.g., $\xi = \xi_a$, $\eta = \pi/2m$), one has:

$$U_m(\xi_a, \pi/2m) \approx \frac{e^{(m-1/2)\xi_a}}{m^{3/2}(Sh\xi_a)^{1/2}}. \quad (19)$$

In eq. (19) the ill-posedness appears as an exponential dependance on the harmonic index m. Because of ill-posedness, a high harmonic content in the data, which could otherwise improve the "resolution" of the solution, will lead to poorer accuracy. The cause of this is readily understood to be that the B.C. on the normal derivative drives the non regular part of the solution, which is divergent on the axis. Any error on the data is thereby amplified and will shadow the regular part, which is the meaningful solution of the elliptic problem.

ON-LINE PLASMA BOUNDARY IDENTIFICATION

Methods of fast identification of the plasma boundary have been successfully implemented in JET and other devices[19]. These are based on the local Taylor-series expansion of the flux function ψ in vacuum. The ill-posedness of the problem is taken into account by a "patchwork" of five regularized local expansions. Near an observation point (R_0, Z_0) the vacuum solution of the G-S equation is written as:

$$\psi(R,Z) = \sum_{\substack{i=0 \\ j=0 \\ i+j\leq 6}}^{6} a_{ij}\left(R^2 - R_0^2\right)^i (Z - Z_0)^j. \quad (20)$$

Imposing the fulfillment of the equation $\Delta^* \psi = 0$ gives 15 linear conditions on the coefficients a_{ik}. Rewriting the expressions of the magnetic flux and tangential field in terms of the remaining coefficients, the fitting of the measurement is imposed as a least-square problem of seeking the minimum of the statistic (cost) function:

$$\chi^2_\alpha = \tfrac{1}{2} \sum_{\text{flux}} w^\alpha_j \left[\psi^\alpha_{mj} - \psi^\alpha_{cj}\right]^2 + \tfrac{1}{2} \sum_{\text{field}} w^\alpha_j \left[B^\alpha_{mj} - B^\alpha_{cj}\right]^2 =$$

$$= \tfrac{1}{2} \sum_{j=1,N_m} w^\alpha_j \left[D_{mj} - \sum_k \psi^\alpha_{mj}(\hat\rho_k,\hat z_k)C_k\right]^2, \quad (21)$$

where the D_{mj} are flux or field measurements. The problem being ill posed, the error in the expansion will rapidly increase far from the region where the data for the least-square fit are located. A sort of uniform asymptotic expansion is obtained by using N (5 in JET) similar expansions and imposing that adjacent expansions agree, in the least-square sense and at a given point, in the region of validity for both, and by adding to the cost function a term of the type

$$w_j \left[\psi_1(R_j, Z_j) - \psi_2(R_j, Z_j) \right]^2 . \tag{22}$$

In this approach the ill-posed problem is effectively regularized and a satisfactory performance can be achieved for implementation of a real-time algorithm on microprocessors or four-port transputer networks. This fast analysis can be eventually completed by on-line evaluation of the global equilbrium properties ($\ell_i, \beta_{p//,\perp}, \mu_I$), as provided by the Zakharov-Shafranov-Shkarofsky current moments and by the Shafranov integrals[10,20,21]:

$$2\beta_{p\perp} + \beta_{p//} + \ell_I - \mu_I = S_1 + S_2 ;$$

$$\beta_{p//} + \ell_I + \mu_I = \frac{R_t}{R_0} S_2 ; \tag{23}$$

$$\beta_{p\perp} - \left(\frac{E^2 - 1}{E^2 + 1} \right) \ell_I - \mu_I = S_3 ,$$

where

$$S_{1,2,3} = \frac{1}{B_p^2(a)V_p} \int_\Gamma ds B_p^2 \left\{ \begin{array}{c} (R - R_0)\underline{e}_R + Z\underline{e}_Z \\ R_0 \underline{e}_R \\ Z\underline{e}_Z \end{array} \right\} \cdot \underline{n}. \tag{24}$$

IMPROVED ANALYSIS WITH NON-MAGNETIC DIAGNOSTICS

As already mentioned, coherently with the criteria given above, improvements in data analysis can be achieved by the use of non-magnetic diagnostics, typically tomography and polarimetry.

Tomography. The J profile can be inferred from a local measurement of the (non circular) shape of the magnetic surfaces, as provided by X-ray tomography or ECE isoemission curves. The geometry of isoemissivity surfaces is given in terms of a well-known parametric representation using elongation and magnetic-axis shift functions[22] of the form:

$$R = R_0 + \Delta(\rho) + \rho\cos\theta + \sum_2^n R_n \cos n\theta;$$

$$Z = \rho E(\rho)\sin\theta + E(\rho)\sum_2^n R_n \sin n\theta. \tag{25}$$

In the parametric representation (for n=1), the relevant metric-tensor elements and Jacobian are:

$$\sqrt{g} = RE\rho \left[1 + \rho\frac{E'}{E}\sin^2\theta + \Delta'\cos\theta \right]; \tag{26a}$$

$$g_{12} = -\Delta'\rho\sin\theta + \rho\left[EE'\rho + E^2 - 1\right]\sin\theta\cos\theta; \tag{26b}$$

$$g_{22} = \rho^2\left[\sin^2\theta + E^2\cos^2\theta\right], \tag{26c}$$

and the G-S equation takes the form:

$$\frac{R^2}{\sqrt{g}}\left\{\frac{\partial^2\psi}{\partial\rho^2}\left(\frac{g_{22}}{\sqrt{g}}\right) + \left[\frac{\partial}{\partial\rho}\left(\frac{g_{22}}{\sqrt{g}}\right) - \frac{\partial}{\partial\theta}\left(\frac{g_{12}}{\sqrt{g}}\right)\right]\frac{\partial\psi}{\partial\rho}\right\} = -[\mu_0 R^2\frac{dP}{d\rho} + F\frac{dF}{d\rho}]/\frac{\partial\psi}{\partial\rho}. \tag{27}$$

A set of moments of eq. (27), taken with weighting factors $R^k\cos(n\theta)$, leads to an algebraic system $\underline{A}\cdot\underline{Y} = \underline{C}$ for the functions

$$\underline{Y} \equiv \left\{-\frac{d\ln\psi'}{d\rho}; \left(\frac{1}{\psi'}\right)^2\frac{dP}{d\rho}; \left(\frac{1}{\psi'}\right)^2 F\frac{dF}{d\rho}\right\}. \tag{28}$$

The determinant of this system,

$$\det[\underline{A}] \cong -\rho(R_0 + \Delta)\frac{E^2 - 1 + 2\rho E'/E}{E^2} - \rho^2\frac{\Delta'\left(1 + 3E^2\right)}{4E^2}, \tag{29}$$

makes clear that non-circularity and non-concentricity of the magnetic surfaces $(E > 1, E', \Delta' \neq 0)$ are essential features for obtaining a solution $\underline{Y}$. In the limits of negligible Shafranov shift and $2\rho E'/E << E^2 - 1$, and in terms of the geometric parameters, one obtains expressions of the type:

$$\frac{d\psi}{d\rho} \cong \psi'_b\frac{\rho}{a}\frac{E^3}{\left[E^2 - 1\right]^2}. \tag{30}$$

Assigning $E_a = E(\rho = a)$ and $\psi'_b = \psi'(\rho = a)$, the current enclosed within a surface r then is

$$\mu_0 I(\rho) = -2\pi\frac{d\psi}{d\rho}\left\langle\frac{g_{22}}{\sqrt{g}}\right\rangle = \frac{\mu_0 I_\phi\rho^2}{\pi a^2}\frac{\left[E_a^2 - 1\right]^2}{E_a^2\left[1 + E_a^2\right]}\frac{E^2\left[1 + E^2\right]}{\left[E^2 - 1\right]^2}, \tag{31}$$

where I_ϕ is the total current. Further, from

$$F\frac{dF}{d\rho} = -\left\{\frac{3E^2 + 2}{2} + \frac{R_0}{a}\frac{3E^2 + 1}{8}\frac{\left(\rho^3\Delta'\right)}{\rho^3}\right\}\frac{(\psi')^2}{\rho E^2}; \tag{32}$$

$$\mu_0 R^2\frac{dP}{d\rho} = \left\{\frac{E^2}{2} + \frac{R_0}{a}\frac{3E^2 + 1}{8}\frac{\left(\rho^3\Delta'\right)}{\rho^3}\right\}\frac{(\psi')^2}{\rho E^2}, \tag{33}$$

both the paramagnetic function $F(\rho)$ and the pressure profile $P(\rho)$ can be obtained. In particular, the safety factor can be expressed as:

$$q(\rho) = \frac{F(\rho)}{2\pi\mu_0 I(\rho)}\left\langle\frac{g_{22}}{\sqrt{g}}\right\rangle\left\langle\frac{\sqrt{g}}{R^2}\right\rangle = q_{cyl}\left(\frac{F(\rho)}{R_0 B_0}\right)\frac{\left[1+E_a^{\,2}\right]}{2}\left(\frac{\left[E(\rho)^2-1\right]}{\left[E_a^{\,2}-1\right]}\frac{E_a}{E(\rho)}\right)^2 \tag{34}$$

with

$$q_{cyl} = \frac{2\pi R_0 B_0 E_a^{\,2} a^2}{\mu_0 I_\phi R_0^2}, \tag{35}$$

or

$$q(\rho) = q(0)g(\rho), \tag{36}$$

with

$$g(\rho) = \left(\frac{F(\rho)}{F(0)}\right)\left(\frac{\left[E(\rho)^2-1\right]}{\left[E_0^{\,2}-1\right]}\frac{E_0}{E(\rho)}\right)^2. \tag{37}$$

In Figs. 1 and 2 a test example is given, correct to within 15%, of reconstruction of the q profile of ITER from "measured" $E(\rho)$ and $\Delta(\rho)$.

Polarimetry. Polarimetric measurements are functionals of combinations of n_e and B. Whenever the shape of the magnetic surfaces is not circular[3,14-17], these functionals are not amenable to simple Abel inversion[14]. In general, the polarization state of a probing wave beam can be dscribed by a unit vector $\underline{s}$, whose components are the Stokes parameters[16]; the vector at the "exit" point is connected to that at the "entrance" point by $\underline{s}_1 = \underline{\underline{M}} \cdot \underline{s}_0$, where the elements of the transition matrix represent Faraday rotation and birefringency effects. This information can be fully included only in a complete numerical MHD reconstruction[16].

Nevertheless, local estimates of the safety factor $q(0)$ on axis can be obtained using only two central polarimetry chords, on both sides of the magnetic axis, assumed to coincide with the electron density maximum. The Faraday-rotation angle is in fact defined as

$$\alpha_F = c_3 \int_{z_0}^{z} n_e(z,R)B_z dz \tag{38}$$

where c_3 is proportional to the square of the wavelength.

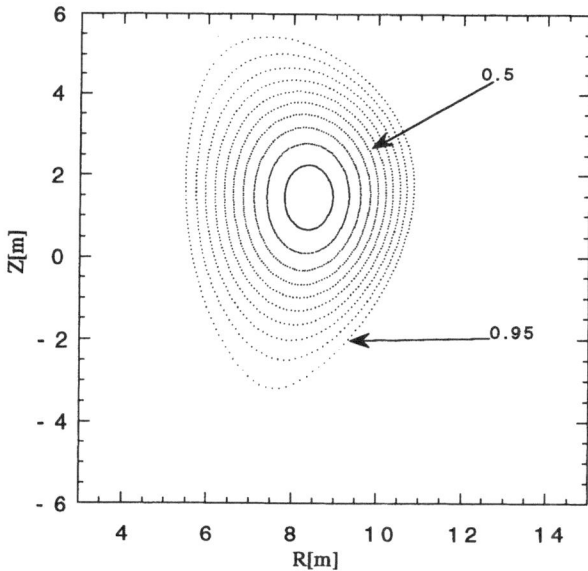

Fig1. Calculated ITER equilibrium with indication of two normalized flux labels.

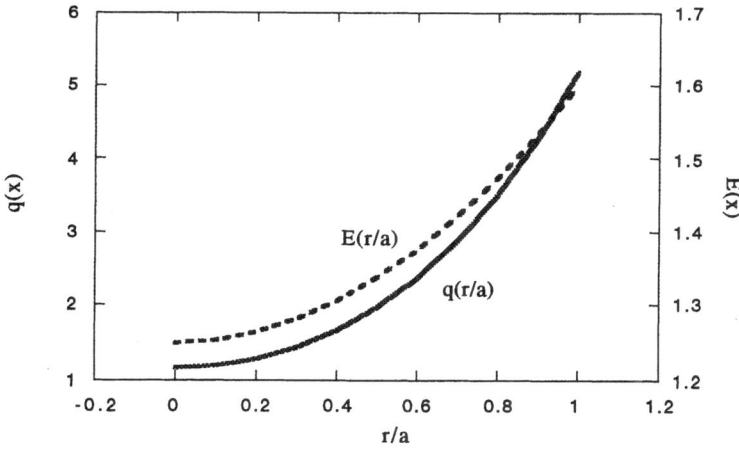

Fig2. Safety factor profile reconstructed from "tomographic measurement" of E(r/a)

The derivative in the direction of the major radius on the magnetic axis is:

$$\left.\frac{\partial\alpha_F}{\partial R}\right|_{R=R_{max(n_e)}} = c_3 \int_{z_0}^{z} dz n_e \left.\frac{\partial B_z}{\partial R}\right|_{R=R_{max(n_e)}} \approx \frac{\alpha_F(R_R) - \alpha_F(R_L)}{R_R - R_L}, \tag{39}$$

where

$$\left.\frac{\partial B_z}{\partial R}\right|_{R=R_{max(n_e)}} = \frac{B_0}{R_m} \frac{1}{q(R_m, Z)}, \tag{40}$$

since $\left.\frac{\partial q}{\partial R}\right|_{R=R_{max(n_e)}} = 0$. Therefore:

$$\left.\frac{\partial\alpha_F}{\partial R}\right|_{R=R_{max(n_e)}} = c_3 \frac{n_e(0)B_0}{R_m q(0)} \int_{z_0}^{z} dz' \frac{f(z')}{g(z')}, \tag{41}$$

where $f(\rho)$ is the density profile function and $g(\rho)$ the q-profile function provided by X-ray tomography or by ECE isoemission curves. Thus, by taking two Faraday-angle measurements on both sides of the magnetic axis an estimate of $q(0)$ is obtained in the form:

$$q(0) \approx \frac{c_3 n_e(0) B_0}{R_m \left.\dfrac{\partial\alpha_F}{\partial R}\right|_{R=R_{max(n_e)}}} \int_{z_0}^{z} dz' \frac{f(z')}{g(z')}, \tag{42}$$

which is weakly dependent on the density and current profiles.

MOTIONAL STARK EFFECT

Due to the motion across $\underline{B}$, a field $\underline{E}_{Stark} = \underline{V}_{neutr} \times \underline{B}$ induced on a neutral atom in its rest frame will cause a wavelength splitting of the emitted (and measured) radiation given by

$$\Delta\lambda_{Stark} = k^2 \left[\alpha_1 B_\phi + \alpha_2 B_p\right]^2. \tag{43}$$

The polarization direction of the emitted radiation gives the local pitch angle of the magnetic field, given by $\tan(\alpha_{Stark}) \equiv (B_z / B_\phi)$. The proof of principle has been given experimentally on JET, TFTR and DIII-D. The diagnostic is delicate but is nevertheless becoming of good accuracy since space resolution can be $\Delta r \sim \pm 2cm$, time resolution $\Delta\tau \sim 50ms$, and the uncertainty on total B field $\Delta B/B < 2\%$.

The local safety factor can be evaluated from motional Stark-effect (MSE) measurements and geometric information (E, Δ) to give:

$$q(\rho) \approx \frac{Z_\theta R^{-1}}{\tan(\alpha_{Stark})} \frac{\left\langle R^{-2}\sqrt{g} \right\rangle}{R^{-2}\sqrt{g}}. \tag{44}$$

This provides the most meaningful piece of information, to be added to the fast and full identification procedures[9].

CONCLUSIONS

The stringent requirements for a diagnostic of the equilibrium in a reactor environment can be best met with the cooperation of full numerical equilibrium reconstruction with tomographic and magnetoptic measurements.

Acknowledgments. I am grateful to U. Tartari for revision of the text and to J. Christiansen, L. De Kock, J. Ellis, D. O'Brien, J. De Haas, S. Nowak, G. Parravicini, M. von Hellerman, S. Segre for the material kindly contributed, which I have freely used for this, by necessity concise, overview.

1. L. De Kock et al., this volume.
2. J. Blum, *Numerical Simulation and Optimal Control in Plasma Physics*, Gauthiers Villars, Paris (1989).
3. J. Blum, E. Lazzaro, J. O' Rourke, B. Keegan and Y. Stephan, *Nucl. Fus.* 30:1475 (1990).
4. F. Alladio and F. Crisanti, *Nucl. Fus.* 25:1421 (1985).
5. A. Tikhonov, V. Arsenin, *Methodes de resolution de problemes mal posees*, MIR Ed., Moscow (1974).
6. M. Bertero, C. De Mol and E.R. Pike, *Inverse Problems,* 1:301 (1985);4:3573 (1988).
7. G.I. Marchuk, *Methods of Numerical Mathematics*, 2nd ed., Springer, N. Y. (1982).
8. W. Feneberg, K. Lackner and P. Martin, *Comput. Phys. Comm.* 31:143 (1984).
9. S.P. Hirshman et al., *Phys.Plasmas* , 1: 2277 (1994).
10. L.E. Zakharov, V.D. Shafranov, *Sov. Phys. Tech. Phys.*, 18:151 (1973).
11. L.L. Lao, H. St.John, R.D. Stambaugh, W. Pfeiffer, *Nucl. Fus.* 25:1421 (1985).
12. J. Bendat, A. Piersol, *Random Data: Analysis and Measurement Procedures*, J. Wiley, N. Y. (1971).
13. G.A. Cottrell, E.S. Fairbanks, R.E. Stockdale, *Rev.Sci.Instrum*, 56:984 (1985).
14. H. Soltwisch, *XI EPS Conf. Contr. Fusion and Plasma Phys.*, Aachen, 7D-I:123 (1983).
15. J. O'Rourke and E. Lazzaro, *XVII EPS Conf. Contr. Fusion and Plasma Phys.*, Amsterdam, 14B-I:343 (1990).
16. S.E. Segre,*Phys.Plasmas* 2:2908 (1995).
17. D. Wrobleski and L.L.Lao, *Rev.Sci.Instrum*, 63:5140 (1992).
18. S.P. Hakkarainen and J.P. Freidberg, MIT Rep. FC/RR-87-22, DOE/ET-501013-244 (1987).
19. J.J. Ellis et al., *IAEA Techn. Meeting on Magn. Diagn. for Fusion Plasmas*, Kharkov, Ukraine (1994)
20. P. Shkarofsky, *Phys.Fluids* 25 :89 (1982).
21. V.D. Shafranov, *Plasma Phys.* , 13: 757 (1971).
22. J.P. Christiansen and J.B. Taylor, *Nucl. Fus.* 22:111 (1982).

FAST AND ACCURATE METHODS OF PLASMA BOUNDARY DETERMINATION IN ITER FROM EXTERNAL MAGNETIC MEASUREMENTS

Yu.K. Kuznetsov, I.V. Yasin

Institute of Plasma Physics, Kharkov, Ukraine

Three methods have been developed and examined in numerical experiments: the variable current filament (VCF) method, the fixed current filament (FCF) method and local field expansion (LFE) method for plasma boundary determination in a tokamak from external magnetic measurements.

CURRENT FILAMENT METHODS

The current filaments methods are based on the approximation of the plasma current by a finite set of current filaments, lying inside the plasma. This model allows to represent any measurable quantity which is related to the poloidal magnetic field generated by the plasma (magnetic field components, fluxes, or their combinations) as:

$$y(r,z) = \sum_{j=1}^{N} G(r,z,r_j,z_j)I_j \qquad (1)$$

where G is known Greens function, r_j, z_j and I_j are co-ordinates and current of the j-th filament. The co-ordinates and currents of the filaments are the parameters of the model (1). The problem of determination of the field generated by the plasma come to calculations of these parameters from the system of algebraic equations of the form:

$$\sum_{j=1}^{N} G(r,z,r_j,z_j)I_j = \overline{y}_i \qquad i=1,...,M \qquad (2)$$

where $\overline{y}_i$ is the value of the i-th measurement, M is the number of measurements. The system of equations (2) for unknown parameters is solved by the least square method, which allows to minimize the functional, its general form being

$$J = \sum_{j=1}^{M} \frac{(y_i - \overline{y}_i)^2}{\sigma_i^2} + \alpha\Phi(a) \qquad (3)$$

Diagnostics for Experimental Thermonuclear Fusion Reactors
Edited by P. E. Stott *et al.*, Plenum Press, New York, 1996

97

where y_i is the value of the i-th measurement, as calculated in accordance to (1), σ_i is the dispersion of the random error of this measurement. The second term in (3) is introduced by necessity to stabilize the solving procedure, a is the vector of the parameters to be searched for, α is the regularization parameter. The need for regularization is connected with an ill-posedness of the problem (2) in the form of a possible instability of the solution procedure with respect to small errors in measurements and calculations. The stability of the system (2) solution is ensured by choosing a sufficiently small number N of unknown parameters (truncation) or by introducing some regularizing term (damping), if the number of parameters is large(too sensitive model).

In the model (1) general case, the unknown could be both filament currents and their co-ordinates. Accordingly, we shall distinguish the VCF method and the FCF method. In the VCF method case, the number of unknown parameters includes the filament co-ordinates, while in the FCF case the filament co-ordinates are fixed. The advantage of the VCF method is in its high universality in respect of plasma shape and position but a large amount of calculations is required. The FCF method is more simple but lacks universality. Therefore an option of some method is possible, depending on the conditions of the experiment.

The total vacuum field and the plasma boundary in the VCF and FCF methods are calculated, supposing that the external current field is known and is produced with the poloidal field coils only. This is valid only if any significant eddy currents are absent. Taking into account the eddy currents needs a model of these currents.

In the VCF method having been developed all the current filaments are considered as known. A good accuracy of the method is available when the current filaments are equal. In the VCF method the Eq. (2) represents a system of non-linear algebraic equations for the filament co-ordinates and is solved by the Newton iteration technique. Numerical experiments have shown, that the convergence of the iteration procedure is weakly sensitive to the initial filament position.

In the FCF method, to show the flexibility of the model (1), we use the combination of neighbouring filaments in filament groups with equal currents. This enables to vary the number of unknown parameters N with a given number of filaments, and thus to optimize the method in its accuracy, depending on the level of measurement errors and the deviation of the configuration from the basic one. The choice of the optimum position of the filaments for the basic configuration is determined by the VCF method with equal values of the current filaments.

The VCF method is characterised by a high universality with respect to variations in the form and position of the plasma, but a considerably larger amount of calculations is required in comparison with the FCF method.

The FCF method provides the faster calculations, but is less universal in comparison with the VCF method.

LOCAL FIELD EXPANSION (LFE) METHOD

The LFE method, we use, is similar to that used on JET [1]. Note, that in this method the external current field may also be considered as unknown, which is an advantage of the method, when significant eddy currents are present.

The FCF and LFE methods are recommended for the steady-state stage of the discharge, while the VCF is recommended for the transition stage.

NUMERICAL EXPERIMENT

In the numerical experiment the equilibrium configurations and sensor signals are calculated by using the EQUS equilibrium code. By numerical codes ROMS the plasma boundary is calculated from the sensor signals and r.m.s. deviation Δ_b of the reconstructed boundary from calculated one:

$$\Delta_b = \left[\frac{\oint_{L_p} (\Delta(l))^2 \, dl}{\oint_{L_p} dl} \right]^{1/2} \tag{4}$$

where l is the distance along the plasma boundary contour L_p, $\Delta(l)$ is the boundary deviation. The errors in determination of special points co-ordinates at the boundary are calculated also. The examinations of the methods developed have been carried out by numerical experiments, with the account of random errors of measurements of signals from the magnetic field sensors. With random errors of measurements taken into account, the experiment is repeated n >> 1 times and the statistical characteristics are calculated for errors of plasma boundary reconstruction.

The results of numerical experiment for divertor configurations of ITER

The VCF method

With the number of filaments increasing, the error initially decreases as a result of model accuracy improvement, but then increases due to deterioration of stability with respect to the measurement errors. N = 5 can be taken as an optimum number of filaments.

The standard errors in determination of boundary and of special point co-ordinates are nearly linear functions of the standard measurement error δ not depending on the choice of special points:

$$\sigma_k = a_k \, \delta \tag{5}$$

The value of the coefficient a_k in (5) depends on a number of factors: the number of filaments, the type, number and position of the sensors, the configuration reconstructed. Besides, this value is different for different points of the plasma boundary.

The dependence of the errors on the number of sensors is well approximated by the relation:

$$\sigma_k = \sigma_k(M_o) \sqrt{\frac{M_o}{M}} \tag{6}$$

The FCF method

The investigations were carried out for a 10-filament set, the optimum position of filaments for the chosen configuration having been found by the VCF method. A deviation from the basic configuration leads to increasing of the method errors in the absence of measurement errors. A higher universality of the method is provided by a more flexible model with $N=N_f=10$. On the other hand, an ill-posedness of the problem manifests itself in sensitivity of errors of the method to random errors of measurements. With N and δ increasing, the error of the boundary determination increases too. Also, a stronger effect occurs, when the numerical iteration procedure of plasma boundary calculation from the values of current filaments becomes unstable. In this case the plasma boundary can not be calculated at all. Thus, in the case of flexible model a stabilization of the solution is required, this stabilization being ensured by introduction of some regularizing functional in the form (3). This subject has been examined in detail. A fairly wide range of the regularization parameter values exists, where an instability of the solution is suppressed, with the universality of the method being not deteriorated simultaneously.

Also, we may conclude from the numerical experiment that the accuracy of the VCF and FCF methods decreases with the sensors being removed from the plasma boundary. Nevertheless, the deterioration of the accuracy is not dramatic, if the sensors are placed on the chamber, comparing with their disposition in the blanket and divertor.

The accuracy of the method is weakly sensitive to the sensor type (local magnetic probes or flux loops).

The LFE method

This method is more sensitive to the sensor distance from the plasma boundary, as compared with the CF methods. Only sensors placed in the blanket and divertor blocks can provide the accuracy required for ITER. Also, the method is more sensitive to the set of sensors.

Some results of the numerical experiments are shown in the Table:

Parameter	VCF	FCF	LFE
Δ_b (cm)	3.2	3.4	2.9
Rin (cm)	4.1	4.7	2.8
Rout (cm)	3.1	3.0	2.2
Rx (cm)	2.2	1.3	1.2
Zx (cm)	3.4	4.5	1.5
Error of measur.	3%	3%	3%

Here the errors of plasma boundary determination are presented for EOB divertor configuration with using full set of sensors proposed by L. de Kock (15 tangential magnetic probes in the blanket; 20 partial flux loops, 15 tangential magnetic probes and 9 two-component magnetic probes in vessel) [2]. In the case of the LFE method 20 two-component magnetic probes in the blanket and 4 full flux loops in vessel were used. All these calculations have been performed for the end-of-burn divertor configuration.

CONCLUSIONS

1) The examinations of the methods developed have been carried out by numerical experiments, with the account of random errors of measurements of signals from the magnetic field sensors.

All the methods provide the accuracy of plasma boundary determination required for ITER. In the absence of measurement errors, the errors of boundary determination are less than 1 cm. The methods are stable relative to measurement errors.

2) The VCF method is characterized by a high universality with respect to variations in the form and position of the plasma, but a considerably larger body of calculations is required in comparison with the FCF and LFE methods.

The FCF and LFE methods provide the most rapid calculations, but yield to the FCF method in the universality.

The FCF and LFE methods give the accuracy required of plasma boundary determination, provided the sensors are placed in the vacuum chamber. To provide the accuracy required, the LFE method needs the sensors to be placed in the blanket and divertor.

3) The studies that have been carried out do not take into account the errors associated with the eddy currents. To make correct conclusions concerning the accuracy of the boundary determination and sensor location, calculations of eddy currents are necessary. For the VCF and FCF methods, taking into account the eddy currents needs a model of these currents. In the LFE method the external current field may also be considered as unknown, which is an advantage of the method, when significant eddy currents are present.

4) The FCF and LFE methods are recommended for the steady-state stage of the discharge, while the VCF is recommended for the transition stage.

ACKNOWLEDGEMENTS

This work was supported by the International Atomic Energy Agency under Research Contract No. 8106/RB.

REFERENCES

1. O'Brien, D.P., Ellis, J.J., Lingertat, J., Nucl. Fusion 33 (1993) 467.
2. Report on the IAEA Technical Committee Meeting on Magnetic Measurements for Large Fusion Devices, Kharkov, Ukraine, 5-7 October 1994. Nucl. Fusion to be published.

A HYBRID MAGNETIC PROBE FOR STEADY STATE MAGNETIC FIELD MEASUREMENTS

Junji Fujita,[1] Kazuo Kawahata,[1] Kiyokata Matsuura,[2] Masataka Sakata,[2]
Setsuya Fujiwaka, [3] Tohru Matoba[4]

[1] National Institute for Fusion Science, Nagoya 464-01, Japan
[2] Daido Institute of Technology, Nagoya 457, Japan
[3] Fujiwaka Office, Toki 509-52, Gifu, Japan
[4] Naka Fusion Research Establishment, Japan Atomic Energy Research
     Institute, Ibaraki 311-01, Japan

INTRODUCTION

The magnetic probe is one of the important basic diagnostic techniques on a tokamak. With an increase in the duration of plasma sustainment, it is urged to develop a technique which can measure the steady state magnetic field, in order to determine the magnitude and the position of the plasma current in a long pulsed tokamak such as ITER. The probe, however, should endure high flux of neutron as well as heat from the fusion plasma. It is also necessary to place the diagnostic equipments as remote as possible.

For this purpose, a hybrid magnetic probe system has been proposed, which is based on a combination of a conventional magnetic probe for the measurement of fast varying magnetic field and a rotating magnetic probe for that of slowly varying field.[1]

It is preferable not to employ any power source of electric as well as magnetic to drive the rotating magnetic coil (rotor) in the vicinity of the tokamak device. It is also preferred not to use any mechanical contacts to pick up the electric signal induced in the rotor from the viewpoint of maintenance-free long life time of the system. From these aspects, a transformer-coupled rotating coil system operated with an air turbine has been chosen to pick up and transmit the magnetic field signal. A circuit has been designed to compensate the change of output signal sensitive to the rotation speed of the coil.

For the purpose of proving the principle of the hybrid magnetic probe, a test system has been constructed. For long term testing, the rotating magnetic probe has been operated continuously for more than one week, and no serious technical problems have been found. Applications to tokamaks under operation will be necessary before designing for ITER.

This work has been conducted partly under 1994 ITER Technology R&D Task Agreement on "Feasibility study of steady state magnetic field measurement"(T64).[2]

Diagnostics for Experimental Thermonuclear Fusion Reactors
Edited by P. E. Stott *et al.*, Plenum Press, New York, 1996

103

OPERATION PRINCIPLE OF THE HYBRID MAGNETIC PROBE

A magnetic probe is widely used for the measurement of plasma current, position, and plasma pressure. In general, an integrating circuit necessary to derive the magnetic field intensity from the magnetic probe signal has an accumulation of a zero-level drift, which limits the use of magnetic probes for a long time discharge. The hybrid magnetic probe system consists of a conventional magnetic probe for the measurement of fast varying magnetic field (hereafter called high frequency system) and a rotating magnetic probe for that of slowly varying one (hereafter called low frequency system).

It is proposed for the high frequency system to have an integrating circuit as is shown in Fig. 1, which has a constant gain (R_2/R_1) for d. c. component below a cross-over frequency corresponding to a time constant of $\tau_H = (CR_2)$, to prevent the accumulation of the drift.

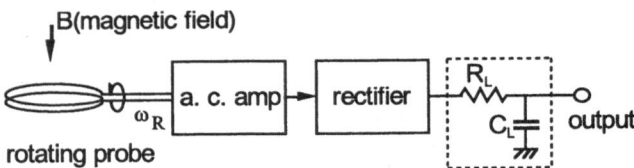

Figure 2. A schematic diagram of a transformer-coupled rotating coil system for magnetic field detection.

The low frequency system has a pick up coil which rotates with an angular frequency of ω_R to measure the magnetic field intensity perpendicular to the rotation axis. The signal is amplified with an a. c. amplifier, rectified and taken out through a low-pass filter with a time constant of τ_L, as is shown in a schematic diagram in Fig. 2.

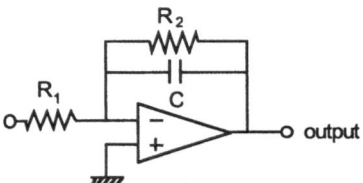

Figure 1. An integrating circuit to be employed for the high frequency system.

The output signal of the rotating magnetic probe V_S is expressed as below when the magnetic field intensity B is present. It changes sinusoidally with the rotational angular frequency ω_R, and the amplitude is proportional to ω_R^2. Here, M is mutual inductance of the transformer, N, the number of turns, and S, the area of the coil.

$$V_s \propto B\,S\,M\,\omega_R^2 \sin \omega_R t. \qquad (1)$$

The frequency transfer function for the synthesis of the outputs from the high and low frequency systems is shown in Fig. 3. If the a. c. voltage amplification gain for the low frequency system b is adjusted to be equal to the integration gain for high frequency system a, and the time constant τ_L of the circuit inserted after the rectifier is to τ_H, the synthesized output for a time varying magnetic field with an angular frequency ω_R is easily obtained from the figure to be simply as $V_0(\omega) = a\,B$.

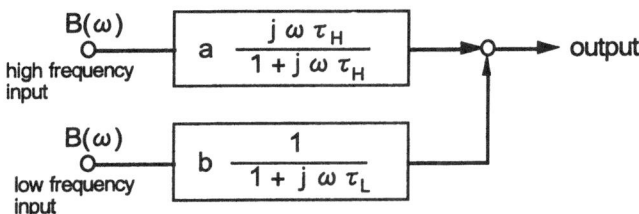

Figure 3. Transfer functions for the synthesis of high and low frequency components to obtain a constant gain in the wide range of frequency for time varying magnetic field.

The magnetic field generated by the plasma current penetrates outside the vacuum vessel with a time delay determined by the electric and magnetic properties of surrounding structural materials. It must be reasonable to assume the delay time to be about one second, and that the direction of the field is vertical when observed at some distance from the plasma current. Therefore, it is possible to place the low frequency system remotely outside the device. This situation eases the problem of neutron irradiation effects on the system.

In order to attain high accuracy of the magnetic field measurement, it is necessary to keep the rotational angular frequency constant and/or employ either a compensation circuit which has a frequency response of $1/\omega_R^2$ by the use of L and C, or electronics to compensate it after A/D conversion, because the sensitivity is proportional to ω_R^2. Effectiveness of this compensation circuit is verified experimentally.

It is recommended to provide a built-in calibration system which generates a known magnetic field for achieving more reliable measurement. Possible errors are predicted, such as temperature dependence of output signal caused by the change of electric conductivity of the rotating coil and transformer with ambient temperature.

MECHANICAL DESIGN OF THE PICK UP SYSTEM

The detection head of the rotating magnetic probe consists of one-turn coil for pick up and 1.5-turn primary coil of a coupling transformer. The rotor is driven by a high speed air turbine operated at an air pressure of 6 atms. An air bearing operated at a pressure of 3 atms is employed for long term operation with high revolution speed as high as 10,000 rpm. The actual rotation speed was 7,800 rpm at an air pressure of 6 atms. The air consumption on the air turbine was 2.6 lit/min, and that on the air bearing was 6.3 lit/min at 3 atms.

It is necessary to keep the air pressure as constant as possible, because the rotation speed is sensitive to the air pressure supplied. A high precision electronically controlled regulator is used to control the air pressure for driving the rotor.

A detail of the mechanical structure is shown in Fig. 4.

LONG TIME TEST

The testing device has been operated for 170 hours without any serious trouble, under the condition of the rotation speed of 7,800 rpm, and a uniform magnetic field intensity of 25 Gauss generated in a Helmholtz coil with a coil current of 1 A. Because no rotation frequency compensation circuit is employed in this stage, the output signal shows a drift of one-day cycle, probably due to the change of rotation speed with the room temperature. This variation is small enough to be cancelled with a rotation frequency compensation circuit.

There found no serious problems, such as abnormal change of rotation speed, heating, noise and others. A typical sensitivity of 0.1 V/Gauss with a resolution of 0.05 Gauss is obtained on this system.

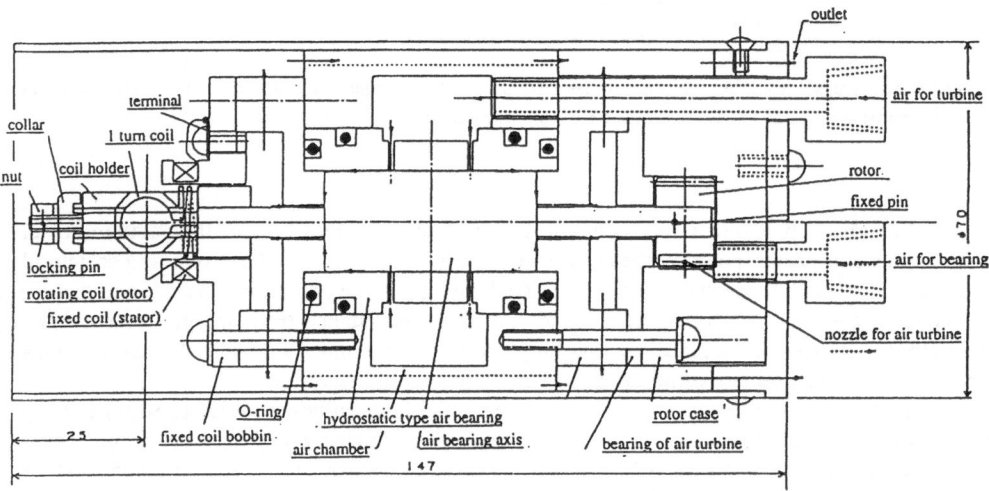

Figure 2. A schematic diagram of a transformer-coupled rotating coil system for magnetic field detection.

CONCLUSION

A hybrid magnetic probe system has been developed for the purpose of measuring steady state as well as time varying magnetic field produced in a long pulse tokamak such as ITER. The output signal from conventional magnetic probe is synthesized with a signal from a rotating magnetic probe so that a flat frequency response can be obtained. A transformer-coupled rotating coil energized with an air turbine was employed. The variation of output signal dependent on the fluctuation of the revolution speed is possible to be compensated with a circuit which has a frequency response of the gain being inversely proportional to the square of rotation frequency. The rotating coil system was constructed and tested for over seven days, that is 170 hours, without any serious problems. A typical sensitivity was 0.1 V/Gauss with a resolution of 0.05 Gauss on this system.

On the basis of the investigation carried out on the hybrid magnetic probe system and the rotation probe, we will be able to design a prototype of the hybrid magnetic probe system for ITER, after finding and solving the problems related to actual applications to the existing tokamaks in operation.

REFERENCES

1. K. Matsuura, M. Sakata, S. Fujiwaka and J. Fujita, A proposal of a hybrid magnetic probe, *J. Plasma and Fusion Research*, 70:397 (1994).
2. K. Kawahata, J. Fujita, K. Matsuura, M. Sakata, S. Fujiwaka and T. Matoba, Feasibility study of steady state magnetic field measurement, *JAERI-Tech* 95-041 (1995).

COMPARISON OF DIFFERENT
REFLECTOMETRY TECHNIQUES

Clément Laviron

Association Euratom-CEA
CEA/Cadarache
13108 Saint-Paul-lez-Durance
France

INTRODUCTION

Reflectometry applied to the measurement of density profiles on fusion plasmas has been subject to many recent developments. For ITER, reflectometry has the great advantage of having only waveguides nearby the vessel, while the windows and the electronics can be at a distance. Different methods have been experimented: linear frequency sweep (broadband or narrowband), dual frequency differential phase, amplitude modulation, pulsed radar, which can be extended to ultra-short pulse, pulse compression, and noise correlation radar. Up to a certain degree, they all encounter the main difficulties limiting the performances, namely the plasma fluctuations and the quality of the transmission lines. Based on actual applications, the implementation of the different techniques is presented, with an analysis of the technological requirements as well as their respective limitations and merits. Expected technical evolution for these techniques, or development of more recent and alternative techniques, is also discussed. Keeping in mind that a whole reflectometry system is based on a method of measurement, but also on the available technique and on the required performances, a comparison is made in prospect of their application in the challenging environment which will be imposed by ITER.

Apart from density profile measurements, reflectometry can also be used for many other applications, like plasma position and control, MHD, plasma rotation, q profile, coupling of heating antennas, fluctuations. Even though the same reflectometry techniques can be used for most of these applications, the advantages and limitations may not be exactly the same as for density profiles. These applications will not be discussed here, and the reader can refer to other and more specific presentations in these proceedings.

Reflectometry relies on the fact that, as an electromagnetic wave propagates through a plasma, its phase is shifted due to the index of refraction being different from vacuum. In O-mode, i.e. when the electric field of the wave is parallel to the magnetic field in the plasma, this index of refraction can be written

$$\mu = \sqrt{1 - \frac{n_e\, e^2}{\varepsilon_0\, m_e\, (2\pi f)^2}} \tag{1}$$

where n_e is the electron density, f the frequency of the wave in vacuum and e, ε_0 and m_e constants. If the density profile is such that there is a layer where the density is large

Diagnostics for Experimental Thermonuclear Fusion Reactors
Edited by P. E. Stott *et al.*, Plenum Press, New York, 1996

107

enough for a specific probing frequency, the index of refraction becomes imaginary and the wave is reflected at the position of that cutoff layer (Figure 1).

After propagation in the plasma, the reflected wave is phase shifted by

$$\varphi(f) = \frac{4\pi f}{c} \int_{r_c(f)}^{r_1} \mu(r) \, dr \; - \; \frac{\pi}{2} \qquad (2)$$

r_1 being the radius of the plasma edge and $r_c(f)$ the radius of the cutoff layer for the frequency f. The effect on the phase is equivalent to a time delay τ after which the echo reflected on the cutoff layer would return:

$$\tau = \frac{1}{2\pi} \frac{d\varphi}{df} \qquad (3)$$

If the phase is measured for many frequencies, the time delay can also be obtained from a transposition between the frequency and time domains through a Fourier Transform:

$$\delta(t - \tau) = FT(\, e^{j\varphi} \,) \qquad (4)$$

τ being known for all frequencies within the range of interest, the determination of the cutoff layer position corresponding to frequency F can be computed from an Abel inversion:

$$r_c(F) = \frac{c}{\pi} \int_0^F \tau(F) \frac{df}{\sqrt{F^2 - f^2}} \qquad (5)$$

The density profile is then extracted from the relation between cutoff density and frequency (using equation (1) with $\mu = 0$).

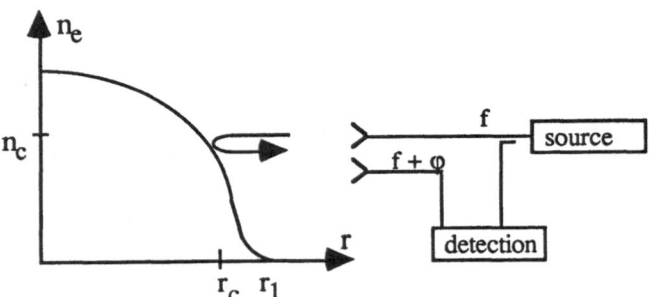

Figure 1. Schematic representation of microwave reflectometry

When many frequencies are used to determine τ, the reflectometer can have some resolution, that is the ability to separate two close reflections (from the plasma or from the transmission system: waveguides, windows, couplers, ...). It can be expressed as

$$\Delta R = \frac{c}{2\Delta F} = \frac{c\Delta\tau}{2} \qquad (6)$$

ΔF being the bandwidth used to determine the time delay from equations (3) or (4) and $\Delta\tau$ the width of the pulse in direct time delays measurements. For a good resolution, a short pulse or a large bandwidth is necessary. Because the position of the reflection layer varies in the plasma with the frequency, the bandwidth cannot be too large and has to be optimised. Moreover, the accuracy of the position of the reflecting layer can be much better than the resolution: it depends on the signal/noise ratio of the detected signal and can be expressed as

$$\sigma \approx \frac{\Delta R}{3\sqrt{S/N}} \tag{7}$$

On the other hand, when only few frequencies are used to determine τ, there is no resolution, and the accuracy is directly related to the precision of the phase difference measurement.

LIMITATIONS ENCOUNTERED BY REFLECTOMETRY DIAGNOSTICS

The determination of the density profile for a given plasma requires measurements of τ or φ for a set of frequencies. In addition to limitations being specific to a given method, two main generic problems have to be considered as they limit the quality of the measurements: the plasma fluctuations and the quality of the transmission lines.

The set of time delays necessary to reconstruct the density profile takes some time to be measured, either through a continuous frequency sweep or through successive measurements at discrete frequencies. If this is too slow, the plasma shape or position may change (macroscopic movements, MHD and microfluctuations) during the measurement and introduce an error. Moreover, even for one specific frequency associated to one reflecting layer, this layer may not behave as a plane reflector, but more likely can be imagined as a lumpy surface made of small sub reflectors. This spatial structure can generate constructive or destructive interferences and give a signal with a strongly varying amplitude. This is intrisic to the plasma, so, for a given frequency at a given instant, and whatever the technique of measurement, the detector would not receive enough signal to make any measurement. But the ability to recover from this loss of signal is not the same for all reflectometry systems. Moreover, and apart from the cutoff layer itself, the fluctuations can also create some scattering of the probing wave during its propagation to and from the reflecting layer.

As it is not possible to place sources and detectors within the vacuum vessel, a waveguide transmission line is necessary to transfer the probing wave from the source to the plasma and, after reflection, back to the detector. For the first reflectometry experiments, the problem was mainly due to the vacuum window as the electronics could be placed close to the vacuum vessel. For specific applications, as probing the high field side or the divertor region, the waveguides configuration inside and nearby the vessel might be complicated. As ITER intends to produce many neutrons, all the electronics and RF components may have to be situated far from the machine, behind the biological shield. This requires long and possibly complicated oversized waveguides which, if not properly optimised, may add parasitic reflections and obliterate the phase measurements. In that case, only the methods using many frequencies can have some resolution and discriminate the parasitic reflections. If, due to multiple reflections, these parasitic reflections overlap the plasma reflection, only the methods allowing a correction procedure from a reference measurement without plasma could be applied.

DESCRIPTION OF THE DIFFERENT REFLECTOMETRY TECHNIQUES

Linear frequency sweep

The first reflectometers were based on the principle of a linear frequency sweep, with a phase measurement. The frequency is swept linearly between two values and the phase difference between a reference arm and the plasma path is measured continuously (Figure 2). τ is computed from applying equations (3) or (4), using a selected bandwidth sliding over the whole frequency band. As no source can be swept continuously over the whole frequency range, the band necessary to probe the plasma is usually split in subbands corresponding to standard waveguides.

The first broadband experiments used BWOs (Back Wave Oscillators), with the frequency range typically swept in few milliseconds. In order to overcome the fluctuations problem, developments have been made on ASDEX-U[1] and DIII-D[2] to use solid state sources (HTOs) which can be swept faster with a high reproducibility (10μs for the whole sweep on ASDEX-U) and make the phase measurements on a "quasi-frozen" plasma. On ASDEX-U, it is combined with a digital frequency discriminator that can extract the phase derivative continuously[3]. Another approach, developed on Tore Supra, is to analyse the measured data by Fast Fourier Transform[4], and to discriminate the plasma reflection from the noise by selective data filtering[5]. For these techniques, because many frequencies can be used for the time delay calculation, there is some resolution and some parasitic reflections can be acceptable from the waveguides. If very reproducible sources are being used, a correction can also be made to cancel out the effects of these reflections (see below, for pulse compression radar). Nevertheless, it is better to use a two waveguide configuration.

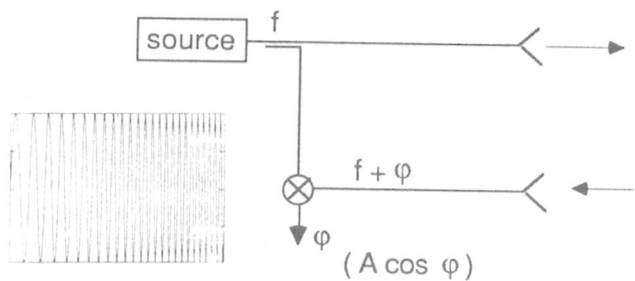

Figure 2. Schematic view of linear frequency sweep technique

An alternative technique has been applied on JET[6,7], with a multichannel narrowband frequency sweep system. A limited number of discrete frequencies are selected and swept around the nominal value. The time delay is determined only for these discrete frequencies, but, as the sweep is made simultaneously for all channels, all the corresponding time delays are determined from the same plasma. This technique has the advantage of simultaneous time delays measurements but the inconvenient of non continuous measurements.

Dual frequency differential phase

Another approach can limit the effects of the fluctuations, by launching two frequencies simultaneously to obtain an instantaneous measurement of the phase difference from the reflection. These frequencies have to be close enough, to reflect from two close cutoff layers, within the correlation length of the fluctuations. So, the effect of the fluctuations are cancelled out in the measured phase difference. Nevertheless, the frequency difference cannot be too small, in order to obtain a phase difference sufficiently above the sensitivity of the detector. As the frequency difference can be controlled through a very stable oscillator, the relative error on the computed time delay is the same as the differential phase measurement error.

This technique has been developed on TFTR by Hanson[8], for edge measurements in X-mode within an ICRH antenna. In these applications, the best compromise for the frequency difference has been found to be of the order of 100 MHz. The instantaneous density profile suffers strong variations, particularly in X-mode at the edge, but there is no loss of data induced by the fluctuations. To obtain only the average density profile, a strong integration time (slow sweep) has been chosen in these applications. But faster sweeps could be used if transients have to be analysed.

The major limitation is that only two frequencies are used to determine a time delay. There is no possibility to differentiate the plasma reflection from any parasitic reflection because the measured phase difference integrates all the contributions from the different reflections. So, only a configuration with two waveguides can be used.

Amplitude modulation

This method has some similarities with the differential phase, as few frequencies are launched simultaneously to cancel the influence of the fluctuations. The amplitude of the probing frequency is modulated (Figure 3), which is in principle equivalent to launch simultaneously three related frequencies:

$$E(t) \ = \ Eo \ (\ 1 + m \cos 2\pi f_m t \ ) \cos 2\pi f t$$

which can also be written:

$$E(t) \ = \ Eo \cos 2\pi f t + \frac{m}{2} Eo \cos 2\pi (f-f_m)t \ + \ \frac{m}{2} Eo \cos 2\pi (f+f_m)t$$

m being the modulation index. The reflected wave carries the phase information for the three frequencies f, $f-f_m$ and $f+f_m$. For a smooth density profile around the cutoff layer corresponding to $f \pm f_m$, and using the first expression, the reflected wave can be written as

$$E(t) \ = \ Eo \ [\ 1 + m \cos 2\pi f_m(t-\tau) \ ] \cos 2\pi f(t-\tau)$$

From this equation, it can be seen that the time delay can be computed from the phase difference for the probing frequency, as for the linear frequency sweep systems, but also from the phase difference, $\Delta\varphi$, between the envelopes of the launched and reflected waves:

$$\tau = \frac{1}{2\pi} \frac{\Delta\varphi}{f_m}$$

This phase difference is related to the phase of the probing frequency by the ratio between the modulation and the probing frequencies.

Amplitude modulation reflectometers have been developed on TJ-1[9], PBX-M[10], T-10[11], Alcator C-Mod[12], FTU[13], and an excellent insensitivity to plasma fluctuations has been reported. This technique is very similar to the differential phase method as the modulation frequency can be very stable, therefore limiting the sources of errors to the phase measurement. But parasitic reflections have also to be avoided and the transmission system needs a two waveguide configuration.

Amplitude modulation, which by principle launches three simultaneous frequencies, can also be extended to more simultaneous frequencies, either by a multiple amplitude modulation or through a sine frequency modulation.

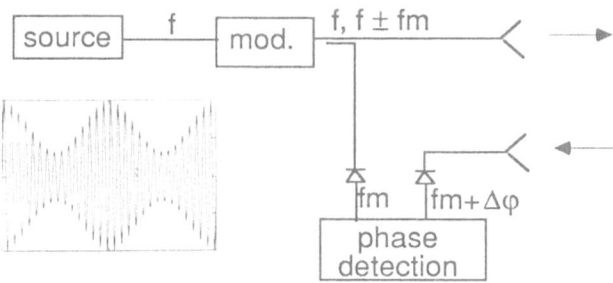

Figure 3. Schematic view of amplitude modulation

Pulsed radar

All the methods presented above determine the time delay through a phase difference measurement. In order to overcome the fluctuations problem, an alternate solution is to launch a short pulse at a given frequency and directly measure the time of flight of the reflected echo (Figure 4). The spectrum sent to the plasma is equivalent to the emission of an infinite number of frequencies in the range $f \pm \Delta f/2$, Δf being the bandwidth of the pulse and equal to the inverse of its duration. In order to measure the time delays for the frequency range necessary to the profile reconstruction, pulses at different frequencies have to be launched.

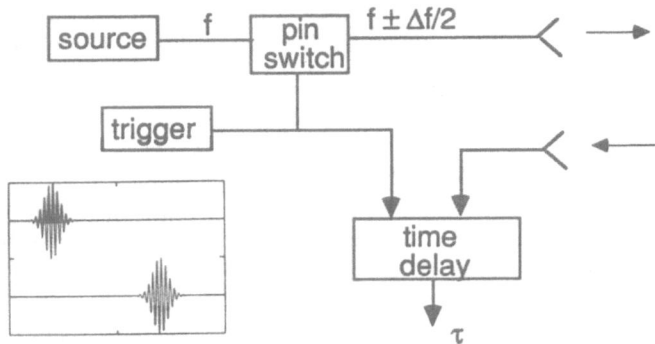

Figure 4 . Schematic view of pulsed radar technique

Although pulsed radar has been used for decades in military and civil applications, technological constraints were not the same as for plasma measurements. To be able to separate in time plasma from false reflections, the pulse length should not excess few ns, and for an accuracy of 1 cm a measurement precision of 60 ps is necessary.

This technique has been implemented on RTP[14], and more recently on T11-M[15] and START[16]. On RTP, measurements are now made on a transmission system using 2 antennas, with a pulse width of 700 ps at 4 different frequencies launched successively. The repetition rate is 500 kHz and the echo detection is made with a time resolution of 70 ps.

Each profile can be reconstructed in a very short time, from few measurements (only one data point per frequency). As the repetition rate of the pulses can be high, the precision of the average density profile can be improved by smoothing the data.

The pulsed radar reflectometer combines the advantages related to the two main limitations stated above. First, each time of flight is a snapshot measurement and the plasma is really frozen. Second, as the time delay measurement can be limited to a predefined time window where the plasma reflection should take place, the parasitic reflections falling out of this window do not corrupt the measurements. Nevertheless, the transmission line should not be too dispersive, as it would broaden the pulse and prevent an accurate time of flight measurement. Another disadvantage in the present systems is that only few discrete points are measured and interpolation has to be made between these points. Instead of using few discrete sources at fixed frequencies, a possible development could be to use a swept source modulated with a broadband pin-switch trigged at a very high repetition rate.

It has been proposed[17] to extend the pulsed radar method, by using ultra-short pulses (few ps), corresponding to a bandwidth covering the whole frequency range. The echo would be spread in time corresponding to the respective delays of all frequencies. The measurement would consist to record the time delays of all the frequency components through a filter bank. By its principle, this method is therefore a way to get the full density profile in one single snapshot. But important technical developments are still necessary to create such a short pulse, before making experiments on a plasma.

Pulse compression radar

Pulse compression radar is also a method which has been used by radarists for a long time. It relies on the fact that a short pulse has a bandwidth, Δf, approximately equal to its temporal width, Δt. By Fourier transform, a pulse can be decomposed into discrete components, each with a particular amplitude and phase. So, a set of discrete frequencies are launched one by one towards the plasma, within a total frequency bandwidth Δf. For each step, the amplitude and phase of the reflected wave are measured and the pulse compression is performed numerically by inverse Fourier transform of the complex vector $A_i e^{j\varphi_i}$ (see equation 4): the full structure of the echo spectrum is then obtained in the time domain. Although it has to be short to limit the adverse effects of plasma fluctuations, the actual measurement duration can be much longer than the reconstructed pulse width, the ratio between them being called the compression factor. This method is very similar to an ultrafast sweep technique with FFT analysis.

This technique has been recently experimented on JET[18,19], showing good density profiles measurements and also its ability to correct for multireflections created in long and complicated waveguides (Figure 5). Even when one waveguide and antenna is used, the plasma profile has been determined with a good accuracy. Using fast frequency synthesizers, one time delay can be obtained in 2 to 5 μs. This value has to be multiplied by the number of probing frequencies necessary to reconstruct the profile, unless measurements are made simultaneously.

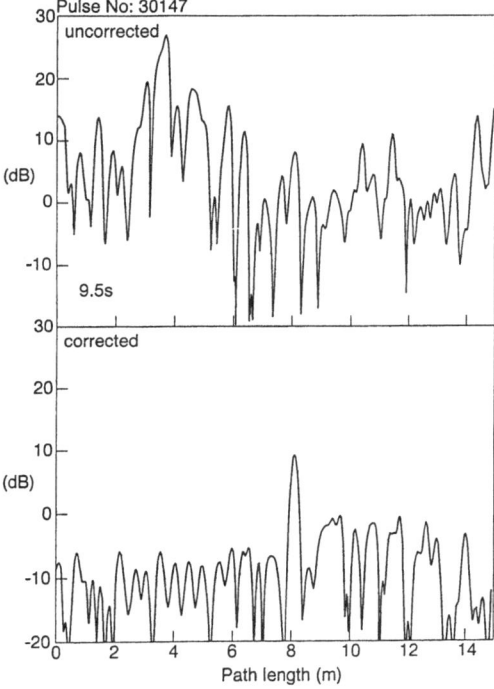

Figure 5. Reconstructed echo structure with pulse compression radar. Before correction (top), the path length of the plasma echo (at ~ 8 m) is 20 dB below many parasitic reflections, including the double windows (at ~ 4 m). After correction (bottom), the plasma echo is 10 dB above noise.

The main advantage of this method is that good measurements can be made through very poor transmission lines, even if the plasma reflection is mixed with parasitic reflections of a higher level. Concerning plasma fluctuations, a compromise has to be found between the number of discrete frequencies used for the profile reconstruction, and the time needed for the whole measurement. Using a synthesizer, the effects of coherent fluctuations can be eliminated by scanning the different frequencies randomly. Alternatively, if fast swept HTOs prove to have reproducible enough frequency and amplitude, they could also be used to increase the speed of measurement, merging the advantages of broadband linear sweep and pulse compression techniques.

Noise correlation radar

Another recent method proposed[20] is to use a generator of stationary broadband microwave signal, around a central frequency. A correlation technique is applied for the determination of the time of flight at this frequency. Because of the high sensitivity, noise sources with a relatively low power can be used. As for the other radar techniques, many frequencies are necessary to reconstruct the full density profile.

Preliminary experiments have been conducted out on the Uragan-2M stellarator[21], using a noise oscillator with a central frequency of 36 GHz and a bandwidth of 300 MHz. On this small plasma (diameter = 20 cm), the first measurements have shown the possibility of observing shifts of the cutoff layer position. Like for pulse compression, noise correlation radar is not a real snapshot measurement. Some time is necessary to sweep and correlate the controlled delay line through the range of times corresponding to the expected plasma time delay. If this system sounds promising, more work is still necessary to fully analyse the advantages and limitations, specifically on the technical point of view.

DISCUSSION

In the above presentation, we have seen that different methods can be applied to measure density profiles in plasma fusion devices. For ITER, in addition to the intrinsic plasma fluctuations, the necessary complexity of the waveguides has to be seriously taken into account. For the temporal fluctuations, two approaches can be made. The first one consists to determine the density profile in a very short time, in order to "freeze" the plasma. If necessary, a large number of successive profiles can then be used to compute an average profile. In addition to slow plasma movements, two different scales have to be taken into account: the MHD with a strong amplitude and a few kHz range, and the microfluctuations in the few 100 kHz range, with low amplitude in the bulk but strong amplitude at the extreme edge. The MHD effects can be limited by doing the measurement in ~100 µs. This is technically possible and has already been implemented for the linear sweep technique (DIII-D, Asdex-U). For the microfluctuations, a much faster or a snapshot measurement has to be considered, but it may not be really necessary in the bulk. The second approach can be applied by the methods which measure directly a phase difference (amplitude modulation and dual frequency). A slow sweep can be made, with a low frequency data filtering and an average profile can be determined if the plasma is stationary. This second approach prevents the analysis of density transients, but might be the only applicable method for highly turbulent plasmas, particularly at the extreme edge in X-mode.

The spatial fluctuations modify strongly the amplitude of the reflected signal which, during a short time, can become too low to be detected. If not properly filtered, fringe counting systems can drift. For reliable measurements, it is better to make independent measurements, as done now in most of modern reflectometers.

In order to limit the effect of parasitic reflections created in the waveguides, it is better to use a two waveguide configuration whenever possible. For some regions of the plasma (divertor, high field side), this may not be possible. If a single complicated waveguide has to be used, only the methods which can discriminate the plasma reflection from any other parasitic reflection can be applied. These are the methods with a good resolution, using many frequencies to compute one time delay, i.e. pulsed radar, pulse compression, noise correlation or linear frequency sweep combined with FFT analysis or other data processing.

The different advantages of these techniques could also be combined, at the expense of a more technical complexity, to create a highly performing reflectometer. By example, the ultra fast linear sweep could be applied to amplitude modulation or differential phase to allow transient measurements. If these sources have also a reproducible enough frequency, a correction technique, as used with pulse compression radar, could be applied to suppress the effects of parasitic reflections. Alternatively, or even combined with it, short pulses could be emitted, combined with a time window detection, instead of having a continuous wave launched toward the plasma.

In the recent years, many developments have been made and there are still good prospects for new improvements. From one frequency at a time, swept slowly in the first applications, we arrive now to systems launching many frequencies, either simultaneously or within a short duration. This is due to new ideas but also to the evolution of the components, with extended characteristics as well as a good reliability and reproducibility. These developments allow reliable measurements of the density profile with strong fluctuations and/or using complicated waveguides.

In ITER, depending of the complexity of the access to the plasma, it will not be necessarily the same technique or combination of techniques for all measurements, in the high or low field sides of the equatorial plane, in the bulk or near the edge, and in the divertor region. Nevertheless, the better the waveguides, the easier it will be to implement the associated reflectometer, and a two waveguides configuration is better wherever possible.

REFERENCES

1. A. Silva, L. Cupido, M. Manso et al, Fast sweep multiple broadband reflectometer for Asdex Upgrade, *17th Simp. on Fusion Techn., Rome* I:747 (1992).
2. K.W. Kim, E.J. Doyle, W.A. Peebles et al, Advances in reflectometric density profile measurements on the DIII-D tokamak, *Rev. Sci. Instr.* 66(2):1229 (1995)
3. A.Silva, L. Cupido, P. Varela et al, Continuous phase derivative evaluation for density profile measurements from FM broadband reflectometry, *Proc. 22nd EPS Conf. on Control. Fus. and Plasma Phys., Bournemouth* in press (1995).
4. M. Paume, J.M. Chareau, F. Clairet and X.L. Zou, Reflectometry experiment on the tokamak Tore Supra, *Proc. of IAEA Tech. Comm. Meeting on Microwave Reflectometry for Fusion Plasma Diagnostics, JET* 21 (1992)
5. F. Clairet, M. Paume and J.M. Chareau, Electron density profile measurements by microwave reflectometry on Tore Supra, *Proc. 21st EPS Conf. on Control. Fus. and Plasma Phys., Montpellier* 18B(III):1172 (1994).
6. R. Prentice, A.E. Costley, J.A. Fessey and A.E. Hubbard, The JET multichannel reflectometer, *Proc. Course and Workshop on Basic and Advanced Diagnostic Techniques for Fusion Plasmas, Varenna* EUR-10797-EN(II):451 (1986).
7. C.A.J. Hugenholtz and A.J. Putter, Source, detection systems and data acquisition for the JET multichannel reflectometer, *Proc. Course and Workshop on Basic and Advanced Diagnostic Techniques for Fusion Plasmas, Varenna* EUR-10797-EN(II):469 (1986).
8. G.R. Hanson, J.B. Wilgen, T.S. Bigelow et al, A swept two-frequency microwave reflectometer for edge density profile measurements on TFTR *Rev. Sci. Instrum.* 63(10):4658 (1992)
9. E. de la Luna, V. Zhuravlev, B. Brañas et al, Density profile measurements by amplitude modulation reflectometry on the TJ-1 tokamak, *Proc. 20th EPS Conf. on Control. Fus. and Plasma Phys., Lisboa* 17C(III):1159 (1993).
10. E. de la Luna, J. Sanchez, V. Zhuravlev et al, Edge density profile measurements on PBX-M using an amplitude modulation reflectometer, *Proc. 21st EPS Conf. on Control. Fus. and Plasma Phys., Montpellier* 18B(III):1180 (1994).

11. V.A. Vershkov, V.V. Dreval and S.V. Soldatov, T-10 plasma investigations with new three waves heterodyne O-mode reflectometer, *Proc. 21st EPS Conf. on Control. Fus. and Plasma Phys., Montpellier* 18B(III):1192 (1994).

12. P.C. Stek and J.H. Irby , Reflectometry on Alcator C-Mod, presented at the *IAEA Tech. Comm. Meeting on Microwave Reflectometry for Fusion Plasma Diagnostics, Princeton* (1994).

13. P. Buratti, M. Zerbini, Y. Brodsky, N. Kovalev and A. Shtanuk, Amplitude modulation reflectometry system for the FTU tokamak, *Rev. Sci. Instrum.* 66(1):409 (1995).

14. S.H. Heijnen, C.A.J. Hugenholtz and P. Pavlo, Pulsed radar; a promise for future density profile measurements on thermonuclear plasmas, *Proc. 18th EPS Conf. on Control. Fus. and Plasma Phys., Berlin* 15C(IV):309 (1991).

15. V.F. Shevchenko, A.A. Petrov, V.G. Petrov and U.A. Chaplygin, Plasma study at T-11M tokamak by microwave pulse radar reflectometer, *Proc. 20thEPS Conf. on Control. Fus. and Plasma Phys., Lisboa* 17C(III):1167 (1993).

16. V. Shevchenko, T. Edlington, M. Gryaznevitch et al, First density profile measurements by multifrequency pulse radar reflectometry in START, *Proc. 22nd EPS Conf. on Control. Fus. and Plasma Phys., Bournemouth* in press (1995).

17. N.C. Luhmann Jr., S. Baang, D.L. Brower et al, Millimiter and submillimiter wave diagnostic systems for contemporary fusion experiments, *Proc. Int. School of Plasma Phys. on Diagnostics for Contemporary Fusion Experiments* 135 (1991).

18. C. Laviron, A. Costley, P. Millot and R. Prentice, Pulse compression radar reflectometry for density measurements on fusion plasmas *Proc. 21st EPS Conf. on Control. Fus. and Plasma Phys., Montpellier* 18B(III):1168 (1994).

19. C. Laviron, P. Millot and R. Prentice, First experiments of pulse compression radar reflectometry for density measurements on JET plasmas, to appear in *Plasma Phys. Control. Fusion* (1995).

20. V.S. Korosteljov, K.A. Lukin, O.S. Pavlichenko et al, New approach to microwave reflectometry: correlation reflectometry via stochastic noise signals, *Proc. of IAEA Tech. Comm. Meeting on Microwave Reflectometry for Fusion Plasma Diagnostics, JET* 236 (1992).

21. K.A. Lukin, O.S. Pavlichenko, Y.A. Alexsandrov et al, Noise radar relectometry: new approach to the fusion plasma density profile measurements, *preprint* (1994).

REFLECTOMETRY APPLICATIONS TO ITER

E.J. Doyle, K.W. Kim, J.H. Lee, W.A. Peebles, C.L. Rettig, T.L. Rhodes, and R.T. Snider[2]

Electrical Engineering Dept. and Institute of Plasma and Fusion Research, University of California, Los Angeles, CA 90095 USA.
[2]General Atomics, P.O. Box 2009, CA 92121 USA.

I. INTRODUCTION

The measurement of many basic plasma parameters on next-step ignition devices such as ITER is a challenge to the fusion community. The ITER environment is different in terms of size, time scale, plasma parameter range, accuracy and reliability needs, etc., such that many current diagnostic techniques will not be directly applicable, while the implementation of other techniques will be significantly different from that employed at present. In this situation, the optimal diagnostic design philosophy for ITER may be quite different from that employed to date, perhaps, for example, by emphasizing measurement packages rather than individual diagnostic systems. This concept is discussed further in Section II below, and it should be noted that in defining the diagnostic requirements for ITER emphasis has been placed on identifying plasma parameters which must be measured, as opposed to selecting specific diagnostics. In general, this situation puts a premium on measurement systems, such as reflectometry, which can be fully demonstrated on current devices, and achieve the required levels of accuracy and reliability.

Reflectometry has emerged as a very attractive candidate for several measurement requirements on ITER, and it is useful to start by listing the primary reasons for this:

1. Flexibility: Reflectometry can measure density profiles, density turbulence, MHD, TAE modes, RF induced waves and associated field structure, etc. Several such measurements can be made simultaneously by sharing a well designed broadband transmission system.
2. Minimal access requirements: Reflectometry can obtain profile measurements from only a single sightline (view), using compact microwave antennas. The transmission system utilizes relatively small microwave waveguides which can incorporate many bends. The minimal access requirements also allow for local measurements, e.g. in RF antenna structures.
3. Compatibility with the harsh ITER environment: Using standard microwave transmission systems, sensitive components such as sources, mixers, amplifiers, etc., can be located outside the high radiation areas. Apart from vacuum windows, all in-vessel components are robust, i.e. waveguide.
4. Demonstrated measurement capability: Measurements are ongoing on current large devices, and these systems have the ability to further develop ITER specific features if necessary.

In this paper a brief overview of potential reflectometry applications to ITER is presented together with a discussion of the unique challenges and requirements posed by operation on ITER. In addition, a detailed description will be presented of specific issues associated with density profile measurements on ITER, many of which are common to other reflectometry applications. A particular objective of the description of density profile measurements is to demonstrate that reflectometry can be applied to ITER with confidence, based on a sound scientific understanding of how to design systems for

Diagnostics for Experimental Thermonuclear Fusion Reactors
Edited by P. E. Stott *et al.*, Plenum Press, New York, 1996

117

optimal performance. To this end, five criteria are introduced by the use of which different profile measurement techniques can be evaluated, and system performance optimized.

The structure of the remainder of this paper is as follows: In Section II, the advantages in the ITER environment of integrated measurement packages are described, and a combined interferometer/reflectometer package for density profile measurements is specifically proposed. Section III describes the cutoff frequencies and accessibility to be expected under ITER reference conditions. An overview of the measurement areas to which reflectometry can contribute, with emphasis on the present status of such measurements and issues associated with their implementation on ITER, is presented in Section IV. In Section V, the issues associated with density profile measurements on ITER are considered in more detail, while a summary is presented in Section VI. Finally, it should be stated that this paper is not a comprehensive review of worldwide contributions to this field.

II. INTEGRATED MEASUREMENT PACKAGES – A DESIGN PHILOSOPHY FOR ITER DIAGNOSTICS

As mentioned above, many standard diagnostic measurement techniques are either not applicable to ITER or will have to be substantially revised. In particular, those diagnostic systems needed for control purposes will have to demonstrate standards of accuracy and reliability exceeding those obtained on current devices. These considerations indicate that it is essential to consider the whole approach to plasma measurements on ITER, specifically the design philosophy employed. Traditionally, diagnostic systems have been designed as individual independent systems. However, substantial benefits can be obtained from an *integrated approach*, where complementary diagnostics are synergistically combined to form a single measurement package. Such an approach should significantly improve overall reliability and accuracy, while also providing superior spatial and temporal coverage, etc. In order to fully realize the synergistic benefits of such integrated packages, however, the constituent diagnostic systems must be *designed from the start* with the idea of producing a composite profile.

As a specific example, we propose that an integrated measurement package comprised of a tangential viewing interferometer/densitometer and a reflectometer should be used for density profile measurements (this is also proposed in Ref. 1). Note that here "densitometer" is used to denote the technique which utilizes the Faraday rotation of a probe beam in a tangential-viewing system to determine the density.[2,3] The superior performance afforded by this combination stems from the excellent edge spatial resolution associated with reflectometry, guaranteed core coverage from interferometry/densitometry, and the application of straightforward inversion techniques that are amenable to real-time analysis. Overall, the proposed combination offers many advantages, is clearly superior to using either diagnostic alone, and ensures that ITER will have available real-time density profile information suitable for control integration.

Utilizing a tangential view on the horizontal midplane of the tokamak entirely circumvents the usual problems associated with inverting line-integrated interferometer data in shaped plasmas, ensures that only a single plasma view is required, and means that a reasonable number of channels, ~3–10, can provide significant information. The data generated by a horizontal array of such beams can be readily inverted to obtain the radial density profile using the straightforward assumption of *toroidal symmetry*. However, multichannel interferometry alone cannot provide density information with the required spatial resolution in the edge plasma, due to insufficient spatial sampling. Since reflectometry can provide excellent radial resolution in the edge plasma, and tangential-view interferometry/densitometry delivers guaranteed core coverage, the integration of the two diagnostics provides a level of reliability and accuracy much superior to that achieved by either diagnostic alone.

III. ACCESSIBILITY AND CUTOFF FREQUENCIES ON ITER

In the absence of core relativistic electron cyclotron absorption effects discussed below, ITER would have very good access for reflectometry. This is due to the aspect ratio of ~3, combined with a relatively high magnetic field strength. This is illustrated in Fig. 1, which shows the O–mode (f_{pe}) and X–mode (f_r, f_l) cutoff frequencies for ITER reference conditions, along with the fundamental and second harmonic cyclotron layers. The X–mode cutoffs were calculated using the vacuum field alone, more accurate calculations using the total magnetic field require more information on the ITER current and beta profiles. The solid and dashed lines for the cutoffs represent the cold plasma and relativistically corrected[4] cutoff frequencies, respectively. The choice of which cutoff to use is influenced by the following considerations: The O–mode cutoff is incompatible with flat profiles and therefore cannot access the core of H–mode plasmas, such as are planned for ITER. O–mode on ITER would consequently be limited to the plasma edge. A frequency range of 20–110 GHz would cover a density range of $0.05–1.5 \times 10^{20}$ m^{-3}, for which sources are easily available. Due to the size of antenna that would be necessary, frequencies below about 15–20 GHz would probably not be used, so the plasma edge at the lowest densities (less than $\sim 5 \times 10^{18}$ m^{-3}) would not be measured. This implies that for profile measurements the position of the plasma edge would not be determined, and the lowest density portion of the profile would have to be assumed.

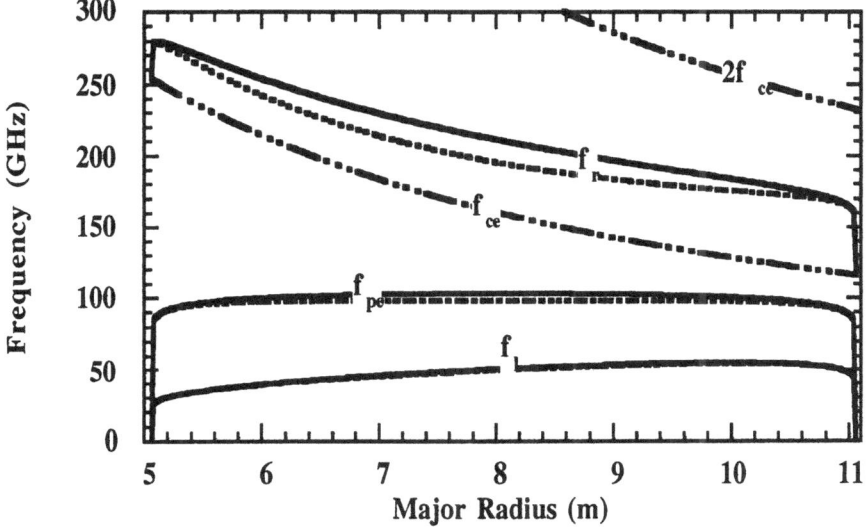

Figure 1. Cutoff and cyclotron frequencies on ITER at reference conditions (5.7 T, 1.25×10^{20} m^{-3} central density, 22 keV central electron temperature). The O–mode cutoff is labeled as f_{pe}, the X–mode cutoffs by f_r, and f_l, and the non-relativistic cyclotron layers by f_{ce} and $2f_{ce}$.

By contrast, the right hand X–mode cutoff is compatible with flat H–mode profiles, and can even "see over" moderate density peaks. X–mode can determine the absolute position of the plasma edge, and measure the entire edge density profile. For full field operation, the frequency range necessary is 116–232 GHz, the lower and upper limits being set by the fundamental and second harmonic edge electron cyclotron (EC) frequencies, respectively. As can be seen from Fig. 1, in the ITER reference case, ignoring EC absorption effects, the right hand cutoff can access the plasma center. If half field operation is catered for the total frequency range necessary increases to 58–232 GHz. This frequency range can easily be accommodated in a single circular corrugated waveguide. At the higher end of this frequency range, broadband sources are a challenge, narrowband less so. At present, fullband solid state sources are available to at least 110 GHz. Advances in solid state source development should further increase this frequency range. However, X–mode polarization requires accurate magnetic field information, a ±1% error on the total edge magnetic field corresponds to a spatial error of ±11 cm on ITER.

A more serious problem is that the unique high electron temperatures (> 20 keV) expected on ITER will seriously compromise the ability of the right hand cutoff to access the core under ignited conditions. At such high temperatures, the electron cyclotron absorption layers will be relativistically broadened and down-shifted, such that the probe beam may encounter an optically thick absorption layer before reaching the plasma core. That such EC absorption could be a problem was first suggested by results such as those presented in Ref. 5, while more recent ITER specific calculations by Drs. D. Bartlett and H. Bindslev of JET confirm that core access is not possible under ITER reference conditions.[6] Fortunately, a solution to this problem of core access exists, inside launch reflectometry using the left hand X–mode cutoff.[7] On ITER, the left hand cutoff has a frequency range of ~20–80 GHz, optimum for millimeter-wave components, and substantially smaller than the frequency of the right hand cutoff. In addition, the relativistic corrections for this cutoff are also substantially smaller than for the other cutoffs, and may be negligible in practice.[7] However, for fixed aperture size, the beam divergence will be larger by the ratio f_r/f_l as compared with the use of the right hand cutoff. The only major problem with implementing such an inside launch system on ITER is the question of access to the inside wall of the torus.

From the above, the conclusions for reflectometry access on ITER (for profile measurements in particular), are as follows: Determining the position of the plasma edge and measuring the low density edge plasma profile requires the use of the right hand cutoff. The right hand cutoff can also measure beyond the steep edge density gradient region in H–mode plasmas, but will not reach the plasma center under high temperature ignition conditions. If guaranteed core access is required then the use of the left hand cutoff and inside launch is necessary. This latter approach is in any case much easier to implement from a microwave technology standpoint.

IV. OVERVIEW OF POTENTIAL REFLECTOMETRY APPLICATIONS ON ITER

Density Profile Measurement: Reflectometer density profile measurements are anticipated to form part of the baseline diagnostic set on ITER, and this is also one of the most developed of the reflectometry applications. Consequently, this subject area has been selected for a more detailed treatment, presented in Sect. V below. The design and implementation of such reflectometer systems is well understood and simultaneous sub-cm spatial precision and sub-ms time resolution have been simultaneously achieved. The main issues for profile measurements on ITER are the effects of relativistically down shifted electron cyclotron absorption on the use of the right hand X–mode polarization, and whether it is possible to use the left hand cutoff for core access. Design studies are also necessary in order to determine optimal transmission system, etc. Continuing advances in computer capability indicate that real-time processing for control purposes will not be difficult.

Divertor Measurements: The divertor region is historically an under diagnosed region of the plasma, but comprehensive diagnostic coverage is essential on ITER where the divertor is crucial to the success of the entire experiment. Reflectometry can provide density profile, peak density and turbulence measurements in the divertor region, and also has the advantage that the transmission system is compatible with interferometry and ECA. In addition, it is essential to monitor tile erosion rates on ITER and reflectometry has the potential to measure the erosion of divertor target plate tiles by simple distance measurements. With the low access requirements of reflectometry, multiple systems could be used for measurements at different heights along the divertor legs. The main issue associated with reflectometry in the divertor region on ITER is the very large density range predicted, up to 10^{22} m^{-3}. Measurement over this full density range implies a very large frequency range for reflectometry, 0–900 GHz for O–mode, and ~255–1036 GHz on the inner leg and ~160–980 GHz on the outer leg for X–mode. Continuous broadband reflectometry measurement over such bandwidths is obviously impractical. In this situation, narrowband, multichannel reflectometry is the best option – a number of channels would be chosen so as measure the peak density and density profile at a range of selected densities. Also, it may not be necessary to measure the entire density range at all locations, or it may be sufficient to know if the peak density is in a particular range, as opposed to a complete profile measurement. Divertor reflectometers are currently under development on JET[8] and DIII–D.[9]

Edge Gap Monitoring System: The edge gap between the plasma separatrix and the wall is controlled by magnetic feedback systems on current machines. On ITER, however, there are concerns that long term drifts in the magnetic systems could lead to the plasma hitting the wall. As a backup system to prevent this, reflectometry could be used to measure the edge profile at a number of positions around the machine and thus validate the magnetic edge gap control. The main issues associated with such a measurement are that an O–mode system is not able to measure the lowest density edge plasma because of minimum frequency limitations, while an X–mode system would require different frequency ranges at different major radii. If less than full field operation is to be catered for, the required X–mode frequency range will become larger. However, there is no doubt as to the ability to perform such measurements; very detailed edge profile measurements have been demonstrated on a number of machines (see for example Fig. 5 later). Continuing advances in computer capability also indicate that real-time processing for control purposes will not be difficult.

Mode Identification System: MHD, TAE, etc., mode numbers are currently measured using toroidal and poloidal arrays of Mirnov loops. However, on ITER the placement of these loops may restrict their frequency response, such that they will be incapable of detecting high frequency MHD and TAE modes, which can have frequencies of several hundred kHz. Reflectometer systems have already demonstrated the ability to accurately detect such modes[10,11] so toroidal and poloidal arrays of reflectometer systems could be used to provide mode number identification on ITER. There are two main issues associated with such measurements. The first is simply one of access and space limitations, but if an edge gap measuring system is feasible, then this should be also, and the two applications could share some waveguides. However, a larger problem is that the requirement on ITER is for mode identification at any radius at any time at any density. This implies that the entire plasma radius would have to be monitored for mode activity at all times at each of the toroidal and poloidal array locations, further implying a very large number of channels over a wide range of frequencies.

Turbulence Measurements: While not necessary for machine operation, turbulence measurements can be crucial for physics understanding.[12-14] Reflectometer turbulence measurements have very good spatial resolution, ~1 cm, coupled with very fast time response, in the microsecond range. These two properties, plus the ability to measure turbulence coherence lengths, make reflectometry a very powerful fluctuation diagnostic. The main issue for ITER applications is whether large aperture imaging optics or phased arrays are necessary in order to obtain quantitative information on turbulence parameters.[14] This is not certain, and requires further ITER specific study; for example, use of the left hand cutoff and inside launch would reduce the microwave frequency by a factor of 4–6 as compared with the normal right hand cutoff, similarly reducing the turbulence induced phase shift, and hence the need for imaging systems. It should also be noted that imaging systems are not necessary for MHD/TAE mode studies, and even qualitative measurements can provide vital information.[12,13] Finally, if an MHD/TAE mode identification system were installed on ITER then this could also easily function as a very powerful turbulence monitoring system.

RF Measurements: If RF systems are used on ITER then reflectometer systems for RF diagnosis should be included as an intrinsic part of the RF package. In particular, reflectometry has a demonstrated ability to measure *local* edge density profiles at RF antennas, and also to measure the internal RF wave structure in the plasma.[9,16,17] The edge density profile is important in determining RF coupling, loading, and the details of the launched spectrum. As the RF itself modifies the density profile, accurate local measurement of the density profile at the launcher structures is vital. Accurate measurements of RF induced profile modification have been performed on TFTR (with differential phase reflectometry) and on DIII–D (with FM reflectometry).[9] In addition, reflectometry can measure coherent density (and in the case of X–mode reflectometry, magnetic) fluctuations associated with the driven RF waves in the plasma. These waves are coherent and are driven at the RF frequency. As the amplitude of these fluctuations is directly proportional to the RF electric field amplitude, measurement of these RF waves provides a way to measure the internal RF electric field profile in the plasma. Such profiles can be used to calculate RF deposition, and also to optimize the RF heating/current drive processes.

Current Profile Measurements: There have been several proposals to measure the internal magnetic field distribution, and hence the current profile, in tokamaks using reflectometry. Usually these proposals rely upon a combination of local measurements using both O-mode reflectometry, which is independent of the magnetic field, and X-mode, which is not. If both O- and X-mode reflectometer beams reflect from the same location, and the source frequencies are known, then the magnetic field is easily determined by calculation. However, as discussed in Sect. III above, because of accessibility constraints it will not be possible to co-locate both O- and X-mode beams on ITER except in the very edge, which is uninteresting. However, it is possible that limited information on the current profile could be provided by measurements of mode activity localized to rational flux surfaces.

L-H Transition and ELM Detector: It is currently anticipated that H-mode operation will be essential if ITER is to meet its design criteria. There is, therefore, a need for a reliable indicator of the L-H transition and H-mode operation. In addition, ELMs will also be important on ITER, both as an impurity removal mechanism and as a potential source of damage to the divertor. Consequently, ELMs and ELM precursors also need to be monitored. On DIII-D reflectometry has been one of the most reliable monitors of the L-H transition, especially in "dithering" transitions, where D-alpha emission signals can be difficult to interpret.[12,13] In addition, reflectometry can routinely monitor both ELM precursors as well as ELM events themselves.[13]

V. DENSITY PROFILE MEASUREMENTS ON ITER

From the list of potential reflectometry applications to ITER density profile measurements have been selected for a more detailed description because this is both the single most important application and also one of the most developed. The goal in this section is to show that the design and optimization of reflectometer systems is well understood, such that designs for ITER can be undertaken in confidence. To date, there has been no published description of how to systematically design a reflectometer system for optimal performance. This need is specifically addressed by introducing five criteria by which different profile measurement techniques can be evaluated and system performance optimized. Reflectometer systems used for profile measurements can be regarded as specialized examples of conventional radar systems used for measuring distance. Radar system design features and optimal data analysis procedures have been fully implemented and incorporated in UCLA reflectometer systems on DIII-D and CCT, and data from these systems will be used to illustrate the design process.

In traversing the plasma, a reflectometer beam suffers a phase shift (or equivalently, a time delay), given by the same basic equation as for interferometry, viz.:

$$\phi\,(f) = \frac{4\pi f}{c} \int_{r_0}^{r_c(f)} \mu(r,f)\,dr - \frac{\pi}{2} \tag{1}$$

where ϕ is the total phase shift in propagating from the plasma edge at r_0 to the cutoff layer at $r_c(f)$, and back, and μ is the plasma refractive index. By varying the probe beam frequency the cutoff layer position can be moved across the plasma radius, or multiple frequencies can be used to reflect from different positions simultaneously. From a data set of phase (or time delay) as a function of frequency, obtained in either of these ways, Eq. (1) can be inverted to recover the density profile. An important point is that as the reflectometer data is gathered along a single line of sight, the *inversion does not require any symmetry assumptions,* as are encountered in the Abel inversion of multichannel interferometer data. In performing reflectometer profile measurements three distinct techniques are employed. The first of these is frequency swept (chirped) reflectometry, where the phase is directly measured as a function of sweep frequency. These systems tend to use continuous (CW) sources, and the technique is usually termed FM or FMCW reflectometry. A second technique, pulse reflectometry, is a direct analog of classical pulse radar measurements, where the time delay of a short (≤ 1 ns) microwave pulse is measured as a function of microwave frequency. The final technique, differential phase reflectometry, does not measure the total plasma phase delay, but instead measures the differential phase shift between two (or more) frequency offset probe beams. As the frequency offset beams are often generated by amplitude modulating a source, this is also often termed AM reflectometry. All three techniques can be implemented in either narrowband or broadband form.

(a) Range Resolution and Precision

Range resolution and precision (accuracy) are two of the most fundamental parameters affecting the design of a radar system[18-21] In a radar system, range resolution is the ability to distinguish (resolve) multiple targets at varying ranges. In this context, the smaller the range resolution the better. The range resolution of a radar system is given simply by

$$\Delta R = c/2B \tag{2}$$

where ΔR is the minimum range difference which can be resolved, c is the speed of light, and B is the radar bandwidth.[18-21] Thus, the range resolution depends *only* on the radar bandwidth, and the resolution is independent of whatever radar technique is employed. It is very important to realize that the range resolution bandwidth in a broadband reflectometer utilizing digital signal processing is variable at will, and is usually much less than the entire sweep bandwidth. For example, if a FMCW reflectometer system is swept over 30 GHz in microwave frequency the mixer IF signal corresponding to this bandwidth can be divided into an arbitrary number of sections (within digitization limits). If each of these data sections is optimally processed then the relevant bandwidth for Eq. (2) is the RF bandwidth corresponding to the length of that data section, not the entire sweep bandwidth. This situation is usually not the case in narrowband systems, or in differential phase reflectometers, where the resolution bandwidth is typically fixed by the hardware system.

In a plasma, of course, the range resolution obtained will differ from that given by Eq. (2) as there is a spatially varying refractive index. As an example, in a pulse reflectometer the plasma changes the pulse-to-pulse separation, and also the pulse length and shape, all three of which contribute to determining the spatial resolution. The amount to which the pulse is affected depends strongly on both the plasma profile shape and the density, so it is hard to generalize as to the actual resolution obtained in a plasma, which has to be determined via numerical simulation. However, the important point is that even in a plasma the only external control parameter on the resolution is still the RF bandwidth.

In a reflectometer system, range resolution is a critical, though often overlooked factor. In a reflectometer system there is only one desired "target," the cutoff being utilized, but there are a multitude of potential false targets; examples of which include spurious waveguide modes, window reflections, multiple reflections from the plasma, reflections from undesired cutoffs, etc. Thus, the range resolution should be utilized to filter out the interfering signals, leaving only the desired signal contribution. However, it is very important to realize that while the ultimate range resolution achievable in a reflectometer or radar system is set by the bandwidth, obtaining this range resolution is not necessarily automatic, but requires data processing. Either filtering/spectral analysis (in a frequency swept reflectometer), or time discrimination (in a pulse reflectometer) are *essential* in order to achieve range resolution. Furthermore, the filtering or time discrimination needs to be *matched* to the resolution bandwidth desired. One important difference from the radar case in a reflectometer is that while in a radar system more bandwidth is always better, in a reflectometer too much bandwidth will cover a distance larger than the desired spatial resolution, smearing the profile which is being measured. This problem is considered in the next sub-section.

The precision of the range measurement in a radar system is a measure of the variation to be expected in a set of measurements under fixed conditions (precision differs from accuracy in that the latter includes systematic effects). For a radar system, the standard deviation on the range measurement is given by:[18,22]

$$\delta r = c/2\beta\sqrt{S} \tag{3}$$

where δr is the rms range error, S is the post detection signal-to-noise ratio, and β is the effective bandwidth, which depends upon the shape of the microwave spectrum being used.[18,22] For a given bandwidth B, the waveform which maximizes β is that of a two frequency CW radar (differential phase reflectometer).[18] Note that the post detection signal-to-noise ratio can be improved by either averaging over a larger bandwidth (in broadband systems), and/or over an ensemble of measurements. With $\beta \sim 1$ GHz, a signal to noise ratio of ~200 yields a range precision of ~1 cm, which is easily achievable. Of course, the range precision will be different in a plasma, particularly as the inversion procedure for the density profile changes the way in which errors propagate. However, the main point is that Eq. (3) identifies the two external controls on the range precision in a reflectometer system, the signal-to-noise ratio, and the effective bandwidth, both of which should be maximized for optimum performance.

An important consequence of the analysis presented in this subsection is that the precision and immunity to false reflections (range resolution) of a reflectometer system do not directly depend on the type of reflectometer utilized, but only on the bandwidths used and the signal-to-noise ratio obtained. For example, a pulse reflectometer using 1 ns pulses (1 GHz bandwidth) has an identical range resolution to a swept reflectometer using a 1 GHz range resolution bandwidth. Consequently, published claims[23] that pulse reflectometers are superior in this respect are incorrect. It should also be noted that both the range resolution and precision do not directly depend on the target distance (range), though the range will affect the SNR, and hence the precision. Finally, in order to obtain range resolution and high precision, it is *essential* that the reflectometer system data be both filtered (for resolution) and averaged (for precision).

(b) Optimization of the Resolution and Smoothing Bandwidths

As shown in the previous subsection, in a radar system more bandwidth is always better for both resolution and precision. In a reflectometer system, however, the cutoff layer changes position with frequency, so the maximum bandwidth which can be used without distorting the inverted profile is a function of the profile itself. The choice of bandwidth should therefore be optimized, as too small a bandwidth will lead to poor system performance (both the ability to remove interfering signals and measurement precision will be poor), while too large a bandwidth will smear out profile details. These considerations indicate that profile measurements via reflectometry are most difficult in large low density devices such as CCT[24] or GAMMA-10.[16,25] This is since the entire profile may correspond to less than 10 GHz in microwave bandwidth, so it is difficult to simultaneously obtain both good spatial resolution and good range resolution and precision. In such cases time resolution must be sacrificed to recover precision via ensemble averaging, while maintaining appropriate spatial resolution.

The optimum solution to this problem is to use bandwidths which *adaptively change with the profile*, i.e. larger smoothing bandwidths can be used in portions of the profile with little structure, while smaller bandwidths are necessary in order to resolve steep density gradients or sharp structures. It should be noted that the density gradient in a plasma can easily change by an order of magnitude, implying that the optimum bandwidths to use should also similarly vary. In broadband swept FM reflectometers where the IF signal is digitized, such adaptive processing is easily achieved as the continuous data can be divided into an arbitrary number of sections (within digitization limits). If each of these data sections is optimally processed, then the relevant bandwidth is the RF bandwidth corresponding to the length of that data section, not the entire sweep bandwidth. An example of a DIII–D H–mode data set analyzed with varying analysis bandwidths is shown in Fig. 2. At the normal analysis bandwidth of ~1 GHz there are ~45 independent data points used in the reconstruction. As the bandwidth is increased to 4 GHz, the number of independent data points decreases to 14, each point now averaging over a wider spatial region. This effectively reduces the available spatial Fourier components. As a result, the inverted profile can no longer resolve the sharp corner at the top of the transport barrier, or the steep density profile itself, as the analysis bandwidth is increased above 1 GHz.

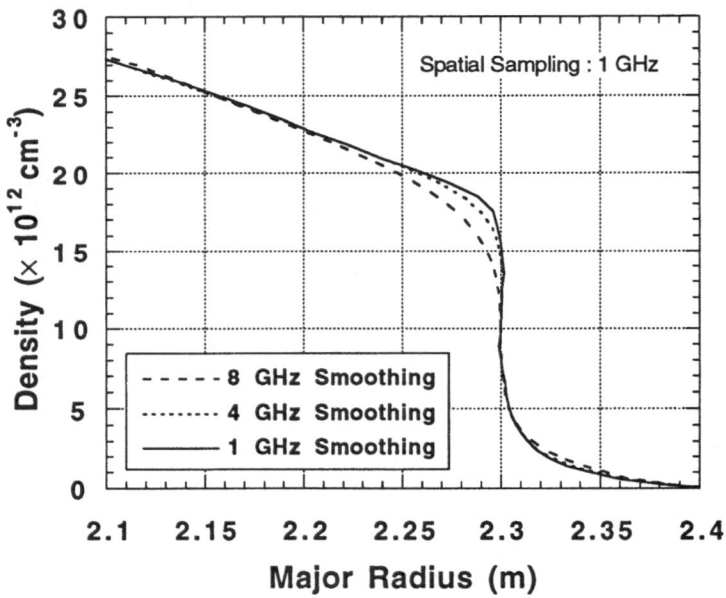

Figure 2. Example of effect of varying smoothing bandwidth on profile reconstruction. 1 GHz bandwidth correctly recovers steep H–mode edge profile, 4 and 8 GHz smoothing bandwidths distort the profile and do not resolve H-mode profile structure.

(c) Spatial Sampling

Mention of profile structure in the preceding subsection introduces another very important (but somewhat neglected) system design parameter, spatial sampling. In measuring a spatial profile containing structure the maximum wavenumber which can be measured without aliasing is $K_{max} = \pi/\delta x$, where δx is spacing between sampling points, while wavenumber resolution is determined by $\Delta K = 2K_{max}/N$, where N is the number of points (assuming equal sample spacing). Thus, avoiding ambiguity due to spatial aliasing requires that δx should be small (large K_{max}), while to simultaneously obtain detailed profiles (small ΔK), N must be large. These requirements are the primary reason for employing broadband reflectometry techniques, as the continuous frequency coverage provides both detailed spatial sampling (small δx), and a large number of samples. By contrast, a limited number of narrowband channels cannot provide the required spatial sampling rates. This is illustrated in Fig. 3, where different profiles have been generated from a single DIII–D data set. The profiles illustrated were obtained by analyzing the data with a fixed resolution and smoothing bandwidth of ~1 GHz, but with a variable gap between the data points utilized during the profile inversion process. By not utilizing some of the available data, the DIII–D data can be made to simulate the performance of a multichannel narrowband or pulse reflectometer system. As expected, spatial resolution is progressively lost as the gap between data sets (channels) is increased, i.e. the profiles become smoother and lose structure. Obviously the smoothing bandwidth and spatial sampling are interrelated. Optimum efficiency and accuracy in profile reconstruction will be achieved through the use of smoothing bandwidths consistent with the existing Fourier components. This can be easily determined iteratively using broadband reflectometry systems that provide the maximum spatial information. Clearly, narrowband systems with a limited number of channels (or spatial samples) are potentially less accurate and smooth the resultant inverted profile.

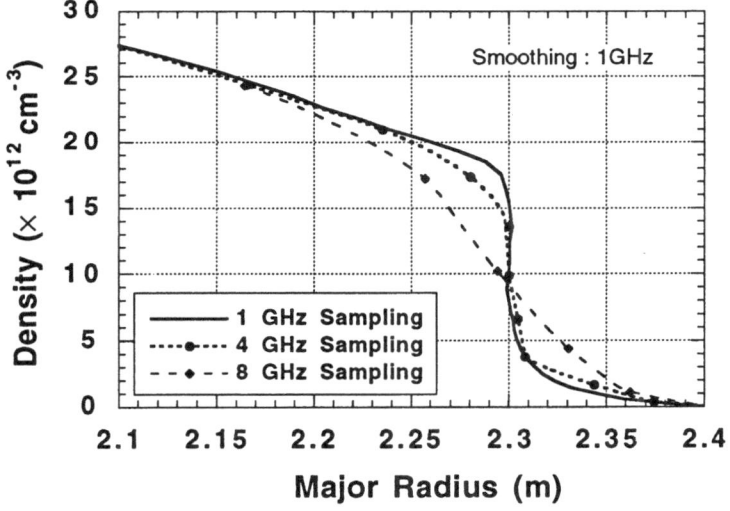

Figure 3. Showing the effect of a varying spatial sampling rate on reconstructed profile quality. A single data set was analyzed with 1 GHz resolution and smoothing bandwidths, but in the first case data every 1 GHz are utilized in the reconstruction, whereas in the other cases, the data are every 4 and 8 GHz, respectively. As can be seen, the lower spatial sampling fails to resolve the "true" profile.

(d) Turbulence Effects on Profile Measurement Techniques

In a profile measurement reflectometer, plasma turbulence may lead to large amplitude and phase modulation of the reflected wave, compromising the ability to measure the density profile. This topic has recently been discussed in Ref. 26 with regard to the relative ability of different reflectometer techniques to work in the presence of turbulence. However, the evaluation performed in Ref. 26 does not include several important effects and cannot be considered as conclusive. In order to qualitatively understand the main effects of turbulence processes a simple analysis is sufficient: If a fixed frequency reflectometer is launched into a plasma it is well known that the return beam will be strongly amplitude modulated (up to 100% in highly turbulent plasmas). It will also be phase modulated due to the movement of the cutoff layer, and Doppler shifts due to plasma rotation can also be observed. Thus, even at fixed frequency the phase modulation is often greater than 2π, making it difficult to track the phase and the reflectometer will accumulate a phase shift which will increase with time. By contrast, in the absence of turbulence, the phase shift from the density profile in a profile system is independent of the time taken to measure the profile.

An equivalent way to consider this problem is from the standpoint of range-Doppler ambiguity in a radar system.[18,19] It is well known that it is impossible for a radar system to simultaneously measure both the exact range and velocity of a target. The Doppler shift induced by a moving target, such as a plasma cutoff layer moving due to turbulence, introduces an unavoidable error or uncertainty in the range. Fortunately, there are (at least) two different approaches to successful density profile measurement in the presence of turbulence: The first is to perform the measurement on a time scale such that the spurious phase shifts induced by the turbulence are negligible. This is the approach pursued in fast sweep FM reflectometers and also in pulse reflectometers, the latter performing measurements on a time scale such that the plasma is entirely frozen. Averaging, both within a sweep or over an ensemble of measurements also helps reduce the effects of turbulence in these cases. Note that in a swept FM reflectometer the fundamental quantity which determines whether the sweep is "fast" is the velocity of the cutoff layer though the plasma, i.e. a "fast" sweep refers not only to the sweep time, but also to the distance the cutoff layer moves. The second approach is to use two or more frequency offset probe beams where the offset is chosen such that the phase difference between the beams is uniquely defined, which requires rms turbulence induced phase shifts of less than one radian. In practice, this requires that the probing beam cutoff layer separation be within a turbulence correlation length, so the probe beam frequency separations are small, ~100 MHz. In this technique a slow sweep can be utilized as the system can track through the turbulence induced phase shifts, and simple data averaging can recover the mean profile. This technique is termed differential phase (or often AM) reflectometry. Both approaches have been applied successfully in practice.profile.

That a fast sweep FM system can successfully cope with plasma turbulence even under ELMing conditions has recently been demonstrated on DIII–D. UCLA has implemented new solid state fast sweep reflectometer systems on both CCT (10 µs full band) and DIII–D (75–100 µs). These systems utilize HTO sources, similar to those employed on ASDEX-U[27] and GAMMA-10.[25] The DIII–D system also has an active solid state frequency quadrupler, delivering 20–80 mW in Q-band (33–50 GHz), with feedback controlled temperature stabilization of both source and quadrupler. It is already apparent that the data quality from the new systems is better than that from the old, i.e. as predicted a faster sweep has improved the quality of the FM data. The new systems are also able to resolve fast density profile changes, a major goal for the new systems. This is illustrated in Fig. 4, showing the time evolution of the edge density profile through a giant ELM on DIII–D. The illustrated data are from *single sweeps*, and this result is the first to demonstrate simultaneous sub-cm spatial precision and sub-ms time resolution in a highly turbulent, rapidly changing plasma regime. profile.

Note that the effects of turbulence cannot be eliminated in *any* system as amplitude modulation of the signal will always be present; even in pulse reflectometers the signal level can be at or below the noise level. However, in flexible software based profile analysis packages this can easily be taken account of in the inversion process by simply not using data portions where the signal amplitude is below a preset threshold level. Consequently, claims that pulse reflectometer systems have superior immunity to a temporary loss of signal[28] are incorrect. It is also interesting to note that turbulence effects and Doppler shifts are also common in classical radar systems; the amplitude of the return signal from aircraft is heavily amplitude modulated, and moving target discrimination via Doppler shifts is a basic radar technique.

(e) Data Processing

It is already apparent from previous subsections that optimal data processing techniques play a crucial role in removing spurious reflections from the signal and in obtaining accurate measurements. These arguments will not be repeated here, instead a brief description of the DIII–D data processing system is presented, along with some comments on software versus hardware data processing implementations. Taking the latter point first, software based analysis implementations have several important advantages as compared with hardware. The most important of these is the ability to adaptively vary resolution, filter and smoothing bandwidths to match the properties of the actual density profile being measured. Such adaptive processing is simply not possible in a hardware based system, where all such parameters are of necessity fixed. Software analysis can also provide much superior phase resolution as compared with hardware zero crossing detectors or peak finders. This position is similar to the situation with modern high performance radars,[19] where the data analysis is invariably digital due to these same considerations of flexibility; by changing the software analysis parameters a single radar system can provide many modes of operation optimized for different needs, i.e. moving target detection, ground mapping etc. The ease of digital signal processing and its relative advantages will continue to increase in the future as computer performance continues to improve at lower cost. The disadvantages of hardware processing extends to the pulse timing circuits used with pulse reflectometer systems. As the range resolution in a pulse system is obtained directly in the time domain it is possible for a hardware timing system to trigger on false signals in cases where the signal is contaminated within the plasma time delay window. Also, the plasma distorts the pulse shape, so there will be systematic profile errors from assigning a single time value to the pulse as opposed to measuring the entire pulse envelope; these systematic errors become progressively more important as the pulse length is shortened.

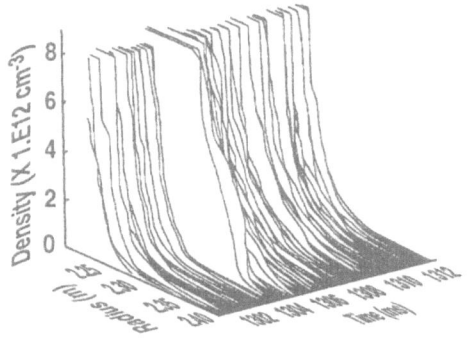
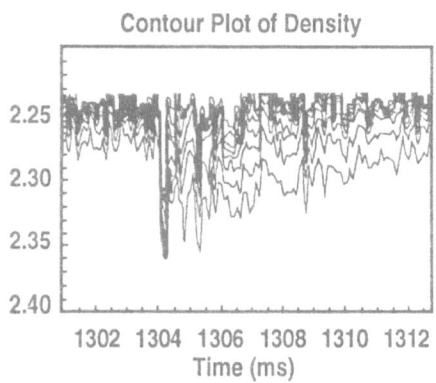

Figure 4. Showing the time evolution of the edge density profile through a giant ELM on DIII–D. The illustrated data are from single sweeps, and this result demonstrates simultaneous sub-cm spatial precision and sub-ms time resolution in a highly turbulent, rapidly changing plasma regime.

The DIII–D data processing system[29] is implemented entirely in software, and is based on classical FMCW radar analysis techniques.[30] The mixer IF signal is digitized, after which the entire data set from the full band sweep is divided into a number of smaller frequency ranges. In each frequency range the IF signal is filtered to remove spurious "targets," the bandwidth of the filter defining the range resolution of the system. The optimum or matched filter case is when the filter bandwidth is simply the inverse of the time duration of the sweep portion being analyzed. The signal phase is then extracted using digital complex demodulation (CDM). CDM is a software implementation of quadrature heterodyne detection, and provides both signal amplitude and phase independently: i.e. it is immune to amplitude modulation. The signal phase is smoothed over a bandwidth chosen to optimize the measurement precision, while maintaining spatial sampling appropriate to the actual profile. The resulting phase versus frequency data are then used to produce the final profiles via a standard numerical inversion algorithm.[31] The system routinely acquires over 1000 profiles per discharge at with a sweep time of 100 μs. Profiles from particular times of interest can now be analyzed between shots. The bandwidth optimization process is also currently being automated; this can be done before profile inversion as the plasma density gradient and profile structure are directly reflected in the phase versus microwave frequency data. Examples of typical profiles obtained on DIII–D using this analysis procedure are shown in Figs. 5 and 6. Figure 5 shows a comparison (by D. Swain, ORNL) of three *independent* edge density profiles, obtained using reflectometry, Thomson scattering and a reciprocating Langmuir Probe. As can be seen, the agreement is excellent. Figure 6 shows an example of the wide dynamic range achievable by a reflectometer; the reflectometer measurements extend all the from the scrapeoff layer to the plasma center.

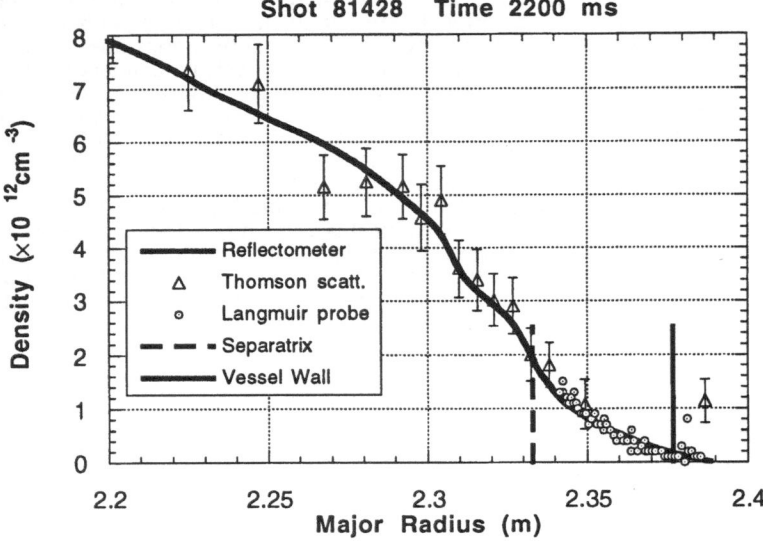

Figure 5. A comparison of three independent edge density profile measurements, obtained using reflectometry, Thomson scattering and a reciprocating Langmuir probe. Langmuir probe data provided by Dr. R.A. Moyer of UCSD.

Recently, an alternate analysis procedure for FMCW reflectometer data has been proposed,[32] based upon pulse compression techniques.[18,19,33] In principle, this is little different than the analysis already applied to DIII–D data, as the RF sweep can be the same, as is the "pulse compression" obtained by filtering the signal so as the obtain the range resolution of a much shorter pulse (hence the term pulse compression). Certainly, the theoretical resolution and performance achievable with the two approaches are identical if both process the data in a properly "optimal" manner. The only significant difference between the pulse compression analysis technique proposed in Ref. 31 and that currently employed on DIII–D is that in the former the range resolution is achieved directly by Fourier transforming the data from the frequency to the time domain, which simultaneously averages the data over the FFT data window. We are currently investigating this approach to see if it offers any advantages in terms of computational efficiency as compared with the current DIII–D approach.

(f) Evaluation of Reflectometer Profile Measurement Techniques

Based upon the criteria outlined in the previous subsections we are now in a position to evaluate the relative merits of the various reflectometer profile measurement techniques. This exercise is useful both in presenting a semi-quantitative evaluation of the various techniques for the first time, and in demonstrating that designs for ITER can be undertaken on a scientific basis. It should be emphasized that the evaluation is based upon current technology and practice, which can be improved, guided in part by evaluations such as this.

Narrowband versus Broadband Reflectometers: Narrowband systems come in several types. The most obvious are the narrowband FMCW systems used on JET[34] and formerly on DIII–D,[35] and the narrowband differential phase system on CMOD.[36] In these systems the microwave bandwidth is typically less than a few hundred MHz. From the discussion presented above, these small bandwidths imply that both the range resolution and precision are roughly an order of magnitude worse than that which can be obtained in broadband FMCW systems. Another major disadvantage of narrowband systems, of whatever type, is poor spatial sampling, resulting in low spatial resolution and potential aliasing effects. By contrast, the continuous frequency coverage employed in broadband systems provides the maximum in spatial sampling capability for a reflectometer system, with consequent superior spatial resolution; simultaneous sub-cm spatial precision and sub-ms time resolution has been demonstrated in practice, see Fig. 4. The superior performance of such broadband reflectometers is analogous to that of modern high resolution, high bandwidth radar systems.[20,21] Low loss multiplexing of multiple narrowband channels is also difficult, and tends to severely limit the number of channels which can be employed. The number of microwave components, such as sources, mixers, couplers, etc., also increases linearly with the number of channels in narrowband systems.

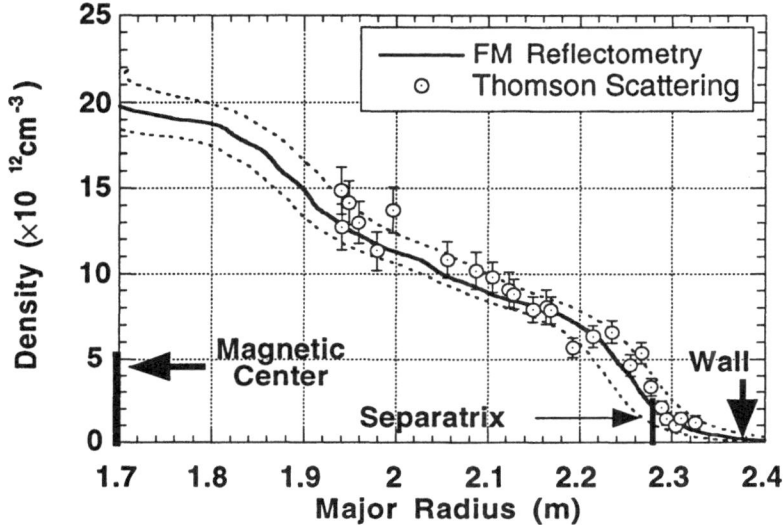

Figure 6. An example of a reflectometer profile measurement extending from the scrapeoff layer to the plasma center. Note the excellent agreement with Thomson data, though the latter does not measure the low density edge, or the plasma center.

Swept differential phase systems are a form of hybrid narrowband/broadband system, in that the range resolution bandwidth is set by the probe beam frequency difference, while the bandwidth used to smooth the data and increase the measurement precision can be varied as in a broadband FMCW system. As the differential frequency offset in broadband swept differential phase systems is typically 100 MHz, or less, these systems are actually "narrowband" with regard to their range resolution bandwidth. Finally, many narrowband systems have their resolution bandwidth fixed in hardware, so that it cannot be optimized for changing plasma conditions as in a broadband system.

Thus, on all the criteria considered, broadband systems are found to be superior to narrowband for general profile measurements. However, narrowband systems may be appropriate on ITER in the divertor region, where continuous frequency coverage is impractical. However, such "narrowband" systems should have a greater resolution bandwidth than is normal at present, at least 1 GHz.

Pulse Reflectometer Systems: It is fundamentally difficult[23] to make a pulse radar system with sub-cm spatial accuracy as this implies time resolution of under 60 ps. This fundamental difficulty is in fact the primary reason for the development of conventional FMCW radars for use in such high resolution close range applications;[37] it is much easier to make frequency measurements to the required accuracy over a longer time period than it is to make the equivalent pulse timing measurement. Nevertheless, pulse measurements are possible, and a 4 channel fixed frequency system has been demonstrated on RTP.[28] Another consideration is that a pulse reflectometer has a duty cycle lower by at least one to two orders of magnitude as compared with CW systems. Thus, for the same source power the total microwave energy used by a pulse reflectometer is also lower by one to two orders of magnitude, with consequent large effects on the achievable signal-to-noise ratios. The disadvantages of this technique with regard to spatial sampling, multiplexing, pulse distortion and hardware implementation have already been described.

The principal advantage offered by this technique is that the plasma is frozen on the time scale of the measurement, so turbulence induced phase errors (time delay errors) are minimized. However, both turbulence induced amplitude modulation (from pulse to pulse) and turbulence induced pulse distortion are still present. The 1 GHz or greater microwave bandwidths employed in pulse reflectometers defines both the range resolution bandwidth, and the bandwidth over which the profile is averaged. The range resolution is consequently very good, though the smoothing bandwidth is not changed to account for profiles where the optimal averaging bandwidth is different. This latter criticism, and the important spatial sampling problem, could be solved by pulse modulating a swept broadband source, so as to achieve a broadband pulse reflectometer. The time resolution of a pulse reflectometer is also potentially very good, though resolution at the level of the pulse repetition frequency is not achievable due to the need to average over pulses. Finally, it is possible to perform pulse reflectometry with ultra short pulses, with very wide bandwidth, which can cover a substantial fraction of the profile with a single pulse.[38]

FMCW Reflectometer Systems: In a plasma environment, FMCW reflectometer systems have to be swept fast, or they don't work.[31] If, however, they can be swept fast enough to avoid turbulence induced errors then broadband FMCW systems are currently superior in performance capability to any of the other techniques. One criterion for how "fast" is fast enough is that the turbulence induced IF frequency broadening should be much less than the IF frequency itself. However, realizing this superior performance capability depends critically upon optimal data processing, and stems from the ability to adaptively vary the range resolution and averaging bandwidth to optimally match plasma conditions. Recent data from fast sweep systems installed on DIII–D and CCT support this contention, showing the ability to track the density profile through ELMs with 100 μs resolution, using *single sweeps* (see Fig. 4 earlier). This demonstrates simultaneous sub-cm spatial precision and sub-ms time resolution in a highly turbulent, rapidly changing plasma regime. Broadband FMCW reflectometers also have a large advantage in having the simplest hardware system, combined with a reduced need for multiplexing and a minimal number of system components.

Differential Phase Reflectometer Systems: As compared with FMCW systems, differential phase systems allow accurate average profile measurements to be made with slow sweep rates. This ability has been demonstrated on several machines.[39,40] However, this is done at the cost of range resolution, which is typically an order of magnitude worse in differential phase systems as compared with broadband FMCW or pulse reflectometer systems. This is the source of the well known sensitivity of AM systems to spurious interference,[30,41] as it is impossible to remove the interference terms by data processing. This has the effect that differential systems require better microwave systems with low spurious signal levels, and also these systems are much more dependent on accurate in-situ calibrations. However, in-situ calibrations cannot account for two potential sources of interference, namely double reflections from the plasma and reflections from undesired cutoffs, both of which can generate 10's of percent of phase error under certain circumstances. Another point is that a single fixed frequency difference is not optimal for all plasma conditions. In current systems the frequency difference is typically small (≤ 100 MHz), a choice driven by the need to keep the phase difference unique under the worst conditions, which occur when the density gradient is shallow. However, this choice results in very small phase shifts when the density gradient is steep; the density gradient scale length can change by an order of magnitude, and the difference frequency has to be chosen to cater for the low end of the range. Thus, a differential phase systems with multiple or variable frequency offsets would offer superior performance.[11] Broadband differential phase reflectometer systems can have a variable smoothing bandwidth and high measurement precision, as in FMCW systems.

Towards an Optimal Reflectometer System: From the above discussion it can be seen that an optimal reflectometer system would combine the range resolution of a broadband FMCW reflectometer system with the turbulence tracking capability of a differential phase reflectometer. Such a system could possibly be achieved by having two simultaneous FM reflectometer channels with a small frequency offset, the data of which, after processing to remove spurious signal components (range resolution), could then be differenced as in a differential phase reflectometer. Such a system would have to have a fast sweep in order for the FMCW section to work. Whether this concept offers practical advantages requires further study, but such a system could easily be implemented by simple modifications to current differential phase systems.

(g) Waveguide Transmission System and Calibration Issues

Many different microwave transmission systems are possible, including fundamental or oversize rectangular waveguide, circular corrugated guide, quasi-optical transmission, etc. While a detailed design study is necessary in order to determine the optimum approach, experience with circular corrugated guide on DIII–D and TFTR has been excellent, and use of this guide has several attractive features.[42] Foremost among these are broadband low loss transmission, and low loss bends, with the ability to propagate arbitrary polarization. This latter point is very important as a single waveguide system can be used to probe all plasma cutoffs, and also the probe beam polarization can be rotated using feedback control so as to match the plasma edge magnetic field pitch angle, minimizing coupling to undesired cutoffs. In addition, circular corrugated guide launches a very pure Gaussian radiation pattern, so separate microwave horns are unnecessary. A bistatic launch and receive arrangement should be used to minimize waveguide modes and spurious reflections. Finally, the waveguide can be constructed from ITER compatible materials, plated if necessary on the inside to provide high conductivity.

In the area of system calibration, ITER presents new challenges. Reflectometer profile systems measure distance from a reference plane; current systems utilize in-vessel calibrations with mirrors during machine vents, and/or in-vessel reference systems such as shutters, pins, or the inside wall, to define the location of this reference plane with respect to the plasma. However, on ITER such techniques may not be sufficient to cater for the anticipated mechanical expansion and movement of the vessel and waveguides during the long ignited discharges. One possible answer to this problem might be a complete additional reference waveguide which would follow the same path through the vessel as the measurement system, but would not view the plasma. However, reflectometer calibration on ITER is still a new and novel challenge requiring a detailed design study.

(h) Proposed Density Profile Reflectometer System for ITER

After the detailed consideration of density profile system design criteria in the previous subsections, we are now in a position to outline the current issues associated with designing an actual system for ITER. At this point in the ITER design process the most important system feature to finalize is the transmission system and the available access. The goal should be to design a *flexible transmission system which can cater for different reflectometer measurement systems and techniques, allow access to all plasma cutoffs, and cater for a wide microwave frequency range.* As described in the previous subsection, broadband circular waveguide is a prime candidate to provide such a system. It is not necessary at this stage in the ITER project to determine which particular reflectometer techniques will be used for any of the measurements; continued advances can be expected before ITER is built. In particular, recent progress on DIII–D, though still preliminary, indicates that accurate, routine operation with sub-cm spatial precision and sub-ms time resolution has been achieved. As currently envisaged, a profile measurement system on ITER would consist of an outboard launch X–mode system for edge profile measurement, with core profiles being provided by an inside launch left cutoff system. These reflectometer systems would be combined into one overall density measurement package along with a tangentially viewing interferometer/densitometer systems, so as to further guarantee accurate density profile coverage under all conditions. Finally, detailed design studies are urgently needed to answer questions relating to inside launch reflectometry, relativistic effects, and system calibration issues.

VI. SUMMARY

Reflectometry is a powerful diagnostic tool with the potential to meet several measurement needs on ITER. In particular, density profile measurements are well understood and profile quality continues to improve. Accurate, routine measurements with sub-cm spatial accuracy, sub-ms time resolution, and high spatial sampling have been demonstrated. Combining reflectometer and toroidal interferometer/densitometer measurements into a single integrated density profile measurement package offers further improved performance in terms of reliability and accuracy over a wide range of conditions. As for other potential applications of reflectometry, some such as RF wave measurements and divertor measurements are currently under development on several machines. Others, such as edge gap control, and MHD/TAE mode identification, need further feasibility studies and experimental demonstration. Finally, although they are not necessary for plasma control, turbulence measurements could prove vital for physics understanding once ITER is operational.

ACKNOWLEDGMENTS

The authors would like to thank Prof. N.C. Luhmann, Jr. of the University of California, Davis for his continuing support, as well as Drs. G. Hanson and J. Wilgen of ORNL for fruitful discussions. Thanks are also due to Dr. R.A. Moyer of UCSD for providing the Langmuir probe data, and to Dr. D. Swain of ORNL for constructing the comparison shown in Fig. 5. This work was supported by US DOE Grant No. DE-FG03-86ER53225 and Contract No. DE-AC03-89ER51114.

REFERENCES

1. R.T. Snider, et al., these proceedings.
2. F.C. Jobes and D.K. Mansfield, Rev. Sci. Instrum. **63**, 5154 (1992).
3. G. Dodel and W. Kunz, Infrared Physics **19**, 443 (1979).
4. E. Mazzucato, Phys. Fluids B **4**, 3460 (1992).
5. M. Sato, et al., J. of the Physical Soc. of Japan **62**, 3106 (1993).
6. D. Bartlett, JET, private communication (1995).
7. E.J. Doyle, et al., Rev. Sci. Instrum. **66**, 1233 (1995).
8. P.R. Thomas, et al., these proceedings.
9. W.A. Peebles, et al., these proceedings.
10. R. Wilson, et al., in *Plasma Physics and Controlled Nuclear Fusion Research 1990*, Proc. of the Washington Conference, IAEA, Vienna (1991).
11. V.A. Vershkov, et al., these proceedings.
12. E.J. Doyle, et al., in *Plasma Physics and Controlled Nuclear Fusion Research 1992*, Proc. of the Wurzburg Conference, Vol. I, 235, IAEA, Vienna (1993).
13. E.J. Doyle, et al., Phys. Fluids B **3**, 2300 (1991).
14. E. Mazzucato and R. Nazikian, Phys. Rev. Lett. **71**, 1840 (1993).
15. R. Nazikian and E. Mazzucato, Rev. Sci. Instrum. **66**, 392 (1995).
16. A. Mase, et al., Physics of Fluids B **5**, 1677 (1993).
17. J.H. Lee, et al., Rev. Sci. Instrum. **66**, 1225 (1995).
18. M. I. Skolnik, *Introduction to Radar Systems*, McGraw-Hill, New York (1962).
19. George W. Stimson, *Introduction to Airborne Radar*, Hughes Aircraft Co., El Segundo (1983).
20. A. W. Rihaczek, *Principles of High Resolution Radar*, McGraw-Hill, New York (1969).
21. D.R. Wehner, *High Resolution Radar*, 2nd edition, Artech, Boston (1995).
22. P.M. Woodward, *Probability and Information Theory, with Applications to Radar*, 2nd edition, Pergamon, Oxford (1964).
23. C.A.J. Hugenholtz and S.H. Heijnen, Rev. Sci. Instrum. **62**, 1100 (1991).
24. T.L. Rhodes, et al., Nuc. Fusion **33**, 1787 (1993).
25. T. Tokuzawa, et al., Jpn. J. Appl. Phys. **34**, L76 (1995).
26. V. Zhuravlev, et al., Proc. 22nd EPS Conference, Bournemouth, UK, (1995).
27. A. Silva, et al., Proc. 21 EPS Conf., Montpellier, France, Vol. 18B Part III, 1188 (1994).
28. S.H. Heijnen, et al., Rev. Sci. Instrum. **66**, 419 (1995).
29. K.W. Kim, et al., Rev. Sci. Instrum. **66**, 1229 (1995).
30. D.K. Barton, ed., *Radars Volume 7, CW and Doppler Radars*, Artech, Boston (1978).
31. E.J. Doyle, et al., Rev. Sci. Instrum. **61**, 2896 (1990).
32. C. Laviron, et al., Proc. 21 EPS Conf., Montpellier, France, Vol. 18B Part III, 1168 (1994).
33. C.E. Cook and M. Bernfeld, *Radar Signals*, Academic Press, New York, (1967).
34. A.C.C. Sipps and G.J. Kramer, Plasma Phys. Control. Fusion 35,1685 (1993).
35. T. Lehecka, et al., Proc. 16th EPS Conf., Venice, Italy, Vol. 13B Part I, 123 (1989).
36. J.H. Irby and P. Stek, Rev. Sci. Instrum. **61**, 3052 (1990).
37. Y. Yamaguchi , et al., IEEE Trans. on Geoscience and Remote Sensing **32**, 11 (1994).
38. C.W. Domier, et al., Rev. Sci. Instrum. **66**, 399 (1995).
39. G.R. Hanson, et al., Rev. Sci. Instrum. **66**, 863 (1995).
40. E. De la Luna, et al., Rev. Sci. Instrum. **66**, 403 (1995).
41. G.A. Ybarra, et al., IEEE Trans. on Microwave Theory and Techniques **39**, 809 (1991).
42. J.L. Doane, Chapter 5, in *Infrared and Millimeter Waves, Vol. 13*, K. Button, ed., Academic Press (1985).

REFLECTOMETRY FOR ITER DENSITY PROFILES

M. Manso[1], D. Bartlett[2], L. Cupido[1], W. Kasparek[3], J. Sanchez[4], P. Stott[2], D. Wagner[3]

[1]Centro Fusão Nuclear, Association Euratom/IST, Lisboa, Portugal
[2]JET Joint Undertaking, Abingdon, England
[3]Inst. fuer Plasma Forschung, Univ. Stuttgart and Euratom/IPP, Germ.
[4]Associacion Euratom/CIEMAT, Madrid, Spain

I. INTRODUCTION

Reflectometry has several features relevant to ITER: it is a robust diagnostic (the waveguides are resistant to radiation and the critical equipment can be placed outside the biological shield); it requires moderate access to the machine and systems can be loss tolerant. Reflectometry is a mature diagnostic that can measure density profiles and local density fluctuations with high spatial and temporal resolutions. Density profiles are, by far, the most demanding in access, transmission line and antennas. These are the main issues at the present phase of EDA and for this reason the study of the EU Home team concentrates on those topics.

After a brief survey (Sec. II) of the background of reflectometry and its actual capabilities, the accessibility conditions are analysed by taking into account the relativistic effects associated with the high temperatures foreseen for ITER (Sec. III). Based on the accessible frequency ranges, a conceptual design is presented of a high performance transmission line compatible with the different reflectometry techniques (Sec. IV). A reference design is proposed for the instrumentation with a modular approach, that can accomodate future developments of the different methods (Sec. V). In section VI the main conclusions are summarized: the application of reflectometry to ITER is already possible with present-day technology but some important problems must be solved, namely the integration of the transmission line and antennas in the inboard blanket.

II. BACKGROUND OF REFLECTOMETRY

1. Reflectometry capabilities

As shown in Table 1, reflectometry has many potential aplications for ITER. Reflectometry can already meet (and even exceed) the requirements for density profile

Diagnostics for Experimental Thermonuclear Fusion Reactors
Edited by P. E. Stott *et al.*, Plenum Press, New York, 1996

measurements in the main plasma. A system is presented based on present technology to measure density profiles in the main plasma, with enough temporal resolution to study also fast profile changes due to MHD, ELMs, Marfes, fishbones and TAE modes.

Table 1. Potential applications of reflectometry for ITER

PLASMA PARAMETERS					
N°	cat (i)	N°	cat (ii)	N°	cat (iii)
2	Plasma position and shape	16	Divertor plate abaltion	47	Density fluctuations
6	Line average density	24	Electron density profile	48	Edge turbulence
8	Locked modes	25	q(r)	52	ICRF Phys. measur.
9	m=2 MHD Sawteeth	27	Fishbones TAE modes		
10	Plasma rotation	41	ne(r) in divertor plasm.		
14	Edge fluct. ELMs, turbul.	45	ICRF control param.		

2. Background

The first density profiles from reflectometry were obtained in the mid eighties with a homodyne slow sweeping FM-CW system[1]. Experiments in hot plasmas revealed that the main problem of profile evaluation was due to plasma fluctuations and in some cases to parasitic reflections of non optimized transmission lines. Much faster FM-CW systems were developed in the nineties[2,3] and new techniques emerged, e.g.: AM[4] and dual differential phase reflectometry[5], pulse radar[6], pulse compression[7]. A great reduction of the effect of fluctuations was obtained by probing the plasma faster ($10~\mu s$ -$100\mu s$, for a complete profile) than the typical period of the plasma perturbations (few kHz for magnetic modes and 100 kHz for turbulence). Disturbances due to spatial fluctuations still remain but some techniques (measuring the differential phase between layers located within the correlation lenght of the fluctuations) seem to be more capable of reducing its effect. It is also possible, with any technique, to make some averaging of the fluctuations by software. The performance of the transmission lines is an important issue. However, current developments in both improved waveguides[3,10] and the possibility that most techniques can make a vector subtraction of the transmission line characteristic, should resolve this problem. A transmission line with high performance is necessary to cope with the methods that are more sensitive to the spurious reflections but less sensitive to the plasma spatial fluctuations. Due to the foreseen advances in technology, mainly driven by other areas (for example computers and telecomunications) significant improvements of the different techniques are expected in the near future. In adition, the evolution of data acquisition systems will enable improved data processing at lower costs.

III. ACCESSIBILITY

In ITER the expected H-mode density profiles have steep gradients at the edge and are almost flat in the core. Temperature profiles are more peaked, as seen in the reference case of Fig. 1. Both the O and X modes can be used to probe the plasma; O mode offers the advantage of depending only on the plasma density while X mode is linked both to the density and magnetic field configuration. Due to the high temperatures foreseen, significant modifications of the cutoffs (see Fig 1.) occur[8]. We have made calculations which show that the X mode upper cutoff (f_{UC}) suffers the largest modifications due to the relativistic effects: f_{UC} decreases and flattens at the outer plasma region. Also, due to relativistic broadening and downshift of the electron cyclotron absorption, the X mode radiation launched from the LFS, is prevented from reaching the core (depending on the exact density and temperature profiles). Therefore, only the steep gradient region of the plasma edge can be probed with X mode, from the LFS. O mode (cutoff at the plasma frequency) is less affected by absorption and can be used to probe the steep gradient regions either from the high or low field sides. The core plasma, however, cannot be probed with O mode when the density profile is nearly flat. The core seems to be accessible from the HFS using X mode, lower cutoff. This is an interesting option because the cutoff is not significantly affected by the temperature and EC absorption is negligible (Fig. 1). We have made ray tracing calculations to investigate the importance

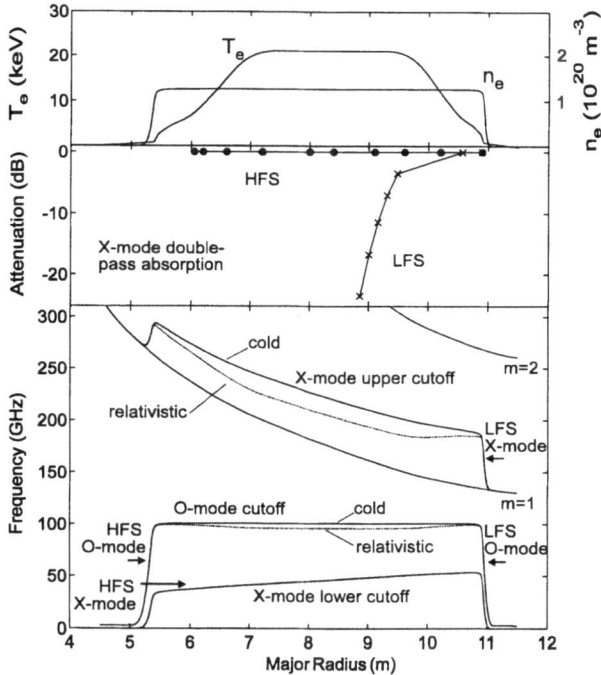

Figure 1. O and X mode cold and relativistic cutoffs and absorption for a ITER reference case with $Te(0) = 21\ KeV, ne(0) = 1.2 \times 10^{20}\ m^{-3}$.

of refractive effects for core probing with X mode. We modelled vertical displacements of the plasma approximately, by launching rays with various angles to the horizontal from a variety of vertical positions. Preliminary results indicate that both the edge and the core should be accessible in the X-mode from the HFS for a range of vertical displacements. However, more extensive calculations are needed to quantify fully the range of equilibria which can be accommodated and the number and characteristics of the required antennas. The frequency range required for O mode measurements is 18-120 GHz, while X mode high cutoff measurements should cover the range 130-190 GHz. The X mode lower cutoff uses frequencies from 18 to 90 GHz.

The frequency range just above the low X cutoff could also be used for delayometry[9], to gain complementary information on the line density as well as on the overall behaviour of the flat part of the density profile (e.g. degree of "hollowness"). Preliminary experiments have been carried out at RTP by using pulse radar but also the other reflectometry techniques (FM, AM...) could be applied.

IV. TRANSMISSION LINE AND ANTENNAS

A solution is proposed for a high performance (low losses and low level of spurious reflections), transmission line and antenna, compatible with the different reflectometry techniques[10]. A possibility is to use corrugated waveguides, operating with the HE_{11} mode, that have several valuable features: the ohmic losses are very low; highly oversized waveguides are broadband (approximately 1 octave); compact mitre bends can be used to change the direction of the lines. Waveguides gaps are possible for mode filtering and allow also for compensation of thermal expansions[10]. Waveguides, however, are sensitive to misalignement, which can lead to mode conversion mainly at the misaligned joints.

1. Transmission line

Separate lines should be employed for transmit and receive in order to minimize the spurious reflections from the waveguides and antennas. The performance of a typical transmission line (diameter D=63.5 mm) with 100 m length and 10 mitre bends has been estimated using a scattering matrix code[10]. A smooth behaviour is observed for frequencies between 56-250 GHz, while for frequencies below 50 GHz the frequency dependence increases due to mode conversion at the mitre bends. An improved characteristic is obtained by including a waveguide gap at each mitred band as shown in Fig. 2. For frequencies below 50 GHz the losses are severe and improved solutions using mitre bends with integrated uptapers or focused reflectors can be implemented (where space is available). Waveguides with smaller diameters (D=31.75 mm) can be also be employed; in this case gradually curved waveguides have a much better performance.

2. Matching of the generators and detection to the transmission line

Due to the excellent coupling (98% efficient) of sources and receivers (TEM_{00} mode) into the corrugated waveguides (HE_{11} mode), quasi-optical techniques are proposed to match the generators and detection systems to the transmission line[10]. An example is shown in Fig. 3 of two microwave sources coupled to the oversized waveguide via a confocal system consisting of two ellipsoidal mirrors with different focal lengths f_1 and f_2. If the ratio f_1/f_2 is chosen to be equal the ratio of the horn output diameter D_1

and waveguide input diameter D_2, the confocal imaging is frequency independent. The summation of the sources is performed by a quasi-optical beam splitter, using a dichroic reflector to minimize the loss. The quasi-optical matching serves also as a mode filter as higher order modes are scattered out due to the finite size of the mirrors. In view of the importance of the inboard launch with severe space constraints, a configuration was studied using only one waveguide for both emitting and receiving. The incident and reflected signals are separated with quasi-optical directional couplers having directivities higher than 60dB[10].

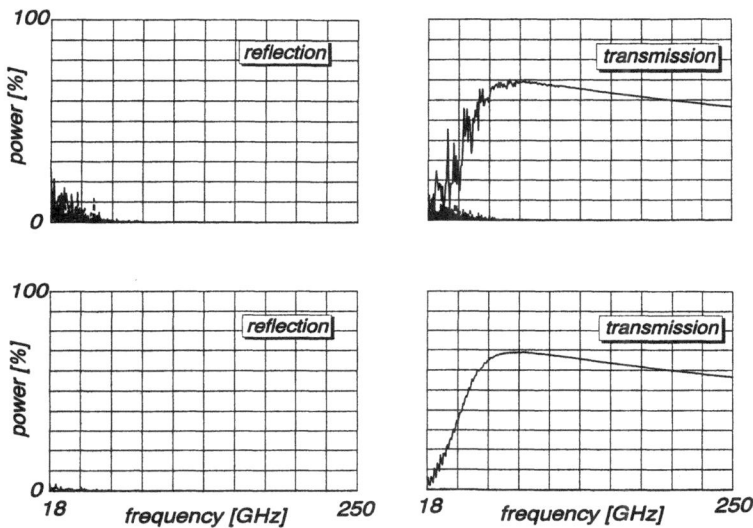

Figure 2. Estimated reflected and transmitted power over a complete line of 100 m length including; (a,b) 10 mitre bends; (c,d) 10 mitre bends and 10 gaps for mode filtering.

3. Vacuum window

For the application of oversized corrugated waveguides over large frequency ranges, a vacuum barrier window consisting of a dielectric disc inclined of the Brewster angle, is proposed[10]. This window will have a broadband transmission for the polarisation parallel to the plane of incidence. Two windows can be combined for polarisation diplexing in a quasi-optical arrangement to ease the integration into the cryostat of ITER. Mode conversion in a window is similar to a gap in the waveguide and therefore tolerable. Some additional losses (few percent) can occur at the lowest frequencies because the HE_{11} mode is not perfectly polarized at low frequencies. The loss terms can be avoided with a quasi-optical transmission of two mirrors in a confocal arrangement. The bandwidth of the vacuum window is essentially given by the dispersion of the material. A suitable material is fused quartz with a Brewster angle of 63 degrees but other materials can be considered (Al_2O_3, BN_3 or BeO). A double window assembly can be employed to increase safety. A dog-leg configuration as shown in Fig. 4 could be installed inside the biological shield of ITER providing simultaneously the vacuum barrier and efficient neutron screening.

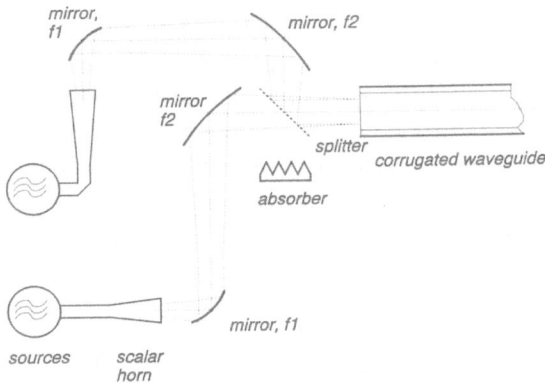

Figure 3. Matching of two generators to the corrugated waveguide via a confocal system with two ellipsoidal mirrors. The summation of the sources is performed by the quasi-optical beam splitter.

4. Antennas

Two options can be considered for the antennas: small antennas with large beam divergence, and large antennas with small beam divergence, that can even be focused. Small antennas have low spatial resolution and large return losses but can be more easily integrated into ITER blanket modules. Large antenna with higher spatial resolution are more sensitive to beam deviations due to plasma movements. The final solution will inevitably be a compromise between available space and required beam width, and will be optimised by means of ray tracing studies. Owing to the good radiation pattern of

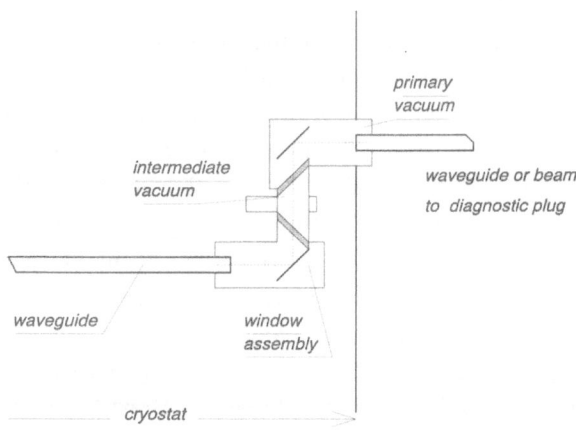

Figure 4. Double Brewster window arrangement and integration into the cryostat of ITER.

the HE_{11} mode, the simple open-ended waveguided would provide a good antenna. The insertion of an up-taper results in a narrower antenna pattern. Transitions to rectangular waveguides can be used to ease the integration, particularly at the HFS. An example of two types of antennas are shown in Fig. 5. The large apperture antenna include metalic reflectors. The quality of the mirror is not very demanding, copper, stainless steel or graphite could be used.

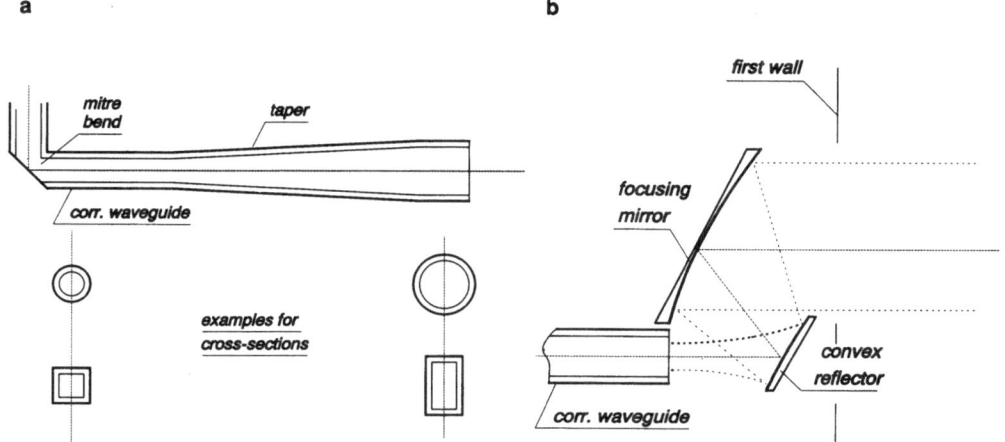

Figure 5. Possible configurations for the antennas: (a) small antenna with a taper using a transition of the waveguide cross-section; (b) large aperture antenna employing a convex and a focusing reflector. Antenna (a) may be used both at HFS and LFS; antenna (b) is proposed for the LFS.

5. Combination with other diagnostics

Other diagnostics may use apertures and/or optics which can be shared by reflectometry. Heating systems like NBI may need large penetrations and therefore it may be possible to take advantage of the NI beam penetration through the first wall. Another example is the LIDAR Thomson scattering system that requires a large metallic mirror ($D\cong0.5$ m) and needs a blanket penetration of 0.3 m. Two solutions were considered for the injection of the microwaves into the optical system[10]: (i) a mirror located at the focal point of the scattering system, with a hole in the center to pass the optical rays; (ii) an optical dielectric coated mirror, with a wedged substrate made from fused quartz. The mirror can be made transparent to microwaves and thus feed the transmission system from an open-ended waveguide (or beam) coming from behind the mirror. The combination with the LIDAR system can be very useful as this diagnostic needs the absolute calibration of density which can be provided by reflectometry.

6. Integration of waveguides and antennas

At the LFS, access is possible through the large main ports near the midplane as well as parallel to the cooling lines coming from the top of the vessel. The waveguides and antennas can be installed within a diagnostic plug. The fact that waveguide connections must only be aligned axially, but gaps are also possible, allows to design waveguide joints without any tight connection which may ease the maintenance of the diagnostic plug.

At the HFS the access is rather limited. The waveguides can be clamped to the back side of the blanket module or they can run parallel to the cooling pipes of the blanket module, using the empty space in the blanket and run until they meet the slit between the blanket segments, close to the midplane. If this is the only space available the horizontal dimensions of the antenna are limited to 2 cm leading to a large divergence in the horizontal direction. The vertical dimension of the antenna can be much larger which opens the possibility to compensate beam broadening effects by a focused beam. Waveguides tapered to appropriate dimensions could be attached on the shielding back plate and end in a mitre bend, directing the power to a corrugated horn located several centimeters behind the first wall. Another solution might be to form the antennas and waveguides in the blanket itself, which would contribute to the cooling of the system. A strong interaction with the ITER engineering is needed to find the best solution for the integration of the HFS channels.

V. REFLECTOMETRY INSTRUMENTATION

1. Emitter-receiver

The emitter-receiver is the starting point for most reflectometry designs but, though important, it would not be wise to give at this time a frozen design of the reflectometry instrumentation for several reasons. Microwave emission and detection technology is evolving very rapidly. The emitter-receiver is placed at the end of the line, outside the biological shield and therefore different systems can be easily installed at later stages in the ITER machine development. Probably the best solution will include complementary or hybrid techniques. Nevertheless, it is important to ensure that, with existing technology and experience, the goals of ITER reflectometry can be met to a high degree.

2. Reference approach

Broadband systems with ultrafast swept operation (10 μs or less) are the preferred solution for density profile measurements because they have the highest spatial sampling rate (which is a key parameter for accurate and detailed density profile reconstruction), as well as good immunity to plasma fluctuations. However, in some circumstances it may be necessary to increase the incident power at the expense of reduced sampling. This may be the case particularly in probing the plasma center of discharges which depart significantly in shape or vertical position from the "standard" equilibria. For these cases, the losses may increase to more than 50 dB, due to refractive effects and a missaligned reflecting layer. With discrete narrow band operation some 15-20 dB amplification of the source power can be obtained. By using networks of multiplexed sources and arrays of multipliers it would still be possible to achieve the required frequency coverage. No matter which technique is used (FM-CW, AM, pulse radar, pulse compression) it will

140

be possible to have emitter/receiver systems with configurations for both broadband operation and discrete frequency probing. All these configurations will be suitable for use with the proposed waveguide/antenna design.

A reference design of a complete system for one specific location is shown in Fig. 6. According to the study of the transmission line and antennas presented in section IV, it may be convenient to use a different design for the line covering frequencies below 50 GHz. Therefore, different channels are foreseen for the low and high frequency ranges. Three transmitting/receiving systems were considered, one for the frequency range 18-50 GHz (Low Frequency Unit) and two for the range 50-190 GHz (Mid and High Frequency units). The Low Frequency Unit may be completely assembled in coaxial technology in a few years time. Each individual transmitter inside one unit shall include one or more generator devices that may be either pulsed, swept or synthetized. The transmitters will probably be of the phase locked type, using a set of schemes to procedure the required signal. All techniques will use sensitive heterodyne receivers with amplitude and phase detection. The modular approach presented above allows for a late decision about the techniques that will suit best each diagnostic aplication, but provides, nevertheless, a solid base for the design of the transmission lines and quasi-optical combiners, which technology and theoretical background is not likely to change significantly in the next years.

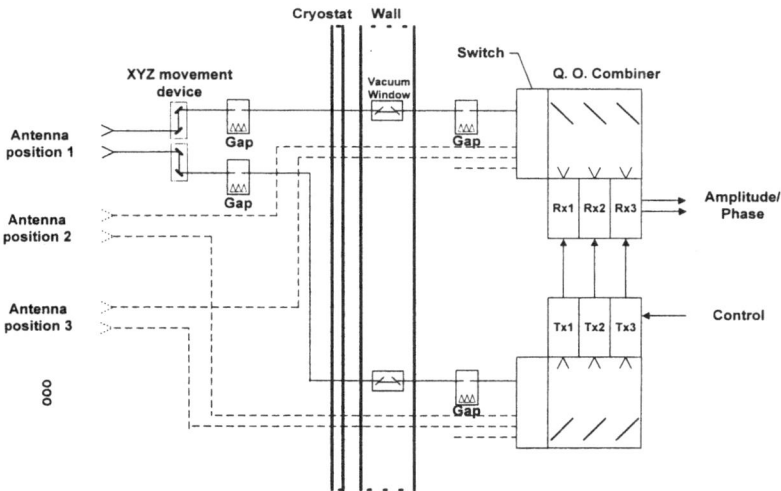

Figure 6. Reference design for the ITER reflectometry system

VI. CONCLUSIONS

Reflectometry is an essential diagnostic for density profile measurements in ITER. At its present status reflectometry is mature, offers good spatial plus time resolution and is already capable of meeting ITER requirements for measurement of the density profile in the main plasma. Access requirements are moderate, specially for the LFS mid plane. Also we have shown that HFS channels are desirable and feasible. We have presented an outline design of a high performance antenna and transmission system that will give good access to the widest range of reflectometer techniques with minimal attenuation and internal reflections. Optimization of the proposed solution is needed to cover the large frequency ranges. Microwave technology in this frequency range is developing rapidly and the final design of the emitter/receiver should be made in a later stage. A system, able to provide base line performance, can be built with state of the art technology. A reference design is given, using different reflectometry techniques.

VII. EUROPEAN REFLECTOMETRY EFFORT

Besides the on-going developments taking place in many EU laboratories, some proposed tasks can give an important contribution for ITER. Further ray tracing and absorption calculations for ITER reference cases are proposed to complete the documentation of the reflectometry study. Also, it is foreseen the test in broadband frequency the corrugated transmission line of the W7-AS with existing dedicated equipment[10].

ACKNOWLEDGEMENTS
This study is the result of many discussions within the EU Home team of reflectometry for ITER and we would like to thank all our colleagues who have contributed, in particular H. Bindslev, A. Donné, H.Hartfuss, S. Heijnen, E. de la Luna, F. Serra, A. Silva and V. Zhuravlov. We also acknowledge the work of Edward Doyle which was presented at an earlier ITER workshop on microwave diagnostics.

REFERENCES
1. F. Simonet, Measurement of electron density profile by microwave reflectometry on toka-maks, *Rev. Sci. Instrum.* **56**, 664 (1985)
2. A. Silva et al., Fast sweep multiple broadband reflectometer for Asdex Upgrade, *17th Simp. on Fusion Techn.*, Rome I:747 (1992)
3. E. Doyle et al., Advances in reflectometric density profile measurements on the DIII-D tokamak, *Rev. Sci. Instrum.* **66**, 1229 (1992)
4. J. Sanchez et al., Amplitude modulation reflectometry for large fusion devices, *Rev. Sci. Instrum.* **63**, 4654 (1992)
5. G.R. Hanson et al., A swept two-frequency microwave reflectometer for edge density profile measurements on TFTR, *Rev. Sci. Instrum.* **63**, 4658 (1992)
6. S.H. Heijnen et al., *Rev. Sci. Instrum.* **66**, 419 (1995)
7. Laviron et al., First experiments of pulse compression radar reflectometry for density measurements on JET plasmas, accepted for publication in *Plasma Phys. Cont. Fusion* (1995).
8. H. Bindslev, Relativistic expressions for plasma cutoffs, *Plas. Phys. Cont. Fusion* **35** 1093 (1993)
9. S. Heijnen et al., proceedings of this conference
10. D. Wagner and W. Kasparek, Transmission line design for reflect. syst. on ITER, *Int. report of Institut fuer Plasmaforschung*, Univ. Stuttgart (1995) and ref. there in

PROPOSAL OF REFLECTOMETRY SYSTEM FOR ITER

V.A. Vershkov

RRC "Kurchatov Institute", Moscow, Russia

INTRODUCTION

Reflectometry is now widely used in tokamak experiment. It may provide the measurements of a number of important plasma characteristics with minimum requirements to the entrance ports. It uses well developed mm wave band and the launched and received waves may be easily transported to and away from tokamak by means of long transmission lines. Significant experience was gained in reflectometry since the first experiments on ATC tokamak[1], but it still in process of development. It means, that there is no final universal version of reflectometry scheme and there is a number of questions connected with the interpretation of reflectometry data. The radial localization of measured phase fluctuations, experimentally appeared to be much more localized then predicted by one or two dimensional calculations[6]. It was shown in[4], that at high turbulence[2,3,4,5] level, the measured radial correlation length may be greatly reduced with respect to the real length and even distort the observed frequency spectrum. The problem of so called "phase run away" is discussed in[7,8,9], which is thought to be connected with the Doppler frequency shift associated with the poloidal rotation of the grating-like cut off layer surface[9]. Thus it is possible to conclude, that reflectometry is powerful diagnostic for ITER, but additional experimental work have to be done to solve some problems.

POSSIBLE GOALS OF REFLECTOMETRY ON ITER

Taking into account previous experience, it is possible to list the following plasma characteristics, which may be measured by means of reflectometry on ITER:

1. The main task of reflectometry is to measure the density profile of the plasma core. It may be done from outside along the major radius in equatorial plane. It is possible to use the frequencies from 110 GHz to 26 GHz for density range $1.5 \cdot 10^{14}$ cm^{-3} to $0.84 \cdot 10^{13}$ cm^{-3} for ordinary mode. For higher cut off extraordinary mode corresponding frequency range (for central magnetic field 5.7 T) is from 123 GHz to 256GHz ($n_e(a)=1 \cdot 10^{13}$ cm^{-3}). Possible absorption at $2\Omega ce$ at the periphery may restrict the access of extraordinary mode to inner half radius. It must be underlined, that taking into account possibility of flat or even hollow density profiles, the use of extraordinary wave is of primary importance. This mode is also preferable for the measuring of turbulence structure with better spatial resolution due to shorter wavelength. The radial accuracy about 5-10 for core and 0.5 cm. for SOL is required. The probing of plasma from high field side with the reflection from lower cut off at frequencies from 26 to 60 GHz seems very promising[10]. In this case it is possible to use low frequency sourses with higer power and to have access to the central part of the plasma without difficulties with the absorption at cyclotron frequencies. The interesting possibility is the measurement of the phase shift, or time delay in interferometer regime, which automatically occurs with O mode at the beginning phase of discharge and with X mode in frequency range

Diagnostics for Experimental Thermonuclear Fusion Reactors
Edited by P. E. Stott *et al.*, Plenum Press, New York, 1996

143

between low cut off and cyclotron frequency.

In principle the phase shift measurements may be done with the accuracy much better, than one tenth of period, which means much less, then 1 mm. for typical conditions. Unfortunately three circumstances usually prevent from using the phase shift at the main frequency. The first is the difficulty of identification of the number of fringes, when this number is big. The second is the difficulty of observation of the phase evolution, when the phase fluctuations become higher, then one fringe. The third is the possibility of wrong phase measurements due to "phase run away". In order to overcome this problem broad or narrow frequency sweep are used, which transfer the measurements to effectively longer wavelength, which enables unambiguous measurements. But T-10 experience showed, as it is seen on Fig. 2, that it is possible to use the phase shift at the main frequency, provided, that phase fluctuations less, then a fringe. Moreover, reducing the fluctuation level by means of narrow band filter, it was possible to use such measurements at least for, registration of the relative phase variation in time. In ITER case such approach definitely can't be used in plasma core measurements, but, in contrary, it seems possible to use in SOL measurements with O mode. As the vacuum probing wavelength at 26 GHz become comparable or even higher, then SOL width, it is difficult to predict the phase fluctuations level much higher, then a fringe. Thus using appropriate narrow band filter it is possible to register the phase variation in time at fixed launch frequency. Special additional experiments of that type are planned on T-10.

2. The evaluation of the plasma shape. In order to do it one needs several poloidal directions of view. Poloidally non-symmetric phenomena like MARFY may cause some difficulties.

3. The measurements of the SOL density and decay length in main plasma and in divertor away and near the plate. The antennas in divertor are needed. It is difficult to measure the densities near divertor plates higher, then $8 \cdot 10^{20}$ m^{-3} with frequencies higher 250GHz.

4. The estimation of internal inductance and plasma energy by measurement of Shafranov shift of magnetic surfaces. Inside and outside equatorial launches is needed.

5. The evaluating of the current profile is possible to make in two ways. The first one is the measurements of radial localization of MHD modes, which gives the q value at a number of radial positions. The example of such measurements on T-10 is presented on Fig. 4. It is possible to use synchronous detection of the reflected wave phase fluctuations with magnetic probes signals. One launch in midplane port is enough for this purpose, but it is possible to obtain wrong results due to modes synchronization. In order to identify independently modes m and n numbers at least two poloidally and two toroidally separated launches are needed. The second way is the measurement of the inclination of magnetic field line with toroidal correlation reflectometry, which was successfully realized on TJ-I[11]. This approach is based on the assumption of long correlation length of turbulence along magnetic field lines. It requires two launches, separated toroidally. Due to large midplane port dimensions it is possible to place them in one port, but in order to increase the accuracy of the measurements it is preferable to have additional launches in adjacent ports. As it will shown below, the search of the magnetic field line position in poloidal direction can be made with the phase array of antennas by means of "holographic scan", which is illustrated by computer simulations, presented on Fig. 8 and 9.

6. The measurement of the rotation of small scale fluctuations, which is in principle the sum of transverse to magnetic field line components of poloidal and toroidal rotation. T-10 experiments show, that it may be qualitative monitor of plasma rotation itself, but may not coincide with it. Poloidal array of antennas is needed at least in equatorial port.

7. The measurements of the amplitudes, the radial position and the m number of long wave modes. Special attention must be paid for the TAE modes exited by fast Alfa particles, which is of primary importance for ITER. High and low field launch is preferable for evaluating ballooning properties of instabilities. Two poloidal and toroidal launch are needed for evaluating m and n number.

8. The measurements of the radial distribution of the amplitude and poloidal structure (poloidal and radial correlation lengths) of small scale turbulence. At least one poloidal array of antenna is needed. Several poloidal launches is needed for the observation of poloidal asymmetry. The typical poloidal wavelength, estimated from relation $k \cdot \rho_i = 0.5 \div 0.1$, where k is the wave number and ρ_i is the ion gyro radius, will be in the range from 6 to 30 cm. It means, that practically for all launched to the core plasma waves, the incident wavelength

much less, then poloidal wavelength of turbulence. Thus the geometrical optics approaches may be used.

It will be shown, that the discussed reflectometry opportunities may be realized with T-10 Three Wave Heterodyne Reflectometer with amplitude modulation and proposed waveguides and antennas configuration.

T-10 THREE WAVE HETERODYNE REFLECTOMETRY SYSTEM WITH VARIABLE FREQUENCY AMPLITUDE MODULATION

It is proposed to use a new type of reflectometer, which was build and tested on T-10[11]. It combines the advantages of amplitude modulation reflectometer in measuring the density profile, with high sensitivity and possibility to measure the density fluctuations of heterodyne one. The scheme of the reflectometer is presented on Fig.1[12]. Two microwave sources are hold by feedback system at the frequency difference of 20MHz. The first one - (S) is used as the source of the launch wave and the second - (L.Q) as the local oscillator. The amplitude of the launch wave is modulated with the p-i-n diode - (M) in the frequency range from 50 to 1000 MHz, thus providing the splitting of the wave into three frequency components, which reflects from the three different minor plasma radius in accordance with there critical densities. The T-10 antenna system consist of 5 horns, each of them can be used as the launcher and two or three of them as the receiver. Thus this system is capable to measure simultaneously the radial and poloidal correlation lengths. The received signal is mixed with the local oscillator and after preamplification is put to a special bloc - (F), which filter separately all three incident frequencies. This bloc is also generates the modulation frequency, which may be changed in 0.5 msec by the control system from 52.5 to 1050 MHz with the step of 52.5MHz. The frequency band of the filter is 2MHz. As those three components are reduced to intermediate frequency of 20MHz, for any modulation frequency from the above mentioned range, they pass through the limiting amplifiers and put after either to the three frequency discriminators, which measure the Doppler frequency shifts of the reflected signals or to the phase detector. It is possible to measure the phase shift either between the reference IF frequency and each of the three reflected waves, or between those three waves. This reflectometer provides simultaneously three kinds of information:

The first is an unambiguous measurement of the time delay of the signal reflected from the plasma. This information is obtained from the phase difference between the three incident

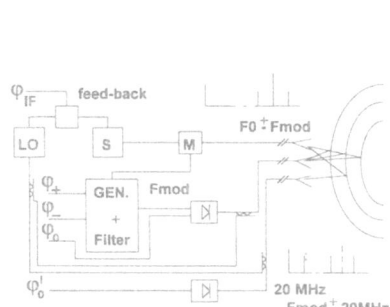

Figure 1. Scheme of Three Wave Reflectometer

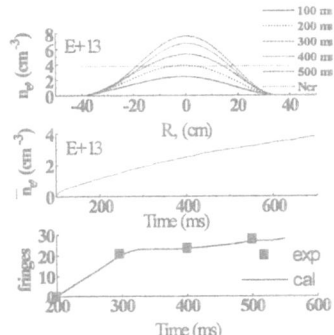

Figure 2. Comparison of reflectometry phase shift on the main frequency with calculated one from interferometry profile measurements.

waves. Due to the fact, that the frequency splitting can be changed, it always may be adjusted to have the phase difference in the range less, then 2π, thus enabling unambiguous measurement of the time delay. For example modulation frequency 50MHz corresponds to effective wavelength of 6m, which is just appropriate for ITER beginning phase and core measurements. In contrary modulation with 1GHz gives effective wavelength (between the two satellite) 15cm, which is just good for the SOL measurements. As the time of variation of modulation frequency is 0.5 msec, both measurements may be done in desired 10 msec resolution time. Thus this system may combine core and SOL measurements. It is natural, that reducing the phase shift to an unambiguous value of one fringe, one looses proportionally the accuracy of measurement, because it can't be much higher, then 1/50 of effective wavelength. Thus it is important to underline, that in proposed system the unambiguous phase measurements between satellites are made simultaneously with the phase measurements at the main microwave frequency with relative accuracy of the some tenth of millimeter. So it is possible to measure the absolute phase shift with the adjusted equivalent radial accuracy between 6-0.5 cm and simultaneously to have relative accuracy of tenth mm, which cover the requirements of above mentioned tasks for ITER.

The possibilities of proposed reflectometry system is illustrated with the experimental data in discharges with slowly rising plasma average density, shown on Fig.2 and 3. They presents the examples of the phase measurements on the main frequency and between the two side band frequencies respectively. The comparison of directly calculated from the experimental interferometry density profile and measured with frequency 55.2GHz reflectometry phase variation is presented on Fig.2. It is seen that the four calculated interferometry points practically overlaps with reflectometry. It was easy to distinguish the phase variation about 0.2 fringe, thus enabling to detect plasma shift about 0.5 mm. Figure 3 presents the phase shift between side bands with the modulation frequencies 157, 630, 997 MHz correspondingly. It is clearly seen, that the phase shift increases with the modulation frequency, thus increasing the accuracy.

The result of the q=2 radial position search is presented on Fig. 4. The first trace is the auto correlation function of the reflectometry phase, taken at the position of maximal m=2 oscillations. The time sequence of 4 msec was analyzed. It is apparent the presence of coherent low frequency oscillations. The Fourier spectrum of this auto correlation function is presented

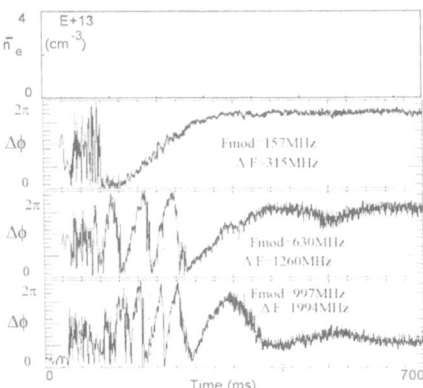

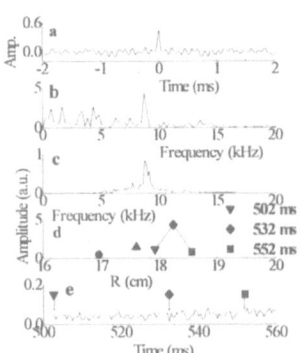

Figure 3. Phase shift measurements between sidebands sitellites with three modulation frequencies.

Figure 4. Measurements of q=2 radial localization with reflectometry. a)- auto correlation function of phase fluctuation, b)- Fourier spectrum of auto correlation function, c)- Fourier spectrum of magnetic probe signal, d)- radial dependence of mode amplitude, e)-time variation of m=2 magnetic fluctuations.

on the second trace and the spectrum of magnetic probe on the third. It is seen the spikes in both spectra at frequency 8.8kHz. The radial dependence of the spike amplitude is presented on the fourth trace. The m=2 oscillations have a very contrast maximum at the radius 18.25 cm., which well agree with estimations of q=2 radial position. The constancy of m=2 in time is proved with trace of m=2 amplitude on fifth trace. The measured width of island about 0.5 cm. Thus the accuracy of q=2 position measurement is about 1.5%.

The results of radial and poloidal correlation measurements for the central parts of the plasma column are presented on Fig. 5 and 6 respectively[13]. All Figures present the result of cross-correlation analysis of the output data of frequency discriminators of two reflectometry channels. For the case of Fig. 5 this channels were radially separated by means of amplitude modulation with frequencies 50, 105, 420 and 840 MHz. The central frequency and the high frequency satellite was used, corresponding to the radial separation of 0.04, 0.08, 0.31 and 0.62 cm. The main frequency was 36.12GHz. The upper trace present the Fourier amplitude averaged over several successive time sequences. The middle trace is averaged cross-phase and the lower trace is the coherency of both channels versus frequency. As amplitude spectrum and cross-phase don't vary with radial separation, only coherence presented for all four cases. It is seen, that the cross-phase is zero at all frequencies, which means, that fluctuations don't propagate radially. Secondly, at low separation coherence is more uniform along the frequency, while at higher separation it deviates from zero only for the frequencies near 100 kHz. This observation suggest the presence of two fluctuation types with different radial correlation length. Poloidal correlation measurements, presented on Fig. 6 give additional support for this suggestion. It is clearly seen on Fig 6 a and b, that coherence also deviates from zero only for some narrow frequency bands. It means, that those two turbulence types have different poloidal correlation length. One can see, that the frequency position of the coherence peak is shifted in two times for Fig. 6 b, although the data were taken in one discharge. The only one difference is that the data of Fig. 6 b were taken after Ne puff. This observation makes some ground to conclude, that turbulence with high correlation length is connected with ion composition of the plasma. Fig. 7 presents the sum of poloidal and toroidal rotations as the function of minor radius of T-10[14]. It is seen, that reflectometry is capable to measure velocity of fluctuations in plasma core and SOL. It is seen even the velocity shear at r=24cm, in agreement with also presented multypine Langmuir probe data. More data about the turbulence properties may be found in[13], while here they just illustrate high potential capabilities of reflectometry in realization the above listed goals in general and possibility to realize them by means of T-10 Three Wave Heterodyne Reflectometry system with variable frequency amplitude modulation. This system also can be upgraded in near future to eliminate the influence of the secondary reflections, which is one of the main problem of the

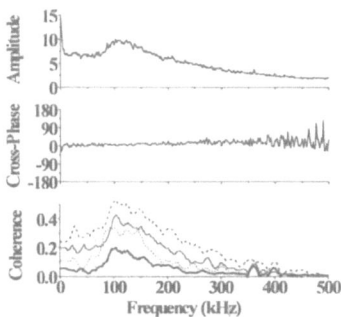

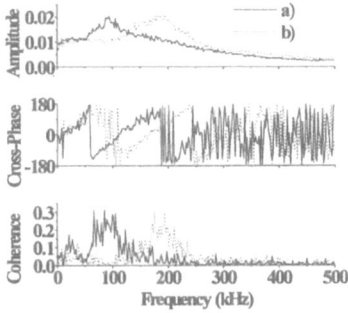

Figure 5. Radial correlation measurements for four values of modulation frequencies. Dashed line- Fm=50 MHz, ΔR=0.04 cm: thin line- Fm=105 MHz ΔR=0.08 cm; dotted line- Fm= 420 MHz, ΔR=0.3 cm; thick line- Fm = 840 MHz, ΔR = 0.6 cm.

Figure 6. Poloidal correlation measurements. Poloidal distance 1.1 cm. Rcr=12 cm. a)- before Ne puff, b)- after Ne puff.

reflectometry. It can be done without big changes of the system, thus conserving all the advantages. In summary it is possible to conclude, that the proposed reflectometry system which already proved in experiment may satisfy all ITER requirements.

HOLOGRAPHY APPROACH IN TURBULENCE MEASUREMENTS

As it was mentioned earlier, the quantitative interpretation of radial localization and absolute amplitudes of fluctuations are still unclear. Moreover it was shown in[14], that at high turbulence level even such qualitative properties, like frequency spectrum may be dramatically distorted. The main reason of this is the interference of reflection from multiple turbulence structures, received by antenna with wide diagram. It was shown previously, that even in this case the use of poloidal or radial correlations may significantly improve the quality of information, but the real solution of the problem is in using wide aperture antennas with good focusing at the measured plasma radius. Such approach was used in W7-AS and was also recently proposed by the authors of[4]. Due to the fact, that it is not possible to realize good focus for every radius, the use of the antenna array as the phase grid was considered in T-10 approach. The example of such array, used for computer simulations[15] is presented on Fig.8. There are one antenna for the launching power. The reflected signal is simultaneously received by the eight antennas. The amplitude and the phase is recorded by ADCs simultaneously for the eight channels, making together 16 signals. Mixing those signals with appropriate phases, it is possible to make mathematical focusing to every needed radius. It is also possible to scan the focus poloidally, directly evaluating the poloidal wavelength. The initial antennas positions for the focusing at the reflection layer and the phase shifts for poloidal scan of the focus were found by means of 2D computer code solving exact wave equation in real geometry[15]. Figure 9 presents the results of the poloidal scan of the focus of the array for the five plasma shapes. For each plasma shape amplitudes and the phases in antennas were found by solving the exact wave 2D equation for the reflection in given plasma configuration. The first reference configuration corresponds to unperturbed circular plasma with linearly rising density. In the four perturbed configurations, presented on the right pictures, the gaussian perturbation was placed at the reflection radius symmetrically and with the shift of 1, 2, and 3 cm. down. The left pictures present the resulting phases, obtained by the poloidal scan of the focus. The trace for the reference unperturbed plasma is also presented on each picture. It is clearly seen, that

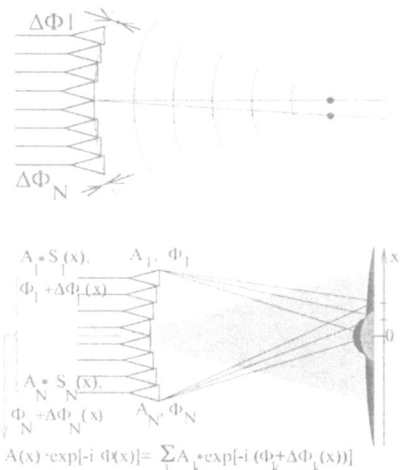

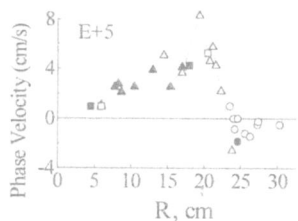

Figure 7. Sum of poloidal and toroidal rotation of fluctuations. Triangles - reflectometry data , circles- Langmuir probe data, squares - magnetic probes.

Figure 8. Principle scheme of holographic measurements.

the position of maximal phase deviation with respect to unperturbed case follow the position of the perturbation. Initial experiments with the array of three antennas are planned on T-10, but for ITER case, additional work is planned to increase the dimensions of computed area.

PROPOSAL FOR ITER REFLECTOMETRY SYSTEM

According to the listed above goals of reflectometry, it is proposed to divide ITER reflectometry system into three parts. They are supposed to be based mainly on T-10 Three Wave system, but to make more reliable measurements it is desirable to use several types of system, such as delayometry or pulsed compression techniques. The first part is the core density profile measurements. The second is the plasma shape evaluation. The third is the plasma turbulence measurements.

Core density profile measurements

Core density profile may be measured with three possible modes: O-mode ; X-mode upper cut off from low field side and X-mode lower cut off from high field side. Due to the access difficulties, connected with flat density profile with O-mode and possible absorption at 2 Ωce for X-mode upper cut off it is proposed to use all three modes simultaneously. The arrangement of core profile system is shown schematically on Fig. 10 and 11. The O-mode and X-mode upper cut of launches are placed in mid plane port and there design seem don't connected with big technical difficulties. The most difficult is the design of high field side launch for X-mode lower cut off. It is connected with design of the waveguides to bring the power to inner part of torus and antennas compatibility with the blanket modules. Apparently it will require to design special blanket section in one poloidal cross-section. Although this is difficult engineering problem, but from one hand the use of lower cut off seems very promising, because it enables the access to the plasma core without the difficulties with absorption and it uses lower frequencies, where high power of the order of 1 w is available and from other hand technical problems have to be solved in any case for plasma shape evaluating

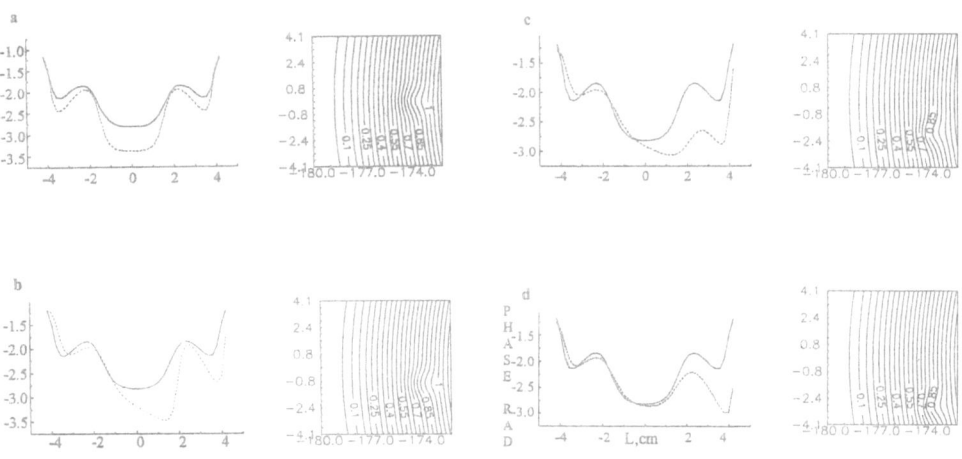

Figure 9. Results of holographic poloidal scan of antenna array focus for four poloidal positions of plasma density peturbation. The reflectometry data were simulated with computer 2D code. Left curves are the phases in radians, right pictures are the perturbed density distributions. Perturbation poloidal position is 0cm (a), 1cm (b), 2cm (c), and 3cm (d). Perturbation amplitude 10% and δx= 0.8cm.

system. The high field launch provides several additional possibilities. It make possible to measure the phase shift with the O-mode in interferometer regime at the beginning stage of discharge with the launch at high field and receiver at low field side. Otherwise it must be done in double pass from low field side, using reflection from inner wall. It is also possible to use the frequency range between lower cut off and cyclotron frequencies in X-mode for the measurements in interferomety regime. Thus the high field side launch may combine the reflectometry with interferometry regime. In order to use this opportunity special receive system for O and X mode low cut off must be placed in mid plane port, as it is shown on Fig. 11. It must be noted, that the reflectometry regime from lower cut off has principle difficulty, due to the fact, that both O and X modes have reflection positions in this frequency range. Thus small error in the polarization of X mode antenna will lead to high level parasitic O mode component, due to the fact, that O mode reflection layer more near to antenna. The natural solution of this problem is the use of pulsed radar technique[16,17] for this launch. Using the antenna polarization at some angle to magnetic field lines it is possible to measure simultaneously both reflection by means of the difference in time of propagation.

It is proposed to use a set of fixed frequency sources. The power of all sources will be amplitude modulated with p-i-n diodes and will be coupled to one launch for each mode. The lowest values of modulation frequencies is set by the requirements of unambiguous measurements, but it may be scanned with period of 10 msec. to higher frequencies in order to increase the accuracy. One receiving antenna is supposed for each mode with the system for decoupling of different frequencies. Three waveguides for each mode must be used. Two of them, launcher and receiver, are ended with antennas and the third waveguide without antenna is directly coupled with the launch waveguide near the antenna and is used for the reference phase measurement. The p-i-n diodes may be made now up to 178 GHz and no principal difficulties are seen to extend them to 250 GHz. Gann diodes are available for fixed frequency from less, then 26 up to 100 - 140 GHz with the power from 1 w to 10 mw (100 GHz). It is possible to use other types of diodes for the higher frequencies up to 250 GHz, either as generators, or in regime of frequency multiplication. The power decreases from 10 mw at 150 GHz to about one at 250 GHz. Backward oscillators up to 250 GHz s is also possible with power about 5 mw.

Plasma shape evaluation

This system is intended for evaluating the shape of plasma column by means of measuring the distance to the given critical density at several poloidal location. As it is seen from Fig. 10, it includes 8 poloidal position in the main chamber, four position in the divertor

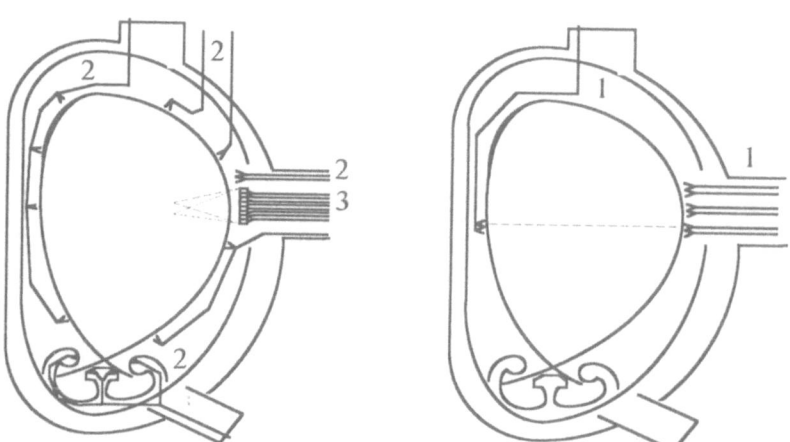

Figure 10. Principal scheme of poloidal positioning of reflectometry systems for ITER. The systems of one poloidal section are presented for convenience on two pictures.
1 - the system for core profile measurements, 2 - the system for shape measurements,
3 - system for turbulence measurements.

and additional supporting system in mid plane port. It is proposed to launch simultaneously three amplitude modulated fixed frequencies in O-mode for channels in main chamber and divertor. The frequencies may be chosen to be 26, 36 and 55 GHz, corresponding for critical densities 0.84, 1.6 and 3.75E19m^{-3}. The possible power from 1 w up to some parts of watt. The modulation frequency may be chosen to be 1 GHz. This corresponds to effective wavelength 30 cm, using the central and one sideband and 15 cm, using measurements between two sidebands. The simultaneous employment of two frequency difference is important, because the first is enables unambiguous measurements up to 15 cm and the last provide the accuracy about 0.5 cm. But the accuracy of that system may be even increased by transition to the phase measurements at the main frequency, as it was discussed above. Estimations show, that for 26 GHz the accuracy is enough to determined unambiguously the total number of fringes, provided, that decay length of the SOL is known only with accuracy of +/- 25 %. Special additional system in the mid plane port is proposed to do it either with the set of fixed frequencies, or slow frequency sweep in discussed frequency range. In this way it seems possible to increase accuracy up to 0.5 mm at 26 GHz. After that, using the known decay length in the SOL, it is possible to determine unambiguously the total number of fringes for 36 and finally for 55 GHz. Thus the accuracy about 1 mm. may be realized at the highest frequency, which correspond to reflection from separatrix.

Plasma turbulence measurements

It is proposed to make this system on the base of phase array of antennas, using extraordinary upper cut off mode. The processing of the data will used the procedure, which was describe above in case of computer model simulation. This system will be placed in mid plane port, as it shown on Fig. 10 and 11. The exact number and type of antennas must be determined with exact computer 2D code for the real ITER geometry. This system supposed also to be based on Three Wave reflectometer system with variable frequency of amplitude modulation to measure the radial correlation length. It is proposed to have in the same port two additional poloidally and two toroidally separated launches for the modes number identification and inclination of magnetic field line measurements. It is desirable to placed two additional launches in adjacent port, to increase the distance and thus the accuracy. The scheme of antennas positions in the adjusent two mid plane ports is shown on Fig. 11. The exact value of poloidal shift of magnetic field line will be determined by means of the holography mathematical scan of the phase array antenna. The data of computer simulation, presented on Fig. 9 show, that the resolution about two probing wavelength may be achieved.

Waveguide System

This system includes the wide band waveguide system inside the chamber and long

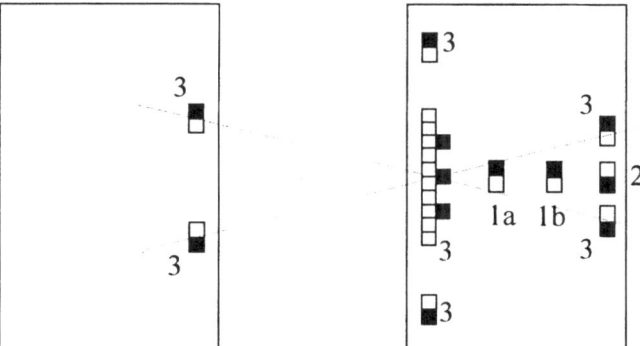

Figure 11. Toroidal view of reflectometry system positions in two adjacent mid ports. The dashed lines are the projections of magnetic field line for two current directions. The antennas phase array for X mode upper cut off at the lines cross-section. Filled and open squares are launch and receive antennas. The notation same as on Fig. 10. 1a and 1b core system for lower cut off and O mode.

transmission line to the registration system. The vacuum waveguide includes antennas, waveguides with several bends and vacuum windows. The transmission line must includes the mixers and filters for a number of different simultaneously used frequencies.

The type of antenna may be different for the three system. Antennas for profile measurements must be chosen in the process of computer modeling. It must be from one hand the compromise between the good amplification and the possibility to loose the signal, if the plasma shape is changed. Antennas with low amplification may be used for the shape measurements, due to small distance to reflected layer. For the turbulence system special antennas must be used, which properties will be found in computer simulations. The simple horn, which is shown on all drawings for simplicity, is not a wide band component and must be changed for reflection type antenna. The final choice must be done on the base of additional computer and experimental work with plasma and compatibility with the ITER design. It is desirable from the signal losses and diffraction divergence considerations to have antenna aperture about 5x5cm for shape and high field side launch measurements.

It is possible to make the vacuum waveguides from rectangular waveguides with dimensions about 2x2cm for plasma shape system and for lower cut off measurements. The calculations show, that the losses may be about 0.15 dB/m. for frequency 60 GHz. The turns in E plane in that case may be made by means of smooth narrowing and bending in E plane. The thickness of the waveguide in turns must be about 3 mm, which increase the losses in factor of 7. If the length of the narrow part 20 cm, the losses will be 0.18dB/turn. Thus the waveguide system of 13m length , which is enough for inner launch, with 6 turns will have the losses for both passes 5 dB. This value seems tolerable, taking into account the possibilities to use the sources power of some hundredth mw. The oversize and corrugated waveguides for mid plane systems are also possible. Vacuum windows may be done in waveguide variant by means of inserting quartz plate in smoothly narrowed in E plane waveguide. The form of the quartz plate and law of waveguide shrinking must be calculated. The quartz plate will be soldered by *Pb* alloy or other material. The estimations show, that for the frequencies up to 300-400GHz long transmission line may be made super conductive. The losses for waveguide 7.4x3.4mm will be of the order of 10^{-3} dB/m, which is negligible for 100m. Corrugated and rectangular oversized waveguides are also possible.

ACKNOWLEDGMENTS

This paper was written after the consultations with many specialists, whose contribution is greatly acknowledged. The author is grateful to Dr. Dreval V.V.,Dr. Soldatov S.V. and Dr. Tsaun S.V., who take part in experimental testing of the reflectometry system and 2D simulations. The developing of reflectometry system was supported by the Russian Research Fond of Fundamental Science Grant # 94-02-06521-a.

REFERENCES

1. E. Mazzucato, Princeton University Plasma Phys. Laboratory Report MATT-1151(1975).
2. Cripwell P.,et al.,Proc. of IAEA Technical Committee Meeting on Reflectometry, JET,1992.
3. Sanchez J, et al, 18th EPS Conf., Berlin (1990) V 15 C, part IV, p 313.
4. Nazikian R.,Mazzucato E., Rev. of Scientific Instr., V 66, No 1, Part II (1995) 392.
5. Baang S., et al, Rev. Sci. Instrum., 61 (1990) 3013.
6. Irby J.H., et al , Plasma Physics and Contr. Fusion, 35 (1993) 601.
7. Sips A.C., Rijnhuizen Report 91-200(19910.
8. Aleksandrov V.O., et al, 19th EPS Conf., Insbruck (1992) V 1, p 111.
9. Bulanin V.V., et al, 20th EPS Conf.,Lisbon (1993) v 17 C, part IV, p 1517.
10. Doyle E.J., et al Rev. Sci. Instrum., v 66 (1995) p. 1.
11. Frances M., et al 22th EPS Conf., Bournemouth, 1995, Abstract S114, page 536.
12. Vershkov V.A. et al, 21th EPS Conf., Montpelier, 1994, V 18, part 3, p. 1192.
13. Vershkov V.A., 22th EPS Conf., Bournemouth, 1995, postdeadline paper.
14. Vershkov V.A. et al, 15th IAEA Conf.(1994), IAEA-CN-60/A2/4-P-8
15. Tsaun S.V.,et all, 22th EPS Conf., Bournemouth, 1995, Abstract S107, page 529.
16. Shevchenko V.F., et al, 20th EPS Conf.,Lisbon (1993) v 17 C, part 3, p 1167.
17. Heijnen S.H., et al, 20th EPS Conf.,Lisbon (1993) v 17 C, part 3, p 1143.

ICRF PHYSICS MEASUREMENTS BY REFLECTOMETRY

A.Mase, H.Hojo, M.Kobayashi, N.Oyama, L.G.Bruskin, E.J.Doyle,[a]
T.Tokuzawa, A.Itakura, M.Ichimura, and T.Tamano

Plasma Research Center,
University of Tsukuba,
Tsukuba 305, Japan

INTRODUCTION

In magnetically confined plasmas, the application of radio frequency (RF) waves is considered to be one of the most promising methods for additional heating in present and future thermonuclear fusion experiments such as ITER. The physics of wave excitation and propagation in the ion cyclotron range of frequency (ICRF) is under intensive experimental and theoretical investigation for efficient heating. The ICRF-wave studies have also been performed for understanding mechanisms associated with a variety of processes such as Alfvén wave current drive,[1] RF-induced radial transport,[2] RF stabilization of plasmas,[3] as well as plasma heating.

We report here the study of ICRF-driven waves in a tandem mirror as measured by reflectometry and cross-polarization scattering (CPS). The reflectometer is a nonperturbing diagnostic method of measuring plasma fluctuations and complements the conventional Thomson scattering method since it has good spatial resolution while lacking wave number resolution. It is currently used to investigate mainly low frequency waves,[4,5] however, it should be suited for coherent waves in RF region, since the high frequency fringes due to the change of cut-off layer is small and good signal to noise ratio is expected. Reflectometry has been applied to the study of RF-driven waves in the GAMMA 10 tandem mirror and DIII-D tokamak.[6,7] The CPS is a new application of electromagnetic wave scattering by magnetic fluctuations.[8-11] This is related to the mode-conversion effect consisting of a polarization difference of a scattered wave with regard to an incident wave. The incident ordinary (extraordinary) wave is transferred to the extraordinary (ordinary) wave by magnetic fluctuations at the position of local cutoff in a plasma. The transferred wave penetrates through the cutoff layer and is detected. Recently, it is considered that magnetic fluctuations as well as density and potential fluctuations are of importance for plasma transport. The CPS is expected as a direct method of measuring internal magnetic fluctuations that remains a long term diagnostic issue.

In linear machine plasmas, a slow Alfvén wave is excited and propagates toward the ion cyclotron resonance layer located near the midplane of the central cell mirror field following the application of ICRF power at the off-midplane of high magnetic field side.

Diagnostics for Experimental Thermonuclear Fusion Reactors
Edited by P. E. Stott *et al.*, Plenum Press, New York, 1996

Because ions are accelerated in the direction perpendicular to the magnetic field line at the resonance layer, pitch angles of the heated ions approach a right angle and the perpendicular plasma pressure becomes high near the midplane and low at the off-midplane. Therefore, the ion velocity distribution function at the midplane tends to be anisotropic. As the plasma pressure increases, instabilities due to the anisotropy of the distribution function are expected to be excited such as the Alfvén ion cyclotron (AIC) wave.[12-14] The energy containment of the ions is considered to be dominated by charge exchange loss and electron drag. During the increase of hot ions the saturation of the energy containment has often been observed in strongly heated plasmas, which cannot be explained by the charge exchange loss and the electron drag. The instabilities relevant to the ICRF heating may play an important role for this saturation.

EXPERIMENTAL APPARATUS

The GAMMA 10 Device

The present experiment is performed in the GAMMA 10 tandem mirror consisting of a central cell, anchor cells attached to both sides of the central cell for MHD stabilization, plug/barrier cells for formation of confining potential, and end cells. It is 27 m in total length, with the central cell vessel 6 m in length and 1 m in diameter. The magnetic field strength at the midplane of the central cell is B_c=0.405 T, and the mirror ratio is 4.9 for the standard operation. It increases up to B_c=0.59 T for the high magnetic field operation. The ICRF power with frequencies of $\omega_{rf}/2\pi$=9.9 MHz (RF1) and 6.3-8.93 MHz (RF2) is employed to build-up a plasma and heat ions, following the gun-produced plasma injection. It is fed by NAGOYA TYPE-III and double half-turn antennas respectively located near the mirror throats of the central cell. The frequency 6.3-8.93 MHz corresponds to the ion cyclotron frequency at the near midplane of the central cell, and 9.9 MHz corresponds to that of the anchor cells. Four gyrotrons providing electron cyclotron resonance heating (ECRH) with a frequency of 28 GHz are then applied to the plug/barrier cells in order to produce confining potentials. The plasma parameters in the central cell are as follows. The density n_c=2-3 $\times 10^{12}$ cm^{-3}, the electron temperature T_e=80-100 eV, and the averaged ion temperature T_i=6-8 keV.

Reflectometry and Cross-Polarization Scattering System

The reflectometers are located at the midplane (z=0 m) and at the west side (z=0.6 m and 1.8 m) of the central cell. The position, z=1.8 m is close to the location of RF-heating antenna. At z=0 m, two systems are installed to observe the horizontal and vertical chords in one shot. The system utilizes a 7-18 GHz and/or 18-26 GHz, 100-150 mW output of a yttrium-iron-garnet (YIG) oscillator as a source. The YIG oscillator is operated in a fixed frequency mode during a plasma shot. A pyramidal horn with an ellipsoidal reflector is used as both transmitter and receiver. The fluctuations of the whole density region of the radial profile can be observed by switching between O-mode and X-mode. The reflected wave is separated from the incident wave by a circulator, and is mixed with the unperturbed local oscillator (LO) wave in a mixer. The homodyne-detected intermediate frequency (IF) signal is amplified by low noise amplifiers and separated into three ports by a power splitter, which is shown in Fig. 1. The one is directly connected to a high speed digitizer with 20 ns sampling time and 4 MB memory, and the others are fed to bandpass filters with the center frequencies of 6.36 MHz and 5.7 MHz that correspond to the RF-heating wave and the AIC wave, respectively, and are rectified through detectors.

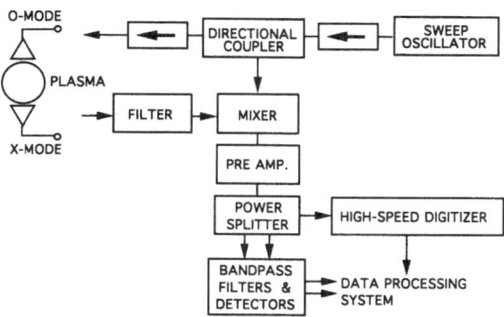

Figure 1. Schematic of the CPS diagnostic system.

The schematic of the CPS diagnostic system is shown in Fig. 1. The system is located at the midplane (z=0 m) of the central cell. A 7-18 GHz, 150 mW output of a YIG oscillator is used as a source. Pyramidal horns with ordinary (O) mode and extraordinary (X) mode are installed in the top and bottom diagnostic windows respectively, and are used as a transmitter or a receiver. The mode-converted scattered wave is mixed with LO wave in a mixer. The IF signals are fed to the similar signal processing system as the reflectometers shown in Fig. 1.

EXPERIMENTAL RESULTS AND DISCUSSION

Frequency Spectra of ICRF Waves

The frequency spectra of the reflectometer signals at two axial positions (z=0 m and 1.8 m) are shown in Fig. 2. The frequency of the YIG oscillator, $\omega_X/2\pi$=16.5 GHz and 18.0 GHz correspond to the same critical density for the X-mode propagation at z=0 m and 1.8 m, respectively. At the position z=0 m, a large peak is seen with frequencies 5.7-5.9 MHz that are lower than the frequency of RF2 (6.36 MHz) applied for the central-cell ion heating. The value of 6.36 MHz corresponds to the ion cyclotron frequency near the midplane (z=0.6 m). On the other hand, the signals with frequencies of applied RF (RF1 and RF2) are dominant at z=1.8 m. These frequencies coincide with those observed with magnetic probes installed in the periphery of the plasma. The low frequency mode with $\omega<\omega_{rf}$ has been identified to be the AIC instability by the magnetic probes.[14]

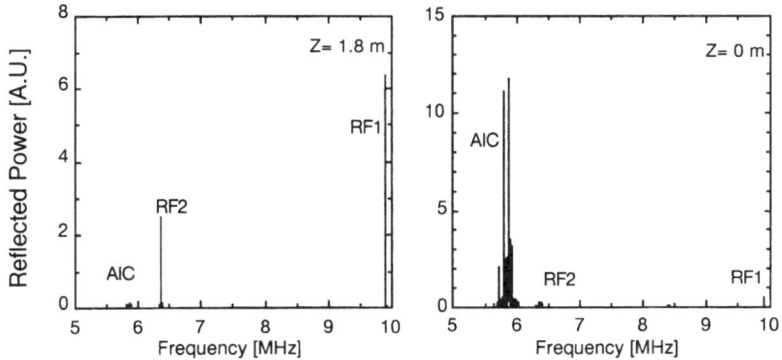

Figure 2. Frequency spectra of the RF-driven waves observed at two axial positions.

In Fig. 3(a) the frequencies of this mode are plotted as a function of the central-cell magnetic field B_c. They are proportional to B_c, and are slightly below the ion cyclotron frequency at the midplane, that is, $\omega \cong (0.90\text{-}0.94)\omega_{ci}$. The dispersion relation of the AIC mode in an infinite and homogeneous plasma is written by[15,16]

$$D(k,\omega) = k^2 c^2 - \omega^2 + \sum_{j=e,i} \omega_{pj}^2 \chi_j(k,\omega) = 0, \tag{1}$$

where k and ω are the axial wave number and the frequency of the AIC mode, c is the speed of light, ω_{pj} is the plasma frequency, and χ_j is the plasma susceptibility. A suffix j means i or e for an ion or an electron term, respectively. The wave number and the frequency are defined as complex numbers of $k_r + ik_i$ and $\omega_r + i\omega_i$. Figure 3(b) shows a typical dispersion relation of the AIC mode obtained from Eq. (1) in the case that the temperature anisotropy is 1/0.09 and 1/0.07, and the averaged plasma beta perpendicular to the magnetic field, $\beta_\perp$ is 0.01. The anisotropy of ion temperature is defined as the ratio of perpendicular to parallel component to the magnetic field line as $T_{i\perp}/T_{i//}$. It is determined from the diamagnetic loops installed in the two axial positions of the central cell.[17] The theoretical dispersion relation shows that the maximum growth is obtained when $k_r c/\omega_{pi} \cong 2.7$. The frequency $\omega/\omega_{ci} \cong 0.92$ at the maximum growth coincides with the experimental results shown in Fig. 3(a).

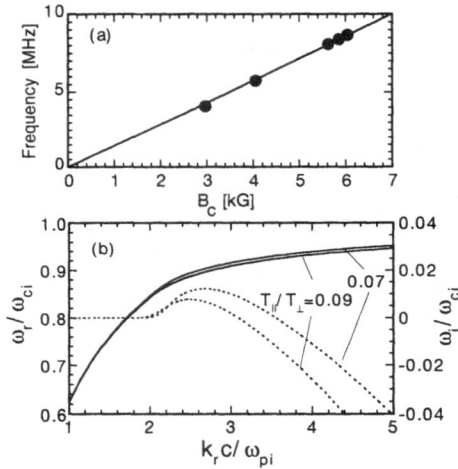

Figure 3. (a) Frequency of the AIC mode as a function of the central-cell magnetic field. (b) Dispersion relation of the AIC mode for $\beta = 0.01$.

The AIC mode is driven by plasma pressure and pressure anisotropy. Shear Alfvén wave couples with free energy derived from the relaxation of an anisotropic population of ion energy state and becomes unstable. It has been observed that the density fluctuation level strongly depends on the AIC driving term, $\beta_\perp (T_{i\perp}/T_{i//})^2$.[18] The dependence is consistent with theoretical predictions of a convectively unstable AIC mode.

Time Evolution of the Waves

Figure 4 shows the time evolution of the detector output as well as the line density and the diamagnetic loop signal at the central cell. The pass band of the filters is selected to be 5.7 ± 0.2 MHz and 6.36 ± 0.15 MHz to observe the AIC instabilities and RF-heating wave,

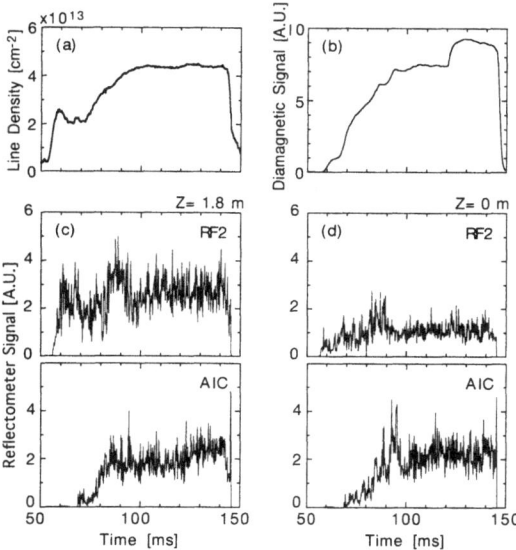

Figure 4. Time evolution of the line density (a) and the diamagnetic loop signal (b) at the central cell, and detected signal via reflectometers at two axial positions (c, d).

respectively. The large fluctuations in the detected signals are caused by the amplitude and phase modulations due to the low-frequency waves. It is noted that the signal level of the AIC mode starts to increase at the later time ($t \geq 70$ ms) after the RF2 power is applied (t=53 ms), begins to saturate, and even decays with time. As mentioned above the growth of the instability is caused by the increase of plasma beta and ion temperature anisotropy. The growth of the AIC mode and the axial plugging by confininig potentials may cause relaxation of the temperature anisotropy and decay of the instability.

In the present experiment, the mixer output of the reflectometer is expressed by a combination of amplitude fluctuation $A(t)$, DC phase change by density profile ϕ_D, and phase fluctuations $\tilde{\phi}_L$ and $\tilde{\phi}_R$ due to the low-frequency and RF waves, respectively, as[19]

$$(1 + A(t))\sin(\phi_D + \tilde{\phi}_L + \tilde{\phi}_R) = (1 + A(t))\left[\sin(\phi_D + \tilde{\phi}_L) + \tilde{\phi}_R \cos(\phi_D + \tilde{\phi}_L)\right] \tag{2}$$

assuming $\tilde{\phi}_R \ll 1$. Since the fluctuation level of the RF waves is estimated to be $\sim 10^{-4}$ and much smaller than that of the low-frequency waves, the contribution of the RF waves to $A(t)$ is negligible. The phase fluctuations are represented by

$$\tilde{\phi}_L = \sum_k A_k \sin\omega_k t,$$

$$\tilde{\phi}_R = A_{RF}\cos\omega_{RF}t + A_{AIC}\cos\omega_{AIC}t, \tag{3}$$

where ω_{RF} and ω_{AIC} show the frequency of RF and AIC waves, respectively. The detector output after each band pass filter is expressed by time integration of

$$\left[(1 + A(t))A_{RF,AIC}\sin(\phi_D + \tilde{\phi}_L)\right]^2, \tag{4}$$

and is given by $A_{RF}^2/2$ and $A_{AIC}^2/2$, respectively. Therefore, the reflectometer output is proportional to the density fluctuation of the waves $(\tilde{n}/n)^2$.

The measurements of damping of the AIC and the RF waves are performed using a high speed digitizer as shown in Fig. 6. It is obvious that the AIC mode starts to decay in the later time after the RF power is turned off at $t=145$ ms. The damping time is typically ~100 μs. The decay of plasma β and the isotropization of ion temperature can be possible candidates for the damping, however, the ion collision time is longer than ~ms. There may be another mechanisms for the wave damping. The damping time of the RF wave ~40 μs mainly corresponds to the RF antenna turn-off time.

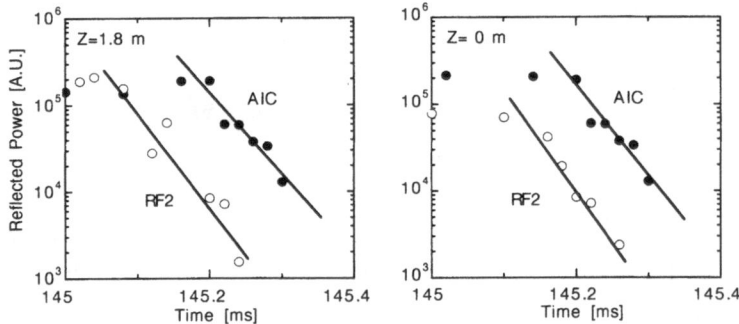

Figure 5. Damping of the waves at two axial positions.

RF-Depostion Profile and Ion Heating

In the experiment, hot ions with several keV produced by ICRF power lose most of their energy to cold electrons with 60-100 eV. The diamagnetic signals increase almost linearly with the ICRF power radiated from the antennas for the low power operation. When the power is further increased, the diamagnetic signals cease to increase and saturate

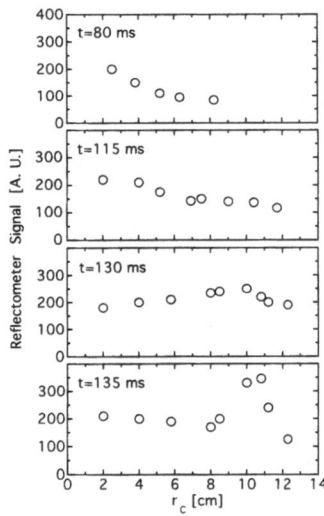

Figure 6. Radial profiles of the RF-driven wave for various times.

as shown in Fig. 4. The RF deposition profile will be one of the key diagnostics for the study of this saturation mechanism. The change in the RF-wave distribution for different times is shown in Fig. 6. This figure is obtained by scanning the incident YIG frequency. It is noted that the RF wave is detected in the core plasma in the early stage of the discharge, however, it is detected more in the edge during the saturation of the diamagnetic pressure. The change in the distribution due to the high ion pressure and the pressure anisotropy may cause the saturation of the plasma energy. The spatial broadening of the ion energy spectrum is observed during the saturation period. The profile of the AIC mode is also observed in the same plasma shot. The AIC mode usually localizes in the core plasma region due to the increase in the plasma beta and the temperature anisotropy.

Cross-Polarization Scattering from RF-Driven Waves

In the previous paper, we have shown the possibility to measure magnetic fluctuations as well as density fluctuations using both X- and O-mode reflectometers.[18] The method can be applied only to a one-dimensional plasma configuration. We will demonstrate here a generalized method to measure the magnetic fluctuations by cross-polarization scattering (CPS). The CPS is a new application of electromagnetic wave scattering by magnetic fluctuations, which is related to the mode-conversion effect consisting of a polarization difference of a scattered wave with regard to an incident wave.

In Fig. 7(a) O- to X-mode CPS (O→X) signal determined from the detector output is plotted for various frequencies of the YIG oscillator. It is seen that the signal level increases during the frequency of 9.5-11.3 GHz. In the present experiment, the on-axis density at the central cell is estimated to be $n_e(0) \cong 2.0 \times 10^{12} \mathrm{cm}^{-3}$, which gives the O-mode cutoff frequency (the electron plasma frequency) f_{pe}=12.7 GHz. Since the magnetic field at the central-cell midplane is 4.05 kG, the pass band of the X-mode wave between the right-hand and left-hand cutoff across the plasma diameter is calculated from the relation $f_{r,l} = \pm f_{ce}/2 + (f_{pe}^2 + f_{ce}^2/4)^{1/2}$ as 8.2 GHz to 11.34 GHz. This value is consistent with the frequency region of the signal enhancement. The incident O-mode approaches a cutoff at

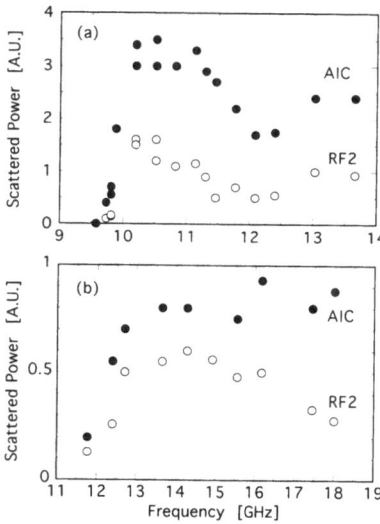

Figure 7. (a) The O→X CPS signal as a function of the incident frequency at t=110 ms. (b) The X→O CPS signal as a function of the incident frequency at t=110 ms.

the local plasma frequency and is reflected, however, a mode-converted X-wave propagates through this cutoff layer and emerges on the other port. The lower limit of the pass band is rather limited by the cutoff frequency of the P-band waveguide, f_c=9.487 GHz.

The X→O scattering is also plotted as a function of the incident frequency in Fig. 7(b). The scattered signal starts to increase at the frequency of ≅12.5 GHz, although the signal of the X-mode reflectometry appears from 11.5 GHz. This will be explained by the overlap of the cut-off frequency between the right-hand X-wave (11.34 GHz) and the O-wave (12.5 GHz). The mode-converted O-wave cannot reach the opposite receiver horn across the plasma during the frequency of 11.34 - 12.5 GHz, and the reflected X-wave is picked up by the receiver horn. In Fig. 7(b), the X-mode incident wave with higher frequency penetrates into the core plasma region until it meets the cutoff layer. The frequency, 18 GHz corresponds to the cutoff density n_e≅1.5x10^{12} cm^{-3}. The AIC mode is stronger in the high β core plasma, however, the RF wave is observed mainly in the edge region. This will cause the saturation of the plasma pressure.

There is no ideal horn antenna which radiates or receives pure O- and X-mode propagations, since it has finite beamwidth and polarization resolution limitation. We have compared the cross polarization scattering (O→X) by magnetic fluctuations with the conventional Thomson scattering due to the spurious X-mode and O-mode incident waves by density fluctuations. It is calculated that the CPS signal is more than factor of two larger than that of the conventional Thomson scattering in the present experimental conditions, that is, the spurious X-mode (O-mode) electric field ~10 %, and the ratio of density fluctuation to magnetic field fluctuation level ~1. Also, the Bragg's relation has to be satisfied for the conventional scattering process. Therefore, the CPS is predominantly observed in the present experiment.

Density and Magnetic Field Fluctuations

The density fluctuation of electromagnetic plasma waves is obtained by integrating the perturbed electron distribution function $\tilde{f}_e$ as $\tilde{n}_e = \int \tilde{f}_e d^3v$. Assuming that the unperturbed distribution function is Maxwellian, and using Maxwell's equation $\partial B/\partial t = -\nabla \times E$ the normalized density fluctuation is calculated as[20]

$$\frac{\tilde{n}_e}{n_e} = -\frac{\omega}{k_z v_{te}} Z\left(\frac{\omega}{k_z v_{te}}\right) \frac{\tilde{B}_z}{B} + i\left[1 + \frac{\omega}{k_z v_{te}} Z\left(\frac{\omega}{k_z v_{te}}\right)\right] \frac{e\tilde{E}_z}{k_z T_e} , \qquad (5)$$

where k_z is the axial component of k, v_{te} is the electron thermal velocity, and Z is the plasma dispersion function. From Eq. (5), we obtain two limiting relationships as

$$\frac{\tilde{n}_e}{n_e} = \frac{\tilde{B}_z}{B} , \qquad \text{for } \omega \gg k_z v_{te} \qquad (6)$$

and

$$\frac{\tilde{n}_e}{n_e} = i\frac{e\tilde{E}_z}{k_z T_e} , \qquad \text{for } \omega \ll k_z v_{te} . \qquad (7)$$

Note that the magnetic field fluctuation $\tilde{B}_z$ and the electric field fluctuation $\tilde{E}_z$ is mot independent each other. The ratio $\tilde{E}_z/\tilde{B}_z$ for the AIC mode can be deduced from the dispersion relation, and the relationship between density fluctuation and magnetic fluctuation is given by

$$\frac{\tilde{n}_e}{n_e} = \left[-\frac{\omega}{k_z v_{te}} Z\left(\frac{\omega}{k_z v_{te}}\right) - \frac{\omega_{ci}}{\omega}\left\{ \frac{T_{i\perp}}{T_{i\parallel}}\left(1 - \frac{\omega}{\omega_{ci}}\right) - 1 \right\}\left\{ 1 + \frac{\omega - \omega_{ci}}{k_z v_{ti}} Z\left(\frac{\omega - \omega_{ci}}{k_z v_{ti}}\right) \right\} \right] \frac{\tilde{B}_z}{B} . \quad (8)$$

Instituting the present experimental parameters into Eq. (8), $\tilde{n}_e/n_e = (0.7 - 1.2)\tilde{B}/B$ is obtained. This value reasonably agrees well with the experimental result of the density and magnetic fluctuation level.[18]

The magnetic fluctuations are also estimated in the CPS measurement. In Fig. 7(b), some amount of signal is observed above the frequency of 12.0 GHz in the O→X system. From theoretical calculations of conventional scattering and CPS with including present experimental parameters, we may also conclude that the similar level of density and magnetic fluctuations (10^{-3}-10^{-4}) are excited in association with the instabilities.

SUMMARY

In summary, the fluctuations associated with ICRF waves are observed to investigate the physics of wave excitation and absorption of RF power using microwave reflectometers. The fluctuations with frequencies of $\omega \leq \omega_{ci}$ have been observed by the X- and O-mode systems. The instability is identified as the Alfvén ion cyclotron (AIC) mode and RF-driven wave. The Fourier amplitude of the AIC instability strongly depends on the plasma β value and the ion-temperature anisotropy, and is stronger in the core plasma region with higher β value. The radial profile of the RF-driven wave is expected to give the deposition profile of the heating power. The change in the RF-wave distribution due to the high ion pressure and the pressure anisotropy may cause the broadening of ion temperature profile and saturation of the plasma energy

The cross-polarization scattering diagnostic method has also been applied for the measurement of ICRF waves. It is observed that the incident wave reaches to the local cutoff layer of the plasma and generates the mode-converted scattered wave which can propagates the cutoff layer across the plasma and is detected by the receiver located at the opposite diagnostic port. This process appears to respond in a similar way for both the O→X and X→O scattering. The level of the magnetic fluctuations ~3x10^{-4} and the similar level of the density fluctuations are evaluated from the CPS diagnostic system as well as from both X- and O-mode reflectometers.

The reflectometric methods are extremely suited for the measurement of coherent waves driven by ICRF heating power. It provides the study of wave excitation and propagation that is necessary for efficient plasma heating. The investigation of magnetic fluctuations as well as density fluctuations becomes important for the transport study of magnetically confined plasmas. The reflectometry is a direct and nonperturbing method for the measurement with good spatial resolution, which will be of importance for the future machine.

ACKNOWLEDGMENTS

The authors thank W. A. Peebles and J. H. Lee for useful discussions. We also thank the members of the GAMMA 10 Group for their collaboration. This work was supported in part by a Grant-in-Aid for Scientific Research from the Japanese Ministry of Education, Science and Culture.

REFERENCES

a) Permanent Address: Institute of Plasma and Fusion Research and Electrical Engineering Department, University of California, Los Angeles, CA 90024

1. T.Ohkawa et al., in *Plasma Physics and Controlled Nuclear Fusion Research,* 1988 (IAEA, Vienna, 1989), Vol.1, p.681.

2. S.Riyopoulos, T.Tajima, T.Hatori, and D.Pfirsh, *Nucl. Fusion* 26:627 (1986).

3. J.J.Browning et al., *Phys. Fluids B* **1**, 1692 (1989).

4. E.J.Doyle et al., *Phys. Fluids B* **3**, 2300 (1991).

5. G.R.Hanson et al., *Nucl. Fusion* **32**, 1593 (1992).

6. .A.Mase et al., *Rev. Sci. Instrum.* **66**, 821 (1995).

7. J.H.Lee et al., *Rev. Sci. Instrum.* **66**, 1225 (1995).

8. T.Lehner, J.M.Rax, and X.L.Zou, *Europhys. Lett.* **8**, 759 (1989).

9. X.L.Zou et al., *Phys Rev. Lett.* **75**, 1090 (1995).

10. L.Vahala, G.Vahala, and N.Bretz, *Phys. Fluids B* 4:619 (1993).

11. L.G.Bruskin, A.Mase, and T.Tamano, *Plasma Phys. Control. Fusion* 37:255 (1995).

12. T.A.Casper and G.R.Smith, *Phys. Rev. Lett.* 48:1015 (1982).

13. S.N.Golovato et al., *Phys. Fluids B* **1**, 851 (1989).

14. M.Ichimura et al., *Phys. Rev. Lett.* **70**, 2734 (1993).

15. G.R.Smith, *Phys. Fluids* 27:1499 (1984).

16. H.Hojo, R.Katsumata, M.Ichimura, and M.Inutake, *J. Phys. Soc. Jpn.* 62:3797 (1993).

17. R.Katsumata et al., *Jpn. J. Appl. Phys.* **31**, 2249 (1992).

18. A.Mase et al., *Phys. Fluids B* **5**, 1677 (1993).

19. J.H.Lee et al., *Proceeding of 11th Topical Conference on RF Frequency Power in Plasmas, Palm Springs,* 1995.

20. H.Hojo, A.Mase, M.Inutake, and M.Ichimura, *J. Plasma Fusion Res.* 69:1043 (1993).

ITER RELEVANT REFLECTOMETRY DEVELOPMENTS ON DIII–D

W.A. Peebles,[1] E.J. Doyle,[1] K.W. Kim,[1] J.H. Lee,[1] G.R. Hanson,[2] T.L. Rhodes,[1] and J.B. Wilgen[2]

[1]Electrical Engineering Dept. and Institute of Plasma and Fusion Research, University of California, Los Angeles, CA 90095 USA.
[2]Oak Ridge National Laboratory, P.O. Box 2009, Oak Ridge, TN 37831.

INTRODUCTION

A number of reflectometer systems are under development on the DIII–D tokamak which are directly relevant to ITER needs and requirements. These include density profile measurement systems, a divertor reflectometer system, an inside launch reflectometer system for core access, and RF wave and edge density profile measurements at an ICRF antenna. For density profile measurements a new solid-state fast sweep FM reflectometer system has been installed on DIII–D. The faster sweep rates, combined with improved/optimal data processing routines developed over several years[1,2] have resulted in more accurate profiles with improved time resolution, as required for routine operation on ITER. Also on DIII–D is an AM reflectometer system developed by ORNL which is being contrasted with the FM technique. A new reflectometer system is now operating in the lower divertor floor and is intended for peak density, density profile, and turbulence measurements. An inside launch reflectometer system has been upgraded for first tests of core profile measurements using the left hand X-mode cutoff. Use of the left hand cutoff and inside launch is necessary for core access for reflectometry on ITER.[3] RF heating is also an option for ITER, and RF coupling/directionality depend on the position and details of the edge density profile. On DIII–D, reflectometry has provided detailed edge density profile measurements at the ICRF antenna. Also, a novel application of reflectometry on DIII–D has been the determination of the internal RF electric field profile in the plasma, which is related to the RF power deposition profile. Design and operational considerations as well as first data from the above systems are presented.

REFLECTOMETRY DENSITY PROFILE MEASUREMENT ON DIII–D

A new fast sweep, broadband FM reflectometer system was installed on DIII–D after a machine vent in early 1995, and is now operational. The system offers new profile measurement capabilities and also represents a substantial improvement over previous backward wave oscillator (BWO) tube based systems. It is intended to fully demonstrate levels of accuracy and reliability necessary for plasma control purposes on current machines and essential for ITER. Initial data are very promising and there is every indication that the system will fulfill its design goals.

In an FM profile reflectometer system the sweep time is of critical importance as density turbulence can distort the measured data at slow sweep speeds, to the point where the data are not interpretable. Fast time response profile measurements are also critically needed to investigate a variety of physics R&D issues associated with the L–H transition and ELMs, where the profile can change on a time scale of 100 μs or less. Advances in solid state sources, amplifiers and multipliers have made it practical to replace the DIII–D BWO sources with an all-solid-state system. As compared with BWO tubes the solid state based system has many advantages: solid state sources are compact, lightweight, rugged and reliable. Magnetic shielding is not required, low voltage DC power supplies can be utilized and, most importantly, they can be swept full band in 5–10 μs. The new DIII–D

Diagnostics for Experimental Thermonuclear Fusion Reactors
Edited by P. E. Stott *et al.*, Plenum Press, New York, 1996

system uses an 8–12.5 GHz hyper-abrupt varactor tuned oscillator (HTO) source, followed by an active frequency quadrupler to cover the frequency range from 32–50 GHz. The present sweep rate is a factor of 5–8 faster than that previously used, and is currently set by digitization limits. It is planned to go to higher sweep rates as faster digitizers are acquired.

Data quality from the new fast sweep system is significantly improved; phase distortions within a single sweep have been significantly reduced and density profiles are routinely produced between discharges. The new system is able to resolve fast density profile changes in highly turbulent plasmas. This capability is illustrated in Fig. 1, where the time evolution of the edge density profile is tracked through a giant ELM. Finally, a system to investigate the viability of using the left hand cutoff for density profiles and fluctuations is under test on DIII–D. Use of the left hand cutoff and inside launch is required for core profile measurements on ITER, and also has the advantage of reduced relativistic corrections.[1,3] The DIII–D system utilizes circular corrugated waveguide and a bistatic launch/receive system from the high field side of the vessel and has a bandwidth of ~40-60 GHz. Initial tests indicate that the system can successfully observe core fluctuations under conditions where the outside launch right hand cutoff is inaccessible due to the n=2 electron cyclotron resonance.

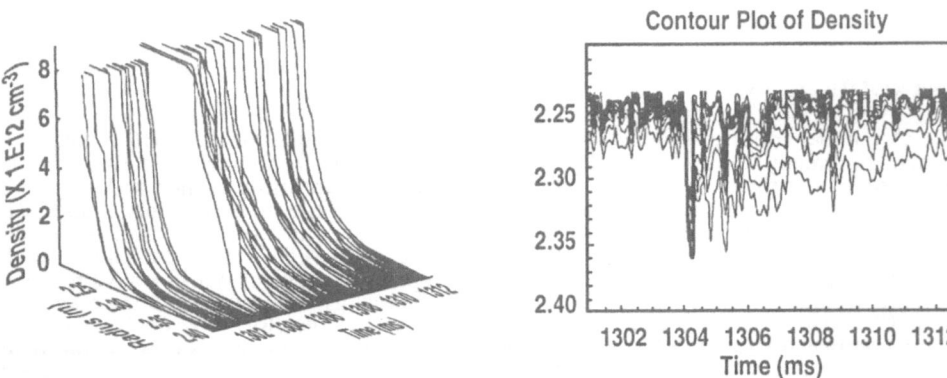

Figure 1. Showing the edge density profile evolution through an ELM. The ELM starts at 1304 ms, time between sweeps is 100 μs.

REFLECTOMETRY APPLICATIONS TO RF OPERATIONS AND MODELING

Over the past year, reflectometer systems on DIII–D have routinely provided detailed edge and core density profiles during RF operation, and have rapidly become the standard density diagnostic in these plasmas. This is due to a) the superior spatial coverage/dynamic range of the reflectometer system, b) the fact that the Thomson scattering system does not access the plasma center, and c) the superior low density edge resolution of the reflectometer system. In addition, the low density plasmas utilized during RF experiments can occasionally present difficulties for the Thomson scattering system; in this case the reflectometer provides the *only* available density profiles. Such profiles have been utilized for code/model validation tests. The profiles have been used as input to the RANT3D code at ORNL[4] in order to calculate a theoretically expected loading resistance, which can then be compared with the experimentally measured loading resistance. The observed comparisons have shown good agreement.[5]

In addition to providing density profile information, reflectometry has also been utilized on DIII–D to determine the radial electric field profile of RF waves within the plasma.[6] Information on internal RF fields is very important as it can be directly related to code and model results providing validation and benchmarking. In addition, mode conversion heating processes can be monitored and correlated with heating results. An upgraded reflectometer system has produced new and interesting results. Profiles of the FW electric field for a VH–mode discharge on DIII–D are shown in Fig. 2. These are the first such internal RF field profiles from a major tokamak. Note that the field decreases towards the edge and that there is significant structure superimposed upon it. This structure may be due to interference effects (due to relatively weak toroidal damping) and is currently under investigation. For comparison two different FW launch directions within the plasma are shown, one towards the reflectometer detection position (co-injection) and one away (counter-injection). There is a clear difference between the two, thereby confirming the FW antenna launch directionality. The non-zero value for the counter-injection is due to the imperfect directionality of the FW antenna, as well as propagation of the FW completely around the tokamak. Comparison of experiment and code are currently underway.[7] Preliminary comparison indicates that the single pass absorption in the plasma is larger than the code predicts. This is in agreement with many other experimental observations. The disagreement with theory may be due to a low assumed value of hydrogen minority content or possibly some unaccounted for damping mechanism.

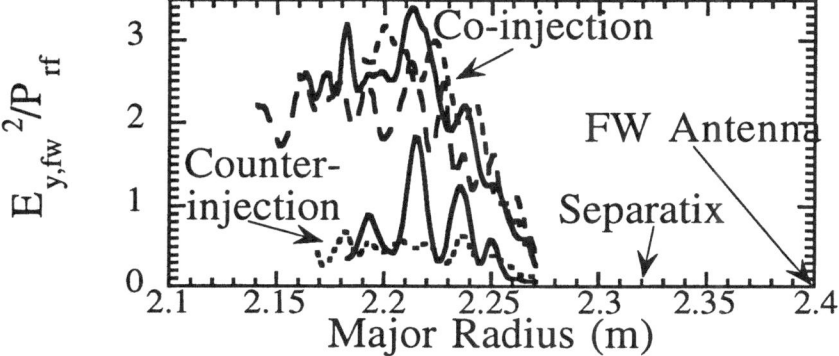

Figure 2. Radial profile of FW electric field from ICRF wave monitor system. Co-injection is launch towards the reflectometer position, while counter is away. Note significant difference in magnitude for the two launch directions. Different curves are from different shots.

DIVERTOR REFLECTOMETER SYSTEM

A divertor reflectometer system was installed on DIII–D during the February 1995 vent. Since ITER has a divertor reflectometer in its proposed diagnostic package the DIII–D system can significantly contribute to ITER R&D needs as well as physics issues. Multiple engineering design problems, similar to those expected on ITER, were solved. These included differential thermal expansion of the vessel with respect to the waveguide fixture, air/vacuum interface in the waveguide, and design of the overmoded waveguide, mode converter, and antenna system within the vessel itself. The system is located at the bottom of the tokamak and will initially probe the plasma from the tiles directly up into the X–point of the divertor. This is currently the only US reflectometer experiment being performed in the important divertor region and thus promises to provide valuable information both regarding the physics in the divertor as well as expertise in the design and operation of a divertor reflectometer. Upgrades to the system will probe the inner and outer divertor legs. The system is designed to allow antenna reconfiguration during a clean vent of the vessel (clean vents occur at a higher frequency than dirty vents, i.e., those that require personnel to enter the vessel).

For first plasma data a 75 GHz fixed frequency microwave source was used. The turbulence data shown in Fig. 3 were obtained during an ELMing H–mode discharge with the system viewing directly up into the X–point of the divertor. Shown in the figure are D-alpha traces with characteristic ELM signatures and the RMS fluctuation power from a 32 GHz outboard mid-plane reflectometer as well as the divertor reflectometer. Note the increase in fluctuation levels on the outboard reflectometer channel while the divertor channel shows a rapid increase just prior to the ELM followed by a decrease in fluctuation level. The significant difference between divertor and midplane may indicate a decoupling (perhaps not unexpected) of the private region fluctuation levels from the rest of the plasma. The decrease in divertor fluctuation level during the ELM may be due to an X-point position change and is currently under investigation. A broadband microwave source (50-75 GHz) has recently been installed and routed to the divertor which will permit density profile measurement. Further tests of the system as well as different antenna configurations are planned for the current and upcoming run periods.

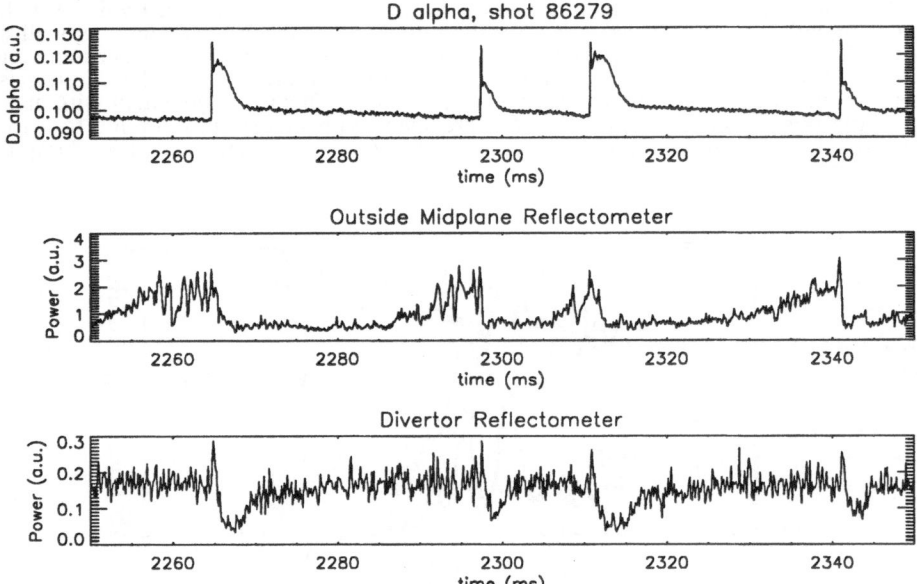

Figure 3. First data from the DIII-D divertor reflectometer. Top trace is D-alpha signal showing ELM timing. Second and third traces are rms power levels from outside launch and divertor reflectometers respectively.

SUMMARY

A number of reflectometry systems have recently been installed on the DIII–D tokamak. These include FM and AM density profile systems, a divertor reflectometer, an inside-launch reflectometer for core access, as well as systems focused on monitoring and optimizing RF coupling. These systems will provide important technical information for possible future diagnostic implementation on ITER. In addition, the physics emerging from DIII–D will also contribute crucial data to address ITER R&D needs. The DIII–D effort represents an invaluable resource to ITER and the overall worldwide fusion effort.

ACKNOWLEDGMENTS

The authors would like to thank Prof. N.C. Luhmann, Jr. of the University of California, Davis for his continuing support, as well as Drs. K.H. Burrell and R.T. Snider of General Atomics for fruitfull discussions. This work is supported by U.S. DOE Grant DE-FG03-86ER53225 to UCLA and Contract DE-AC03-89ER51114 to General Atomics.

REFERENCES

1. E.J. Doyle, *et al.*, these proceedings.
2. K.W. Kim, *et al.*, Rev. Sci. Instrum. **66**, 1229 (1995).
3. E.J. Doyle, *et al.*, Rev. Sci. Instrum. **66**, 1233 (1995).
4. M.D. Carter, *et al.*, "Three-dimensional modeling of ICRF launchers for fusion devices," accepted for publication in Nuclear Fusion.
5. D.W. Swain, *et al.*, Bull. Am. Phys. Soc. **39**, 1649 (1994).
6. J.H. Lee, *et al.*, Rev. Sci. Instrum. **66**, 1225 (1995).
7. F. Jaegar, ORNL, private communication.

STATUS OF THE RTP PULSED-RADAR REFLECTOMETER AND PROSPECTS FOR ITER

S.H. Heijnen, C.A.J. Hugenholtz, H.J. van de Meiden and A.J.H. Donné

FOM-Instituut voor Plasmafysica 'Rijnhuizen', Association Euratom-FOM,
P.O. Box 1207, 3430 BE Nieuwegein, The Netherlands

INTRODUCTION

At the Rijnhuizen Tokamak Experiment (RTP) in the Netherlands, a four-channel pulsed-radar reflectometer is developed and installed. Carrier frequencies of 29, 33, 36 and 39GHz are modulated with sub-nanosecond pulses and transmitted in the O-mode. This means that these pulses, once transmitted into the plasma, will reflect from a critical density layer where the plasma frequency equals the probing frequency. The time of flight of these pulses is detected with an accuracy of 70ps, corresponding to an accuracy of 1cm when reflections from a metal mirror in vacuum are detected.

SYSTEM DESCRIPTION

The four channels of the RTP pulsed-radar system are time multiplexed. This means that at every step in a sequence only one probing frequency is handled by the system. The pulse repetition frequency is 2MHz, yielding a time resolution per channel of $2\mu s$. Using heterodyne detection techniques, the system can cope with a round-trip loss of 45dB while maintaining a signal-to-noise ratio high enough for an accuracy of 70ps in the time-of-flight measurements. During a plasma pulse, however, the signal-to-noise ratio decreases due to plasma emission, resulting in an accuracy of 150ps. Combined with the high sampling frequency, this is enough to detect effects on the density profile caused by MHD oscillations, pellet ablation and disruptions[1]. Density profiles found with the pulsed-radar system match well with those found with a 180-point Thomson-scattering system and a 19-channel FIR interferometer, see Fig. 1. An example of an MHD oscillation as detected with pulsed-radar reflectometry is shown in Fig. 2. In (a) the measured flight times are shown after filtering with a 20kHz low-pass filter. The reference value 0ns represents reflection from the vacuum vessel in the absence of plasma. In (b) the positions of the reflecting layers are shown. For the inversion, Abel inversion techniques are used. The frequency of the oscillations, derived from pulsed-radar data, was 8kHz.

Diagnostics for Experimental Thermonuclear Fusion Reactors
Edited by P. E. Stott *et al.*, Plenum Press, New York, 1996

167

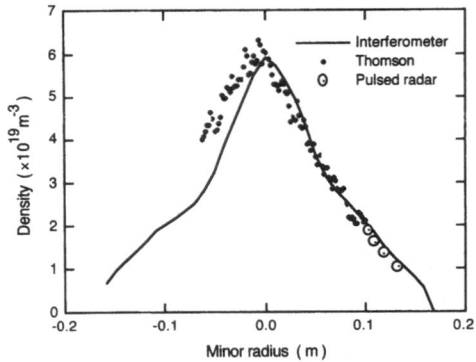

Figure 1. Comparison of density profiles from pulsed-radar reflectometry with density profiles from Thomson scattering, as a function of the vertical position without Shafranov shift, and interferometry, as a function of R including Shafranov shift.

PULSED-RADAR REFLECTOMETRY FOR ITER

One of the advantages of a pulsed-radar reflectometer over conventional reflectometers is that the interpretation of the signal does not depend on the time history. This means that even after the signal has been lost for some time, the density profile can still be calculated. This has obvious advantages in the long plasma pulses, of up to 1000s as foreseen for ITER.

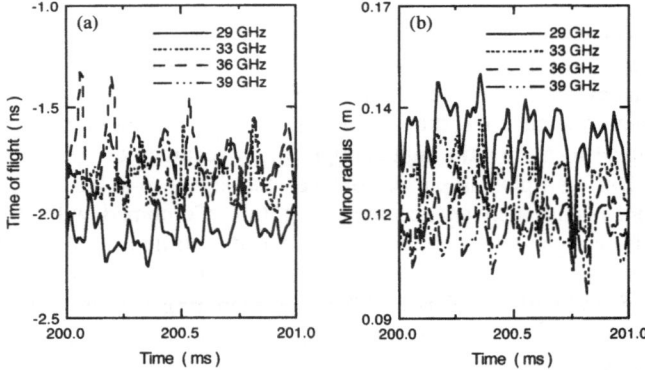

Figure 2. MHD oscillations (m/n = 2/1) as detected with pulsed-radar reflectometry. (a) before and (b) after inversion of the measured flight times to positions of the reflecting layers.

An other advantage is the possibility to separate reflections from different positions. This means that reflections from the plasma can be separated in time from reflections from e.g. the vacuum window. As also reflections from X and O-mode critical layers can be separated in time, the alignment of the antenna to the pitch angle of the magnetic field is not critical. Combined interpretation of measurements in X and O-mode could reveal information on the magnetic field structure in the plasma and thus reveal information on the current density profile[2].

The maximum time of flight measured in the RTP tokamak with O-mode probing is less than 4ns. This means that the spatial accuracy, relative to the plasma diameter, is rather poor. In ITER however, the expected flight time will be much larger, up to 60ns for the O-mode and up to 200ns for the X-mode. Therefore, the relative accuracy will be much larger than on RTP. This is shown in Fig. 3, where the flight time is calculated for ITER with O-mode and X-mode probing. In X-mode, reflections from the lower cut-off layer and the upper cut-off layer are shown. For the density profiles, a parabolic shape is assumed with a maximum density of 1×10^{20}m^{-3}, while for the current density profile a fourth-power profile is used with a total integrated current of 10MA. Fig. 3 shows that for a fixed accuracy in the time-of-flight measurements, the accuracy in the position will improve for reflections deeper into the plasma.

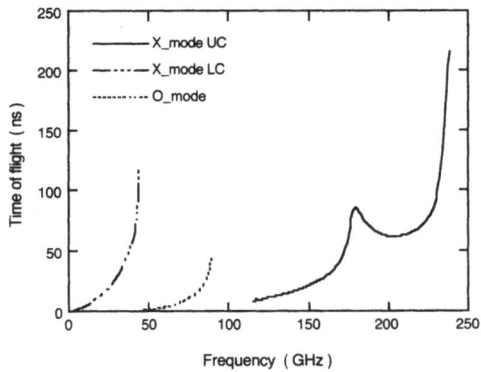

Figure 3. Flight time for ITER O- and X-mode probing. In the X-mode, the time of flight for reflection from both the lower as well as the upper cut-off frequency are shown.

TRANSMISSION MEASUREMENTS FOR ITER

In low-density plasmas in ITER it will be possible for pulses, with a carrier frequency above the maximum plasma frequency and emitted in the O-mode, to pass through the plasma instead of being reflected by the plasma. In this case, the flight time will carry information on the line-integrated value of the refractive index. It has been shown[3] that this can give additional information on the shape of the density profile in the centre of the discharge. Normally, this region is inaccessible by O-mode reflectometry. In ITER, it will not be possible to perform these low-frequency transmission measurements in high density discharges due to electron cyclotron absorption. Above about 110GHz (corresponds to a critical density of 1.5×10^{20}m^{-3} for transmission in the O-mode), the absorption is complete, while at 3THz still 10% of the injected power is absorbed. At 3THz, the incremental time of flight above the vacuum situation is only 30ps. Therefore, this mode of operation is not regarded as a real option for ITER. However, transmission measurements in low-density discharges, done with probing channels intended for reflection measurements in high-density discharges, should be used to derive the central part of the density profile. For X-mode probing, a window for transmission measurements will exist between the maximum value of the lower cut-off frequency and 110GHz above where absorption becomes dominant. This will increase the window in central density values where transmission measurements can be done, but the interpretation will become more complicated as the refractive index for the X-mode depends on both the local electron density and the local magnetic field strength.

CONCLUSIONS

It has been shown that the pulsed-radar reflectometer developed for the RTP tokamak works reliable and gives density profiles comparable to those from Thomson scattering and interferometry. A space resolution of 1cm is achieved with a temporal resolution of 5×10^{-5}s. Therefore, it is believed that pulsed-radar reflectometry is an option for measuring the electron-density profile in ITER and for studying fluctuations in the density caused by e.g. MHD oscillations.

In general, it is not believed that pulsed transmission measurements will be a good solution for measuring the line-integrated density although in low-density discharges, transmission measurements can add valuable information for determining the central part of the density profile.

ACKNOWLEDGEMENTS

This work was performed as part of the research programme of the association agreement of Euratom and the 'Stichting voor Fundamenteel Onderzoek der Materie' (FOM) with financial support from the 'Nederlandse Organisatie voor Wetenschappelijk Onderzoek' (NWO) and Euratom.

REFERENCES

1 S.H. Heijnen, Thesis University of Utrecht, September 1995.

2 D.L. Grekov, O.S. Pavlichenko and A.I. Skibenko, ITER-IL-PH-07-0-56 (1990).

3 S.H. Heijnen, A.J.H. Donné, C.A.J. Hugenholtz, Th.F.M.M. Maas, H.J. van der Meiden, F.C. Schüller and the RTP-team, Proc. 22th EPS Conf. on Contr. Fusion and Plasma Phys., Bournemouth (1995). To be published

MILLIMETER-WAVE DIAGNOSTIC SYSTEMS FOR ITER

N.C. Luhmann, Jr., C.W. Domier, R.P. Hsia, W.R. Geck, and B. Deng

University of California
Davis, California, U.S.A. 95616

INTRODUCTION

To date, reliable measurements of electron temperature fluctuations in major tokamak experiments have remained elusive. Since electron heat transport in tokamaks is anomalously large, and appears to be driven by turbulent fluctuations, measuring spatially resolved electron and density temperature fluctuations is key to understanding both transport and the resultant turbulent confinement. Reflectometry has shown itself to be extremely sensitive to density fluctuations, and is a proven technique for making accurate electron density profiles. Electron cyclotron emission (ECE) imaging is a method whereby 2-D images of large amplitude coherent temperature fluctuations as well as 2-D electron temperature profiles of tokamak plasmas may be obtained. In this paper, we discuss the present status of each of these UC Davis diagnostic research areas, with particular emphasis on how the diagnostics may best be implemented on ITER.

ULTRASHORT PULSE REFLECTOMETRY

Pulsed radar techniques offer considerable advantages in determining density profiles on next generation tokamaks such as ITER. Such time-of-flight radar systems function by reflecting short pulses of electromagnetic radiation. By collecting double-pass time delay data at many distinct frequencies, it is then possible to invert the time delay data to generate plasma density profiles [1]. Ultrashort-pulse reflectometry (USPR) is a revolutionary technique pioneered at UC Davis in which an extremely short pulse (or chirped waveform) is used that contains frequency components spanning the desired plasma density profile [2]. In this manner, it is therefore possible to obtain a complete density profile with a single source and a single set of measurements.

The first multichannel USPR system was designed and fabricated by the UC Davis Plasma Diagnostics Group, and successfully tested on the CCT tokamak at UCLA in late 1994. Experimental results showed good agreement with both theory and with other CCT diagnostics. An improved 8 channel USPR system (see Fig. 1) has since been fabricated, capable of acquiring high resolution (25 psec corresponding to < 4 mm spatial resolution)

Diagnostics for Experimental Thermonuclear Fusion Reactors
Edited by P. E. Stott *et al.*, Plenum Press, New York, 1996

multichannel time delay data at high repetition rates (≥ 100 kHz). This new system is scheduled to run on CCT in early October, with two primary goals in mind. The first is to make detailed profile comparisons with a conventional 8-18 GHz swept-FM reflectometer system. The second is to explore if USPR can serve as a simultaneous density profile and density fluctuation diagnostic (limited to frequencies below that of the pulse repetition rate of the ultrashort-pulse system, which in principle can be in excess of 500 kHz).

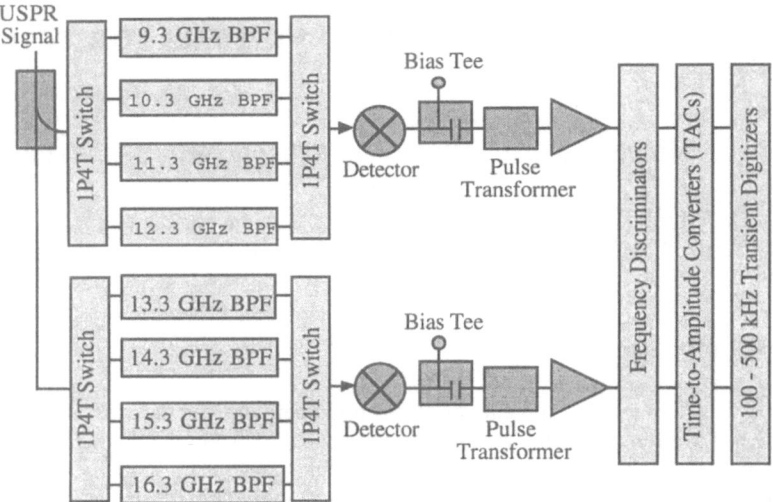

Figure 1. Multichannel ultrashort-pulse reflectometry IF receiver system.

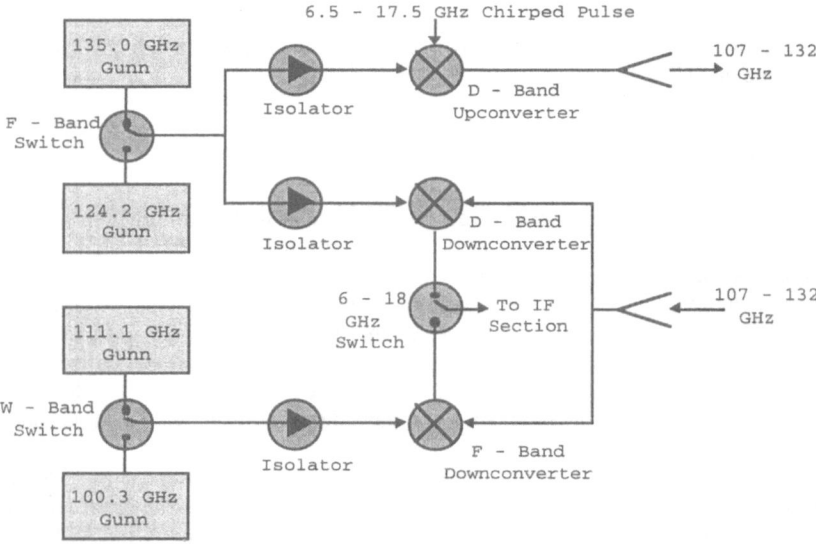

Figure 2. X-mode USPR system for use on ITER.

Higher frequency extensions of the basic USPR system are now in development. The present 8 channel CCT system is limited by the 16.3 GHz frequency of the highest frequency channel. A pair of wideband 2-26 GHz mixers have been purchased to upconvert the present system to 25.3 GHz. Similar frequency extensions have been designed to implement X-mode USPR on ITER (see Fig. 2) utilizing relatively low-cost commercially-available millimeter-wave hardware. Eventually, however, we hope to utilize nonlinear transmission lines (NLTLs) as extremely fast (~few ps) risetime sources. Such NLTLs are presently under development at UC Davis, and would simplify the overall system tremendously by eliminating the need for frequency upconversion as the extremely wide bandwidth inherent in such fast risetime sources would be sufficient to span the desired ITER X-mode profile.

ECE IMAGING

Low cost, planar arrays are under development to image electron cyclotron emission (ECE) from tokamak plasmas. Conventional multichannel waveguide mixer array implementations require expensive hardware including individual mixers, horns, waveguide, and couplers for each channel. A planar array implementation, on the other hand, integrates all of these elements in a much simpler and low cost manner. ECE imaging systems such as these are especially well suited to physically large tokamaks such as ITER, as a single viewing port can be utilized to image much of the plasma area. The high temperatures expected in these devices, however, leads to significant harmonic overlap at the second and third X-mode harmonics typically utilized in most present-day tokamaks. Nevertheless, spectra calculations for ITER [3] indicate that second harmonic X-mode systems may probe the outer plasma regions (r/a > +0.4) corresponding to frequencies in the 240 GHz to 290 GHz range. Still more of the plasma is accessible (-0.4 ≤ r/a ≤ 1.0) with the collection of fundamental O-mode radiation corresponding to the 120-190 GHz frequency range.

Wideband 20 channel hybrid and monolithic mixer antenna arrays have been designed and fabricated, with initial tests at frequencies between 90 and 110 GHz. The imaging system consists of a pair of high density polyethylene focusing lenses which image the plasma electron cyclotron emission onto the array (see Fig. 3). The spot size within the plasma for these imaging optics has been experimentally determined and varies from channel-to-channel over a 1.0-1.7 cm range, with an interchannel spacing of 1.1-1.5 cm. First tokamak tests of the arrays will take place on TEXT-U in September, 1995.

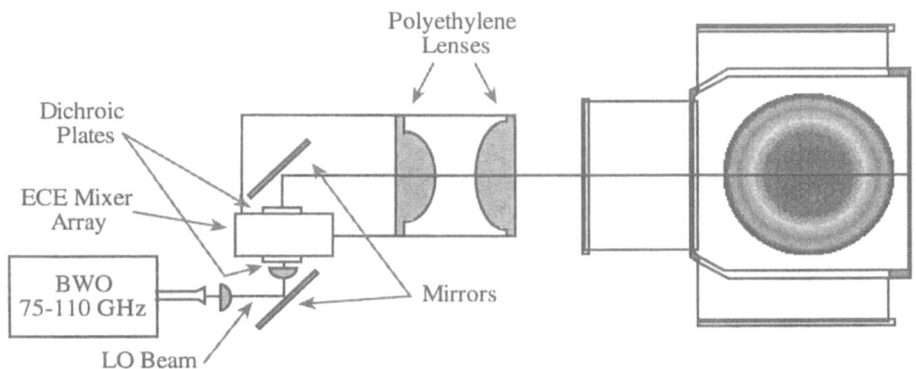

Figure 3. ECE imaging implementation on TEXT-U.

ACKNOWLEDGMENT

The authors wish to thank Drs. D. Brower, G. Cima and K. Mizuno for their support and advice. This work is supported by the U.S. Department of Energy under contracts No. DE-FG03-95ER54295 and W-7405-ENG48.

REFERENCES

[1] S.H. Heijnen, M. de Baar, A.J.H. Donne, M.J. van de Pol, C.A.J. Hugenholtz, and the RTP team, Review of Scientific Instruments **66**, pp. 419-421, 1995.

[2] C.W. Domier, N.C. Luhmann, Jr., A.E. Chou, W-M. Zhang, and A.J. Romanowsky, Review of Scientific Instruments **66**, pp. 399-401, 1995.

[3] D.K. Akuluna, Proc. Course and Workshop "Diagnostics for Contemporary Fusion Experiments," **ISPP-9**, Varenna (1991) 199.

EXPERIMENTAL RESULTS OF AMPLITUDE MODULATION REFLECTOMETRY ON HIGH DENSITY TOKAMAK PLASMAS

M. Zerbini,[1] P. Buratti[1] and P. Amadeo[2]

[1]Associazione EURATOM-ENEA sulla Fusione
  Centro Ricerche Frascati, CP 65, 00044 Frascati, Rome, Italy.
[2] ENEA guest

INTRODUCTION

The measurement of round trip group delay vs. frequency for electromagnetic waves reflected at a cutoff layer in the plasma can be used to reconstruct electron density profiles[1]. In Amplitude Modulation (AM) reflectometry[2,3] the group delay is obtained from the phase delay ϕ of the modulating envelope as $\tau_g = \phi/\Omega$, where $\Omega/2\pi$ is the modulating frequency; phase measurement is unambiguous provided that $\Omega\tau_g < 2\pi$;

The AM reflectometry system of FTU tokamak[4] consists of two reflectometers and three vacuum waveguide systems. The microwave frequency range is 52-90 GHz, corresponding to critical densities for the ordinary polarization $0.34 \div 1 \times 10^{20}$ m^{-3} and can be varied by jumps between values manually preselected in eight sub-bands. The modulating frequency is 200 MHz. In the phase delay region close to 2π the linearity of conversion was observed to degrade, due to the internal delays in the electronics. To compensate for this effect, an additional phase detector (PD) was incorporated in the reflectometer scheme, with a reference signal shifted by π relative to the first PD. This arrangement allows to use only the linear part of the voltage vs phase characteristic. The PD accuracy is better than 0.02 rad, corresponding to a distance of 2.4 mm in vacuum. In this paper some experimental reflectometry data collected on the FTU tokamak will be shown and discussed.

EXPERIMENTAL RESULTS

The PD output has been absolutely calibrated using the signal reflected by the FTU vacuum wall; the calibration remained stable during the three months of reflectometer operation. Fig. 1 shows the time evolution of the AM phase as measured in two FTU discharges, with the reflectometer operating at fixed probe frequency; as described in the caption, the general features of the signals agree with the plasma density evolution as measured by a DCN 5-chords interferometer.

The expected AM phase delay for a given plasma density distribution has been calculated assuming for simplicity a parabolic density profile.[1] Fig. 2 shows the comparison between the time evolution of the measured and calculated AM phase: the agreement is good, the differences should be due to the non-parabolic features of the actual density profile.

The reflectometer capability of detecting "fast" plasma fluctuation are illustrated by the

Diagnostics for Experimental Thermonuclear Fusion Reactors
Edited by P. E. Stott *et al.*, Plenum Press, New York, 1996

175

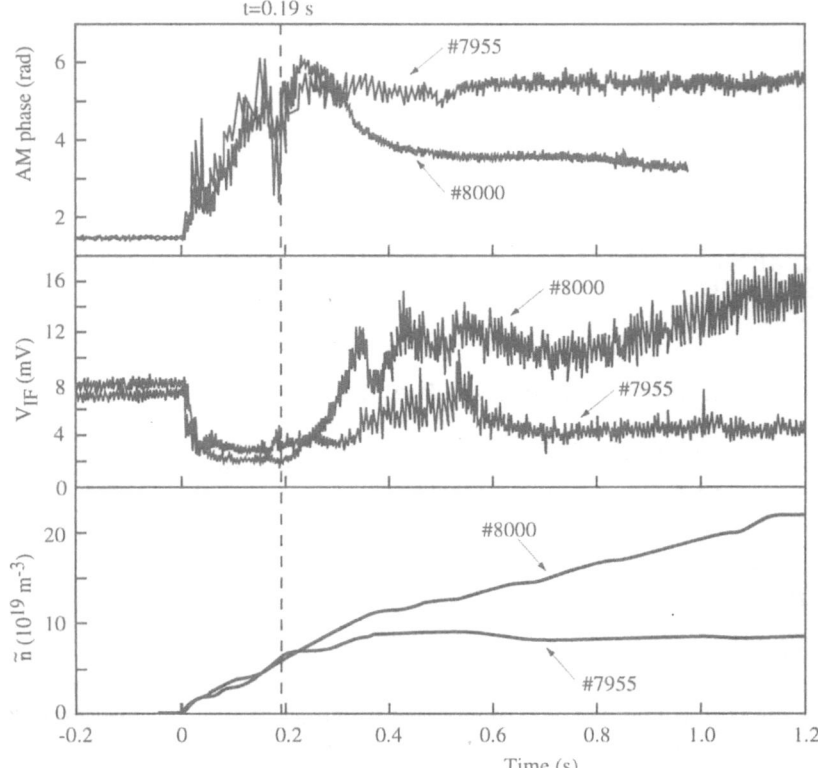

Figure 1. a) AM phase as measured during two FTU shots, with 85.39 GHz probe frequency; b) reflected signal amplitude detector; c) DCN line average density. For t<0 the plasma is not present and the phase delay is due to the reflection on the vacuum wall; for 0<t<0.19 s the maximum plasma density is below the cutoff value; for t >0.19s the cutoff layer reflects the radiation; after this time, for the shot #8000, the density increases and the reflecting layer moves toward the antennae, so that the phase delay reduces and the reflected signal amplitude increases; for the shot #7955 plasma density and consequently reflectometer signals remain relatively stable.

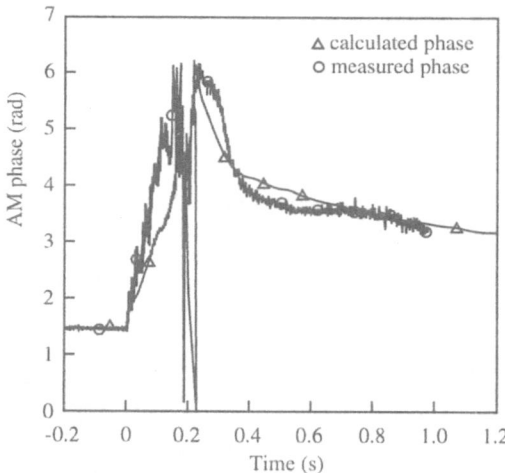

Figure 2. Comparison between the time evolution of the measured and calculated AM phase during shot #8000.

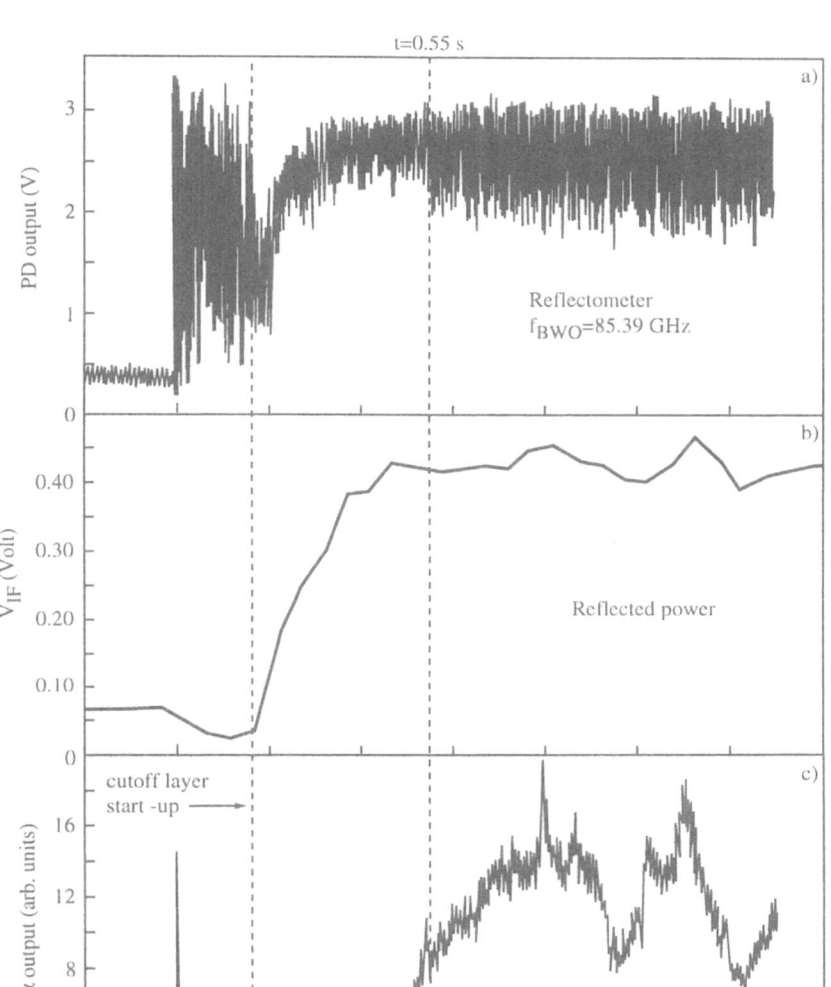

Figure 3. a) PD signal sampled at 10 kHz during shot #7997. b) Reflected power detector signal, averaged over 10 ms. c) D$_\alpha$ signal.

data in Fig. 3: the noise on the AM phase increases for t > 0.55 s, when also the D$_\alpha$ signal increases indicating a higher level of particles recycling, that should be explain the observed turbulence. The level of the reflected signal does not decrease at this time, then the phase fluctuations cannot be attributed to low signal effects on the phase detector operation.

In the reflectometric measure-ments at FTU, the PD output has been sampled at frequencies between 10 kHz and 100 kHz. The recorded signal showed usually a relevant noise, which can be attributed to plasma fluctua-tions. The noise can be greatly reduced by time averaging (1 ms was used in Figs 1 and 2), provided that the fast fluctuating signal is well within the PD output linear range.

The plasma density profile obtained with Abel inversion of frequency scan measure-

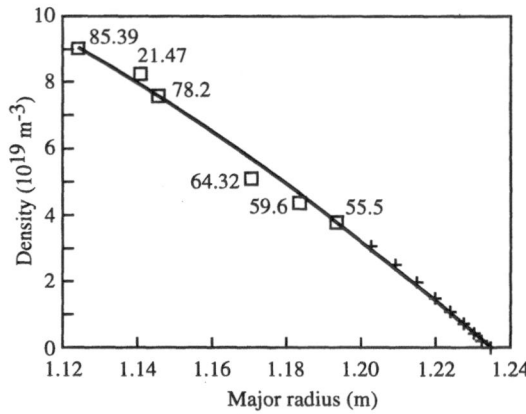

Figure 4. Squares: Abel inversion of reflectometry data (shot #8093), the probe frequencies in GHz are indicated; solid line: parabolic density profile with line average from DCN interferometer;

ments is shown in Fig. 4 in comparison with parabolic one normalized by a DCN line average measurement. The unprobed external part of the profile has been assumed parabolic.

CONCLUSIONS

The FTU AM reflectometry system operated successfully for three months in routine tokamak operation. The PD output has been calibrated to give the absolute group delay: this allowed to perform the study of plasma low-frequency phenomena and the density profile inversion. Spurios reflections effects are not very large as shown by the agreement between the calibrations with and without plasma in the vacuum chamber. The detection of plasma fluctuations indicates a diagnostic potential in this field.

REFERENCES

1. J. L. Doane, E. Mazzucato, and G. L. Schmidt, Plasma density measurements using FM-CW millimeter wave radar techniques, Rev. Sci. Instrum. **52**, 12 (1981).
2. V.A. Vershkov and V.A. Zhuravlev, Plasma diagnostics experiments on the T-10 tokamak by means of a reflected microwave signal, Sov. Phys. Tech. Phys. **32**, 523 (1987).
3. J. Sanchez, B. Branas, T. Estrada, E. de la Luna, and V. Zhuravlev, Amplitude modulation reflectometry for large fusion devices, Rev. Sci. Instrum. **63**, 4654 (1992).
4. P.Buratti, M.Zerbini, Y.Brodsky, N.Kovalev, A.Shtanuk, Amplitude Modulation Reflectometry Systemfor the FTU Tokamak, Rev.Sci.Instrum. **66** (409), 1995.

TWO DIMENSIONAL EFFECTS OF TURBULENCE ON DENSITY PROFILE MEASUREMENTS BY REFLECTOMETRY

V. Zhuravlev, J. Sanchez, E. de la Luna

Asociación EURATOM-CIEMAT
28040 Madrid, Spain

INTRODUCTION

Reflectometry is a very attractive diagnostic for measurements of density profile and plasma turbulence in reactor conditions, but the problem of interpretation of the results is not solved utterly. The reason for this is the plasma turbulence which leads to interference of signals, reflected simultaneously on different turbulent waves and back scattering of signal. Other source of perturbations is the multiple reflection of the signal between plasma and chamber wall. These effects can be simulated completely only in a 3-d full-wave code, though simulation of effects of scattering and interference may be reduced to 2-d taking advantage of the homogeneity of the plasma in the toroidal direction. Full-wave 2-d codes have been developed in recent time [1], but the use of this codes is limited due to the very complicated computations involved (a realistic simulation of a density profile measurement involves typically the calculation of the E-field detected at the antenna for a set of 2000 frequency values). In this work, a simple WKB 2-d code was used for qualitative simulation of effects of interference in situations where the turbulence size is larger than the signal wavelength.

2-DIMENSIONAL WKB CODE

The code is based on the quasi optic equations for the wavefront propagation[2]. This equations may be used only when the size of the turbulence Λ is much larger than the wavelength of signal λ. In the 2-d case the condition of applicability depends also on the angle of reflection. But in real experiments the signal, reflected with big angles usually do not reach the receiver and the condition of applicability does not differ significantly from $\lambda \ll L$. The wavefront is propagating in the 2-d space with variable refraction index. Amplitude **a** and phase ϕ of the reflected signal are computed as the result of interference of parts of reflected wavefront, coming to the receiving antenna. The main question for 2-d WKB applicability is the spatial size of the real plasma turbulence. Computation with the 2-d code with fixed frequency of microwave signal have shown that when the poloidal size of turbulence is less than the size of reflectometer antennas the spectrum of phase oscillations is completely noisy[3].

Experimental spectra of oscillations of the phase of reflected signal are usually not noisy. That is the real turbulence has rather big size and the WKB code is applicable for it.

The model for the turbulent density profile consists of two parts: 2-d unperturbed density distribution and a set of interchanging hills and depressions due to turbulence. Amplitudes of any hill or depression and distances between them in both directions were varying randomly inside +/- 50% on average values. For this structure the average size of

Diagnostics for Experimental Thermonuclear Fusion Reactors
Edited by P. E. Stott *et al.*, Plenum Press, New York, 1996

179

turbulence **L** is the average distance between tops of hills. Poloidal and radial average sizes were equal. Different turbulent structures with the same average size **L** and amplitude **A** could be created by modifying an initial random parameter. For all results below sizes of turbulence exceeded the vacuum wavelength of the launched waves by a factor 10-15 in order to keep applicability of the WKB solution.

SIMULATION OF DENSITY PROFILE MEASUREMENT

It is commonly accepted that the instantaneous density profile shows complicated structures produced by the turbulence. On the other hand, the averaged density profile is of high physical interest. In the following section we will use the simulation code to evaluate the ability of the different reflectometer techniques to perform this kind of average.

Simulations were performed for swept homodyne, AM and pulse reflectometers. The phase delay $\phi(f)$ and amplitude $a(f)$ of the reflected signal were computed as a function of the carrier frequency for a frozen turbulent structure (situation equivalent to a sweeping rate much faster than the turbulence time scale). From these data the measurements of each reflectometer were simulated.

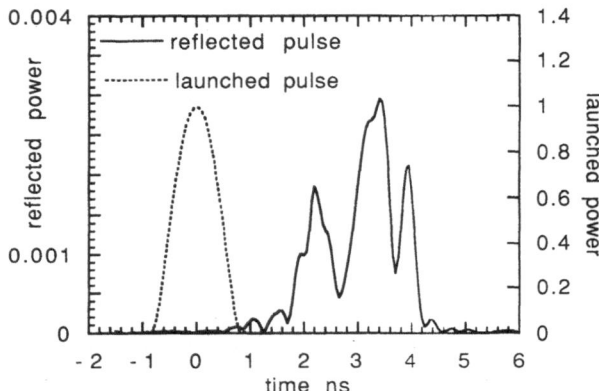

Figure 1. Launched and reflected pulse simulation for a typical turbulent profile. W7AS parameters

For the pulse radar, pulses were reconstructed by backward Fourier transform of the spectrum of reflected pulse, including the effects of phase and amplitude changes. The pulse width was about 1 ns, its frequency components were reflecting in the plasma within a radial extension shorter than the correlation length. The time delay was defined as the time between maximum power of reflected and launched pulses. The maximum power of the reflected pulse suffered variations more than 20 dB due to turbulence. Shapes of many reflected pulses were distorted, as shown in fig.1, due to the interference of pulses reflected in different points of plasma.

For the AM reflectometer the time delay was computed from the phase differences in the 3-component spectrum, with 200 MHz distance between components. Results were averaged over carrier frequency sweep so that the frequency resolution of the AM reflectometer was close to that of the pulse radar. In real AM reflectometers this vector averaging is done by IF filter. This averaging eliminates short splashes of the measured time delay due to interference, accompanied usually by falls in the signal amplitude. Results of averaging over 1 and 3 periods of turbulence for fixed carrier frequency and rotating turbulent structure are shown in fig.2.

For the sweeping homodyne reflectometer the signal $a \times \sin(\phi)$ was reconstructed (see fig.3. Standard filtering (rejection of too small or too large beat frequencies) and fringe detection (zero crossing) were used to compute the time delay. The strong fluctuations in the amplitude advise the use of heterodyne receivers able to measure both phase and amplitude of the reflected beam.

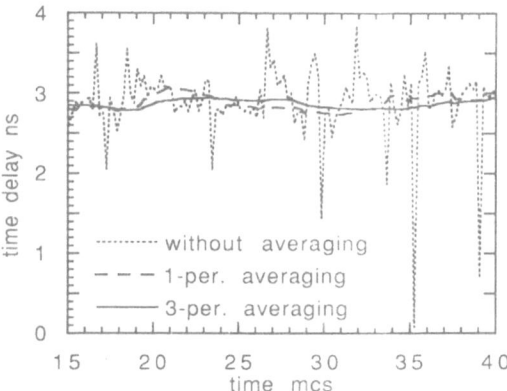

Figure 2. Simulated time delay from AM reflectometer with increasing level of vector average. (W7AS parameters)

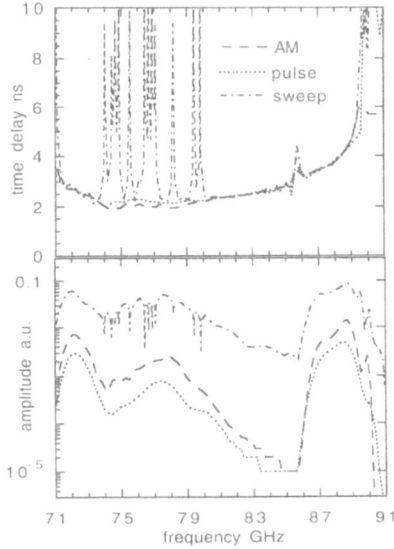

Figure 3. Time delay and amplitude of the measured signal for simulated AM, Pulse radar and swept FM operation. (W7AS parameters)

Results were simulated for the geometries of W7AS and LHD stellarators operating in the X-mode. Turbulence fluctuation level was about 2.5% and 5 cm correlation length for W7AS and 0.7% level and 6 cm length for LHD. Turbulence with this parameters leads to strong interference and to a completely noisy spectrum of phase oscillations. The 2-d effects are stronger in LHD, suggesting effects of the larger distances involved (5 times larger in LHD than in W7AS for the same wide antenna pattern). That would mean a more severe problem in the case of ITER which would show still 3-5 times larger distances than LHD for the innermost probed points. Attempts to simulate the ITER conditions failed due to problems of precision. Improvements in the code are being introduced to cope with the problem.

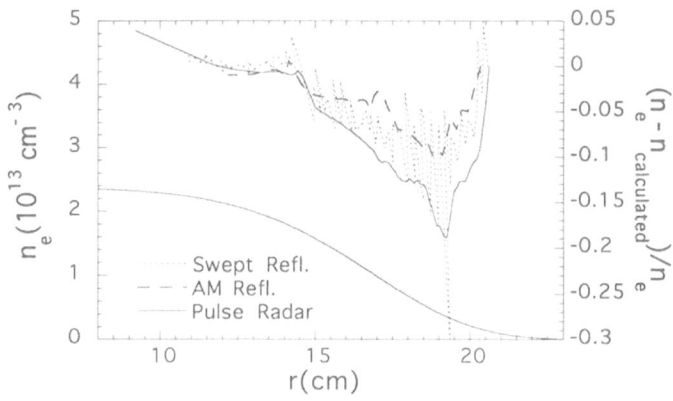

Figure 4. Simulation of original and relative error of the reconstructed density profiles by AM, Pulse radar and FM reflectrometry. FM results improve with proper filtering of the homodyne signal.

In spite of the turbulence, the reconstruction of the density profile from the simulated signals leads to a very good approximation to the actual average one (see fig.4). The key point is the vector average that the systems perform: the E-field reaching the antenna after reflection at the plasma is the result of an interference process, it has low amplitude and random phase when the interference is destructive and high amplitude with well defined phase for constructive interference. The vector average along several frequencies (or time instants) includes mainly the good values, whereas the random ones carry very little weight because of the low amplitude. This selective weighted average would be a powerful tool for application to ITER if the signals were too turbulent. Pulse Radar and AM tend to include the average in the detection process. For the FM systems it is usually performed by post-detection signal processing [4].

CONCLUSIONS

The 2-d WKB code, able to compute interference effects, was used for qualitative investigation of the perturbations of turbulence in reflectometry measurements. The code is applicable when the size of turbulence is clearly larger than the wavelength of the signal.

Turbulence causes strong, more than 20 dB, modulation in the power of the reflected signal of the AM reflectometer and pulse radar but the average density profile can be obtained to good approximation. For the homodyne sweeping reflectometer the good profile was obtained after proper filtering of the beat signal. The vector average of the received E-field might be the best solution for the determination of the average density profile in ITER.

REFERENCES

1 E. Mazzucato and R. Nazikian , Rev. Sci. Instrum. 66(1) (1995)
2 V.L. Ginzburg *The propagation of electromagnetic waves in plasmas* Pergamon
    press, Oxford (1970).
3 V.Zhuravlev, J.Sanchez, E.de la Luna *Two-dimensional effects of turbulence in density profile
    measurements by reflectometry*. Proc. 22nd EPS Conf. on Controlled Fusion, Bournemouth, 1995
4 A. Silva, l.Cupido et al *Continuous phase derivative evaluation for density profile measurements from FM
    broadband reflectometry* Proc. 22nd EPS Conf. on Controlled Fusion, Bournemouth, 1995

PHYSICS ISSUES OF ECE AND ECA FOR ITER

D V Bartlett

JET Joint Undertaking, Abingdon Oxon, OX14 3EA, UK

and members of the European Home Team for Microwave Diagnostics *

ABSTRACT

The parameters predicted for the core of ITER plasmas are sufficiently different to those of existing tokamaks that a re-evaluation of the physics basis of ECE measurements is required. This is particularly important in the light of the detailed, and demanding, targets for the measurement of these parameters which have been set in the ITER design. In this paper we examine quantitatively the localisation of ECE T_e measurements in the core and edge plasma, as well as the problem of access to the plasma centre at high T_e. It is concluded that while ECE will be able to provide good measurements of T_e for a wide variety of plasma conditions, the limitations need to be analysed in detail. We also look briefly into the possible potential of ECA as a divertor plasma diagnostic.

INTRODUCTION

The physics issues addressed in this paper are those concerned with the localisation of T_e measurements and the access for those measurements to the plasma centre, since these are the most likely areas of change from the ECE T_e measurements on current tokamaks. The applications of ECE measurements and other ECE physics related issues, such as the consequences of emission from current drive electrons, or the need for an accurate knowledge of the plasma internal magnetic fields, are not discussed here. The experimental issues are considered in another paper [1] at this workshop.

The most important factor determining the behaviour of the core plasma Electron Cyclotron Emission (ECE) in ITER will be the high electron temperatures (>20 keV). Also significant will be the small aspect ratio (which causes harmonic overlap and so limits access to the plasma centre) and the possibility of high electron densities (which might limit access for low frequencies). The large plasma volume (which increases the emission in the high frequency, optically thin, region) will influence the overall form of the emission spectrum. A preliminary investigation of these effects was made during the ITER CDA [2]. In the present work, the consequences of these effects have been evaluated by running a

Diagnostics for Experimental Thermonuclear Fusion Reactors
Edited by P. E. Stott *et al.*, Plenum Press, New York, 1996

183

number of ECE simulation codes for realistic ITER parameters and measurement scenarios. This allows the accessibility (limited by re-absorption in other harmonics) and the spatial resolution (set by spectral broadening) to be determined.

Edge T_e measurements have the same physical basis as the core measurements, but the different plasma conditions (particularly the strong gradients of density and temperature) result in different behaviour for the spatial resolution. This is discussed in relation to the demanding spatial resolution requirements set by ITER

Electron Cyclotron Absorption (ECA) may be able to provide useful Te related information over a reasonably wide range of divertor plasma parameters. The accessible parameter range will be described and the possible performance of a measurement system, based mainly on extrapolation of present experiments, will be discussed.

The paper begins by recalling briefly the method by which ECE spectra are simulated. An example spectrum for predicted ITER conditions is shown. A qualitative explanation of the behaviour of ECE at high T_e is used to explain the features of this spectrum and to illustrate the problem of access to the plasma centre and localisation of the measurements in the core and edge plasmas. The possibility of reducing harmonic overlap by using near vertical sightlines through the plasma is discussed. Quantitative results from the simulations are compared with the requirements for T_e measurements which have been established as part of the ITER design. Finally, the physics basis and likely performance of a divertor ECA diagnostic are discussed.

SIMULATION METHOD AND PLASMA PARAMETERS

In this paper we are concerned with emission from a Maxwellian plasma, and the following simple analysis is appropriate. For each radiation frequency of interest it is necessary to solve the radiation transport equation [3]:

$$dT_{RAD}(s)/ds = j(s) - T_{RAD}(s).\alpha(s), \tag{1}$$

where s is a variable describing the distance along the ray path through the plasma, the radiation intensity is represented as radiation temperature T_{RAD} (defined as the temperature of a black-body which would radiate the same power at the given frequency) and j and α are the emission and absorption coefficients respectively. In a Maxwellian plasma we can use Kirchhoff's law to replace $j(s)$ by $\alpha(s)$ and the electron temperature:

$$dT_{RAD}(s)/ds = \alpha(s).\left(T_e(s) - T_{RAD}(s)\right). \tag{2}$$

This equation has this solution for the intensity emerging from the plasma:

$$T_{RAD} = T_o.e^{-\tau} + \int_{IN}^{OUT} G(s).ds, \tag{3}$$

where $T_o e^{-\tau}$ represents the intensity of the radiation entering the plasma by wall reflections and attenuated by the total optical depth:

$$\tau = \int_{IN}^{OUT} \alpha(s).ds. \tag{4}$$

The integrations in Equations 3 and 4 are carried out along the ray path, from where it enters to where it leaves the plasma. The integrand in Equation 3 is

$$G(s) = T_e(s).\alpha(s).e^{-\tau(s)}, \tag{5}$$

where $\tau(s)$ is the cumulative optical depth:

$$\tau(s) = \int_{S}^{OUT} \alpha(s') . ds . \tag{6}$$

In the present case, we are interested in local temperature measurements so we restrict the discussion to those frequencies for which the total optical depth of the plasma is large, that is, $e^{-\tau} = 0$. This means that T_O is not important and the localisation of the observed emission is determined entirely by the spatial width of the function $G(s)$.

There is a large body of theory in the literature concerned with electron cyclotron interactions in tokamak plasmas (see for example [4] and [5]), and a wide variety of approximations, appropriate to the calculation of emission and absorption coefficients in different circumstances are available.

Also, a large number of ECE simulation codes exists, each code either specifically including or excluding various physics ingredients, and using a variety of techniques to obtain the absorption coefficient. The initial feasibility study for ITER by the European Home Team [6] included a quantitative comparison of six of the codes available in European laboratories. While some differences between the codes were apparent, the level of agreement is certainly good enough to give confidence that predictions made for conditions within the limits of validity of each code are reliable. Most of the results presented here are from the JET code which uses a formulation of the absorption coefficient due to Bornatici [5], which is valid for propagation perpendicular to the magnetic field, high electron temperatures, and includes the "finite density" effects. The limitation to perpendicular propagation is not a problem for the illustrative results presented here, but more general calculations, valid for arbitrary directions, will be needed to complete this study.

Figure 1 shows some results from the JET code which illustrate the behaviour of the quantities $\alpha(s)$, $\tau(s)$, $G(s)$ and the cumulative integral of $G(s)$ for the second harmonic extraordinary mode (E-mode) with ITER-like plasma parameters. The cumulative integral of $G(s)$ is defined in the same way as that for the optical depth (Equation 6), and for convenience we denote it by $\int G(s)$. The shape of the $\alpha(s)$ curve arises from the relativistic expression for the cyclotron resonance frequency, for propagation perpendicular to $\underline{B}$ (ie without Doppler broadening):

$$\omega = m . \frac{e.B}{m_e}, \tag{7}$$

where m is the harmonic number (m=2 in Figure 1), e is the electron charge and m_e is the relativistic electron mass. The width of $\alpha(s)$ is due to the Maxwellian distribution function giving rise to a range of electron masses, the higher masses requiring a stronger magnetic field (ie. smaller R) to radiate at the given frequency. The vertical line on each plot of Figure 1 is at the location of the non-relativistic cyclotron resonance, obtained by substituting the electron rest mass, m_0, in Equation 7.

A significant feature of Figure 1 is that although the absorption coefficient profile is very wide (> 2m), the width of $G(s)$ is only about 0.1m. This is because the optical depth is large and there is almost total re-absorption of radiation emitted over most of the line width. For observations from the outboard side, only the outer surface of the resonance layer is seen. To quantify the spatial localisation of the observed emission, we use the fact that, according to Equation 3, $\int G(s)$ determines the intensity reaching a measurement antenna. We therefore use the three levels marked on the $\int G(s)$ curve: 90% of the observed intensity originates from the region between the 5% and 95% points, while the 50% point may be considered as the mean radius from which the observed emission originates.

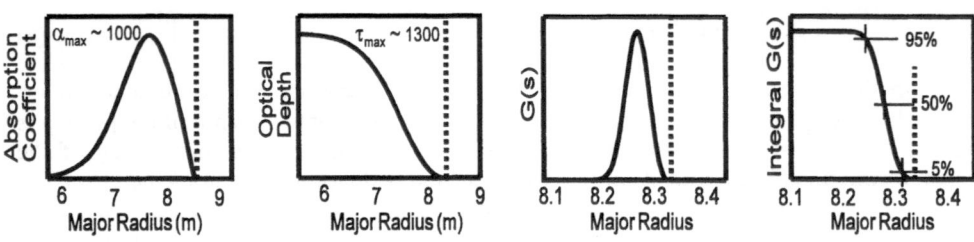

Figure 1: From left to right the curves are: the absorption coefficient for the second harmonic E-mode for typical ITER conditions, the optical depth obtained by integrating the absorption coefficient (Eq. 6), the function $G(s)$ (Eq. 5) and its cumulative integral. The radii corresponding to the levels marked on the $\int G(s)$ curve are used to characterise the localisation of the emission (see text).

ITER PLASMA PARAMETERS AND AN EXAMPLE ECE SPECTRUM

To date, two sets of plasma parameters provided by ITER [7] have been used in the simulations. These parameters have been derived from transport code simulations of ITER reference ELMy H-mode conditions. They exhibit the typically flat density profile of H-modes. One set (denoted "CASE1") simulates a sawtoothing discharge and has a flat region at the centre of the T_e profile, while the other ("CASE2") is sawtooth free and has a peaked T_e profile. The full magnetic equilibrium of these discharges is used in the ECE simulations. Figure 2 shows the n_e and T_e profiles for these two cases. The flux surfaces for CASE1 can be seen in figure 7.

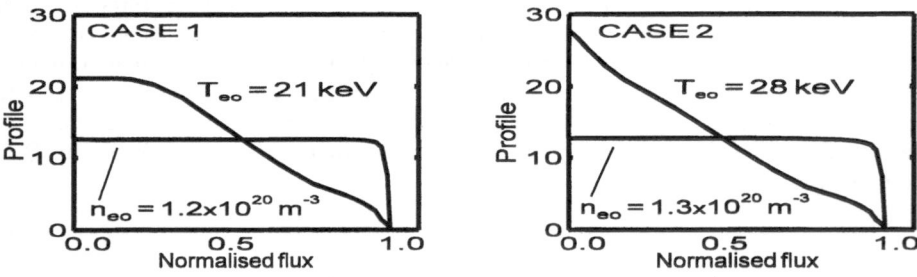

Figure 2: ITER density and temperature profiles used in the ECE simulations. The sawtoothing discharge on the left is referred to as CASE1, while the sawtooth free plasma on the right is denoted CASE2.

Figure 3a shows an example of an E-mode ECE spectrum calculated for CASE1 conditions, using the method described above. A simple model of wall reflections has been included, but this affects only the very low and very high frequency regions. The spectrum appears to consist of the usual series of overlapping harmonics, from which it should be possible to obtain localised measurements of the electron temperature profile. Using the second harmonic region of the spectrum and the $|B|$ profile shown in Figure3b, the usual, simple, analysis which uses the non-relativistic resonance condition (Equation 7 with $m_e = m_0$), gives the result shown in Figure 3c. There appear to be three problems in this deduced T_e profile: excess harmonic overlap (in a region where the geometry of the magnetic field indicates that overlap should not occur), an outward shift of the deduced profile on the outboard side, and an overestimate of T_e in the edge region.

A qualitative understanding of the behaviour of the deduced temperature profile in Figure 3 can be obtained from the sketches of the absorption coefficient in Figure 4. Frequency f_1 corresponds to second harmonic E-mode emission from close to the plasma

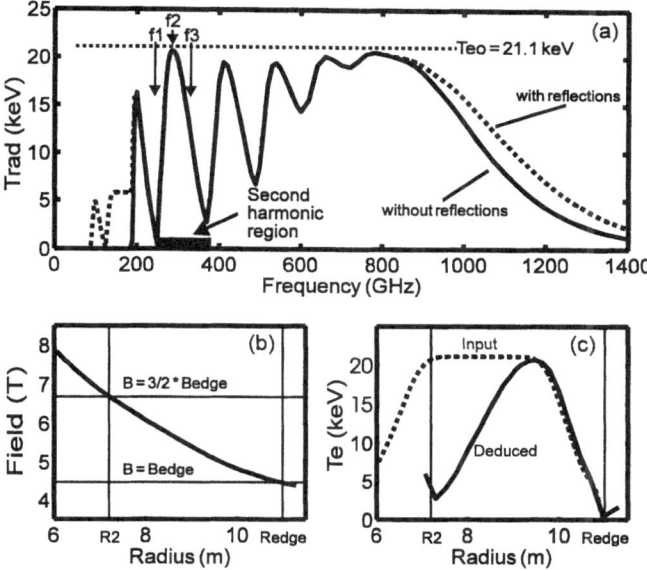

Figure 3: Part (a) is an example of a simulated extraordinary mode ECE spectrum, both with and without wall reflections. The frequencies labelled f1, f2 and f3 are referred to in Figure 4. Part (b) shows the spatial profile of $|\underline{B}|$ used in the simulation, and used to deduce a T_e profile from the second harmonic region of the spectrum. The radius and field at which third harmonic at the outer edge starts to overlap the second harmonic is indicated. The deduced profile, in part (c), is obtained using the non-relativistic resonance condition. It shows some differences compared to the profile which was used as input to the calculations.

edge: unless the edge optical depth is large, the tail of $\alpha(s)$ (which is in a region of much higher temperature) can dominate the observed intensity. At frequency f_2 the second harmonic has large optical depth and a well localised value of T_e is obtained. At even higher frequencies, f_3, the tail of $\alpha(s)$ due to the third harmonic enters the plasma (even though the non-relativistic resonance of the third harmonic may be well outside it) and begins to obscure the second harmonic emission.

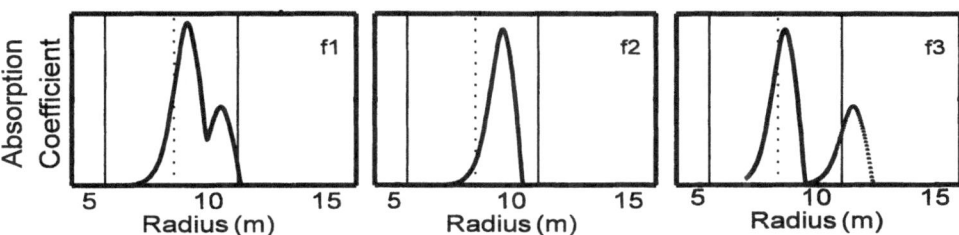

Figure 4: Sketches of the absorption coefficient for the three frequencies around the second harmonic region which are indicated on Figure 3. The solid vertical lines mark the inner and outer edges of the plasma.

LOCALISATION AND ACCESS TO THE PLASMA CENTRE

From the above discussion it is clear that it is not possible to obtain information about the localisation of the emission simply by looking at spectra, such as the one shown in Figure 3 . A better way to present this information is to plot the location and width of the function $G(s)$ as a function of frequency, on a graph of frequency vs. radius. Examples are shown in Figure 5, for both the E-mode and the ordinary mode (O-mode), for an antenna on the outboard side on the plasma midplane, and for the plasma conditions of CASE1. The horizontal bars link the radii corresponding to the 95% and 5% levels of $\int G(s)$, while the thicker central line marks the 50% level. At each frequency, the intensity which would be measured is the average of T_e across the radial width of the bar. These bars and the central line are drawn over the frequency range for which the plasma is optically thick. This range is indicated by the vertical bar next to the frequency scale. Also shown are the locations of the non-relativistic cyclotron resonances (Equation 7 with $m_e = m_0$, and using the spatial variation of $|\underline{B}|$ given by the equilibrium data), and the cutoff locations for each polarisation mode.

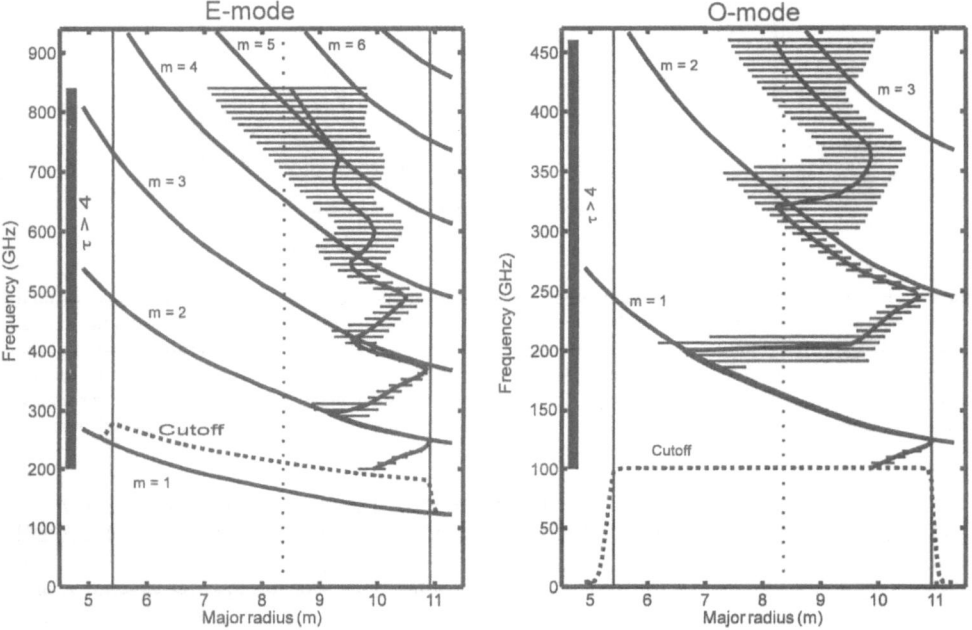

Figure 5: Horizontal bars representing the width and location of the emitting layer on a frequency vs. radius plot. The intensity which would be measured at each frequency by an outboard side antenna on the plasma midplane is the average temperature across the bar. The left hand plot is for the E-mode, and the O-mode is on the right. In each case the non-relativistic cyclotron harmonics are shown as well as the appropriate cutoff locations. The vertical bar next to the frequency axis shows the frequency range for which the plasma is optically thick.

Using the data in this figure the behaviour of the deduced T_e profile, explained qualitatively in Figure 4, can be made quantitative. Over some part of the second harmonic E-mode (and most of the first harmonic O-mode) the emission is localised to a narrow band just behind the non-relativistic cyclotron harmonic. The inward shift, relative to the non-relativistic resonance, is not a serious problem since its magnitude can be estimated quite accurately from an approximate knowledge of n_e and T_e. At higher frequencies, emission which is downshifted from the next higher harmonic starts to obscure the well localised emission and T_e measurements are not possible. At lower frequencies, close the plasma edge, there is not sufficient optical depth to re-absorb downshifted emission from the same harmonic which is emitted deeper in the plasma. Higher harmonics (third harmonic E-mode and second harmonic O-mode) show some moderately localised emission over a small radius range. For these plasma conditions, there are no frequencies for which there is well localised E-mode emission from near the plasma centre, but localised T_e measurements at radii well inboard of the centre are possible in the O-mode.

Repeating these calculations with either the n_e or T_e profiles multiplied by a scaling factor gives an indication of how the localisation and central access will vary with plasma conditions. For T_{eo} below about 15 keV, access to the plasma centre is possible using the second harmonic E-mode. First harmonic O-mode emission from the plasma centre can propagate to the outboard side for T_{eo} up to about 30 keV and (limited by cutoff) densities up to about 3×10^{20} m^{-3}. Over this parameter range the width of localisation in both polarisation modes does not vary very much. In addition, these conclusions are not strongly affected by changes of density, apart from the problem of cutoff.

NEAR VERTICAL SIGHTLINES TO REDUCE HARMONIC OVERLAP

A possible way of reducing the harmonic overlap, while maintaining propagation almost perpendicular to the magnetic field, is to use near vertical sightlines which are directed towards the plasma centre from a vertical port. The radiation path through the plasma covers a smaller range of radius, and therefore experiences a smaller range of $|\underline{B}|$, which reduces harmonic overlap.

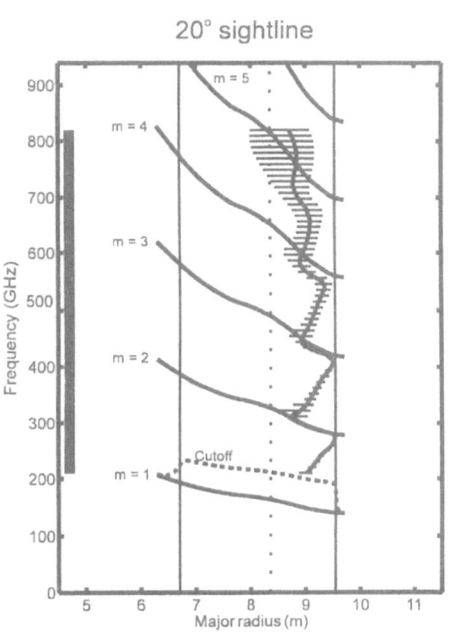

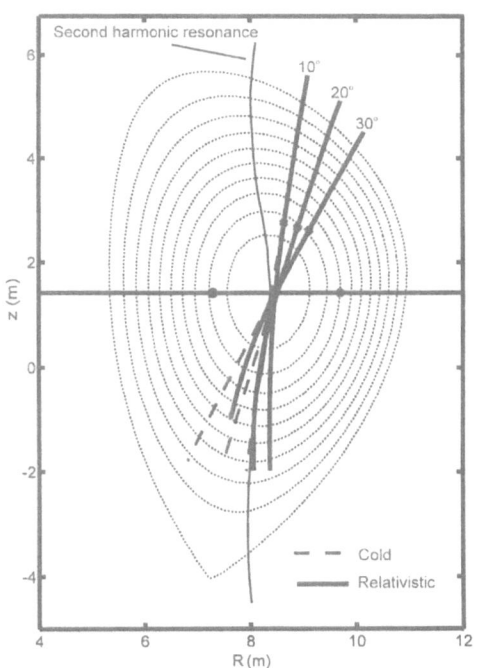

Figure 6: Localisation for the E-mode along a near vertical sightline, showing that the limiting radius in the second harmonic is slightly closer to the plasma centre than for the horizontal sightline of Figure 5.

Figure 7: Relativistic ray tracing for the E-mode along three near vertical sightlines. The deflection near the plasma centre is due to anomalous dispersion, and may prevent good access to the centre.

Figure 6 shows the localisation of the second harmonic E-mode for CASE1 plasma conditions, and a sightline 20° from the vertical. Although the overlap in the second harmonic region is reduced compared to the plasma midplane sightline of Figure 5, it is still not possible to reach the centre. Moreover, there is a further effect which becomes important at these high temperatures. Figure 7 is the output from a relativistic ray tracing code [8] for three E-mode rays, launched at angles of 10°, 20° and 30° from the vertical, and calculated for a frequency equal to the central second harmonic. Near the resonance layer, relativistic effects cause anomalous dispersion: a kink in the refractive index whose spatial width increases with T_e. The effect of this perturbation is to deflect the rays which are propagating near vertically away from the resonance layer. This effect is complex to characterise, since both the magnitudes and profiles of n_e and T_e play a role, but it appears that at the high temperatures where harmonic overlap is a problem the deflection will generally be large enough to prevent the plasma centre from being observed.

SUMMARY AND COMPARISON WITH THE ITER REQUIREMENTS

Data of the type plotted in Figure 5 can be used to give a simple summary of the spatial resolution which can be achieved with ECE measurements. Table 1 summarises the results for a horizontal sightline, viewing the plasma from the outboard side and passing through the plasma centre.

Table 1: Spatial resolution of T_e measurements for propagation nearly perpendicular to $\underline{B}$.

	Core plasma, with T_{eo} = 10 keV	Core plasma, with T_{eo} = 20 keV	Plasma edge
First harmonic, O-mode	0.10 m	0.13 m	0.04 - 0.06 m*
Second harmonic, E-mode	0.08 m	0.10 m	0.02 - 0.04 m*

* See note in text.

It is apparent that the ITER target resolution for core plasma measurements of 0.10 m can be achieved. The resolution target of 5 mm set by ITER for edge T_e measurements has not been achieved in these simulations. It should be noted that the edge resolutions given in the table apply to the particular edge profiles used in these calculations. They can only be taken as typical values, since this resolution is strongly profile dependent. For very steep edge T_e profiles the edge resolution is somewhat worse, but it is unlikely that the 5 mm target will be reached for any plasma conditions.

Access to the plasma centre is limited by re-absorption at high temperature (for the second harmonic extraordinary mode) and by cutoff at very high density (for the first harmonic ordinary mode). The results, again for an outboard horizontal antenna with a sightline through the plasma centre, are summarised in Table 2.

Table 2: Access to the plasma centre for T_e measurements.

	Max. T_e for access to centre*	Max. n_e for access to centre*	Inner R limit for CASE1
First harmonic, O-mode	~ 30 keV	~ 3×10^{20} m^{-3}	7.2 m
Second harmonic, E-mode	~ 15 keV	~ 3×10^{20} m^{-3}	9.5 m

* See note in text.

It should be noted that the values for maximum n_e and T_e are dependent on the equilibrium which is used in the calculations: changes in profile shapes would modify the values slightly, while a reduced toroidal field would significantly reduce the density limits.

The only sightlines which might give some improvement in access are those which pass near vertically through the plasma centre. However, the calculations presented here indicate that such improvements are in fact not achieved.

PRINCIPLE AND PROJECTED PERFORMANCE OF AN ECA DIAGNOSTIC

It is very unlikely that the plasma conditions in the divertor will be such that a localised measurement of T_e using ECE will be possible. The principle problem is that the magnetic field strength in the divertor region is the same as that in the core plasma, while the electron temperatures should be a factor of 100 or more lower. The intense ECE from the core plasma will reach the divertor region by wall reflections, and a high optical depth ($\tau \sim 8$) in the divertor plasma would be required to attenuate it to a level much less than that of the local emission. The density and temperature predicted for the ITER divertor plasma give optical depth values of order one.

Since the optical depth is too low for useful ECE measurements, the alternative is to determine the optical depth itself by measuring the absorption of radiation from an external source. Such a diagnostic has been constructed for the JET divertor [9]. It must overcome a variety of serious technical difficulties, related to access to the plasma, long waveguides to transport the radiation to and from the divertor, and standing waves which can cause spurious modulation of the transmitted power. Since the experience of the JET system is that these problems can be overcome, we examine here the plasma parameter range over which such a system might be able to operate in the ITER divertor.

The theoretical basis of a calculation is much the same as that presented above for ECE, although simpler in this case, for two reasons. First, we are concerned only with transmission of radiation through the plasma, that is, we obtain the optical depth from an experimentally measured transmission, then use Equation 4 to deduce plasma parameters. Secondly, the spatial width of the absorption coefficient is very small (~ 1 mm) for the low temperatures in the divertor. It can be approximated by a delta function, making the integration in Equation 4 unnecessary. There are, therefore, relatively simple expressions [4] for the spatially localised optical depth as a function of frequency. In the second harmonic E-mode the optical depth is proportional to the density - temperature product. The second harmonic E-mode is the best choice for such a measurement since it has greater optical depth than the first harmonic O-mode and a higher cutoff density.

Taking into account the other ITER parameters, and assuming that transmission can be reliably measured over the range 10% to 90%, the range of $n_e.T_e$ which can be determined is approximately the following for the second harmonic E-mode:

$$2 \times 10^{20} \text{ eV.m}^{-3} < n_e.T_e < 4 \times 10^{21} \text{ eV.m}^{-3}. \tag{8}$$

This equation breaks down at high densities ($\sim 6 \times 10^{20}$ m^{-3}), close to cutoff. The result is illustrated graphically in Figure 8, which plots several transmission contours in $n_e.T_e$ space.

Although the parameter range shown above is wide, it does not cover the whole of the potential range given by current predictions for the ITER divertor (1 to 100 eV and 10^{19} to 10^{22} m^{-3} near the target). A more extensive analysis, including other factors (such as refraction) in the calculations, and using realistic 2-D plasma profiles will be required before more precise predictions can be made.

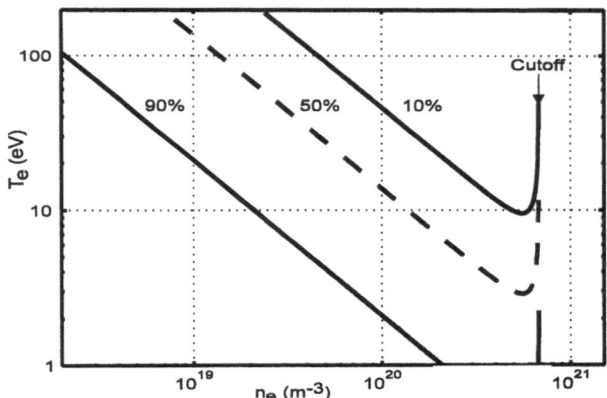

Figure 8: Contours of constant transmission through the ITER divertor plasma in $n_e.T_e$ space, for the second harmonic extraordinary mode. At constant transmission the $n_e.T_e$ product is constant, except near the cutoff.

SUMMARY AND CONCLUSIONS

This paper has shown that the plasma conditions predicted for the core of ITER plasmas are sufficiently different to those of existing tokamaks that the behaviour of electron cyclotron emission is qualitatively different. A re-evaluation of the characteristics of ECE in such plasmas is required. The first stages of this investigation have shown that ECE will be able to provide measurements of electron temperature for a wide variety of plasma conditions. The spatial resolution requirements for such measurements set in the ITER design can be met. There is a problem of access to the plasma centre at high T_e for the second harmonic extraordinary mode, but the this can be overcome by using the first harmonic ordinary mode, except at very high densities.

The potential of ECA as a divertor plasma diagnostic has been examined briefly. A simple theoretical analysis indicates that it should be possible to measure the electron density - electron temperature product over a wide range of the predicted operating space. However, results from current experiments, such as the one at JET, need to be evaluated before reliable predictions, leading to the design of hardware, can be made.

ACKNOWLEDGEMENTS
The author wishes to thank Dr H Bindslev for his considerable efforts in modifying his ray tracing code to operate with the ITER equilibria, and Dr D Boucher for providing the equilibria used in the calculations presented here.
* The members of the European Home Team for Microwave Diagnostics who have participated in this work are: A Airoldi, P Buratti, L Cupido, T Donné, G Giruzzi, H Hartfuss, E Joffrin, M E Manso, S Nowak, R-L Meyer, G Ramponi, J Sanchez, B Schokker, A Silva, P Stott, V Tribaldos, G Waidmann

REFERENCES
[1] H Hartfuss, this workshop.
[2] A E Costley and D V Bartlett, Proc 8th Joint Workshop on ECE and ECRH (Gut Ising, 1993), IPP report IPP III/186.
[3] G Bekefi, "Radiation processes in plasmas", Wiley (1966).
[4] M Bornatici et al, Nuc Fusion **23**, 1153 (1983).
[5] M Bornatici, F Engelmann and U Ruffina, Sov J Quantum Elec, **13**, 68, (1983).
[6] Report by the EU Home Team for ITER Task Agreement S 56 03 94-9-27-F.
[7] D Boucher, private communication.
[8] H Bindslev, Proc 9th Joint Workshop on ECE and ECRH (Borrego Springs, 1995).
[9] R J Smith et al, Proc 8th Joint Workshop on ECE and ECRH (Gut Ising, 1993), IPP report IPP III/186.

ECE DIAGNOSTIC ON ITER

Marino Bornatici,[1] and Umberto Ruffina[2]

[1]Physics Department, University of Ferrara, Ferrara, Italy,
and Physics Department "A. Volta", University of Pavia,
Pavia, Italy
[2]Department of Nuclear and Theoretical Physics, University of Pavia, Pavia, Italy

ABSTRACT

On the basis of a detailed numerical investigation of the ECE spectra for flat as well as peaked model profiles for temperature and density, it is shown that for ITER conditions with central temperatures $\gtrsim 20 keV$ the first harmonic ordinary mode is the most suitable choice for localized measurements of the temperature over a significant portion of the plasma cross–section. For the plasma core and flat profiles the corresponding spatial resolution is 10 cm, typically, improving somewhat for peaked profiles. The second harmonic extraordinary mode, instead, is more strongly affected by harmonic overlap, which spoils its utilization for measurements of the central temperature profile, its diagnostic potential being restricted at best to the outboard plasma region.

INTRODUCTION

At the relatively high temperatures expected in ITER, the diagnostic potential of ECE measurements has to be re–assessed in respect of both the relativistic and Doppler effects, which, relative to the low temperature case, will lead to an increase in the width, as well as to a displacement in the location, of the layer wherefrom the measured radiation originates.[1] Here the feasibility of ECE measurements to provide the electron temperature (T_e)profile for ITER conditions is investigated in terms of the **radiation temperature**, which, for observation from the outboard side of a plane Maxwellian plasma slab of thickness $2a$, is

$$T_{rad}^{(i)}(\omega, N_\parallel) = \int_{-a}^{a} dx T_e(x)\alpha^{(i)}(\omega, N_\parallel, x)e^{-\int_{x}^{a} dx' \alpha^{(i)}(\omega, N_\parallel, x')}$$

$$\equiv \int_{-a}^{a} dx G^{(i)}(\omega, N_\parallel, x) \tag{1}$$

with $\alpha^{(i)}$ the EC absorption coefficient of the i.th mode ($i = O, X$, respectively for the ordinary and extraordinary mode), and $N_\parallel$ the parallel refractive index ($N_\parallel = 0$ for perpendicular observation). The effects of wall–reflections are accounted for on multiplying the r.h.s. of (1) by $[1 - \rho e^{-\int_{-a}^{a} dx' \alpha^{(i)}(\omega, N_\parallel, x')}]^{-1}$, with ρ the wall-reflectivity (taken equal to 0.6 in the numerical evaluations).

NUMERICAL RESULTS AND DISCUSSION

The device parameters of ITER used in the numerical evaluation of the radiation temperature (1) are:[1] major radius $8.1m$, minor radius $3.0m$; (central) toroidal field $5.7T$, for which the EC frequency is $f_c(0) = 159.6GHz$; central density $10^{20}m^{-3}$, so that $(\omega_p(0)/\omega_c(0))^2 = 0.315$, and central temperatures in the range 15 to $25keV$. As for the profiles used as input to the calculations, along with the vacuum toroidal field profile,

Diagnostics for Experimental Thermonuclear Fusion Reactors
Edited by P. E. Stott *et al.*, Plenum Press, New York, 1996

193

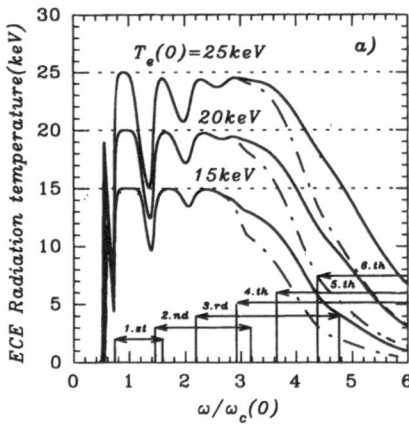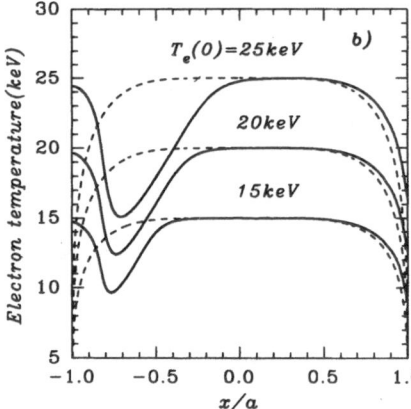

Fig.1: a) Spectra of the radiation temperature of the perpendicular ($N_\parallel = 0$) **O–mode**, for $T_e(0) = 15, 20, 25 keV$, both with (continuous curves) and without (dot–dashed curves) wall–reflections. Trapezoidal–like profiles for both T_e and n_e are used as input to the calculations. The different frequency ranges are determined on the basis of the cold EC resonance

$$\omega = n\omega_c(x) = n\omega_c(0)/(1 + \epsilon\frac{x}{a}), \quad \epsilon = 0.37, \quad -a \leq x \leq a.$$

b) Temperature spatial profiles deduced from the radiation temperature of the **1.st** harmonic O–mode, on using the cold EC resonance, i.e., $x/a = (1/\epsilon)(\omega_c(0)/\omega - 1)$, $0.73 \leq \omega/\omega_c(0) \leq 1.59$. The model T_e profile is also indicated (dashed curves).

flat $\left(\sim [1 - (x/a)^5]^{0.25}\right)$ profiles, to be referred to as trapezoidal–like profiles, for both T_e and density (n_e), as well as peaked $\left(\sim [1 - (x/a)^2]^\beta\right)$ profiles, with $\beta = 3/2$ for T_e, to be referred to as (3/2)–profile, and $\beta = 1$, i.e., parabolic profile, for n_e, are considered.

The radiation temperature of the **O–mode** is shown in Fig.1a as a function of frequency (normalized to the central EC frequency $\omega_c(0)$), for observation along a major radius ($N_\parallel = 0$), up to the 6.th harmonic.

It appears that i) a narrow feature occurs at frequencies downshifted w.r.t. the 1.st harmonic (cold) resonance at the outboard edge of the plasma, $\omega_c(x = a) \simeq 0.73\omega_c(0)$; ii) a feature with a plateau–like maximum occurs throughout the low–frequency part of the 1.st harmonic frequency range, with $T_{rad}(\omega) = T_e(0)$ in the plateau, i.e., the corresponding emission is that of a blackbody at temperature $T_e(0)$, along with a pronounced dip in the frequency range where the 1.st and 2.nd harmonic overlap; iii) in the 2.nd harmonic frequency range, $T_{rad} \lesssim T_e(0)$, for $T_e(0) \geq 20keV$, with a pronounced dip due to the re–absorption at the downshifted 3.rd harmonic; iv) the high frequency part of the spectra is characterized by a continuum, for which $T_{rad}(\omega) < T_e(0)$ and enhanced by the effect of wall-reflections.

The diagnostic potential of an ECE spectrum can be assessed on examining the temperature spatial profile obtained in correspondence to a chosen frequency range of the spectrum. In Fig.1b the T_e profiles deduced from the 1.st harmonic portion of the ECE spectra of Fig.1a are shown: it appears that the deduced profiles match up the corresponding model profiles except in the inboard side of the plasma cross–section where harmonic overlap with the 2.nd harmonic is substantial. Note that the outward shift of the deduced profile relative to the model profile on the outboard side is an artifact due to the use of the cold EC resonance. On considering, instead, the 2.nd harmonic part of the ECE spectrum of the O–mode, the deduced T_e profile is found to deviate significantly from the model profile as a result of the substantial overlap with (mainly) the 3.rd harmonic over most of the plasma cross–section, which precludes the utilization of the 2.nd harmonic O–mode for T_e measurements.

To ascertain the diagnostic potential of the 1.st harmonic O–mode, it is important to evaluate the **spatial resolution** of the envisaged T_e measurements. Referring to the G–function required for the evaluation of the radiative temperature, cf. Eq.(1), from Fig.2a it appears that such a function is a narrow, Gaussian–like function for the frequency range for which the deduced T_e profile reproduces the model profile. The spatial resolution of T_e measurements is just the spatial width of the G– function, i.e., $\Delta x(\omega, N_\parallel) \equiv \int_{-a}^{a} dx G(\omega, N_\parallel, x)/G(\omega, N_\parallel, x = x_{max})$, where x_{max} denotes the maximum of the G–function. Δx, normalized to the minor radius, is shown in Fig.2b for core plasma measurements.

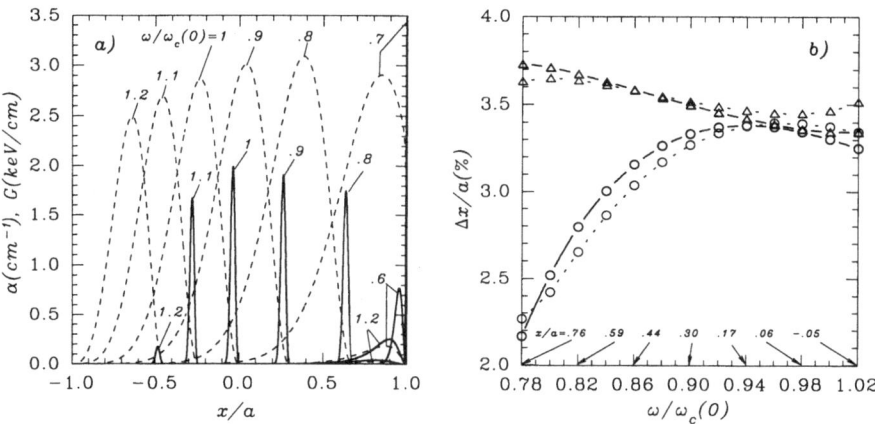

Fig.2: a) The spatial profile of both the absorption coefficient $\alpha(\omega, x)$ (dashed curves) and the function $G(\omega, x)$ (continuous curves), cf. Eq.(1), for the perpendicular O–mode in the 1.st harmonic frequency range, for trapezoidal–like model profiles and $T_e(0) = 20 keV$. Note that for $\omega > \omega_c(0)$, in addition to a narrow Gaussian–like component on the inboard side, the G–function comprises a broad component on the outboard side which accounts for ECE from superthermal electrons at the downshifted 2.nd harmonic.
b) Spatial resolution of core plasma T_e measurements based on ECE of 1.st harmonic O–mode, for both perpendicular ($N_\parallel = 0$, continuous curves) and oblique ($N_\parallel = 0.17$, i.e., $\theta \equiv \angle(\mathbf{k}, \mathbf{B_o}) = 80°$, dotted curves) observation, for trapezoidal–like (triangles) as well as 3/2–profile for T_e and parabolic profile for n_e (circles). $T_e(0) = 20 keV$. For trapezoidal–like profiles, it is $\Delta x/a \approx 3.5\%$, i.e., $\Delta x \approx 10.5 cm$, which satisfies the ITER target resolution for core plasma measurements.[1]

Numerical calculations of ECE spectra have been performed on varying the input T_e and n_e profiles. A sample of this analysis is shown in Fig.3, and other cases can be found in Ref.s 2 and 3.

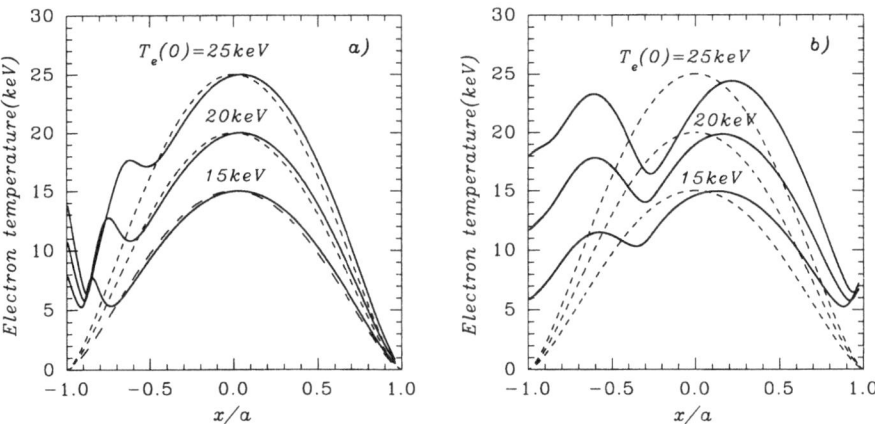

Fig.3: a) Temperature profiles deduced from the 1.st harmonic ECE of the perpendicular O–mode, for parabolic n_e profile and (3/2)–profile (dashed line) for model T_e. With respect to the case of trapezoidal–like model profiles shown in Fig.1b, the deduced and model T_e profiles match up over a

more extended part of the plasma cross–section, the spatial resolution of T_e measurements being also superior, as it appears from Fig. 2b.
b) Same as Fig.3a on using the **2.nd** harmonic O–mode. The strong harmonic overlap prevents T_e measurements.

As for the **X–mode**, ECE spectra are shown in Fig.4a for trapezoidal–like profiles as input to the calculations,and the T_e profiles deduced from the 2.nd harmonic frequency range of the ECE spectra are shown in Fig.4b.

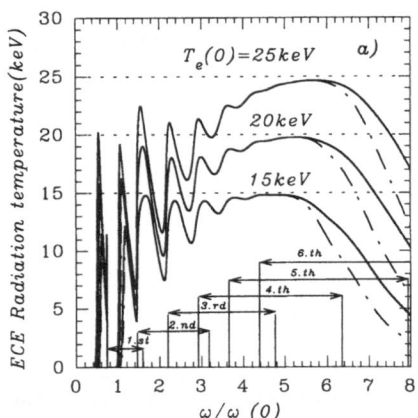 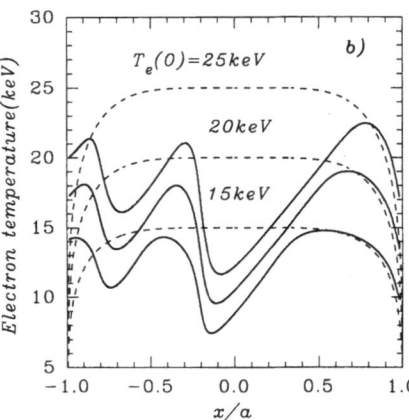

Fig.4: a) Same as Fig.1a for the **X–mode**.
b) The reconstructed T_e profiles (continuous curves) derived from the **2.nd** harmonic frequency range of the ECE spectra of the X–mode; the dashed curves refer to the model T_e profile.

From Fig.4b it appears that the diagnostic potential of the 2.nd harmonic X–mode is restricted to the outboard plasma region,[2] the core plasma being not accessible due to the strong re–absorption at the downshifted 3.rd harmonic, cf. the pronounced dip around $\omega = 2\omega_c(0)$ in the ECE spectra of Fig.4a. For parabolic model profiles for both n_e and T_e, the feasibility of T_e measurements by means of the 2.nd harmonic X–mode improves somewhat,[3] being still inferior to the 1.st harmonic O–mode.

In addition to the results given here, we have carried out a detailed comparative analysis of the ECE spectra of the O and X–mode which leads to the conclusion that the **1.st harmonic O–mode** is the most suitable choice for localized T_e measurements in ITER, confirming earlier investigations.[2,3] Concerning the density limit for propagation in the O–mode, the cold plasma cut–off for the fundamental perpendicular O–mode gives a limit of $3.2 \times 10^{20} m^{-3}$, for a central toroidal field of 5.7T. Such a limit is somewhat increased by relativistic effects,[4] the enhancement in the density cut–off for the O–mode being $\Delta n_e/n_e \approx 4.9 \times 10^{-3} T_e(keV)$, approximately, so that for $T_e(0) = 20 keV$ the density limit is increased to $3.5 \times 10^{20} m^{-3}$. Finally, for the parameters considered here, $[\omega_p(0)/\omega_c(0)]^2 = 0.315$ and refraction effects are expected to be weak.

REFERENCES

1. A.E. Costley, *Diagnostic Requirements for ITER: the Potential Role for ECE and ECA Measurements*, in Proceed. 9th Joint Workshop on ECE and ECRH, Borrego Springs, California, January 22–26 (1995).
2. A.E. Costley, and D.V.Bartlett, Proceed. 8th Joint Workshop on ECE and ECRH, Gut Ising, Germany (1992) (Max Planck Institut für Plasmaphysik, Garching (1993), IPP III/186, Vol.1, p.159.
3. M. Sato, S. Ishida, and N. Isei, *J. Phys. Soc. Japan*, **62**:3106 (1993).
4. E. Mazzucato, *Phys. Fluids B*, **4**:3460 (1992); H. Bindslev, *Plasma Phys. Controlled Fusion*, **34**:1601 (1992).

INSTRUMENTATION OF ECE FOR ITER

H.J. Hartfuss

Max-Planck-Institut für Plasmaphysik, EURATOM Ass.,
85748 Garching, Germany

and the Microwave Diagnostic Working Group of the EU Home Team

INTRODUCTION

Simulations of the electron cyclotron emission spectra have shown that the ITER requirements on spatial resolution of electron temperature measurements (0.1 m) in the plasma core can be met by ECE diagnostics measuring the cyclotron emission along a major radius in the horizontal midplane of the machine.[1] The parameters used in these simulations are 20 keV and 1×10^{20} m^{-3} central electron temperature and density, central field $B_0 = 5.7$ T, and major and minor radii $R_0 = 8.1$ m and $a = 3.0$ m.

The simulations showed that the emission spectrum extends from about 100 GHz into the 1000 GHz range of frequencies. The maximum intensity is emitted around 800 GHz. A great part of the spectrum is governed by harmonic overlap and relativistically down-shifted emission. However, the radiation temperature of the first harmonic emission in ordinary mode (O1) and second harmonic emission in extraordinary mode (E2) polarization can directly be related to the electron temperature in the emission volume considered, at least for parts of the temperature profile.

The paper first describes the requirements in terms of the spectral ranges to be measured and the spectral resolution necessary to meet the fundamental spatial resolution determined by ECE physics. It then discusses the applicability of the spectrometers in use on existing fusion experiments to ITER. Both quasioptical instruments and heterodyne radiometers are being considered. Radiation collection and transmission will be discussed next. Horn antennas and oversized waveguides are considered as well as quasioptical collection and transport systems. A typical sightline is being discussed in more detail on the basis of Gaussian beam optics.

To make optimum use of the measurements a hierarchy of calibration and stability control measurements is being outlined and the proposal of instruments summarized.

Diagnostics for Experimental Thermonuclear Fusion Reactors
Edited by P. E. Stott *et al.*, Plenum Press, New York, 1996

DIAGNOSTIC REQUIREMENTS

The spectral range of electron cyclotron emission lies between 100 and about 1200 GHz at a central temperature and density of about 20 keV and 1×10^{20} m^{-3} respectively, with the upper limit strongly depending on the intensity of the high cyclotron harmonics which is determined by both the electron temperature and density.

For an overview of the emitted spectrum, for comparison with simulations and to verify that the spectra are thermal, the total spectrum needs to be measured. A resolution of about 5 to 10 GHz seems appropriate. However, electron temperature information can only be deduced unambiguously from the measurement of the O1- and E2-emission. Their maximum spectral ranges are determined by the central B-field and the aspect ratio of the machine. It turns out that there is only little harmonic overlap of the first harmonic emission by the second one at the high field side plasma edge, but that there is strong overlap of the second harmonic emission by the third.[1,2] Because the plasma is optically thick at the third harmonic for the projected ITER temperatures and densities, the second harmonic emission is completely reabsorbed for frequencies above 360 GHz. So most of the temperature profile is accessible by O1 emission (if density cut-off allows, $n_e < 2 \times 10^{20}$ m^{-3}) but less than one half (R > 7.5–8 m) of the temperature profile is accessible via E2 emission. The spectral ranges of interest for electron temperature evaluation are therefore 120–240 GHz in O-mode polarization and 240–360 GHz for the E-mode.

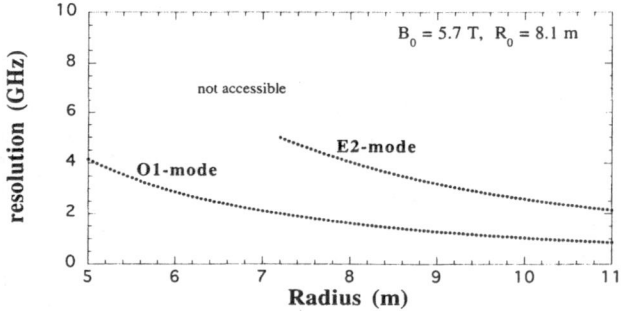

Figure 1. The estimated instrumental resolution as a function of location for spatial resolution of 0.1 and 0.08 m for O1 and E2 emission respectively.

The spatial resolution which can be reached at the projected ITER parameters is 0.1 and 0.13 m using the O1 emission and 0.08 and 0.1m using the E2 emission for 10 and 20 keV central temperatures respectively as determined from the simulations referenced before. The resolution can be worse if the viewing divergence angle in toroidal direction is larger than 5 degrees due to increased Doppler broadening. To reach these values the frequency resolution of the ECE spectrometers must be appropriate. For an estimate, the df/dR dependence of the emission frequency on the location has been used to calculate the corresponding frequency resolution for dR = 0.1 m and 0.08 m for O1 and E2 emission. Figure 1 gives the results for the frequency ranges accessible. The frequency resolution should not be worse than 1–4 GHz for the O1 spectrometers and 2–5 GHz for those measuring the E2 emission depending on the spectral (radial) ranges of interest.

INSTRUMENTATION

A number of different instruments are in use at existing fusion experiments. They can be divided into two categories, quasioptical instruments and classical microwave heterodyne receivers. Quasioptical instruments are the Michelson interferometer, the grating polychromator and the Fabry-Perot filter. The radiation detection is accomplished by wideband detectors in the input band of frequencies. Heterodyne receivers are categorized by their bandwidth and the radio frequency (RF) sidebands they are sensitive to. Double sideband (DSB) types with low and narrow intermediate frequency (IF) and single sideband (SSB) type instruments with wideband IF and filterbank are distinguished. Radiation detection occurs at the intermediate frequency using diode detectors.

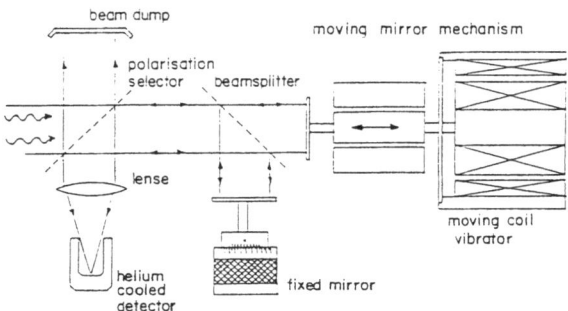

Figure 2. Simplified Martin-Puplett type arrangement of a Michelson interferometer.[3]

Michelson interferometers were first used in infrared spectroscopy[4] and are now widely used in plasma diagnostics[5]. An interferogram is generated by moving a mirror over a distance large compared to the average wavelength of the spectrum to be resolved. The mirror is electromechanically or pneumatically driven. A total scan lasts typically 10 ms. A reference of constant path length and the beam reflected at the moving mirror interfere at the detector. The spectral intensity is gained by Fourier transform of the interferogram. In this way the spectrum is measured continuously with a resolution depending on the maximum path difference of the interfering beams. Typical values are 4–10 GHz in the range 100 to 1000 GHz. The sensitivity is mainly determined by the detector responsivity which is mostly a He–cooled InSb crystal. The overall sensitivity results in a noise equivalent temperature of the order of 10 eV.

Michelson interferometers rely on a well developed technology which is unlikely to evolve radically on the time scale of ITER. Time resolution may be increased by a factor of 3 to 10 by mechanical refinement: rotating helicoidal rather than oscillating reflectors.[6] Other characteristics will probably not change too much. However characteristics of existing devices already meet some of the ITER requirements. Absolute calibration is possible using large area blackbody sources and averaging the interferograms over a few hours. The system responsivity is stable on long term periods.

One or two Michelson interferometers are therefore an obvious choice for overview measurements of the whole ECE spectrum and for T_e profiles. Because the spectral intensity in ITER is very high, peaking at the high cyclotron harmonics in the 800 to 1000 GHz range of frequencies, it might be necessary to use different instruments, optimized for different spectral coverage, for overview and T_e profile measurements.

Besides two-beam interferometers, multibeam types like the Fabry-Perot can be used to realize single frequency filters with high resolution to measure the spectral intensity as a function of time. Unfortunately transmission occurs also in higher orders which makes octave-band filtering in front of the instrument necessary for unambiguous measurements. These limitations of Fabry-Perot interferometers are such that they do not need further detailed consideration. One spectral channel per waveguide, filtering of higher orders and in addition wing leakage due to the Lorentzian instrument transmission function are severe handicaps for an application to ITER.

Measurement of the temperature at fixed radii with high temporal resolution is of great importance for any kind of fluctuation and transient phenomena analysis. A quasioptical multichannel instrument fulfilling this demand is the grating polychromator. Continuous and simultaneous electron temperature measurements are possible at a number of radial channels. Its time resolution is determined by the detectors used and is typically of the order of 1 μs. An Echelette grating is often used as the dispersive element, He-cooled InSb crystals as radiation detectors.[7] The resolution and the sensitivity obtained are similar to those of Michelson interferometers. Long term stability is comparable as well. Grating polychromators can be absolutely calibrated. The problems with this kind of instrument are similar to those of the Fabry-Perot interferometers. Low pass filtering is necessary.

Grating polychromators could certainly work at the required frequencies with adequate spectral coverage. However, the well known problems of stray light and sensitivity to higher diffraction orders will be much more severe on ITER than on the present tokamaks. It appears very likely that they will be entirely superseded by heterodyne instruments.

The good properties of microwave heterodyne receivers concerning frequency resolution and sensitivity are now extending into the sub-mm range. Heterodyne radiometers especially of the broadband multichannel type[8] have proven their value for ECE diagnostics on a number of current fusion devices.[5] Figure 3 gives the block diagrams of the main components of the types in use.

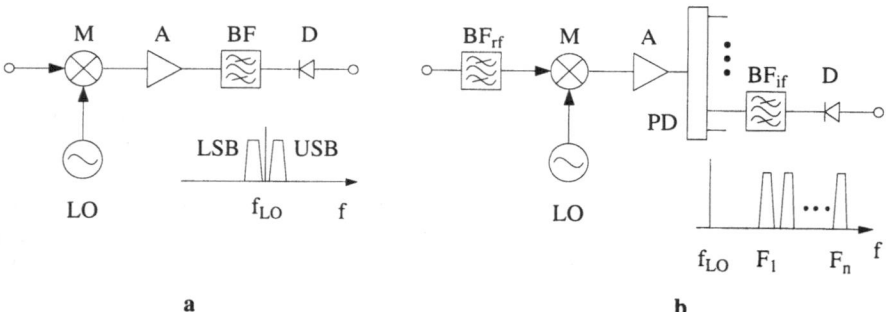

Figure 3. Block diagram of a narrowband double sideband receiver (a) and a broadband multichannel receiver (b). M= mixer, LO=local oscillator, A= IF-amplifier, BF=bandfilter, D=detector, PD=power divider or diplexer.

The input radio frequency (RF) signal is mixed with a local oscillator (LO) signal in a mixer where it is downconverted into the IF frequency range, further amplified, filtered and detected. In this way both high frequency resolution determined by the predection IF filter bandwidth and high time resolution determined by postdetection bandwidth are accomplished. Without RF filtering the instrument is sensitive to both the upper (USB) and the lower (LSB) sidebands corresponding to sum- and difference-frequencies of the IF and LO signals. They are called DSB receivers. SSB receivers are those where an RF filter suppresses one of the sidebands, typically the lower one. The mixer is sensitive then only to the upper sideband. Technical development made very wideband mixers possible with an instantaneous RF and IF bandwidth of several 10 GHz. With these mixers multichannel radiometers became possible which use a filterbank in the broad IF band for the simultaneous measurement of the intensity at a number of discrete frequencies similar to a grating polychromator but with much higher selectivity and sensitivity and with negligible cross talk between the individual channels.

As shown in table 1, very braodband mixers are available up to frequencies of about 220 GHz, in near future to 250 GHz. This is just the frequency range of the O1 emission of ITER. Five mixers can cover the whole range of interest. The situation is worse for the E2 emission range 240–360 GHz. At the moment only narrowband DSB mixers are available in this band. About 15 to 20 are necessary to cover the undisturbed E2 emission range with a sufficient amount of channels.

Table 1. Instantaneous input and output bandwidths of broadband mixers

RF-range (GHz)	IF-range (GHz)
100–120	6–26
120–150	6–36
150–180	6–36
180–220	10–50
210–250	10–50 (near future)

To complement Michelson interferometers, heterodyne radiometers seem to be the best choice for detailed transient T_e studies on ITER. They can provide continuous and simultaneous measurements at a large number of radial locations with the necessary spatial and temporal resolution. The sensitivity is high. Calibration with blackbody sources of a few hundred Kelvin can be done within relatively short times.

First and second harmonic measurements with heterodyne receivers are feasable with current technology. Substantial technical improvements as expected on the timescale of ITER may further reduce the number of individual mixers. But this development is not essential to this proposal based on an extensive use of heterodyne receivers.

RADIATION COLLECTION AND TRANSMISSION

The large distances from the diagnostic flange to the plasma center (> 3 m) and the large plasma diameter (> 6 m) call for antennas for radiation collection with minimum divergence angles. Due to their diffraction limited operation it makes large apertures necessary and structures of some depth as well. These collectors can only be mounted in the plug of the diagnostic ports. Antennas small enough to be located in the blanket structure are unlikely to meet the antenna pattern requirements. Only sightlines from the low field side are therefore possible and are being discussed in this paper. The value of others than horizontal viewing lines is discussed elsewhere.[1] As turned out, the most important ones are indeed those viewing the plasma along a major radius. Sightlines in the horizontal midplane as well as those above and below it are of equal importance. The access of all these sightlines is similar. Therefore only the midplane sightline is being discussed in more detail.

Three different designs are possible: 1. oversized waveguide horns, feeding directly into oversized waveguides. 2. Non-imaging concentrators and 3. Gaussian beam collection optics.

1. Oversized horns are simple and robust. For ITER horns with low beam divergence are necessary. Good values of divergence are of the order of $\pm$ 2–3 degrees, resulting in a spot size in the plasma center of about 0.3 x 0.3 m^2 . Small sidelobes are unavoidable. Oversized transmission lines coupled to these horns could go through the plug. Doglegs with mitre bends are necessary to provide sufficient shielding. The part between vessel and cryostat can be overcome with straight sections and a periscope-like part to compensate for mechanical movements. Because of the large size of the ports, a large number of viewing lines can be accommodated.

Although it looks simple, experiments indicate that it may not be possible to achieve an antenna performance providing sufficient spatial resolution in toroidal and poloidal direction to fulfill the ITER requirements.

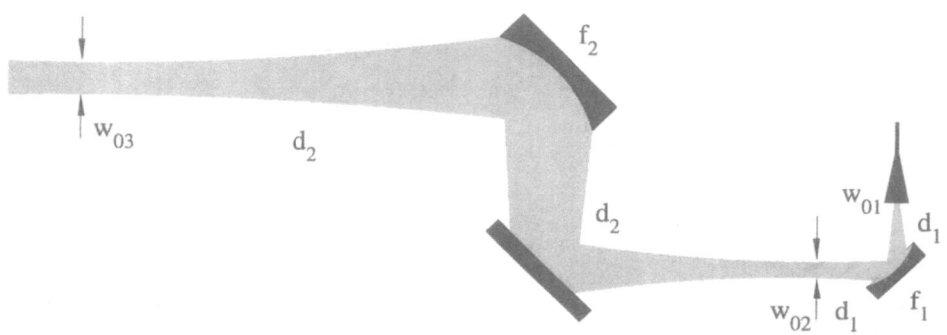

Figure 4. Gauss beam telescope. Two mirrors with focal lengths f_1 and f_2 separated by $d_1 + d_2$; $f_1 = d_1$, $f_2 = d_2$. In this case beam waist location and dimension w_{03} is independent of wavelength.

2. Non-imaging concentrators[9] could substitute the horn antennas as radiation collectors. The principle is to collect radiation by an oversized waveguide and to limit the collection angle by a quasioptical system. Small divergence angles of $\pm$ 2 degrees could be realized resulting in a beam FWHM of 0.14 m at a 4 m distance. The overall length is 0.45 m. With the concentrators efficient and robust coupling to the oversized waveguides can be achieved. Because of their well defined and narrow antenna pattern and the good coupling properties into an oversized waveguide as transmisson line to the spectrometers, non-imaging concentrators might be a suitable choice for wideband measurements with the Michelson interferometers.

3. Gauss beam collection optics probably offer the best means of achieving suitable antenna pattern control. A collection and transmission section from the plasma through the cryostat and the shield are briefly sketched in the following.

In a most simple arrangement the divergent beam of a corrugated feedhorn which has to a good approximation a Gaussian field amplitude distribution transverse to the beam is focused with an elliptical off-axis mirror to a beam through the plasma. This arrangement is only possible if the spectral bandwidth is clearly below one octave. For higher bandwidths the Gauss beam telescope offers a possibility to form beams whose dimensions are to first order independent of wavelength. The basic arrangement is given in Figure 4. Two focusing mirrors are placed at a distance equal to the sum of their focal lengths and the waist of the input beam is located in focal length distance of the first. The location of the waist of the output beam and its dimension is then independent of the wavelength. The ratio of output to input beam waist is determined by the ratio of the focal lengths of the two focussing mirrors.[10]

For an ITER sightline the focal length of the plasma facing mirror is chosen equal to the distance to the plasma center which is about 3.6 m, if it is mounted in the plug about 0.5 m behind the blanket structure. To avoid unnecessary large mirrors the beam waist at the plasma center is chosen to be $\cong$ 0.06 m. This determines the focal length of the second mirror via the size of the waist of the feedhorn to about 0.5 m. To overcome the whole distance through plug, cryostat and shield, the two mirrors with focal lengths as determined are not enough. Additional telescopes can be added to match the geometrical requirements. We introduce only one additional mirror with the same focal length as the plasma facing one to overcome the distance. Figure 5 gives the resulting design. The wavelength dependence introduced by this measure can be compensated by choosing feed antennas with different beam waists which handle the coupling to the waveguide section outside the cryostat. The resulting beam diameters as function of distance for feedhorns with different waists between 5 and 8 mm at frequencies between 350 and 90 GHz are given in figures 6a and 6b.

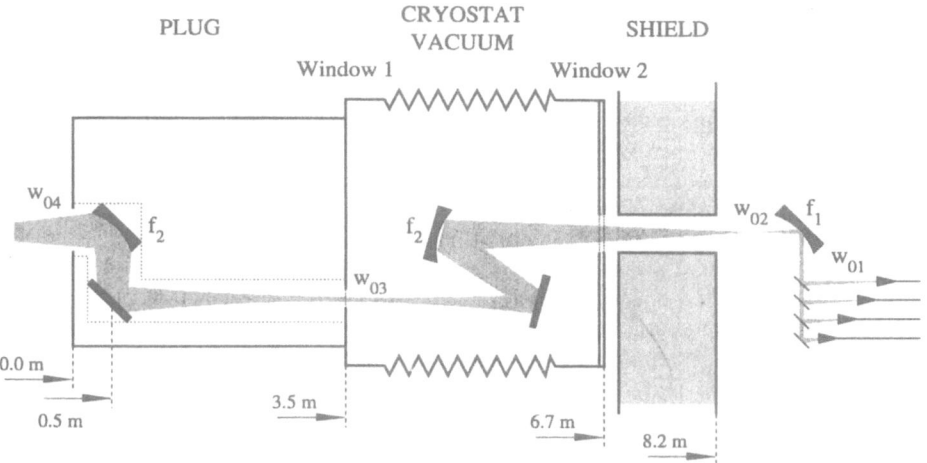

Figure 5. Gaussian beam collection and transmission optics through plug, cryostat and shield. f_2 = 3.6 m, f_1 = 0.5 m, w_{01} = 2–8 mm depending on frequency between 90 and 350 GHz. The beam waist w_{04} in the plasma center is 0.06 m. (Only dimensions in radial direction to scale.)

To use frequency dependent beam waists of the feedhorns is not a severe restriction because corrugated antennas always have finite bandwidth. In addition the heterodyne radiometers to be coupled through these horns have also restricted bandwidth which makes more than one antenna necessary.

A number of beam splitters must be introduced with integrated high pass filters to separate the spectral band segments as sketched in figure 5. For wideband use in combination with Michelsons, the solution with an additional telescope must be chosen. Direct coupling of the Gauss beam to oversized waveguides is then preferable and seems feasable. It seems advantageous to make the whole beamline monomode.

Two windows are necessary one at the outside of the plug and one to seal the secondary vacuum. The first one is most important because it has to withstand 5 bar. Probably wedged or tilted quartz windows can fulfill the requirements on stability and bandwidth. Radiation exposure might due to the folded beam path be not a problem. Preferably the various beam waists are located close to the windows to keep their size minimum. The waist dimension w03 of figure 5 is between 17 and 63 mm for frequencies between 350 and 90 GHz.

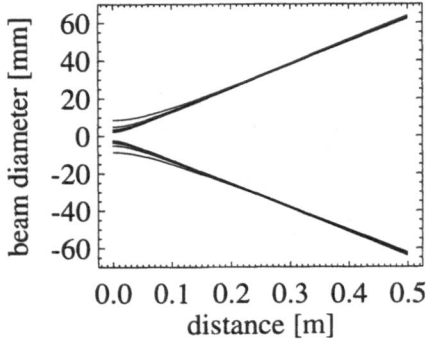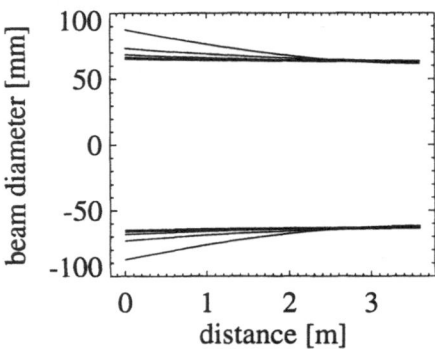

Figure 6. (a) Beam diameter from the various feedhorns to first mirror. Due to the different feedhorn waists chosen, the mirror is completely illuminated in the whole 90-350 GHz range of frequencies. (b) Beam diameter from the front of the plasma facing mirror to the plasma center. The different lines correspond to frequencies within 90 to 350 GHz.

For optimum beam quality the plasma facing mirror must have a sufficient size. The diameter is chosen four times the beam diameter resulting in almost 0.5 m which makes a hole of about 0.35 m in the plug necessary. Smaller diameters are of course possible resulting in larger beam diameters in the plasma.

Although the design discussed is only a crude estimate it shows that Gaussian beam collection optics is the best means to achieve a well controlled beam with low divergence. The mirrors can be fabricated of metal which makes the construction mechanically stable and robust. The folded beam path allows for shielded windows. The beam splitters need some extra consideration. But even if these are delicate constructions they are mounted outside the cryostat and the shield and can be adjusted and maintenanced easily. The mechanical movement has not been taken into account yet. But it seems, that if all optical components are mounted at the plug, during mechanical movements of the torus, the beam moves only outside the shield where readjustment is less problematic.

CALIBRATION

Absolut calibration of ECE spectrometers is difficult. But the experience shows that it can be done according to the accuracy (10%) as required by ITER. The procedure is to use a blackbody source of known physical temperature, to illuminate the plasma facing antenna and to measure the temperature sensitivity of the arrangement end-to-end. Attention must be focused on the linearity of the detectors and the supposition of blackness of the calibration source.

The main difficulty for the ITER ECE diagnostic is certainly that an end-to-end calibration can probably not be done in the radioactive environment in later operation phases of the machine. A three-level hierarchy of calibration measurements is therefore proposed.

During the time of commissioning of the machine an end-to-end calibration should be made using the standard large area blackbody sources in the empty vacuum vessel. This procedure should be used during any vacuum openings before the machine becomes too activated. After the activation remote handling of the calibration sources is necessary. If this is not feasable other methods must be applied. A blackbody source as close as possible to the torus should be provided. Figure 7 gives an idea how this could be realized. If the first mirror is made rotatable it could be switched into a position as given in the figure. It could then be illuminated by a heated blackbody source mounted just outside the plug. In this way the full transmission line is being included in the calibration with only two additional plane mirrors to form a dogleg in the plug because of shielding reasons. This allows to control and to check the calibration at any time. Another possibility could be to use the heated torus itself as a calibration source. A reference temperature then may be realized by a cooled plate just opposite the beams.

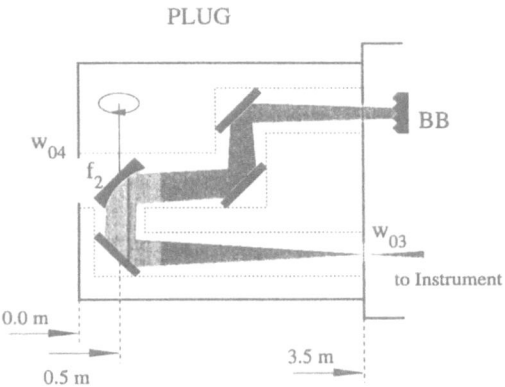

Figure 7. Front end of the transmission line. A rotatable first mirror allows end-to-end calibration under almost unchanged diagnostic conditions. BB=blackbody calibration source.

Finally, calibration stability checks with more sources close to the instruments are required. This is especially necessary for the heterodyne radiometers because their IF amplifiers are not feedback–controlled and are therefore temperature dependent. On this basis a strategy for making optimum use of the measurements combining them into a full system calibration procedure which is periodically updated can be developed.

SUMMARY

The cyclotron emission spectrum between 100 and 1200 GHz expected for ITER can be registered with existing techniques. Best suited are the well developed Michelson interferometers with adaption either for overview measurements or for T_e profile evaluation from first and second harmonic emission. Both total spectrum and profile information are available at least every 10 ms, meeting the ITER requirements on time resolution of electron temperature measurements. Michelsons using an oscillating helicoidal mirror could provide even better time resolution. For detailed studies of transient phenomena, simultaneous and continuous measurements of the temperature at a number of radial positions is necessary.[11] Heterodyne radiometers of the broadband SSB type are the best choice to cover the spectral range of the first harmonic O-mode emission, by which most of the temperature profile information is accessible for a wide range of ITER parameters. They offer the spatial and temporal resolutions required. The spectral range of second harmonic emission can only be covered with narrowband DSB mixers. A number of 15–20 is necessary. Technical improvements expected may reduce this number significantly.

Well controlled antenna patterns are best realized applying Gaussian beam optics. Simple estimates with imaging on the basis of beam telescopes showed that with the dimensions given well collimated narrow sightlines through the plasma are basically possible for the horizontal sightlines from the low field side. Only those are possible due to the geometrical constraints of ITER. All focusing elements can be made of metal resulting in high beam quality and low losses (compared to dielectric lenses). Folded beams allow for shielded windows. For more detailed studies of the optical collection and transmission system full wave treatment of the beam path is necessary.[12] It seems possible to design beamlines resulting in spot sizes in the plasma of a few cm^2 suited even for temperature fluctuation measurements which call for high poloidal resolution to avoid extensive averaging of the fluctuation wavenumber spectrum.[13]

Due to the size of ITER and its shielding extremely long transmission lines to the spectrometers are necessary. Oversized waveguides with mitre bends are a possible choice. Careful adjustment is necessary to avoid mode conversion at the bends and the transitions. Again quasioptical systems like the one proposed may give transmission with lowest attenuation. The whole beam path could be enclosed in a pipe under vacuum or filled with dry air to avoid water vapour absorption.

The accuracy of electron temperature measurements required can be reached on the basis of a strategy making optimum use of system calibarations either end-to-end or of the individual components depending on the status of the machine. If blackbody calibration sources can be mounted near the machine as sketched in the paper, overall calibration is possible at any time.

REFERENCES

1. D.V. Bartlett, Physics issues of ECE and ECA for ITER, *These Proceedings*
2. A.J.H. Donne and B.C. Schokker, Advantages and limitations of microwave diagnostics in ITER, *These Proceedings*
3. T.C. Hsu et al., A novel fast scan Michelson interferometer for ECE diagnostic applications on Alcator C-MOD, in: *Proc. of Joint EC-7 and IAEA Techn. Comm. Meeting,* Hofei, China (1989)
4. D.H. Martin, E. Puplett, Polarised interferometric spectroscopy for the millimetre and submillimetre spectrum, *Infrared Phys.* 10:105 (1959)
5. D.V. Bartlett et al., Overview of JET ECE measurements, in:*Proc. of EC-6,* Oxford, UK (1987)
6. P. Buratti, M. Zerbini, Fourier transform spectrometers employing rotating helicoidal reflectors, in: *Proc. of EC-9,* Borrego Springs, USA (1995)
7. A. Cavallo et al., Twenty-channel grating polychromator for millimeter wave plasma emission measurements, *Rev. Sci. Instrum.* 59:889 (1988)
8. H.J. Hartfuss, M. Tutter, Fast multichannel heterodyne radiometer for electron cyclotron emission measurements, *Rev. Sci. Instrum.* 56:1703 (1985)
9. P. Buratti, M. Zerbini, Non-imaging Concentrators, *These Proceedings*
10. P.F. Goldsmith, Quasi-optical techniques at millimeter and submillimeter wavelengths, in: *Infrared and Millimeter Waves Vol. 6,* K.J. Button, ed., Academic Press, New York (1982)
11. N.J. Lopes-Cardozo, Perturbative transport studies in fusion plasmas, *Plasma Phys. Control. Fusion* 37:799 (1995)
12. T.C. Hsu et al., Quasioptical transmission system for ECE measurement on Alcator C-Mod, in: *Proc of EC-8,,IPP report III/186 Vol.2,* Gut Ising, Germany (1993)
13. S. Sattler and H.J. Hartfuss, Experimental evidence for electron temperature fluctuations in the core plasma of W7-AS stellarator, *Phys. Rev. Lett.,* 72:653 (1994)

ROTATING REFLECTOR SPECTROMETERS FOR ECE DIAGNOSTIC IN LARGE FUSION DEVICES

P. Buratti and M. Zerbini

Associazione EURATOM-ENEA sulla Fusione
Centro Ricerche, Frascati, CP 65, 00044 Frascati, Rome, Italy

INTRODUCTION

Fourier transform spectrometers (FTS) give a unique possibility to perform a spectral overview of electron cyclotron emission (ECE) by a single, absolutely calibrated instrument; furthermore, detailed measurements of the electron temperature profile can be obtained, provided that the spectral resolving power is adequately large. Since the latter is proportional to the maximum path difference introduced in the two-beam interferometer which multiplexes the spectrum, the time resolution for a given spatial resolution is limited by the time taken to scan the path difference.

The introduction of path difference scanning devices based on rotating helicoidal reflectors[1] allowed a significant improvement of the time resolution, in fact the prototype developed for the FTU tokamak was able to scan a path difference of 4 cm in a minimum time of 1.2 ms. The prototype also demonstrated other advantages of rotating helicoids, in particular the high reliability (no maintenance was required in five years of routine operation), and the possibility to sample the interferograms at constant time intervals, due to the fact that the path difference is scanned linearly in time.

The prototype helicoidal reflector was optimised for high scan speed; recently a new modulator with increased size has been realised[2] in order to optimise the bandwidth and the throughput of the FTS and to perform four-channel simultaneous measurements with a single rotating modulator. A time resolution of 3 ms was easily obtained with the new instrument.

The limits on achievable spectral performance and time resolution depend on the non-planar shape of the helicoidal surface and on the strength of materials subjected to the centrifugal force respectively. Such limits have been translated into well established design constraints owing to the experience gained with the first rotating modulators.[2]

Two-beam interferometers measure the difference between the spectra entering at two input ports.[3] For the flat-mirrors polarising optical configuration adopted with the first helicoidal reflectors one of the input ports coincides with the detector port. Since cryogenic detectors have to be used to achieve an adequate sensitivity, a relatively large signal is measured with the main input at room temperature. This background signal is not a serious problem for the diagnostic installed on FTU, but it would become intolerable if the calibration signals were depressed by a long and complex transmission line. In a polarising interferometer the background spectrum can be suppressed if roof top mirrors that rotate the polarisation by 90° are used as reflectors.[3] For this reason a new helicoidal reflector which is equivalent to a roof top mirror has been developed.

In this paper the laboratory tests performed in order to demonstrate the feasibility of a FTS based on roof top rotating helicoidal reflector will be described, and the parameters of a rotating helicoid that could fit ITER requirements will be shown.

Diagnostics for Experimental Thermonuclear Fusion Reactors
Edited by P. E. Stott *et al.*, Plenum Press, New York, 1996

THE ROOF TOP MIRROR HELICOIDAL REFLECTOR

The helicoidal surface equivalent to a plane mirror (Fig.1) is determined by the requirement that the normal direction remains unchanged during rotation. In this case the normal to the surface is parallel to the optical axis and its inclination with respect to the rotation axis determines the path difference excursion Δ.[1] The other parameters of the reflector are the distance r_0 between the optical axis and the rotation axis, the radial half width ρ of the helicoidal surface, the number N of helicoidal sectors, and the rotation period τ (time resolution is $\delta\tau=\tau/N$). In the helicoidal surface equivalent to a roof top mirror the optical axis is placed as before, while the normal lies on the plane defined by the radial direction and by the optical axis, and is inclined with respect to the latter by 45° if $r<r_0$ and by -45° if $r>r_0$.

In order to perform the feasibility tests, one of the sectors of the four-channel FTS has been temporarily substituted by a helicoidal roof top reflector, while one of the interferometric channels has been properly reconfigured and equipped with a conventional roof top mirror as fixed reflector. The parameters of the four-channel FTS are $r_0=15$ cm, $\rho=1.5$ cm, $\Delta=4$ cm and N=4. The roof top sector has been realised by joining two separately machined and polished parts (extending from $r_0-\rho$ to r_0 and from r_0 to $r_0+\rho$ respectively). In this experimental configuration the rotating reflector was only operated at 150 ms rotation period due to the mechanical unbalance introduced by the roof top sector.

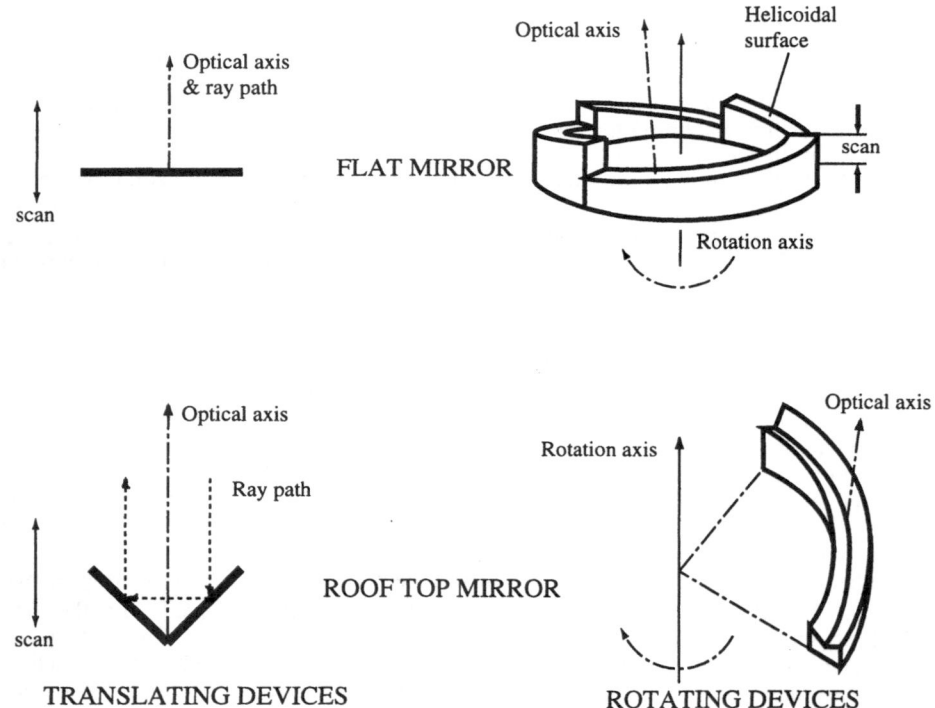

Figure 1. Illustration of rotating mirrors concepts. For the rotating helicoid equivalent to a flat mirror an assembly of four sectors is shown; during a complete rotation the surface remains perpendicular to the optical axis and the path difference is scanned four times. For the helicoidal sector equivalent to a roof top mirror a single sector is shown.

EXPERIMENTAL TESTS

The reflecting surface was well polished, so that it was possible to align carefully the interferometer using a HeNe laser; the wire grid polarizers were replaced for alignment with thin mylar films. After the alignment procedure the signals obtained from each arm of the interferometer with a high pressure mercury arc lamp turned out to be well balanced.

A first check of the FTS interferometric performance has been obtained by measuring the broadband fringe visibility of the instrument, defined as[1] $V=(I_{av}-I_{zpd})/I_{av}$, where I_{av} is the sum of the signals independently obtained from each arm, or equivalently the signal measured when the path difference is large, while I_{zpd} is the residual signal at zero-path-difference (zpd). The orientation of the output polarizer is chosen such that the interferogram has a minimum at the zpd position, so that V gives the incoming power fraction that the interferometer is able to multiplex ($V=1$ for an ideal interferometer). Besides common effects like imperfect alignment, helicoidal reflectors have a specific cause of imperfect modulation, in fact the reflecting surface intrinsically deviates from a planar one, and modulation becomes poor at wavelengths comparable to such deviation.

The high pressure mercury lamp has been used for these measurements; the interferogram baseline has been obtained by chopping the lamp radiation. A visibility $V=0.74$ has been measured with the lamp faced to the interferometer without any filtering, while, using a 300 GHz optical low pass filter to reduce the source bandwidth, the visibility increased to $V=0.9$. For the full-bandwidth measurement an upper frequency of about 1200 GHz can be assumed, because the detector sensitivity drops at higher frequencies.

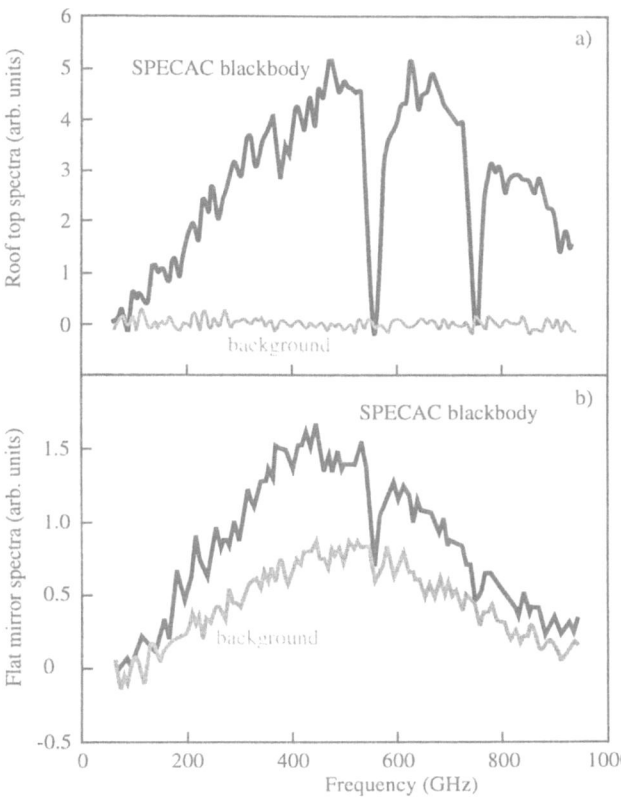

Figure 2. Comparison between the SPECAC hot source spectra (solid lines) and the corresponding background (shaded lines); (a) roof top rotating mirror FTS, 10000 scans average; (b) flat rotating mirror FTS, 100000 scans average.

These results qualitatively agree with the predictions[1] for a conventional helicoidal reflector with the same geometric parameters, in fact for the latter the interferometric modulation would be 98% at 300 GHz and 80% at 1200 GHz.

The background suppression has been checked by comparing the background spectrum with the one measured on a large-area 600 C black body source (SPECAC Ltd) used for the absolute calibration of the ECE diagnostic on FTU (Fig. 2a). Results obtained in calibration runs with a standard helicoidal reflector are shown for comparison in Fig. 2b: in this case the background signal amounts to about one half of the hot source signal.

Fig 2a also shows that the bandwidth of the FTS with roof top reflectors extends up to 1000 GHz with a standard InSb detector, so that an extension to 1500 GHz or more can be obtained introducing magnetically enhanced InSb detectors.

THE HELICOIDAL ROOF TOP REFLECTOR FOR ITER DIAGNOSTIC

Having demonstrated that the performances of roof top helicoidal reflectors are equivalent to the ones obtainable with conventional rotating reflectors, we can use the expressions given in Ref. 2 to determine the parameters of a modulator that could fit the requirements for ECE measurements on ITER, i.e. a radial space resolution of 8 cm, a wavelength range extending from 3 mm to 0.2 mm, and detector limited étendue. The results shown in Table 1 refer to a single FTS covering the whole spectral range; time resolution could be improved if two spectrometers with different parameters were employed.

Table 1. Parameters of the helicoidal reflector for ITER

Maximum path difference	Δ	10.5 cm
Average radius	r_0	21 cm
Radial half width	ρ	1.5 cm
Number of sectors	N	3
Attainable time resolution	$\delta\tau$	4 ms

REFERENCES

1. P. Buratti and M. Zerbini, A Fourier transform spectrometer with fast scanning capability for tokamak plasma diagnostic, to be published on Rev.Sci.Instrum (August 1995).
2. P. Buratti and M. Zerbini, Fourier transform spectrometers employing rotating helicoidal reflectors, in Proceedings of Ninth Joint Workshop on Electron Cyclotron Emission and Electron Cyclotron Resonance Heating, Borrego Springs (USA), January 1995.
3. D. H. Martin, in *Infrared and Millimeter Waves*, edited by K. J. Button (Academic, New York, 1982), Vol. **6**, p. 79.

MULTIMODE LIGHT COLLECTION SYSTEMS FOR ECE DIAGNOSTICS

P. Buratti and M. Zerbini

Associazione EURATOM-ENEA sulla Fusione
Centro Ricerche Frascati, CP 65, 00044 Frascati, Rome, Italy

INTRODUCTION

The spatial localisation required for ECE temperature measurements on ITER gives rise to two main constraints for the light collection system:[1] the antenna pattern width has to be smaller than $\Delta z = 15$ cm, and, in order to avoid excessive Doppler broadening, very little power has to be accepted outside the angular range $\pm \theta_B$, with $\theta_B = 5°$ in the toroidal direction.

The required antenna pattern characteristics can be achieved over a broad spectral range using fundamental Gaussian beam collection optics;[2] in this case the étendue E at a wavelength λ is limited by $E \leq \lambda^2$. This is not a real restriction as far as coherent detection is used, but for Fourier spectrometers the detector étendue may be $E_D \gg \lambda^2$, and its full exploitation (which is important to reduce the calibration time) requires to collect and transport a multimodal beam with a number of modes $N = E_D / \lambda^2$.

The use of light pipes for the transport of multimodal beams gives some advantages, namely low losses, small pipe size, and relative insensitivity to misalignment. Simple light pipes however do not allow to define a narrow field of view; for this reason hybrid systems, i.e. combinations of light pipes and optical elements have to be considered. This scheme has been successfully used for ECE diagnostics on the FTU tokamak employing imaging optical elements.[3] Substitution of the latter by non-imaging flux concentrators[4] may give significant improvements, in particular unwanted radiation can be turned back at the front end instead of being absorbed by a field stop; furthermore the use of concentrators allows to design robust systems that do not require any in-situ alignment.

In this paper the design principles for hybrid light collection systems will be outlined, and the main elements of a system which could fulfil ITER requirements will be described.

DESIGN PRINCIPLES

The front end of a hybrid light collection system is characterized by the aperture size a and by the acceptance angle θ_e; rays incident on the aperture are assumed to be transmitted to the detector if they form an angle $\theta < \theta_e$ with the optical axis and rejected if $\theta > \theta_e$. In this section we will give expressions for the antenna pattern width W and for the étendue E of the light collection system, and determine the ranges of a and θ_e values which are compatible with the following requirements: a) $W \leq \Delta z$; b) $\theta_e \leq \theta_B$; c) $E \geq E_D$.

The antenna pattern can be defined as the ratio between the power received from a point source and the one that would be received from the same source with a flat angular response, i.e. for $\theta_e = \pi/2$. A lower limit W_g for the half width at half maximum can be readily determined in the geometric optics limit: at large distances D from the aperture there is a

Diagnostics for Experimental Thermonuclear Fusion Reactors
Edited by P. E. Stott *et al.*, Plenum Press, New York, 1996

211

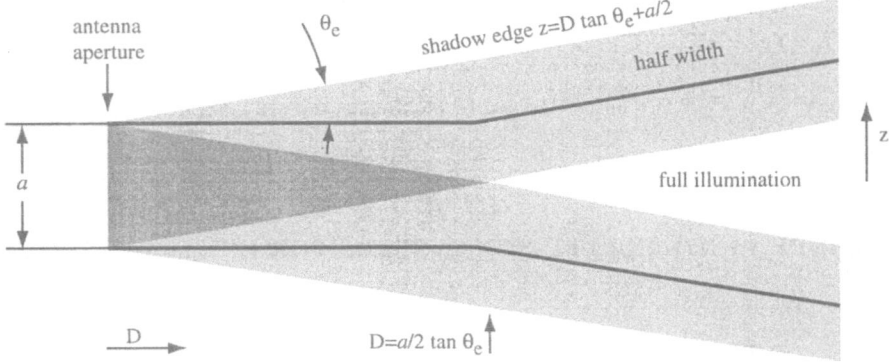

Figure 1 Geometric antenna pattern for an antenna aperture with diameter a and acceptance angle θ_e. The edge of the full illumination region is given by $|z|=D\tan\theta_e - a/2$. The illumination is linearly tapered in the shadow region.

shadow region limited by $|z|=D\tan\theta_e \pm a/2$ (Fig. 1), and $W_g=D\tan\theta_e$, while for $D<a/(2\tan\theta_e)$ the shadow region invades the boresight direction and $W_g=a/2$. In both cases we have:

$$W_g = \max\{a/2, D\tan\theta_e\}. \tag{1}$$

The real antenna pattern width will exceed the geometric one W_g in consequence of finite wavelength effects. The latter can be accounted for by assuming that the angular cutoff at $\theta=\theta_e$ is smoothed by diffraction at the input aperture;[3,5] in this way we obtain:

$$W = \left[W_g^2 + (D\lambda/a)^2\right]^{1/2}. \tag{2}$$

The étendue of a light collection system with revolution symmetry (in this case a is the diameter of the circular aperture) is given by:

$$E = \left(\pi a \sin\theta_e/2\right)^2. \tag{3}$$

Using equations (1)-(3), requirements a) and c) can be translated into:

$$\max\left\{\frac{2E_D^{1/2}}{\pi\sin\theta_e}, \frac{\lambda}{\left((\Delta z/D)^2 - \tan^2\theta_e\right)^{1/2}}\right\} \leq a \leq 2g\Delta z, \tag{4}$$

where $g=(0.5+0.5(1-(D\lambda)^2/(\Delta z)^4)^{1/2})^{1/2}$.

The optimum θ_e value is the one which minimizes the lower limit on the antenna size subject to requirement b); neglecting terms $O(\tan^2\theta_e)$ we obtain:

$$\tan\theta_e = \min\left\{\tan\theta_B, \frac{\Delta z}{D}\left(1 + \frac{\pi^2\lambda^2}{4E_D}\right)^{-1/2}\right\}. \tag{5}$$

For ITER we have $\theta_B = 5°$, $\Delta z = 15$ cm and $\lambda \leq 3$ mm; furthermore we assume $E_D = 0.4$ cm^{2}sr, which is a typical value for the étendue of InSb detectors, and D=3 m, i.e. the distance between the low field side antenna and the plasma center. The latter assumption is done in consideration of the fact that the antenna pattern width has little influence on spatial resolution on the high field side, where the temperature is nearly constant in the vertical direction. Substituting into equation (5) we find that the optimum acceptance angle for ITER is $\theta_e = 2.3°$, and the antenna diameter must be in the range $10 \leq a \leq 27$ cm.

CONTROL OF THE ACCEPTANCE ANGLE

In the light collection system realised for the FTU tokamak[3] the acceptance angle is determined by a lens-stop combination, while coupling with the oversized waveguide that transports collected radiation to the spectrometers is accomplished by a second lens confocal with the first one. This scheme allows to obtain good spatial resolution and efficient coupling with the oversized waveguide, but its application could be difficult in a hostile environment because absorbing foam is needed to realise the stop and to baffle stay light. For this reason a self-baffling system has been considered, in which the lens-stop combination is replaced by a nonimaging flux concentrator[4] which transmits rays incident at the entry aperture with inclination $\theta < \theta_e$ and reflects back rays with larger inclination.

The simplest flux concentrator is the compound parabolic concentrator (CPC). In a meridian section the CPC profile (Fig. 2) is a parabola having the focus at the edge of the exit aperture, the axis inclined by θ_e and the tangent at the entry aperture parallel to the optical axis. The 2D concentrator generated by translation of the profile in the direction perpendicular to the plane of Fig. 2 has an ideal (i.e. with a sharp cutoff) transmission-angle diagram; 3D concentrators, realised either by revolution around the optical axis (not around the parabola axis) or by combining two 2D CPCs at right angles (lower polarisation scrambling is to be expected in this case), approach very closely the ideal behaviour.[4] The ratio between the diameters of the exit and entry apertures of the CPC is $a'/a = sin\theta_e$, and the overall length is:

$$L_{CPC} = a(1 + sin\theta_e)/2\tan\theta_e. \tag{6}$$

For $\theta_e = 2.3°$ and a=10 cm the CPC is quite long, in fact $L_{CPC} = 130$ cm.

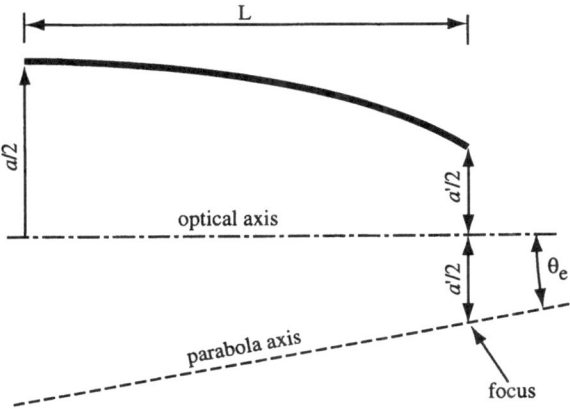

Figure 2 CPC profile drawn for $\theta_e = 30°$.

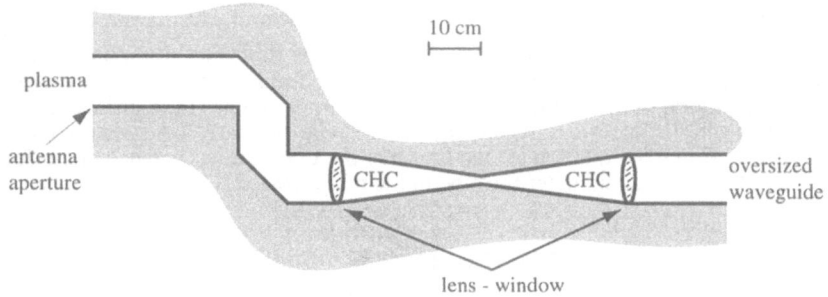

Figure 3. Schematic of a possible double-window hybrid light collection system for ITER; three sections of straight lightpipe with two mitre bends avoid direct plasma exposure of the first window.

The length of the concentrator can be reduced by introducing a lens at the entry aperture; this leads to the design of the compound hyperbolic concentrator (CHC), in which the mirror profile is a hyperbola with a focus placed as in the CPC and the other one at the position where extreme rays are focused. Lens aberrations can be corrected by modifying the mirror profile[5] provided that the lens focal ratio $f/a \geq 4$. The CHC length is given by

$$L_{CHC} = f \frac{1 + sin\,\theta_e}{1 + \dfrac{2f}{a} tan\,\theta_e};$$ (7)

Substituting θ_e=2.3°, a=10 cm and assuming f=4a, we obtain from (7) L_{CHC}=31 cm.

The need to introduce a lens at the entry aperture of the CHC can be an advantage, in fact, if the light collection system is properly designed, the lens can be used as a vacuum window; in this way the problem of etalon effect in the window is naturally solved.

The exit aperture is quite small for both CHC and CPC (a'=4 mm for the sample values used before), but a good matching with the oversized waveguide can be obtained by a second reversed concentrator (Fig. 3) which is the equivalent of a collimator.

CONCLUSIONS

Non-imaging concentrators can be used to realise light collection systems with narrow antenna pattern and large étendue as required for Fourier transform spectrometers. The requirements for ECE measurements on ITER can be fulfilled by employing a CHC with an aperture diameter a=10 cm, an acceptance angle θ_e=2.3° and a length L_{CHC}=31 cm. Efficient coupling with the oversized waveguide can be obtained by combining two CHC's joined at their small apertures; such a system also offers a natural solution to the problem of realizing double vacuum windows with low etalon effect.

REFERENCES

1. D.V. Bartlett, private communication.
2. H.J. Hartfuss, private communication.
3. P. Buratti, O. Tudisco, and M. Zerbini, A broadband light collection system for ECE diagnostics on the FTU tokamak, *Infrared Phys.*, 34:533 (1993)
4. W.T. Welford and R. Winston, *The Optics of Nonimaging Concentrators*, Academic Press, New York (1978).
5. J. Keene, R.H. Hildebrand, S.E. Whitcomb, and R. Winston, Compact infrared heat trap field optics, *Appl. Opt.* 17:1107 (1978).

ADVANTAGES AND LIMITATIONS OF MICROWAVE DIAGNOSTICS IN ITER

A.J.H. Donné and B.C. Schokker

FOM-Instituut voor Plasmafysica Rijnhuizen
Associatie EURATOM-FOM
P.O. Box 1207
3430 BE Nieuwegein
The Netherlands

I. INTRODUCTION

Microwave diagnostics are widely applied to study the parameters of, in particular, the electron distribution in magnetically confined plasmas. Thanks to their high accuracy, their good spatial and temporal resolution as well as some other advantageous properties, microwave diagnostics are considered as candidate measuring techniques for a large number of different parameters of the ITER plasma, including many category I parameters for machine protection and performance control. Microwave techniques that appear on the list of ITER candidate diagnostics are electron cyclotron emission (ECE) and -absorption (ECA), reflectometry, interferometry, polarimetry and ion (collective) Thomson scattering.

In evaluating the specific merits and drawbacks of microwave diagnostics for ITER, one should carefully take into account the expected range of plasma parameters, as well as the specific geometry of ITER. Both differ substantially from present day devices and therefore, the application of many diagnostics is not as straightforward as one should naively expect. The aim of this paper is to discuss the specific advantages and limitations of microwave diagnostics for ITER.

In this paper we will limit the discussion to electron cyclotron emission and -absorption (Sec. II), reflectometry (Sec. III) and interferometry/polarimetry (Sec. IV). Hence, collective scattering for diagnosing density fluctuations and fast ions is not considered here. For this topic one is referred to another paper in these proceedings.[1] For each of the microwave diagnostics we will first give a brief explanation of the principle as well as a short discussion of the merits and drawbacks of the technique when applied to present confinement devices. After that we will focus on the specific issues encountered when diagnosing ITER. In Sec. V some general remarks will be made about more technical aspects of microwave diagnostics. Because of the relatively large number of different microwave techniques as well as of the length restrictions of these proceedings it is not possible to enter into details. For more information on the various diagnostics the reader is referred to other papers in these proceedings that focus onto the specific microwave diagnostic techniques.

Diagnostics for Experimental Thermonuclear Fusion Reactors
Edited by P. E. Stott *et al.*, Plenum Press, New York, 1996

215

II. ELECTRON CYCLOTRON EMISSION AND -ABSORPTION

II.A. Merits and Drawbacks at Existing Confinement Devices

The electrons in the plasma gyrate around the magnetic field lines and as a result emit electromagnetic radiation at the electron cyclotron frequency $\omega_e = eB/m_e$ and its harmonics, with e and m_e the electron charge and -mass and B the magnetic field strength. See references [2] and [3] for details on the theory and experiment, respectively. For optically thick plasmas the electron cyclotron emission (ECE) is directly related to the electron temperature, T_e, via the Rayleigh-Jeans approximation. Because the magnetic field strength varies in a known way across the plasma cross section, the frequency of the emission is a function of the position. Therefore, it is possible to determine the spatial profile of the electron temperature, $T_e(r)$, by measuring the spectrum of the emission. This can be done with good temporal resolution ($\leq 5 - 10$ µs) and with an absolute accuracy of about 10%, but with a much better relative accuracy. The interfacing of ECE diagnostics with the machine is fairly small: in principle one waveguide with an antenna at the low field side is sufficient, although often separate antennae are used for the O-mode ($\vec{E} /\!/ \vec{B}$) and the X-mode ($\vec{E} \perp \vec{B}$). Typical ECE frequencies for present day plasmas are in the range 60 - 600 GHz. The ECE emission is sensitive to the presence of small populations of suprathermal electrons and the absolute calibration is rather cumbersome.

In optically thin plasmas the radiation temperature measured by ECE cannot directly be converted to an electron temperature, because of the unknown amount of reflected radiation. In this case it is possible to perform active measurements with electron cyclotron absorption (ECA). From the transmitted part of the radiation one can determine the value of the optical thickness. In the special case that the launched radiation is in the second harmonic X-mode, the optical thickness is approximately proportional to the electron pressure, $p_e = n_e T_e$. If the full absorption spectrum is measured one obtains the spatial profile of the electron pressure. In ECA the frequency of the microwave source is usually swept over the full bandwidth of the receiver (typically in ≈ 1 ms for the full sweep). The interpretation of the data is rather complex because of refraction, scattering and mode conversion of the microwaves in the plasma. Moreover, cut-offs and harmonic overlap in the plasma can limit the applicability. As a result of the requirement that the plasma should be optically grey (typically $0.05 \leq \tau \leq 3.0$), the applicability of ECA is limited to small-sized magnetic confinement devices and to divertor plasmas of large plasma devices.

II.B. Additional limitations at ITER

A variety of physics effects will limit the measurement capability of ECE. Some of these effects are already encountered in existing experiments: harmonic overlap will limit the range of major radius that can be investigated, antenna patterns and refraction will limit the spatial resolution, cut-offs in the refractive index may occur at too high a density, and shine through will limit measurements in the edge region. Additionally, the high temperatures expected in ITER will have consequences for the spatial resolution and radial coverage of the T_e measurements because of relativistic effects. Moreover, the relatively large poloidal magnetic field in ITER will have also an effect on the spatial resolution.

To obtain a first indication of the measurement capabilities, the Microwave Diagnostics Group of the European ITER Home Team decided to benchmark the different available ECE simulation codes for two typical ITER discharges, one with $T_e = 10$ keV, $n_e = 5 \times 10^{19}$ m^{-3} and the other with $T_e = 20$ keV, $n_e = 10^{20}$ m^{-3}. Both profiles featured very flat profiles $(1-\rho^5)^{0.25}$, with $\rho = r/a$. The benchmarking was performed for an ITER plasma with a circular geometry, but with standard parameters $R_0 = 8.1$ m, $a = 3.0$ m, and $B_0 = 5.7$ T. The poloidal field was set to zero by taking $q_a = 1000$. The above mentioned effects have been taken into account in the simulations. As an overall conclusion of the benchmarking one can state that the different codes are in good agreement, although minor differences between the various simulations exist. The result of the simulations will be described extensively by Bartlett elsewhere in these proceedings.[4]

Nevertheless, to make some quantitative statements about the specific limitations of ECE for ITER we will have a closer look at the results of the NOTEC-code.[5] In Fig. 1 the full ECE spectra are shown for the O-mode and the X-mode for the two typical ITER cases mentioned above. The line structure in the spectra is determined by harmonic overlap and relativistically downshifted absorption. For each frequency one can determine the effective emission radius and the spread of the profile. These are shown in Fig. 2 for the first harmonic O-mode, as well as the second harmonic O- and X-mode, for the two typical ITER cases.

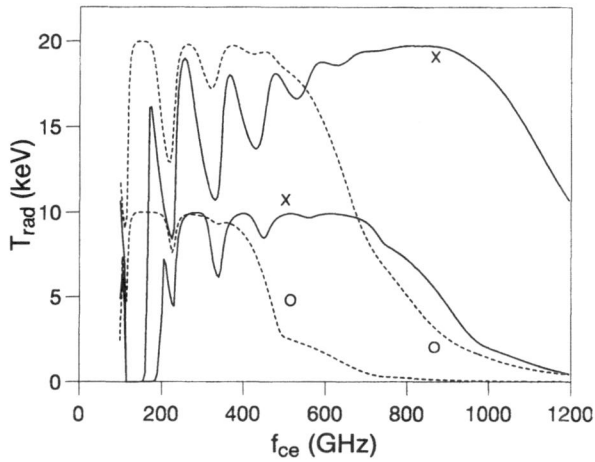

Figure 1. ECE spectra in the O-mode (dashed lines) and X-mode (solid lines) for the two ITER scenarios.

In the 10 keV case the first harmonic O-mode can be used to diagnose the electron temperature at major radii in the range 7 - 11 m, with a localization of about 12 cm, which is within the requirements set by the ITER team. At low temperatures also the second harmonic X-mode can be applied with an even better localization of 9 cm, however, only in a limited range from 9.3 to 11 m. At high temperatures the first harmonic O-mode can be used to diagnose nearly the full low field side of the profile from 8.5 - 11 m, with a localization of 16 cm, which is still within the requirements. The applicability of second harmonic X-mode is limited to the very plasma edge: 10.5 - 11 m.

The ranges for which ECE measurements are feasible are indicated in Table 1. The radial range for which measurements can be performed has been translated also in the required frequency range of the instrumentation. For the first harmonic O-mode, frequencies up to 185 GHz have to be applied, which is well feasible. The heterodyne technology for measurements up to 278 GHz needed for the second harmonic X-mode also exists, but has not yet been applied to plasmas.[6] The temperature and density at the plasma edge are too low to fully absorb the (relativistically downshifted) emission from further inside the plasma. This shine-through effect makes the interpretation of ECE measurements from the outermost part of the plasma very difficult. The precise thickness of the edge layer for which shine-through effects are important, depends on the detailed electron density and temperature profiles, but amounts typically to 20 - 50 cm. Shine-through effects were not taken into account in determining the ranges in Table 1.

The simulations have been performed for a single antenna in the equatorial plane at the low field side. For resolving toroidal MHD modes it is required to implement more ECE systems at different toroidal positions. Furthermore, a non-standard viewing ECE systems (e.g., from the top) should be considered to obtain spectral information on down-shifted ECE.[7]

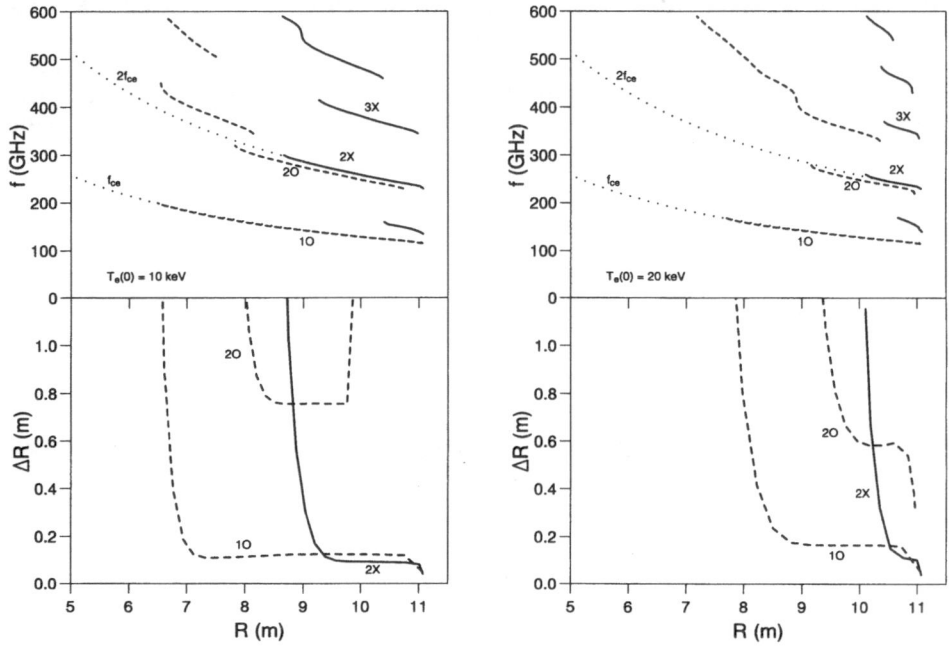

Figure 2. Resonance curves and localization as a function of the major radius for the O-mode (dashed lines) and the X-mode (solid lines). As reference also the non-relativistic resonances are indicated (dotted lines). The left graphs refer to the 10 keV plasma, the right ones to the 20 keV case.

Table 1. Ranges for which ECE measurements in ITER are feasible.

mode	$T_e(0)$ keV	R m	f GHz	ΔR cm
First O-mode	10	7 - 11	118 - 185	12
	20	8.5 - 11	118 - 152	16
Second X-mode	10	9.3 - 11	235 - 278	9
	20	10.5 - 11	235 - 246	10

The optical depth of the divertor plasma is expected to be too low for ECE to be feasible. Instead ECA should be employed to measure the electron pressure profile across the divertor leg (in second harmonic X-mode). If the electron density in the divertor leg is measured by other techniques, e.g. interferometry or reflectometry, then the electron temperature can be determined. It is difficult to extrapolate the experience with ECA on present systems to the ITER divertor. The only ECA system employed thus far in a divertor is at JET.[8] Preliminary measurements have shown that the level of absorption seems to be very low, and possible explanations for this are still being investigated. At the RTP tokamak, ECA is routinely used to measure the electron pressure profile absolutely at 20 positions.[9] Standing waves which might occur in the long and complex waveguides could deteriorate the measurements, and special emphasis should be put on avoiding them. ECA systems are suited for opaque plasmas with optical thicknesses in the range 0.05 - 3.0 (between 5% and 95% absorption), corresponding to an electron pressure in the range 5×10^{19} - 3.0×10^{21} eVm^{-3}. Opposing antennae at both sides of the divertor leg are needed.

Precise simulations for ECA in the ITER divertor have not been performed yet. Since the divertor plasma is cold, many problems which are encountered in the plasma core are not present in the divertor. The spatial resolution is mainly determined by the bandwidth of the instrument and the gradient of the magnetic field, and will be approximately a few cm. The time resolution depends on the precise technique used (many cw narrowband sources or a swept source).

III.   REFLECTOMETRY

III.A. Merits and Drawbacks at Existing Confinement Devices

In reflectometry, microwave radiation is launched into the plasma along the gradient of the electron density and reflected at the layer where the electron density equals a critical density value, $n_e = n_c$.[3,10] Either the phase change of the reflected radiation with respect to a reference beam, or the time-of-flight of short microwave pulses reflected at the critical density layer is measured. To obtain the density profile, $n_e(r)$, the plasma is usually probed in the O-mode at many different frequencies, either by sweeping the source frequency or by using many sources at fixed frequencies. The density profile is obtained by Abel-inversion of the measured data. For the present generation of fusion devices the frequencies for O-mode reflectometry are in the range 15 - 100 GHz. The measurements with reflectometry are relatively localized and the time resolution can be very high (≤ 10 μs). Like with ECE only one waveguide in principle suffices to carry out the measurements. In most cases though, separate waveguides and antennae are used for launching and receiving the microwaves. For measuring the full density profile, antennae at both sides of the plasma are needed (unless the plasma is assumed to be symmetric). Even in this case it is not possible to probe the flat central part of the density profile by O-mode reflectometry because of refraction. Furthermore, one has to make assumptions about the undiagnosed part of the density profile at the edge (below the lowest probing frequency) which can have a large impact on the total density profile. Finally, difficulties can arise from density fluctuations in the plasma.

The plasma can also be probed by X-mode reflectometry, using either the lower or upper cut-off layer as critical density layer. If the density is known from other diagnostics, the total magnetic field may be derived from these measurements or vice versa. X-mode reflectometry could be used to probe the core of the plasma, contrary to O-mode reflectometry, albeit at the cost of larger interpretational difficulties.

III.B. Additional limitations at ITER

Various physics effects will limit the measurement capability of reflectometry in ITER.[10] First, the probing beam could be absorbed due to the presence of relativistically down-shifted ECE absorption in front of the cut-off layer, with the result that part of the density profile is not accessible.[11] Second, refraction effects are large in the vicinity of the cut-off layer. Moreover, because of refraction it is necessary to keep the direction of propagation to within ≈2° of the density gradient. Third, filter techniques or time averaging are required to cope with density fluctuations, which has an adverse effect on the time resolution.

To study the accessible ranges for reflectometry quantitatively, simulations have been performed with the NOTEC code for the 20 keV case which was also used for ECE simulations.[5] A major limiting effect for reflectometry appears to be absorption of the microwaves by the relativistically down-shifted electron cyclotron absorption. Reflectometry is possible if the power losses in the plasma are less than about 99%, corresponding to $\tau < 5$. It turned out that, O-mode reflectometry is not limited by ECA, even at the highest temperatures expected in ITER. X-mode reflectometry on the other hand is limited to a frequency of about 170 GHz at 20 keV, implying that only the outer 1 - 1.5 m is accessible, depending on the precise profile shapes. For more peaked density profiles it will be possible to probe further into the plasma. At $T_e = 10$ keV the upper cut-off frequency can be used to probe the full outer half of the plasma ($R > 6.5$ m). X-mode reflectometry at the lower cut-off is not hampered at all by the presence of ECA. The cut-off densities belonging to the maximum probing frequencies are somewhat larger than the values calculated with the classical formula for the critical density. This is because at the high temperatures present in ITER relativistic effects become important causing an apparent reduction of the critical density.[12]

Although the O-mode radiation can penetrate fully towards the centre of the ITER plasma, reflectometry measurements in O-mode will be limited to the outer part of the plasma where the density gradient is significantly different from zero. In the plasma centre the density profile is expected to be very flat such that the phase or time-of-flight information of the reflected signals becomes deteriorated and reflectometry becomes unfeasible. Also X-mode reflectometry at the lower cut-off observing from the low field side is limited to the density gradient region. However, observed from the high field side, the lower cut-off frequency is continuously (but slowly) increasing for $R < 10$ m. Detailed feasibility studies have to be performed to investigate whether this cut-off layer can be used for reflectometry measurements in the centre of the plasma. If this turns out to be possible, then one is still dependent on another diagnostic, since the total magnetic field must be known to extract the density information from the plasma. The ranges for which reflectometry measurements are feasible are indicated in Table 2. The accessible ranges for O-mode and X-mode lower cut-off are almost independent of the plasma temperature. The range for the X-mode upper cut-off strongly depends on the plasma temperature as well as on the shape of the density profile. As far as the resolution is concerned one can state that at the plasma edge it will be roughly similar to that in existing devices, typically 1 cm. However, in the centre of the plasma, most cut-off layers are rather flat which has a negative effect on the resolution. Precise feasibility studies must be performed to investigate the localization of reflectometry measurements in the plasma centre.

To probe the full electron density profile by means of reflectometry it is required to have antennae in the equatorial plane at both the low field side and high field side. Especially the high field system is complicated and research should be performed with the aim to find a possible routing of the waveguides through the blanket without deteriorating the required screening of the super-conducting coils.

Table 2. Ranges for which reflectometry measurements in ITER are feasible.

mode	$T_e(0)$ keV	R m	f GHz
O-mode cut-off	< 20	5 - 6 & 10 - 11	< 90
X upper cut-off	20	9.5 - 11	120 - 170
	10	6.5 - 11	120 - 210
X lower cut-off	< 20	5 - 11	< 50

Many different techniques have been developed for reflectometry including broadband and narrowband swept, fixed multiple frequency, AM, pulsed radar, pulse compression radar, noise radar, etc.[13] Although all techniques have the same underlying principle, they have different characteristics and specific advantages and drawbacks. Most of the techniques are still under development and their characteristics are being continuously improved. It is therefore too early to make a definitive choice which reflectometry system to use. This implies that the waveguide systems should be designed as versatile as possible such that it is compatible with most of the techniques in the future; the optimum reflectometer system could well be a combination of several existing techniques.

For diagnosing the density profile in the divertor legs and in particular for monitoring the position of the flame front, various reflectometer channels need to be installed along the divertor leg. Most probably the optimum diagnostic for monitoring the density in the divertor plasma is a combination of reflectometer and interferometer, using an array of opposing antennae along the divertor leg, very similar to the comb reflectometer/interferometer at JET.[8] The spatial resolution is in this case very much determined by the distance between successive antennae along the divertor leg (this could be 20 - 25 cm), whereas the resolution across the leg is strongly determined by fluctuations in the divertor plasma. Without having more detailed information about the parameters of the divertor plasma, it is impossible to make any accurate assessment. It is doubtful though, whether the resolution of 3 mm across the leg, as required by the ITER team can be obtained with reflectometry.

IV. INTERFEROMETRY/POLARIMETRY

IV.A. Merits and Drawbacks at Existing Confinement Devices

In interferometry the phase change profile of a set of (mostly parallel) far-infrared probing beams is measured and Abel-inverted to obtain the electron density profile.[3,14] By simultaneously measuring the change of polarization of the probing beams with polarimetry, one determines additionally the poloidal magnetic field profile, from which the important profile of the safety factor, $q(r)$, can be derived. The optimum frequency for interferometry/polarimetry in present plasmas is 600 GHz - 3 THz, corresponding to a wavelength range of 500 - 100 μm. So strictly spoken, interferometry/polarimetry is a far-infrared rather than a microwave diagnostic. Nevertheless, since the underlying techniques are very similar to those of reflectometry and ECE, interferometry/polarimetry is often categorized into the group of microwave diagnostics. Interferometry yields values for the line-integrated density with high accuracy ($\approx 1\%$) and high time resolution (≤ 5 μs). Polarimetry usually features a much lower time resolution (≥ 5 - 10 ms) whereas an accuracy of 0.1 - 0.2° is feasible. The disadvantage of interferometry/polarimetry is that the measurements are line-integrated, so it is required to perform an inversion procedure to obtain $n_e(r)$ and $q(r)$, which is only permitted in quiescent symmetric plasmas. For measuring the full profiles, many smaller access ports (or a few larger ones) are needed at two opposite sides of the plasma to make a fair number of parallel probing chords possible. Instead of launching the beams at one side of the plasma and collecting it at the other side, one can also (retro-)reflect the beam such that it passes the plasma twice.

III.B. Additional limitations at ITER

The wavelength range for interferometry/polarimetry is on the one hand limited by refraction (≤ 200 μm), ellipticity (≤ 100 - 150 μm) and the huge EC background emission (see Fig. 1), whereas vibrations on the other hand give a lower limit to the wavelength (≥ 5 μm). So the optimum wavelength for interferometry/polarimetry is in the range 5 - 150 μm, which implies that only CO_2-lasers (10.6 μm) and FIR-lasers (e.g., at 118 μm) can be applied as a source. The optical tolerances of the reflecting surfaces should be better than typically $\lambda/20$.[14] Especially for systems with a smaller wavelength this implies that the quality of the measurements slowly degrades if the reflecting surfaces are subject to blistering (and to a lesser extend swelling)

The optimum source and geometry depends very much on the specific parameter one wants to measure. For density feedback a tangential polarimeter has been proposed.[15] Since the vacuum magnetic field is well known it can be unfolded from the measurements to yield the density information. An advantage above standard interferometry, which is used in many existing tokamaks for density feedback, is that polarimetry is rather robust. The method is not sensitive to signal losses because fringe counters are not used. A fan of tangential polarimeters could be envisaged for measuring a rough electron density profile. The beam could be introduced via one of the diagnostic plugs at the low field side and reflected by retro-reflectors mounted in other low field side diagnostic plugs at various toroidal locations.

Measuring the electron density and poloidal magnetic field profiles in ITER in the same way as it is presently done on various existing devices is impossible, or at least extremely difficult, because of the large number of diagnostic access ports needed at the top and bottom of the device. Nevertheless, polarimetry could well be the only existing method to determine the q-profile unambiguously, especially at the high field side of the tokamak. Namely, it can be doubted whether the penetration of neutral beams is large enough to use the motional Stark effect in this region. The only other alternative is combining a full density profile measured with a tangential polarimeter, with a reflectometry measurements in the X-mode (lower cut-off) from the high field side, but this has not been demonstrated yet. Instead of a vertical set of beams, one could build a polarimeter system with a set of (close to) horizontal chords and reflect from the back wall. When FIR-lasers are employed as source, it is probably possible to reflect the beam simply from the blanket tiles at the back wall, without necessity to mount retro-reflectors. Since probably only chords not too far above and below the midplane can be employed, the best information one can expect from such a polarimeter system is a value for $q(0)$. Of course it is not too difficult to upgrade this polarimeter system to a full interferometer/polarimeter yielding the central part of the density profile (in addition to the edge density profile from reflectometry).

In case ITER will be equipped with reflectometry antennae at both the high and low field side, it is also possible to extract line-integrated information on the density, but strongly weighted towards the plasma centre, by measuring the delay of microwave pulses with a frequency just above the O-mode cut-off (and of course below the down-shifted EC absorption). As has been shown in ref.[16] this new diagnostic technique, called delayometry, is a sensitive method to extract additional information about the central part of the density profile.

Interferometry can also be used for diagnosing the divertor plasma. Probably the best choice here is to combine the divertor interferometer with the reflectometer, using the same waveguide system (see Sec. III.B).

V. GENERAL CONSIDERATIONS

Microwave radiation can be guided to and from the tokamak by means of waveguides (see Fig. 3). Hence, the sensitive equipment can be located remote from the plasma in an area which is accessible during tokamak operation, such that routine maintenance can be carried out. Possible choices for the waveguide system are open, quasi-optical, systems or closed waveguide systems.[6] For the closed waveguide systems one can make a further choice between fundamental and oversized. The final choice which waveguide is most suited depends very much on the specific diagnostic application. All types of waveguide systems are compatible with multiple bends, which are necessary to prevent any direct lines from the tokamak plasma to the outside world.

For launching (receiving) the radiation into (from) the plasma one can use mirror systems (e.g., two mirrors in a Z- or X-configuration), standard microwave antennae or parabolic concentrators.[17] Especially the mirror systems are compatible with quasi-optical waveguides. According to gaussian beam optics it is possible to focus the microwave beams to a waist of 10 cm diameter in the plasma, using mirrors in the diagnostic plugs with diameters of about 30 cm. It should be studied whether viewing lines could be shared by different diagnostics. Reflectometry and ECE systems can possibly share the same oversized or quasi-optical waveguides, and even possibilities exist to share viewing lines with LIDAR Thomson scattering. Because of the limited number of waveguides that is possible in the divertor region one will be forced to make shared use of them. The waveguides and antennae for the divertor system could be made quite narrow such that they can be located in the small gap between adjacent divertor cassettes.

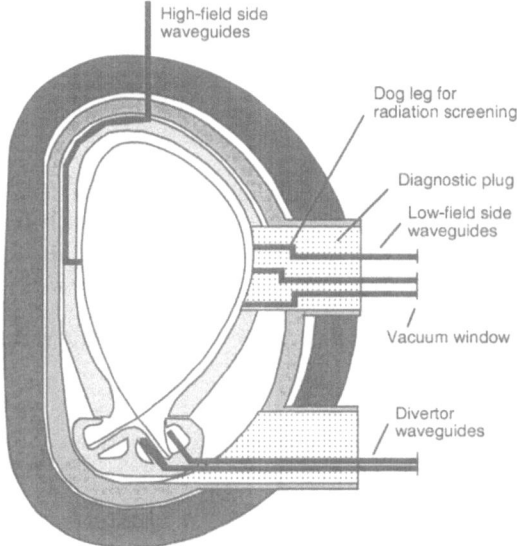

Figure 3. Schematic diagram of a poloidal ITER cross section. Indicated are some typical waveguides for microwave diagnostics. The antenna at the high field side is especially dedicated to reflectometry. The antennae at the low field side can be used (simultaneously) for ECE and reflectometry. The waveguides contain a dog leg for radiation screening. Divertor waveguides are envisaged for both divertor legs. (For keeping the figure clear they are only indicated for one leg.) The vacuum windows are positioned outside the diagnostic plugs.

Some applications (e.g., reflectometry) would greatly benefit from an antenna at the high field side. Fundamental waveguides seem to be the preferable option here, since they can contain sharp bends while still having a good transmission. The drawback is that many separate fundamental waveguides are needed for a broadband diagnostic system. Moreover, some specific microwave diagnostics are incompatible with long lengths of fundamental waveguide.

The vacuum windows can be positioned relatively far from the plasma, outside a direct viewing line with the plasma. Therefore, the neutron load on these mirrors is much reduced, compared to windows facing the plasma directly. Much effort has to be put on the design of window materials which maintain good transmission properties during long periods of ITER operation. The waveguides seem to be less critical than the vacuum windows. Nevertheless, their front-ends are possibly located very close to the plasma and are subject to strong irradiation and mechanical stresses. Therefore, also attention should be devoted to the development of waveguides that maintain good transmission characteristics under these conditions.

VI. CONCLUSION

Microwave diagnostics are standard techniques on existing confinement devices to measure the parameters of the electron distribution with good accuracy and with high spatial and temporal resolution. The application of these diagnostics to the ITER plasma is more limited as in present devices. The reason for this is especially the higher electron temperature in ITER. Furthermore, because of the required shielding the number of diagnostic access ports will be strongly limited and the waveguides to transport the radiation should contain many bends. In spite of all these additional difficulties, microwave diagnostics are indispensable for ITER since they promise to yield spatially and temporally resolved information on many important parameters over a substantial part of the poloidal cross section. By combining two separate microwave diagnostics (e.g. a tangential interferometer and a reflectometer operating at the lower cut-off from the high field side, to yield the current density profile) information can be extracted from the plasma, which is hard to obtain by any of the individual diagnostics.

ACKNOWLEDGEMENTS

The contents of this paper is based on discussions in the Microwave Diagnostics Group of the European ITER Home Team and we would like to thank the members of this group for their active participation at the various meetings. This work was performed as part of the research programme of the association agreement of EURATOM and the 'Stichting voor Fundamenteel Onderzoek der Materie' (FOM) with financial support from the 'Nederlandse Organisatie voor Wetenschappelijk Onderzoek' (NWO) and EURATOM.

REFERENCES

1. U. Tartari, Collective microwave scattering applications to ITER, *These proceedings.*
2. M. Bornatici, R. Cano, O. De Barbieri and F. Engelmann, Electron cyclotron emission and absorption in fusion plasmas, *Nucl. Fusion* 23:1153 (1983).
3. I. Hutchinson, "Principles of Plasma Diagnostics", Cambridge University Press, Cambridge (1987).
4. D.V. Bartlett, Physics issues of ECE and ECA for ITER, *These proceedings.*
5. B.C. Schokker, Feasibility study of ECE measurements at ITER, *submitted to Plasma Phys. Contr. Fusion.*
6. H. Hartfuss, Instrumentation of ECE for ITER, *These proceedings.*
7. Y. Michelot et al., Electron temperature measurements by microwave transmission in Tore Supra, *Proc. 21st EPS Conf. on Contr. Fusion and Plasma Phys.*, Montpellier (1994) Vol. III, p. 1208.
8. R. Prentice et al., New microwave measurements of electron density and temperature in the JET divertor, *Proc. 22nd EPS Conf. on Contr. Fusion and Plasma Phys.*, Bournemouth (1995) to be published.
9. J.F.M. van Gelder et al., Diagnosis of nonthermal ECE by means of LFS ECE, HFS ECE and ECA measurements, *Proc. 22nd EPS Conf. on Contr. Fusion and Plasma Phys.*, Bournemouth (1995) to be published.
10. M.E. Manso, Reflectometry in fusion devices, *Plasma Phys. Contr. Fusion* 35:B141 (1993).
11. E. Doyle, *private communication.*
12. H. Bindslev, Relativistic expressions for plasma cutoffs, *Plasma Phys. Contr. Fusion* 35:1093 (1993).
13. C. Laviron, Comparison of different reflectometer techniques, *These proceedings.*
14. A.J.H. Donné, High spatial resolution interferometry and polarimetry in hot plasmas, *Rev. Sci. Instrum.* 66:3407 (1995).
15. R. Snider, Interferometry applications to ITER, *These proceedings.*
16. S.H. Heijnen et al., Measuring the line-averaged density by the time delay of short microwave pulses, *Proc. 22nd EPS Conf. on Contr. Fusion and Plasma Phys.*, Bournemouth (1995) to be published.
17. P. Buratti, O. Tudisco and M. Zerbini, A broadband light collection system for ECE diagnostics on the FTU tokamak, *Infrared Phys.* 34:533 (1993).

APPLICATION OF INTERFEROMETRY AND FARADAY ROTATION TECHNIQUES FOR DENSITY MEASUREMENTS ON ITER

R.T. Snider,[1] T.N. Carlstrom,[1] C.H. Ma,[2] W.A. Peebles[3]

[1]General Atomics, P.O. Box 85608, San Diego, CA 92186-9784
[2]Oak Ridge National Laboratory, P.O. Box 2009, Oak Ridge, TN 37831
[3]University of California at Los Angeles, CA 90024-1597

ABSTRACT

There is a need for real time, reliable density measurement for tokamak plasma density control, compatible with the restricted access and radiation environment on ITER. Line average density measurements using microwave or laser interferometry techniques have proven to be robust and reliable for density control on contemporary tokamaks. In ITER, the large path length, high density and high density gradients, limit the wavelength of a probing beam to shorter then about 50 µm due to refraction effects. In this paper we consider the design of short wavelength vibration compensated interferometers and Faraday rotation techniques for density measurements on ITER. These techniques allow operation of the diagnostics without a prohibitively large vibration isolated structure and permit the optics to be mounted directly on the radial port plugs on ITER. A beam path designed for 10.6 µm (CO_2 laser) with a tangential path through the plasma allows both an interferometer and a Faraday rotation measurement of the line average density with good density resolution while avoiding refraction problems. Plasma effects on the probing beams and design tradeoffs will be discussed along with radiation and long pulse issues. A proposed layout of the diagnostic for ITER will be presented.

INTRODUCTION

The measurement of the electron density in large tokamaks play a critical role in the operation and understanding of the plasmas produced. Active feedback control of the density is routinely used in present day tokamaks to optimize the performance of the plasma, to avoid density limit disruptions and to perform systematic studies with the density as an independent variable. In the next generation of machines, such as the International Thermonuclear Experimental Reactor (ITER), density measurements and active density control will take on an even more important role due to the increased seriousness of disruptions and the added element of burn control. Many of the present critical areas of concerns for the ITER design team, such as fueling and disruption avoidance, assume very accurate and reliable density measurements.[1]

Line average density measurements using microwave or laser interferometry techniques have proven to be robust and reliable for density control in tokamaks and can provide the accuracy needed for ITER. Another density measurement technique proposed by Jobes[2] determines the density by measuring the Faraday rotation of a probing beam tangent to the toroidal field. While the Faraday rotation method has yet to be demonstrated on a tokamak it has some advantages over interferometer techniques particularly in long pulse operation since it does not rely on keeping track of fringe shifts. A density interferometer and a Faraday rotation density measurement can both use the same optics path and port interface and can probably use the same probing beam allowing both measurements to be made simultaneously. There are distinct advantages in using both techniques in terms of complimentary information and improved reliability.

ITER presents significantly new and difficult challenges to interferometer and Faraday rotation density measurements due to the large plasma path length, long pulse length, high

Diagnostics for Experimental Thermonuclear Fusion Reactors
Edited by P. E. Stott *et al.*, Plenum Press, New York, 1996

225

electron density, severely restricted access and the nuclear environment. A beam path designed for 10.6 μm (CO_2 laser) with a path tangential to the toroidal field at the midplane through the plasma allows both a vibration compensated interferometer and a Faraday rotation measurement. These diagnostics address ITER relevant issues without penetrations or disturbance of the ITER shield modules other than the radial port plugs. This design has minimal impact on the ITER machine design and uses well-understood proven techniques that pose little technical risk.

FARADAY ROTATION AND INTERFEROMETRY TECHNIQUES FOR ELECTRON DENSITY MEASUREMENTS

Laser interferometry is a well established method for measuring electron density in high temperature plasmas going back to the first demonstration by Ashby and Jephcott in 1965.[3] In a high temperature plasma the change in the refractive index is dominated by the electron inter-actions and, in a tokamak plasma can cause significant displacement, deflection and phase lag of a probing beam. The phase lag caused by the electron density (n_e) is given approximately by $\phi = (\pi/\lambda n_c) \int n_e d\ell$ where n_c is the cutoff frequency [$n_c = m_e \, \varepsilon_0 \, (2\pi c^2/e\lambda)$], e is the electron change, c is the speed of light in vacuum, ε_0 is the dielectric constant, and m_e is the electron mass. Many techniques have been developed to measure this phase lag but the most common technique now used on tokamaks is to split the laser beam into a plasma probing beam and a reference beam with the reference beam Doppler shifted, either by moving mirrors,[5] moving gratings[5] or acoustic optic cells.[6] The beams are combined after the plasma probing beam passes through the plasma. The resulting mixed beam has a beat frequency approximately equal to the Doppler frequency shift, due to interference between the plasma probing beam and the Doppler shifted reference beam. The plasma induced phase shift is then measured by measuring the difference between the period of the beat frequency (by measuring the time between zero crossings) and the period of the Doppler shift. The change in the line average density during each cycle of the beat period is the quantity measured and a running total of these phase shifts gives the total line average density. Any loss of this running total, or total plasma induced phase shift during a plasma discharge results in the loss of the density measurement.

Mechanical motion of the mirrors in the direction of beam propagation produce phase shifts ($\Delta\phi=2\pi \, \Delta l/\lambda$) in the plasma probing beam. Vibration isolation structures are impractical on a machine the size of ITER, particularly with the restricted port access. Further, at the short wavelengths appropriate for ITER even a vibration insolated structure cannot reduce vibrations to an acceptable level. However a second wavelength interferometer can be used to compensate for any motion of the optical components.[5] This allows the optical components, such as the retrore-flectors and relay mirrors, to be directly mounted onto the vacuum vessel or port structures.

A measurement of the Faraday rotation of a probing beam that is tangential to the toroidal field at the midplane can also be used to monitor the plasma density. This technique was pro-posed for ITER by Jobes[2] and has several advantages over interferometry. The measurement does not depend on the past history of the plasma discharge (the absolute measurement of the Faraday rotation is all that is required as long as the total Faraday rotation is less than 2π) and vibration of the optics does not affect the measurement. A linearly polarized beam propagating in the direction of the magnetic field experiences a rotation in the polarization of $\Omega = [\lambda^3 \, e^3/(8 \, \pi^2 \, c^3 \, \varepsilon_0 \, m_e^2) \int n_e B_\| \, d\ell.]$ where $B_\|$ is the magnetic field parallel to the beam propagation. At the midplane $B_\|$ is just the toroidal field ($B_T=B_0 R_0/R$ where B_T is the toroidal field, B_0 is the toroidal field at the center of the discharge, R_0 is the major radius at the center of the discharge and R is the radial distance from the centerline of the tokamak) and $\Omega \propto \int (n_e/R) \, d\ell$. If the beam is off midplane, as would almost certainly be the case in ITER, the Poloidal field would have some component along the beam line and would have some effect on the polarization. This effect is small and antisymmetric and should not cause a significant error. An Abel inversion is required in order to extract the density from the data (i.e. a single channel does not give the line average density as with an interferometer), however as discussed earlier because of the toroidal symmetry a rough Abel inversion can be made with few channels. Further, it is likely that the density control function can be done with a single channel $\int n_e B \, d\ell$. measurement just as the line-average density is used in existing tokamaks for density control and not a density profile or the peak density.

The tangential geometry proposed for a density measurement on ITER allows the possibil-ity of making both an interferometer measurement and the Faraday rotation measurement using the same optics and indeed the same laser beam. Further the measurements are complementary in that the interferometer for a given wavelength has better density resolution given the current demonstrated detector technology, while the Faraday rotation does not have the long pulse problem of keeping track of fringe shifts. Also the two methods give different information (i.e. $\int n_e d\ell$ and $\int (n_e/R) \, d\ell$) that could be used to improve the density profile information from a given set of beamlines.

226

DIAGNOSTIC GEOMETRY

The port access on ITER will be severely limited and will limit the lines of sight available for probing beams. As many as six horizontal lines of sight at the midplane, tangential to the toroidal field, can be realized in ITER using the existing radial ports without modifying the shield/blanket modules or any other machine structure other than the radial port plugs as shown in Fig. 1. Penetrations or modifications to the shield/blanket modules are highly undesirable and

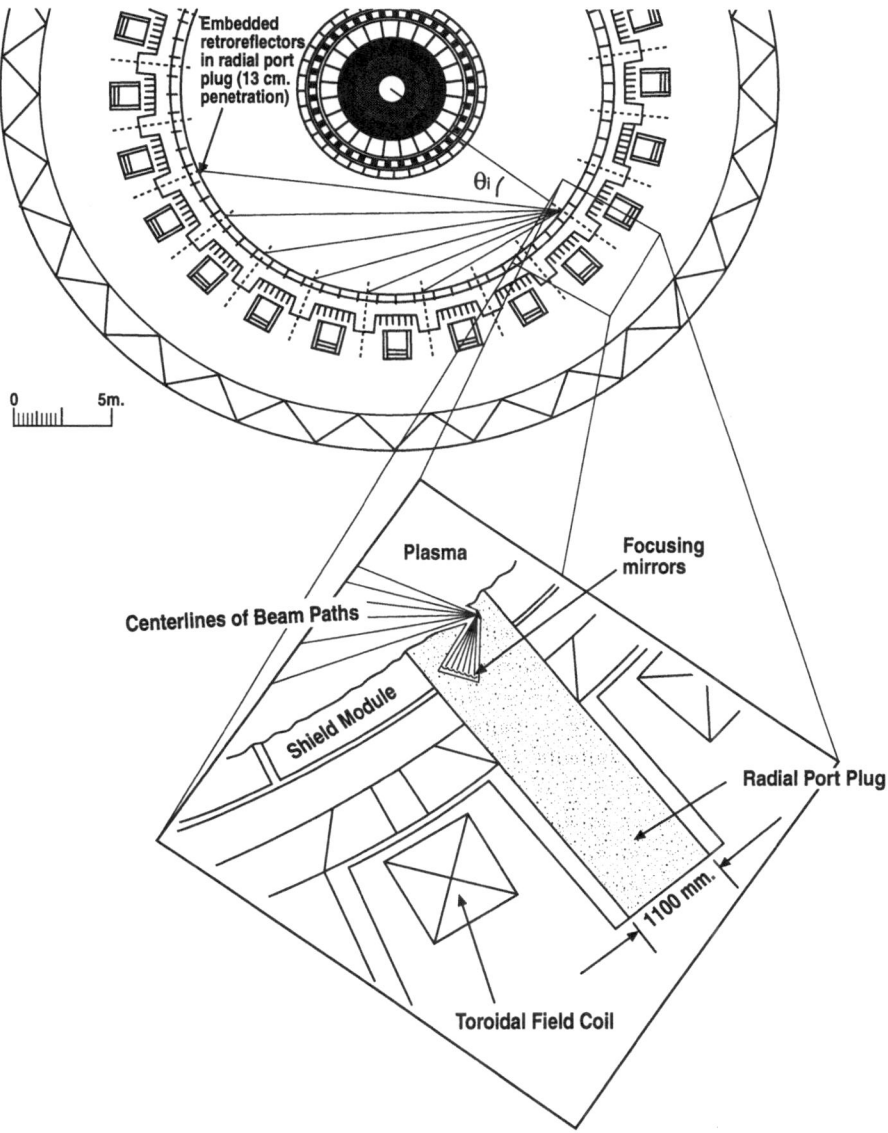

Figure 1. The proposed layout uses a single radial port for the entrance and exit beamlines for all of the sight lines and retroreflectors embedded in other radial port plugs. Relay optics in the radial port transport the beams through a shielding labyrinth and are then focussed onto and returned by the embedded retroreflectors.

227

costly due to the complexity and the remote handling requirements of the modules. Further, penetrations through the shield/blanket modules (either inboard or outboard) compromise the nuclear shielding of the super conducting coils. These considerations restrict or perhaps eliminate geometry other than those similar to the proposed layout, such as radial lines of sight withretroreflectors mounted in the inboard shield or vertical lines of sight with retroreflectors mounted in the lower part of the outboard shield/blanket module in line with the vertical port.

The proposed tangential layout has other advantages over either radial or vertical sight lines: 1) The symmetry about the center post allows a straight forward Abel inversion of the line integrated data yielding rough density profiles from relatively few channels. While the six sight lines proposed could not give fine, detailed density profiles, the profiles produced would be adequate for density profile control measurements and for most physics analysis, such as confinement time studies. 2) The tangential sight lines always sample the entire profile (in particular the center) regardless of the Shafranov shift. 3) The tangential geometry allows the option of using the Faraday rotation of the probing beam as a density monitor.

CONSTRAINTS ON WAVELENGTHS

In considering possible wavelengths for a probing beam for a plasma density measurement (either an interferometer or a Faraday rotation measurement several effects must be taken into account.[7] The refraction and diffraction of the probing beam favor short wavelengths while density resolution, optical surface quality of plasma facing components, and in the case of interferometry, mechanical vibrations favor longer wavelengths. Tradeoffs between these competing effects drive the usable wavelengths for ITER into a very narrow range, between roughly 1 μm and 50 μm.

The variation of the angle of deviation (α_m) of a 119 μm beam as a function of tangential angle of incidence (θ_i defined in Fig. 1) due to refraction, is shown in Fig. 2 for an ITER-sized plasma with a central electron density of 1.1×10^{20} m^{-3} for four density profile shapes.[8]

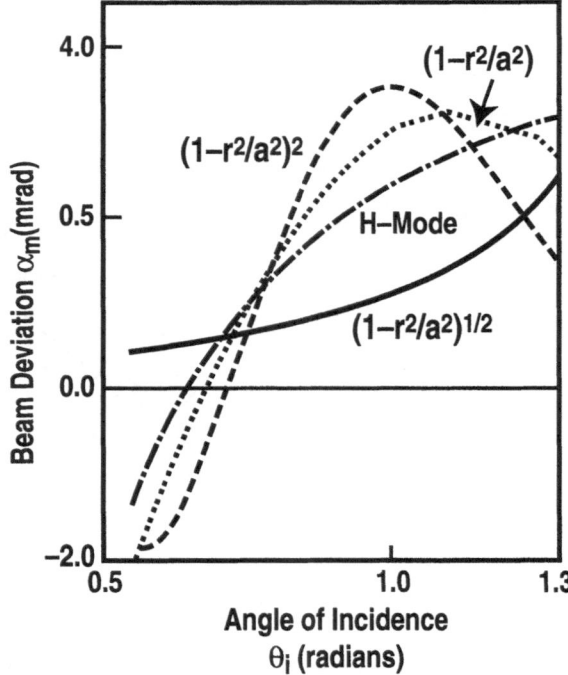

Figure 2. The variation of the angle of deviation (α_m) of a 119 μm beam as a function of the tangential angle of incidence (θ_i) for an ITER-sized plasma with a central electron density of 1.1×10^{-20} m. The characteristic H–mode density profile has an edge density gradient of 3.21×10^{21} m^{-4} with constant density across the central region.

For a wide variety of density profile shapes α_m has a similar range of values, all between -2 mrad and 4 mrad for a 119 μm beam. The refractive bending of the probing beam results in a translation of the beam at the location of the return retroreflector ($d\ell$) which is approximately proportional to $n_0\lambda^2$ for small deviations and is plotted in Fig. 3 as a function of wavelength for a family of tangential angles of incidence for a parabolic density profile. A beam with wavelength of 10.6 μm has a maximum deflection of less then 1 mm and is an order of magnitude smaller then typical beam diameters and can be ignored. At 50 μm the deflection is about 1 cm and at 100 μm the deflection is several centimeters. The large displacements caused by refraction above 50 μm make alignment of the beam line difficult.

The divergence and the diameter of the probing beam and the path length determine the dimensions of the plasma facing optics and the opening in the port plugs and thus play a large role in determining the long wavelength bounds on the wavelength. The path length between the retroreflector and the turning mirror (Z_0) shown in Fig. 2 is about 20 m. We want to minimize the size of the retroreflector in order to cause the smallest perturbation in the port plugs that house the retroreflectors. For a Gaussian beam, the laser spot size that contains 95% of the power at the beam waist (set to be at the retroreflector) is $d_0 = 2\eta\,(2\lambda Z_0/\pi)^{1/2}$ where η is a factor representing the increase in the divergence of the laser beam above the defraction limit (for example, commercially available CO_2 lasers are available with $\eta \cong 2$). A focusing mirror near the plasma would be required and the laser spot at that mirror would be $d_m = \sqrt{2}\,d_0$.

With all of these factors included, $d_0 = 4\eta\,(2\lambda Z_0/\pi)^{1/2}$ for 10.6 μm, the retroreflector would be at least 9.2 cm (assuming $\eta = 2$) and the turning mirror would be 13 cm (for a path length of 20 m corresponding to $\theta_i = 0.77$). As can be seen in Fig. 1 the path length changes as a function of θ_i and thus the optics sizes change somewhat for the various paths. The opening in the port plug through which all six beams pass, would be 13 cm in the poloidal dimension and 25 cm in the toroidal dimension.

The density resolution for an interferometer is determined by the total phase shift caused by the plasma and the detector/electronics resolution of the phase shift. The density resolution for a Faraday rotation density measurement is likewise determined by the Faraday rotation caused by the plasma and the detector/electronics resolution of that polarization shift. Table 1 summarizes relevant parameters for various wavelengths including the total number of fringe shifts and the Faraday rotation. The last two columns in Table 1. list the density resolution for both types of measurements assuming detector/electronics with 1/100 of a fringe resolution for the interferometer and 1/1000 of a radian resolution of the polarization for the Faraday rotation measurement. These numbers are consistent with demonstrated detector/electronics,[4,7,9] and improvements are probable. It should be noted that the density resolution for a vibration compensated interferometer is reduced because of the use of the second wavelength interferometer.

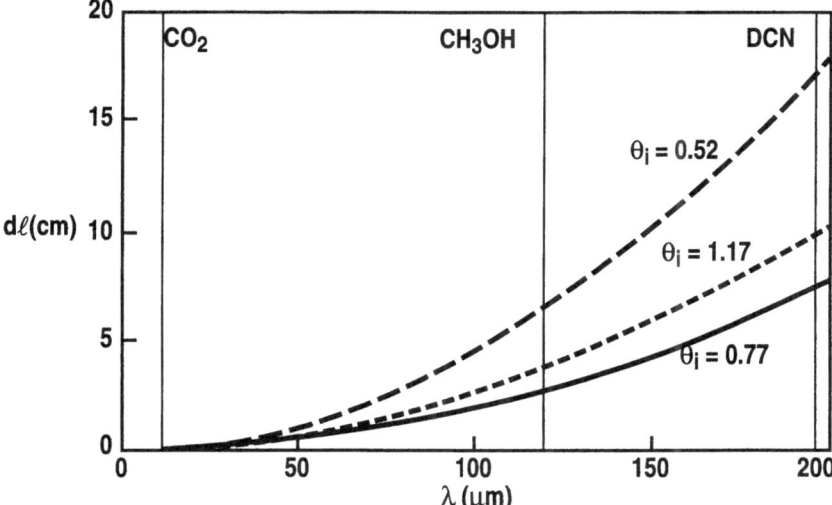

Figure 3. The translation of a beam at the location of the return retroreflector due to refraction as a function of wavelength for ITER geometry for $n_e = 1.3\times10^{20}$ m^{-3} $(1 - r^2/a^2)^{1/2}$.

Table 1. Summary of relevant parameters for various wavelengths for a tangantial ($\theta_i = 0.77$) double pass beam in ITER. The density resolutions assume $\Delta\phi$ resolution of 1/100 radian and polarization resolution of 1/1000 radian.

λ (µm)	Port Penetration Size (cm) Retroreflector/ Mirror	θ_i n_e (o) ($\times 10^{19}$ m^{-3})	Total Phase Shift (rad)	Faraday Rotation (rad)	Interferometer $\Delta\bar{n}_e/\bar{n}_e$ ($\times 10^{-2}$)	Faraday Rotation $\Delta\overline{n_eB}/n_eB$ ($\times 10^{-2}$)
195	74/91	13	1969	73	0.0005	0.0014
		1	151	5.6	0.006	1.8
119	44/57	13	1202	27	0.008	0.0037
		1	92	2.1	0.01	0.048
50	22/30	13	505	4.8	0.002	0.02
		1	39	0.37	0.026	0.027
10.6	10/13	13	107	0.21	0.0093	0.48
		1	8	0.016	0.12	6.2
3.39	6/8	13	34	0.022	0.029	4.5
		1	2.6	0.017	0.38	60
1	3/4	13	10	0.0019	0.1	52
		1	0.78	0.0005	1.3	666

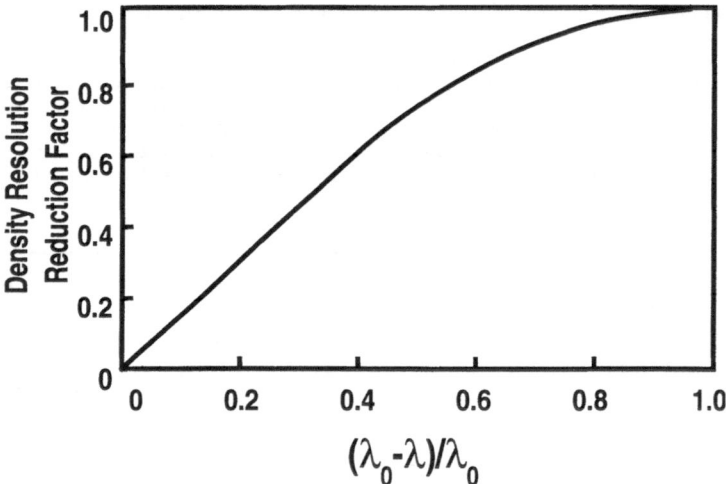

Figure 4. Reduction of the density resolution of a two color interferometer as a function of the normalized differences between the interferometer wavelengths. λ_0 is the primary interferometer wavelength and λ is the second interferometer wavelength.

The density resolution decreases as the difference between the two interferometer wavelengths decreases by the factor shown in Fig. 4. With existing technology a 10.6 µm interferometer would give line average density resolution of better than 0.2%, while the Faraday rotation measurement would give a nB resolution of about 6% at low densities (see Table 1).

All of the optical surfaces will require flatness of $\lambda/4$ across the diameter of the component and surface roughness to be less then about $\lambda/10$ for the life of ITER. The optical quality of metal front surface mirrors should not experience degradation due to neutron damage, however the plasma facing optical surfaces will be subjected to particle bombardment from the plasma. The particle flux can cause sputter damage and coating of the surface. While it is not clear at this time how badly the optical surfaces will be degraded in ITER from the particle flux it is advantageous to use as long a wavelength as possible to minimize the effect on the density measurement. From existing tokamak experience, primarily the short wavelength interferometer used on DIII–D, visible light is problematic due to coating from the plasma on optical components close to the plasma. While the coating and damage that optical components will experience on ITER will be different (different wall material, radiation fluxes and fluence and energy distribution), it is likely that the optical surface degradation due to coating and damage will be worse on ITER than on existing tokamaks. This then argues for long wavelengths, at least longer than visible light.

CO_2 LASER AS OPTIMUM CHOICE OF WAVELENGTH

Satisfying the constraints on the wavelength of a probing beam we propose a 10.6 μm (CO_2) system for ITER including both a Faraday rotation measurement and a vibration compensated interferometer using 3.39 μm (HeNe) as the second wavelength. This choice has many practical advantages. Commercially available lasers with adequate power are available, detectors and window material are well developed and a interferometer using these wavelengths has been operational on the DIII–D tokamak for several years.[10] Figure 5 shows representative time traces for a DIII–D discharge including the line average density and the vibration of the vacuum vessel measurement by the two color interferometer. This demonstrates the existing technology associated with this wavelength choice. The penetrations through the port shield plugs of ITER are

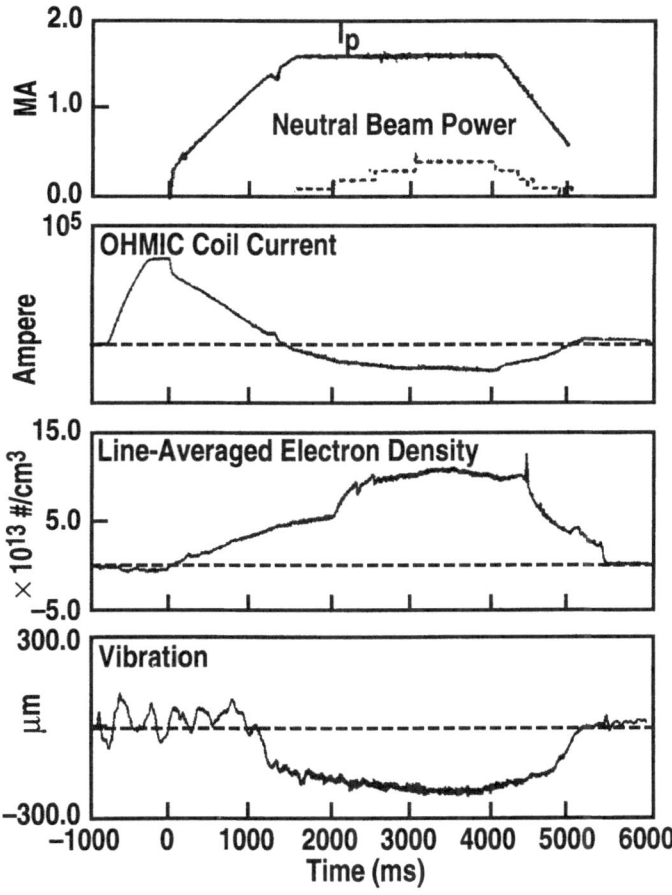

Figure 5. A vibration compensated interferometer using lasers at 10.6 and 3.39 μm and a retroreflector mounted directly on the vacuum vessel has been build and operated on the DIII–D tokamak. (Taken from Ref. 10).

manageable (~13 cm x 25 cm), and the density resolution for the interferometer is much better than required by ITER. Faraday rotation density measurements at 10.6 mm can provide adequate density resolution, provides additional profile information and, most importantly, is an instantaneous measurement and is thus inherently more reliable than an interferometer. Overall the proposed 10.6 μm and 3.39 μm system represents a minimum of technical risk satisfies the mechanical and geometrical constraints and can easily meet the specifications required by ITER.

RADIATION EFFECTS

The plasma facing optical components will experience neutron fluxes and fluences that will approach the first wall levels. The recessed, tangential geometry shown in Fig. 1 allows some shielding of the retroreflectors and the final turning mirrors, but the neutron fluence experienced by those components will be close to 5×10^{17} n/cm^2. Radiation testing of optical components indicate that first surface, metal mirrors will survive these fluence levels. However, coated mirrors and refractive optics almost certainly will not survive. Very little radiation testing of infra-red (IR) optic elements have been done at the ITER neutron levels, but the radiation testing of visible refractive optics indicate serious problems with radiation darkening and other damage.[11] For an interferometer or Faraday rotation measurement darkening is less of a problem then intensity sensitive diagnostics since only the phase or polarization information is used, not the absolute intensity. This means that the first several optical elements closest to the plasma will have to be metal mirrors that relay the beams through a shielding labyrinth (shown in Fig. 1) to a place where the neutron levels are low enough for refractive optics to survive. This implies that the focus of radiation testing of IR refractive optics should be to determine the neutron flux and fluence that the element can survive which then defines how close to the plasma in the beam path the element can be placed.

INTEGRATION OF DIAGNOSTIC DESIGNS

The limited access and high machine interface costs of ITER makes it imperative that the diagnostic suite as a whole be optimized and that any synergistic advantages that exist among diagnostic systems be exploited. An example of such a complimentary set of diagnostics is the combination of reflectometry, toroidal interferometry and Faraday rotation measurement. The strengths and weaknesses of the systems are complimentary. The interferometry and Faraday rotation measurements provide reliable accurate core measurements in real time that can be simply Abel inverted, the addition of data from a reflectometer would further increase the accuracy of an inversion particularly by defining the edge density. Determining edge density, particularly the scrape-off density, is one of the biggest sources of error for a density inversion of line integrated density, while there are difficulties in making a central density measurement using reflectometry alone in ITER.[12] Integrating the data from these systems to produce a composite profile is straightforward, such that together, a much more accurate, reliable, and spatially detailed profile can be produced with guaranteed core and edge measurements with good resolution. Combining the diagnostics as a single system reduces the technical risk of each and reduces the total penetrations required and hence the total cost. To fully exploit this synergism requires that the concept of using the diagnostics together be considered during the design of these systems, not just combining the data after the systems are built.

CONCLUSIONS

The need for a reliable density measurement for plasma control on ITER within the considerable limitations imposed by the ITER environment has lead us to propose a limited array (six spatial channels or less shown in Fig. 1) of probing beams, tangential to the toroidal field, at the midplane of ITER employing wavelengths of 10.6 μm and 3.39 μm. Both an interferometer and a Faraday rotation density measurement can be made at these wavelengths using the same beam path and even the same laser. The choice of tangential geometry and in particular the layout shown in Fig. 1 has minimal impact on the ITER device and requires no modification to the critical shield/blanket modules. Due to the large physical size and the harsh environment of ITER the choice of wavelength is severely limited by refractive and diffractive effects and by optical surface quality/survivability of the plasma facing components. Practical limits on the wavelengths are roughly 1 μm and 50 μm. The choice of 10.6 μm and 3.39 μm is driven by laser and detector availability and by nearly optimum compromise of constraints. The proposed system is straightforward (with much of the technology already demonstrated), has minimal impact on the ITER design, has a proven tokamak track record of reliability, and can provide the density measurements needed for control of ITER plasmas.

The combination of an interferometer and a Faraday rotation measurement using the same diagnostic probing beam has synergistic advantages since the measurements are complimentary and provide different information that can be exploited to improve any density profiles measurements. The integration of diagnostic designs to exploit other synergisms (such as reflectometry and interferometry) between diagnostic should be aggressively explored due to the restricted access and high cost of machine interfaces inherent in ITER.

REFERENCES

1. ITER Report TAC-95-15; TAC-JCT informal technical reviews.
2. F.C. Jobes and D.K. Mansfield, *Rev. Sci. Instrum.* 63: 5154 (1992).
3. Ashby, D.E. T.F. and Jephcott, D.F. Appl. Phys. Lett., 3, 13: (1963)
4. D.R. Baker and S.-T. Lee, Rev. Sci. Instrum. 49: 919 (1978).
5. G. Dodel and W. Kunz, Infrared Physics 18, 773 (1978)
6. T.N. Carlstrom, D.R. Ahlgren, and J. Croshie, Rev. Sci. Instrum. 59: 1063 (1988).
7. D. Viron, *Infrared and Millimeter Wave*, edited by K.J. Button, (Academic, New York, 1979), Vol. 2.
8. P. Gohil, et al., *Phys. Rev. Lett.* 61: 1603 (1988).
9. D.P. Hutchinson, at this conference
10. R.T. Snider and T.N. Carlstrom, *Rev. Sci. Instrum,* 63: 4979 (1992).
11. E.H. Farnum et al., J. Nucl. Mater. 219: 224 (1995).
12. E.S. Doyle, T.L. Rhodes, J.L. Doane and W.A. Peebles, *Rev. Sci. Instrum.* 66: 1233 (1995).

DEVELOPMENT OF DUAL CO$_2$ LASER INTERFEROMETER FOR LARGE TOKAMAK

Akira Nagashima, Yasunori Kawano, Takaki Hatae and Soichi Gunji

Department of Fusion Plasma Research,
Naka Fusion Research Establishment,
Japan Atomic Energy Research Institute,
Naka-machi, Naka-gun, Ibaraki-ken, 311-01, Japan

INTRODUCTION

To measure electron density of plasmas with sufficient reliability is one of the key issues of tokamak fusion research. Also in ITER (EDA) diagnostics, reliable electron density monitor is required as for absolute density calibration of Thomson scattering. The laser interferometry will play an important role in the future large tokamak device. This report presents a new technical progress about CO$_2$ laser interferometer, which has been developed in JT-60U. The choice of the laser is expected to eliminate refraction effect by dense and large plasma, Faraday rotation due to the large magnetic field especially in the case of a tangential chord, and the problem of optical transmission darkening which affects seriously for mirrors and windows in case of utilizing a light source of short wavelength around visible. The wavelength 10.6μm (λ_1) of CO$_2$ laser is not too long as FIR and not too short as visible which are sensitive to the Faraday rotation and the optics darkening, respectively. We consider the selection is very suitable to neglect the problems and to consider the commercial availability. For the interferometry, however, on large geometry machine with such wavelength, another wavelength (λ_2) is necessary to monitor the optical path difference. The wavelength is selected ordinarily not too close to the fundamental one ($x \equiv \lambda_1/\lambda_2 \gg 1$, or $\ll 1$), to keep the density resolution, because the signal to noise ratio is shown as equation (1), where ω_p is the plasma frequency, ω_1 the CO$_2$ laser frequency, L$_p$ and r$_f$ correspond to plasma length and fringe resolution.

$$\frac{S}{N} = \frac{1}{2} \frac{\omega_p^2}{\omega_1^2} \frac{L_p}{\lambda_1} \frac{1}{r_f} (1 - \frac{1}{x}) \tag{1}$$

For the interferometer of JT-60U, we have initially examined the 10.6μm and

Diagnostics for Experimental Thermonuclear Fusion Reactors
Edited by P. E. Stott *et al.*, Plenum Press, New York, 1996

235

3.39μm combination system for a toroidally tangent line of sight,[1,2] which is very similar to the ITER (CDA) proposal.[3] This configuration was introduced and successfully operated for the central density measurement of the " high βp H mode " plasmas in JT-60U, which made the world record of the fusion triple product of 1.1×10^{21} m^{-3}•s•keV.[4]

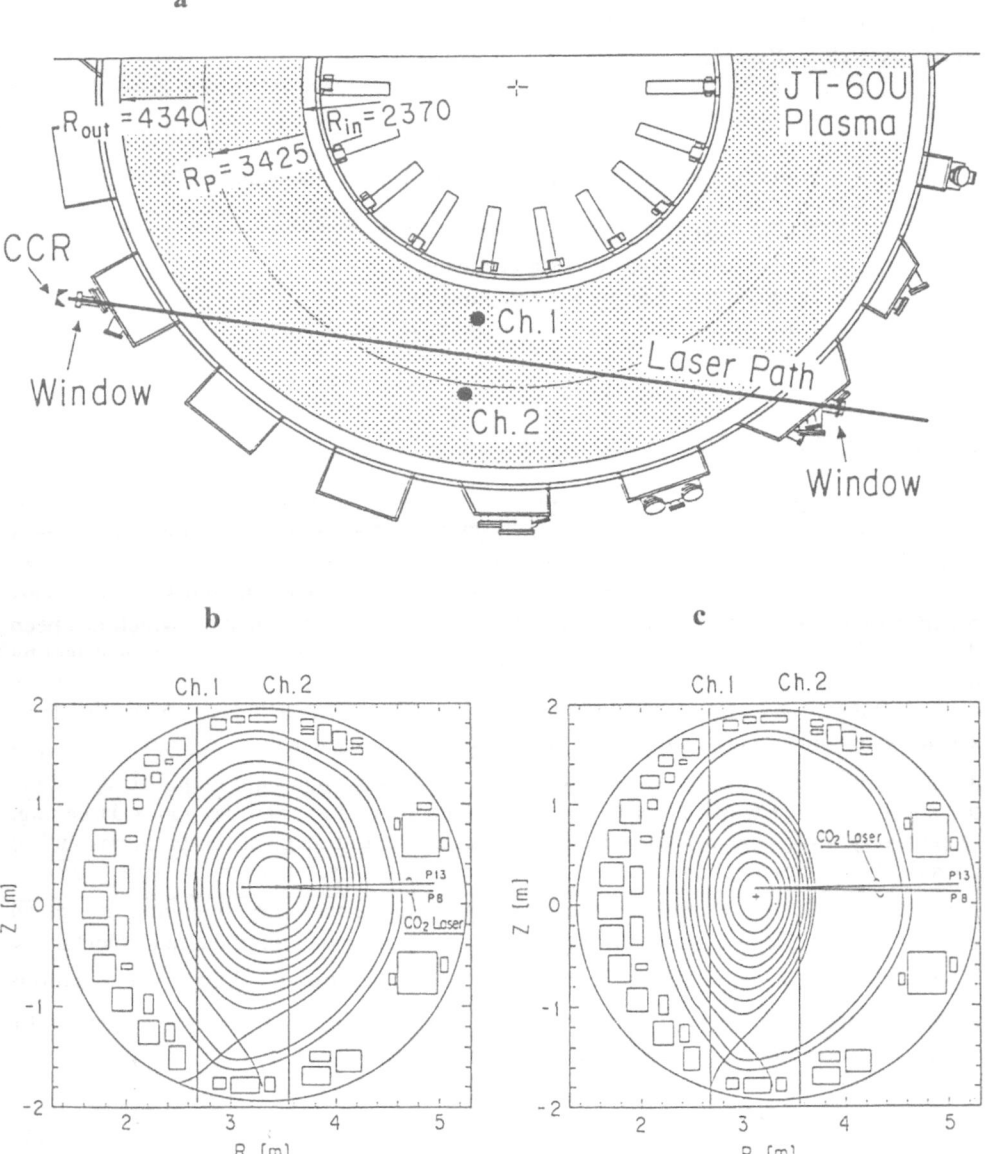

Figure 1. Laser beam line in the vacuum vessel, (a) top view of the JT-60U vacuum vessel with the laser path in the case of a large plasma configuration, (b) cross sectional view in the case of a large plasma configuration, (c) in the case of the high βp configuration, respectively. Vertical chords of ch. 1 and ch. 2 show the beam line for the 119μm alcohol laser interferometer.

However, there were some practical problems in the operation, when performing the high plasma current operation or rapid change of the plasma position. (1) Fringe loss on IR-HeNe was sometimes detected. (2) The effect of darkening of windows and mirrors were much sensitive for IR-HeNe comparing to CO_2. (3) The difficulty in alignment for IR-HeNe. (4) Independent modulation frequency used for each AOM (Acoust Optic Modulator) of the two colors resulted to baseline drift on phase detection. These problems must be resolved for the practical use as monitor. To resolve the problems, we introduced a dual CO_2 laser two color system which uses close wavelengths of 10.6μm and 9.27μm.[5] The system has advantages on practical problems such as mentioned above; robust to the darkening of mirrors and vacuum windows, better capability for large mechanical vibration, and easiness for laser beam alignment. Further more, the layout of optical components could be simplified by using close wavelengths. A new technique of common use of one AOM for the two colors could remove another serious problem of asynchronous effect concerning to the AOM drive frequencies. By these improvements, we have succeeded in very reliable measurement of the line electron density. The new configuration enabled us to investigate even a fast major disruption, which has been almost impossible. The observed density resolution, however, was relatively poor as was expected. In order to resolve the problem, we are developing an ultra high resolution phase comparator, with $1/10^4$ fringe resolution and a time response of 1μs. The resolution is a hundred times better than our present phase comparator, and will resolve the difficulty in detecting fine density change.

SCHEMATIC OF INTERFEROMETER

The dual CO_2 laser interferometer consists of seven major parts; CO_2 laser oscillators, frequency shifters, relay optics, retardation optics, vacuum windows, detectors,

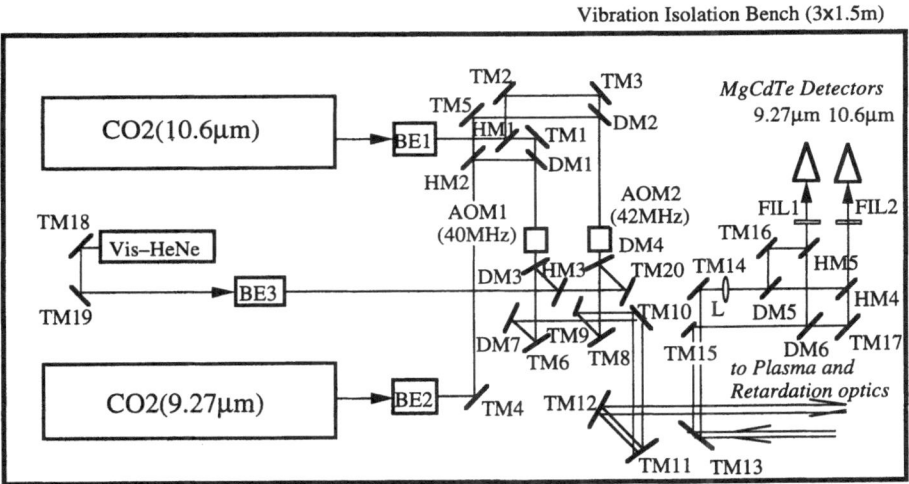

Figure 2. Layout of optical components on the vibration isolated bench. Optical components are indicated as BE: beam expanders, TM: total reflection mirror, HM: half mirror, DM: dichromatic mirror, FIL: filter for splitting/combining 9.27 μm and 10.6 μm, L: lens, AOM: acoust-optic-modulator.

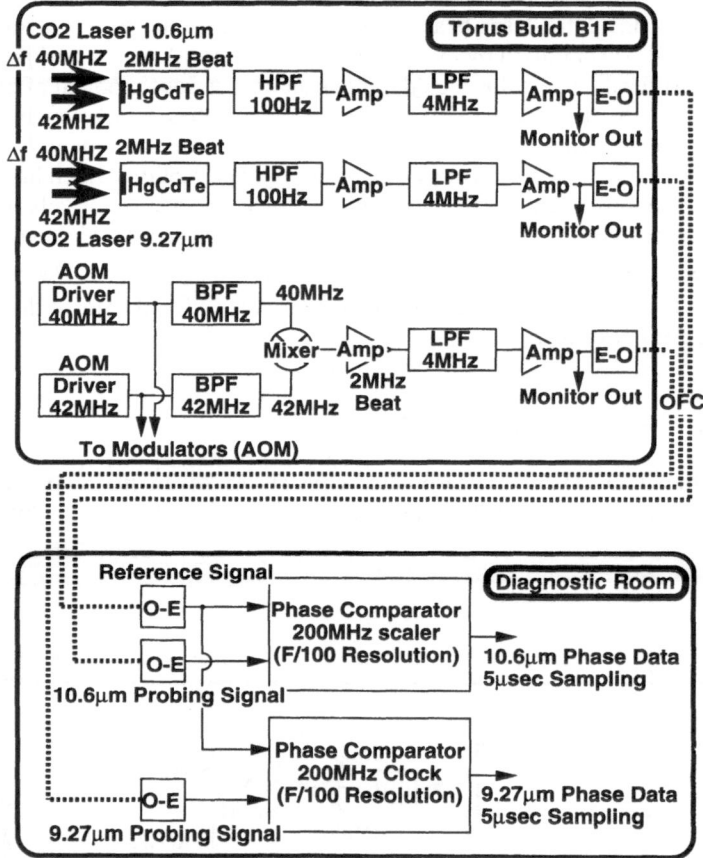

Figure 3. Schematic of the signal detection and the data acquisition. Signal detection part consists of two same probing branch and a reference branch. Each interference light is detected by a HgCdTe detector. Reference beat signal is generated from 40 and 42MHz AOM driver signals by an electrical frequency mixer.

and data acquisition. Figure 1(a) shows the laser beam line in the vacuum vessel of JT-60U. The laser beam is launched into the vacuum vessel tangentially through ZnSe windows at an equatorial port. A CCR (Corner Cube Reflector) is located outside of the window. Laser path length in the plasma is about 6m (for one way) for large plasma configuration shown in figure 1(a). 10W single mode CW CO_2 laser oscillators, frequency modulators and HgCdTe detectors are laid out on a vibration isolated bench. Figure 2 shows the configuration. The laser oscillators are tuned and stabilized to the different lasing branches 10P(20) of 10.6μm and 9R(20) of 9.27μm. Each laser beam is divided into two parts. One is for the probing which propagate into the plasma and another is for the phase reference for heterodyne detection. The center beat frequency of 2MHz to utilize the existing phase comparators is produced by the differential use of 40 and 42MHz AOM (ISOMET model 1206B-7) for each CO_2 laser beams. Figure 3 shows the schematic of signal detection and data acquisition. The differential frequency of AOM and two probe beat signals of 10.6 and 9.27μm from interferometer are transmitted to the processing area by optical fiber line. The phase comparator can measure the phase shift with 1/100 fringe resolution by a 200 MHz scale clock. The phase data are sampled by every 5μs at muximum speed over 15sec of the plasma duration time, and processed in the main frame computer (Fujitsu M780/10).

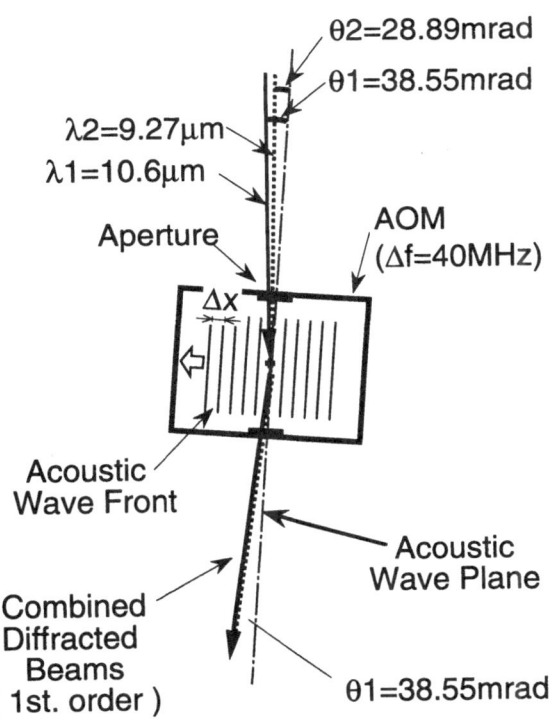

Figure 4. Alignment of the common use of an AOM for different wavelength CO_2 beams. Incident angles are adjusted so that output beams of 10.6μm and 9.27μm are coaxially propagated.

SIMULTANEOUS TWO COLOR FREQUENCY MODULATION BY AN AOM

Even very small difference of two color beat frequencies results in base line drifts on the calculated density trace. The effect was difficult to remove when use of independent AOM pairs for two color modulation. To resolve the problem, we developed a new technique of two color simultaneous modulation by single AOM. The very close wavelengths of the new interferometer enables us to make use of the new technology. Figure 4 shows the schematic of two laser beams for 40MHz frequency upshift. A medium of the AOM is made of single crystal germanium and the Bragg's angles of the 10.6μm and the 9.27μm are 38.55 and 33.72mrad, respectively. For 10.6μm, incident angle is set with normal condition, that is at its Bragg's angle. For 9.27μm, incident beam is adjusted at an angle slightly sifted from the exact angle of it own, that is 28.89mrad (=. 38.55mrad x 2 - 33.72mrad) so that the two color output angle be the same. Though the incident angle of 9.27μm is not optimum for its diffraction condition, sufficient diffraction power is available.

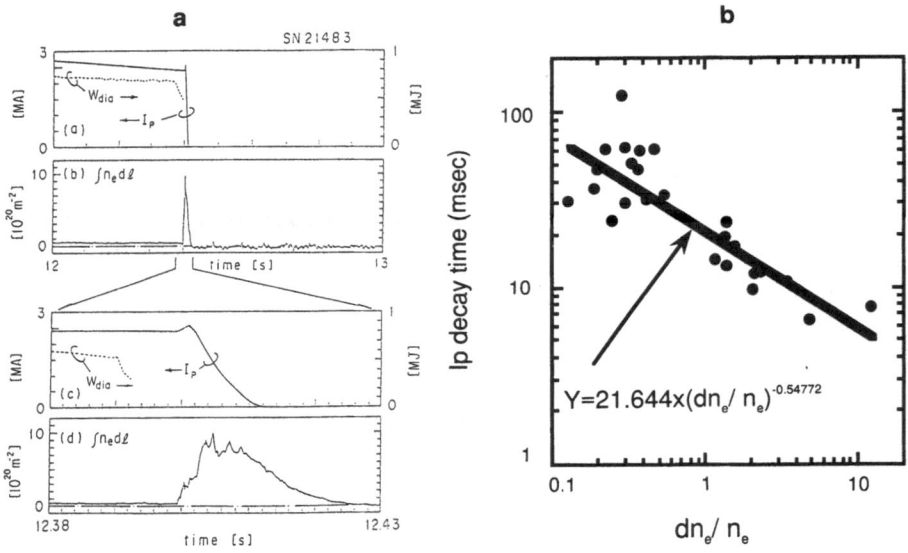

Figure 5 Electron density behavior of a high li disruption (a) typical waveform of disruption and (b) obtained relation between the current decay time and the density change at disruption.

OPERATION RESULT IN MAJOR DISRUPTION

At the major disruption, it is often difficult to measure the electron density by conventional far infrared alcohol laser interferometers. A rapid change of the electron density or the large density gradient prevents reliable phase detection. The previous 10.6μm and 3.39μm combination system is also used to suffer from the fringe loss of the 3.39 mm signal due to a large mechanical vibration and displacement of mirrors at the disruption. These problems become more serious with an increase in the disruption speed.

In contrast to above systems, the dual CO_2 combination is more reliable and it succeeded in measuring the density of fast disruptions. Figure 5(a) shows a typical waveform of major disruption, and (b) a relation[6] can be seen between the rapid change of electron density and the current decay. The behavior is obtained at a high li disruption with a characteristic decay time of plasma current defined by $Ip/(dIp/dt)$ of 11ms.[7] The electron density is $0.32\times10^{20}m^{-2}$ before the energy quench. It rises to 3.4 times larger value as $1.1\times10^{20}m^{-2}$ just before the start of current decay. Finally, it reaches the highest value of $9.6\times10^{20}m^{-2}$ after multiple density spikes.

ULTRA HIGH RESOLUTION PHASE COMPARATOR

The phase resolution of the present comparator is 1/100 fringe, which corresponds to a line electron density of about $2\times10^{18}m^{-2}$. The effective density resolution, however, is reduced significantly in the dual CO_2 combination due to the very close wavelengths. The reduction factor for a two color scheme is $(1-1/x)$. Therefore the reduction factor is 0.125 for dual CO_2 system $(x=10.6/9.27)$, which means that the effective phase resolution for the density measurement is 1/12.5 fringe. This corresponds to a relatively poor resolution of line density about $1.6\times10^{19}m^{-2}$, which agrees with the observed resolution. To improve the density resolution, a ultra high resolution phase comparator has been developed. The phase resolution is originally designed as $1/10^4$ fringe for 2MHz signal, which corresponds to the time resolution of 50ps.[3] This specification depends on the discrimination accuracy of the high speed phase measuring module in the comparator. A test signal generated by a synthesized oscillator is divided into two signals. One signal is for direct input as a reference signal and another is an input as a probing signal via a variable electrical delay line. When the delay line lengths is changed by a step, the measured time difference change between both signals indicates the accuracy of the time discrimination. Measured time differences are plotted against the delay line length in Figure 6. The test frequency is 1MHz and the number of sampling is 1000, respectively.

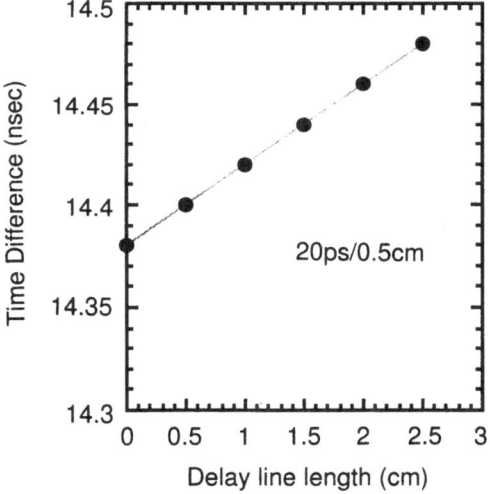

Figure 6. Measured time differences against the delay line length. Each number of sampling is 1,000. Time difference of 20ps is well discriminated.

One step of the electrical delay line length is 0.5 cm which corresponds to 20ps. It is found that the time difference is well discriminated for each case. The jitters of measured times are also an important index. The histogram of independent measurements for 10,000 sampling shows the standard deviation of 30ps. From these results, it is concluded that the accuracy of the high speed phase measuring module is determined by the jitters with the standard deviation of 30ps, which satisfies the designed value. The new phase comparator has been assembled and an initial measurement for an actual plasma is planned in 1995. This new comparator will enable more precise analysis of the fast density change.

DISCUSSION

As mentioned above, two AOMs provide a 2MHz beat signal. The ultra high resolution phase comparator was examined about the capability over the beat frequency range from 10kHz to 50MHz. Therefore even 40 MHz beat signals is directly acceptable, and only one commercial AOM is sufficient for the heterodyne detection. An electrical frequency converter from 40MHz to 2MHz can also reduce the number of AOM and complexity of optics. We expect that the interferometer will become more stable by this way of simplification of the AOM system.

A design of a multi-chordal interferometer for future large devices must face a port limitation issue. Vertical or tangential ports which have a pair of windows at each side of a vacuum vessel are suitable to minimize the influence from the darkening and the vibration of reflection mirrors. If reflection mirrors must be installed inside of the vacuum vessel to provide sufficient number of diagnostic chords, the dual CO_2 combination is one of the possible candidates. Its unique ability of the vibration compensation at the infrared radiation range is an advantage point. The relatively high lasing power, the simple configuration and easiness of commercial availability to obtain the required hardwares are also attractive.

SUMMARY

The dual CO_2 laser interferometer has been developed for large tokamaks. The combination of $10.6\mu m$ and $9.27\mu m$ two colors are utilized to simultaneous measurement of the electron density and the optical path components. A single AOM is commonly used for different wavelengths as the key technique to stabilize the system.

Several data show the basic performance of the dual CO_2 interferometer under the different operational modes. The dual CO_2 system can measure the electron density of 3MA plasmas in the case of JT-60U. The observed effective density resolution is $2 \times 10^{19} m^{-2}$, which is determined by the phase resolution of the present phase comparator. It is presented that the stabilized laser mode with retardation optics is suitable for long duration discharges. The electron density behavior of fast major disruptions is successfully measured without a fringe loss.

The ultra high resolution phase comparator has been developed to improve the density resolution of the dual CO_2 laser interferometer. The phase resolution is designed as $1/10^4$ fringe for the 2MHz signal which corresponds to the time measurement accuracy of 50ps, which is a hundred times better than that of the present one. The achieved accuracy of discrimination is 30ps. The study of a fast density change can be significantly improved by use of the dual CO_2 system.

Through this work, the feasibility of the dual CO_2 interferometer is well demonstrated for future large machines such as ITER.

ACKNOWLEDGMENTS

The authors acknowledge useful discussions with Drs.T.Matoba, T.Fukuda, S.. Ishida, H.Shirai, H.Yoshida and R.Yoshino. The authors appreciate Y.Endo, T.Kakizaki, M.Ohzeki, H.Sunaoshi, S.Uno, M.Shitomi and M.Uramoto for their technical cooperation. The authors wish to thank Drs. M.Mori, M.Kikuchi, H.Ninomiya, M.Shimada, M.Nagami, A.Funahashi, M.Azumi, H.Kishimoto and Y.Tanaka for their continuous support and encouragement.

REFERENCES

1 Y.Kawano, A.Nagashima, S.Ishida, T.Fukuda and T.Matoba, CO2 laser interferometer for electron density measurement in JT-60U tokamak, *Rev. Sci. Instrum.* 63:4971 (1992).

2 Y.Kawano, T.Hatae, A.Nagashima, T.Fukuda and T.Matoba, First Operation Result of CO2 Laser Interferometer, *JAERI-M report* 93-057, 351 (1993) .

3 ITER Tokamak Device, ITER Documentation Series 25 (IAEA, 1991) Vienna.

4 M.Mori, S.Ishida, T.Ando, K.Annoh, N.Asakura, M.Azumi, A.A.E.van.Blokland, G.J.Frieling, T.Fuii, T.Fujita, T.Fukuda, A.Funahashi, T.Hatae, M.Hoek, M.Honda, N.Hosogane, N.Isei, K.Itami, Y.Kamada, Y.Kawano, M.Kikuchi, H.Kimura, T.Kimura, H.Kishimoto, A.Kitsunezaki, K.Kodama, Y.Koide, T.Kondoh, H.Kubo, M.Kuriyama, M.Matsuoka, Y.Matsuzaki, N.Miya, M.Nagami, A.Nagashima, O.Naito, H.Nakamura, M.Nemoto, Y.Neyatani, H.Ninomiya, T.Nishitani, T.Ohga, S.Ohmori, M.Saidoh, A.Sakasai, M.Sato, M.Shimada, K.Shimizu, H.Shirai, T.Sugie, H.Takeuchi, K.Tani, K.Tobita, S.Tsuji, K.Ushigusa, M.Yamada, I.Yonekawa, H.Yoshida, R.Yoshino, Achievement of High Fusion Triple Product in the JT-60U High βp H Mode, *Nucl. Fusion* 34, 1045 (1994).

5 Y.Kawano, A.Nagashima, T.Hatae and S.Gunji, Development of Dual CO2 Laser Interferometer for Large Tokamak, *JERI-Research* 95-023 (1995)

6 Y.Kawano, et.al., to be published

7 R.Yoshino, Y.Neyatani, N.Hosogane, Y.Neyatani, N.Hosogane, S.W.Wolf, M.Matsukawa, H.Ninomiya, The softning of current quenches in JT-60U, *Nucl. Fusion* 33, 1599 (1993).

INFRARED LASER DIAGNOSTICS FOR ITER

D. P. Hutchinson,[1] R. K. Richards,[2] and C. H. Ma[2]

[1]Instrumentation and Controls Division
[2]Physics Division
 Oak Ridge National Laboratory
 Oak Ridge, TN 37831

ABSTRACT

Two infrared laser-based diagnostics are under development at the Oak Ridge National Laboratory (ORNL) for measurements on burning plasmas such as ITER. Our primary effort is the development of a CO_2 laser Thomson scattering diagnostic for the measurement of the velocity distribution of confined fusion-product alpha particles. This diagnostic utilizes small-angle collective scattering of infrared light from the electron cloud surrounding the alpha particles. Key components of the system include a high-power, single-mode CO_2 pulsed laser, an efficient optics system for beam transport and a multichannel low-noise infrared heterodyne receiver. A successful proof-of-principle experiment has been performed on the Advanced Toroidal Facility (ATF) stellarator at ORNL utilizing scattering from electron plasma frequency satellites. The diagnostic system is currently being installed on Alcator C-Mod at MIT for measurements of the fast ion tail produced by ICRH heating. A second diagnostic under development at ORNL is an infrared polarimeter for Faraday rotation measurements in future fusion experiments. A preliminary feasibility study of a CO_2 laser tangential viewing polarimeter for measuring electron density profiles in ITER has been completed. For ITER plasma parameters and a polarimeter wavelength of 10.6 μm, a Faraday rotation of up to 26° is predicted. An electro-optic polarization modulation technique has been developed at ORNL. Laboratory tests of this polarimeter demonstrated a sensitivity of $\leq 0.01°$. Because of the similarity in the expected Faraday rotation in ITER and Alcator C-Mod, a collaboration between ORNL and the MIT Plasma Fusion Center has been undertaken to test this polarimeter system on Alcator C-Mod. A 10.6 μm polarimeter for this measurement has been constructed and integrated into the existing C-Mod multichannel two-color interferometer. With present experimental parameters for C-Mod, the predicted Faraday rotation was on the order of 0.1°. Significant output signals were observed during preliminary tests. Further experiment and detailed analyses are under way.

COLLECTIVE THOMSON SCATTERING

Our primary effort is in the development of a CO_2 laser Thomson scattering diagnostic for the measurement of the velocity distribution of confined fusion-product alpha particles.[1] Collective Thomson scattering measurements at the CO_2 laser wavelength generally requires operation at small scattering angles. For fusion relevant plasma conditions, this scattering angle between the source and receiver is typically around one degree or less. Such small angles present problems with beam alignment and stray light rejection. A proof-of-principle test of the diagnostic system has been performed on ATF with a measurement of an electron resonance feature,[2] which examined these problems and successfully demonstrated the capability of this diagnostic. As an application for this diagnostic, the equipment used in the proof-of-principle test has been upgraded and moved to the Alcator C-Mod tokamak for the measurement of an ICRH-produced ion tail.

Diagnostics for Experimental Thermonuclear Fusion Reactors
Edited by P. E. Stott *et al.*, Plenum Press, New York, 1996

245

The receiver system makes use of heterodyne detection for reduction of stray light and improvement in the signal-to-noise ratio. The post heterodyne signal-to-noise ratio, S/N, is given by:

$$S/N = \frac{P_S}{P_s + P_N} \sqrt{1 + \overline{B\tau}} \tag{1}$$

where B is the receiver bandwidth, τ is the integration time, P_S is the scattered signal power, and P_N is the system noise. A large signal-to-noise ratio requires a large product of bandwidth time integration time. The bandwidth is set by the Doppler broadening of the ions to be measured and the integration time is limited to the laser pulse length. For the alpha particles in ITER the full spectral width is expected to be around 10 GHz, which will limit the bandwidth of a spectral channel used in measuring the spectral shape to around 1 GHz. According to Eq. (1) with a laser pulse length of 1 microsecond (as used in the proof-of-principle test), the signal-to-noise ratio will have a maximum value of 32. For the experiment at Alcator C-Mod (which also has a 1 GHz bandwidth), the pulse length has been extended to 5 microseconds to permit a signal-to-noise ratio of up to 71. For ITER, the laser pulse length must be further increased to permit larger signal-to-noise ratios.

The source laser is a modified Lumonics TEA-103. It consists of four cells, unstable resonator optics, and is controlled by a low power cw injector laser which sets the wavelength, polarization, and lengthens the pulsewidth. A small quantity of tripropylamine is added to the gas mix which allows the discharge to remain stable with increased nitrogen levels, which produces longer pulse lengths.

For this experiment, the velocity is determined by measuring the scattering amplitude in k space (i.e., in scattering angle) rather than conventional frequency space. For ITER the system will be expanded to include measurements of scattering in frequency and time.

INTERFEROMETER/POLARIMETER

The second diagnostic under development at ORNL is an interferometer/polarimeter system for phase shift and Faraday rotation measurements in future fusion experiments. The feasibility of the system for ITER has been investigated both theoretically and experimentally. Theoretical analyses have been carried out to study the wave propagation in ITER for both vertical and tangential viewing systems. Computer codes have been developed and have been used to calculate the phase shift, ϕ, Faraday rotation angle, θ_f, ellipticity, ε, and the angle of beam refraction. The results are summarized in Table I. The values corresponding to Alcator C-Mod plasma are also given for comparison.

Table I. Summary of plasma effects on phase shift, Faraday rotation angle, ellipticity, and angle of refractions for various wavelengths using parabolic electron and current distributions.

	C-Mod	ITER							
		Viewing vertically				Viewing tangentially			
Wavelength (microns)	10.6	1.0	3.39	10.6	119	1.0	3.39	10.6	119
Major radius (m)	0.65	8.06							
Minor radius (m)	0.2	3.01							
Central density ($10^{20}/m^3$)	10	1.27							
Plasma current (ma)	3	25.0							
Toroidal field (t)	9	5.7							
Faraday rotation (deg)	1.4	0.018	0.2	2	248	0.23	2.7	26	3277
Ellipticity (e)	$7.6 \cdot 10^{-6}$	$9.6 \cdot 10^{-11}$	$4.3 \cdot 10^{-8}$	$1.3 \cdot 10^{-5}$	$0 \le e \le 1$	$1.3 \cdot 10^{-10}$	$5.7 \cdot 10^{-8}$	$1.7 \cdot 10^{-5}$	$0 \le e \le 1$
Phase shift (fringes)	2.5	0.46	1.6	4.8	55	1.4	4.6	14.4	162
Angle of refractions (deg)	$3 \cdot 10^{-3}$	$6.5 \cdot 10^{-6}$	$7.5 \cdot 10^{-5}$	$7.3 \cdot 10^{-4}$	$9.2 \cdot 10^{-2}$	$9.6 \cdot 10^{-6}$	$1.1 \cdot 10^{-4}$	$1.1 \cdot 10^{-3}$	$1.4 \cdot 10^{-1}$

An electro-optic polarization modulation technique has been successfully developed at ORNL to achieve the high sensitivity and time resolution required for the measurement of the Faraday rotation. The polarimeter is a modification of a previous system developed for CIT.[3] Because of similarities in the expected phase shift and Faraday rotation between ITER and Alcator C-Mod (Table I), a collaboration between ORNL and the MIT Plasma Fusion Center has been undertaken to test this polarimeter system on Alcator C-Mod. A 10.6 μm polarimeter for this measurement has been constructed and integrated into the existing C-Mod multichannel two-color interferometer. Detailed analyses and experiments of the previous systems have been reported.[4] Only a brief description of the modified system is therefore presented in the following.

The two-color heterodyne interferometer consists of ten CO_2 laser (10.6 μm) channels and three HeNe laser (0.6328 μm) channels viewing the plasma vertically with a Michelson geometry. The main components are: a CO_2 laser, a germanium Bragg cell, and a HgCdTe thermoelectrically cooled detector array. The acoustic optic Bragg cell diffracts ~50% of the CO_2 laser power into a reference beam. This cell also introduces a frequency shift of 40 MHZ in the diffracted beam. The undiffracted probing beam is passed through a CdS quarterwave plate, a CdTe electro-optic modulator and a mechanical polarization rotator. Emerging from the polarization modulator, the Gaussian beam is expanded to an elliptical beam with a profile of 1×20 cm via cylindrical mirrors in order to view the plasma cross section accessible by the viewing ports. The return beam in the two-pass system is decollimated and is passed through an analyzer. The reference beam is also expanded and is guided to the signal detector array to mix with the probing beam.

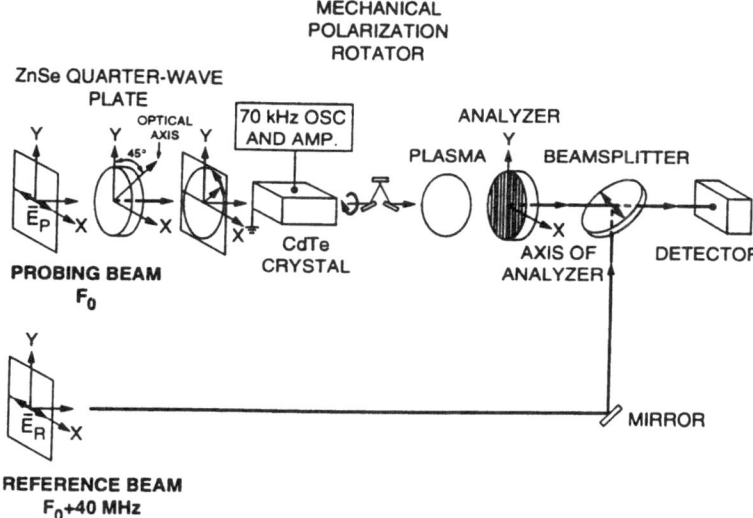

Figure 1. Illustration of the electro-optic polarization-modulation technique used in the CO_2 laser polarimeter on Alcator C-Mod.

Figure 1 shows a simplified sketch of the arrangement of the polarimeter. Both the probing beam and the reference beam are initially linearly polarized with their electric fields in the x direction. The quarter-wave plate is placed with its optical axis at 45° to the y-axis, thereby changing the probing beam into a circularly polarized wave. The modulator is driven by a 70 kHz oscillator-amplifier system with the electric field applied in the y direction. The mechanical polarization rotator consists of three mirrors mounted on a rotatable frame such that the polarization of the probing beam changes by twice the angle of the frame rotation. Emerging from the plasma, the probing beam is passed through an analyzer. The analyzer is oriented with its axis in the x direction, so that only the x component of the probing beam is mixed with the reference beam at the detector. The output of the signal detector V_s can be expressed by the following relation:

$$V_s = RP_pJ_1(\theta_m)\sin(2\theta)\sin(\omega_m t) + R(P_pP_r)^{1/2}[\cos(\theta)\cos(\Delta\omega t + \phi)$$

$$+ \sin(\theta)\sin(\Delta\omega t + \phi + \phi_m)] + \textit{terms of dc and other frequencies,} \qquad (2)$$

where R is the responsivity of the detector; θ is the sum of the polarization rotation angle due to the mechanical rotator θ_b, and the Faraday rotation in plasma θ_f; ω_m is the modulation frequency; $\Delta\omega = 40$ MHz; $J_1(\theta_m)$ is the Bessel function of the first kind with order 1; and P_p and P_r are the power of the probing and reference beam at the detector, respectively. The modulation of the phase shift between x component and y component of the electric field ϕ_m is sinusoidal and can be written as: $\phi_m = \theta_m \sin(\omega_m t)$, where θ_m is the amplitude of the modulation angle in degrees and is related to the modulation voltage V_m and the half-wave voltage of the CdTe crystal V_π by: $\theta_m = 180 \, V_m/V_\pi$. The detector signal at the modulation frequency is synchronously detected by a lock-in amplifier. For small θ_f and by setting the mechanical polarization rotator at $0°$ position ($\theta_b = 0°$), the output voltage of the amplifier is then $V_{out} = ARP_pJ_1(\theta_m)\sin(2\theta_f)$, where A is the voltage gain of the amplifier. The values of the individual components in this equation may be lumped into a single calibration constant V_0 so that the equation becomes

$$V_{out} = V_0 \sin(2\theta_f), \qquad (3)$$

where $V_0 = ARP_pJ_1(\theta_m)$. The value of V_0 can be obtained by setting the mechanical polarization rotator at a calibration angle θ_c of a few degrees ($\leq 2°$) away from its original $0°$ position and measuring a calibration voltage V_c at the output of the lock-in amplifier before the beginning of the plasma discharge. The voltage V_0 is related to V_c and θ_c by: $V_0 = V_c/\sin(4\theta_c)$. The V_0 as determined by this technique calibrates the polarimeter in a manner that does not require the absolute knowledge of the laser power, the detector responsivity, the modulation angle, or the gain of the amplifier. For small rotation angles, $(\theta_f + 2\theta_c) \leq 5°$, V_{out} is a direct measure of θ_f, since $\sin[2(\theta_f + 2\theta_c)] \approx 2(\theta_f + 2\theta_c)$, and the above equations are reduced to the form

$$\theta_f = 2\theta_c(V_{out}/V_c - 1). \qquad (4)$$

The outputs of the lock-in amplifiers are digitized for computer storage and processing. The values of the Faraday rotation angles along the ten vertical chords are calculated according to Eq. (4).

A test bench of the polarimeter system was set up at ORNL to determine the sensitivity of the polarimeter. The calibration experiment was carried out with the mechanical polarization rotator oriented at $1°$ position. Under this condition, a Faraday rotation of $2°$ was simulated. The output voltage of the lock-in amplifier was measured to be about 350 mV. With both beams blocked, an output voltage in the range of 1-2 mV was observed, thus a polarimeter sensitivity better than $0.01°$ was achieved. The polarimeter has been added in the existing two-color interferometer system on Alcator C-Mod. With present experimental parameters for C-Mod, the predicted Faraday rotation was on the order of $0.1°$. Significant output signals were observed during preliminary tests. Further experiment and detailed analyses are under way.

This work was sponsored by the Office of Fusion Energy, U.S. Department of Energy, under Contract No. DE-AC05-84OR21400 with Lockheed Martin Energy Systems.

REFERENCES

1. R. K. Richards, C. A. Bennett, K. L. Fletcher, H. T. Hunter, and D. P. Hutchinson, *Rev. Sci. Instrum.* 59:1556 (1988).
2. R. K. Richards, D. P. Hutchinson, C. A. Bennett, H. T. Hunter, and C. H. Ma, *Appl. Phys. Lett.* 62:28 (1993).
3. C. H. Ma, D. P. Hutchinson, and K. L. Vander Sluis, *Rev. Sci. Instrum.* 59:1629 (1988).
4. C. H. Ma, D. P. Hutchinson, R. K. Richards, J. Irby, and T. Luke, *Rev. Sci. Instrum.* 66:376 (1995).

A THOMSON SCATTERING SCHEME FOR OBTAINING
T_e AND n_e PROFILES OF THE ITER CORE PLASMA

C Gowers[1], P Nielsen[1], F Orsitto[2], F J Pijper[3], H Salzmann[4], B Schunke[1]

[1] JET Joint Undertaking, Abingdon, Oxon, OX14 3EA, UK
[2] Ass. EURATOM-ENEA sulla Fusione, CRE Frascati, 00044 Frascati, I
[3] FOM Inst. voor Plasmafysica, Rijnhuizen, 3430 BE Nieuwegein, ND
[4] MPI für Plasmaphysik, 85748 Garching, D

ABSTRACT

The hostile environment, the very restricted access geometry, the required spatial resolution and the target accuracy all impose severe constraints on the design of many techniques proposed for spatially resolving ITER plasma parameters. Despite the constraints it appears feasible using LIDAR Thomson scattering to make measurements of the electron temperature and density profiles in the ITER core plasma. A generic scheme using reflective optics for the front end is adopted. The scheme uses a folded mirror system positioned inside a shielded labyrinth located in a standard radial port. The same mirror surfaces and vacuum window are used for both the laser input and the collection path. Radial and tangential sight lines are considered. The sight line uses a single 30 cm diameter blanket penetration and an exit window of <20 cm diameter. The specified spatial resolution of 30 cm can be met using existing laser and detector technology. Calculations of the expected accuracy of the temperature measurement are presented for different laser/detector combinations. The results are based on extrapolations from the existing JET LIDAR system and are therefore bounded by the constraints of a realistic optical system. The background to the design will be reviewed and some details of the front end optical design will be presented. Possible methods of obtaining some of the necessary calibration data will also be discussed. The critical outstanding problem for the design is to identify suitable materials which can simultaneously withstand the heat, gamma, neutron and laser beam fluxes.

Diagnostics for Experimental Thermonuclear Fusion Reactors
Edited by P. E. Stott *et al.*, Plenum Press, New York, 1996

249

INTRODUCTION

In the following the applicability of Thomson scattering to the ITER core plasma is investigated. This is not done to the extent of a complete design study. The main scope of the work presented is to demonstrate the access requirements and to show that the requirements for the measurement of electron temperature and density profiles, i.e. spatial resolution, parameter range and accuracy, can be fulfilled.

GENERAL CONSIDERATIONS

In this section a few general considerations regarding the applicability of different scattering schemes to the ITER geometry and plasma parameters are discussed. From this discussion some guidelines for the design of suitable diagnostics are derived.

LIDAR vs Conventional Scattering

There are two ways to provide spatial resolution for a scattering system. Either the laser beam is crossed with the beam of the collection optics ("conventional" set-up) or the origin in space of the scattered light is determined by a time-of-flight method (LIDAR). In the conventional set-up the spatial resolution can easily be made high, being limited only by the reduced signal-to-noise ratio (SNR) resulting from the fact that the number of scattered photons decreases with the length of the scattering volume. In the LIDAR scheme the time-of-flight differences and thus the spatial resolution are maximised for a backscattering arrangement and go to zero for small scattering angles (forward scattering). For back-scattering, the spatial resolution is given by

$$\Delta x = \frac{c}{2} \sqrt{ (\tau_{laser}^2 + \tau_{detection}^2) }$$

where c is the speed of light, τ_{laser} is the laser pulse duration and $\tau_{detection}$ is the risetime of the detection system.

In the LIDAR scheme, a good spatial resolution requires a short, high-energy laser pulse and a high-speed detection system. Provided the spatial resolution is sufficient, these drawbacks are however more than compensated for by the following factors

i)   the plasma background radiation added to the scattered signal is very small due to the high-speed detection, thus improving the SNR
ii) in the backscattering arrangement the collection and laser optics are coupled mechanically, making alignment both stable and simple
iii) only one access route/window to the plasma is required,
iv) all the spatial channels are transmitted via one line of optics. This means that simpler relay optics (and much less space close to the machine) are required than for the wide field-of-view collection optics needed in a conventional system,
v)   all spatial channels are measured with one single detection system. This has the advantage that the absolute (density) calibration of the system need not be performed separately for each spatial channel, comparison to a density line integral is sufficient (see 'Density Calibration' section).

These arguments, especially those pertaining to the access problem, lead to the conclusion that a LIDAR system is to be preferred, as long as the spatial resolution is sufficient.

Choice of Laser Wavelength and Dynamic Range of the Measurements

For the LIDAR backscattering scheme the choice of laser wavelength is based on the following criteria:

i) high-speed detectors with a sufficiently large sensitive area (to cope with the large étendue, dF*dΩ, of the collection optics) are presently available only in the visible spectral range (MCP photomultiplier tubes, streak tubes). The laser should therefore yield scattering spectra which cover this spectral range for the electron temperatures of interest.

ii) the 'blue' wing of the scattered spectrum should be located in the accessible spectral range since this wing varies more strongly with temperature (at high temperatures) than the 'red' one.

iii) there should be a possible laser candidate already with today's laser technology.

Fig. 1a shows Thomson scattered spectra for a laser wavelength of 1.06 μm. It shows that for temperatures above 2 - 3 keV, spectral power measurements in the wavelength range 400 nm - 800 nm could well resolve the temperature.

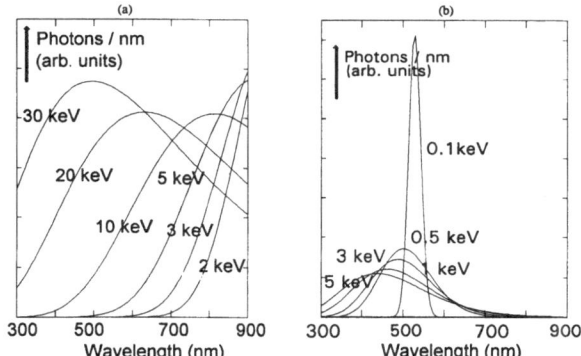

Figure 1a & 1b Scattered spectra for Nd fundamental and 2nd harmonic wavelengths respectively

For temperatures below that value the use of the harmonic laser wavelength seems advantageous, as can be seen from Fig.1b. For this purpose, frequency doubling of the laser is required which can be done with an efficiency well above 50%. So as not to confuse the measurements at both wavelengths, use can be made either of the different polarisation of the fundamental and the harmonic and thus of the related scattered light or of a delay of one of the laser pulses by about 100 ns.

However, account must be taken of the spectral response of commercially available photocathodes. Their quantum efficiency tends to decrease at both the red and blue edge of the measurements. The combinations of different laser wavelengths with different types of photocathodes has been investigated and the results are summarized in Fig.2. The figure shows the dynamic range for electron temperature measurements. The use of GaAs photocathodes extends the long wavelength limit of the accessible spectral region. The error curves are derived by calculating the expected signal in 20 equally sized spectral channels over a fixed wavelength range. The spectra are convoluted with simplified quantum efficiency curves; i.e. a straight line between the wavelength extremes. GaAs is taken to have a constant 10 % quantum efficiency between 600 - 900 nm; MA-1 (extended red tri-alkali) is 10% at 400 nm and 1% at 800 nm.

The numbers for the error bars in Fig.2 are scaled to match the actual performance of the JET LIDAR system. All the curves are then calculated using fixed geometry (F#), fixed laser energy and fixed transmission. As can be seen, measurements with the Nd fundamental wavelength can cover a dynamic range (defined here as <20% statistical error) from 2 keV up to 50 keV when using GaAs detectors. The dynamic range for measurements with the frequency doubled laser radiation can extend up to 6 keV. Thus the simultaneous use of these two wavelengths with some GaAs detectors will indeed cover the interesting temperature range without a gap. Without using GaAs photocathodes the accuracy is inadequate in the 5-10 keV range. It is interesting to note that this analysis shows that a laser at 790 nm (e.g. Alexandrite) can cover the full temperature range using extended red photocathodes.

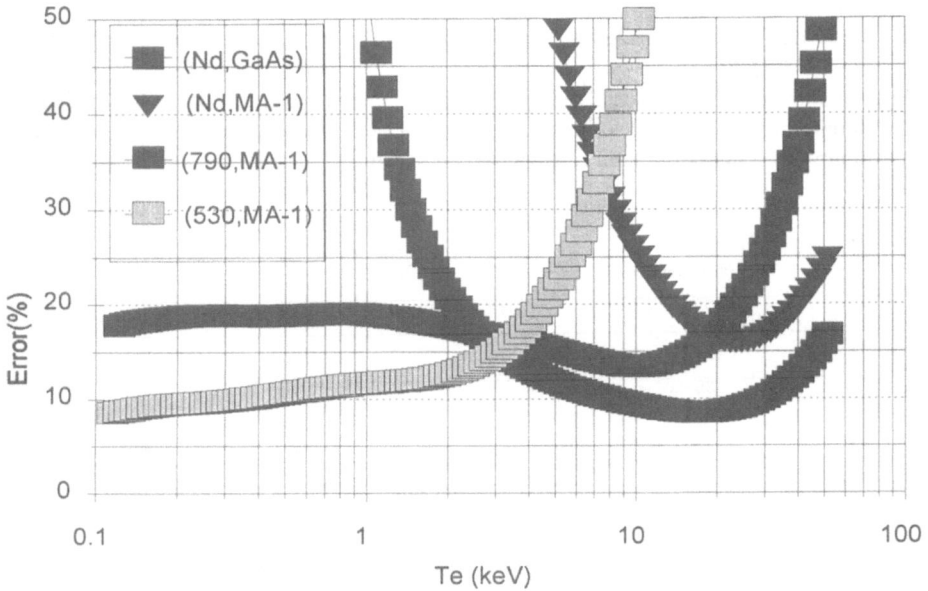

Fig.2 Relative error in temperature from Thomson scattering in five cases:
(Nd,GaAs) - Nd laser, 1064 nm, detector quantum efficiency $\eta = 0.1$ from 400 to 900 nm
(Nd,MA-1) - Nd laser, 1064 nm, detector $\eta = 0.1$ at 400 nm dropping to 0.01 at 800 nm
(790,MA-1) - 790 nm laser, same detector
(530,MA-1) - Nd laser 2ω, 530 nm, detector $\eta = 0.1$ at 400 nm dropping to 0.01 at 800 nm

Access

Access will be provided through a large horizontal port. Since the size of the window(s) on this port should be as small as possible, the front collection optics must be located inside the port and the window(s) should be positioned at a waist of the beam(s) of collected light. In addition, it is advisable to have this/these window(s) shielded from the flux of streaming neutrons. We also believe that the first optical element, which has to face the plasma, should be a mirror, preferably a metallic one.

These considerations lead to a folded optical arrangement within the port. In order to make the output window as small as possible, these optics should image the blanket penetration onto this window, demagnifying the image as much as possible.

In the following, we will consider only optics and lines of sight which can be realised using a standard radial port. It seems advisable that the optics located within the area of the horizontal port and the cryostat, should all be contained within the cross section of the port flange.

The idea of introducing optics into the vessel structure immediately raises the question of alignment between the laser and the collection optics. We intend to solve this problem by transmitting the laser via the same optical components as those being used for the collection optics. Thus the alignment is guaranteed for all time and allows us to use non-adjustable optics.

However, metallic mirrors like those intended for the front optics have a lower laser damage threshold than dielectric ones. This requires that a rather large area of the collection optics must be filled with the laser beam which could then lead to 'vignetting' (see below) when viewing the laser beam with the collection optics.

"Vignetting"

When collecting scattered light in a backscattering arrangement the solid angle of collection unavoidably decreases with distance from the collection optics. This represents no problem as long as the laser energy is great enough to ensure a sufficient signal-to-noise ratio for the most distant spatial points.

Another effect which has to be considered carefully, is that of vignetting. In optics, the usual meaning of vignetting is the variation of the solid angle of collection across the field of view. This effect occurs when stops exist in the collection optics which are located outside either the plane of the collecting lens or the image plane. In a Thomson scattering set-up vignetting across the laser beam renders density measurements difficult if the laser beam intensity profile changes from shot to shot. A special case exists when this is not a problem. If we assume a perfect relay system then our collection optics may be defined by two apertures. We may select these apertures to correspond to the blanket penetration and the first collection mirror. These two apertures are in turn imaged on the final focusing lens and the detector front surface. The two apertures define a cone with apex at P0 (Fig.3). Scattering that takes place within this double cone shows no variation in solid angle across the cone axis. We obviously still have the 1/R dependence of the F# as we move away from the apertures, but the solid angle of collection for points in the region of P1 is defined solely by the collecting mirror. Conversely the solid angle of collection in the region of P2 is defined solely by the blanket penetration. From this it is clear that the scattered signal does not depend on the location or the distribution of the laser beam, as long as it is confined within this double cone. In the design we take advantage of this and make the laser fill as much of the collecting mirror as required by damage considerations. The laser must of course be focused at the apex point P0.

Lines of Sight

Of course, it is desirable to realise the maximum possible number of optical chords for T_e and n_e profile measurements.

One line of sight is essential, namely a line of sight through the plasma centre. Further chords passing through the plasma inclined with respect to the horizontal will yield information about plasma movement in the vertical direction. Up to two additional chords seem to be feasible.

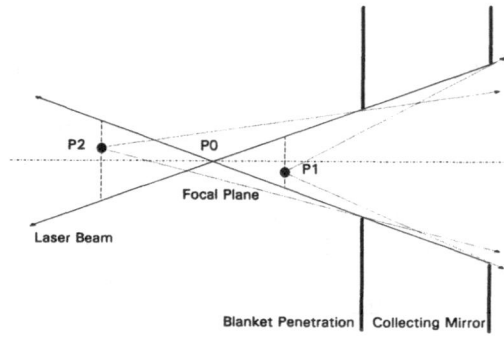

Fig. 3 Vignetting (Schematic)

Sharing with other diagnostics

From the point of view of density calibration of the Thomson scattering diagnostic alone, sharing the optical chord with another diagnostic is advisable. This diagnostic could be

    i)   a single interferometer chord collinear with the LIDAR line of sight (possible for both radial and tangential LIDAR geometry),
    ii)   a single polarimeter chord collinear with the LIDAR line of sight (only for a tangential LIDAR geometry)
    iii) a reflectometer collinear with the LIDAR line of sight (only for a radial LIDAR chord, normal to the outer flux surfaces). This diagnostic would be compatible with the intended use of metallic mirrors for LIDAR.

In addition, the LIDAR collection optics could also be used for spectroscopic diagnostics in the visible and infra-red and ECE.

Optical Components

The environment faced by the components inside the ITER vacuum vessel is the major challenge to making a Thomson scattering system. This is particularly true for the first component (mirror) facing the plasma. This mirror has to withstand a high 14 MeV neutron flux in addition to a possible high temperature (400 $^{\circ}$C) and possibly plasma erosion.

This environment precludes the use of refractive optics near the front end where bulk damage will be highest. As the albedo for neutrons is high the neutrons in the beginning of any labyrinth tend to slow down, but the neutron flux remains high. We should therefore keep refractive elements as far down the labyrinth as possible. It should be noted that Thomson scattering requires a high light throughput which is incompatible with a tight labyrinth.

We have conducted neutron damage tests of conventional laser mirrors, ie. dielectric coatings on quartz substrates. High Z compounds in a few layers survived best. However, these tests together with earlier tests at high temperatures suggest that we cannot use this type of element.

An alternative is to use all metal mirrors, possibly with some overcoat. Tests conducted with silver mirrors on quartz with an overcoat of sapphire show very high reflectivity over the wavelength of interest and good laser power handling performance. Tests at high temperature showed some promise. We would like to propose silver mirrors on e.g. Be substrates, possibly with sapphire protection. Tests on this and other types of mirrors are outstanding.

"Temporal Resolution" , Repetition Rate of Measurements

The target resolution with respect to temporal resolution could be confusing since no distinction is made between the integration time for a single measurement and the repetition rate of the measurements, e.g.: the target measurement resolutions and accuracies for ITER are given in task agreement S 56 TD 01 94-09-27 FE. The time resolution of the electron temperature and density profile measurements is given as 10 ms. It should be noted that the temporal resolution of the Thomson scattering measurement of these quantities is about 50 ns (nanoseconds). The repetition rate of the measurements, as far as the laser technology of the 2000's can be foreseen, will be of the order of 100 pps. Thus, fast events such as ELMs can be frozen and temporally resolved by sampling several events.

CORE

The target specifications for the core measurements foresee a spatial resolution of 30 cm, which is a factor 3 worse than that already achieved with a LIDAR system on JET using yesterdays technology. Thus, according to the discussion in the section on 'LIDAR vs Conventional Scattering' above, for the core plasma LIDAR is the optimum technique for Thomson scattering.

Multichord System

To achieve the maximum possible number of chords the lay-out of the front optics inside the radial port needs to be optimised. This has not yet been completed but to demonstrate that a multichord system is feasible, Fig. 4 shows as an example a possible set-up with 3 tangential chords. In this set-up the angles between the chords are still small enough (10°) to allow sharing of a single blanket penetration and a single output window and the chords pass close to the inner wall. A fan of rays is shown which originates from the inner plasma boundary (the laser focal plane for this set-up). The folding optics within the port consist of sets of plane and spherical mirror pairs, imaging the centre of the blanket penetration (30 cm dia.) onto a common window (17 cm dia.). All the optics have not yet been optimised with respect to imaging quality. This example of optics is shown just to demonstrate that the port volume is sufficiently large to allow the introduction of a multichord system. Such a system may use tangential or radial lines of sight.

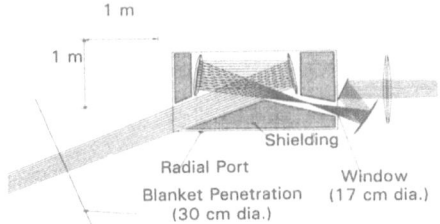

Fig. 4a Tangential multichord system, top view

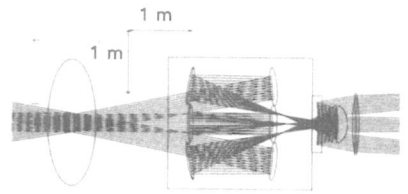

Fig. 4b Tangential multichord system, side view

Tangential vs Radial Lines of Sight

Two lines of sight, radial and tangential, are possible giving different advantages and disadvantages. The current LIDAR system at JET is radial. A radial system cannot measure close to the inner wall due to an intense stray light signal resulting from the laser beam hitting the inner wall. A tangential view, tangent to the plasma at the inner wall avoids this problem and additionally offers

a) the possibility of better spatial resolution near the inner plasma edge (limited to the diameter of the laser beam as it passes the inner wall)

b) the possibility in principle of measuring temperature and density on the second pass through the plasma.

In the case of an inclined beam this gives additional flux surface information. The tangential view also allows lines of sight through the plasma for calibration measurements, e.g. interferometry.

However, the ability of the tangential system to provide better spatial resolution is subject to a number of restrictions. The length over which the spatial resolution is actually better than that obtained in the radial set-up is limited to only 3 or 4 spatial points and in any case cannot meet the 5 mm requirement. The location of this restricted high resolution region may not actually coincide with the location of the plasma region of interest. Also improved spatial resolution is not in fact realised if the error bars are too large. As the scattering volume is further removed from the collection mirror we may find that the necessary laser energy to achieve the required accuracy is too high. In addition one loses spatial resolution at the *outer* boundary because we propose to fill the optics with the laser beam and unlike the radial case the beam is not normal to the flux surfaces so different points across the beam diameter will be on different flux surfaces.

SNR, Required Laser Energy

We have adopted a design criterion which avoids "vignetting" problems. The laser beam is allowed to fill the collection optics and the étendue is fixed by the detector. Aiming at a spatial resolution of 10 cm, a maximum sensitive detector area of 20 mm is assumed to ensure the required detector risetime. For the optics that concentrate the collected light onto the detector, the maximum possible F number, F/1, is chosen.

In the tangential design the laser must be focused at the inner edge, whereas we focus the laser at the plasma centre in the radial design. The size of the first collection mirror and the size of the blanket penetration is then given and we can calculate the resulting vignetting curve in the two cases (F# vs. major radius), see fig 5.

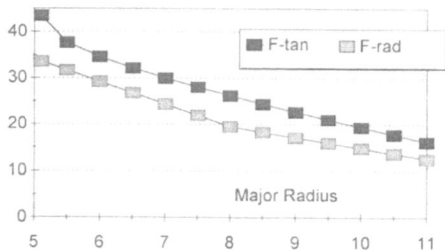

Figure 5 Collection F# versus major radius

Using JET experience we know that we need ~1 Joule at F/12. The required energy in the two ITER designs becomes 2.5 J and 5 J for the radial and tangential designs respectively.

The energy required to make good measurements at the inner edge is ~13 Joules, and the concept of the second pass of the plasma is clearly not realisable. It is worth noting that if we choose a larger, slower detector (~ 40 mm dia.) we can manage with a laser energy of less than 1 Joule and still meet the criterion of 30 cm spatial resolution in the radial design.

Spectral Calibration

The relative spectral sensitivity of the detection system, which must be known to infer the electron temperature from the scattered signals, is measured by illuminating with a calibrated, tuneable, monochromatic light source. This light source is usually placed outside the vessel and thus the transmission of the window and of optical in-vessel components is not measured, (except at the very beginning before the start-up of plasma experiments). However, it has been observed on a number of experiments, during plasma operation absorbing layers build up on in-vessel optics. These can lead to chromatic absorption of the transmitted light. For ITER, effects of neutron and gamma irradiation are expected to add to this problem. There are potentially several solutions:

i) Rayleigh scattering from a gas filling using a tuneable laser or lasers at a number of different emission wavelengths.
ii) Observation of plasma radiation at different wavelengths in the visible with known intensity ratios and bremsstrahlung on standard Ohmic discharges

Investigation of the feasibility of these schemes has to be part of a detailed design study.

Density Calibration

Density calibration of a Thomson scattering arrangement is usually done by performing either Rayleigh or Raman scattering from a rather high pressure (up to 1 bar) gas filling of the discharge vessel. Of course, such a procedure will not be possible on ITER after the initial start-up. However, the absolute (density) calibration could be obtained by taking advantage of the fact that a LIDAR system measures the different spatial points with the same set of detectors, i.e. with constant sensitivity. This is the method currently used at JET. The variation of the solid angle of collection along the line of sight is purely geometrical in nature, it does not vary with time and can either be measured before the start-up of the experiment by Raman scattering or can be calculated sufficiently precisely by optical ray-tracing. Thus the LIDAR system measures directly a relative density profile along the chord. The unknown ratio to the absolute density, which is constant for all spatial points, can be inferred either from comparison to

i) a line integral of the electron density n_e obtained from an interferometer
ii) a line integral of $n_e * B_{||}$ obtained from a polarimeter
iii) a comparison with the density at a single point measured by another diagnostic, such as a single fixed frequency reflectometer channel.

Item i) is a well-established method on JET[1]. In item ii) a polarimeter replaces an interferometer along a tangential line. Since the magnetic field along the tangential path is known, the density can be determined from the measured polarization change. [In fact this could evolve into a density profile measurement diagnostic in its own right.] iii) simply requires a direct comparison of the LIDAR profile with the position of the cut-off density measured by the reflectometer. This sets the LIDAR absolute level for the whole profile.

There is a minor caveat for the applicability of methods i) and ii) to the proposed LIDAR scheme: Since, as we have seen above, it may be difficult to cope with the whole temperature range of an ITER plasma using a single laser wavelength, the relative density profile along the chord may have to be fitted together from two profiles: one for the high temperature core and one for the 'low temperature' (< 3 keV) outer edge.

Spatial Resolution

The LIDAR system at JET has a spatial resolution of 13 cm. Using a 300 ps laser and a 20 mm diameter MCP photomultiplier with a response time of 450 ps and a 1 GHz recorder a spatial resolution of 10 cm is achievable. This is all based on performance figures of commercially available equipment, much of which is already in use in the JET LIDAR system. However, since we are only asked for 30 cm resolution we can relax all these parameters. A large MCP photomultiplier, e.g. 40 mm diameter, would improve the light gathering power and reduce the required laser power. A longer laser pulse improves the damage limit and is obviously easier to make. We could therefore aim for something like a 1.5 ns laser and a 1 GHz recorder (commercial portable oscilloscope), the larger photomultiplier would have a response time of ~1 ns. The resultant resolution of this combination is still within the 30 cm.

SUMMARY AND ACTIONS

The required specifications can be met by a LIDAR system without difficulties once it is established that the front optics can withstand the ITER conditions. It should be possible to provide three lines of sight, one crossing the plasma centre and two inclined ones. This will yield information on the plasma vertical position. At the rather low level of laser energy necessary for a spatial resolution of 30 cm, a repetition rate of the measurements of 100 Hz seems possible. This would enable the possibility of using this diagnostic for control purposes.

The actions required can be listed as follows:

i) Laser damage, thermal and irradiation tests should be conducted for metal mirrors, e.g. Be mirrors both with a high reflectivity silver coating and an aluminium coating (each with and without a dielectric protective coating) in the following sequence:
   a) laser damage threshold tests
   b) reflectivity and laser damage threshold after heating to relevant temperatures
   c) reflectivity and laser damage threshold after irradiation

ii) Conduct a design study for the layout of the front optics inside the port and of the relay optics between port and cryostat, thereby continuously assessing (in discussion with ITER) the required blanket penetrations and their effect on shielding. Determine the levels of heat, gamma and neutron flux at the position of the optical components.

References

1. Salzmann et al, JET Report JET-R(89)07, 1989
2. ITER task agreement S 56 TD 01 94-09-27 FE

THE CONCEPT OF *ITER* DIVERTOR PLASMA DIAGNOSTICS BY MEANS OF *LIDAR* TECHNIQUE

G.T.Razdobarin, A.Daavittila*, V.K.Gusev, E.E.Mukhin,
R.Salomaa*, S.Yu.Tolstyakov

IOFFE Phisico-Technical Institute,
Politechnicheskaya, 26, St.Petersburg, RUSSIA
*Departent of Technical Physics, Helsinki University of Technology
FIN-02150 Espoo, FINLAND

ABSTRACT

The general considerations of Thomson and Rayleigh scattering diagnostic methods using a time-of-flight (**LIDAR**) principle in **ITER** divertor are discussed regarding the dynamic gas target divertor operation. Compatibility with specific plasma conditions and divertor operation scenario is examined to define the guidelines of a design. In order to validate the concept a lot of experimental activity and technological developments is to be fulfilled. Some of them that have recently been carried out are demonstrated and discussed.

CONTENTS:

1. GENERAL CONSIDERATIONS

1.1. The Guidelines of a Design

As is commonly recognized, the ITER program implementation relies upon the ability of a divertor problems successful solution. In a condition of strongly enhanced power production and under the transition to burning it is of primary importance to address the problems of power handling

Diagnostics for Experimental Thermonuclear Fusion Reactors
Edited by P. E. Stott *et al.*, Plenum Press, New York, 1996

259

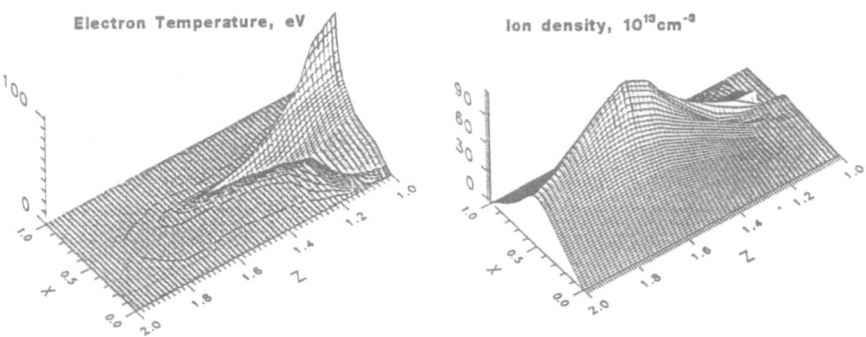

Figure 1. Divertor detached plasma parameters. Level lines correspond to Te range 0.5-5.0 eV.

and impurity locking divertor capability. The moderate steady-state power handling is envisaged by the project to insure the acceptable life time of divertor target plates. This goal task looks feasible under the use of a so-called dynamic gas target operation scenario[1] which is the alternative of a well known high recycling regime. According to the conception a volume power exhaust due to radiation and the multiple charge exchange interactions is resulting in a reduced power load onto a target surface. The existence of a detached plasma space predicted by simulations (see Fig.1) is the crucial point to validate this conception. As is seen from Fig.1, the ionization front is just a surface that separates the up- and downstream plasma regions of essentially different parameters. The upstream region including the plasma in X-point vicinity is distinguished by relatively large temperature of a few hundreds eV and electron density within the range of 10^{13}-10^{14}cm^{-3}. It represents discharge conditions which are somewhat experienced in the moderate size tokamak studies. Nevertheless, in spite of the valued diagnostic experience gained on a large number of existing machines a lot of practical constraints arising from the harsh radiation environment and limited access routes is to be encountered here. At the downstream region of rather cold, high density and strongly impurity contaminated plasma besides of practical constraints, the physical obstacles are foreseen for the application of a number of well-established techniques. In the following we shall give a survey of those laser-aided diagnostic techniques that appear to be satisfactory for divertor applications. A set of diagnostic arrangements falls into two separate categories. The first one that is mostly confined by ITER operation physics phase is intended to explore a divertor detached plasma formation. The electron temperature as well as the electron and neutral particle density distributions up- and downstream of ionization front are to be delivered by means of Thomson and Rayleigh scattering techniques. The matter of concern is also the monitoring of abnormal energy deposition events during power excursions and their influence on the divertor plasma properties. A reasonable divertor operating windows compatible with core plasma scenarios should be examined and specified. The second diagnostics category is suggested for ITER technology phase. It is founded on the most safety access routing that uses a reduced number of front plasma viewing optical components with a local shielding. The axis that comes through the X-point seems to be the most appropriate route. A Thomson scattering shared with Balmer lines spectroscopy is refered to as a favorable candidate.

The measurements of T_e, n_e distributions in a plane of X-point location are beneficial to monitor SOL plasma parameters in a region of a significant flux expansion. The measurement of neutrals distribution behavior nearby a divertor baffle limiter is advantageous to monitor a baffling efficiency and the retention capability of recycling neutrals and impurities. The interrelationship between both diagnostic categories is to be learnt about in a course of experiments for the establishment of a robust optimal diagnostic technique. It should provide a discharge conditions control and monitoring during the high neutron flux and fluence period of the technology phase.

1.2. The Plasma Background Radiation and Selfabsorption Estimates

According to the available specifications[1] the downstream plasma parameters are expected to be as follows.

Electron temperature in a steep gradient ionization front zone falls from more than 100eV down to nearly 1eV. The total density of neutrals and ions is within the range of 10^{14}-10^{15}cm^{-3} at the extend of nearly one half a meter and exceeds 10^{15}cm^{-3} in close vicinity of a target plate. In these conditions the plasma is expected to become optically thick at least at the strong resonance optical transitions. An absorption cross-section in central part of Doppler broaden profiles $\sigma=(\pi e^2 f_{12}/m_e c^2)\lambda_{12}^2/\Delta\lambda_D$ is of the order of $10^{-4}\lambda_{12}^2$ provided the oscillator strength $f_{12}=1$, and spectrum line Doppler width $\Delta\lambda_D$ ~0.1nm. Thus, the reabsorption condition σnl~1 in VUV spectrum range (λ~100nm) is satisfied for a plasma layer length l>10cm and a ground state atomic density n~10^{13}cm^{-3}. This condition is favorable to a diffusion-dominated divertor plasma where the ground level population may keep many orders of magnitude above the coronal equilibrium value.

For the extension of reabsorption condition onto the visible lines as well (e.g. D_α of deuterium) the enhanced optical thickness (σnl~10) is required to intervene the collisional-radiative coupling between the bound states[2]. Under a strong reabsorption the line radiance is saturated to a black body limit for a fitted temperature which may strongly depart of kinetic one (e.g. to our estimates, less than 1eV for D_α, D_β transitions of deuterium in nonequilibrium diffusion-dominated plasma). This saturated line background radiance (nearly one watt from cm^2 in sr within Doppler bandwidth of D_α and/or D_β line profiles) represents a serious constraint diagnostic factor. An effective rejection notching of a background line emission is of primary importance for divertor Thomson scattering. The perfect notching is the necessary but not still a sufficient condition. The line-free windows appropriate for Thomson scattering are really occupied by rather strong bremsstrahlung and recombination emissions. As opposed to spectrum lines, these windows are optically transparent (due to a rather small photoionization cross-section, see appendix I). Nevertheless a bremsstrahlung radiance in a visible spectrum band remains still significant, e.g. in the range 10^{-3}-10^{-4}W/(cm^2 nm sr) for T_e, n_e like in Fig.1 and a length of sight l~100cm. This estimate that follows immediately from eq.3 looks dangerous for divertor Thomson scattering especially on account of significantly broad band spectrum channels 0.5-5.0nm for λ~530nm. It seems very likely that no one conventional Thomson scattering system instead of LIDAR could be applied to explore a down-stream divertor plasma.

This conclusion is supported additionally by the expectation of emissions from multiplets of low-Z light impurities and metallic constituents of a target plate that could sometimes dominate over the bremsstrahlung radiance[3].

The plasma emission and selfabsorption restrictions urge to revise not only a Thomson scattering but some of the established techniques based on localized measurements of excitation and radiation events.

2.THOMSON SCATTERING IN THE DIVERTOR OF ITER

2.1 Sensitivity Considerations

A high resolution LIDAR Thomson scattering meets the requirement to diagnose the electron parameter variation in anticipated range and is the most beneficial of study strongly inhomogenious plasma regions such as the X-point and/or ionization front vicinity. The upgrading of the diagnostic spatial resolution to nearly 3-5cm became possible with a use of high speed streak-camera for scattered light recording[4]. Because of the photocathode dimension and the screen sweep length restrictions the laser beam image spot on a tube entrance window usually does not exceed of 2mm. This obstacle and a large scattering length ~1m are responsible for a limited value of the laser beam diameter ~30mm in plasma and light collection cone ~F/20. The sensitivity of the LIDAR Thomson scattering diagnostics can be evaluated consequently under the following assumptions:

-spatial resolution along the laser beam axis is 5 cm,
-light collection solid angle is $2*10^{-3}$sr,
-total transmission of spectrometer and collection optics is 10%,
-photocathode quantum efficiency is 10%,
-resolved spectrum band is 0.15 of the scattered spectrum profile halfwidth.

If so, a 3J laser output is needed to yield nearly one hundred photoelectrons per spectrum channel at the lower limit of the electron density range $\sim 10^{13}$ cm^{-3}. This provides nearly 10% accuracy of temperature measurement respectively.

The given above estimate looks somewhat optimistic at least for the purpose of detached plasma studies inside the divertor legs. In spite of a significant density increase resulting in more than 10^3 photoelctrons per single channel the sensitivity fall is expected because of the plasma background continuos radiation. The bremsstrahlung and recombination contributions can produce according to eq.3 a total number of detected photoelectrons of a few tens (up to hundred) per spectrum channel. An extra worry are the multiplets of low-Z impurities that may overlap the spectrum recording diagnostic channels. On account of these it is normal to multiply the bremsstrahlung estimate at worst by 100. On account of plasma emission influence a 3J laser output is considered only as a low limit estimate.

2.2 Access Requirements.

The LIDAR diagnostics inherent feature is that the laser equipment and recording apparatus are installed far from the reactor~ 50m beyond the biological shield. For this reason the problem of radiation resistance to neutron and X-ray radiation loads refers only to small number of optical units nearby the reactor(optical windows and mirrors). The use of plasma-viewing mirrors at the divertor chamber immediate vicinity is likely not a strong constraint factor at least for the metallic polished surfaces that are essentially resistive to fusion reactor conditions.

Divertor Legs Access Considerations. A lot of engineering effort has been made to evaluate the impact of different diagnostic access routes on ITER design. The ITER divertor and divertor legs volume in particular are still continue to be the most sophisticated issue of a design. Just up to now we have no a detailed view to cope with this problem. It is hardly to be relied upon the underneath divertor assembly of reflective mirrors. The more reliable route suggested by the JET team is a downward looking by means of a front optics while sharing a single blanket penetration with a core plasma Thomson scattering. Another problem is the plasma location in proximity (near one half of meter) of a target plate. First of all, it is strongly desirable to keep a viewing axis away of a target energy deposition spot that may be a source of strong black-body emission[5].

Secondly, a stray light originating from the area that a probe beam strikes upon represents a strong diagnostics constraint. This may become a major problem since the scattered in plasma light is usually below by the orders of magnitude. A fine technology is required to discriminate a low intensive useful pulse against an extremely large spurious one that is delayed by only a portion of ns.

Special purpose demagnifying streak cameras are now under development[6] to insure the property of an instant and a deep stray light dumping under the enhanced number of independent measurements over a chord length. Preliminary investigations of a blooming resistance capability and a dynamic range of the streak-camera option were carried out at IOFFE Institute using a LIDAR Rayleigh scattering in an air. The experimental arrangement layout is shown in Fig.2. It consists of QUANTEL Nd:YAG laser (1), spherical F=800mm mirror collector (2), streak-camera (3), supplied with neutral filters and F/2 imaging objective (4) and CCD readout camera (5). The laser output at second harmonics was less than 0.1J. The laser pulse duration in these experiments was nearly 0.1ns which corresponds to a laser beam "pencil" (7) length approximately of 3cm.

The different kinds of target surfaces including white paper, stainless steel and carbon plates have been located in proximity after the gas scattering volume.

The streak speed was taken as 1ns/cm. This means the detector response time be nearly 0.25ns provided the 2.5mm image spot sweeping across the screen. On account of a negligible short laser pulse duration the resolved scattering length along a probe axis was restricted by $\Delta x = c\tau/2 = 4$cm.

As is seen from Fig.3(a) a very strong stray light signal associated with a beam diffuse backreflection from a target surface occupies a well-defined blurring spot area. The signal inside a blurring spot composed of more than 10^6 photoelectron counts is strongly dumped due to a MCP intensifier saturation effect. Because of this significant technological advantage the detection of an extremely weak scattered signal represented by the a discrete photoelectron counts on a streak trace becomes possible just at a distance of no more than 10cm from a target surface. As is seen, the

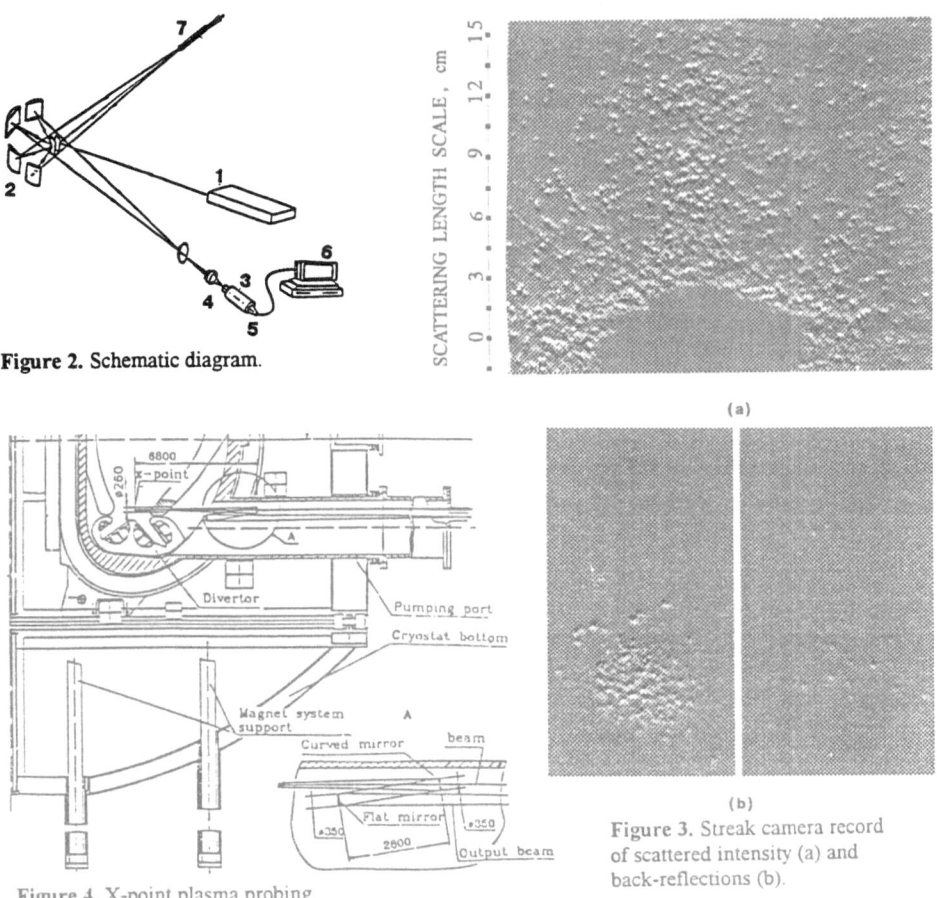

Figure 2. Schematic diagram.

SCATTERING LENGTH SCALE, cm

(a)

(b)

Figure 3. Streak camera record
of scattered intensity (a) and
back-reflections (b).

Figure 4. X-point plasma probing

streak-camera gain is quite sufficient to detect the individual photoelectron marks on CCD unit.

In Fig.3(b) the results of a streak-camera dynamic range testing are demonstrated. Each of 2.5mm image spot patterns corresponds to the back-scattered signals from target of different density. A small intensity pattern is composed of discrete photoelectron marks and is suitable for a count data processing. A large intensity pattern that contains nearly 10^4 photoelectrons marks is suitable only for analogous data processing. Under the combination of both data processing approaches the resulting dynamic range for a cell of 2.5mm size is probably 3-4 orders that satisfies the LIDAR diagnostic requirement.

The X-point Plasma Probing. The routing to this particular plasma region is more favorable through one of the pumping ducts inside the cryostat. The engineering design sketch is sampled in Fig.4. The penetration duct is ended outside the cryostat with a pair of input and output beam guide pipes of 100mm and 400mm in diameter respectively which are welded in the pumping tank door. Each of these circular straight pipes is dumped with a vacuum optical windows beyond the bulk shielding. The input and output pipe axes are tracing in parallel on either vertical or horizontal planes. A gap between the input and output pipe axes is taken as 430mm that approximates the diameters of collimated light beam (or plasma-viewing mirror) with a few cm reservation.

The plasma-viewing mirror that is removed from X-point by 6.8m is a round disc of 350mm in diameter with a curved surface shaping of curvature radius exceeding 9m. The mirror is installed opposite to the diagnostic channel passing through blanket to direct the line of sight via the X-point.

A slit-like diagnostic channel is taken as large as 260mm in length to allow a line of sight scanning up and downward nearby the X-point region by means of the curved mirror angular adjustment. The curved mirror is supplied with a central hole of 50mm in diameter for the laser beam emerging. The use of a narrow straight penetration over the whole laser beam path from entrance window to X-point seems to be reasonable in avoid of the severe hazardous phenomena at the reflective mirror surface resulting from mutual action of the hostile plasma environment and the laser beam radiological loads. A supplementary periscopic flat round mirror of a same diameter 350mm is installed at 2.8m apart from a curved one with a rear side facing the plasma. The rear sides of both curved and flat mirrors are protected with a local shielding to reduce the neutron fluxes at the axis of horizontal straight penetration. A curved mirror is intended to image the probe beam section from the X-point onto a conjunction plane of output optical window at a distance more than 13m. This helps to minimize the vacuum window useful area to a diameter limited by 300mm. The input and output vacuum windows are made of fused silica. To secure against tritium leaks all windows are double assemblies with separate pumping of the interspace. Outside the vacuum flange the probe and scattered light beams are transported beyond the biological shield by means of quartz prism units and the mirrors incorporating dielectric coatings.

3. SHARING WITH OTHER DIAGNOSTICS

3.1. Lidar Rayleigh Scattering Near the Strong Optical Transitions

The application of near-resonant Rayleigh scattering technique for density measurements of neutral hydrogen and impurities in plasma experiments has been outlined in a number of recent publications[7,8,9]. This absorption-free plasma diagnostics is especially advantageous for a divertor detached plasma region. The important advantage as well is the compatibility in principle with a time-of-flight LIDAR technique. In order to predict the diagnostic sensitivity it is reasonable to compare the Rayleigh and Thomson scattering emission.

Within a Born approximation that is valid for nondegenerated lower and upper atomic levels a Rayleigh scattering cross section can be presented in the terms of oscillator strengths as:

$$\sigma_R = \sigma_{Th}\{f/2 \cdot \lambda/\delta\lambda\}^2 \tag{1}$$

where $\sigma_{Th} = r_o^2$ is a Thomson scattering cross section, f is an oscillator strength of adjacent transition, λ and $\delta\lambda$ are the probe laser wavelength and the laser-to-transition wavelength detuning respectively.

Unfortunately, the advantage of strong cross-section increase with small detuning does not mean the gain in sensitivity (as opposed to Thomson scattering) because of the saturation effect. In order to avoid the saturation of the scattering intensity the laser power flux density is not allowed to exceed the upper limit $\Phi_{sat} = (\delta\lambda)^2 \pi hc^2/r_o f\lambda^5$ which follows from a condition that Rabi frequency is got the laser-to-transition frequency detuning.

A variety of fixed frequency probe lasers can be adopted for the Rayleigh scattering plasma diagnostics using the harmonics generation with a help of frequency mixers and Raman shifters. The choice of a laser option is obviously wider for a larger laser-to-transition wavelength detuning allowance. Moreover, the increase of detuning is helpful to enjoy a large gathering power of a spectrometer that is used to resolve the wavelengths of a Rayleigh and the adjacent transition contributions. The detuning off resonance is really confined by the instrumental and physical reasons. First of all, in order to conserve the scattering signal intensity the cross-section fall in a proportion to $(\delta\lambda)^2$ should be compensated by the respective probe energy increase.

Secondly, as is follows from equation (1), to overcome a Thomson scattering signal the admitted detunings should not exceed of a few nanometers for a visible light provided the scatter particle concentration be nearly 0.1% relative to electron density.

In order to verify the diagnostics operation within the predicted range of laser-to-transition wavelength detunings the Rayleigh scattering on Li atomic beam with a density of ground state atoms a few times 10^{10}cm^{-3} was performed. For this reason a tunable dye laser wavelength was a shot-to-shot scanned over a 5nm spectrum band. A scattered light collected from nearly 10mm length beam section into $2*10^{-2}$ sr solid angle was analyzed by monochromator under 1.3nm/mm reciprocal dispersion. The laser output was nearly 1mJ. Fig.5 shows the scattered spectrum recording with a dye laser tuned at 0.6nm apart from 2P-2S resonance line λ=670.8nm. Each datum point in a recorded spectrum (Fig.5) represents an average over a few tens shots. Two spectrum components were found as is predicted by theory. The component that occurs at the laser wavelength is the Rayleigh scattering. The shifted component centered on a transition line is a collisional-induced fluorescence emission. The widths of these both components are 0.26nm in according with monochromator slit openings.

As is seen, a fluorescent emission overcomes a Rayleigh scattering contribution by several times. On the other hand, when the laser was tuned by more than 3nm apart of resonance a fluorescence contribution was not observed at all.

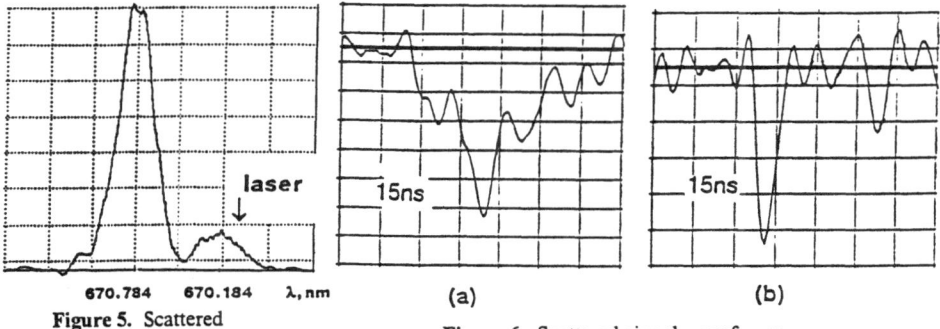

Figure 5. Scattered spectrum recording.

Figure 6. Scattered signal waveforms.

The oscillograms (a) and (b) in Fig.6 show in separate the waveforms of fluorescent and Rayleigh scattering signals respectively. The fluorescence pulse duration of nearly 30ns is well related to a spontaneous decay time for 2P-2S transition of Li. A small duration of a Rayleigh scattering signal of nearly 10ns is related as is to be, to a laser pulse length. Such an accurate matching of the both probe and Rayleigh scattering signal durations is in accordance of LIDAR detection principle. In order to get a fine spatial resolution of nearly 3cm in LIDAR diagnostics both matched pulse durations are to be of nearly 0.15ns. The most essential result of this experimental run is the evidence of diagnostics low detection limit.

The feasibility of LIDAR Rayleigh scattering relies on the powerful laser source. In a considered Li experiment a good option is Nd:YLF λ=523.5nm second harmonics combined with H_2 Raman shifter. The wavelength detuning 1.79nm calls for the saturation power flux density of nearly 200MW/cm^2. For LIDAR device under a laser pulse duration 0.15ns and probe beam diameter 3cm the laser output is restricted by saturation limit 0.2J. Under a given above performance of a LIDAR detection system this yields 10^3 photoelectron counts provided the Li atomic density being $\sim10^{10}$cm^{-3}.

Table I

Element	λ nm	Laser options	$\Delta\lambda$ nm	σ_R/σ_T 10^4	E^{sat} J	n^{lim} 10^{10}, cm^{-3}
Li $^{2S-2P}$	670.8	1047nm(λ/2)$\rightarrow$523.5nm(H_2)$\rightarrow$669nm	1.8	2.0	0.2	1.0
D_α $^{2P-3D}$	656.1	1064nm(λ/2)$\rightarrow$532nm(HD)$\rightarrow$659.4nm	3.3	0.4	1.0	1.0
		1315nm (λ/2)$\rightarrow$657.5nm	1.4	2.0	0.2	1.0
He $^{2P-3D}$	667.8	1047nm(λ/2)$\rightarrow$523.5nm (H_2)$\rightarrow$669nm	1.2	4.0	0.1	1.0
He $^{2P-3D}$	587.6	1064nm(CF_4)$\rightarrow$1177.8nm(λ/2)$\rightarrow$588.9nm	1.3	2.0	0.3	1.0
He $^{2S-3P}$	501.6	1047nm(CH_4)$\rightarrow$1507.4nm(λ/3)$\rightarrow$502.5nm	0.9	0.2	1.0	2.0
Ne $^{3S-3P}$	640.2	1064nm(O_2)$\rightarrow$1275nm(λ/2)$\rightarrow$637.5nm	2.7	0.3	1.0	1.0
Ar $^{4S-4P}$	763.5	1047nm(D_2)$\rightarrow$1524.4nm(λ/2)$\rightarrow$762.2nm	1.3	0.4	0.2	3.0

A list of the important elements and suitable transitions for a divertor plasma study are sampled in Table I. The corresponding population density detection limit that holds for 10^3 photoelectron counts is illustrated in the right column. To get a probe beam and the transition wavelengths coupling the laser options were restricted by Nd:YLF, Nd:YAG with use of Raman shifting and frequency doubling technique. The iodine laser second harmonics 657.5nm is additionally included as the most straightforward for scattering on deuterium [7]. As is seen from the table right column, the population density detection limit in neglection of plasma background radiation is kept from one to a few times $10^{10}cm^{-3}$ (10^3 photoelectron counts respectively). The corresponding ground state atomic density will depend on a plasma equilibrium condition. On account of resonance line selfabsorption that expected for a dense divertor plasma, the excited atomic states are to be overpopulated relative to a coronal equilibrium distribution. As for example, the deuterium second level population may be nearly a percent of ground state atomic density [10]. Thus, a Rayleigh scattering near D_α transition is capable to detect down to $10^{12}cm^{-3}$ of ground state deuterium atoms. This provides the sensitivity reservation to allow the plasma selfemission influence. The accurate predictions for the impurities are hardly to be made just now because of the essential uncertainty of atomic densities. In any case, the enhanced population of the existing metastable levels is in favor of the noble gas impurities (the last three rows of a table).

3.2. Balmer Line Emission Spectroscopy (Along the Axis Coming Through the X-point), Laser Induced Ionization

The optical chord coming through the X-point could probably be shared with spectroscopic diagnostics in the visible. The detection of Balmer deuterium lines near the bottom of baffle unit is consistent with a purpose to investigate a divertor locking capability and a particle confinement time behavior.

According to a gained experience [11] the routine spectroscopic diagnostics may be modified to provide the line-of-sight profile studies. It looks really valuable because of quite different plasma physical conditions near the X-point and at a baffle vicinity. Besides of physics aspects, it is an obvious technological reason to share not only the LIDAR path collection optics but the LIDAR probe laser and the detection system as well. This new invented diagnostic method is based on laser induced ionization of excited atomic states populated in plasma. The respective population depletion is recorded as a ramp down of plasma selfradiation during the laser action. Under the complete photoionization the measured burn out signal is to yield a local number of excited deuterium atoms inside the given resolved volume. Note, that unlike a complicated fluorescence diagnostics there is no necessity of exact tuning of the laser to the transition frequency with the only allowance that photon energy exceeds the upper level ionization potential.

To exhibit the diagnostic operation in FT-1 tokamak discharges (T_e=200-500eV, $<n_e>~10^{13}$ cm^{-3}) the conventional $90°$-laser scattering geometry with the probe neodymium laser output of nearly one hundred joules at $1.055\mu m$ was employed. The use of crossed probing and viewing beams provided the signal recording from a well-defined spatial region of a few cm^3 resolved volume. The local population density of the excited hydrogen on the level with a principal quantum number n=4 was determined and the respective number of ground-state atoms n_1 was calculated on a basis of the population distribution tabulated by L.C.Johnson and E.Hinnov [10]. Measurements of hydrogen density were carried out for the torus cross-sections with and without gas puffing. Figs.7 demonstrates the ground state population $n_1(r)$ density distributions as a function of radius r in both cross sections.

For a reason of more close fitting to the LIDAR experimental layout a proof-of-principle experiment was carried out. The emission of tokamak plasma at the H_β line was observed along the probe beam at the extend of a minor diameter of 30cm length. The time integrated signal has shown a pulsed decrease of spontaneous emission to an infinitely small value during the laser action.

In order to extrapolate the available data onto the LIDAR experimental conditions two points should be accounted for. The first one is the atomic photoionization rate. It should be large enough to insure a full ionization during the laser action. The ionization time arrives to τ=100ps when a probe power flux density $\Phi=h\nu_{las}/\sigma_{ion}\tau\cong160MW/cm^2$. This value is well related to the LIDAR standards.

The second point is a spatial resolution that follows from the time-of-flight detection of photoionization front propagation induced by a laser pulse flight. The time derivative of a swept output signal holds the plasma line emission spread along the beam direction and the respective atomic density profile. To our experience, the plasma emission scan starting from the plasma

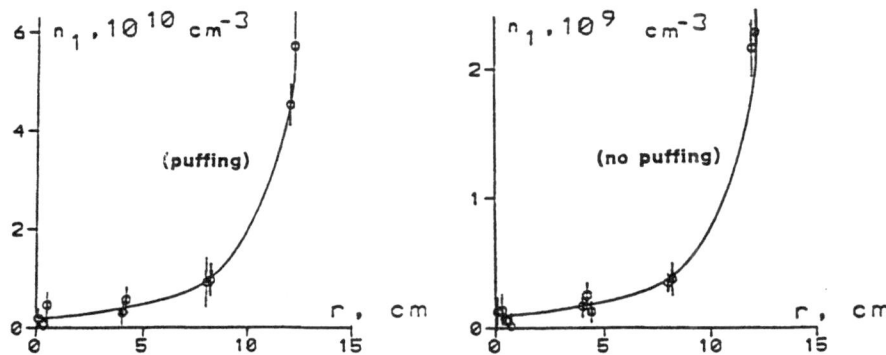

Figure 7. Hydrogen ground state density profiles.

periphery used to cover density falling range by nearly one order of magnitude. Under the detection arrangement specifications like in a section 2.1 approximately 10^3 photoelectron counts during a resolved time 0.15ns are predicted for D_β line emission sweep apart from photoionization. To make the estimate a ground state neutral density was considered to drop from 10^{13}cm^{-3} at the extend of tens of cm from the baffle. Thus, it looks well-founded to incorporate this new developed diagnostics in a LIDAR scattering device.

4. CONCLUSIONS

A LIDAR Thomson scattering is really a favorable diagnostics of an ITER divertor plasma for both physics and technology phases and there is no seen an unsolved obstacle for its implementation. The strongly appreciated sharing with the appropriate laser-aided plasma diagnostics looks possible as well. Two of them LIDAR Rayleigh scattering and Laser Induced Ionization which are tightly compatible with Thomson scattering are intended to obtain the atomic species density profiles in the divertor legs and near the baffle region. Their application relies upon a presently gained experience. At least two urgent actions are necessary at this moment. The first one is to get a clear understanding of the laser diagnostics impact on a divertor design. The second one is to continue extensively the researches in order to get more precise fitting of laser diagnostics to ITER divertor.

APPENDIX I

The off-line spectrum distributed volume emission ε_λ can be defined in terms of black-body radiance B_{black} regarding a Kirgoff law $\varepsilon_\lambda = B_{black}\kappa_\lambda$. The corresponding absorption coefficient κ_λ can be estimated for the high excited hydrogenic terms according to Unsold[12] as:

$$\kappa_\lambda = (1-e^{-hc/\lambda kT}) \sum_z \sum_{i=p}^{\infty} \{n_i \sigma_i(\lambda)\}_z \qquad (2)$$

where n_i the Saha equilibrium population density of hydrogen and/or of impurity particle energy level with a quantum number i, $\sigma_i(\lambda)$ is the respective Kramers photoionization cross-section for hydrogen excited atomic level, $1-e^{-hc/\lambda kT}$ is a term regarding the influence of a photon stimulated emission. The summing is carried out over the variety of ionic species n_z and quantum numbers i.

Both bound-free and free-free transition contributions are included in (2). The wavelength coupling condition for a series limit wavelength λ_i exceeding λ is satisfied by means of the corresponding choice of a lower i-number summation limit. Namely, the lower limit holds $p=z(\chi_H\lambda_p/hc)^{1/2}$, where the transition wavelength λ_p is the shortest one to satisfy $\lambda_p \geq \lambda$, $\chi_H=13.6\text{eV}$.

On allowance for nonsignificant difference between the summing and integration over i and after the summing over z in formula 2:

$$\varepsilon_\lambda = B_{black}Z_{eff}n_p<\sigma>_p \qquad (3)$$

where $n_p=n_e^2h^3(2\pi m_e kT_e)^{-3/2}(\chi_H\lambda_p/hc)e^{hc/\lambda kT}$ is a Saha equilibrium population density of a hydrogenic coupled level with a principal quantum number $p=(\chi_H\lambda_p/hc)^{1/2}$, $<\sigma>_p=5*10^{-3}\lambda^3(hc/\chi_H\lambda_p)^2(kT_e\lambda_p/hc)(1-e^{-hc/\lambda kT})$ is a net averaged cross-section which includes the absorption and stimulated emission for the both bound-free and free-free transition contributions ($<\sigma>_p$ is given in units of cm^2, λ in cm, Gaunt factor is taken as unity).

The off-line plasma volume emission in a low temperature range $kT_e<hc/\lambda_p$ is dominated by bound-free, otherwise by free-free transition contributions. In the range of $kT_e \geq hc/\lambda_p$ the cross-section $<\sigma>_p$ is weakly depended on temperature and is of the order of $10^{-17}cm^2$ for $\lambda_p \sim \lambda = 500\text{nm}$.

REFERENCES

1. G. Janeschitz, K.Borrass, G.Federici et al., ITER divertor concept, *ITER JCT, Joint Worksite Garching* (1994).
2. H.W.Drawin and F.Emard, Influence of laser radiation on the collisional-radiative recombination and ionization coefficients, *Z.Physik*, 266:257 (1974).
3. K.Kadota, M.Otsuka, J.Fujita, Spase-and time-resolved study of impurities by visible spectroscopy in the high density regime of JIPPT-II tokamak plasma, *Nucl. Fusion*, 20:209 (1980).
4. H.Fajemirokun, C.Gowers, P.Nielsen et al, A high-resolution LIDAR Thompson scattering diagnostic for JET, *Rev.Sci.Instr.*61:2843 (1990).
5. E.E.Bogdachenkov, U.A.Djushembiev, I.K.Konkashbaev, I.S.Landman, Interaction of plasma flows with a divertor during the termal phase of a disruption, *Phizika Plasmy*, 19:963 (1993).
6. A.V.Dech, G.G.Feldman, V.V.Grebenschikov et al., Research and design study of a streak camera for LIDAR Thompson diagnostics, *Plasma Devices and Operation*, 2:301 (1994).
7. V.A.Belyakov, A.A.Besshaposhnikov, V.K.Gusev et al., Laser diagnostics of the electron and nuetral plasma components in ITER divertor with LIDAR system, *Plasma Devices And Operations*, 1:227 (1991).
8. D.A.Shcheglov, A.B.Berlizov, I.V.Moskalenko, Application of near-resonant laser scattering to plasma diagnostics, *Proc.5 Int.Symp. on Laser-Aided Plasma Diagnostics*, p.205 (Bad-Honnef, Germany 1991).
9. B.Schunke, P.Nielsen, C.Gowers, Discussion of the possible combination of the LIDAR technique with Rayleigh scattering, *JET Papers Presented to Course and Workshop on Diagnostics for Contemporary Fusion Experiments*, p.273 Varenna, Italy (1991).
10. L.C.Johnson, E Hinnov, Ionization, recombination and population of excited levels in hydrogen plasmas, *J. Quant. Spectrosc.Radiat. Transfer*, 13:333 (1973).
11. V.I.Gladuschak, V.K.Gusev, M.Yu.Kantor et al, Development and application of photoionization method for measuring the density of hydrogen atoms in a tokamak plasma, *Tech. Phys. Lett.* 19:362 (1993).
12. A.Unsold, Uber den kotinuierlichen absorptionskoeffizienten, und das spectrum einer sternatmosphare, welche nur aus wasserstoff besteht, *ZS.f.Ap.* 8:32 (1934).

THOMSON SCATTERING AT THE EDGE OF ITER

D. W. Johnson[1], T. Carlstrom[2], B. Grek[1], and R. Snider[2]

[1]Princeton Plasma Physics Laboratory, Princeton, NJ, USA
[2]General Atomics, San Diego, CA, USA

INTRODUCTION

Parameters at the plasma edge away from the divertor are found to correlate with performance parameters in other regions. For example, sharp edge pressure gradients in H-mode profiles may limit global confinement by exceeding the ballooning MHD stability limit. Edge temperatures and densities determine the efficiency of impurity radiation from the plasma boundary, which affects power flow to the divertor. Thus, ITER will benefit from a good characterization of the edge region. Futhermore, ITER's success in operating within various limits in parameter space may depend on the ability to control edge parameters in real time.

Among the most important of these edge parameters are the electron temperature T_e and density n_e. This study addresses the feasibility of measuring these parameters at the edge of ITER using Thomson scattering.

Thomson scattering diagnostics are also proposed for measuring profiles of T_e and n_e in the plasma core and in the divertor region. Because of advantages in required access and in plasma light rejection offered by the LIDAR Thomson scattering technique, this is the technique of choice for the core and divertor systems (ITER Expert Group Report, 1995).

The technical challenges facing Thomson scattering at the edge of ITER are formidable. As described in the following section, ITER requires high spatial resolution measurements. This generally means that the scattering length and hence the signal strength for a given laser energy is small. This problem is further compounded by the fact that, due to the harsh ITER radiation environment, the most compatible design will be one which minimizes the blanket penetration size for the collection optics. Finally, for imaging systems, the high bremsstrahlung background, which is much brighter on ITER than on existing devices, causes severe contrast problems.

To solve these problems we propose imaging concepts which achieve the desired specifications in a somewhat modified sense, with technology clearly on the horizon in the commercial market.

Diagnostics for Experimental Thermonuclear Fusion Reactors
Edited by P. E. Stott *et al.*, Plenum Press, New York, 1996

269

TARGET MEASUREMENT SPECIFICATIONS

We summarize below the requirements for measuring T_e and n_e at the edge as defined in the ITER Expert Group Report (ITER Expert Group Report, 1995). The reference ignition regime for ITER is the ELMy H-mode. Radial profiles of the electron temperature and density expected at the outer midplane of ITER are shown in Figure 1.

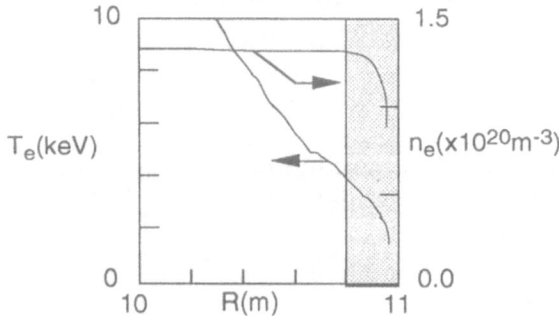

Figure 1. Expected profiles of T_e and n_e at the outer midplane of ITER for ignited plasma in H-mode.

The outer edge profiles have large radial gradients in T_e and particularly n_e. The edge profile measurements should also extend into the scrape off layer (SOL). T_e is expected to be in the range of 0.3-1 keV in the SOL, and n_e will vary from $(0.3-1)10^{20}m^{-3}$. The SOL width is expected to be 0.5-2 cm, again implying large radial gradients. The edge Thomson scattering system would need to resolve these gradients, implying that both the optical resolution and the sensitivity needs to be high.

Target measurement specifications for T_e and n_e in the various regions are given in Table 1 below:

Table 1. Measurement specifications as recommended by ITER Expert Group Report for edge electron temperature and density.

region	param.	range	spatial res.	time res.	accuracy
Edge	T_e	0.05-10 keV	0.5 cm	10 ms	10%
Edge	n_e	$(0.05-3)10^{20}m^{-3}$	0.5 cm	10 ms	5%

These specifications represent a balance between anticipated ITER needs and anticipated technical capability. Since T_e and n_e are expected to vary by more than the required accuracy over the required spatial resolution, it would be desirable to make these specifications more demanding. As we will see, this would be particularly difficult at the lower density range. The authors concur with the targets for the edge Thomson measurements as determined by the Expert Group on Diagnostics.

Currently electron temperature and density measurements are listed as potentially needed for plasma control. We feel that it is appropriate to evaluate the feasibility of real time control using the edge Thomson measurements.

DESIGN CONSTRAINTS

Design concepts for edge Thomson scattering must consider the ITER machine configuration. Shown in plan view schematically in Figure 2, the separatrix of the ITER plasma is about 30 cm from the first wall of the blanket section of the port plug. The 0.5 m thick blanket consists mainly of stainless steel with a dense array of water cooling channels. Currently these cooling channels are oriented primarily in the poloidal direction. Detailed guidance on the design of penetrations to the blanket does not exist, however, the size and complexity of these penetrations will be minimized. Moving out from the blanket is the rest of the port plug, consisting mainly of shielding material. Figure 2 shows a viewing penetration with a 10 cm aperture at the back of the blanket. This aperture size, which represents ~f/10 optics, is a reasonable target for this design, and is probably within a factor of 2 in area of what will be feasible from a radiation standpoint.

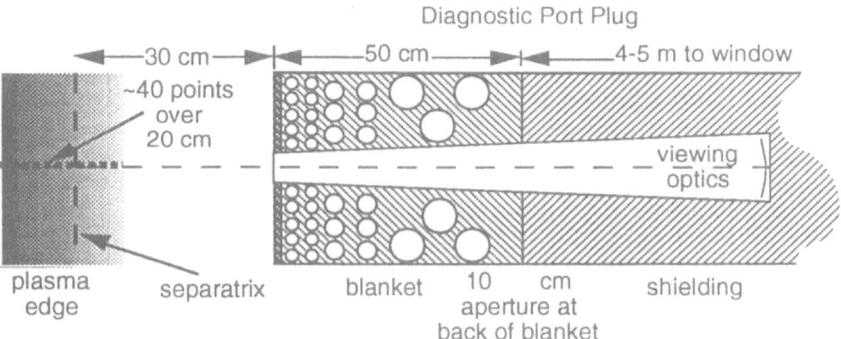

Figure 2. Schematic of the ITER edge region showing approximate distances relevant to edge Thomson scattering.

At a distance > 1 m from the first wall, a large reflective optical element would be located for viewing the scattering volume. The image would be relayed through the port plug via reflective and/or refractive optics (possibly including fiber optics) through a vacuum window at the outside of the port plug.

It is beyond the scope of this report to consider the optical design of the various options discussed, except to give a judgement on the relative difficulty involved. These optical designs depend not only on the penetration constraints, but also on the effects on the primary mirror of loading due to thermal radiation, neutrons, gammas, and charge exchange neutrals. Because of the severity of this loading, it is important to note not only the size but the orientation of the viewing penetration. The loading on mirror with a line of sight through the plasma core will be more severe than that with the same size penetration which views the edge region.

Some options considered involve penetrations which are not normal to the first wall. We assume this is more feasible if the axis lies in a poloidal plane so that the penetration would not block water flow paths.

LIDAR EDGE MEASUREMENTS

A LIDAR Thomson scattering approach, such as that currently used on JET, offers many advantages for the core profile system on ITER. These include relatively simple access, an optical system with coupled laser and collection optics to insure alignment, and short integration times to avoid noise due to plasma background light. On JET with an f/12 optical system, a ruby laser energy of 1 J is needed to achieve a resolution of 10 cm with an accuracy of 10% (Salzmann, 1987)

An ITER edge LIDAR system looks feasible, following a proposal by Molvik, et al at LLNL (Molvik, 1995). Using a chirped pulse amplification laser and a streak camera with cylindrical electrostatic focussing, the required 15 ps time response is within reach.

For a system with the same optical collection and detector efficiency as that in use on JET, a laser energy of 20 J would be necessary to achieve the same accuracy with 5 mm spatial resolution. However, a dedicated radial system would have to view only the 20 cm region at the edge as shown in Figure 2. Thus it is conceivable that a faster f/6 optical system could be used, considering that the penetration could narrow near the first wall, providing better radiation shielding for the first mirror. Furthermore, since the edge system would be measuring lower temperatures, a laser in the green or blue could be used, a wavelength region where the photocathode efficiency is 2-3 times higher than that at the ruby wavelength. Taking credit for both of these effects, and considering that there are fewer photons/J at the shorter wavelength, a laser energy of ~3 J would be adequate to achieve current JET capability on ITER. However, to achieve the target resolution at the lower density limit listed in Table 1, we estimate that 8-10 J would be needed.

While the streak camera detector does exist, a 20 ps, 3 J, 100 Hz laser in the blue or green capable of steady state operation would require considerable development. For example, laser options which are buildable today include (Molvik 1995):

1)	20 ps	4.0 J	527 nm	3-5 Hz	1.5 M$ (1994)
	Flashlamp pumped APG-2 slab amplifier				
2)	10 ps	0.6 J	800 nm	100 Hz	2.2 M$ (1994)
	Diode pumped Nd:YAG pumping Ti-Sa. disk amplifier				

However, these lasers are not tested, and the development needed to obtain reliable operation of a laser appropriate to an ITER edge system is uncertain.

Power densities can be extreme for these laser pulses, and care must be taken to avoid component damage. Another concern is beam breakup and filamentation due to nonlinear phase retardation in refractive optics. Mirrors must be used to transport the beam. However, vacuum windows cannot be avoided. Since this effect scales linearly with intensity and thickness, one approach is to make the beam and the window larger, or alternately, make the window thinner. However, both of these may be incompatible with pressure loading requirements for vacuum windows.

We conclude that an edge LIDAR diagnostic is feasible, in terms of achievable spatial resolution. However, to obtain the desired sensitivity in a radial system, lasers far beyond what is available are needed. The possibility of a LIDAR measurement tangent to the edge is considered in the discussion section below.

IMAGING OPTIONS FOR EDGE THOMSON MEASUREMENTS

The Capabilities Of Existing Thomson Scattering Systems

Multipulse systems with high repetition rate based on Nd:YAG technology, developed first on ASDEX (Rohr 1987), come closest to satisfying all of the ITER edge T_e and n_e measurement requirements. For example, the DIII-D Thomson scattering system has 160 Hz capability with resolution along a vertical line at the plasma edge of 1 cm (Carlstrom 1992). Because of flux expansion the effective midplane radial resolution for this system is near the 0.5 cm specified above. With f/5 optics on DIII-D, this system is adequate for the specifications listed in Table 1. However, such high throughput systems are not likely to be feasible on ITER. In addition, while plasma background light is not a serious problem on DIII-D, it will be much more of an issue on ITER.

Rejection of Background Plasma Light in Imaging Sytems

Ignited operation on ITER will require much higher line average densities than attainable on existing devices. The chord-integrated bremsstahlung brightness will be a factor of 10-20 higher. One means of minimizing this problem would be to orient the sightline to minimize the chord length through the plasma. However, with the available access, it may be hard to avoid looking through the bulk of the plasma. Noise due to plasma background light can seriously degrade the accuracy of these measurements. Roughly speaking:

$$\Delta T_e/T_e \; \alpha \; [(\, S + 2\, B)/S^2]^{0.5} \tag{1}$$

where   S······# of detected photoelectrons from scattered light
              B······# of detected photoelectrons from background

In order to achieve an accuracy of $\Delta T_e/T_e = 10\%$ in the absence of background light, $S \sim 1\text{-}2 \times 10^3$ photoelectrons. For an imaging system operating at the fundamental Nd:YAG wavelength, with avalanche photodiode detectors with a 50 ns integration time for the ITER background, B/S=0.5-1.0. This results in significant degradation in accuracy, particularly for low density edge measurements. As described later, this may lead to design tradeoffs to improve the contrast ratio.

The Spatial Resolution of an Imaging System

Before considering possible scattering geometries for imaging systems, we review the factors which affect spatial resolution and sensitivity. When the spatial resolution specification is comparable to the laser beam width, as it is in this case, several geometrical effects need to be considered. The lower limit on the spatial resolution is given by:

$$dR \cdot F = dl \cdot \sin\theta_{lf} \; + \; w \cdot \sin(\theta_{lv} - \theta_{lf}) / \sin\theta_{lv} \tag{2}$$

where

     dR·····spatial resolution referred to the midplane outer edge
     F·······flux expansion from midplane to measurement position
     dl··scattering length θ_{lf}··angle laser to flux surface
     w··laser beam width θ_{lv}··angle laser to viewing axis

Notice that unless the viewing axis is tangent to the flux surface ($\theta_{lf} = \theta_{lv}$), the spatial resolution is ultimately limited by the laser beam width. For most geometries the laser is focussed with a focal length of several meters. With typical beam divergence, this results in a beam width of ~2 mm, so the second term in (2) can be important for dR = 5 mm.

Equation (2) ignores the degradation in resolution due to the measurement error determined by photon statistics and background light subtraction, and considers only the scattering geometry. This degradation is minimized by maximizing the scattering length dl at fixed dR. This can be done by increasing F or decreasing $\sin\theta_{lf}$.

For the case where the laser is tangent to the flux surface, the first term in (2) diverges, because this equation also neglects the curvature in the flux surfaces. In this case, the following equation applies:

$$dl = [8R(dR \cdot F - w)]^{0.5} \tag{3}$$

where

R ⸱⸱⸱⸱⸱⸱⸱radius of curvature of flux line.

For ITER, at the outer midplane, where R=11m and F=1, dl=420mm for w=3mm and dR=5mm.

Options For Edge Imaging Thomson Scattering Systems

Imaging Option 1 - Radial Laser Beam Normal to Outer Flux Surfaces. Shown schematically in Figure 3, the scattering angle for this geometry is ~150° ($\theta_{lv}=30°$, $\theta_{lf}=90°$). The scattering length for 5mm spatial resolution is degraded according to the beam width according to equation (2). This is shown in Table 2, which compares the scattering length for all of the imaging options we will consider.

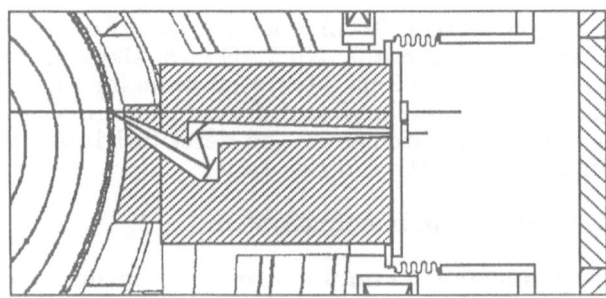

Figure 3. Scattering geometry for imaging Option 1

Because of the small scattering length for this option, the scattered signal is too small to achieve the desired precision with realistic assumptions about available technology. This is illustrated in Table 3, which includes the results of a prediction analysis (Wohlberg, 1967) of the measurement precision. For this table, the f-number of the collection optics was varied in

Table 2. Comparison of scattering lengths for various imaging options.

Option	Laser Configuration	θ_{lv} (deg)	θ_{lf} (deg)	R (mm)	dl (mm) for w = 1.5mm	2mm	3mm	equation
1	normal to outer midplane surfaces	150	90		2.4	1.53		2
2	angled to outer midplane surfaces	90	22		9.63	8.39	5.92	2
3	tangent to outer midplane surfaces	90	0	11000	555	514	420	3

the analysis to achieve the measurement accuracy specified in Table 1 for low density conditions. For example, in the geometry of Option 1, at a laser energy of 1 joule, f/2 optics are needed. Also shown in the table are the number of photoelectrons detected in the scattered and background components for the f/10 optical design target.

Table 3. Comparison of the f-number requirements of the imaging options to achieve $\Delta n/n = .05$ and $\Delta T/T = 0.1$ for $n_e = 5 \times 10^{18} m^{-3}$ and $T_e = 100 eV$ in a single laser pulse. Also shown are detected signal components at target design of f/10 in a 100nm bandwidth .

Option	θ_{scat} (deg)	dl (mm)	w (mm)	f-number for spec. precision	scattered p.e. at f/10	background p.e at f/10
1	150	2.4	1.5	1.9	120	69
2	90	9.6	1.5	3.2	560	270
3	90	300	3	10	1.7×10^4	5.6×10^4

Assumed Parameters

laser energy (1064nm)	1.0 J	integration time	50 ns
transmission	10%	T_{brem}	10000 ev
T_e	100 ev	n_{brem}	1.3×10^{20} m^{-3}
n_e	$5 \times 10^{18} m^{-3}$	Z_{eff}	1.5
excess noise	1.5	path length	5.0 m
spectrally dependent QE		multiplier	1

The values chosen in Table 3 attempt to realistically represent the best that technology will have to offer to achieve the target resolution and repetition rate along with the target precision. Note that the calculations include the effects of bremsstrahlung background, as parameterized by the assumed values listed in Table 3.

To achieve the specified precision at f/10 with a single laser pulse, the options 1-3 would require laser energies of 17, 6.5, and 1 J, respectively.

Imaging Option 2 - Laser Angled to Outer Flux Surfaces. Figure 4 shows this geometry.

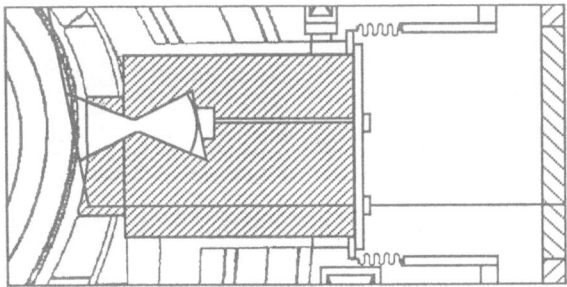

Figure 4. Scattering geometry for imaging option 2

In this case the scattering angle varies around 90°. Since the laser is more aligned with the flux surfaces, the scattering length is longer for the target spatial resolution. However, as indicated in Table 3, the required access for the penetration is still large (≈ 30 cm diameter) to be feasible with 1 J laser energy to achieve the specified accuracy on a single laser shot. However, as discussed below, data can be averaged from multiple lasers or a high repetition rate laser to improve accuracy.

Imaging Option 3 - Laser Tangent to Outer Flux Surfaces. This concept, shown schematically in Figure 5a, utilizes a laser tangent to the flux lines to measure a single radial position at the edge with each laser pulse. The ability to measure a full profile per laser pulse is sacrificed in order to achieve maximum sensitivity. This sensitivity is high due to the long scattering length that results from this geometry, limited only by how long a slot can be made in the blanket (we have assumed a 30 cm scattering length for the calculation in Table 3). As shown in Table 3, this results in greatly reduced size requirements for the viewing penetration.

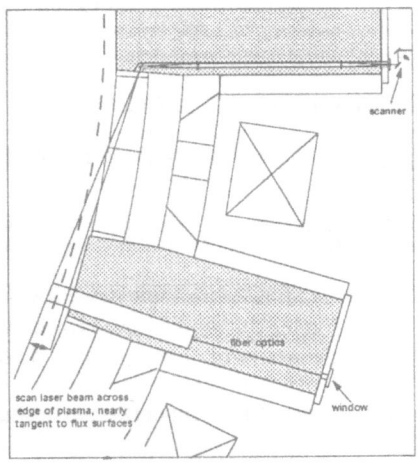 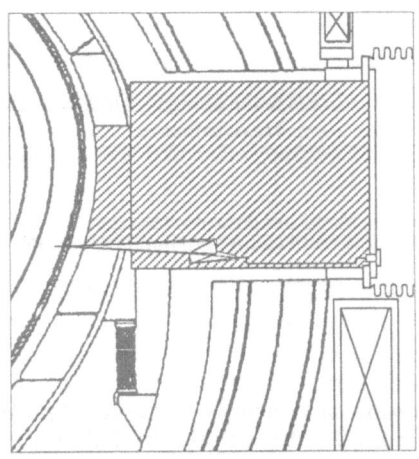

Figure 5.   a) Plan view of imaging option 3   b) Elevation view of 3

To obtain a profile, multiple lasers could be used firing at different times, or a single high repetition rate laser could be scanned across the edge. Thus the spatial information would be temporally encoded with a single analysis system.

The optical design of options 1 and 2 above are made difficult by the small values of dl and w in Table 3. However, for option 3, good resolution is required in only one dimension, and even that is relatively less severe. This should ease the design of the relay optics which carry the light to the analyser. The calculations of the background brightness for option 3 assume an aperture area of dl*10cm/(f-number), consistent with the need to accomodate the large depth of field of ± 10 cm.

As shown in the figure, this option requires access for laser beams through more than one blanket module. It may be possible to take advantage of the gap between the port plug blanket module and the blanket module above or below it for this narrow slot as shown in Figure 5b. It also requires a horizontal slot in a different port plug for viewing, perhaps through a modification of the top or bottom edge of the port plug blanket module.

Imaging Options at the Top. There is a variant of option 3 utilizing a top port for access. Using a laser deflected from the outer part of the port, it is possible to scan the beam tangent to the top edge flux surfaces, which in this location have expanded by ~x3 from their spacing at the midplane (F=3). However, the required depth of field is expanded by a corresponding amount. One advantage over the midplane versions of this concept is that the laser and viewing components are on the same port plug. In addition, the slots for the laser and the viewing are in the poloidal direction, parallel to the coolant paths in the blanket. Similarly, there is a variant of imaging option 2 with the laser intersecting the top edge at a shallow angle. For both options, the viewing sightline has a shorter chord through the plasma than the midplane views, reducing plasma background light.

DISCUSSION

Given the design constraints that we achieve the specified accuracy with a laser capable of high repetition rate (implying an energy E~1J), and with an optical system with an f-number of 10 or higher for compatability with blanket design, we are unable to find a solution capable of providing a full profile in a single laser pulse. To do this would require a high repetition rate laser capable of ~ 10J/pulse.

However, we believe both options 2 and 3 can be considered if one is willing to acquire a profile over many 1J laser pulses, perhaps from multiple lasers. Figure 6 shows the measurement precision at one position for options 2 and 3 . Since to acquire a full profile at full resolution with option 3 would take ~ 40 laser pulses, we show for comparison the accuracy expected for averaging over 40 pulses for option 2, which also clearly meets specifications. It is better than the option 3 case shown because of its better background rejection. Averaging over 12 1J laser pulses in option 2 is equivalent to the option 3 case shown.

There are a number of issues with both of these options. First, to aim the laser beam, both options require a mirror close to the back surface of the blanket as shown if Figures 4 and 5. This mirror is shielded from the core of the plasma, and can be fairly small ~ 4 cm.

The aiming of the multiple beams or scanning of the single laser beam over ~ 3° in imaging option 3 would be difficult because of the need to keep a fairly tight focus for efficient background rejection, and the impossibility of using moving optics at the final mirror position.

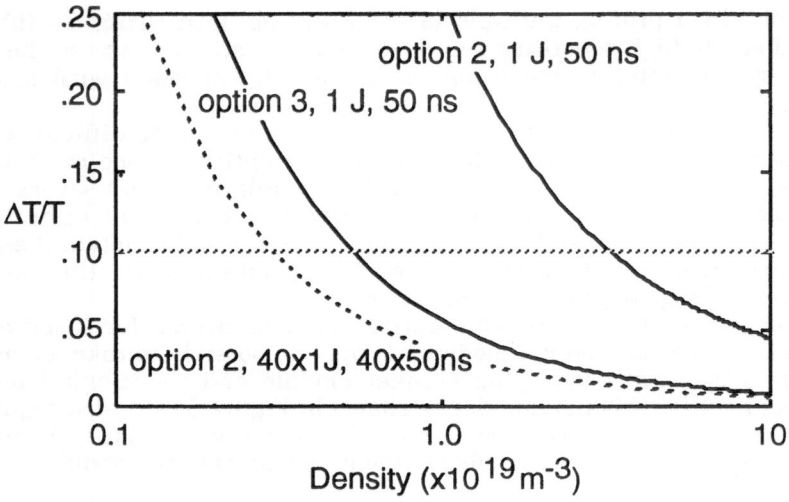

Figure 6. Measurement accuracy as a function of density for options 2 and 3, and also for option 2 with averaging of 40 laser pulses.

Although in both options 2 and 3, the scattered signal could be used to adjust the alignment transverse to the viewing axis, knowing the location of the scattering volume along the viewing axis to 5 mm is also a problem. Perhaps a small penetration at the laser beam strike point could be instrumented to provide a positioning fiducial. Since the laser beam position is fixed in option 2, obtaining absolute position stability would be easier in this case.

The size and shape of the blanket penetration for option 3 is definable assuming f/10 optics. At the back of the blanket module, it would consist of a 10 cm diameter hole. At the first wall, the penetration would be a slot with approximate dimensions 5 cm x 20 cm. For option 2 the first wall slot would need to be 5 cm x 35 cm.

The laser beam impinges at a glancing angle on another part of the blanket. Depending on the angle and the beam diameter at this point, a special laser dump may be needed to avoid ablation damage to the first wall.

The choice of laser wavelength is not clear. With current laser technology, operating at the Nd:YAG fundamental with 50 ns integration time provides larger signals but poorer contrast than operation at half that energy at the second harmonic with a gate time of 5 ns. At this point, without getting into more details on the background assumptions and detector efficiencies, operation at the fundamental appears marginally more favorable. This choice also has the advantage that optics are less sensitive to coating and damage at this wavelength.

Depending on the number and repetition rate of lasers used, detector and digitizer recovery time may limit the speed with which a full profile can be acquired with these techniques.

The choice between these options may become more clear as ITER decisions are made on how these measurements can be most useful. For example, if high precision measurements of n_e or T_e or their gradients at a particular position are needed to correlate with other measurements or for feedback control, option 3 is clearly superior. On the other hand, if there is a willingness to sacrifice accuracy at low density in order to follow the evolution of the full profile, option 2 is preferable.

We have considered the possibility of using LIDAR tangent to the edge. However, to achieve sufficient signal with a 1J laser at 532 nm would require f/10 optics even with a 30 cm scattering length. Since the access is such that the collection mirror would be ~ 3-4 m away from the scattering volume, the required blanket penetration would be too large (30 - 40 cm diameter), and a mirror would be needed near the plasma, which is not desirable.

REFERENCES

Carlstrom, T.N., *et al* , 1995, Design and operation of the multipulse
  Thomson scattering diagnostic on DIII-D, *Rev. Sci. Instrum.* 63:4901.
Barth, C. J. *et al* , 1992, PROMT: a multiposition Thomson scattering system
  for RTP, *Rev. Sci. Instrum.* 63:4947.
Molvik, A.W., Hooper, E.B., Lerche, R.A., and Perry, M.D., 1995, in Review
  of Initial Definition and Specification of the ITER Diagnostic System,
  ITER/95/PH-07--27, 11.
ITER Expert Group Report, 1995, Minutes of the Second Meeting of the
  ITER Physics Expert Group on Diagnostics, S CX MI 3 95-03-23 F 1.
Rohr, H., Steuer, K.-H., Murmann, H., and Meisel, D., Periodic
  Multichannel Thomson Scattering on ASDEX,IPP III/121 B, 1987.
Saltzmann, H. *et al* , 1987, *Nuclear Fusion*, 27:1925.
Wohlberg, J.R., 1967, *Prediction Analysis*, D. Van Nostrand,New York.

ACTIVE SPECTROSCOPIC DIAGNOSTICS FOR ITER UTILIZING NEUTRAL BEAMS

Earl S. Marmar

MIT Plasma Fusion Center
Cambridge, MA 02139 USA

INTRODUCTION

The use of neutral beams, combined with visible spectroscopy, is widely applied to the diagnosis of tokamak plasmas. Local plasma parameters which have been measured with these techniques include low Z impurity densities (e.g. Isler, 1977; Boileau, et al., 1989; von Hellerman, et al., 1992) and impurity transport coefficients (e.g. Synakowski, et al., 1990), ion temperature and rotation (e.g. Fonck, et al., 1984), internal magnetic field, (and thus current density and safety factor (Levinton, et al., 1989; Wrobleski and Lao, 1992; Wolf, et al., 1993)), and electron density fluctuations (eg. Durst, et al., 1992). Very recently, the first measurements of the energy distribution of slowing-down D-T fusion α particles have been reported (McKee, et al., 1995). An examination of the proposed measurement requirements for ITER (ITER 1995), as developed by the ITER JCT in consultation with the ITER Diagnostic Experts Group and diagnosticians from around the world, reveals several obvious candidates for application of diagnostic neutral beams (DNB's). Of these, the one for which there appears to be no alternative whatsoever, is measurement of the thermalized helium density profile in the confinement region. Since helium ash accumulation is one of the key physics issues to be tested in long pulse (> 1000 second) operation, the inability to measure this quantity accurately in real time would be a significant handicap and could conceivably compromise some of the main ITER missions. A diagnostic beam, combined with charge exchange recombination spectroscopy (CXRS) will be able to provide this measurement.

Assuming appropriate spectroscopic and neutral particle diagnostics can be designed to observe the DNB, additional enhanced profile measurement capabilities that will become available include ion temperature, plasma rotation, and low Z impurity concentrations, including Be, C, O and Ne. Ion temperature and rotation measurements in the plasma core can (and should) be available using high resolution X-ray spectroscopy (Hill, et al., 1992; Bitter, et al., 1993; Widman, et al., 1995). However, these are inherently chord integral measurements, and may require seeding the plasma with a medium Z impurity, such as argon or krypton. The low Z impurity density profiles probably cannot be obtained by any means other than the DNB/CXRS com-

Diagnostics for Experimental Thermonuclear Fusion Reactors
Edited by P. E. Stott *et al.*, Plenum Press, New York, 1996

281

bination. While detailed knowledge of beryllium, carbon, oxygen and neon profiles is not quite as important as the α ash question, knowing these quantities will be extremely useful, and the only alternative information in the core plasma will come from profiles of radiated power and Z_{eff}. Two additional important uses of DNB's, on presently operating tokamaks, are to measure internal magnetic field (and thus q profiles) via the Motional Stark Effect (MSE) and to measure fluctuations via Beam Emission Spectroscopy (BES). In both of these cases, it is the radiation resulting from excitation of beam neutrals which is observed. For CXRS, the cross section for population of the desired upper level in the H-like ion is a strongly decreasing function of interaction energy. As a result, even though beam attenuation will be very significant, the optimum energy for CXRS is in the range from 100 keV/amu to 150 keV/amu for hydrogenic beams. The excitation cross sections for MSE and BES are weaker functions of neutral energy than for charge exchange recombination, so energies in the range of 400 to 600 keV/amu, or perhaps higher, are advantageous for those applications. As a result, a DNB which is optimized for CXRS will not be so well suited for MSE; assuming NBI heating is available on ITER, it is possible that one of the heating beams (with time modulation) would be better suited to MSE. Given the need for high time resolution, it is likely that BES would be most useful in measuring fluctuations near the periphery of ITER plasmas; BES will not be analyzed quantitatively here.

CHARGE EXCHANGE RECOMBINATION SPECTROSCOPY

Alpha Ash Measurements

As the measurement of the α ash density profile is the primary motivation for wanting to add a diagnostic neutral beam to ITER, we focus our considerations first on that application. As has been previously pointed out (von Hellerman and Summers, 1993), beam attenuation in the large, high density ignited ITER plasma will be severe. This, combined with the rapid decrease of charge exchange population of excited He II as the neutral energy increases above 50 keV, implies that the signals from the core of the plasma will be relatively weak. In all cases presented in this paper, the plasma has been modeled as follows (ITER Diagnostic Group, 1995): $R = 8$ m, $a = 3$ m, $T_i = T_e = 20 \times [1 - (r/a)^2]^2$ keV, $n_e = n_0 \times [1 - (r/a)^2]^{.5}$, $n_{He} = .1 \times n_e$, $Z_{eff} = 1.5$.

Figure 1 shows the results of penetration calculations for a range of neutral hydrogen beam energies, assuming perpendicular injection along the midplane. The plasma parameters used are those of the high density, fully ignited case, with central density $n_0 = 1.4 \times 10^{20}$ m^{-3}. For the beam attenuation calculations, ionization due to electron impact (Janev, et al., 1987), proton impact (Fite, et al., 1960; Gilbody and Ireland, 1964) and charge exchange (Freeman and Jones, 1974) have been taken into account. The emissivity in the H-like He line of interest is given by:

$$E = \frac{n_\alpha \, n_{H_0} \, \langle \sigma v \rangle_{cx}}{v_{H_0} \, \Delta \lambda_D} \quad \text{ph/m}^3/\text{s/nm}, \tag{1}$$

where $\langle \sigma v \rangle_{cx}$ is the effective rate for emission of the $n = 4 \to 3$ transition at $\lambda = 468.6$ nm, taken from (Fonck, et al., 1984), v_{H_0} is the neutral velocity and $\Delta \lambda_D$ is the doppler width due to the ion thermal distribution. The neutral beam is assumed to deliver an equivalent current density of 10^3 amp/m^2 (before plasma attenuation), independent of beam energy, with a cross-section of $.2 \times .2$ m. At 125 keV, this corresponds to a beam power of 5 MW. For a detection geometry where the optics

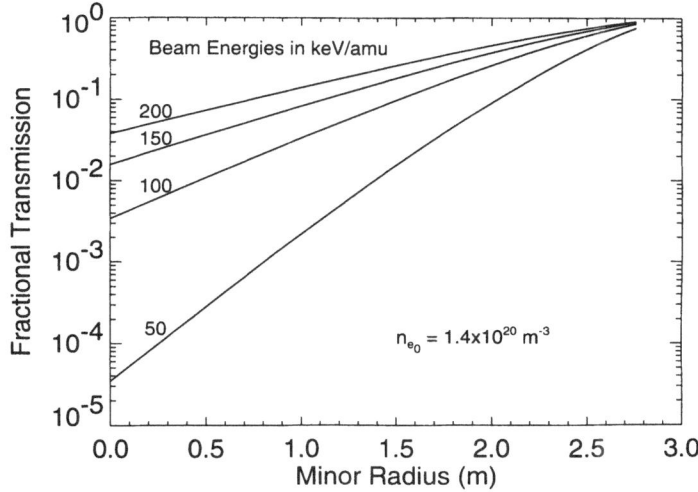

Figure 1. Neutral hydrogenic beam transmission as a function of minor radius for the high density, fully ignited scenario. The curves are parameterized by beam energy in keV/amu.

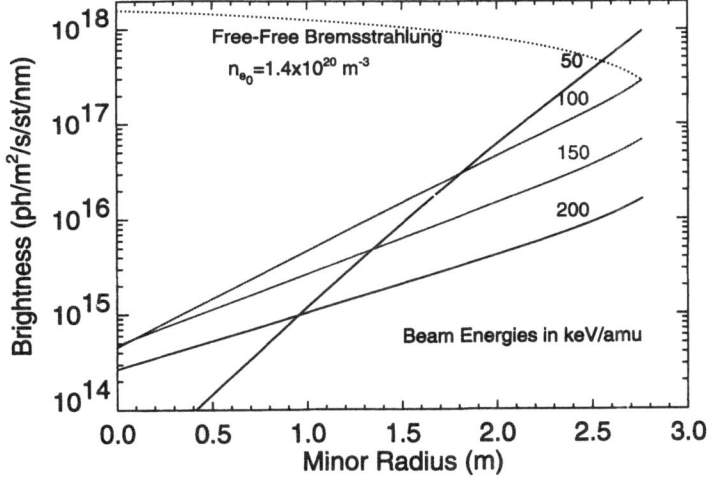

Figure 2. Brightness in the He$^+$ ($n = 4 \rightarrow 3$) line at $\lambda = 468.6$ nm, as a function of minor radius, for 4 different neutral energies. The continuum due to free-free bremsstrahlung at the same wavelength is shown by the dashed curve.

view the beam tangentially (at r = 0), the line brightnesses which would result are shown in figure 2. For comparison, the continuum due to free-free bremsstrahlung (Kadota et al., 1980), is also shown. Figure 3 shows the same data, plotted to give the ratio of the bremsstrahlung to the CXRS signal, as a function of beam energy for several radial locations. In the center of the plasma ($r/a = 0$), the continuum will be more than 3 orders of magnitude brighter than the CXRS signal, even for the optimum beam energy of about 125 keV. In order to pull the signal out of this large background, it will be necessary to modulate the DNB.

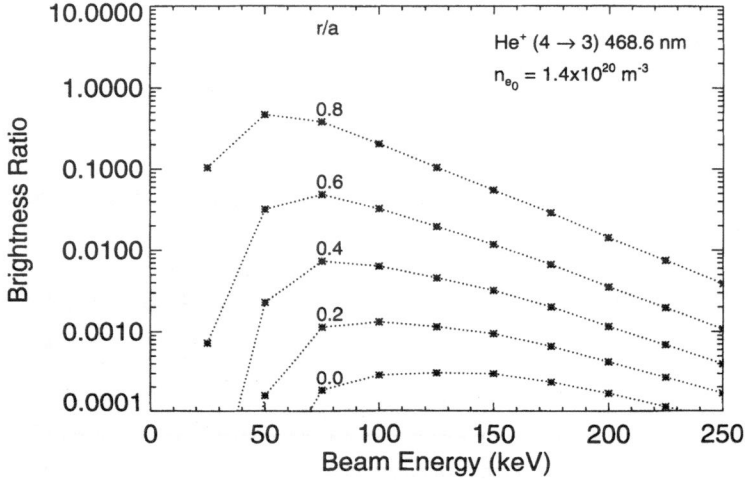

Figure 3. Ratio of the bremsstrahlung continuum divided by the charge exchange signal for He^{+}, as a function of neutral energy. The 5 curves are for different values of fractional minor radius (r/a).

The signal to noise ratio of the measurement is fundamentally limited by the fluctuations in the continuum signal (von Hellerman and Summers, 1993). The minimum level of these fluctuations is determined by photon statistics. In order to estimate this, it is assumed that the optical system has an etendue of 2×10^{-7} m$^2 \cdot$ st (determined by the spectrometer), and an overall transmission and detection efficiency (mirrors, fibers, lenses, detectors) of 1% (R. Fonck, private communication). To measure the density of thermalized helium in the discharge, we integrate over the doppler broadened spectrum of the line; at $T_I = 20$ keV, the full width at half maximum will be ~ 2 nm. With these assumptions, for the high density case, and a beam energy of 125 keV, the detected photon count rate will be $\sim 6 \times 10^9$ ph/s. Figure 4 shows the signal which would result, for a beam which is modulated at 5 Hz (0.1 seconds on, 0.1 seconds off) with 1 msec time averaging. In general, the signal to noise ratio (SNR) is given by:

$$\text{SNR} = \frac{S_{cx}\sqrt{t}}{\sqrt{2(2S_{vb} + S_{cx})}} , \tag{2}$$

where S_{cx} is the charge exchange signal, S_{vb} is the continuum signal (both in photon/s) and t is the integration time. Taking the difference between two successive 0.1 second averages ($t = 0.2$ s) would, in this case, yield SNR ≈ 5, as shown in figure 5.

In applying these calculations, an immediate concern is that, while counting statistics will certainly determine the minimum contribution to the noise, this minimum may not be achieved in the real experiment. It is heartening, in this regard, that CXRS measurements of the slowing down spectrum of D-T produced α particles have been recently reported from TFTR (McKee, et al. 1995). In the TFTR experiment, spectra out to about 0.6 MeV have been measured, even though the signal level at these energies was 1% or less of the bremsstrahlung continuum. In addition, time modulation of the beam during a shot was not available, so the data were taken on successive shots, one with the beam on, the second with it turned off. It is also worth noting that monitoring the continuum intensity simultaneously with the charge exchange measurements was crucial for the TFTR measurements. Similarly, on ITER it will be necessary to monitor the continuum in order to detect slow changes in the bremsstrahlung intensity. The presently planned ITER diagnostic set includes such a system, to measure Z_{eff}. It might be convenient to use the same tangential view for both the CXRS and visible bremsstrahlung front end optics.

Ion Temperature Measurements

Measuring ion temperature profiles, as well as those of rotation velocity, is in some ways simpler than measuring impurity density, since the absolute density of the donor neutral particles in the observation region need not be known, nor is it necessary to measure absolute brightness. However, because the desired information is contained in the spectral shape, there is a requirement to increase the spectral resolution of the measurement, by about a factor of 10. Thus, instead of the 2 nm which is required for the α ash density measurement, ~ 0.2 nm resolution, along with the need for multiple spectral channels, is required. From equation 2, it is apparent that, for the same SNR, a factor of ~ 10 will be lost in time resolution. Figure 6 summarizes the signal to noise for the thermal ash density measurements, which can be achieved with beam energy of 125 keV, as a function of integration time for several minor radii. All of these cases are calculated for the high density plasma ($n_0 = 1.4 \times 10^{20}$ m^{-3}).

Assuming that SNR of 10 is sufficient, an integration time of about 2 seconds will be required for the central α ash density measurement; the situation improves rapidly as the radius of observation increases (due to the decrease of beam attenuation), so that at $r/a = .5$, a similar SNR can be achieved with 0.1 second time resolution. The results for ion temperature measurement are similar, except for the factor of 10 increase required. It would thus appear that, at least for the highest density operation, central T_I (as well as rotation) measurements will require very long integration times (10 seconds or longer), but profile information for $r/a > 0.4$ should be available with much higher time resolution. Things also improve rapidly if the plasma density is lowered, for two reasons: first, the beam transmission is roughly $\propto e^{-(a/\lambda_i)}$, where λ_i is the beam ionization mean free path, which is inversely proportional to density; second, the bremsstrahlung emissivity is $\propto n_e^2$. One further consideration, with regard to measurement uncertainties, has been pointed out by von Hellerman (1993). Because of the exponential dependance of transmission on ionization, uncertainties in the ionization rates themselves lead to increasing uncertainties in the beam density as the attenuation increases: the fractional uncertainty in transmission is equal to $(a/\lambda_i) \cdot [\delta(\lambda_i)/\lambda_i]$, where $\delta(\lambda_i)/\lambda_i$ is the fractional uncertainty in λ_i.

The use of very high power, short pulse beams, based on ion-diode technology, has also been proposed (Rej, et al., 1992; Bartsch, et al., 1995). The goal would be

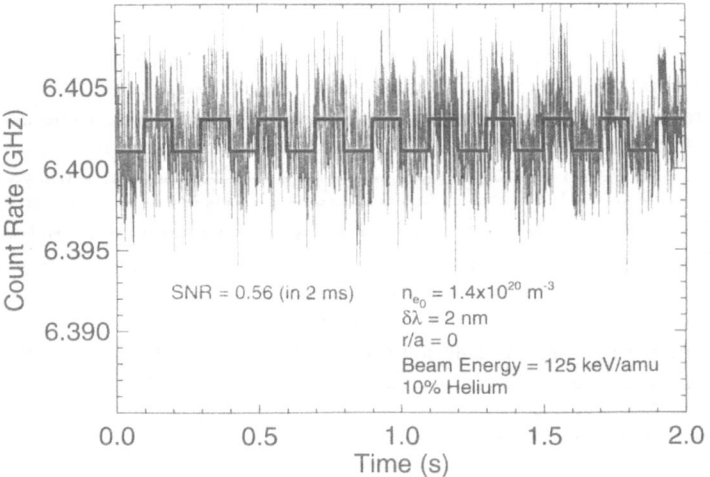

Figure 4. Simulation of signal which could be expected if the noise is due solely to photon counting statistics. The square wave shows the expected signal if there were no noise. The 2 nm spectral width is chosen to look at thermalized He density, not temperature or rotation.

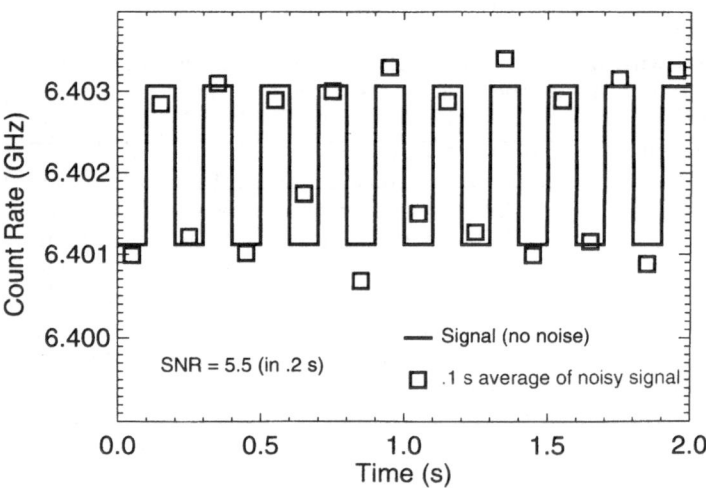

Figure 5. Results of averaging the simulated data of figure 4. The boxes indicate 0.1 second averages, from successive times with the beam first off and then on.

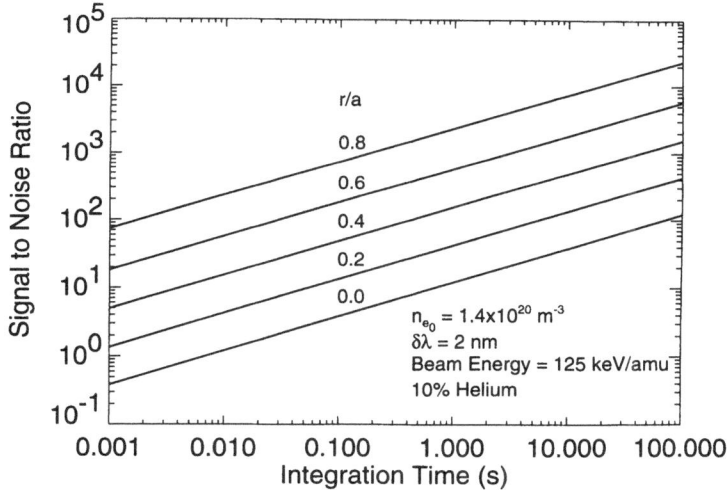

Figure 6. Signal to noise ratio, as a function of measurement time resolution (total of beam-on plus beam-off), for different values of normalized minor radius. These calculations are relevant to the thermal α ash density measurement ($\delta\lambda = 2$ nm). As discussed in the text, SNR for ion temperature measurements is reduced by one order of magnitude.

to develop beams with 5×10^9 W power and pulse length in the microsecond range. The beam would be pulsed repetitively, at up to 30 pulses per second. Compared to the 5 MW conventional beam, the high power beam would yield signal increases of 10^3 for CXRS measurements. In this way, the signal to noise ratios obtained with a $1\,\mu s$ beam pulse would be comparable to those obtained with the conventional beam in 1 second.

CXRS measurements on Be, C, O and Ne will also be possible (and desirable). The effective cross sections increase with the Z of the target. At 125 keV/amu, the effective rate for C VI emission at $\lambda = 529.2$ nm is about a factor of 5 greater than that for He II at 468.6 nm (Fonck, et al., 1984; Zinov'ev and Korotkov, 1989). Since the expected helium density is more than a factor of 10 greater than that of C, the signal to noise ratios for C will be, if anything, slightly smaller than those shown for He.

Fast Alpha Distribution

As already mentioned above, the spectrum of slowing-down D-T fusion produced α's ($E_\alpha \leq 0.6\,\text{MeV}$) was recently measured on TFTR, using CXRS (McKee, et al., 1995). (Higher energy α spectra were also obtained on TFTR (Fisher, et al., 1995), using a lithium pellet charge exchange technique, observing outgoing neutrals produced in the pellet ablation cloud.) In principle, similar CXRS measurements could be undertaken on ITER. The signals rapidly decrease with increasing α energy, and the expected density of tail α's also decreases with energy. It is likely that CXRS measurements of the core non-thermal α distribution become impractical for energies greater than that where the density in a 0.2 MeV bandwidth is about 10% of the thermal density.

MOTIONAL STARK EFFECT

The final diagnostic application of DNB's considered here is the measurement of internal magnetic field, and thus q profiles, by taking advantage of the motional stark effect (MSE). As the high energy neutrals traverse the plasma, crossing the magnetic field, the atoms, in their rest frame, experience an electric field, $\mathbf{E} = \mathbf{v} \times \mathbf{B}$. The resulting stark effect shifts the energy levels in the atom, and line emission (e.g from Balmer-α) is split into multiple components. The various lines are also polarized, depending on the relative directions of $\mathbf{v} \times \mathbf{B}$ and the optical view (von Hellerman and Summers, 1993; Wroblewski and Lao, 1992; Levinton et al., 1989).

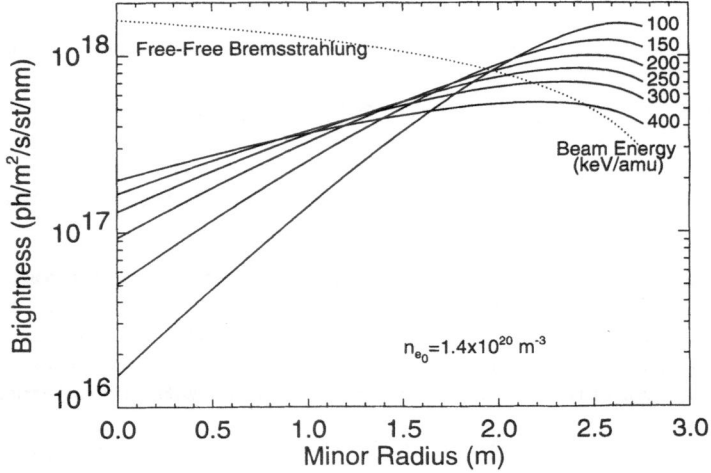

Figure 7. Balmer-α brightness in the unshifted σ stark component, (polarized perpendicular to the electric field) as a function of minor radius for several beam energies. For comparison, the bremsstrahlung continuum brightness profile is indicated by the dotted line.

In order to see a component of the poloidal magnetic field, as well as to provide a doppler shift to move the lines spectrally away from the edge neutral emission, it is necessary that the view of the beam not be tangent to the toroidal direction. Since beam attenuation remains as an issue for MSE, we again assume perpendicular injection along a major radius in the midplane, with a viewing angle of 45°. Figure 7 shows the results of a simulation, similar to those done for CXRS, giving the brightness in the central, unshifted σ component of H_α. The spectral width is determined by the doppler broadening; here we assume this is dominated by the viewing cone for each spatial channel, taken to be ±0.8°. The excitation cross-section, which for these energies is dominated by proton impact, does not fall as rapidly with increasing neutral energy as the CXRS rates. Consequently, because of the improving beam penetration, it is advantageous to go to higher energies. For the highest electron density case considered ($n_o = 1.4 \times 10^{20} \mathrm{m}^{-3}$), the on-axis signal reaches its maximum for an energy slightly in excess of 400 keV. Again the bremsstrahlung continuum is brighter than the signal of interest, but only by about a factor of 10 at $r/a = 0$. The measurement itself requires splitting the light into multiple polarization components, so for a given total intensity, the ultimate signal to noise ratio will be somewhat smaller than for CXRS. However, given the intensities in figure 7, it should be possible to measure poloidal field profiles with excellent time resolution (< 0.1 s).

CONCLUSIONS

A number of possible applications for diagnostic neutral beams on ITER have been considered. The optimum beam energy for charge exchange recombination spectroscopy is ~ 125 keV. Using such a beam, with about 5 MW injected power, CXRS should yield quantitative measurements of the thermalized α ash density profiles. Concurrently, the availability of such a beam, and the associated spectroscopic diagnostics, will allow for measurements of other low Z impurity density profiles (Be, C, O, Ne) as well as ion temperature and rotation profiles. For temperature and rotation, on-axis measurements will be difficult, and long integration times will be necessary (> 10 s); for $r/a > 0.4$, all of these measurements should yield accurate results with excellent time resolution. It is crucial, with respect to all these measurements, that the spectroscopic diagnostics collect as much light as possible, since the signal to noise ratio improves with the square of the number of detected photons. For the same reason, the time resolution which can be achieved, for a given SNR, improves rapidly with decreasing density, since the bremsstrahlung continuum emissivity is proportional to n_e^2.

DNB's can also be used to measure internal magnetic field, through the motional stark effect. In this case, the optimum beam energy is closer to 500 keV, requiring a separate beam from that used for CXRS. Consideration should be given to using one of the heating beam sources (operating in hydrogen, with modulation) for this purpose.

ACKNOWLEDGEMENTS

The author is pleased to acknowledge helpful discussions with G. McCracken, J. Terry, J. Goetz, M. Greenwald, M. von Hellerman, R. Fonck, B. Stratton, C. Barnes, P. Thomas, M. Sasao, A. Costely, V. Mukhovatov, and K. Young. This work is supported by U.S.DoE Contract No. DE-AC02-78ET51013.

REFERENCES

R.R. Bartsch, et al., 1995, *Rev. Sci. Instrum.* **66** 306.

M. Bitter, et al., 1993, in *Atomic and Plasma-Material Interaction Processes in Controlled Thermonuclear Fusion*, R.K. Janev and H.W. Drawin (editors), Elsevier Science Publishers, Amsterdam, page 119.

A. Boileau, et al., 1989, *Plasma Phys. Contr. Fusion* **31** 779.

R.D. Durst, et al., 1992, *Rev. Sci. Instrum.* **63** 4907.

R.K. Fisher, et al., 1995, *Phys. Rev. Lett.* **75** 842.

W.L. Fite, et al., 1960, *Phys. Rev.* **119** 663.

R.J. Fonck, et al., 1984, *Phys. Rev. A* **29** 3288.

E.L. Freeman and E.M. Jones, 1974, *Atomic Collision Processes in Plasma Physics Experiments I*, UKAEA Report CLM-R137, Culham Laboratory, Abingdon, England.

H.B. Gilbody and J.V. Ireland, 1964, *Proc. Roy. Soc.* **A-277** 137.

K.W. Hill, et al., 1992, *Rev. Sci. Instrum.* **63** 5032.

R.C. Isler, 1977, *Phys. Rev. Lett.* **38** 1359.

ITER, 1995, Report S CX MI 3 95-03-23 F 1, *Minutes of the Second Meeting of the ITER Physics Expert Group, Naka, Japan.*

ITER Diagnostic Group, 1995, Report S 55 RE 1 95-07-11 F 1, *ITER Information: Updated Background Information relating to Diagnostic Task Agreement.*

R.K. Janev, et al., 1987, *Elementary Processes in Hydrogen-Helium Plasmas*, Springer-Verlag, Berlin.

K. Kadota, et al., 1980, *Nucl. Fusion* **20** 209.

F. Levinton, et al., 1989, *Phys. Rev. Lett.* **63** 2060.

G. McKee, et al., 1995, *Phys. Rev. Lett.* **75** 649.

E.J. Synakowski, et al., 1990, *Phys. Rev. Lett.* **65** 2255.

D.J. Rej, et al., 192, *Rev. Sci. Instrum.* **63** 4934.

M.G. von Hellerman and H.P. Summers, 1993, in *Atomic and Plasma-Material Interaction Processes in Controlled Thermonuclear Fusion*, R.K. Janev and H.W. Drawin (editors), Elsevier Science Publishers, Amsterdam, page 45.

K. Widman, et al., 1992, *Rev. Sci. Instrum.* **66** 761.

R.C. Wolf, et al., 1993, *Nucl. Fusion Lett.* **33** 663.

D. Wroblewski and L.L. Lao, 1992, *Rev. Sci. Instrum.* **63** 5140.

A.N. Zinov'ev and A.A. Korotkov, 1989, *JETP Lett.* **50** 307.

SPECTROSCOPY FOR IMPURITY CONTROL IN ITER

N J Peacock,[1] R Barnsley,[2] N C Hawkes,[1] K D Lawson,[1] M G O'Mullane[3]

[1] UKAEA (Government Division–Fusion), Euratom/UKAEA Fusion
   Association, Culham Laboratory, Abingdon, Oxfordshire, OX14 3DB, UK
[2] Department of Physics, University of Leicester, UK
[3] Department of Physics, University College Cork, Ireland

INTRODUCTION

Bremsstrahlung losses from the common low Z impurities play a much larger role in ITER than in present experiments due to the larger plasma dimensions and relatively better energy confinement of ITER. Control of the level of medium and high Z impurities inside the main plasma is also of utmost importance for achieving ignition in ITER. In this paper we outline the need for spectroscopic diagnostics not only covering the main plasma but also the radiation mantle, the scrapeoff layer (SOL) and the divertor. The total wavelength coverage is ~1Å to 1μm. We also examine the effect of the particle diffusion parameters on the spatial emission and fractional abundances of impurities. Transport coefficients appropriate to a wide range of operating regimes are adopted. The spatial distribution of the ion charge states determines the wavelength coverage and sensitivity of the diagnostic.

Preliminary designs exist for visible spectroscopic diagnostics (ITER, 1995). All of these visible systems, which have been proposed so far, rely on reflectors as the first element in a labyrinthine optical line-of-sight and are exposed directly to the plasma emission. They are appropriate to passive spectroscopy of the bulk plasma continuum and influxes of low charge states near the edge as well as active, atomic beam spectroscopy of more highly charged ions in the bulk plasma (Marmar, 1995; von Hellerman et al., 1995). X-ray systems are also under active consideration (Barnsley et al., 1995a; Hill et al., 1995).

A somewhat neglected region is the VUV and XUV region (25Å < λ < 2000Å). Results from present tokamaks are presented which show the importance of this spectral region in diagnosing the mantle, SOL and divertor plasmas. The main technical problems to be overcome are the durability and performance of the first reflector, the lack of a tritium barrier and induced neutron noise in the detectors. We discuss the JET experience with

Diagnostics for Experimental Thermonuclear Fusion Reactors
Edited by P. E. Stott *et al.*, Plenum Press, New York, 1996

291

neutron noise levels and present calculations of expected noise ratios in ITER. Some radical solutions to the first reflector problem are also advanced.

EMISSION FROM ITER

A reasonable operating scenario of an ignited ITER is 300MW of α power, P_α, released into the bulk plasma with an 100MW of additional heating (Janeschitz, 1995). In Figure 1 we illustrate the levels of radiation expected from the different plasma regions with their characteristic electron parameters. Up to 100MW could be radiated from the bulk as bremsstrahlung and synchrotron radiation. The central impurity fraction should be limited to keep $Z_{eff} \le 1.65$ with a maximum He concentration of 14% giving $P_{brem} \sim 0.35P_\alpha$. Consequently 50MW can be radiated in the mantle as a result of Ar gas puffing. Engineering constraints allow a maximum of 50MW onto tungsten divertor strike plates. The remaining 100MW must be radiated in the SOL and in the divertor. In these outer regions, atomic processes of recombination and charge exchange will necessarily need to be taken into account. The diagnostic challenge is to monitor the full spectral spread of this radiation.

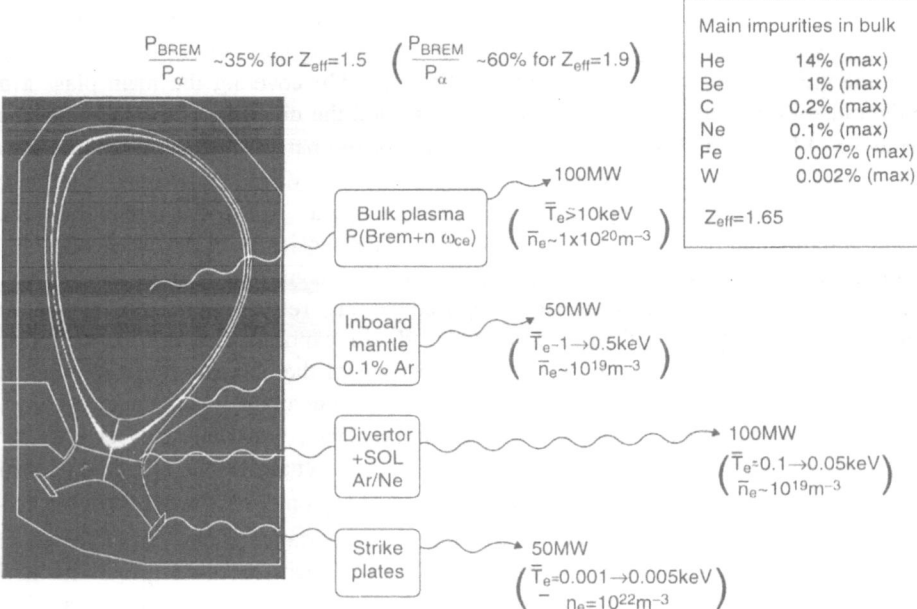

Figure 1 Illustration of putative impurity abundances (relative to n_e) and impurity emission from different regions of the ITER plasma.

Spatial Emission in Different Transport Regimes

The detailed spatial distribution of impurities depends on the transport (confinement regimes) and on the temperature and density profiles. The blanket and vacuum vessel walls are expected to be fabricated from stainless steel with Be tiles as the first wall material (Janeschitz, 1995). A radiation mantle of argon is also likely during operation. The spatial variation of these elements, in various transport regimes, has been modelled using the 1.5-D impurity transport code, SANCO (Lauro-Taroni, 1995), supported by the atomic data acquisition system, ADAS (Summers, 1994). Plasma profiles are taken from the ITER–EDA documents, (ITER, 1995). We present illustrative calculations of the impurity emission profiles of iron, one of the more likely medium Z impurities. The maximum concentration of Fe allowed for $Z_{eff} = 1.65$ is 0.007% (see Fig.1). There is a large uncertainty in the shape of the temperature and density profiles in an ignited 1.5GW ITER H-mode plasma. The 'nominal' reference density and temperature profiles (Becker, 1995a), appropriate to ELMy H-mode ignition and the 'extreme' L-mode profiles (Becker, 1995b), have both been used in the present simulations and are shown in Figure 2.

The choice of impurity transport coefficients is crucial to the calculation of the radial distributions of the impurity ionisation stages. Determining these coefficients is an empirical procedure, being derived from observations on existing tokamaks. Since these coefficients seem to be applicable over a wide range of conditions, we extrapolate their use to ITER conditions. Three situations are modelled, 'normal' H-mode, H-mode with extra outward convection, and L-mode. The coefficients adopted are shown in Figure 2. Experience of atomic beam induced H-modes in JET indicates hollow impurity profiles, i.e. the impurities accumulate off-axis (Lauro-Taroni et al., 1994). This necessitates the introduction of a positive outward convection (O'Mullane et al., 1995c) in the flux formalism of

$$\Gamma_z(r) = -D(r)\nabla \cdot n_z + V(r)n_z$$

assuming steady-state values. An amendment to the 'usual' convection profile (Pasini et al., 1990) to account for this outward flow is shown in Figure 2.

Calculations of the total radiation profiles for Fe contamination are shown in Figure 3. With an outward convection pinch, particles accumulate off-axis. The internal 'barrier' is sufficient to prevent a large number of particles penetrating towards the centre thus reducing core radiation. ELMs will be required to relieve the monotonic build-up of particles. The plasma profiles are not a sensitive parameter in this confinement mode. In the more 'normal' H-mode, Fe^{25+} radiation from the core region dominates and bremsstrahlung emission is seen at the plasma centre. In L-mode the absence of any transport barrier permits strong radiation both in the core and also in the outer part of the plasma. The electron profiles in this situation have more of an effect, a characteristic of the 'reference' profile being the relatively intense radiation from the edge. Similar results have been calculated for argon.

The main lesson from this exercise is the futility of relying on a *single* line-of-sight (LOS) chord for interpreting impurity emission or even for controlling impurities. Multiple LOS's essential. If off-axis accumulation is strong, perhaps due to core refuelling or temperature gradient screening, emission from the core may be insufficient to off-set the alpha power, if this is required for burn control.

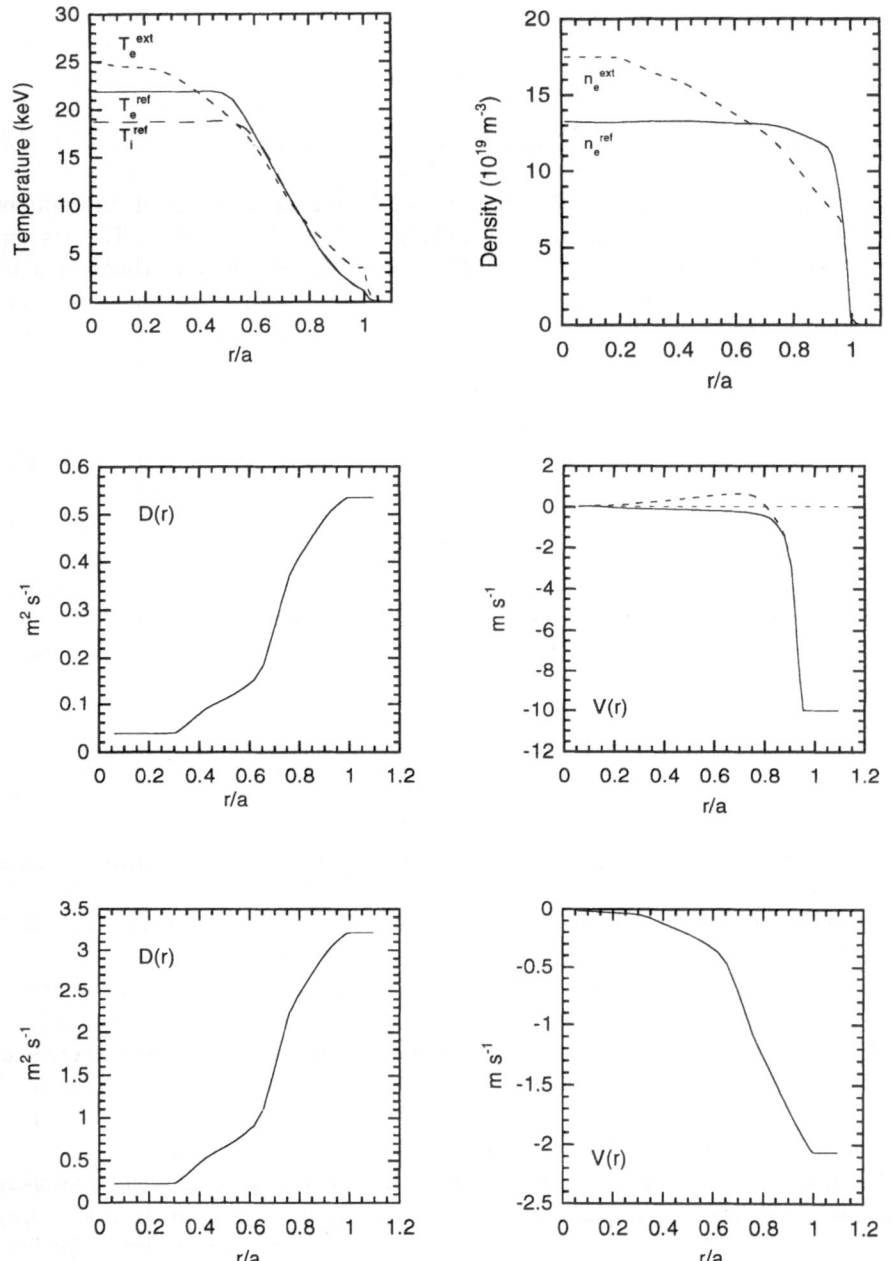

Figure 2 Electron density and temperature profiles (top, L and R). 'Ref.' and 'ext.' refer to the reference and extreme profiles. Impurity coefficients used in transport simulations – 'standard' H- mode transport coefficients with (dashed line) and without (solid line) an outward convection term are shown (middle) with L- mode coefficients at the bottom.

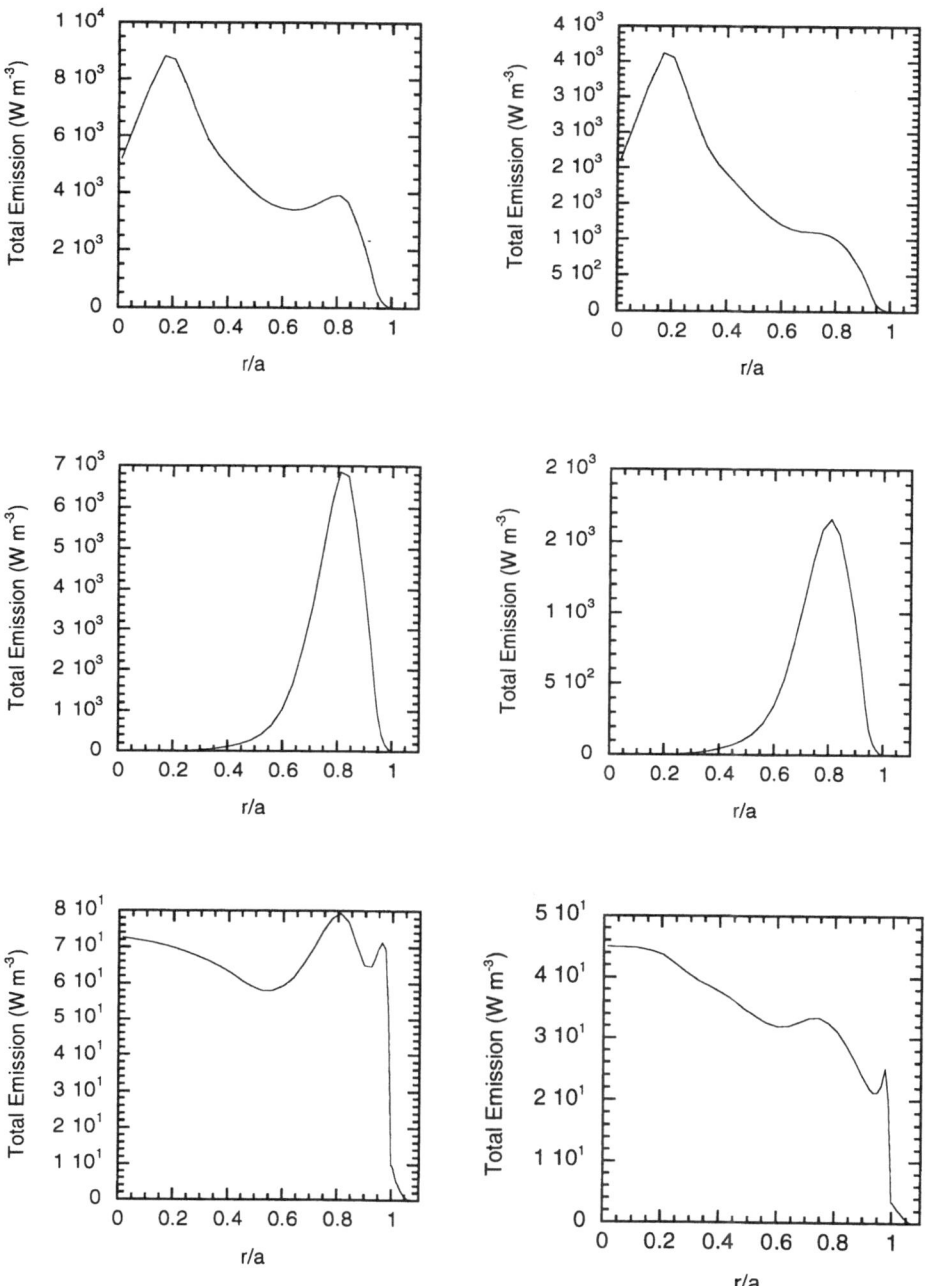

Figure 3 Total emission from Fe in ITER. The profiles using the 'reference' data of Figure 2 are in the left hand column with the 'extreme' in the right hand column. From the top there is normal H mode (note the extreme sensitivity to weak outwards convection in the core), then H- mode with an outward convection term required to model off-axis accumulation (Lauro-Taroni , 1995) and L- mode at the bottom. The impurity influx is the same in each simulation and has not been adjusted to give the same concentration in each plot, hence the absolute emission values are not directly comparable.

SPECTROSCOPIC SYSTEMS FOR PLASMA CONTROL

Using visible spectroscopy it is possible to monitor $H\alpha$, He^+, influxes of low ionisation states in the SOL, ionisation balance in the divertor region, ELM's, the existence of MARFES, and $Z_{eff}(r)$. To maximise the spatial information and monitor the common impurities deeper into the plasma, a diagnostic atomic beam system is recommended (von Hellerman et al., 1995; Marmar, 1995). From the generic design studies by deKock et al. (1995) multiple LOS chords with dog-leg optical paths and F/6 aperture appear possible. Degradation of the first reflector, see later, and of refractive optical components may be monitored using in-situ calibration techniques involving reference optical paths as discussed by Morgan et al. (1993). The dispersion and detector systems for the visible present no great problems and may be low resolution elements such as a thin film filter for bremsstrahlung measurements (Mandl et al., 1994) or an interferometer for resolving the $D\alpha$ and $T\alpha$ isotope abundances (Skinner et al., 1995).

Similarly in the VUV, XUV and X-ray regions, line emission from the radiation mantle and bulk plasma may be monitored for impurity or burn control. Since crystal diffraction is essentially monochromatic, however, spectral survey presents some difficulties in the X-ray region. On the credit side, the X-ray region can offer, in common with the visible and VUV down to 1200Å (Grützmacher et al., 1995), robust windows which will support atmospheric pressure and act as a tritium barrier.

In general, the need to obtain sufficient photon statistics for analysis is the factor that limits the cycle time of a real-time spectroscopic diagnostic. Possible exceptions to this are certain bright visible transitions, $H\alpha$ and Bragg crystal spectrometer measurements (Barnsley et al., 1992). A simple computation of the intensity of a number of selected spectral lines could readily be achieved within a spectrum integration time, say ~ 20ms, and the output of this calculation combined with that from other spectroscopic systems which give P_{RAD} and Z_{eff} , to provide a real-time assessment of the different impurities in the plasma (Lawson, 1995). This information might be included within the control loop which, for example, controls the isotope fuel mix, the admission of the divertor radiating species or the build up of He ash. Real-time measurement of divertor ion temperatures could be used for divertor protection or control of detachment, while charge exchange and X-ray line profiles would yield information about ion temperatures deeper inside the plasma. Such information, augmenting neutron yield data, might be used to control the thermonuclear burn. It is difficult to visualise, however, that control of the spatial distribution of the radiated power would not be better managed by real time analysis of the data from the bolometric arrays.

XUV AND VUV EMISSION

In this section we examine the information which would be lost by ignoring the technically difficult XUV/VUV region. This waveband (25Å $< 1 < 2000$Å) contains the $Ly\alpha$ lines of the most important elements in the plasma viz., D, He, Be and C. It also contains the most intense lines arising from charge exchange collisions between impurity nuclei and diagnostic atomic beams. The $n=2-2$, $n=3-3$ transitions of the common metals and $n=4-4$ transitions in heavy metals occur in this region, making it also the best region for monitoring high and medium Z contamination. Analyses of the spectra from present day tokamaks are cited as evidence that this region contains much of the information associated with the plasma performance. In particular, we illustrate the importance of broad band VUV/XUV spectroscopy in characterising the edge plasma and in monitoring the abundances of the impurity elements in JET.

SOL and Divertor Analyses

In JET a grazing incidence XUV spectrometer (Schwob et al., 1987) and a VUV (SPRED) survey spectrometer (Fonck et al., 1982) view the plasma through the horizontal mid-plane port while a further SPRED instrument, located on a top port, looks vertically downwards into the divertor.

In a sequence of experiments to investigate the SOL emission, the XUV spectrometer was tilted to view the lower part of the vessel with a line-of-sight passing from the midplane to the lower X-point. Thus, in the initial limiter phase of the discharge, it views the SOL. When the X-point is formed the LOS view includes part of the bulk and the divertor. Figure 4 indicates the differences in the H-like and He-like carbon and nitrogen emission lines during these phases of the discharge. When viewing the bulk plasma, the spectrum appears normal. However, in the earlier limiter phase, the Lyman series decrement appears distorted with Lyγ becoming relatively intense. Furthermore the carbon G ratio (He-like intercombination line / resonance line) is grossly inverted, as shown in Figure 4. Recombining carbon nuclei from the bulk plasma diffusing across the last closed flux surface cause the inversion. Charge exchange with background neutral hydrogen must be included in order to account for the observed ratios. A SOL neutral density of 10^{17} m^{-3} has been deduced (O'Mullane et al., 1995b).

During radiative divertor experiments, nitrogen is puffed to provoke detachment from the strike plates. Intense radiation is localised in the divertor region. The vertically viewing VUV spectrometer indicates that NIV and NV are the dominant radiators with NIII increasing throughout the radiative divertor phase. Ionisation balance temperatures ~20eV are deduced. In the case when a radiation-induced instability produces a detached plasma and finally an isolated region or MARFE (O'Mullane et al., 1995c), the XUV spectrum shows only a recombining continuum and the single NVI intercombination line, see Figure 5. The absence of the corresponding resonance line confirms the low temperatures in the MARFE ~ 10eV. Figure 6 shows the XUV spectrum of the BeIV and BeIII resonance lines in JET. The intercombination line of BeIII at 101.69Å can be used together with the 3724Å line in the UV region for branching ratios calibration of the XUV spectrometer (Hawkes et al., 1992). Note also the tantalum (Z=73) in this discharge. This spectral region will therefore indicate potential tungsten (Z=74) contamination of the main plasma. It is not possible to achieve the requisite spectral resolution to deduce the above information on edge physics using diffractors.

Elemental Concentrations

A method has been developed (Lawson, 1995), to derive impurity concentrations using representative VUV lines of the significant impurities and normalising their intensities to the bolometric measurements of the total radiation $P_{rad}(r)$ or even $Z_{eff}(r)$. The semi-empirical coefficients thus derived are used to relate the intensities of these representative lines to the elemental contributions to the total radiated power. In general, the relative concentration of each impurity varies throughout the discharge. However, in limiter discharges, there is

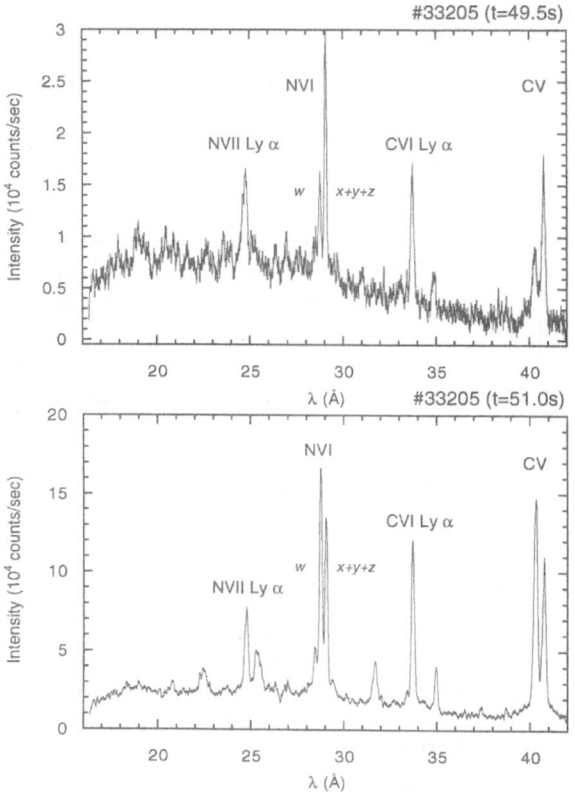

Figure 4 XUV spectra during a JET discharge showing the difference between limiter configuration (top) and the X-point configuration (bottom).

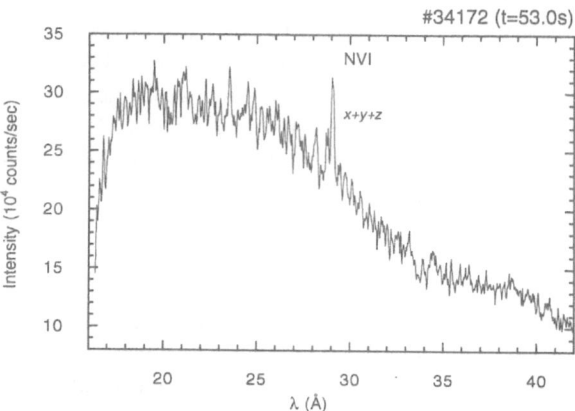

Figure 5 XUV spectrum of a MARFE during radiative divertor experiments at JET

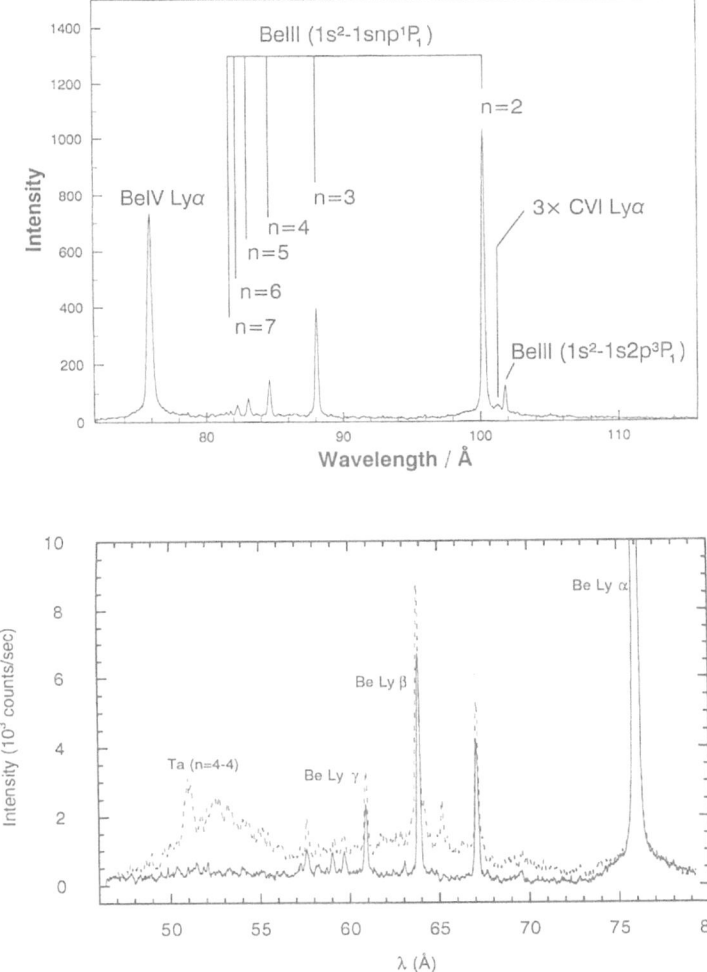

Figure 6 Survey of part of the XUV spectrum of JET (top) showing the Be resonance line emission, including the $1s^2$ -$1s2p$ $^3P_1$ intercombination line at 101.69Å which can be used together with the $1s2s$ $^3S_1$ - $1s2p$ $^3P_1$ line at 3724Å for branching line ratios calibration. The lower spectrum is from a different region of the XUV showing the Be resonance line and a tantalum emission feature following laser ablation.

excellent agreement between the total power as measured by bolometry and the spectroscopic method. Figure 7 shows the total radiated power from a typical JET discharge with the separate elemental contributions which together make up the total bulk plasma radiation. The semi-empirical components have also been scaled such that their sum matches the measured Z_{eff}. With the long ITER discharge times, this could be implemented as an active feedback method. More recent analyses, based on data from 'crystal' diffractors has been successfully applied to the derivation of elemental concentrations in JET (Barnsley et al., 1995b). This technique, however, is not inherently broad band and requires either moveable diffraction elements or multiple LOS viewing, neither option being easy to implement on ITER.

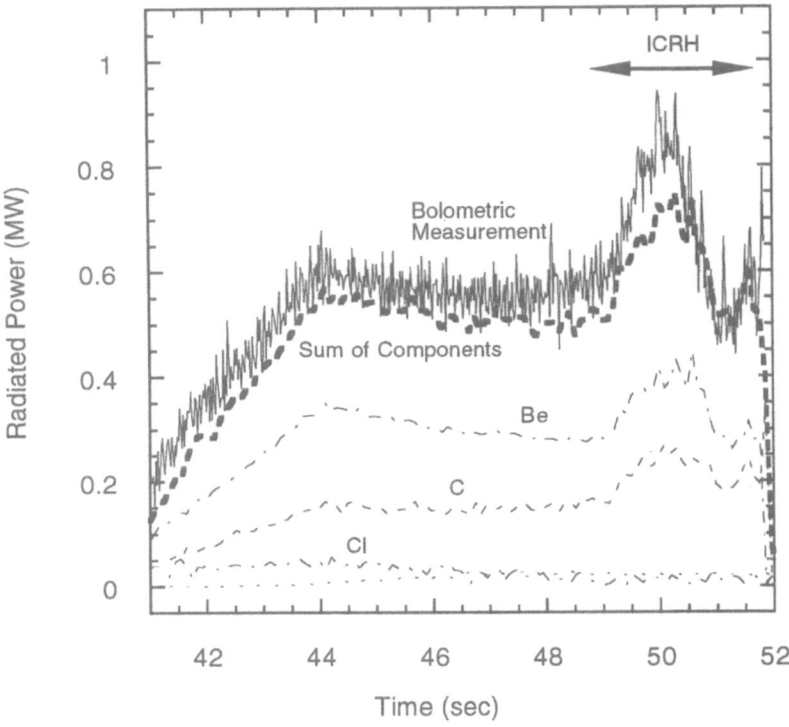

Figure 7 Semi-empirical reconstruction of $P_{rad}(Z)$ from VUV and XUV line emission.

RADIATION INDUCED NOISE IN THE VUV SPECTRUM

The VUV (SPRED) spectrometers on JET have been operated while exposed to the radiation from variable D-D reaction rates. Of relevance to ITER is the performance of the SPRED detector which is a Si-diode array coupled to a front-end, microchannel plate (MCP) via a phosphor screen and fibre optics. Figure 8 shows the neutron noise at different flux levels during a JET discharge with the detectors located 5.65m from the plasma centre on an unshielded, horizontally viewing SPRED. It is significant and encouraging that the VUV spectrometer can operate in neutron fluxes much above the $\sim 1 \times 10^7$ n/cm²s level that is predicted by the generic studies of dog-leg shielding with 90^0 bend angles (ITER, 1995). Experiments on JET using a steel shield around a SPRED detector (15cm thick in the direct LOS to the plasma with 5cm thickness elsewhere), indicated a reduction by a factor of eight in radiation induced noise (Wolf et al., 1995). The variation of the background noise with the MCP voltage indicates that the ionising radiation interacts with the MCP itself rather than the phosphor or other detector components.

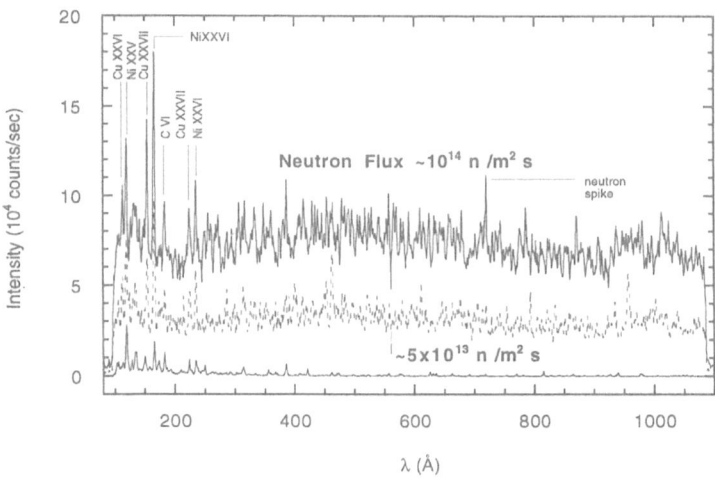

Figure 8 VUV spectra taken with an unshielded spectrometer at three different times during a JET discharge. Neutron induced noise increases with the neutron flux. The 'background' signal in the high flux spectra is due to interactions with the MCP – additional spikes are ascribed to interactions with the Si-diode array or associated readout electronics.

DETERIORATION OF VUV REFLECTORS ON JET

In a series of experiments on JET, the horizontally viewing SPRED viewed the plasma via a gold-coated mirror at a bend angle of 30^0. The mirror was placed 2.5m inside a 150mm diameter tube whose entrance aperture was 1.5m from the plasma edge. After 9673 discharges, equivalent to approximately 50 ITER shots, discolouration of the mirror was accompanied by a drop in sensitivity by a factor of 5 at the 118Å NiXXV line. At wavelengths longer than about 160Å there appeared to be no significant loss in reflectivity.

NEUTRON SHIELDING AND SIGNAL/NOISE IN VUV USING DOG-LEG LABYRINTH

The ITER generic dog-leg with 45^0 angle of incidence gives reflection only above 300Å, seriously limiting the spectral coverage. Shorter wavelengths can be accessed using mirrors at grazing incidence. This, however, reduces the effectiveness of the neutron shielding. In this section we present an illustrative calculation of the signal / (radiation induced noise) ratio experienced by an XUV/VUV detector of the type described above. The calculations assume a line source intensity of 10^{15} ph.cm^{-2}s^{-1} and a spectrometer inverse sensitivity of 1.5×10^{10} ph.cm^{-2}/cnt.pixel^{-1} with an integration time of 20ms, values typical of SPRED. For simplicity, the spectrometer sensitivity is assumed constant with wavelength. The light is ducted to the spectrometer via the generic, double reflection dog-leg, using the tabulations of Williams and Howells (1979) to calculate reflection losses from the gold surfaces.

The noise calculation assumes an ITER first wall flux, given in the ITER EDA report (1995), of 6×10^{13} γ cm^{-2}.s^{-1} incident on the aperture leading to the dog-leg and a detector sensitivity to gammas of 2.2×10^{-2} counts per γ, (Timothy and Bybee, 1979). On the basis of the available literature, the detector is more sensitive to gammas than neutrons. However, the corresponding neutron flux is 1.4×10^{14} $cm^{-2}.s^{-1}$ and with a sensitivity of 1.7×10^{-3} counts per $^1n_0$, (Medley and Persing, 1981), the neutron flux would induce rather less noise in the detector. The attenuation of the light throughput via the dog-leg is calculated using an equation based on the 'Simon-Clifford' technique, (Schaeffer et al., 1973), and normalised to agree with the ITER-EDA calculations in the 90^0 bend angle case. These calculations may be too pessimistic at shallow grazing angles. The detector pixel area is taken as $25\mu m$ x $2.5mm$ and the same integration time, 20ms, is used.

Figure 9 shows the number of counts per pixel during the 20ms integration time from each source. The ITER-EDA report and Orlinsky (1995) highlight the strong dependence of the neutron flux on aperture diameter and so two apertures, 20cm and 10cm, are considered. Inspection of Figure 9 indicates that with a bend angle of 20^0 (10^0 grazing angle) the shortest wavelength which can be resolved above the noise level is ~ 70Å, long enough to measure the Lyman lines of Be^{3+}, He^+, and D, T and the most intense charge exchange lines of the other light elements. These calculations are only 'illustrative', probably erring on the pessimistic side. They should serve as a starting point for more accurate Monte-Carlo modelling.

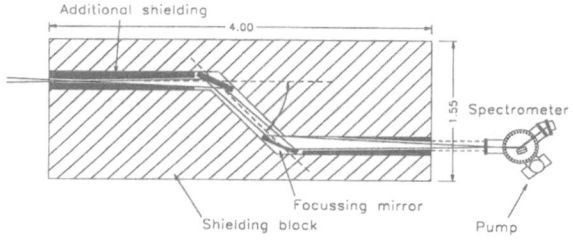

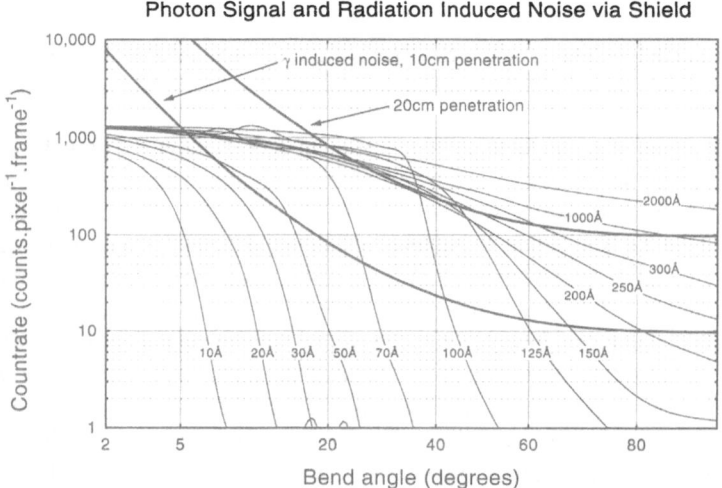

Photon Signal and Radiation Induced Noise via Shield

Figure 9 Illustrative calculations of signal and radiation induced noise for fixed line intensity and variable photon wavelength. The spectrometer views the ITER plasma using mirrors in a dog-leg labyrinth with bending angle as shown. Radiation induced noise is calculated for 20cm and 10cm apertures in the stainless steel shield. For details of the assumptions, see text.

Finally we present some engineering considerations in accommodating the XUV and VUV spectrometers. Figure 10 shows the use of a (removable) focusing mirror system to narrow the aperture in the shielding blanket, thus improving the signal / noise in the VUV region still further. In Figure 11 we introduce the possibility of double diffraction system, using multilayers, which essentially acts as a narrow band tuneable filter for XUV grating spectroscopy. Finally, in Figure 12 we suggest an optical layout for sharing the beam line between VUV and XUV spectrometry.

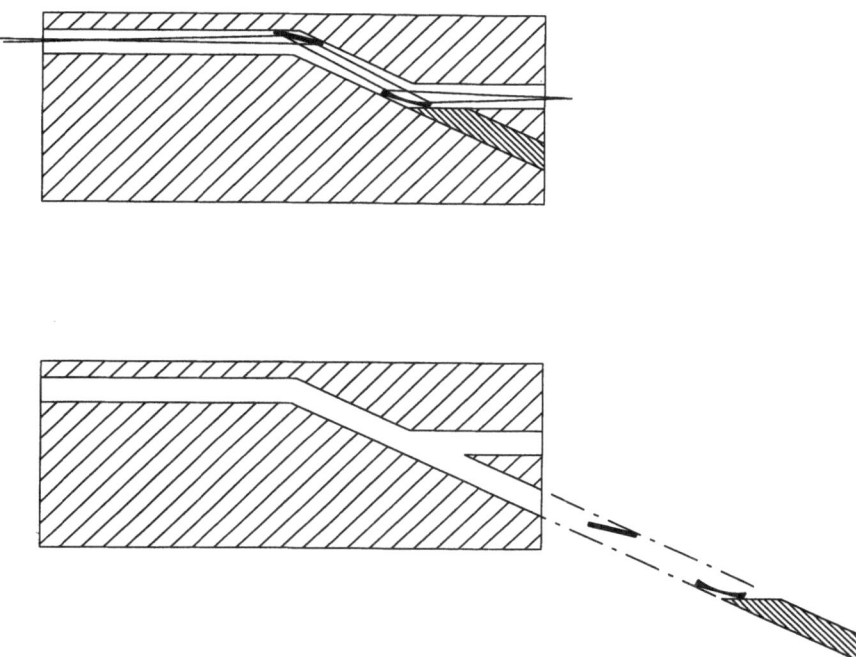

Figure 10 A removable focusing VUV mirror system in the shielding blanket.

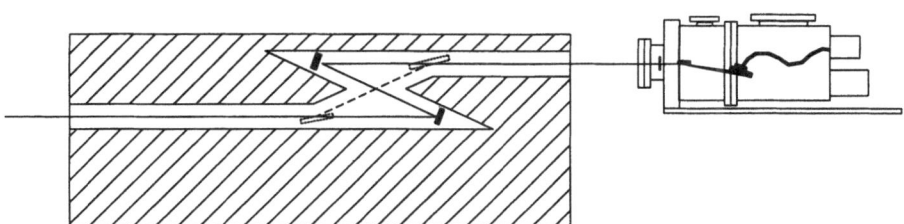

Figure 11 A double diffraction system which acts as a narrow band tuneable filter for XUV grating spectrometers.

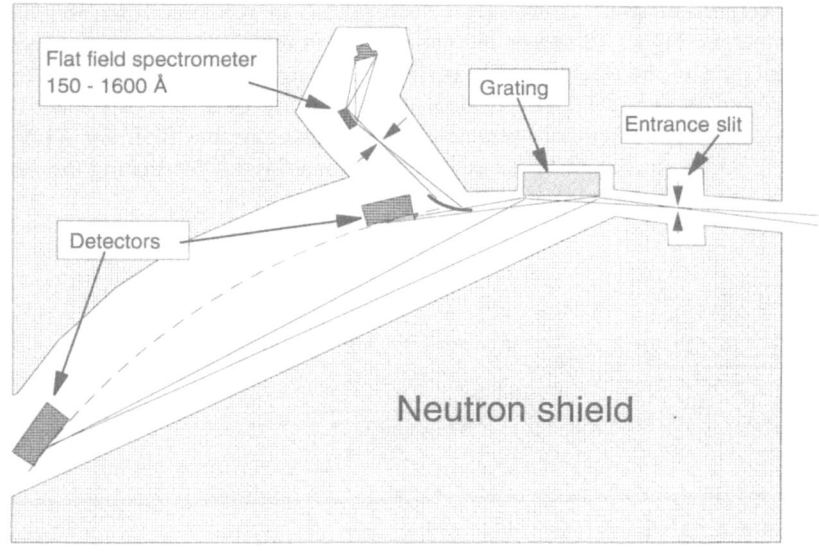

Figure 12 An optical layout with the VUV and XUV sharing a common beamline.

SUMMARY

Spectroscopic diagnostics on existing tokamaks can be adapted to ITER using reflectors in labyrinthine optical paths. The greatest obstacle to this approach is the performance and integrity of the first reflector. However, with this approach to neutron screening, radiation induced noise can be reduced to acceptable levels. There is no single white light reflector (WLR) with complete wavelength coverage between 1Å and 1μm. Above ~ 300Å, mirrors can give adequate reflectivity and screening against neutrons and γ-rays. In this long wavelength region there is a possibility of multi-chord arrays from a single reflector with F/6 aperture. In the X-ray region ($\lambda < 25$Å), no ideal WLR exists but several limited-bandwidth regions each of $\Delta\lambda/\lambda$ ~0.1, covering important wavelength regions, are feasible. This implies the installation of several viewing chords or a single chord with moving diffractors or reflectors. Innovative techniques using $2d$-graded diffractors and capillaries are being investigated (Barnsley et al., 1995a; Hill et al., 1995). Nevertheless, a broadband survey using a single static element is not possible in the X-ray region.

The XUV and VUV region presents the most technically demanding problems. It should not be neglected since this region holds essential information on the mantle and divertor plasmas. The best compromise is a first mirror with a bend angle of 20°-30° in a dog-leg labyrinth. This gives a low-λ cut-off of ~70Å with reasonable signal-to-gamma noise. A minimum of three line-of-sight cords is recommended viewing the bulk, divertor and edge plasmas. Windows to isolate tritium contamination cannot be used, so a continuously pumped cavity containing the relay optics and spectrometers must be employed. A radical approach would be to remove all windows and use multiple straight-through, vacuum beamlines similar to and possibly sharing diagnostic functions with the neutron camera, (Marcus et al., 1995). This would introduce new problems of shifting the tritium barrier to the housings for the diagnostic spectrometers – a radical but feasible option.

REFERENCES

Barnsley, R. et al., 1992, *Proc. 10th Colloquium on UV and X-ray Spectroscopy of Astrophysical and Laboratory Plasmas.* (Berkeley, USA).

Barnsley, R. et al., 1995a, This conference.

Barnsley, R. et al., 1995b, *22nd EPS Conference on Controlled Fusion and Plasma Physics,* (Bournemouth, UK), To be published.

Becker, G., 1995a, *Nucl. Fusion* 35:39.

Becker, G., 1995b, Transport simulations of ITER with broad density profiles and high radiative fraction, To be published in Nuclear Fusion.

Fonck, R. J. et al., 1982, *Applied Optics* 21:2115.

Grützmacher, K. et al., 1995, This conference.

Hawkes, N. C. et al., 1992, *Proc. 10th Colloquium on UV and X-ray Spectroscopy of Astrophysicl and Laboratory Plasmas.* (Berkeley, USA).

Hill, K. W. et al., 1995, This conference.

ITER, 1995, ITER–EDA, Section 5.5.E (S 55 RE 95–07–11 F 1).

Janeschitz, G., 1995, *22nd EPS Conference on Controlled Fusion and Plasma Physics,* (Bournemouth UK), To be published.

Lauro-Taroni, L. et al., 1994, *21st EPS Conference on Controlled Fusion and Plasma Physics,* (Montpellier, France), Vol.18B, Part I:102.

Lauro-Taroni, L., 1995, Private communication, JET.

Lawson, K. D., Private communication, UKAEA Culham Laboratory.

Mandl, W. et al., 1994, *21st EPS Conference on Controlled Fusion and Plasma Physics* (Montpellier, France), Vol.18B, Part III:1276.

Marcus, F. et al., 1995, This conference.

Marmar, E. S., 1995, This conference.

Medley, S. S. and Persing, R., 1981, *Rev. Sci. Instr.* 52:1463.

Morgan, P. D. et al., 1993, *Proc. 17th Symposium on Fusion Technology,* (Rome, Italy) 1:722.

O'Mullane, M. G. et al., 1995a, *22nd EPS Conference on Controlled Fusion and Plasma Physics,* (Bournemouth, UK), To be published.

O'Mullane, M. G. et al., 1995b, *5th European Conference on Atomic and Molecular Physics* (Edinburgh, UK).

O'Mullane, M. G. et al., 1995c, *Proc. 11th Colloquium on UV and X-ray Spectroscopy of Astrophysical and Laboratory Plasmas.* (Nagoya, Japan), To be published.

Orlinski, D. V., 1995, This conference.

Pasini, D. et al., 1990, *Nucl. Fusion* 30:2049.

Schaeffer, N. M., 1973, *Reactor Shielding for Nuclear Engineers,* Publ. USAEC, Office of Information Services.

Schwob et al., 1987, *Rev. Sci. Instr,* 58:1601.

Skinner, C. H. et al., 1995, *Rev. Sci. Instr.* 66:646.

Summers, H. P., 1994, *The Atomic Data and Analysis System (ADAS),* JET Report IR(94)06.

Timothy, J. G. and Bybee, R. L., 1979, *Rev. Sci. Instr.,* 50:743.

von Hellermann, M. et al., 1995, This conference.

Williams, G. P. and Howells, M. R., 1979, Brookhaven National Laboratory Report BNL 26121.

Wolf, R. et al., 1995, submitted to Review of Scientific Instruments.

EDGE n_T/n_D BY TWO-PHOTON INDUCED Ly-α FLUORESCENCE

K. Grützmacher,[1] M.I. de la Rosa,[1] J. Seidel,[1] A. Steiger,[1]
G. Fußmann,[2] W. Bohmeyer,[2] D. Voslamber[3]

[1] Physikalisch-Technische Bundesanstalt (PTB), Berlin, Germany
[2] Max Planck Institute of Plasma Physics (IPP), Berlin Branch,
   Ass. EURATOM
[3] Ass. EURATOM-CEA, DRFC-STPF, CEN Cadarache, France

ABSTRACT

Recent theoretical and experimental investigations had promising results concerning the potential of Doppler-free two-photon induced Lyman-α fluorescence (2γ Ly-α) as a means for selective direct measurements of atomic densities of hydrogen isotopes in magnetically confined fusion plasmas. Even under tokamak conditions, all the two-photon absorption profiles of H, D, and T exhibit sharp, nearly unshifted central components with typical widths fitting well with the bandwidth of pulsed single-mode laser spectrometers which will be able to deliver the required power and quality needed for tokamak diagnostics. A first 2γ Ly-α fluorescence signal was observed at the plasma generator PSI-1 of IPP Berlin with high spatial resolution (Ø 3mm, 8 mm length) in a pure deuterium plasma beam, Ø 100 mm, $n_e \approx 5 \cdot 10^{12}$ cm^{-3}, $T_e \approx 5$ eV. We are therefore convinced, that measurements of the n_T/n_D ratio from 0.1 to 10 will be possible in the edge plasma of ITER, with single shot accuracy better than 20% and with high spatial and temporal resolution at a laser repetition rate of 50 Hz. The detection limit for each isotope is estimated to be 10^9 cm^{-3}.

TWO-PHOTON ABSORPTION LINE PROFILES

In general, Doppler-free two-photon induced Ly-α fluorescence is essentially a three-step process: Hydrogen atoms are first excited from the 1s ground state to the 2s state by absorption of two photons (one from each of two counterpropagating laser beams), the excitation is then transferred to the 2p states by collisions, and finally the Ly-α fluorescence is emitted during radiative decay from 2p to 1s. The basic theory of Doppler-free 2γ Ly-α fluorescence for hydrogen atoms in magnetically confined fusion plasmas was developed by

Diagnostics for Experimental Thermonuclear Fusion Reactors
Edited by P. E. Stott *et al.*, Plenum Press, New York, 1996

307

Voslamber.[1] The two-photon absorption line profile was found to consist of two broad side components (due to the motional Stark effect in the electric field $E = v \times B$ as seen by an atom moving with velocity v through a magnetic field B), but also has always a sharp, un-shifted central component, regardless of atomic velocity. This central component can be used for effective Doppler-free two-photon excitation. Moreover, since the corresponding un-shifted upper state is a mixture of 2s and 2p in this case, due to the motional Stark effect, di-rect emission of fluorescence radiation from this state occurs even without collisional redistri-bution of the excitation.

Further consideration showed, however, that the theoretical treatment had to be extended[2] to account for the fine structure of the upper energy level (and, in addition, even for the hy-perfine splitting of the lower level) as well as the second order relativistic Doppler effect. Acting alone, each of these two effects would lead to a distinctly larger line width (accompanied by a line shift) of the central component, but acting together, they partly cancel one another (Fig. 1), and the width of the central component (which remains very nearly un-shifted) does not exceed the order of 1 GHz (FWHM at the frequency of the laser used for excitation) under tokamak conditions (Fig. 2). This matches the laser line width of about 500 MHz rather well, since the atomic excitation rate (and, hence, the fluorescence intensity) is roughly inversely proportional to the larger of the two widths. Therefore, improved theoretical calculations[2,3] still predict the possibility of efficient Doppler-free two-photon excitation of hydrogen Ly-α fluorescence as a means of measuring ground-state hydrogen atomic densi-ties.

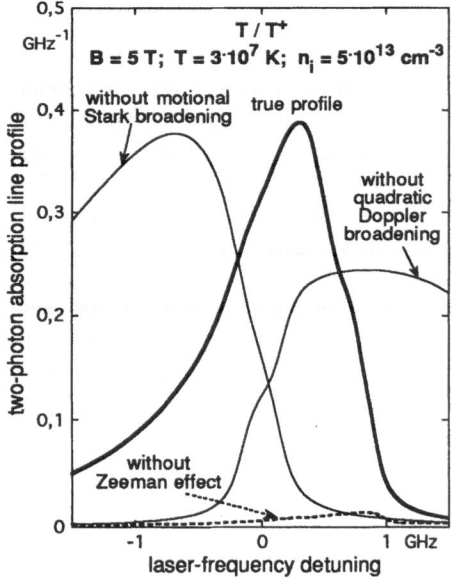

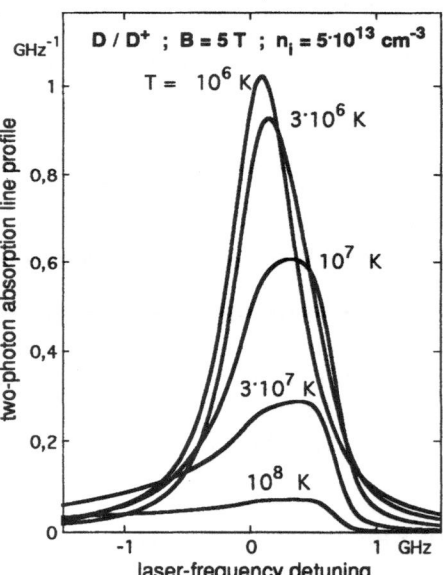

Figure 1. Two-photon absorption line profiles (cen-tral part) for tritium in a tritium plasma under fusion conditions: Fictitious profiles resulting from the neglect of motional Stark broadening and quadratic Doppler broadening, respectively, and the true profile comprising these effects. All profiles are normalized to unit area, inclusive of the broad side components which are not shown.

Figure 2. Two-photon absorption line profiles (cen-tral part) for deuterium in deuterium plasmas under tokamak conditions. Profiles are area-normalized as in Fig. 1.

SINGLE-LONGITUDINAL-MODE
SOLID-STATE LASER SPECTROMETER

Because of the small two-photon absorption probability, the method requires tunable laser radiation of high spectral and beam quality at 243 nm with much more power than commercial laser systems can presently provide. Therefore, we developed a new type of solid-state laser spectrometer based on a high-power injection seeded Q-switched Nd:YAG laser, an OPO/Ti:Sapphire system pumped by the second harmonic of the Nd:YAG, and a BBO crystal for the final generation of 243 nm radiation by frequency mixing. At present, this table-top system delivers 243 nm pulses of 2,5 ns duration with energies up to 50 mJ, a bandwidth of less than 500 MHz, and very good spatial laser beam profile at a repetition rate of 10 Hz. The experience with this system allows to define the design parameters for a laser system especially suited for tokamak diagnostics.

FIRST DEMONSTRATION OF 2γ Ly-α
AT THE PLASMA GENERATOR PSI-1

Since 1992, IPP and PTB are working in close cooperation in order to develop 2γ Ly-α fluorescence as a means for tokamak diagnostics. First investigations at the plasma generator[4] PSI-1 were to demonstrate the capability of the new method and to prove theoretical predictions. To provide Doppler-free excitation, two counterpropagating circularly polarized laser beams (8 mJ each in 2.5 ns) were directed into the target chamber of PSI-1 perpendicular to the plasma beam. The Ly-α fluorescence was observed with high spatial resolution (Ø 3mm, 8 mm length) at a pure deuterium plasma of Ø 100 mm, $n_e \approx 5 \cdot 10^{12}$ cm^{-3}, $T_e \approx 5$ eV. Two important properties could be verified:
– Within the laser bandwidth of 500 MHz, the two-photon resonances in the plasma and in a reference cell occur at the same laser wavelength, and
– the pulse duration of the 2γ Ly-α fluorescence signal is less than 10 ns in agreement with theoretical predictions.

Current work concentrates on the determination of kinetic temperatures and number densities in a wide range of plasma parameters for different hydrogen-deuterium mixtures and on improvements of the detection limit. Testing and optimization of novel techniques concerning the Ly-α fluorescence detection system have to be continued.

Future application at ASDEX Upgrade is under discussion. We expect that already now measurements of the variation of isotope densities should be possible down to 10^8 cm^{-3} in tokamak plasmas at temperatures up to 10^7 K. Corresponding kinetic temperatures can be measured if the counterpropagating laser beams are crossing under a very small angle.

APPLICATION TO ITER DIAGNOSTICS

Doppler-free two-photon excitation for the measurement of neutral density ratios, e.g. for D and T, requires two laser systems, each providing a pair of counterpropagating laser beams, at the corresponding resonance frequencies (which are separated by the isotope shift). The interaction volumes of the beams for D and T will be identical, so that the n_T/n_D ratio can be measured with a single Ly-α detection system. The spatial resolution is determined by the diameter of the laser beams in the interaction volume (Ø 1 cm is reasonable for a pulse energy of 100 mJ in each beam), while the spatial filter of the Ly-α detection system defines the observed length, typically a few cm.

The laser pulses have a temporal duration of 3 ns, and the 2γ Ly-α fluorescence occurs nearly instantaneously within less than 5 ns. If the time separation of the laser pulses for D and T is set to 50 ns, for example, it is sufficient to sample the signal from the Ly-α detection system for a period of 100 ns with GHz time resolution to obtain the Ly-α background flux from ITER and the two contributions of the 2γ Ly-α fluorescence from D and T.

Normalizing the 2γ Ly-α fluorescence signals obtained for D and T with respect to the energies of the two laser pulses results in the atomic density ratio n_T/n_D. It should be noted that this technique does not require calibration of the Ly-α detection system.

In the edge plasma of ITER, measurements of the n_T/n_D ratio from 0.1 to 10 will be possible with single shot accuracy better than 20%, with ns time resolution at a laser repetition rate of 50 Hz. The detection limit for each isotope is estimated to be below 10^9 cm^{-3}. Furthermore, a multichannel detection system will allow for measurements of the n_T/n_D ratio with spatial resolution of about 1 cm^3 in the plasma edge region.

EXPERIMENTAL SET UP AT ITER

The two laser systems will be located at large distance from ITER, generating radiation in beam diameters of about 3 cm. Final beam preparation, beam splitting, delay line etc. will be located near ITER.

Four colinear beams (∅ 1 cm), which have to enter at a Mid Horizontal Port, will be reflected to the upper Vertical Port and retro-reflected there. In this arrangement the laser beams cross the edge region under an angle of about 60° which provides excellent spatial resolution. The alignment of the counterpropagating beams will be controlled outside ITER.

The two-photon excitation can be set to occur in the edge-plasma either near the Mid Horizontal Port or near the Vertical Port. Measurements at both positions will be possible with two Ly-α detection systems, one at the Mid Horizontal Port and one at the Vertical Port, simply by changing the time delay of the laser beams.

REFERENCES

1. D. Voslamber, Determination of neutral particle density and magnetic field direction from laser-induced Lyman-a fluorescence, II - Two-photon excitation, Report EUR-CEA-FC-1387 (1990)
2. D. Voslamber, Some novel concepts for spectroscopic diagnostics in tokamaks, in: *Spectral Line Shapes, Vol. 8* (AIP Conf. Proc. 328), A.D. May, J.R. Drummond, E. Oks, eds., American Institute of Physics, New York (1995)
3. J. Seidel and D. Voslamber, to be published
4. H. Behrendt, W. Bohmeyer, L. Dietrich, G. Fussmann, H. Greuner, H. Grote, H. Kammeyer, P. Kornejew, M. Laux, E. Pasch, in: *21st EPS Conference on Controlled Fusion and Plasma Physics*, E. Joffrin, P. Platz, P.E. Stott, eds., Europhysics Conference Abstracts Vol. 18 B, part III, p. 1328

DEVELOPMENT OF LUMINESCENT DETECTORS FOR HOT PLASMAS

B. Zurro[1], A. Ibarra[1], A. U. Acuña[2], R. Sastre[3] and K. J. McCarthy[1]

[1]Asociación EURATOM/CIEMAT para Fusión. E-28040 Madrid
[2]Inst. Quimica-Física Rocasolano, CSIC. E-28006 Madrid
[3]Inst. Quimica-Física Polímeros, CSIC. E-28006 Madrid

INTRODUCTION

Among the main requirements which a broadband radiation detector for high temperature plasmas should possess include: a known or traceable response over as broad a wavelength range as possible; a quick response time; immunity to electromagnetic interference and moreover, in fusion devices the capability to cope neutron producing plasmas; resistance to radiation and operation with a limited access to the machine. Since most of the present designs lack sufficient radiation hardness, research is needed to find a solution.

The luminescent properties of phosphor materials are commonly used in radiation detection and measurement. The energy absorbed by the phosphor is partly converted into light whose wavelength range can be tailored by appropriate material and impurity content. Potentially the main advantage for a fusion plasma is that only a thin film of the phosphor plus radiation filters need to be close to the plasma, whereas the most sensitive parts of the system, which receive the luminescent signal via metallic optics and fibres, can be withdrawn to a safer distance. Here we summarise the effort invested over the last years at CIEMAT to develop broadband plasma radiation detectors with spatial, temporal and energy resolution. Particular attention is given to the extrapolation of such detectors to the harsh environment of an ITER-like device.

LUMINESCENT DETECTOR DEVELOPMENT AT CIEMAT

At CIEMAT we have developed various types of detectors based on the luminescence of different materials. Our first development[1] took advantage of the luminescence produced by plasma radiation impinging on a quartz window, where the impurities present in the quartz were the main source of luminescence. This idea can be extrapolated to a fusion device to build a global radiation monitor as long as the luminescence produced by the plasma radiation can be separated from the background produced in the window and fibres. Heated fibres

Diagnostics for Experimental Thermonuclear Fusion Reactors
Edited by P. E. Stott *et al.*, Plenum Press, New York, 1996

311

seem promising for this task while a blind channel should provide a background measurement.

A further step in this work has been the construction of a radiation detector, based on the luminescence of the well known phosphor sodium salicylate (ss), which has spatial temporal and spectral resolution[2]. It involves optically scanning the fluorescence of this phosphor which converts into visible light any plasma radiation, from x-rays to 350 nm, which falls onto it after passing through a collimator and a selectable broadband filter. The fluorescence is relayed via a mirror and a quartz fibre bundle to a filtered photomultiplier located several meters away. By repetitive scanning of the fluorescence during a long tokamak discharge we can get radiation profiles within a selected spectral band. When operated without scanning, the system can follow the time evolution of a spectral VUV band along a particular plasma chord. This system has several useful features like compactness, portability, inherent freedom from ground loops and good sensitivity, which make it very attractive for the harsh environment of a hot plasma device. Therefore, it may prove to be a useful diagnostic for certain experimental tasks, which are usually solved by bolometer or X-ray detector arrays and by high resolution VUV spectrometers. Typical results obtained in the TJ-I tokamak with the system are shown in Figs. 1 and 2. Fig. 1 depicts a repetitive scan of the phosphor fluorescence produced by plasma radiation in the range 50-220 Å, while Fig. 2, shows a high resolution profile of the plasma radiation (100 to 900 Å) where significant flattenings and hums can clearly be observed. They appear at approximately symmetric radii and their positions do not change significantly for two successive profiles.

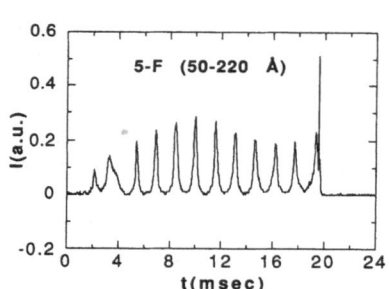

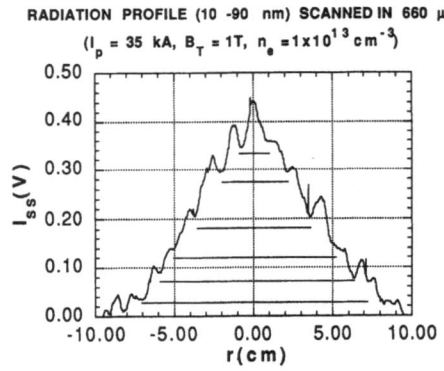

RADIATION PROFILE (10 -90 nm) SCANNED IN 660 μs

(I_p = 35 kA, B_T = 1T, n_e =1x10^{13} cm^{-3})

Fig. 1. Space-time evolution of TJ-I plasma radiation in the spectral range 5-22 nm.

Fig. 2. High-resolution radiation profile in which flattenings and hums are observed at symmetric radii.

Finally, we have developed a simple radiation monitor based on the luminescence of P-45 phosphor, which operates in the ECR heated plasma of the TJ-UI torsatron. It has the advantage over standard bolometers and standard X-ray detectors, apart from its extremely high sensitivity, of not being influenced by the strong microwave field disturbing other more standard detectors like thermal bolometers and pyroelectric detectors. Fig. 3 shows typical signatures of this monitor in discharges where the microwave power was modulated at different frequencies. It should be noticed that fluctuations close to the Ghz range could be monitored using a much faster phosphor than P-45 , as long as the plasma respond to such.

LUMINESCENT MATERIALS

In order to find a phosphor better than ss for high temperature plasmas, we have studied[3] the X-ray and VUV response of several luminescent materials to make comparisons with the response of the standard VUV phosphor sodium salicylate. The selected phosphors include a powdered and a polymeric version of sodium salicylate, several well known inorganic phosphors [Y_2SiO_5:Ce (P-47), Y_2O_3:Eu (P-22), Y_2O_2S:Tb (P-45)] and two crystalline-like materials, namely spinel ($MgAl_2O_4$:Mn^{++}) and ruby (Al_2O_3:Cr^{3+}). These materials were chosen because of properties which make them interesting for plasma applications. For instance, ruby and P-22 have high quenching temperatures while P-45 is the most sensitive phosphor for soft X-ray detection. In addition, some of them fluoresce in the red (P-22 and Ruby) where high sensitivity avalanche photodiodes are available. With regard to the temporal response, P-47 is the only phosphor here whose ns lifetime is comparable with that of sodium salicylate, 5-10 ns, whereas other promising materials, like ruby, P-22 and P-45, whose fluorescence emission lies at longer wavelengths, have luminescence lifetimes of around 1 ms.

The relative response of several phosphor samples to the radiation emitted by the TJ-I tokamak, is shown in Fig. 4 for the range 150 to 800 Å. It should be noted that only two of the phosphors, namely P-22 and P-45, have a better response than sodium salicylate in this range. The relative response of different samples in the X-ray range (5-50 keV), obtained by exciting them with a standard X-ray tube, are shown in Fig. 5. The excellent behaviour of the crystalline materials like spinel and ruby over this energy range should be noted, especially in light of their higher tolerance to radiation damage as compared with standard luminophors.

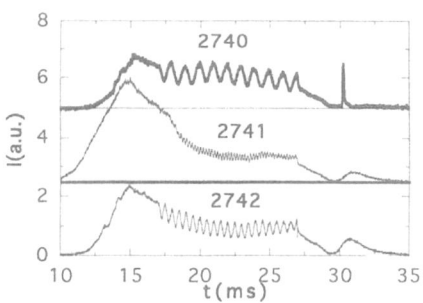

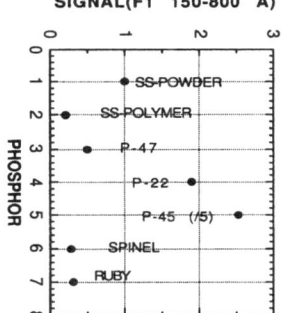

Fig. 3. Modulation seen in the P-45 phosphor monitor when the ECR heating power was modulated at different frequencies in the TJ-UI torsatron. Faster phosphors should be more appropriated for this task.

Fig. 4. Relative response of selected luminescent materials to plasma radiation centred in the range (150-800 Å). Note that only two phosphors exhibit better response than ss in this spectral range.

A LUMINESCENT DETECTOR FOR ITER

In order to extrapolate the previous design to an ITER-like device several relevant points need to be considered. The radioluminescent material which must be radiation hard and have a high quenching temperature, should be placed inside the vacuum vessel with shielding provided to minimise the level of gamma rays. The problem of background signal separation which then arises may be overcome by monitoring not only the phosphor fluorescence exposed to the plasma radiation but also that produced by background neutrons and gammas only. In addition, tuning of the phosphor thickness can help to maximise its response to the plasma radiation while minimising its response to the background radiation. In this respect phosphor deposited as thin films may help. Since the luminescence must pass through an optical window which can attenuate the signal or contribute to it by its own radioluminescence, we must take account of the best transmission range; i.e. for sapphire[4] this seems to be from 500-600 nm, thus the luminescent material should ideally be tailored to emit in this range. However, some very resistant luminescent materials like BaF_2 and

YAP:Ce emit far below this range, whereas spinel which emits in the appropriate range has a longer lifetime than ideally wished.

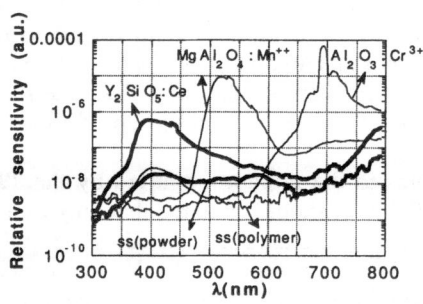
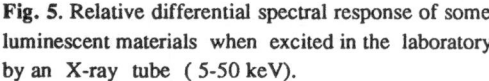

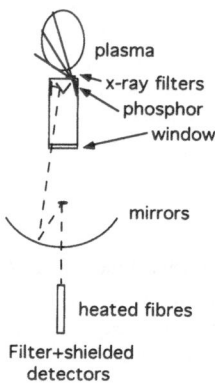

Fig. 5. Relative differential spectral response of some luminescent materials when excited in the laboratory by an X-ray tube (5-50 keV).

Fig. 6. Sketch of a phosphor detector for an ITER-like plasma. Only the thin phosphor layer and filters must withstand the harsh environment.

Unfortunately, the radiation hardness of other potential phosphors is unknown at present. If the luminescent material is used with no filter in front, then plasma UV and visible radiation in conjunction with an operating temperature of a few hundred ºC, might contribute to in situ annealing of the phosphor and window against the darkening produced by the neutron and gamma fluxes. Note; the luminescence process is unaffected until very high radiation levels, however the latter can trigger absorption mechanisms, but these can be reduced in high purity materials. Therefore, progress in this area may help to alleviate part of the problem.

Fig. 6 shows the sketch of a possible scheme to monitor the VUV and X-ray radiation of an ITER-like plasma with spatial, temporal and energy resolution. Not shown in this figure is the possible labyrinth between plasma and phosphor needed to separate them. Phosphor deposited on a cylindrical surface receives the plasma radiation selected by a filter through a collimator. Using a set of three metallic mirrors, as shown in the figure, the phosphor luminescence is relayed away from the plasma and collected by a set of heated, high-purity silica fibres. The signal is picked up by a set of detectors, a filter tuned to the phosphor luminescence is used in front of the end detectors to optimise the plasma light rejection of the system. Non illuminated parts of the phosphor are used to monitor the background. More quantitative considerations would need the development of numerical codes which take into account the detailed geometry, plasma source emission and multiple effects produced in the luminescent material and transmission window by stray background radiation, to assess the potential of these type of detectors for an ITER-like environment. However, these detectors cannot be ignored for fusion plasma diagnostics in light of the significant improvements in luminescent materials brought about by their interest in high-power lasers, high energy calorimeters, medical applications and x-ray and gamma detection, which allows us to foresee that some of the present limitations can be overcome by future developments.

REFERENCES

1. B. Zurro, C. Pardo and J.L. Alvarez-Rivas, *J. Phys. D: Appl. Phys.* 19: 1895 (1986).
2. B. Zurro and B. García-Castañer, *Rev. Sci. Instrum.* 65: 2580 (1994).
3. B. Zurro, A. Ibarra, K.J. McCarthy, A.U. Acuña and R. Sastre, *Rev. Sci. Instrum.* 66: 534 (1995).
4. E. R. Hogdson, Personal Communication.

MULTILAYER MIRROR BASED MONITORS FOR IMPURITY CONTROLS IN ITER

S.P. Regan, M.J. May, V. Soukhanovskii, M. Finkenthal, and H.W. Moos

Department of Physics and Astronomy
The Johns Hopkins University
Baltimore, MD

ABSTRACT

Multilayer Mirror (MLM) based monitors are compact, high throughput diagnostics capable of extracting XUV emissions (the wavelength range including the soft-x-ray and the extreme ultraviolet, 10 Å to 304 Å) of impurities from the harsh environment of large fusion reactor type devices. Since 1989 the Plasma Spectroscopy Group at Johns Hopkins University has investigated the application of MLM based XUV spectroscopic diagnostics for magnetically confined fusion plasmas. MLM based monitors have been constructed for and extensively used on DIII-D, Alcator C-Mod, TEXT, Phaedrus-T, and CDX-U tokamaks to study the impurity behavior of elements ranging from He to Mo. On ITER MLM based devices would be used to monitor the spectral line emissions from Li I-like to F I-like charge states of Fe, Cr, and Ni, as well as 'extractors' for the bands of emissions from high Z elements such as Mo or W for impurity controls of the fusion plasma. In addition to monitoring the impurity emissions from the main plasma, MLM based devices can also be adapted for radiation measurements of low Z elements in the divertor. The concepts and designs of these MLM based monitors for impurity controls in ITER will be presented. The results of neutron irradiation experiments of the MLMs performed in the Los Alamos Spallation Radiation Effects Facility (LASREF) at the Los Alamos National Laboratory will also be presented. These preliminary neutron exposure studies show that the dispersive and reflective qualities of the MLMs were not affected in a significant manner.

INTRODUCTION

The spectral range under consideration is that of the soft-x-ray and extreme ultraviolet emissions, 10 Å to 304 Å. In this range the characteristic emissions of H I and He I-like low Z impurities (B, C, N, O) are emitted (15 Å -70 Å), as well as most of the L-shell emission of intermediate Z impurities, Ti to Ni (80 Å - 200 Å). Moreover, if higher Z elements such

Diagnostics for Experimental Thermonuclear Fusion Reactors
Edited by P. E. Stott *et al.*, Plenum Press, New York, 1996

315

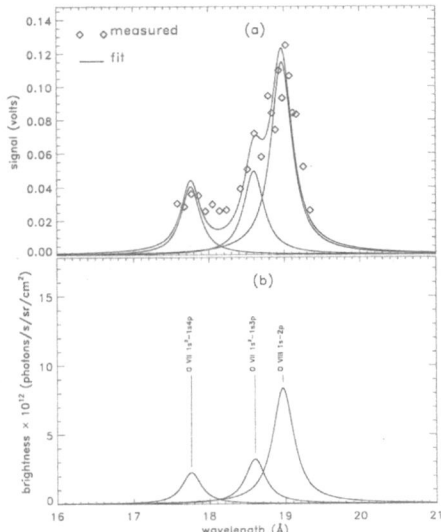

Figure 1. Measured wavelength scan around O VIII Lyman α emission at 19 Å in (a), and calibrated spectral fit in (b)

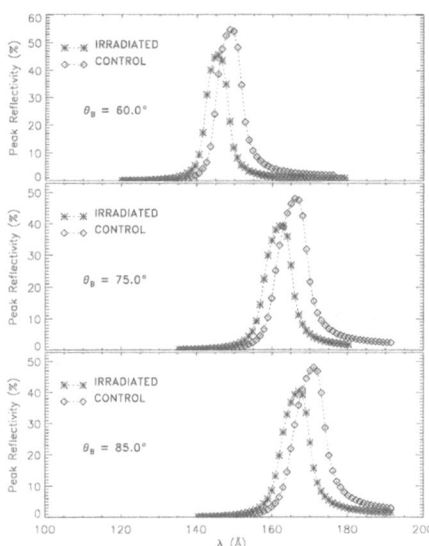

Figure 2. Measured Peak reflectivities of neutron irradiated MLM and control MLM (Mo/Si, 2d=186 Å, N=100)

as Mo or W will be used in reactor type devices, M and N-shell emission is also emitted in this spectral range.[1, 2] In a plasma with $T_e(0) = 10$ keV, the light impurity emission will exist at the very edge and the intermediate Z, L-shell emission will be confined at $r/a \approx 0.7 - 0.8$. Monitoring these emissions with relatively high time resolution is important not only for power loss estimates, but also because of the potential use of these signals in feedback routines for impurity control procedures. In a radiative divertor scenario for ITER, the gas target ('coolant') temperature may be as high as 300 eV in the divertor region. In addition to the above mentioned, intrinsic light impurity emissions, Ne or Ar emission will also have to be measured. Argon, will emit strong Ne I-like lines in the 40 Å - 50 Å region.

The XUV spectral range has been traditionally observed with grazing incidence spectrometers utilizing gratings as dispersive elements. These instruments have very good spectral resolution[3]; however, their low throughput and their complex viewing geometry makes their use on ITER impossible. On the other hand, the relatively high reflectivity of the multilayer mirrors (MLMs) and their narrow bandwidth below 50 Å ($\Delta\lambda \approx 1$ Å or less) make them suitable 'filters' for the characteristic emission of light impurities. Moreover, our work has shown that in spite of an increasing bandwidth at longer wavelengths, meaningful information can be extracted insofar as high Z emission is concerned.[4, 5]

Both flat and curved MLM based 'extractors' should be considered for monitoring the ITER emission in the divertor as well as in the main plasma regions. In most of our experiments performed to date we have used flat MLMs; however, large f number, curved mirrors can be produced. We have constructed an imaging camera for 150 Å O VI emission, and operated it on the Phaedrus-T tokamak.[6]

Radiation damage is one of the major concerns for any plasma facing component. As shown in the next section, the results of the neutron irradiation of W/C and Mo/Si multilayer mirrors at levels comparable to those of ITER, are very encouraging.[7]

MULTILAYER MIRRORS

The properties and physics of multilayer mirrors (MLM) have been presented extensively in the literature.[8] We shall only mention here that the most frequently used MLMs are those with W/C and Mo/Si bilayers; their 2d values range from 50 Å to 200 Å and the number of layers can be as high as 100. Figure 1 presents the spectrum recorded with a polychromator using MLMs as dispersive elements (in the range 17 Å - 20 Å). The Lyman alpha (Ly_α) emission of O VIII is well separated form the He I-like O VII lines. The Lyman beta (Ly_β) line (not shown in the figure) is also well separated form adjacent iron emission, thus making possible an accurate estimate of the temperature sensitive Ly_α/Ly_β ratio.[9]

Figure 2 shows the effect of neutron irradiation on a Mo/Si MLM. One observes a small shift in the Bragg peak, indicating a change in the 2d value, and a reduction of the reflectivity by about 20 %. These changes are probably due to a 'swelling' of the layers as well as to a small degradation in their smoothness. However, the changes are small enough not to significantly affect the spectroscopic measurements.

RESULTS OBTAINED WITH MLM BASED SPECTROSCOPIC DEVICES

During the last few years we have operated MLM based devices on DIII-D[10], Alcator C-Mod[4], Phaedrus-T[9], and most recently on CDX-U. Most of these devices use flat MLMs as dispersive elements. In general, XUV emission is collimated on the MLM; the Bragg reflected beam falls on a phosphor which transforms the XUV light into visible. The visible emission is coupled optically (either through fibers or lenses) onto a photo-detector. The emission line profile can be obtained in one single shot by scanning the Bragg angle. Radial distributions can be measured at a fixed wavelength, again, either shot by shot or during a single tokamak discharge if a multi-MLM system is used. On the Phaedrus-T tokamak, the MLM based polychromator achieved both spectral and spacial scans on a shot-by shot basis. On Alcator C-Mod, the Mo monitor measures simultaneously the core and 'intermediate' (i.e. $r/a \approx 0.5$) emission. The comparison of time histories of the MLM based Mo monitor emission and that of a 2.2 m grazing incidence spectrometer at 128 Å (Mo XXXII) shows that in spite of the large bandpass, the time history of the monitor signal is identical to that of the grazing incidence device. The temporal resolution of the MLM device is 0.2 ms, while that of the grazing incidence spectrometer is 8 ms. At 75 Å the Mo XXIV-XXVI band of emission is recorded. The signals from these two channels in conjunction with accurate atomic physics modelling yields a quantitative estimation of the power losses in the plasma. An imaging device at 150 Å using two curved MLMs has also been built (see figure 3). In the context of the ITER application, the interest is in the increased light gathering capability of a system using curved reflecting surfaces. The principle of operation is the same as for the flats; the XUV light is transformed into visible and read out by a conventional photo-detector.

FUTURE DEVELOPMENTS OF MLM BASED DEVICES

A 'periscope' type device, which is shown schematically in figure 3, is in the process of being constructed for Alcator C-Mod in order to extract 'local' information from the region around the X-point. This device addresses both the problem of local measurements in the divertor area, and the efficiency of the extraction of XUV signals through multiple reflections on flat MLM surfaces. For ITER, visible and IR signals will be extracted in a similar way.

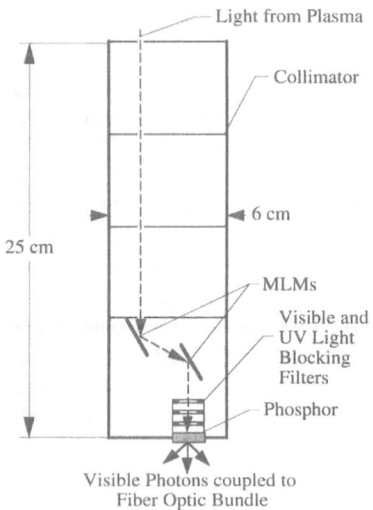

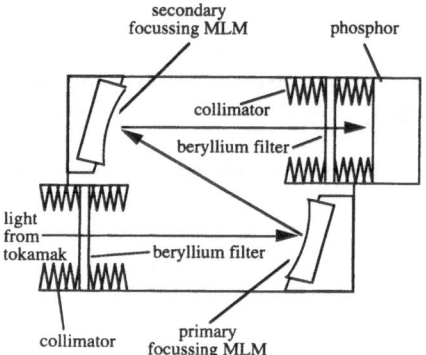

Figure 3. Schematic of MLM periscope for the Alcator C-Mod tokamak

Figure 4. Schematic of 150 Å high throughput imaging device

One option for the extraction of the soft-x-ray signals is to use an optical path similar to the one envisaged for the visible light collection, and replace the visible mirrors with MLMs. At 150 Å, the MLM reflectivities will be only 50% (compared with almost 100% of the visible mirrors); however, the higher brightness of the XUV signals will compensate for the lower efficiency (resonance transitions of L-shell ions will have brightness between one and two orders of magnitude larger than the $\Delta n = 0$ transitions among excited states). A device using a combination of gratings and MLMs for radiative divertor studies on DIII-D is under construction.

In conclusion, we propose to monitor the XUV emission of impurities from both the divertor and central part of the ITER plasma, with devices based on MLM as dispersive elements (filters). Both the 'periscope' and focussing systems are considered. The mechanical constraints will ultimately decide which of the two is appropriate. At the present level of technology a three reflection system leads to measurable signals beyond the last shielding volume, if $Z_{eff} = 1.5$ is assumed.

ACKNOWLEDGMENTS

This work is supported by DOE Grant DE-FG02-86ER53214.

REFERENCES

1. M.W.D. Mansfield, N.J. Peacock, C.C. Smith, M.G. Hobby, and R.D. Cowan, "The XUV spectra of highly ionised molybdenum", J. Phys. B: Atom. Molec. Phys. 11:1572 (1978).
2. M. Finkenthal, L.K. Huang, S. Lippmann, H.W. Moos, P. Mandelbaum, J.L. Schwob, M. Klapisch, and the TEXT group, "Soft x-ray bands of highly ionized W, Au, and Pb emitted from the TEXT tokamak plasma", Phys. Lett. A 127:255 (1988).
3. J.L. Schwob, A. Wouters, S. Suckewer, and M. Finkenthal, "High resolution duo-multichannel soft x-ray spectrometer for tokamak plasma diagnostics", Rev. Sci. Instrum. 58:1601 (1987).

4. M.J. May, M. Finkenthal, S.P. Regan, H.W. Moos, J.L. Terry, M.A. Graf, K. Fournier, and W.L. Goldstein, "Measurements of molybdenum radiation in the Alcator C-Mod tokamak using a multilayer mirror soft x-ray polychromator", Rev. Sci. Instrum. 66:561 (1995).

5. A.P. Zwicker, S.P. Regan, M. Finkenthal, and H.W. Moos, "High throughput multilayer mirror based soft x-ray spectrometer for metallic emission from tokamaks", Rev. Sci. Instrum. 61:2786 (1990).

6. L.K. Huang, S.P. Regan, M. Finkenthal, and H.W. Moos, "Laboratory test of a LSM-based narrow bandpass and high throughput camera for tokamak plasma imaging between 100 and 200 Å", Rev. Sci. Instrum. 63:5171 (1992).

7. E.H. Farnum, R.W. Clinard, Jr., S.P. Regan, and B. Schunke, "Neutron damage to diagnostic mirrors", Proceedings of American Ceramic Society's Fourth Symposium on Fabrication and Properties of Ceramics for Fusion Energy, Indianapolis IN, (1994).

8. W. Moos, A.P. Zwicker, S.P. Regan, and M. Finkenthal, "Layered synthetic microstructures for soft x-ray spectroscopy of magnetically confined plasmas", Rev. Sci. Instrum. 61:2733 (1990).

9. S.P. Regan, "Soft-x-ray spectroscopy on the Phaedrus-T tokamak", Ph.D. thesis, The Johns Hopkins University (1995).

10. A.P. Zwicker, M. Finkenthal, L.K. Huang, S.P. Regan, M.J. May, and H.W. Moos, "A soft x-ray multilayer mirror scanning monochromator for magnetically confined fusion plasmas", Rev. Sci. Instrum. 63:5035 (1992).

FEASIBILITY OF QUANTITATIVE SPECTROSCOPY ON ITER

M.G. von Hellermann[1], W.G.F. Core[1], A. Howman[1], C. Jupén[2],
R.W.T. König[1], M.F. Stamp[1], H.P. Summers[3], P.R. Thomas[1], K-D. Zastrow[1]

[1]JET Joint Undertaking, Abingdon UK, OX14 3EA,
[2]University of Lund, Sweden
[3]University of Strathclyde, Scotland

ABSTRACT

The options and needs for a comprehensive spectroscopic diagnostic package on ITER are discussed. Recent JET results illustrate the role of key parameters leading to acceptable levels of data consistency. A high level of accuracy in bulk plasma data, magnetic surfaces and the inclusion of an active beam diagnostic appears to be indispensable for a prospective quantitative spectroscopy on ITER.

INTRODUCTION

Powerful spectroscopic techniques have been developed during the last five decades of fusion research and the understanding of atomic processes in plasmas close to thermo-nuclear conditions has considerably advanced. The main progress has been achieved by the concerted use of several complementary plasma diagnostic systems and the development of appropriate evaluation procedures. The need for consistency checks involving the entire plasma environment has been recognised to play a major role. This is very demanding in terms of data range, quality and integration. In particular, the plasma edge and divertor region is still difficult to assess, so that local or semi-local spectroscopic measurements leading to absolute particle densities rather than incompletely analysed photon counts, are a major challenge.

A crucial part of any quantitative spectral analysis is the deduction of a *local ion velocity distribution function* from a measured spectrum. The experience gained at present fusion devices has clearly demonstrated that *temperature* and *density* are inseparable quantities, and intensity-only or Doppler-width-only measurements can not survive thorough consistency checks. In order to deduce local impurity densities from measured absolute

Diagnostics for Experimental Thermonuclear Fusion Reactors
Edited by P. E. Stott *et al.*, Plenum Press, New York, 1996

321

spectral intensities the *background plasma density* (consisting of electrons, ions and neutrals) and their respective *temperatures* needs to be unambiguously established. In the case of active charge exchange spectroscopy[1,2] the local neutral beam strength is required for the deduction of impurity densities. A precise reconstruction of *magnetic surfaces* enabling the mapping of all plasma data on common flux co-ordinates is an essential part of this procedure.

SPECTROSCOPIC DATA NEEDS

Two examples, the deduction of the *kinetic plasma energy content* and the prediction of *thermal-thermal and beam-thermal neutron rates* are used to highlight key input data necessary for an overall consistency check. The following table summarises the required data and the interdependence of measured and calculated quantities.

Table I Reconstruction of kinetic plasma energy and neutron yield from key experimental data. ◆ indicates direct dependence and ○ weak or indirect dependence. Cross-reference to magnetic field B indicates the need of mapping of data to a common flux grid. B is the result of an equilibrium calculation.

	B	n_e	T_e	T_i	Ω	n_z	$n_{d,th}$	P_{NB}	E_{NB}	v_{crit}	τ_s	ζ	S	$n_{d,fast}$
B		○	○	○	○	○	○							
n_e	◆													
T_e	◆													
T_i	◆				○	○								
Ω	◆			○		○								
n_z	◆	◆	○	○	◆	○		◆	◆			◆		
$n_{d,th}$	◆	◆	○	○	○	◆		○	○	○	○	○	○	◆
W_{th}	◆	◆	◆	◆	◆	◆	◆	○	○	○	○	○	○	○
v_{crit}	◆	◆	◆	○	○	◆	◆	○	○			○		
τ_s	◆	◆	◆											
ζ	◆	◆	○	○	○	◆			◆				○	
S	◆	○	○	○	○	○		◆	◆				◆	
W_{fast}	◆	○	○	○	○	○	○	○	◆	◆	◆	○	◆	
$n_{d,fas}$	◆	○	○	○	○	○	○	○	◆	◆	◆	○	◆	
Y_{dd}^{th-th}	◆	○	○	◆	○	○	◆	○	○	○	○	○	○	○
Y_{dd}^{b-th}	◆	○	○	◆	○	○	◆	○	◆	○	◆	○	◆	

For an ITER plasma with $<Z_{eff}>=1.5$ deuteron densities can be deduced from simultaneous CX measurements of the dominant low-Z impurities (carbon and helium) and electron density data. The dilution is: $d = \dfrac{n_d}{n_e} = 1 - \sum_{z>1} Z_z \dfrac{n_z}{n_e}$ (1) and its error: $\dfrac{\delta d}{d} = \dfrac{1-d}{d} < \dfrac{\delta c}{c}>$ (2). For dilution factors d close to 1 and estimated absolute uncertainties in local impurity concentrations $c_z = \dfrac{n_z}{n_e}$ of 30% the actual error in dilution values is quite low. For d>0.5 the deduction of deuteron densities from charge neutrality and CX measurements of the main impurities is to be preferred to a direct measurement of deuteron densities[3]. The errors are based on the assumption that electron density profiles can be established reliably by a combination of Thomson scattering (LIDAR) providing precise radial shapes and far infrared interferometry providing absolute fringe shifts and thus line integrated electron densities with an accuracy of less than 10%.

Kinetic plasma energy

The thermal ion energy density and the volume integrated kinetic energy content is determined by a CX measurement of radially resolved ion temperature profiles and the main light impurities respectively and the electron pressure measured by Thomson Scattering.

$$W_{thermal} = \frac{3}{2}\int (p_{ion} + p_{electron})dV \text{ with } p_{ion}(\rho = r/a) = n_{ion}(\rho)\cdot T_{ion}(\rho) \qquad (3)$$

$$\frac{n_{ion,thermal}}{n_e} = (1 - \sum_{z>1} Z\frac{n_z}{n_e}) - \frac{n_{d,fast}}{n_e} + \sum_{z>1}\frac{n_z}{n_e} \quad (4); \qquad n_d = n_{fast} + n_{thermal} \qquad (5)$$

The fast deuteron density is determined by the source rate S of neutral beam injection. For JET two injector boxes with eight neutral beams each contribute to the source rate:

$$S_{tot}(\rho) = \sum_{p=1}^{8}\sum_{k=1}^{3}\frac{f_{k,p}\cdot P_{k,p}}{e\cdot E_{k,p}}\frac{d\zeta_{p,k}(\rho)}{dV(\rho)} \qquad (6)$$

$f_{k,p}$: power fraction, $P_{k,p}$: beam power, $E_{k,p}$: beam energy. Indices k and p refer to energy fractions and beam number, $\rho = r/a$ is the minor radius. The neutral beam attenuation factor ζ depends on atomic stopping cross-sections and electron and ion densities:

$$\zeta(\rho) = \exp\left\{-\int ds(\rho)\cdot n_e(\rho)\sum_j \sigma_{stop,j}c_j(\rho)\right\} \qquad (7)$$

The fast deuteron density is:

$$n_{fast}(\rho) = \frac{1}{3}\tau_s(\rho)\sum_{p=1}^{8}\sum_{k=1}^{3}S_{k,p}(\rho)\ln[1 + (\frac{v_{b,k,p}}{v_c(\rho)})^3] \qquad (8)$$

with the slowing-down time:

$$\tau_s(\rho) = \frac{3\pi^{3/2}\varepsilon_0^2}{e^4\ln\Lambda}\frac{m_e m_\alpha}{n_e(\rho)Z_\alpha^2}\left(\frac{2T_e(\rho)}{m_e}\right)^{3/2} \qquad (9)$$

and critical velocity:

$$v_c^3(\rho) = \frac{3}{4}\sqrt{\pi}\left(\frac{2T_e(\rho)}{m_e}\right)^{3/2}\sum_j\frac{n_j(\rho)}{n_e(\rho)}Z_j^2\frac{m_e}{m_j} \qquad (10)$$

The total fast energy content is calculated from volume integration of the fast particle pressure profile:

$$P_{fast}(\rho) = \frac{1}{2}\tau_s(\rho)\cdot\sum_{p=1}^{8}\sum_{k=1}^{3}S_{k,p}(\rho)E_{k,p}\left[1 - \frac{2}{3}(\frac{v_c(\rho)}{v_{b,k,p}})^2\left\{\begin{matrix}\frac{1}{2}\ln\frac{v_{b,k,p}^2 - v_{b,k,p}v_c(\rho) + v_c^2(\rho)}{v_{b,k,p}^2 + v_{b,k,p}v_c(\rho) + v_c^2(\rho)}\\ +\sqrt{3}\tan^{-1}(\frac{\sqrt{3}v_{b,p,k}}{2v_c(\rho) - v_{b,p,k}})\end{matrix}\right\}\right] \qquad (11)$$

Thermal and beam-thermal neutron rates

In a beam-heated JET plasma the observed neutron rates are the sum of: Thermal-thermal, beam-thermal and beam-beam neutron rates. The thermal-thermal can be derived from CX ion temperatures and thermal deuteron density profiles:

$$Y_{th-th}(\rho) = \frac{1}{2}n_{thermal}^2 <\sigma v_{dd}(T_i(\rho))> \qquad (12)$$

The thermal-thermal d-d reaction rates are calculated using a modified Padé expansion as described in [4]. The beam-thermal neutron rate depends on neutral beam energies, impurity and electron density profiles as well as thermal target density, ion temperatures, toroidal angular rotation frequency Ω. The latter may have a considerable effect on the effective collision energy: $E_{col} = \frac{1}{2}M(v_b - R_{imp} \cdot \Omega)^2$ and hence the beam-thermal reaction rate. For JET hot-ion modes the rotation induced energy correction may reach values up to 35keV (of 140 keV). The beam-thermal neutron reactivity and its emissivity profile is:

$$<\sigma v>_{b-th} = \int_0^{v_{b,p,k}} \frac{v^2 dv}{v^3 + v_c^3(\rho)} \cdot \frac{2}{m \cdot v} \cdot P(\frac{1}{2}mv_{th}^2 u_1^2) \sqrt{\frac{u_1^3}{u_1^3 + \alpha}} \cdot \exp\left\{-\left[(u_1 - \frac{v}{v_{th}})^2 + \frac{\alpha}{u_1}\right]\right\} \quad (13)$$

with $\alpha = \sqrt{\frac{2B^2}{mv_{th}^2}}$ and $2u_1^2(u_1 - \frac{v}{v_{th}}) - \alpha = 0$

P is the Padé function and B a constant [4].

$$Y_{b-th}(\rho) = n_{thermal}(\rho) \cdot \tau_s(\rho) \cdot \sum_{p=1}^{8}\sum_{k=1}^{3} S_{p,k}(\rho) \cdot <\sigma v>_{b-th}(\rho, p, k) \quad (14)$$

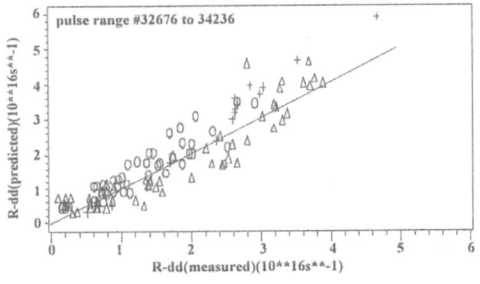

Fig.1 Comparison of predicted (see table I and equations above) and measured neutron yield

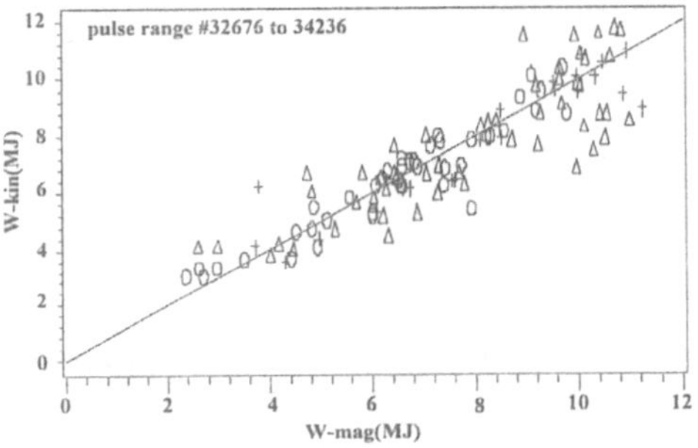

Fig.2 Comparison of kinetic (eqn.(3)) and diamagnetic energy content

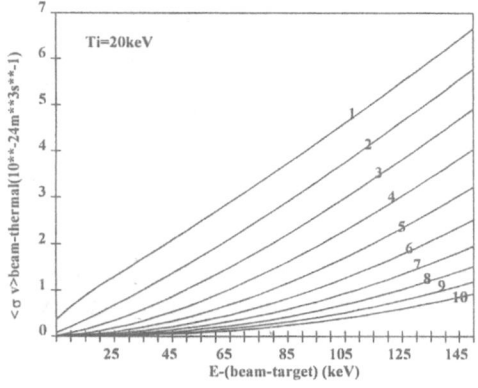

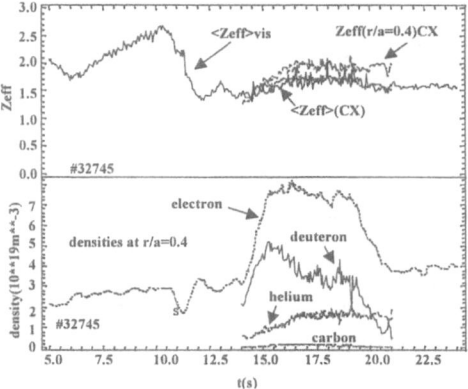

Fig. 3 Beam-thermal averaged reaction rate (eqn.13). Numbers 1 to 10 refer to critical velocity in steps of $5 \cdot 10^5$m/s.

Fig.4 a) <Zeff> from bremsstrahlung and profile averaged CX data in a radiative divertor plasma b) impurity and bulk plasma densities

324

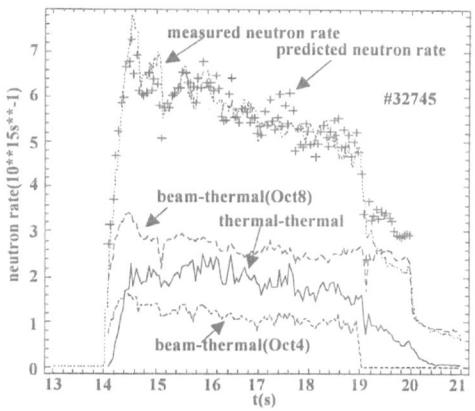

Fig. 5 Measured and predicted (eqns.(12, 13, 14)) thermal-thermal and beam thermal neutron rate in a radiative divertor plasma

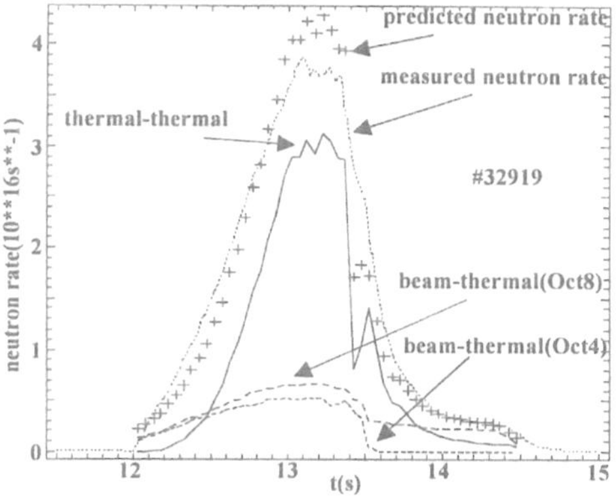

Fig. 6 Measured and thermal-thermal and beam thermal neutron rate in a hot-ion mode plasma

FEASIBILITY OF AN ACTIVE BEAM DIAGNOSTIC ON ITER

The chapter addresses three specific problems related to the feasibility of an active beam diagnostic on ITER. One is the role of *background spectral lines* emitted at the plasma edge, the second is the *detection of weak CX spectra* with amplitudes comparable to fluctuations of the continuum background radiation, and finally, what must be considered a fundamental bottle-neck, the *error propagation of the neutral beam attenuation* calculation.

Background spectral emission

This is the most difficult part to assess in advance of an ITER operation. Experience at JET and other tokamaks have shown that viewing lines looking onto wall sections close to high recycling areas may lead to complex passive emission spectra with appreciable signal levels. As a rule of thumb, if neighbouring edge line intensities exceed that of the active CX signal by more than order of magnitude any quantitative analysis (that is unambiguous extraction of a velocity distribution function) is put at risk. This applies in particular to the analysis of the CX HeII spectrum where intense CIII, CIV and BeII lines affect the analysis. The high quantum shell edge lines of CIII, CIV and BeIV are currently investigated [5]. Only sparse data exist on edge emissions close to CX spectra in the UV wavelength region [6] but present results indicate an even greater complexity of passive background line emissions. A UV CX spectroscopy for ITER appears therefore not to be a true alternative.

Signal-to-Noise considerations

In the case where fluctuations of the continuum radiation signal is the dominant noise source in an observed charge exchange spectrum, and passive line emission is not significant, the signal-to- noise can be expressed by, cf. [7] :

$$\{\frac{S_z}{N}\} = \frac{I_n c_z \sigma_{CX} \exp\{-\int dr n_e \sum_z c_z \sigma_{s,z}\}}{\pi w_\perp e \sqrt{Z_{eff} G_{ff} L_p B_c}} \sqrt{\frac{c \cdot v_e}{v_z^2}} \sqrt{\Delta t \frac{\Delta\lambda}{\lambda}} \, \mathfrak{R} \qquad (15)$$

I_n : neutral beam current, $I_{neutral} = e \cdot n_{beam} \cdot v_{beam} \cdot A_{beam}$

A_{beam}	: beam area, $A_{beam} = \pi \cdot w_\perp \cdot w_{l.o.s}$
n_z, $c_z{=}n_z/n_e$	: impurity ion density and its relative concentration
Δt	: integration time
$\Delta\lambda$	: spectral wavelength interval
$\mathfrak{R}$	: detection efficiency ($\mathfrak{R}{=}t{\cdot}\eta{\cdot}\varepsilon$ is the number of counts per radiance, with t : optical transmission, η : detector quantum efficiency, ε : spectrometer étendue $\varepsilon{=}\Delta\Omega{\cdot}A_{sp}$
σ_{cx}	: CX emission cross section.
σ_{sz}	: effective neutral beam attenuation stopping rate, including electron and ion impact ionisation as well as charge exchange processes
G_{ff}	: free-free Gaunt factor
L_p	: effective length of the continuum line-of-sight integration:

$$\frac{<n_e^2(0)> L_p}{T_e^{1/2}(0)} = \int ds(r) \frac{n_e^2(r)}{T_e^{1/2}(r)},$$

$$B_c = \left[\frac{e^2}{4\pi\varepsilon_0}\right]^3 \frac{16}{3hc^4 m_e^2}\sqrt{\frac{\pi}{3}}$$

Z_{eff} :effective plasma ion charge:

$$Z_{eff} = 1 + \sum_{z>1} c_z \cdot Z_z \cdot (Z_z - 1) \tag{16}$$

Fundamental limits for a quantitative CX diagnostic

A crucial part in the deduction of a local impurity densities from measured CX photon fluxes is the ability to provide accurate local neutral beam density data. The local beam strength is determined by stopping cross-sections (Fig.7) as well as electron, impurity densities and temperature profiles. The error in the deduced density introduced by an error in the attenuation is given by:

$$n_z = \frac{4\pi\phi_{cx}}{n_b L_{cx}\langle\sigma v\rangle_{cx}}$$

$$n_b = n_b(0)e^{-\int dn_e \sum_z \sigma_{sz} c_z} \tag{17}$$

$$\frac{\Delta n_z}{n_z} = \frac{\Delta n_b}{n_b} \approx \frac{\Delta(\sigma_{sz}, n_e)}{(\sigma_{sz}, n_e)} \cdot \sum_z \sigma_{sz} c_z \cdot \int\limits_{plasma-boundary}^{cx-volume} ds \cdot n_e$$

A convenient approximation for the relative error at 100 keV/amu (neglecting in a first approximation the contributions of impurities) is given by:

$$\frac{\Delta n_z}{n_z} \approx \frac{\Delta n_b}{n_b} \approx 2 \cdot \frac{\Delta x}{x} < n_e L / 10^{20}\,m^{-2} >_{cx-pl.bdy.} \tag{18}$$

For example, a relative error $\dfrac{\Delta x}{x}$=0.1 of the main input data (primarily the electron density) implies that, due to the exponential decay of the beam strength along the penetration path into the plasma, at an attenuation factor of $\zeta{<}0.05$ the errors in beam strength exceed 50%. A further option, cf. [3], which is currently explored, is the use of active Balmer Alpha beam emission spectroscopy for an independent deduction of local neutral beam strength. This alternative method might remove errors related to the exponential error propagation but needs further extensive tests.

326

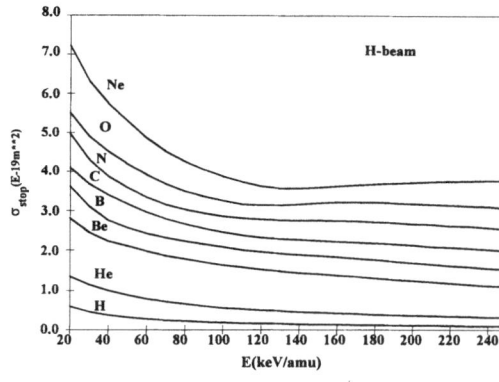

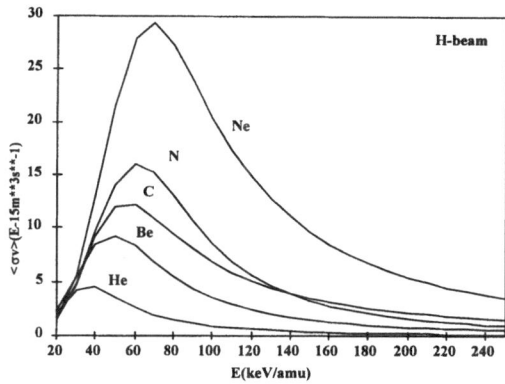

Fig.7 Total stopping cross-sections for a hydrogenic neutral beam (H^0, D^0) interacting with the main light impurities in a plasma. The total cross-section refers to the sum of electron and ion ionisation and charge exchange interactions.

Fig.8 Effective emission rates for CX transitions He(n=4→3, 4685.25Å), Be(n=6→5, 4658.55Å), C(n=8→7, 5290.5Å), N(n=9→8, 5671Å) and Ne(n=11→10, 5249.7Å) for a hydrogenic neutral beam (D^0 or H^0) as an electron donor.

DIAGNOSTIC BEAM OPTIMISATION

The signal-to-noise (equation (15)) depends approximately linearly on the impurity concentration c_z since its occurrence in Z_{eff} leads only to a weak modification ($1 < Z_{eff} < 2$). For an optimised operation of a fusion reactor the helium ash level is constant and needs to be kept below approximately 10% of the electron density. The signal-to-noise ratio is therefore primarily determined by the energy dependencies for CX excitation rates and beam stopping cross-sections respectively. Most of the CX excitation rates for low-Z impurities (cf. Fig.8) reach a maximum at about 50 keV/amu (HeII(4→3)@37keV/amu, CVI(8→7)@49keV/amu and NeX(11→10)@63keV/amu) whereas beam-stopping cross-sections continuously decay with increasing beam energy. If there were no limits to power supplies, neutral beam designs etc. one could consider an apparent gain by pushing the beam energies to higher levels. However the excitation rates for charge exchange recombination radiation drop more rapidly than the gain in beam penetration (Fig.7 and Fig.8).

The following beam optimisation study is dictated by technical constraints given by the ITER environment. Neutral beam ion sources must be shielded from magnetic stray fields. This implies a location of the source at a considerable distance away from the toroidal field coils. At the same time a high neutral current density (minimisation of $w_\perp$) must be maintained in the plasma centre for an optimisation of the signal to noise ratio. This leads to the requirement of a low-divergence neutral beam. Present day neutral beam technologies based on the development of negative ion sources promise a gain in beam divergence compared to positive ion sources. High power negative ion beam sources, which are presently under construction cf. [9] have specifications coming close to data estimated for spectroscopic optimisation studies as presented in this paper.

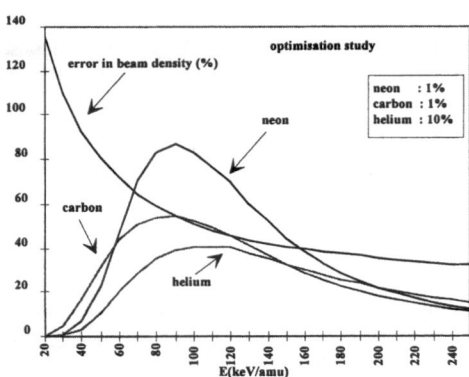

Fig.9 Figure of merit (a.u.). for the detection of CX spectra of helium (10%), carbon (1%) or neon (1%). The line integrated density along the neutral beam path is $2 \cdot 10^{20} m^{-2}$. The error in beam density is calculated from the attenuation factor: $-10*\ln(\zeta)$. The beam neutral current is assumed to be constant over the entire energy range so that eqn. (15) applies.

Table II Comparison of JET CX spectroscopy related diagnostic beam data and extrapolation to ITER

	JET	ITER
ion type	positive ion source	negative ion source
effective beam power for CX diagnostic (MW)	2	10
beam species	T^0, D^0	H^0
energy (keV/amu)	53 or 65	120
neutral beam current (full energy)	30 A	80 A
beam divergence	0.8°	0.2°
distance ion source plasma centre (m)	11.0	30
beam dimensions in plasma centre (m)	$w_\perp$=0.15, $w_{l.o.s.}$=0.15	$w_\perp$=0.1, $w_{l.o.s.}$=0.3
central target density (m^{-3})	$5 \cdot 10^{19}$	$10 \cdot 10^{19}$
beam path length (m)	1.2	2.8
line integral to centre (m^{-2})	$6 \cdot 10^{19}$	$20 \cdot 10^{19}$
impurity composition	D(74%),He(10%),C(1%)	D(74%),He(10%),C(1%)
emission rates ($10^{15} m^3 s^{-1}$)	3.6 12.2	0.56 5.09
Z_{eff}	1.26	1.26
beam attenuation	0.12	0.013
error in beam attenuation	21%	43%
integration time	0.05s	0.5s
detection sensitivity ($\frac{counts/s}{photons/s/sr/cm^2}$)	10^{-6}	$5 \cdot 10^{-6}$
path length for continuum radiation (m)	6	12
figure of merit (CVI spectrum)	55	43

COMPLEMENTARY ION TEMPERATURE DIAGNOSTICS

Any passive ion temperature diagnostic, e.g. X-ray emission spectroscopy, neutral particle energy analysis or neutron spectroscopy represents a line-of-sight integrated, and therefore *non-localised* measurement. This necessitates local values to be reconstructed either by tomographic techniques, which requires a substantial number of viewing lines, or the calculation of location and radial width of emission shells. Recent results at JET, cf. [8], have demonstrated that by choosing an off-axis X-ray line-of-sight for the collection of NiXXVII spectra, a comparatively narrow emission shell contributing to the signal may arise and consequently a reduced uncertainty in the radial position is gained. The method is based on the local emissivity, which is calculated from electron density and electron temperature profile data, providing thus a weighting function which leads to an *expectation value* and its *variance* of the magnetic flux surface at which the line of sight integrated X-ray spectrum is dominantly emitted.

$$<\rho> = \frac{\int \rho \cdot w(\rho) d\rho}{\int w(\rho) d\rho} \quad \text{with} : w(\rho) = f_{NiXXVII}(T_e(\rho)) \cdot n_e^2(\rho) \cdot \varepsilon(T_e(\rho)) \cdot g(\rho) \tag{19}$$

f is the local fraction of helium-like nickel, ε the emissivity, and g a geometry factor. The width of the emission layer:

$$\sigma_\rho = \sqrt{<\rho^2> - <\rho>^2} \tag{20}$$

is calculated by the second moment of the same weighting function. For the present geometry of the JET divertor phase typical variances are $\sigma_\rho = \pm 0.2$, that is, x-ray results represent an average over 20% of the minor radius. In theory it would be a possible option for ITER to design a passive ion temperature diagnostic system consisting of a fan of x-ray lines-of-sight each dedicated to a specific ion stage and to reconstruct a finite number of ion temperature values and their respective emission shell positions.

ALTERNATIVE LOW-Z ION DENSITY DIAGNOSTICS ?

There is to the present knowledge no alternative to active beam based charge exchange spectroscopy which may provide *local density data* of low-Z impurities (fully stripped helium, carbon, beryllium, boron etc.) with sufficient accuracy. If, for example, helium ash densities are to be deduced from (visible) continuum radiation measurements and of line averaged Z_{eff} values, this would have to rely on a known impurity composition (e.g. 1% C, 1%Be). For a 10% level of helium changes in Z_{eff} of the order 0.2 have to be measured:

$$<Z_{eff}> = 1 + \sum Z(Z-1)\frac{n_z}{n_e} \text{ or } <\frac{n_{He^{2+}}}{n_e}> = \frac{<Z_{eff}>-1}{2} - 15 < \frac{n_{C^{6+}}}{n_e}> \tag{21}$$

This implies that $<Z_{eff}>$ values have to be measured with an absolute accuracy of better than 0.05 which puts unrealistic demands on the accuracy of radial electron density and temperature profiles. Moreover, this is already the optimum accuracy of the absolute calibration of the measured bremsstrahlung.

The deduction of radial helium profiles from multi-chord bremsstrahlung measurements is even more doubtful. Uncertainties in Abel inversion procedures, in particular for the case for flat or hollow impurity profiles, exceed by far required data accuracies. Similar arguments apply for the use of tomographic reconstruction of x-ray emission profiles where accurate electron density and temperature data and impurity compositions will be needed for the deduction of a single low-Z impurity.

ABSOLUTE AND CROSS-CALIBRATION OF SPECTROSCOPIC INSTRUMENTS

There is one important aspect related to the measurement of line integrated continuum radiation and the deduction of $<Z_{eff}>$ which plays a central role for the deduction of absolute densities. This is the provision of an established cross-calibration source in addition to primary absolute calibrations. In most of the present fusion experiments, and this applies even more for ITER, in situ absolute calibrations during extended periods of operation are difficult, if not entirely impossible. Provided Z_{eff} is a constant on a magnetic flux surface and each spectroscopic line of sight can be characterised by its impact minor radius ρ_{imp} then cross-calibrations of spectroscopic instruments with arbitrary lines of sight can be established at any time.

$$< Z_{eff}(\rho_{imp}) > = \frac{\int\limits_{\rho_{imp}}^{1} I_{cont}(\rho)\frac{ds}{d\rho}d\rho}{B_c \int\limits_{\rho_{imp}}^{1} d\rho \frac{ds}{d\rho} \frac{n_e^2(\rho) \cdot g_{ff}(1,\rho)}{\lambda \cdot \sqrt{T_e(\rho)}}} \tag{22}$$

The constant B_c is defined in eqn. (15) and the integration ds refers to the path along the line-of-sight and $d\rho$ its corresponding increment in magnetic flux co-ordinate or normalised minor radius.

SUMMARY

The experiences at the JET experiment and also other major fusion devices have shown that the measurement of high accuracy ion temperature profiles is central for the control and analysis of plasmas close to thermo-nuclear conditions. Active beam techniques provide to the present knowledge the only access to profiles with the required accuracy ($\frac{\Delta T}{T} < 0.1$ and $\frac{\Delta r}{r} < 0.05$). A prerequisite for a quantitative CX deduction of impurity and deuteron densities is the availability of accurate data on electron density and temperature and also the provision of magnetic data. The deduction of local ion densities with errors less than 30% should be feasible. Adequate neutral beam sources with sufficient energy, power etc. for an active beam diagnostic on ITER do exist and should not be a problem. However, fundamental limits for a CX diagnostic dictated by errors in calculated beam attenuation factors, corresponding to an upper line density along the beam path of $2 \cdot 10^{20}$ m^{-2}, emphasise the need for additional passive techniques, for example, X-ray spectroscopy or neutron spectroscopy providing complementary, less-localised ion temperatures values. There is no alternative for thermal alpha particle or, in general, for low-Z impurity concentration measurements except by CX spectroscopy.

REFERENCES

1. R.Isler, Plasma Physics and Contr. Fusion, **36**, 171(1994)
2. M. von Hellermann and H.P.Summers, Atomic and Plasma Material Interaction Processes in Controlled Thermonuclear Fusion, Ed. Janev, 'Elsevier Science Publishers 1993',135-164
3. W.Mandl, R.Wolf, M von Hellermann, Plasma Phys. Contr. Fusion,**35**,1373, 1993
4. H.S.Bosch, G.M. Hale, Nuclear Fusion 32,611(1992)
5. C.Jupen, I. Martinson, M. Tunklev, University of Lund, Sweden, to be published
6. H.W.Morsi, R.W.T. König, H. Schröpf, M.G. von Hellermann
   Accepted for publication in Plasma Physics and Controlled Fusion, JET-P(95)24
7. M von Hellermann, W.G.F.Core, J.Frieling, L.D.Horton, R.W.T. König, W.Mandl, H.P.Summers, Plasma Phys. Contr. Fusion,**35**,799, 1993
8. M. von Hellermann, P. Breger, W.G. Core, U. Gerstel, N.C.Hawkes, A. Howman, R.W.T. König, C.F. Maggi, A.C. Maas, A.G. Meigs, P.D. Morgan, J. Svensson, M.F. Stamp, H.P. Summers, R.C. Wolf, K.-D. Zastrow, Proc. ICPP 1994, Foz do Iguacu, Brazil, Nov. 94, JET-P(94)58
9. Iiushi et al., Proc. ICPP 1994, Foz do Iguacu, Brazil, Nov. 94

LASER-INDUCED FLUORESCENCE

K. Muraoka,[1] K. Uchino,[1] T. Kajiwara,[1] S. Kuroda,[1] T. Okada,[2] and M. Maeda[2]

[1] Kyushu University, Kasuga Fukuoka 816, Japan
[2] Kyushu University, Hakozaki Fukuoka 812, Japan

1. INTRODUCTION

Laser induced fluorescence (LIF) is the technique in which the laser frequency ν is tuned to match a transition of a group of atoms or molecules, whose energy difference is E_{12}, by a relation $h\nu = E_{12}$ (where h is the Planck's constant), and the resultant fluorescence is observed. By the resonant nature of the excitation, the fluorescence intensity is many orders of magnitude larger than other sources of scattering, such as Rayleigh, Raman, or Thomson scattering, if the relevant number densities are the same order of magnitude. Therefore, the LIF technique is very selective of species to be detected, sensitive and/or yields high spatial and temporal resolution. In addition, because electric and magnetic fields affect E_{12}, the values of these field strength should, in principle, be obtainable using LIF.

The availability of the tunable source dictates the usable ranges of the technique. Usually, the technique is useful for excitation from the ground levels of neutral or lowly ionized particles, otherwise excitation using multi-photon processes or from the excited levels is necessary.

In this presentation, the principle of LIF and tunable lasers, its key hardware, are briefly reviewed in Section 2, followed by the explanation of experimental results of LIF applied to high-temperature studies in Section 3. Then, proposals of LIF experiments for ITER diagnostics are described in Section 4, followed by conclusion in Section 5.

2. PRINCIPLE OF LIF AND LASER SOURCES

There are excellent textbooks on LIF (e.g., Demtröder, 1982). Also, reviews were written on LIF for plasma measurements up to 1986 (Bogen and Hintz, 1986), and up to 1993 (Muraoka and Maeda, 1993) and interested readers may refer to these publications. Here, we give a brief description of the principle of the technique for explanation at the later chapters.

Figure 1 illustrates, as a typical example, an energy level diagram of atomic hydrogen for LIF. The first requirement for an LIF experiment is to match the laser wavelength to transitions of the atoms, molecules and ions to be detected. Usually, this requirement sets the limit of applicability of LIF to detection of various species. The subsequent optical, spectroscopic and detection system is common to usual optical emission or scattering experiments of plasmas.

The relationship between the laser wavelength λ [nm] and energy difference among the transition of species to be detected, E_{12} [eV], is given by

$$\lambda = 1239.8 / E_{12} \tag{1}$$

Diagnostics for Experimental Thermonuclear Fusion Reactors
Edited by P. E. Stott *et al.*, Plenum Press, New York, 1996

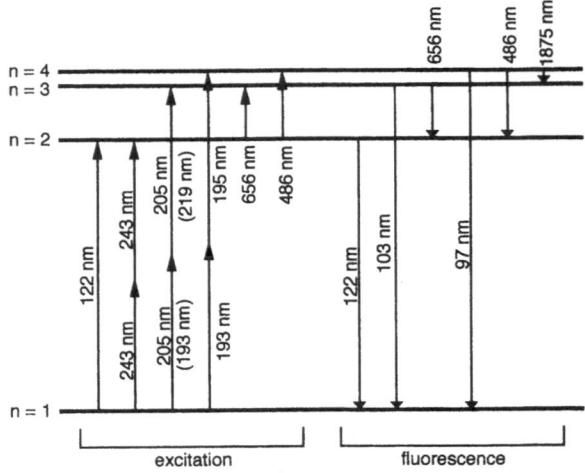

Figure 1. Low-lying energy levels of a hydrogen atom and wavelengths of some transitions used for LIF.

Because atoms and ions in high temperature plasmas (and molecules in low temperature plasmas) are predominantly at the ground level, excitation from there is necessary. E_{12} for alkali atoms is around 2 eV, for most metal atoms around 5 eV, for light atoms (H,O, C, and others) around 10 eV and for ions substantially higher than the respective atoms. The corresponding excitations wavelengths are, from Eq. (1), in the visible, ultraviolet (UV) and vacuum ultraviolet (VUV) regions, respectively.

The tunable sources developed most extensively for the above purpose are dye lasers in these wavelength regions. By the proper choice of organic dyes, spectral regions of visible, near infrared and UV wavelengths can be covered. A wide tunable range can be covered by the pumping of a high-power, short-pulse laser such as excimer, Nd:YAG (SHG,THG) and N_2 lasers. The pulse durations of these lasers are around 10 ns with a repetition rate of 10~100 Hz. Flashlamp-pumped dye lasers can operate over longer pulse widths, typically around 1 μs, in the visible region. CW operation is possible by the pumping of an argon ion laser. The coherence of them is excellent, although the wavelength access over a wide spectral region is poor.

In order to cover a wider spectral (mainly shorter wavelength) region than obtainable by dye lasers, as required for detection of metal atoms and light atoms in plasmas, various nonlinear frequency conversion techniques have been developed. Figure 2 illustrates such a scheme, where tunable sources based on an excimer laser are shown. The visible region is covered by the excimer laser pumped dye lasers, while the near UV spectral region can be covered by second harmonic generation (SHG) in nonlinear optical crystals, such as KDP, ADA, and BBO. In particular, efficient UV generation down to 205 nm has become very easy by the development of the new crystal BBO. Excimer lasers themselves have tunable range of around 1 nm centred on respective lasing wavelengths, such as 308 nm for the XeCl laser, 248 nm for the KrF laser and 193 nm for the ArF laser. In order to cover the VUV region, third harmonic generation (THG) and four-wave mixing techniques in gaseous media have been developed. Although the conversion efficiency from dye lasers by THG and four-wave mixing is extremely low at around 10^{-5}~10^{-6}, tunable VUV sources usable for LIF have become available such as for detection of C (165 nm) and H (121 nm). Also, Raman shifting of excimer laser outputs has proved useful for certain applications. These high power tunable sources are also useful for experiments of multi-photon excitation.

Lately, new solid-state tunable sources such as Ti:sapphire laser and optical parametric oscillator (OPO) are rapidly being developed. Because these sources have wider tunable range, higher efficiency, and better reliability than those of dye lasers, they are most promising tunable sources in the next generation. The necessary laser power is discussed from the viewpoint of saturation among energy levels concerned. It is easy to obtain the saturation

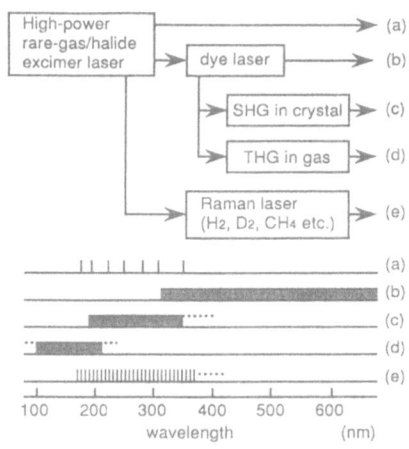

Figure 2. Wide range tunable sources based on a high power excimer laser.

condition for the visible or UV regions using these tunable sources, while it is usually difficult for the VUV region because of the small available laser power.

Coarse tuning of tunable lasers is made using gratings or birefringent filters. For measurements of spectral line profiles due to Doppler, Zeeman and Stark effects, a laser spectral line width of around 1 pm is sometimes required. For such a fine tuning, insertion of an intra-cavity etalon is used. In this respect, the RAFS (rapid frequency scan) laser (Honda *et al.*, 1987) is noteworthy, where the spacing of intra-cavity etalon is piezoelectrically driven to yield a spectral scan in a few μs. This is useful for observation of a spectral line shape in a transient condition.

3. EXPERIMENTAL RESULTS OF LIF APPLIED TO HIGH-TEMPERATURE PLASMA STUDIES

Applications of LIF to particle measurements in plasmas are classified by the wavelength of the used tunable lasers as visible ($\geq$ 400 nm), UV (400~200 nm), and VUV (< 200 nm). This is because there are distinct differences in the excitation source for each region. In addition, the optical and detection components are mostly decided by the excitation wavelength, and the factors to be considered are accordingly different.

3.1 LIF Using Visible Lasers

Tunable sources in this wavelength region are readily available commercially as shown in Section 2. However, from Eq. (1), visible lasers correspond to E_{12} < 3 eV, and only atoms and ions of alkali metals and rare earth metals have transitions in this range of E_{12} from the ground levels. Because of this, basic plasma physics researches have been carried out in vapours and plasmas of these atoms (e. g., Bowles *et al.*, 1992). Otherwise, such atoms have to be injected into high-temperature plasmas. A scheme to measure a magnetic field in a tokamak using an injected Li beam combined with LIF was tested on the TEXT tokamak (West *et al.*, 1987).

Another possibility is to use a transition from an excited level to higher levels. The most notable among various such researches is the atomic hydrogen measurement in high temperature plasmas. This is described below.

Measurements of Hydrogen Atoms. The energy level diagram of atomic hydrogen was shown previously in Fig. 1. For the detection of hydrogen atoms in magnetically confined plasmas for understanding particle behaviour and density control, tunable sources tunable to the

Lyman series have recently been developed and LIF experiments performed, as discussed in Section 3.3 . However, the detection limit of the atomic hydrogen density n_H by the scheme is around 10^{17} m^{-3}, and n_H in the confinement region in high-temperature plasmas is below this value. Therefore, LIF among excited levels, usually the Balmer series, has been performed. The measurements yield, after calibration of the optical system using the Rayleigh scattering, a population at an excited level. n_H is calculated by a collisional-radiative model, which relates the ground and excited level densities using independent measurements of election temperature and density.

The first successful measurements using the scheme was performed on FT-1 Tokamak at Ioffe Institute (Burakov *et al.*, 1977). Subsequently, the technique was applied to a mirror machine to study radial loss of hydrogen by charge exchange (Muraoka *et al.*, 1985), and to a stellarator, and then to Heliotron E (Muraoka *et al.*, 1990) and CHS (Compact Helical System) (Uchino *et al.*, 1992) to study the detailed particle behaviour there. The results of the last experiment is briefly described.

The experimental arrangement for LIF on CHS (l=2/m=8 helical device having a major radius R=1 m, average plasma radius a=0.2 m and the magnetic field B=0.9 T with a peak density of 3.4×10^{19} m^{-3}, and the electron temperature of ~300 eV at the plasma centre and ~30 eV at the edge) is shown in Fig. 3. A flashlamp-pumped dye laser with an oscillator and an amplifier was used, which yielded 100 mJ at the Balmer-alpha wavelength (H$_\alpha$, 656.3 nm) with a pulse duration of 1 μs FWHM (output power of 100 kW). The size of the observation volume was 30 mm in depth, 4 mm in width and 50 mm in height. The detection solid angle was 2×10^{-2} sr.

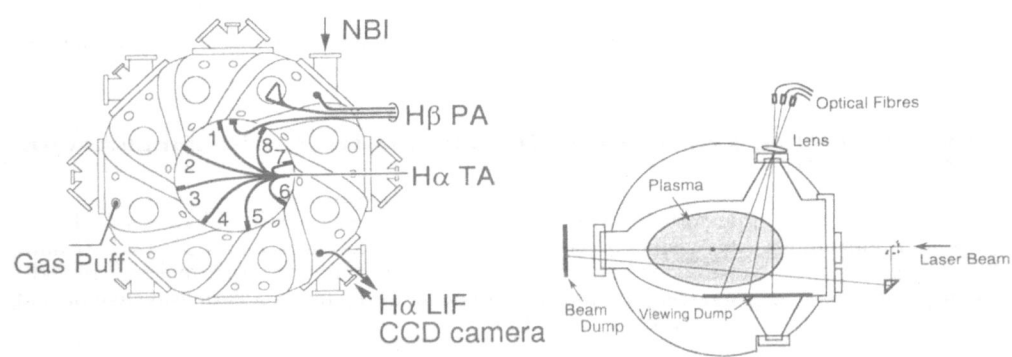

Figure 3. Experimental arrangement for measurements of atomic hydrogen densities on the CHS device.

The other diagnostic system for atomic hydrogen behaviour consisted of arrayed detectors of H$_\alpha$ and H$_\beta$ emissions and a CCD camera. The LIF system yielded local atomic hydrogen densities (n_H) at six points in a poloidal cross section of CHS. Each of toroidally arrayed H$_\alpha$ emission detectors observed the plasma at the same configuration at each of the eight sections of CHS. There were three sets of poloidally arrayed H$_\beta$ emission detectors. The CCD camera having an H$_\alpha$ interference filter observed, for a small toroidal extent, the toroidal change of recycling due to the helical change of the magnetic configuration.

Figures 4 (a) and (b) show n_H profiles measured by LIF tuned to H$_\alpha$, at a cross section where the plasma is horizontally elongated, for different chords as shown in the insets. Also shown are calculated profiles from the DEGAS code (Heifetz, 1986). The calculations were performed for the cases of a localized wall source (local.) at the divertor traces on the wall and

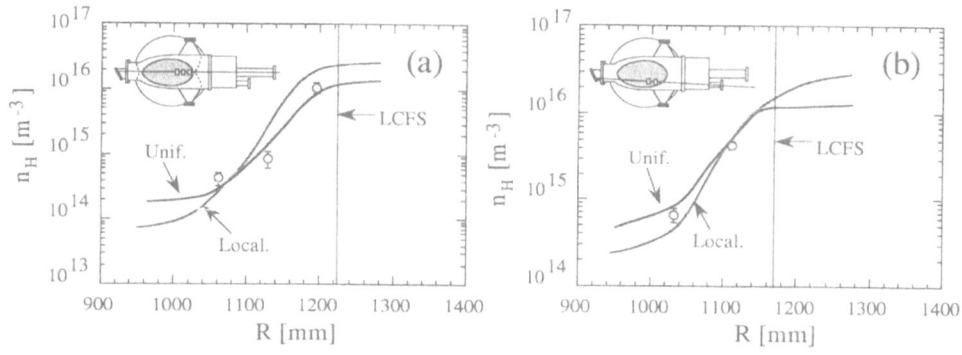

Figure 4. n_H profiles measured by Hα LIF for (a) a central chord and (b) a lower chord, as shown in the insets. Also shown are calculated profiles from the the DEGAS simulation code for cases of a localized wall source (local.) at divertor traces on the wall and of a uniformly distributed wall source (unif.).

of a uniformly distributed wall source (unif.), with the atomic energy set as eV (Muraoka *et al.*, 1990). Because the calculated profiles of n_H were in relative units, they were adjusted to best fit the experimental data. The fittings for Figs. 4 (a) and (b) are not independent but related, and we have to adjust the two profiles of the simulation simultaneously to the measured data. As can be seen from the figures, the fitting to the experimental points is better for the case of the uniform wall source.

By combining the above results with measurements of absolute emission intensity at the Balmer alpha and beta lines at various toroidal and poloidal cross sections, particle behaviour was discussed (Takenaga *et al.*, 1995), such as origins of hydrogen atoms, the resulting hydrogen recycling at the wall, the overall hydrogen supply, the diffusion coefficient, and particle confinement.

3.2 LIF Using UV Lasers

Tunable sources in this wavelength region are also readily available, where dye laser outputs are frequency up-converted (wavelength down-converted) using SHG, as shown in Section 2 with a typical conversion efficiency of 10%. Usually, the available laser power is above the saturation of transitions mentioned in Section 2 and the LIF experiments can be performed almost as easily as in the visible regions. $\lambda > 200$ nm imposes, from Eq. (1), $E_{12} < 6$ eV, which means that transition from the ground levels of metal atoms are in this wavelength region.

The first attempt of LIF for high temperature plasma studies was indeed intended for metallic impurity detection in the TEXTOR tokamak (Bay and Schweer, 1984), when the machine was envisaged as the one to study plasma-surface interactions, in particular the reduction of impurities by various means, such as surface modifications and a pumped limiter. Subsequently, similar experiments were performed on the ASDEX tokamak (Schweer *et al.*, 1982), and other machines.

Because optical setup and calibration procedures are very similar to those for the visible LIF as shown in Section 3.1, we do not go into further details here.

3.3 LIF Using VUV Lasers and Multi-Photon Excitation

As tunable sources in this wavelength region are not commercially available, we have to develop the sources themselves for each measurement. However, by various novel ideas, notably a recent development of a multi-photon excitation scheme, detection of light elements, such as H, C and O and various ions, has become feasible.

The most important information to be obtained in this wavelength region is atomic hydrogen behaviour. The energy level diagram of atomic hydrogen has already been shown in Fig. 1. Because a tunable source for the Lyman alpha transition (121.6 nm) is available by THG of a dye layer output (364.8 nm) in Kr/Ar gas mixture, the first measurements (Mertens and Bogen, 1989) were performed using the two-level system for the transition. From the elaborate arrangement on the TEXTOR tokamak, an excimer laser-pumped dye laser light was guided into the tripling cell installed on the wall of the machine. The resulting VUV light was irradiated hydrogen atoms at the edge region. The fluorescent light thus yielded was first converted to spectral profiles and then the atomic hydrogen density using the Rayleigh scattering calibration. The resulting hydrogen temperature was 0.7 eV, and a density of 4×10^9 cm^{-3} per velocity interval of 10^6 cm/s was found at the maximum.

Although the above scheme is straightforward and it is easy to convert the fluorescence intensity to atomic hydrogen density through calibration using the Rayleigh scattering, it is usually difficult to have a special optical system to reduce stray light necessary for the two-level system on large fusion machines. Therefore, a different scheme has recently been devised. As shown in Fig. 1, the next longest wavelength in the Lyman series after the Lyman alpha is the Lyman beta (102.6 nm), which is beyond the cutoff of a LiF window. However, two-photon excitation is feasible, because a very large laser power is available at 205.1 nm, which is twice the Lyman beta wavelength. The emission can be generated by Raman shifting of an ArF laser emission at 193 nm with D$_2$ gas, and the wavelength can be tuned over 1 nm at 205 nm.

Figure 5 shows the experimental arrangement of the system on Heliotron E (Kajiwara et al., 1991). The laser characteristics were 10 mJ of output energy in an 8 ns FWHM pulse with a spectral width of about 30 pm. The spectral width was optimized for the density measurements of atomic hydrogen at a temperature of several eV. The laser beam was focused at the observation point by a lens having a focal length of f=810 mm. Fluorescence signals were observed by a photomultiplier through a pinhole having a diameter of 0.4 mm and a Hα filter. Even with this coaxial arrangement, the spatial resolution was found to be about 20 mm, and the system was calibrated through the comparison between perpendicular and coaxial observations to the laser beam. The measurements were performed at 30 mm outside the separatrix.

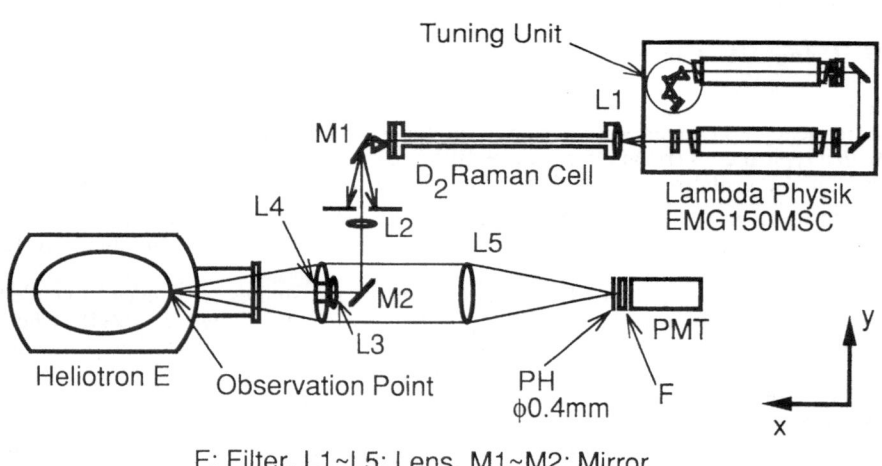

F: Filter, L1~L5: Lens, M1~M2: Mirror,
PH: Pinhole, PMT: Photomultiplier tube

Figure 5. Experimental arrangement of two-photon excited LIF for atomic hydrogen detection on Heliotron E.

Figure 6 shows typical traces of fluorescence signals for the two timings. At t=4 ms after the ECH initiation [Fig. 6 (a)], the hydrogen density at the observation point was high enough for the fluorescence signal to be recognized above the background Hα fluctuation.

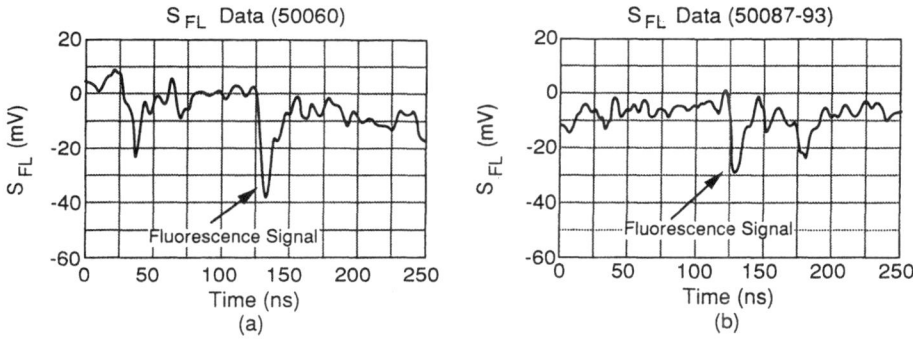

Figure 6. Typical traces of fluorescence signals. (a) was obtained at t=4 ms after the initiation of the ECH plasma. (b) was obtained at t=12 ms, the integrated data of seven observations.

However, at $t=12$ ms, the fluorescence signal was not large enough to be clearly observed in a signal observation. Therefore, data from seven shots were accumulated and the result is shown in Fig. 6 (b). These results were put into the absolute calibration to yield atomic hydrogen densities of 3×10^{16} m^{-3} for the former and 1×10^{16} m^{-3} for the latter.

Previous measurements of atomic hydrogen densities in core plasmas in Heliotron E were combined with the DEGAS computer code written for the Heliotron E configuration to yield behaviour of hydrogen atoms and molecules in the scrape-off layer as well as in the plasma core. The atomic hydrogen densities estimated for the present experimental condition, namely at the position of observations for the ECH sustained plasmas at the two times from the initiation of the discharge, were consistent with the above values within a factor of 2.

The SN ratio and the detection limit of this system were estimated, where the dominant noise source was the fluctuation of the background Hα radiation, caused by the shot noise of the photomultiplier during the integration time needed to observe the fluorescence (10 ns). The results yielded the detection limit, determined by the SN ratio, to be $(n_H)_{min}=2 \times 10^{16}$ m^{-3} for a single shot measurement, and $(n_H)_{min}=1 \times 10^{16}$ m^{-3} for the case of 7 shot accumulation. These values agreed with the above observation.

The detection limit is good enough for application to measurements at large machines, because the repetition rate of commercial lasers can be as high as 250 Hz, opening up even possibility of measuring Doppler profiles by observing fluorescence while scanning the frequency of a laser whose output has been spectrally narrowed.

4. PROPOSALS OF LIF EXPERIMENTS FOR ITER DIAGNOSTICS

During the CDA phase of the ITER design activities, the Kurchatov group studied possibilities of applying LIF for measurements of densities and velocity-distribution functions of helium, carbon, boron, beryllium and lithium atoms around the ITER divertor. There, no quantitative description of the possibility of the signal detection was made in the literature (Mukhovatov *et al.*, 1990) except the phrase, " Using a laser wavelength close to an atomic transition will result in Rayleigh scattering with cross sections about 4 orders of magnitude greater than the Thomson cross section. Ground state densities of hydrogen and helium of at least 10^{16} - 10^{17} m^3 will be detectable". Subsequently, they further studied the conventional LIF among exited levels of helium atoms (I. V. Moskalenko and D. A. Shcheglov, 1994a).

In the following, these and the authors' proposals (Uchino *el al.*, 1995) are described of density measurements of atomic hydrogen and helium, the neutral elements by far the most important around the divertor region.

4.1 Hydrogen Detection

For measurements of hydrogen atoms around the ITER divertor, the laser induced fluorescence technique could be applied using three schemes, namely (1) among excited levels as described in section 3.1, (2) the Lyman alpha transition, and (3) the multi-photon excitation from the grand level, both (2) and (3) being described in section 3.3.   Among these, the scheme (1) may be difficult to apply because the laser and fluorescence are at the same wavelength, and the technique of stray light suppression would be difficult to implement at the ITER divertor, and for the scheme (2), the VUV optical system, including relaying of the VUV laser light and of detection with the stray light suppression system, looks difficult to be implemented at the ITER divertor.   Therefore, the scheme (3), namely the method based on two-photon excitation (two-photon LIF) from the ground level to the n=3 level, looks the most appropriate. Fluorescence observation is at Hα and the ground level population is directly probed without using VUV light. In addition, the problem of stray light is eliminated. An excitation source for the two-photon LIF is obtainable using a ArF-excimer laser in combination with a D_2 Raman-cell. In order to get absolute values of hydrogen densities from the fluorescence signals, the detection system is calibrated using a hydrogen flow tube and a titration procedure with NO_2 gas.   Expected SN ratios have previously been estimated for measurements of atomic hydrogen densities around divertors of JET, DIII-D and JFT-2M tokamaks, and were found to be sufficient to determine the values.

The same estimation procedures were followed for the ITER divertor conditions, by the arrangements shown in Fig. 7.   Because of the lack of reliably predicted values of plasma parameters in the ITER divertor, a model was established for profiles of n_e and T_e there and hydrogen penetration was evaluated against this plasma.   The plasma radiation at the Balmer alpha was compared with the fluorescence at the same wavelength resulting from the two-photon laser excitation. The SN ratio thus estimated is shown in Fig. 8 as a function of atomic hydrogen density at the observation point.   Here, two arrangements were envisaged, namely vertical and coaxial, as shown in Fig. 7.   From Fig. 8, it is evident that much of the ITER divertor could be probed by this scheme.

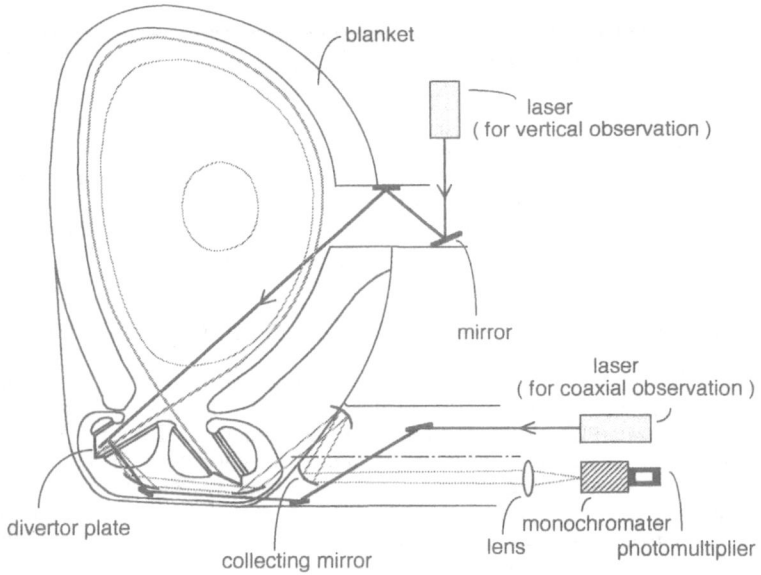

Figure 7. Schematic arrangement of two-photon excited LIF on the ITER divertor.

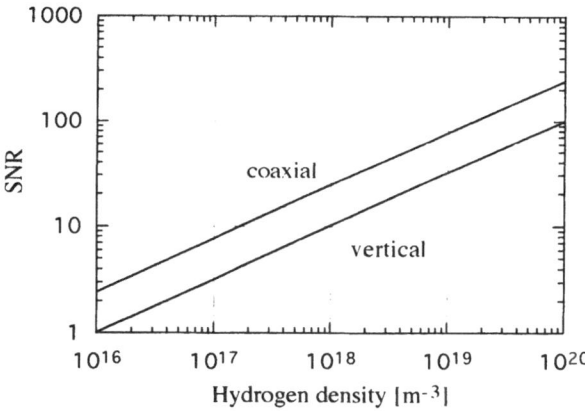

Figure 8. SN ratio plotted against atomic hydrogen density n_H, for vertical and coaxial observations (cf. Figure 7).

4.2 Helium Detection

For helium atom measurements, only excited levels can be probed by laser radiation. In this case, the Kurchatov group (Moskalenko and Shcheglov, 1994b) proposes to excite at 388.9 nm ($2^3S \rightarrow 3^3P$) and to observe 587.6 nm ($3^3D \rightarrow 2^3P$) or 706.5 nm ($3^3S \rightarrow 2^3P$), the transition among the upper levels being due to collisional transfer processes. The fluorescence observations at different wavelengths from the laser wavelength are to try to avoid the stray light from swamping signals. Because no estimate was made of the expected SN ratio for this configuration, we describe an alternative proposal of the laser induced ionization (LII) method (Uchino *et al.*, 1993) which yields a local atomic density without elaborate techniques such as frequency tuning and suppression of stray light.

When a strong laser beam ($\sim$10 MW/cm^2) whose wavelength is shorter than 660 nm is injected into the plasma, the emission from the n=3 to n=2 transition of the helium atom is reduced due to photo-ionization of the n=3 level. This can be detected quantitatively. Thus, the helium atomic density at the excited state can be obtained and atomic density at the ground state is calculated using a collisional-radiative model for known values of electron temperature and density. Actual measurements will be performed at a laser power density of the order of MW/mm^2, which is a power level at which Thomson scattering is possible if the electron density is larger than 10^{18}m^{-3}. This means that electron temperature and density can be determined simultaneously with the LII signal, and so the atomic helium density at the ground state can be obtained without depending on other diagnostic systems.

Proof-of-principle experiments were performed on ECR discharge plasmas. LII signals and the saturation characteristics were clearly observed, as was expected from pre-experiment estimations.

SN ratios were estimated for the LII applied to the ITER divertor. The optical arrangement is similar to the vertical configuration of Fig. 7. If the atomic helium density there is larger than 10^{17} m^{-3}, SNR > 10 is expected after the data accumulation over 100 laser shots.

5. CONCLUSION

After brief descriptions of principle of LIF and laser sources, its key hardware, examples of its application to high temperature plasmas were shown. Further, proposals of its application to the ITER divertor studies were described. The application of LIF to plasma diagnostics is in its juvenile stage, in that the study has been seriously pursued for less than a decade, and that it has not yet become indispensable for plasma diagnostics. It has, however, started to show potential for future promise. The last comment is particularly true for studies of hydrogen behaviour in the scrape-off layer of high temperature plasmas. This is because these studies can not be pursued with other diagnostic means.

For these measurements to become routine for plasma studies, maintenance-free operation of the tunable sources is highly desired. In this respect, recent developments of tunable solid-state sources, such as alexandrite and Ti:sapphire lasers and OPO are noteworthy. High-power semiconductor lasers and new solid-state lasers pumped by them will also change basic structure of the laser system in the near future.

Combining new developments of tunable lasers with measurements of physical quantities of plasma state which are not possible by other means, LIF will, hopefully, evolve into indispensable diagnostic technique for plasma studies during the next decade.

The authors are indebted to many people, Dr. T. V. Moskalenko in particular, who supplied them useful information.

REFERENCES

Bay, H. L., and Schweer, B., *J. Nucl. Mater.* 1984, **128&129**, 257.
Bogen, P. and Hintz, E., 1986, in Physics of Plasma Wall Interactions in Controlled Fusion (Eds. Post, D. E. and Behrisch, R., Plenum, New York) p. 211.
Bogen, P., and Mertens, Ph., 1989, *Proc. 4th Int. Symp. Laser-Aided Plasma Diagnostics* (Kyushu Univ.) p.117.
Bowles, J., McWilliams, R. and Rynn, N., 1992, *Phys. Rev. Lett.* **68**, 1144.
Burakov, V. S. *et al.*, *JETP Lett.*, 1977, **26**, 403.
Demtröder, W., 1982, Laser Spectroscopy-Basic Concepts and Instrumentation (Springer, Berlin).
Heifetz, D. B., 1986, in *Physics of Plasma-Wall Interactions in Controlled Fusion* (Eds. Post, D. E. and Behrisch, R., Plenum, New York) **17**, 86.
Honda, C., Maeda, M., Muraoka, M., and Akazaki, M., *Rev. Sci. Instrum.*, 1987, **58**, 759.
Kajiwara, T., Shinkawa, T., Uchino, K., Masuda, M., Muraoka, K., Okada, T., Maeda, M., Sudo, S., and Obiki, T., 1991, *Rev. Sci. Instrum.* **62**, 2345.
Kajiwara, T., Takeda, K., Kim, H. J., Park, W. Z., Okada, T., Maeda, M., Muraoka, K., and Akazaki, M., 1990, *Jpn. J. Appl. Phys.* **29**, L154.
Mertens, Ph. and Bogen, P., 1989, *Proc. 16th Europ. Conf. Contr. Fusion Plasma Phys.*, 1989, **III**, 983.
Moskalenko, I. V, and Shcheglov, 1994a, *Proc. ITER Workshop* (Garching).
Moskalenko, I. V, and Shcheglov, 1994b, *Minutes Moscow. ITER Prog. Meeting.*
Mukhovatov, V. *et al.*, 1990, *ITER Diagnostics* (ITER Documentation Series, No, 33, IAEA, Vienna)
Muraoka, K., Uchino, K., Itsumi, Y., Hamamoto, M., Maeda, M., Akazaki, M. et al, 1985, *Jpn. J. Appl. Phys.* **24**, L59.
Muraoka, K., Uchino, K., Maeda, M., Kajiwara, T., Matsuo, K., Okada, T., Honda, C., Suehiro, Y., Yano, N., Takeda, K., Hagiwara, H., Akazaki, M. *et al.*, 1990, *J. Nucl. Mater.* **176&177**, 231.
Muraoka, K., and Maeda, M., 1993, Plasma Phys. Contr. Fusion **35**, 633.
Schweer, B., Bogen, P., Hintz, E., Rusbüldt, D., Goto, S. and Steuer, K. H., 1982, *J. Nucl. Mater.* **111&112**, 71.
Takenaga, H., Nakao, T, Uchino, K., Muraoka, K., Maeda, M., Iguchi, H., Ida, K., Yamada, I., Okamura, S., Yamada, H., Morita, S., Takahashi, C., and Matsuoka, K., 1995, Nucl. Fusion **35**, 107.
Uchino, K., Takenaga, H., Kajiwara, T., Okada, T., Muraoka, K., Maeda, M. *et al.*, 1992, *J. Nucl. Mater.* **196-198**, 210.
Uchino, K., Takeshita, M., Muraoka, K., and Maeda, M., 1993, *Proc. 6th Symp. Laser-Aided Plasma Diagnostics*, 61.
Uchino, K., Kuroda, S., Kajiwara, T., Muraoka, K., Okada, T., and Maeda, M., 1995, *This workshop*.
West, W. P., Thomas, D. M., deGrassie, J. S., and Zheng, S. B., 1987, *Phys. Rev. Lett.* **58**, 2758.

CONCEPTS AND REQUIREMENTS FOR ITER X-RAY DIAGNOSTICS

K. W. Hill, M. Bitter, and S. von Goeler

Princeton University
Plasma Physics Laboratory
P. O. Box 451
Princeton, NJ 08543

I. INTRODUCTION

X-ray diagnostics are used on present-day tokamaks to measure a variety of plasma parameters.[1,2] Some x-ray diagnostics are not directly adaptable to ITER because of the high sensitivity of the detectors to noise and damage induced by the radiation background from fusion neutrons.[3,4] Those diagnostics which use compact, highly efficient solid-state detectors are particularly vulnerable. This work addresses mainly the issues involved in adapting a subset of the existing x-ray instrumentation to ITER, mainly by use of additional x-ray optics to deflect the x-ray beam out of the path of the neutron beam.[5,6,7] The topics discussed will be the required x-ray intensities and energy ranges, the necessary reduction in radiation background, and the x-ray optics and geometries best suited to accomplishing these requirements. Preliminary work on this subject appears in Refs. 5-7. We will be focusing on the x-ray energy range around 5 - 40 keV, and somewhat higher in some cases. The main diagnostics discussed are the XIS (X-Ray Imaging System) or SX arrays, the broad-band PHA (Pulse-Height-Analyzer) continuum and line measurement spectrometer, and the high energy resolution XCS (X-ray Crystal Spectrometer).[2] Other diagnostics suitable for ITER have been discussed by Barnsley.[8,9]

One of the most useful applications of the XIS is the measurement of radial profiles of fast fluctuations in x-ray emissivity due to electron temperature (T_e) and density (n_e) fluctuations, but it also diagnoses slower fluctuations and MHD phenomena, enables determination of the sawtooth inversion radius, and can be used for Z_{eff} and electron temperature (T_e) measurement.[10] The T_e and n_e fluctuations can be distinguished by use of foils. The XIS is based on measurement of the total x-ray intensity over a broad band, from 2 to 20 keV on TFTR. On ITER this band should cover about 5 to 40 keV to enable T_e - fluctuation measurements at T_e values of 20 - 30 keV. The XCS is most important for its measurement of ion temperature (T_i) and plasma rotation, but can also provide information on impurity charge-state distributions, T_e, and atomic physics of highly charged ions.[1,11] It is based on very high energy resolution ($E/\Delta E \sim 7000$) spectroscopy over a bandwidth of order 0.1 keV at x-ray energies ranging from 5 to 17 keV. The PHA is a convenient diagnostic of metallic impurity content,[12] but can also measure T_e and Z_{eff} and diagnose nonthermal electron distribution functions.[13] It typically has a resolving power ($E/\Delta E$) of

Diagnostics for Experimental Thermonuclear Fusion Reactors
Edited by P. E. Stott *et al.*, Plenum Press, New York, 1996

about 30 at 6 keV and requires a bandwidth of a few keV in the region 5 - 17 keV for impurity monitoring, or broader band operation (4 - 100 keV) for T_e measurement. Higher-energy versions, covering 25 to ~ 500 keV are used for studying the nonthermal electron populations generated by RF current drive.[2,13]

The motivations for trying to adapt the x-ray diagnostics to ITER are several fold. First, the diagnostics measure a range of plasma parameters and have been developed and tested on existing tokamaks. Secondly, other techniques for measuring the same parameters on present tokamaks may suffer technical problems on ITER. For example, the charge-exchange recombination spectroscopy for measuring T_i may suffer from excess beam attenuation. Also, the Thomson scattering diagnostic for T_e may suffer from radiation damage to its window, and the ECE diagnostic may have problems at the high magnetic fields in ITER. Recent increases in the count-rate capability of PHA detectors by a factor of 10 make that diagnostic more attractive as a temperature diagnostic.[14] A high level of redundancy in diagnostic measurements would be prudent for a complicated device like ITER. Thirdly, the solid state detectors used in the XIS and PHA are highly efficient and compact, the absolute calibration is known to good precision, and the signal processing is simple. Thus, relatively small penetrations through the ITER blanket are required, resulting in relatively low fluxes of neutrons streaming to the outside. Fourth, these systems do not necessarily require a vacuum path extending through the biological shield since a thin metallic x-ray window can be used near the tokamak. Fifth, at the high temperatures and densities of ITER the x-ray emission will be intense, and it serves as a good basis for diagnosis of core plasma parameters.

The basic limitation of both the XIS and PHA, with regard to use on ITER, is that the solid-state detectors would be quickly damaged by the incident neutron fluxes, even if the neutron aperturing were the same as that of the x rays.[3] For the XIS, a second problem would be frustration of measurement of fast fluctuations due to large spikes or grass on the detector signals due to neutron interactions.[4] Thus, we consider the use of x-ray deflectors as "neutron filters" to deflect the x-ray beam out of the direct neutron beam from ITER. Although the XCS, if properly shielded, could operate without additional deflectors, since the detector is already well removed from the neutron beam, x-ray deflectors could also be used between the ITER blanket and the biological shield to reduce neutron streaming through penetrations in the biological shield and onto the XCS apparatus.

In section II the bandwidth, deflection angles, reflection efficiencies, and radiation-noise reduction capabilities of potential x-ray optical elements are discussed, as well as their general applicability to the x-ray diagnostics discussed in this report. In sections III, IV, and V, the XIS, PHA, and XCS are discussed from the perspective of the signal levels required, the tolerable radiation background noise and detector-damage, and finally a calculation of the aperture and mirror sizes and deflection geometries for a channel of each instrument which could be deployed on ITER.

II. X-RAY DEFLECTOR CHARACTERISTICS

The concept of the x-ray deflector is simple. The x-ray and neutron beams are narrowly collimated onto an x-ray deflecting element. Shorter multiple deflectors, as are used for x-ray telescopes, are more practical than one long mirror. The lateral x-ray deflection must be large enough so that the neutron beam does not enter the instrument or exit collimator. The minimum spread of the neutron beam is determined by the intensity requirements for the x-ray measurement. The lateral deflection achievable depends on the type of deflecting element chosen and decreases with increasing x-ray energy, as shown in Fig. 1 for grazing-incidence (GI) x-ray mirrors and natural or (synthetic) multilayer (ML) Bragg diffractors. Nuclear radiation can enter the exit collimator by three general processes to cause background noise. These are (1) scattering from the deflector and its support, (2) scattering from the end of the mirror entrance collimator or any mass that the instrument views, and (3) scattering of the direct neutron/gamma-ray beam from the impact point on the wall into the mouth of the exit collimator. It is desirable to have a long exit collimator leading the x rays to the measuring instrument to minimize the effects of process 3. A long path length from the mirrors to the diagnostic, to enable a large lateral x-ray deflection, and a narrow angular spread of the neutron beam are desirable.

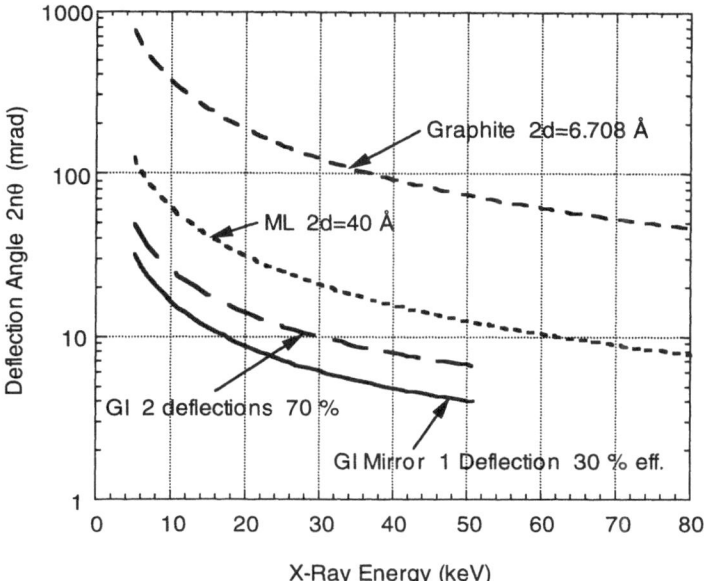

Figure 1. Deflection angles 2θ times the number of deflections n for grazing-incidence (GI) mirrors, a multilayer (ML) diffractor, and a graphite crystal as a function of x-ray energy.

To illustrate the possible application of such x-ray mirrors to ITER, the locations of the shields surrounding ITER should be discussed. A 2-m thick biological shield is located about 6 m from the ITER blanket. The outer shield wall of the diagnostics or instrument area is 10 m beyond the biological shield. The x-ray deflector can be placed in either of two locations, (1) just outside the biological shield or (2) between the blanket and the biological shield, which we shall refer to as "inside" (radially) the biological shield. The location outside the biological shield is preferable technically because of the long deflection arm, about 8 - 9 m; in location 2 the deflection arm could be only 3 - 4 m because a collimation length of 2 - 3 m would be required between the blanket and the mirror to minimize the spread of the neutron beam. Location 2, however would be preferable from the point of view of minimizing the neutron flux penetrating the biological shield.

As an example of the achievable background reduction from a grazing-incidence (GI) x-ray mirror, a MCNP calculation of the scattering of 14-MeV neutrons from ITER by the mirror into the instrument collimator was performed.[5] The mirror dimensions were 1 x 30 x 200 cm³, based on an early concept of the size needed to provide high x-ray fluxes. The material was silicon to simulate either a glass or silicon substrate, the incidence angle was 5.25 mrad, and the distance from ITER to the mirror and from the mirror to the exit collimator were both 10 m. The ITER aperture size was 1 x 30 cm². The calculated 14-MeV neutron flux at the exit collimator position, displaced 10.5 cm laterally from the axis of the incident neutron beam, was smaller by a factor 1.6×10^{-10} than the flux onto the ITER wall, which was 4.4×10^{13} n/cm² s for 1 MW/m² of neutron wall loading. This flux onto the detector of about 7×10^3 was also reduced by 1/15000 relative to that incident onto the x-ray mirror (which would also be the flux incident onto the detector for a direct plasma viewing

XIS channel). This was a very conservative calculation, because both the area and the thickness of the mirror were much larger than necessary; a factor of 30 - 50 reduction should be possible in the total mirror scattering volume, resulting in negligible scattered neutrons at a deflection angle 2θ of 10.5 mrad and a distance of 10 m. However, as we reduce the deflection angle, as may be necessary for extending the instrument to higher x-ray energies (20 - 40 keV), the scattering will increase. Also the detector signal may become dominated by scattering of the direct neutron/gamma-ray beam from the impact point on the wall into the mouth of the exit collimator.

The choice of x-ray deflector to be used depends on its reflection characteristics, i. e. bandwidth and deflection angle, and the spectral bandwidth, intensity, and degree of isolation from nuclear radiation required by the diagnostic. For example, GI x-ray mirrors can reflect a broad range of x-ray energies, from low energies up to 40 or 50 keV with high efficiency, although the incidence angles, relative to the surface must be very small (~ a few mrad) at the higher energies.[15,16] These elements are, thus, preferable for use with both the XIS and the PHA diagnostics. Bragg deflectors, either natural crystals or synthetic multilayer (ML) elements, can provide much larger deflection angles than the GI mirrors, but usually pass only a narrow bandwidth, E/ΔE ~ 100 for multilayers and natural crystals with the broadest rocking curves.[17,18] Thus, they may be well suited for use with narrow-bandwidth spectrometers, but not with diagnostics which require broad bandpass, such as the PHA, and they are less attractive for use with the XIS because of its requirement for high power throughput. Graded-d multilayers can provide reflections over a much broader bandwidth (e. g. 20 - 60 keV) than can the single-d multilayers (or natural crystals), but the incidence angles are small (2 - 2.5 mrad), and the reflectivities are somewhat small (10 - 30 percent) at the higher energies.[19,20] These are called either "supermirrors" for very broad bandwidth elements or "super-band" reflectors for more limited energy bandwidths. These elements may be suitable for use with a broad-band continuum spectrometer because of the smaller solid angle requirements and smaller background-noise-reduction requirements of the PHA. It is expected that a reflectivity of 25 percent can be achieved at 60 keV at an angle of 3 mrad, and around 15 percent up to 45 keV at 6 mrad.[21] Further optimizations of these elements may be possible for specific x-ray diagnostic applications. A deflector suitable for use with the hard x-ray spectrometer for monitoring nonthermal electron populations has not been discovered, since the critical angle for efficient reflection becomes extremely small at high energies. Cauchois transmission crystal spectrometers have been operated in the 100+ keV range; further analysis should be done to investigate their suitability for diagnosing tokamak hard x-ray continuum spectra.

III. THE X-RAY IMAGING SYSTEM

The x-ray signal requirements of the XIS are dictated by the desired time resolution for fluctuation measurements. To measure small fluctuations with one percent statistics at a sampling period of 1 microsecond requires about 10^{10} x-ray counts/second. At an average photon energy of 7 keV, this corresponds to a signal in the silicon surface-barrier detector of 3 μA, if the detection efficiency is 100 percent. Indeed, the central detector of the TFTR horizontal XIS typically registers signals of 1 - 15 μA, and the edge channels register 0.1 - 1 μA. This corresponds to an AΩ product (emitting area of plasma x solid angle of detector) of about 2×10^{-4}. For less demanding physics studies, such as measurement of lower frequency MHD fluctuations or the sawtooth inversion radius, the required signal levels could be reduced an order of magnitude, resulting in smaller x-ray optic and aperture sizes. In order to be sensitive to electron density fluctuations, the ITER XIS should respond in the x-ray energy range of at least 10 - 20 keV. To measure electron-temperature fluctuations, however, at a central temperature of 20 - 30 keV, the XIS region of sensitivity should extend to higher energies; an upper energy of 30 keV would be adequate, but 40 keV would be preferable.

The tolerable radiation-generated background level in the XIS detectors, again, depends on the frequency of the fluctuations to be measured and the accuracy desired. A study of the frequency-domain power spectral densities of both XIS signals and DD nuclear-radiation-induced background concluded that at a 1 nA DD neutron/gamma-ray background current the high frequency (>20 kHz) is ~ 1/10 the level associated with a strong plasma signal (in these shots about 1 μA); thus studies of such fluctuations would become impossible at radiation background currents much above 1 nA. Other studies, such as characterization of lower frequency and more intense fluctuations and location of the sawtooth inversion radius could be done at higher background levels, such as 10 nA or higher. The radiation noise characteristics in the TFTR XIS Si detectors has been presented in Ref. 4. The background noise level from neutrons in an XIS-type Si detector viewing the central chord of ITER with a neutron $V\Omega$ product ($A\Omega$ x plasma diameter) similar to that of the TFTR XIS is estimated to be about the same order as the signal level, or of order 10 μA. The lifetime of a Si surface-barrier detector in this case would be about one hour of ITER operation at 3000 MW if the damage limit is 10^{13} n/cm^2. Thus a reduction of the neutron flux by a factor of order 10^3 is desirable to insure both no loss of high frequency, low level fluctuation information and survival of the detectors for of order 10^3 hours of operation. The detector-damage requirement can be mitigated by use of radiation hard detectors, such as fast, gridded gas ionization chambers[22] or diamond detectors, but these detectors are not as optimal for x-ray measurements as Si detectors, and further development would be required.

In principle, the x-ray deflector most suitable for use with the XIS is the GI mirror because of its broad bandwidth and high x-ray power throughput. Both calculations[7] and measurements with a ML-based prototype XIS channel[23] on TFTR suggest that the $A\Omega$ product of an ML-based deflector system would need to be of order 100 times larger than that of the GI mirror for the same power throughput, because of the small spectral bandwidth of the ML element. On the negative side, however, is the small deflection angle of the GI

Table 1. Parameters of an ITER XIS channel using a grazing-incidence x-ray mirror just outside the biological shield, and a multilayer mirror between the ITER blanket and the biological shield.

Parameter	Units	GI Mirror	ML Mirror
Distance ITER to mirror	cm	700	250
Dist. mirror to exit coll.	cm	1000	350
First aperture width	cm	0.8	2
First aperture height	cm	10	15
Second aperture width	cm	1.8	3
Mirror height	cm	10	15
Mirror length	cm	40	30
Number of mirrors		5	4
Neutron-beam half width	cm	2.8	5
X-ray energies	keV	6 - 22	17 21 27 39
Incidence angles	mrad	3.7 - 5.5	8 - 18
Lateral x-ray deflection	cm	8.8 - 9.6	18 - 18.4
$V\Omega$ for TFTR	cm^3 sr	0.028	0.028
$V\Omega$ relative to TFTR		1.5	110
Fusion neutron power	MW	3000	3000
Neutron brightness x 5	(cm^2 s sr)$^{-1}$	2.5 x 10^{14}	2.5 x 10^{14}
Neutron flux at mirror	cm^{-2} s^{-1}	3 x 10^9	9 x10^{10}
Neutron-damage limit	n cm^{-2}	10^{18}	10^{18}
Mirror lifetime	hr	8 x 10^4	2.7 x 10^3

mirror, and in some situations, such as between ITER and the biological shield, the larger deflection angles available with the ML mirror may more than compensate for the lower hroughput. These ideas are illustrated in Table 1, which summarizes the characteristics of an XIS channel using a GI mirror located outside the biological shield, and one using a ML mirror radially inside the shield. The aperture and mirror sizes for the second case are chosen much larger than for the first case to compensate for the narrow bandwidth of the ML mirror. As a consequence, the neutron flux onto the reflective element is 30 times higher, and the lifetime shorter. However, even with the smaller deflected-path length available inside the biological shield, the lateral deflection of the x-ray beam relative to the neutron beam is 13 cm, compared to 6 cm for the GI mirror outside the shield, and the maximum x-ray energy reflected by the ML is about 40 keV, compared to 22 keV for the GI mirror.

Another interesting possibility for enabling XIS measurements is the x-ray "light pipe",[24] which is also based on grazing-incidence reflection. This is a 2-cm diameter bundle of small ID (280 μm) lead glass tubes up to 80 cm long. When embedded in a solid neutron/gamma-ray shield these elements provide a window for x rays that is, in effect, more penetrating or larger than for neutrons and gamma rays because of the larger $A\Omega$ for x rays due to reflection at small angles from the walls of the glass tubes or "capillaries". Laboratory measurements of the effective "gain" ($A\Omega$ for x rays relative to the geometrical $A\Omega$) for 70-cm long bundles showed values ranging from 100 to 150 at energies ranging from 4.5 to 10 keV. The length of 70 cm was calculated to result in a neutron background current of 1 nA in a Si detector when exposed to 30 MW of DT fusion neutrons from TFTR. A prototype device has been partially tested on TFTR. The good news is that no signature of the neutron noise has been seen at DT fusion powers up to 10 MW. Unfortunately, no x-ray signal has been seen above the noise level of the detectors. It is hoped that this failure is due to misalignment, since the angular pointing must be precise, and 90 percent of the viewing area originally assigned to the diagnostic has since been blocked by installation of another diagnostic. More sensitive photon counting detectors will be used to ascertain the x-ray throughput and whether other blockages have occurred.

IV. THE X-RAY PULSE-HEIGHT ANALYSIS DIAGNOSTIC

The x-ray signal levels required for the broad-band continuum and line-radiation spectrometer PHA diagnostic are very small relative to the surface brightness of x rays. The maximum acceptable detector count rates are $\sim 8 \times 10^4$ for the present TFTR system and $\sim 8 \times 10^5$ for new detectors with lower FET capacitance. Thus, the time resolution for measurements can be reduced from about 100 ms to 10 ms, making the system more attractive. The PHA, thus, has a very small $A\Omega$ product and would require only small penetrations in the biological shield; on TFTR a value of 10^{-8} is used for supershots and a maximum value of 2×10^{-7} is available.

The acceptable neutron background noise levels in the PHA can be estimated from TFTR data during DT shots. The measured neutron-noise level in the PHA Si(Li) detectors is about a factor of 20 too high for comfortable metal impurity and T_e measurements in DT supershots producing about 6.3 MW of fusion power (Q=0.2) or 1.8×10^{18} neutrons/s (Fig. 2). The neutron aperture, however, has an area of about 100 times that of the x-ray aperture used for these shots. Thus, good quality measurements could be made if the neutron aperture hole diameters were reduced from the present fixed values of 1.5 cm to values about twice that of the x-ray aperture diameters, which are 0.15 cm. To understand these conclusions we look at Fig. 2. These are raw spectra from three detectors of the PHA system integrated over 0.4 s during DT pulse 73268. The beam heating power was about 30 MW. The detectors had filter foils with e^{-1} transmission values E_c at 23.3, 12.2, and 8.6 keV for detectors 1, 2, and 3, respectively. The shaded regions are the regions of validity of the data, ranging from the energy E_{02} corresponding to an x-ray transmission of 2 percent to an energy 1.7 times that of the e^{-1} cutoff value (to avoid pulse-pileup distortion which is a maximum near twice E_c. The dashed curves represent the purely neutron/gamma-ray noise

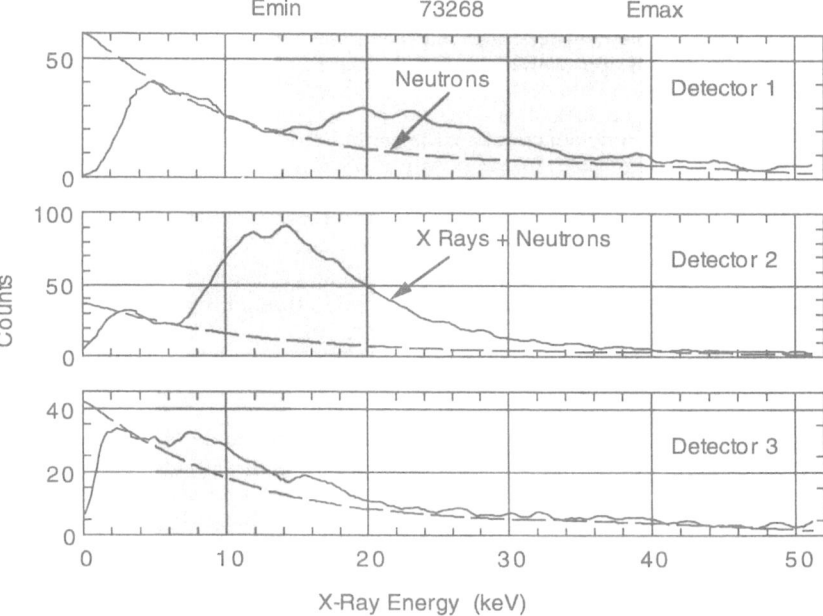

Figure 2. X-ray plus neutron spectra (solid curves) and approximate neutron-generated background (dashed curves) measured in 3 of the 6 TFTR PHA Si(Li) detectors during a shot producing 1.8 x 10^{18} 14-MeV neutrons per second.

level measured during a DT shot with the gate valve closed, adjusted to the level of the spectra below E_{02}; in non-DT shots no x-ray counts are seen in this region. Because of low statistics the noise reference spectrum was taken as an average of the signals from the six detectors. The levels for the dashed curves agree approximately with those resulting from scaling of the neutron-noise background spectrum by the total DT neutron production rate of shot #73268 relative to that of the reference shot, and for individual detectors making an adjustment for the known dead time of the amplifiers as a function of detector count rate. In fact, the measured count rate from neutrons during the background reference shot in Fig. 2 agrees with an $A\Omega$ product about 150 times that used for the x rays. Thus, we conclude from the spectra in Fig. 2 that at Q=0.2 in TFTR if the neutron and x-ray aperture sizes were identical, the background noise levels in Si(Li) detectors would be easily tolerable.

On ITER, however, directly viewing pulse-height spectrometry with Si(Li) and HPGe (high purity germanium) detectors is not feasible, possibly because of excessive noise but certainly because of detector damage. This statement may not be so clear for potential alternative detectors which may be more radiation hard. The measured response of HPGe detectors, which would be necessary in the 50 - 100 keV range, to 14-MeV neutrons is about a factor 1.5 to 4 higher than that of the Si(Li) detectors used on TFTR. Considering this higher response plus the higher neutron reactivity at ignition and the expected enhancement of neutron and gamma-ray fluxes in ITER due to multiple scattering of neutrons, it is likely that the signal-to-noise ratio in a directly viewing HPGe PHA channel in ITER would not be acceptable. One obvious, simple technique for reducing the neutron-generated background, which could be tested on TFTR, would be to locate a neutron/gamma-ray detector behind each x-ray detector and inhibit those counts in the x-ray detector which are coincident with a neutron or gamma-ray count. The test should be done on TFTR to provide a better measure of the maximum tolerable neutron count rate, regardless of the fact that directly viewing Si and Ge detectors are not applicable to ITER, because it would also be applicable to other potential alternate radiation hard detector types which may be more suitable for use on ITER.

It is estimated that the detector lifetimes of a Si and Ge based PHA-type system viewing ITER directly, e. g. if coincidence techniques could reduce the noise to acceptable levels, would be less than one hour at a fusion power of 3000 MW. Thus x-ray optical techniques would be necessary to adapt the present PHA to ITER. The neutron damage threshold (factor of about 5 increase in detector capacitance and leakage current) for Si(Li) and HPGe detectors is about 10^{11} and 10^{10} neutrons/cm^2, respectively.[3] In the case of TFTR, it is determined that as of August 15, 1995 the machine had produced an estimated 3 x 10^{20} 14-MeV neutrons in just those shots having neutron source strength values greater than 6 x 10^{16} n/s. The fluence of 14-MeV neutrons onto the PHA Si(Li) detectors is, thus, estimated to be about 3 x 10^9 cm^2, or about 3 percent of the Si(Li) damage threshold and 30 percent for HPGe (which are not presently used on TFTR). The fluence of DD neutrons is roughly the same level.

It is possible that alternate detectors, such as mercuric iodide, CdZnTe, GaAs, NaI, or x-ray microbolometers[25] might have higher damage thresholds than Si or Ge; an improvement of at least 3 orders of magnitude would be necessary to make the detectors attractive. Also, the signal to noise characteristics and the energy resolution must be considered to ascertain the applicability of alternate detectors. To monitor metal impurities such as Ti, Cr, Fe, and Ni, an energy resolution less than 250 eV FWHM is needed. To simply measure the continuum spectrum at energies from 10 to 100 keV for T_e and nonthermal electron studies, cruder resolution is acceptable.

Calculations were made for a PHA-type diagnostic using GI mirrors located both just outside the biological shield and between the shield and the ITER blanket. The throughput or $V\Omega$ product assumed was that provided by the largest aperture used on the TFTR PHA, which is 20 times larger than that used for supershots. The aperture and GI mirror dimensions, neutron beam halfwidth, lateral x-ray beam deflection, and neutron flux incident onto the mirror were calculated for various angles of incidence, assuming 3000 MW of fusion power and a total neutron flux 5 times higher than the virgin 14-MeV neutron flux. Due to space constraints, we will only briefly summarize the results. The aperture sizes and mirror widths are very small; e. g. for the "outside" configuration they are 0.4 x 1.0 cm^2 and 1.0 cm, respectively. The resulting neutron flux at the mirror is also small, resulting in a very long estimated lifetime (~ 10^6 hours). The neutron beam halfwidth at the wall is 0.8 cm, and the lateral x-ray beam deflection ranges from 7 to 10 cm for the cases illustrated, with maximum energies, respectively, of 25 and 17 keV for one deflection, and 35 keV for 2 deflections totaling 10 cm. Inside the biological shield the available path lengths are shorter, as are the aperture dimensions. It is much more difficult to achieve a reasonable deflection of the x-ray beam from the neutron beam; for example a lateral x-ray deflection of only 4.2 cm was calculated at 28 keV for two reflections.

V. THE X-RAY CRYSTAL SPECTROMETER

The required signal levels and radiation noise background rates for the XCS are also determined from experience. The XCS detectors can usually accept count rates up to 10^6 s^{-1}. Such a rate would provide a time resolution of 1 or a few ms in ion-temperature measurement. A background count rate of less than 5 percent of the x-ray rate should be a design goal, although measurements can probably be made at higher rates. Calculations of direct neutron scattering from the crystal, as well as experience with a large focal length, well shielded XCS at the JET tokamak,[9] suggest that the XCS could operate successfully with the crystal directly viewing the ITER plasma if the detector is well shielded. Charge-state distribution calculations with the MIST impurity transport code suggested that the impurity element krypton, which could be easily introduced into the plasma in controlled amounts, would be suitable for spectroscopic ion-temperature measurement of the plasma center at electron temperatures of 20 to 30 keV. Simulations of the resonance-line x-ray emissivity from H-like and He-like charge states show that with a relatively small concentration of Kr ($n_{Kr}/n_e \sim 10^{-4}$), modest sized apertures at ITER (1 x 7 cm^2) and through the biological shield and small crystal sizes (1 x 12 cm^2) would be adequate for detector count rates of a few times 10^5. Focusing errors are acceptably small for the case studied. Due to space limitations these results will not be reproduced here; the reader is referred to Ref. 6. Also high resolution measurements of Kr x rays on TFTR have shown that this element can be used for T_i measurement, and that the radiative cooling agrees with MIST simulations.[26,27] Of course, it would be wise to allow for initial operation at lower x-ray emissivities by providing for exchangeable aperture plugs of different sizes in the biological shield and the shield block outside ITER. More thorough studies of the detector background levels with realistic geometries and using MCNP codes would be desirable.

The use of x-ray optical elements to reduce the neutron flux striking the XCS crystal[28] and of doubly focusing crystals to maximize the XCS solid angle and sensitivity to low signal levels while minimizing the size of the x-ray and neutron beam[29] have been considered.

VI. DISCUSSION AND CONCLUSIONS

Virtually all the conventional tokamak x-ray diagnostics can be adapted to the harsh radiation environment of ITER, with limitations in some cases. For the XCS this adaptation means careful collimation and restriction of the neutron/gamma-ray beam falling onto the crystal and careful collimation and shielding of the detector from ambient and streaming radiation; for the XIS and PHA systems, it means addition of suitably designed x-ray optical elements or deflectors and careful collimation of the neutron beam and the viewing region of the detector or spectrometer, as well as shielding of the detector. The penetrations required in the radiation shield near ITER to view the x-ray emission are relatively small, ranging from less than 1 cm^2 for the broad-band photon-counting spectroscopy to about 1 x 10 cm^2 for a chord of the XIS and XCS systems. The radiation levels at the x-ray optical elements for the aperture sizes required for useful signals are quite acceptable. The estimated lifetimes at full power range from several thousands of hours for ML mirrors located inside the biological shield used with the XCS to 10^6 hours for GI mirrors used outside the biological shield with PHA-type systems.

The adaptation of the x-ray systems to ITER has some limitations. For example, the broad-band continuum spectroscopy will suffer an upper limit on x-ray energy somewhere in the 35 - 50 keV limit with GI-type mirrors, simply because the grazing angles become too small for substantial lateral deflection of the x-ray beam. This same limit makes it difficult to extend the XIS sensitivity beyond 25 - 30 keV, as would be helpful to optimize sensitivity to T_e fluctuations at a central T_e value of 30 keV. Other technical problems will have to be addressed. One is the design, fabrication, alignment, and maintenance of alignment of long, neutron-collimating blocks with narrow penetrations, as well as alignment of 20 - 40 cm long mirrors at grazing angles of order 3 - 30 mrad. The location and radiation-damage characteristics of the x-ray transmitting vacuum-boundary window have not yet been considered. Beryllium is the material usually used for these windows, but the radiation-damage characteristics have not been considered. If Be should be easily damageable then the window should be extended outside the ITER port cover plate to a location within or beyond a substantial thickness of neutron shielding to extend its lifetime.

Additional research and development would be helpful in defining the systems discussed here. Further studies of the radiation scattering into the instrument entrance aperture from the point of impact of the neutron beam on the face of the collimator should be done, especially at small distances between the neutron beam and the aperture. Background reduction by coincidence with neutrons or gamma rays measured in detectors behind the XCS and PHA detectors should be studied. The search for alternative detectors with higher radiation tolerance should continue, and further optimization of x-ray microbolometers and supermirrors should be pursued.

ACKNOWLEDGEMENTS

The authors gratefully acknowledge the continuing support and helpful discussions with K. M. Young and helpful discussions with L.-P. Ku, and K. Joensen. This work was supported by the U. S. Department of Energy under contract number DE-AC02-76-CHO-3073.

REFERENCES

1   K. W. Hill, M. Bitter, et al., Tokamak Physics Studies Using X-Ray Diagnostic Methods, in *Proceedings of the Course and Workshop on Basic and Advanced Fusion Plasmas, Diagnostic Techniques*, (Varenna (Como), Italy, September 1986), (Monotypia Franchin, Citta' di Castello, Italy and Office for Official Publications of the European Communities, Luxembourg, Belgium, 1987), edited by P. Stott, D. K. Akulina, G. G. Leotta, E. Sindoni, and C. Wharton, Vol. I, pp. 169-200

2   K. W. Hill, P. Beiersdorfer, et al., Tokamak X-Ray Diagnostic Instrumentation, *Ibid.*, pp. 201-226.

3   S. von Goeler, K. W. Hill, et al., X-Ray Diagnostics for TFTR, Proceedings of the Course, Diagnostics for Fusion Reactor Conditions, Varenna, Italy, September 1982, Vol. I, pp. 69-85

4   K. W. Hill, H. Adler, et al., Analysis of nuclear-radiation-induced noise in spectroscopic and x-ray diagnostics during high power deuterium-tritium experiments on the tokamak fusion test reactor, *Rev. Sci. instrum.* **66**, 913 (1995).

5   K. W. Hill, K. M. Young, et al., ITER X-Ray Diagnostic Studies, *Rev. Sci. Instrum.* **63**, 5032(1992)

6   K. W. Hill, M. Bitter, et al., Design Studies for ITER X-Ray Diagnostics, Princeton University, Plasma Physics Laboratory Report PPPL-3034 (January 1995)

7   K. W. Hill, M. Bitter, et al., ITER X-Ray Fluctuation Diagnostic Possibilities, Princeton University, Plasma Physics Laboratory Report PPPL-3008 (February 1995)

8   R. Barnsley et al., Bragg spectroscopy of impurities during the JET preliminary tritium experiment, *Rev. Sci. Instrum.* **63**, 5023 (1992)

9   R. Barnsley et al., *X-ray spectroscopy for ITER*, presented by N. Peacock at the ITER Progress Meeting on Spectroscopic Systems, Kurchatov Institute, Moscow, Russia, Nov. 29 - Dec. 1, 1994.

10   K. McGuire, R. J. Colchin, et al., Diagnostic Applications of the TFTR XIS System, *Rev. Sci. Instrum.* **57**, 2136 (1986)

11   M. Bitter, S. von Goeler, et al., Doppler-Broadening Measurements of X-Ray Lines for Determination of the Ion Temperature in Tokamak Plasmas, *Phys. Rev. Lett.* **42**, 304 (1979)

12   K. W. Hill, M. Bitter, et al., Studies of Impurity Behavior in TFTR, *Nucl. Fusion* **26**, 1131 (1986)

13   S. von Goeler, J. E. Stevens, et. al., Angular Distribution of the Bremsstrahlung Emission During Lower-Hybrid Current Drive on PLT, *Nucl. Fusion* **25**, 1515 (1985)

14   N. Madden, Lawrence Berkeley National Laboratory, private conversation.

15   D. H. Bilderback and S. Hubbard, X-Ray Mirror Reflectivities from 3.8 to 50 keV, Part II - Pt, Si, and Other Materials, *Nucl. Instrum. Methods*, **195**, 91 (1982)

16   Martin Elvis, Daniel G. Fabricant, and Paul Gorenstein, Grazing Incidence Imaging from 10 to 40 keV, *SPIE* Vol. **830**, 296 (1988)

17   A. F. Jankowski and D. M. Makowiecki, Manufacture, Structure, and Performance of W/B4C Multilayer X-Ray Mirrors, in *X-Ray Multilayers for Diffractometers, Monochromators, and Spectrometers*, Ed. Finn E. Christensen, (SPIE, Bellingham, WA 1988), Vol. 984, p. 64

18 Anthony Burek, Crystals for Astronomical X-Ray Spectroscopy, *Space Science Instrum.* **2**, 53 (1976)

19 K. D. Joensen, F. E. Christensen, et al., Medium-Sized Grazing-Incidence High-Energy X-Ray Telescopes Employing Continuously Graded Multilayers, Proc. SPIE **1736**, p. 236 (1992)

20 Peter Hoghoj, Eric Ziegler, et al., Broad-Band Focusing of Hard X-Rays using a Supermirror, in *Physics of X-Ray Multilayer Structures*, 1994 Technical Digest Series (Optical Society of America, 1994), Vol VI, p. 142

21 K. D. Joensen, P. Hoghoj, et al., Multilayered supermirror structures for hard x-ray synchrotron and astrophysics instrumentation, in *Multilayer and Grazing Incidence X-Ray/EUV Optics II*, SPIE proc. **2011**, (1993)

22 M. A. Goldman, K. W. Hill, et al., A Gridded Ionization Chamber for Detection of X-Ray Wave Activity in Tokamak Plasmas, *Rev. Sci. Instrum.* **56**, 349 (1985)

23 J. A Penkethman, Layered Synthetic Microstructure X-Ray Mirror Focusing Instrument: Bent Silicon Wafer Substrate, *Optical Engineering* **27**, 99 (1988)

24 D. Parsignault and A. S. Krieger, X-Ray Fiber Optics from 60 eV to 10 keV, in *X-Ray Detector Physics and Applications*, edited by R. Hoover, (SPIE, Bellingham, WA 1992), Vol. 1736

25 S. H. Moseley, J. C. Mather, D. McCammon, Thermal Detectors as X-Ray Spectrometers, *J. Appl. Phys.* **56**, 1257 (1984)

26 M. Bitter, H. Hsuan, et al., Spectra of Heliumlike Krypton from Tokamak Fusion Test Reactor Plasmas, *Phys. Rev. Lett.* **71**, 1007 (1993)

27 M. Bitter, H. Hsuan, et al., X-Ray Spectra of Heliumlike Krypton as a Potential Ion-Temperature Diagnostic for the International Thermonuclear Experimental Reactor (ITER), *Atomic and Plasma-Material Interaction Processes in Controlled Thermonuclear Fusion*, (Elsevier Science Publishers, 1993), edited by R. K. Janev and H. W. Drawin, pp. 119-133

28 V. A. Bryzgunov, A. B. Gil'varg, and A. N. Svetchkopal, X-Ray Optics for ITER Spectrometer, Russian Research Center Kurchatov Institute report number IAE-5645/14 (Moscow 1993), presented at the Sixth National Topical Conference on High Temperature Plasma Diagnostics, St. Petersburg, May 26 1993

29 M. Bitter, B. Fraenkel, K. W. Hill, H. Hsuan, and S. von Goeler, Numerical Study of the Imaging Properties of Doubly-Focusing Crystals, paper 3.22, presented at the 10th Topical Conference on High Temperature Plasma Diagnostics, 8-12 May, 1994, Rochester, N. Y, *Rev. Sci. instrum.* **66**, 530 (1995)

APPLICATIONS OF X-RAY SPECTROSCOPY TO ITER

R Barnsley[1], R M Giannella, K D Lawson[2], N J Peacock[2],

JET Joint Undertaking, Abingdon, OX14 3EA, UK.
[1]Leicester University, Leicester, LE1 7RH, UK.
[2]UKAEA Culham Laboratory, Abingdon, OX14 3DB, UK.

INTRODUCTION

The soft x-ray spectral region between 0.1 nm and 10 nm includes the peak of the radiated power and contains a wide range of diagnostic information about the temperature and density of electrons and impurity ions. Spectroscopic coverage of this band will be essential for ITER.

Two main applications of x-ray spectroscopy will be important. Firstly, impurity species monitoring for impurity control, wall protection and the derivation of Prad and Zeff components. This requires broad-band coverage between about 0.1nm and 10nm. Secondly, derivation of ion temperature and plasma rotation, which requires high-resolution Doppler broadening and shift measurements of one or more narrow spectral reqions. At JET, these two functions are performed by independent instruments, which operated successfully during the preliminary tritium experiment, and which together form the basis of an x-ray spectroscopy installation for ITER. The main requirement is for a vacuum beamline leading to a shielded bunker outside the cryostat. In such a location the problems of crystal life and detector shielding will be manageable.

Spatial resolution will be particularly difficult to achieve on ITER, and a system cannot be specified until port access is more closely defined. Several options will be discussed, with multiple direct lines of sight being preferred.

ITER REQUIREMENTS

ITER diagnostic requirements[1] are divided into three measurement categories; (i) machine protection and control, (ii) performance evaluation and optimisation, and (iii) additional measurements for physics understanding. Parameters derivable from x-ray measurements are shown in table 1, which distinguishes between central-chord measurements and those requiring spatial resolution.

All impurities have lines below about 10 nm, and can be monitored by a broad-band x-ray spectrometer of medium resolving power ($100 < \lambda/\delta\lambda < 1000$). Doppler line width and shift measurements require high resolving power of about 10^4, and are limited to below about 1 nm (and hence to $Z \geq 10$) by the properties of available crystals. The measurements requiring low or no resolving power can be made with pulse-height analysis and energy integrating systems respectively, or by a hybrid of the two.

Diagnostics for Experimental Thermonuclear Fusion Reactors
Edited by P. E. Stott *et al.*, Plenum Press, New York, 1996

For spectroscopy between 0.1 nm and 10 nm, a direct vacuum view of the plasma is required. Vacuum isolation windows are possible, but must be thin below about 1 keV, 1 μm of polypropylene being typical. All the spatially resolved measurements can be made at energies above about 5 keV, allowing the use of relatively thick windows of materials such as beryllium.

Table 1. Measurement category, and parameters derivable from x-rays.

Central chord		E/δE	Spatially resolved		E/δE
(i)	Plasma rotation	H	(i)	Plasma position	N
(i)	Impurity monitor	M	(ii)	T_e profile	L
(i)	Z_{eff} (line av.)	L	(ii)	Z_{eff} profile	L
(ii)	Central T_i	H	(ii)	T_i profile	H
(ii)	Impurity density	M	(iii)	Core MHD	N
			(iii)	T_e fluctuations	L

(Required resolving power E/δE: **High, Medium, Low, None.**)

EXPERIENCE AT JET

JET has three x-ray spectroscopy beamlines, all the instruments being situated in secondary shielding enclosures outside the torus hall. An energy-dispersive semiconductor detector[2] monitors the spectrum between about 1 keV and 50 keV, two different crystal spectrometers share a second beamline to survey impurity line-radiation between about 0.1 nm and 10 nm, and a 25 m curved-crystal spectrometer[3] makes high-resolution Doppler broadening and line-shift measurements. The impurity survey beamline will be discussed in some detail because it exemplifies most of the features required for ITER.

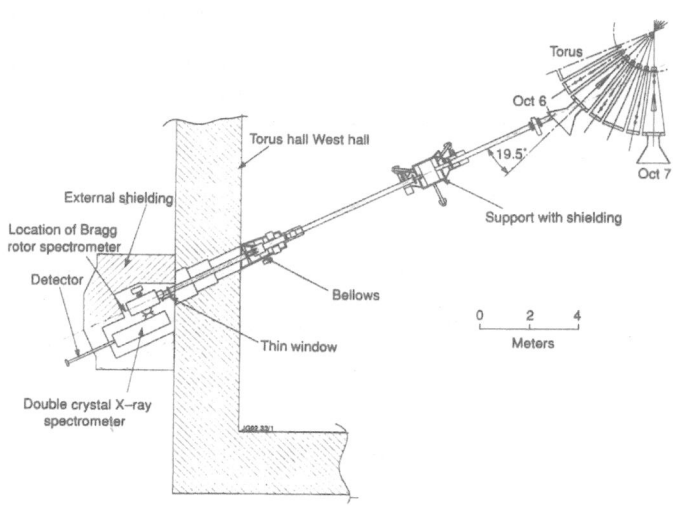

Fig.1 The JET vacuum beamline, bunker, and x-ray survey spectrometers.

The JET x-ray survey spectroscopy beamline

The beamline (fig.1), the spectrometer chamber, and their respective vacuum systems, are all designed to contain tritium. There is an isolation valve at the torus end, but no window, this being located inside the secondary shielding bunker. The window of 1 μm polypropylene is supported on both sides by a fine mesh and steel ribs. A specimen has been tested to failure at a differential pressure of 3 bar, though the design could be modified to withstand a much higher pressure. Under normal operating conditions its purpose is to isolate the torus vacuum from the spectrometer vacuum and to prevent free flow of tritium into the spectrometer during D-T discharges. Two instruments share the available aperture: a single-reflection Bragg rotor spectrometer[4], and a double-reflection shielded spectrometer[5].

The Bragg rotor spectrometer contains two independent sections. An almost complete spectral survey between 0.1 nm and 10 nm is provided by a selection of crystals and multilayers mounted on a hexagonal rotor, the diffractors being scanned sequentially. The spectrum (fig.2) is relatively simple, consisting mainly of H-like and He-like lines of low and medium-Z impurities, well separated by regions of continuum. An exception is the region between about 0.75 nm and 1.5 nm, which is crowded with L-shell lines of medium-Z metal impurities such as Fe and Ni. Because of the wide spectral coverage, unexpected impurity influxes can be monitored. The scan is programmable, and can cover the full spectrum, reciprocate over a narrower band, or be monochromatic.

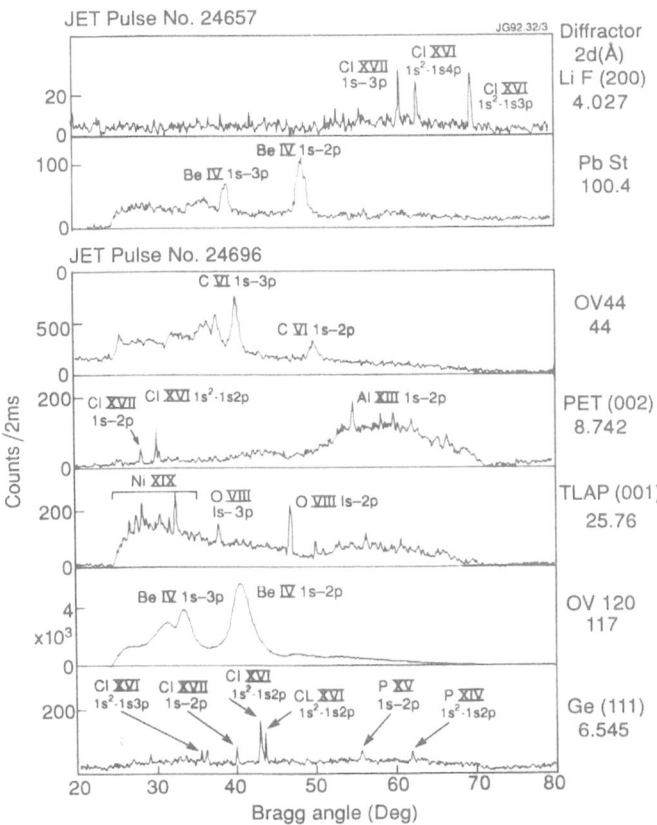

Fig.2 Survey spectra from JET, from a range of crystals and multilayers.

The monitor section of the instrument uses a side-by-side array of four crystals and multilayers, each with its own detector channel, to monitor representative lines of the main intrinsic impurities. An important use of the data from the Bragg monitor channel has been to fit line intensities to the total radiated power[6]. An example of such a fit is shown in figure 3, where the impurity radiated power components have been derived for a reference discharge. During the discharge the plasma was moved to different positions in the vessel to evaluate the wall conditioning.

The double-crystal spectrometer achieves a high degree of shielding by using two reflections from identical crystals in parallel, resulting in a labyrinth optical path between the beamline and a fixed detector.

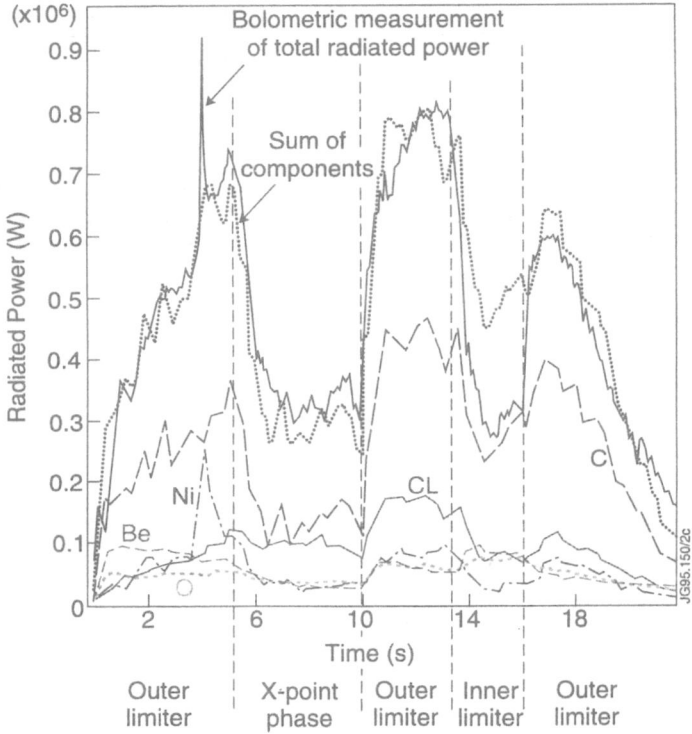

Fig.3 Empirical fit of x-ray line intensities to the total radiated power for a JET reference discharge.

JET tritium experiment

The Bragg rotor spectrometer and the double-crystal spectrometer together monitored all the main impurities throughout the JET preliminary tritium experiment[7] (PTE). Procedures were developed for the operation and subsequent clean-up of the beamline and spectrometer chamber in the presence of tritium.

The double-crystal instrument showed excellent shielding from nuclear radiation, even though the full shielding was not installed. For a D-T neutron production rate of 6.10^{18}/s, the peak neutron-induced noise was about 10^3 count/s, with about 400 count/s in the discriminated energy band, compared with peak count-rates of 10^5 count/s in the main lines. Figure 4 shows the detector pulse-height spectrum for D-D and D-T discharges, and illustrates the excellent low-noise properties of a double reflection system where, for D-D discharges, only diffracted photons reach the detector. The prompt background (fig.5) closely follows the neutron production, followed by a delayed background of similar intensity. The background showed broad agreement with design calculations for shielding[8], from which it is estimated that installation of the complete shielding will give an improvement of a factor 200. Optimisation of the detector aperture and pulse processing should give a further factor 5 improvement, resulting in a shielding performance adequate for ITER.

The Bragg rotor spectrometer detector was very close to the direct line of sight to the plasma, resulting in a peak background count-rate of 3.10^5 count/s, making measurement of the weaker lines marginal. Some improvements, probably a factor 10, can be made, but it is unlikely that such an instrument with a close-coupled detector will be suitable for the D-T phase of ITER. No pulse-height background measurements were made for the 25 m Johann spectrometer, but no increase in background was detectable in the wavelength spectrum during the tritium burn.

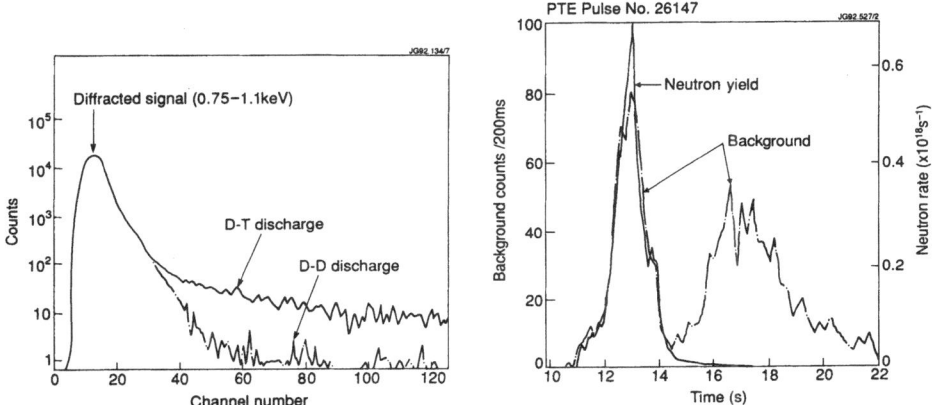

Fig.4 (left) Pulse-height spectra from the double-crystal spectrometer.
Fig.5 (right) The double-crystal spectrometer background in the pulse-height window above 5 keV, and the neutron production, for a JET D-T discharge.

X-RAY SPECTROSCOPY BEAMLINE FOR ITER

To fulfil the ITER requirements, two x-ray spectrometers are required; a medium-resolution survey instrument with full coverage between about 0.1 nm and 10 nm, and a high-resolution instrument with a narrow coverage in the range between about 0.1 nm and 0.2 nm.

A vacuum beamline is required, with integrated high-resolution and survey instruments sharing the aperture in a secondary shield outside the bio-shield (fig.6). All the mechanisms and x-ray optics, including the windows, are located outside the bioshield. As well as being the most reliable and maintainable, this scheme is the most adaptable to future changes in optics and detectors. The installation is compatible with the clearances available on the equatorial level of ITER, and requires an aperture of 180 x 100 mm^2 at the first wall, the height being determined by the free bandwidth of the curved-crystal instrument. The beamline is tilted down at 5° to save space outside the bioshield.

The use of VUV/XUV instruments is considered difficult, due to their need for a windowless sight-line and relatively wide angular view. However, if combined with doubly-curved mirrors, they could operate with pinhole apertures, or with astigmatic focusing and differentially-pumped orthogonal slits. Grating spectrometers might then be acceptable if, as with the x-ray instruments, they were made part of the tritium containment. Given the

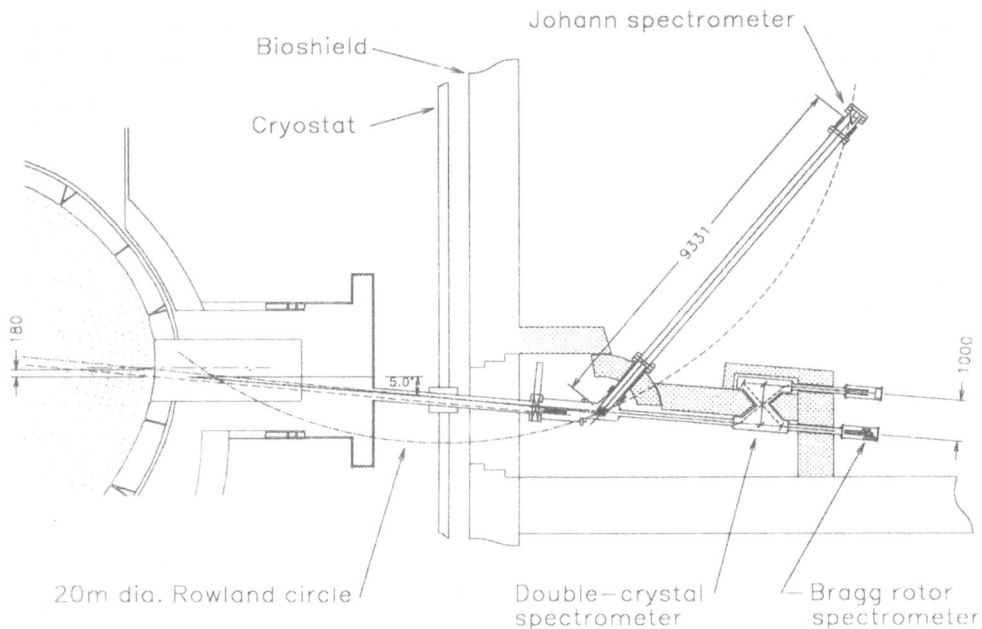

Fig.6 Integrated vacuum spectroscopy beamline for ITER.

importance of the VUV/XUV region[9], this design provides aperture for VUV and XUV grating spectrometers, preceded by gold and multilayer mirrors respectively.

A study of neutron damage to multilayer mirrors[10] has shown that they were degraded but functional after a neutron fluence of $1.1 \cdot 10^{19}/cm^2$. Depending on the beamline geometry, neutron fluxes up to about $10^{11}/cm^2 s$ can be expected at the first diffractor of an instrument sited outside the bioshield, implying a comfortable safety margin. Further studies[11] are required of radiation damage to crystals, multilayers and mirrors, but it is assumed here that outside the bioshield crystal damage will not be a serious problem.

X-ray survey spectrometer

The double-reflection spectrometer (fig.7) incorporates several changes relative to the JET double-crystal instrument. Four channels are necessary for full coverage between 0.1 nm and 10 nm without changing crystals. For the multilayer channels, the parallelism constraints are coarse, and a simple pantograph-like mechanism will suffice, instead of a sophisticated servo-system as on the JET instrument. By eliminating the high-resolution collimator, and by using crystals with relatively broad (~1-10 arcmin) diffraction profiles, the positioning constraints can be relaxed from the present arc-second level without affecting the survey capability of the system. This relaxation will permit a faster scan time, the aim being less than 250 ms for a full Bragg angle survey (1 s at present) and 10 ms for monitoring line-scans.

The total aperture of 100 x 100 mm^2 is shared between four channels, each 25 x 100 mm^2 and independently covering the Bragg angle range between $22.5°$ and $67.5°$. The gas flow proportional counter, with window and gas depth optimised for each channel, has a count-rate capacity exceeding $10^7/s$ in each channel. A suitable diffractor arrangement is shown in table 2.

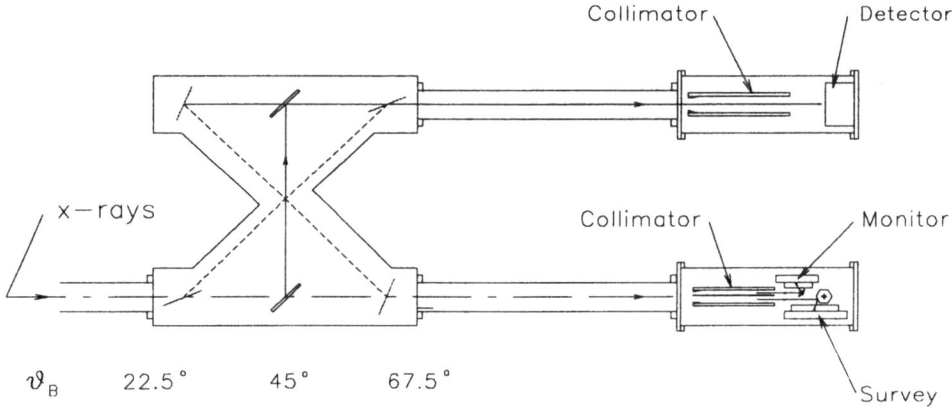

ϑ_B 22.5°   45°   67.5°

Double—crystal spectrometer   Bragg rotor spectrometer

Fig.7 X-ray survey spectrometers for ITER.

Table 2. Diffractors for the double-reflection x-ray survey spectrometer.

Diffractor	2d (nm)	λ_{min} (nm)	λ_{max} (nm)	$\lambda/\delta\lambda$	Impurities
Multilayer	9.0	3.44	8.3	20	BeIII, BeVII.
Multilayer	4.5	1.72	4.15	40	CV, CVI, NVI, NVII.
TAP(001) (002)*	2.576	0.98 0.49	2.38 1.19	600	OVII, OVIII. L-shell Cr, Fe, Ni.
PET(002) (004)*	0.8742	0.33 0.17	0.81 0.40	700	Al, Si, S, Cl, Ar. Ar, K, etc.

* Higher orders discriminated and stored separately.

Multiple reflection techniques, while providing good shielding, inevitably restrict sensitivity, bandwidth and time resolution. Therefore, for the non-active phase of ITER, a simple Bragg survey spectrometer should view the plasma directly, using that portion of the aperture later to be used by the double-reflection spectrometer (fig.7). The superior sensitivity and flexibility of this instrument will be important for plasma commissioning, and in the achievement of a low Z_{eff}.

High-resolution x-ray crystal spectrometer

This instrument (fig.6) is very similar to the one at JET. The use of a relatively low central Bragg angle of 27.5°, adjustable between 25° and 30°, reduces the dispersion and gives a wider bandwidth for a given beamline divergence. Smaller movements are then required when changing wavelengths, and the detector can be further back from the bio-shield. Within this range, H-like and He-like lines of Cr, Fe, Ni and Kr can be observed using first and second order reflections from a single Ge(220) crystal (table 3). At the chosen Bragg angle, a crystal radius of 20 m is about the largest which can be accommodated in the space outside the bioshield, giving a mean crystal-detector distance of 9.5 m. This geometry places the Rowland circle virtual entrance slit about 1 m away from the plasma, inside the shielding plug, which is sufficient to ensure that the whole crystal is illuminated at all wave-lengths. The beamline is shorter than at JET, allowing a wider divergence of

0.9° which, combined with the reduced dispersion, gives an increase in the spectral window from 0.5% to 3%, thereby encompassing a complete He-like Fe spectrum. For a crystal aperture of 120 x 60 mm^2, and an Fe concentration of $10^{-4}.n_e$, count-rates of several MHz are estimated in the Fe Ly$\alpha_{1,2}$ lines.

The detector is a sealed, Xe-filled multi-anode proportional counter with 1-D position resolution of 0.5 mm. Each anode has an individual amplifier-discriminator channel capable of count-rates exceeding 10^6/s.

Table 3. Settings for the high-resolution x-ray crystal spectrometer.

	Ge(220) 2d = 0.400 nm		
Transition	λ (nm)	θ$_B$	
Fe XXV "w"	0.18503	27.55°	
Fe XXV "z"	0.18689	27.84°	
Fe XXVI Lyα_1	0.1778	26.39°	
Fe XXVI Lyα_2	0.1784	26.49°	
Kr XXXV "w"	0.0946	28.23°	(2nd order)
Kr XXXV "z"	0.0955	28.52°	..

For a central T$_e$ of about 20 keV, there will be significant emmission from H-like ions of the medium-Z metals such as Fe and Ni, and therefore routine Kr puffing may not be necessary solely to measure T$_I$.

It is possible to isolate the curved crystal from a direct view of the plasma by using a graphite or multilayer preselector, either of which could pass the necessary band-width of a few percent. However, given that for this instrument crystal life and detector noise are not expected to limit performance, it is better to accept some activation of the crystal and alignment mechanism rather than the losses of sensitivity and flexiblility incurred by using a preselector. At JET the crystal focus is checked regularly, but has not needed adjustment since about 1990.

SPATIALLY RESOLVED X-RAY MEASUREMENTS

In spite of the importance of spatially resolved x-ray measurements, it is not possible to make firm proposals until it is known what degree of blanket and cryostat penetration will be feasible. Here we discuss some options.

Multiple direct-view chords

The preferred option for sensitivity and flexibility is to have multiple chords with direct views of the plasma. As shown in figure 8a, a wide range of views can be achieved via small penetrations to the cryostat and bioshield. Vertical views would be particulary useful if they can be achieved. One option is to provide a common fan-array for several diagnostics, including high-resolution x-ray spectroscopy, x-ray imaging and PHA, and a neutron camera[12].

Multiple reflections within shielded plug

A useful range of chordal views is possible using a variation of the proposed visible/IR shielding plug[13] (fig.8b). Multiple Bragg reflections within such a plug would transmit a narrow bandwidth of about 3%, with transmission between about 1% and 10%. A similar principle has been demonstrated by the double-crystal spatial-scan mechanism at JET[14]. The above-mentioned study of neutron damage to diagnostic mirrors suggests that the first diffractor in the shielding plug would have a life equivalent to about 1000 D-T discharges.

Single reflection from curved diffractor panel

An option which gives a full view of the plasma via a quasi-parallel sight-line, but with no moving parts, is shown in figure 8c. A curved graphite panel close to the plasma relays a different wavelength for each chord. Graphite[15] is a good candidate for such a close-coupled diffractor, because it has a relatively broad diffraction profile, and its lattice spacing (2d=0.671nm) gives access to transitions from ions in widely different temperature regions, ranging from He-like Cl to H-like Ni. Provided overlapping diffraction orders are avoided, the broadband spectrum from a pulse-height-analysis system will contain two energies for each chord, allowing the T_e profile to be derived.

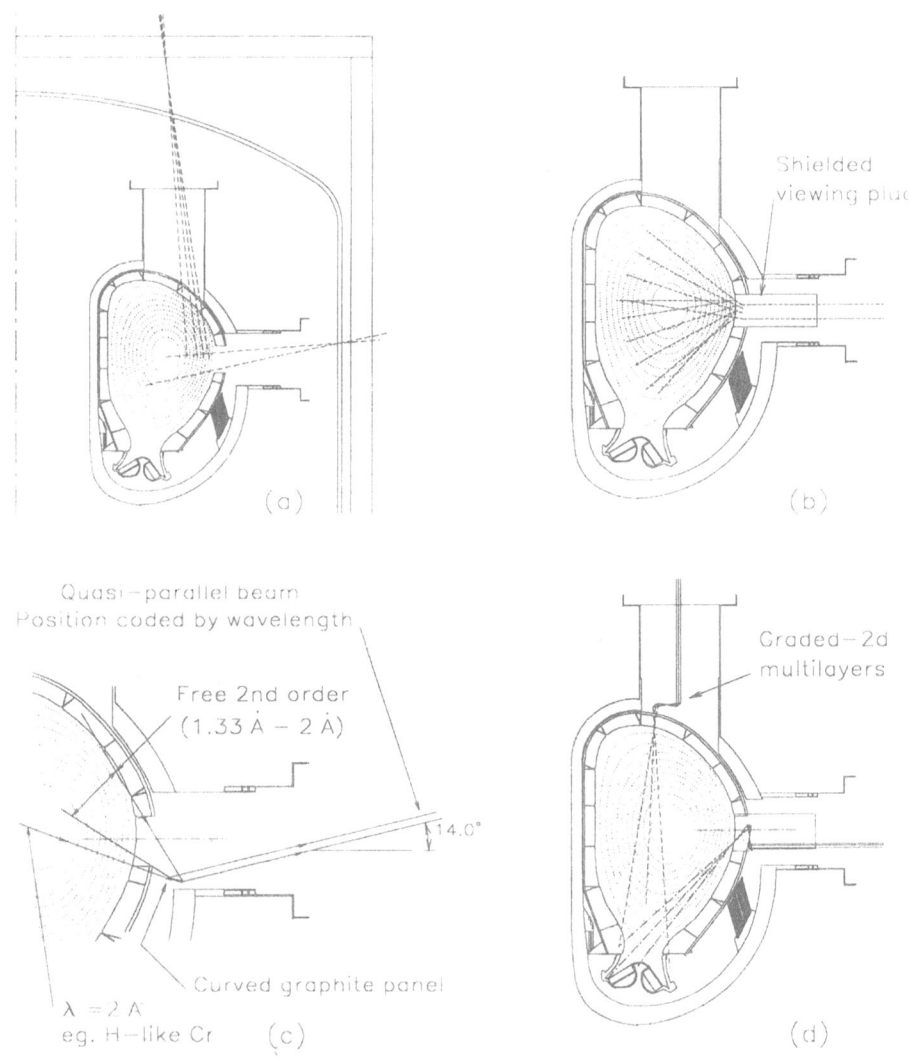

Fig.8 Options for spatial resolution; a) Multiple direct views,
b) Double-reflections within shielded plug, c) Curved graphite panel,
d) Divertor imaging with graded-2d multilayers.

DEVELOPMENTS

Development and testing is required in several areas, including neutron damage to diffractors and thin windows, advanced pulse processing for improved background rejection, and real-time data reduction. The ITER system must be flexible enough to accommodate future developments in x-ray optics and detectors, including such areas as multilayer gratings and cryogenic x-ray bolometers.

Developments at JET

The JET instruments are being upgraded to improve their spectral coverage, time resolution and signal-to-noise ratios for future tritium experiments. It is planned to fit two crystals and detectors to the high-resolution spectrometer to allow simultaneous observation of H-like and He-like Ni. The monitor section of the Bragg rotor spectrometer will be fitted with a more remote and better-shielded detector, and its spectra will be analysed and displayed in real-time.

Detectors

Several x-ray detector developments are being driven by the needs of x-ray astronomy. X-ray CCD arrays are being developed with larger areas, and improved low and high energy response[16]. X-ray bolometers[17,18] and superconducting tunnel junctions[19] are being developed, and should be capable of 10^{-3} energy resolution, sufficient to act as impurity monitors. They may also be useful as ex-vessel high-sensitivity bolometer reference channels.

An array of proportional counters would combine the functions presently filled by diode arrays for x-ray tomography, and by semiconductor detectors for energy dispersive spectroscopy. Gas proportional counters, extensively used for pulse-height spectroscopy in x-ray astonomy[20,21], can have high count-rate capability ($>10^7$/s), useful energy resolution (~0.1), good background rejection, and are intrinsically radiation hard.

X-ray optics

Multilayer mirrors[22,23] are a major development area, and have high reflectivity, with bandwidths suitable as preselectors either with crystal or grating instruments. In x-ray astronomy, crystal arrays will be coated with multilayers to obtain two or more simultaneous pass-bands[24], a technique which may be necessary on ITER if sight-lines are at a premium. Graded-2d multi-layers[25], where the 2d varies across the diffractor surface, have been proposed as constant deviation spectrometers, and could be used as pre-selectors for an XUV grating spectrometer. Multilayer mirrors have obtained monochromatic soft x-ray images of the sun[26], and could similarly image the divertor, as illustrated in figure 8d.

Required developments

Further studies are needed of neutron and other damage to diffractors, particularly those which may be located close to the plasma. Ray tracing is usually necessary to determine the sensitivity and bandpass of spectrometers and relay optics which use multiple Bragg reflections.

CONCLUSION

Since the ITER CDA study[27,28], progress has been made at JET toward the design of a soft x-ray/xuv system for ITER. All the features necessary for the central-chord soft x-ray system have now been demonstrated, and these are incorporated in the proposed x-ray spectroscopy beamline. The instrumention is completely located outside the bio-shield – an important feature for category (i) measurements.

It is not possible to specify a spatially resolving system until the vessel access constraints are finalised. Several options have been considered, with multiple direct views being much preferred, possibly shared with other diagnostics such as a neutron camera. Apert from the multiple direct view solution, the spatial-view options contain aspects which have yet to be demonstrated, and these major implications for the in-vessel design.

Significant developments can be expected in x-ray optics and detectors during the life of ITER, and any x-ray diagnostic system must be adaptable enough to accommodate relevant improvements .

References

1 A E Costley et al, ITER plasma diagnostics, to be publ. in: *Controlled Fusion and Plasma Physics,* Proc. 22nd EPS conf. Bournmouth, July 1995.

2 D Pasini et al, The JET x-ray pulse-height analysis system, *Rev.Sci.Instrum,* **49** 693 (1988).

3 R Bartiromo et al, *Rev Sci Instrum.* 60 237 (1989).

4 R Barnsley, K D Evans, N J Peacock, N C Hawkes, Bragg rotor spectrometer for tokamak diagnostics, *Rev. Sci. Instrum.* **57**(8) 2159 (1986).

5 R Barnsley, U Schumacher et al, Double crystal spectroscopy at JET, *Rev. Sci. Instrum.* 62(4) 889-898 (1991).

6 K D Lawson et al, Proc. 17th EPS Conf. *Controlled Fusion and Plasma Plasma Physics*, Amsterdam. Vol III 1413 (1990).

7 R Barnsley et al, Bragg spectroscopy during the JET preliminary tritium experiment, *Rev. Sci. Instrum.* 63(10) 5023 (1992).

8 R B Thom, Radiation levels in the vicinity of the KS1 spectrometer, AEA Winfrith rept. RPD/RBT/1107 (1985).

9 N J Peacock et al, this conference.

10 E H Farnum et al, Neutron damage to diagnostic mirrors, *J. Nucl. Materials,* **219** (1982) 63.

11 N J Peacock et al, Two-axis goniometer for reflectivity measurements of x-ray diffractors used in fusion research. *Rev. Sci. Instrum.* 66(2) 1175-1179 (1995).

12 F Marcus, this conference.

13 ITER Diagnostic Group report, S 55 RE 95-07-11 F 1.

14 U Schumacher et al, Continuously space-resolved x-ray spectroscopy at JET, *Rev. Sci. Instrum.* 60(4) 562-566 (1989).

15 K D Evans et al, Calibration data for the Ariel 5 Bragg spectrometer, *Space Sci. Instrum.* 2 313-323 (1976).

16 A D Holland et al, The MOS CCDs for the european photon imaging camera, *SPIE* **2006/3** (1993).

17 S H Mosely et al, X-ray microcalorimeters - principles and performance, in: *Proc. ESA Symposium on Photon Detectors, Nordwijk, Netherlands, Nov 1992.* ESA SP-356 (1992).

18 D McCammon et al, Thermal calorimeters for high resolution x-ray spectroscopy, *Nucl. Inst. Meth.* **A326** 157-165 (1993).

19 P Hübner et al, Superconducting tunnel junctions as photon counting detectors, *SPIE* **2006,** 308-323 (1993).

20 G W Fraser, *X-ray Detectors in Astronomy,* Cambridge Univ. Press (1989).

21 M J L Turner et al, *Publ. Astron. Soc. Japan.* **41** 345-372 (1989).

22 W Moos et al, Rev. Sci. Instrum. 61(10) (1990) 2733-37.

23 T W Barbee Jnr. AIP conf. proc. 75. Low energy x-ray diagnostics. Eds. D T Attwood and B L Henke (1981).

24 F E Christiansen et al, Objective crystal spectrometer (OXS) for the Spectrum-X-γ satellite, in: *EUV, X-ray, and Gamma-Ray Instrumentation for Astronomy.* SPIE **1344** 14-22 (1990).

25 D J Nagel, J V Gilfrith, T W Barbee, *Nucl. Inst. Meth.* **195** 63 (1982).

26 M E Bruner et al, Soft x-ray images of the sun using normal incidence optics, *Journal de Physique,* **49** C1-115 (1988).

27 R Barnsley, R Giannella, M Stamp, P Thomas. Visible, grating VUV/XUV and crystal spectroscopy on ITER. ITER-IL-PH-7-0-80, (1990).

28 ITER Diagnostics (CDA) V Mukhovatov, H Hopman et al, ITER Documentation Series, No 33 (1990).

DIAMOND DETECTOR BASED SXR ARRAY FOR ITER

Andrey G.Alekseev, Vladimir N.Amosov, Vladimir S.Khrunov[1], Anatoli V. Krasilnikov, Dmitry V. Portnov, Albert Yu. Tsoutskikh, Evgeniy G. Utjugov

Troitsk Institute of Innovating and Fusion Researches
Troitsk, Moscow region, Russia
[1]Institute of Physical and Technical Problems, Dubna, Moscow region, Russia

INTRODUCTION

Silicon surface barrier detector based Soft X-Ray (SXR) imaging systems have been successfully used in tokamak experiments for plasma shape, sawtooth and other MHD oscillations measurements. But the high fluxes of 14 MeV neutrons and γ-rays make it impossible to use silicon detectors in such system in ITER deuterium-tritium experiments. The most important problems are detector's radiation resistance and low signal (due to SXR flux) to noise (due to neutron and γ fluxes) ratio[1]. Natural Diamond Detectors (NDD)[2,3] have essentially higher radiation resistance and lower current sensitivity to DT neutrons and gammas. Here we will discuss the possibility to use NDDs for ITER SXR-imaging.

NATURAL DIAMOND DETECTOR PROPERTIES

In comparison with Si and CdTe semiconductor detectors, NDDs proved to be preferable due to their outstanding features: low leakage currents (~ 1 pA at low radiation background), high radiation damage level ($\sim 5*10^{14}$ n/cm^2), etc. NDD's spectral response can be obtained in the range from UV to 10 keV photons depending upon chosen entrance window. Lower cut-off energy of the order of 5.5 eV is equal to diamond band gap (at the absence of any window and use of planar contacts). Due to absorption in a contact NDDs with 20 nm thick gold layers covered both sides of diamond plate have lower cut-off energy of the order of 500 eV. The upper edge of spectral curve is limited by the sensitive layer thickness (~ 0.3 mm) and is about 10 keV. The use of NDD with two diamond layers of the same thickness provides "two-color" sensitivity of the detector (see Fig.1).

Diamond, silicon and CdTe SXR detectors with similar dimensions have been comparatively tested at the T-10 tokamak[4]. In the case of NDD application the essentially higher signal to noise ratio (by more than order of magnitude) approximately at the same value of current sensitivity have been reached. For application at more severe operation conditions (high neutron and gamma backgrounds, high electromagnetic fields and acoustic

Diagnostics for Experimental Thermonuclear Fusion Reactors
Edited by P. E. Stott *et al.*, Plenum Press, New York, 1996

365

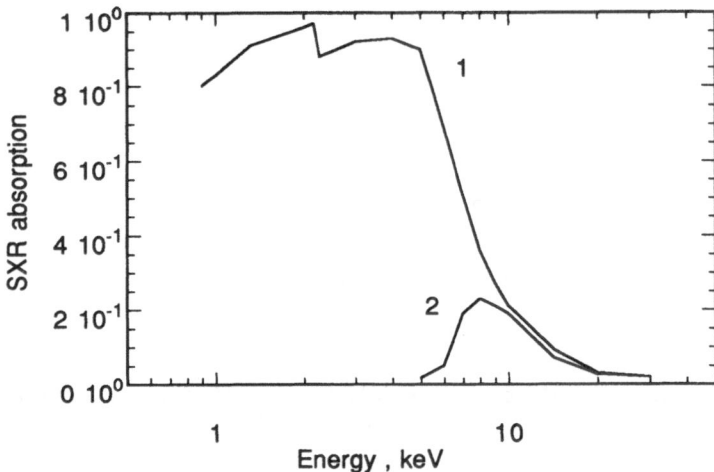

Figure 1. Diamond detector response function: 1 - first 300 μm diamond layer with 20 nm gold contact , 2 - second 300 μm diamond layer of "two-color" sensitive detector.

noises) further NDD based detector module development have been done. It was divided into two parts: remote detector head including the NDD with entrance window (beryllium foil as usual) and first preamplifier stage (low noise FET or nouvistor, if high radiation damage level is needed) and basic amplifier unit with the system of dc bias, leakage current and/or thermal output voltage drift compensation, frequency and noise correction and NDD power supply filtering. The NDD, developed for SXR flux measurements, specification is presented in Table 1.

Table 1. Natural diamond detector specification

Spectral range	0.006 - 10 keV
Sensitive area	0.2 cm^2
Dark current	~ 1 pA
Dynamic current range	10^{-12} - 10^{-6} A
Response time	< 10^{-8} s
Detector voltage supply	- 100 V

The responses of first and second (second diamond layer in "two-color" sensitive detector) NDDs without any entrance window are shown in Fig.1. Such response is typical for detector with narrow, in comparison with plasma electron temperature, spectral range. The electron temperature contrast of narrow spectral band detector can be estimated by

$$\frac{dJ}{J} \Big/ \frac{dT_e}{T_e} \approx E/T_e - 1$$

where: J is SXR flux and E - average energy of detector spectral response.

So NDD can be used for SXR imaging of plasma in the spectral range up to 8 keV. More detail description of natural diamond (IIa) and NDD properties can be found in ref. 5.

POSSIBILITIES FOR NDD APPLICATION IN ITER SXR MEASUREMENTS

Current sensitivity of best NDDs to 14 MeV neutrons and 4 Mev gammas are of the order of $3*10^{-18}$ (A/mm^3)/(n/cm^2/s) and $1.2*10^{-18}$ (A/mm^3)/(γ/cm^2/s). These values are ~ 2 and 7 times lower than corresponding current sensitivities of silicon surface barrier detectors. This gives possibility to increase signal to noise ratio in the case of NDD application for SXR flux measurements in the conditions of high neutron and gamma background fluxes. But this increase still not enough for ITER conditions. So, to improve NDD shielding, we propose to use curved glass capillary bundles. Spatial resolution will be limited by the angle of SXR reflection (~ 5 mrad) in capillary bundle and the distance from plasma to bundle entrance edge (~ 3 m). In the case of use ~ 2 m length bundles NDDs can be removed from straight neutron and gamma flux at a distance of about 25 cm, and so will be good shielded. Table 2 shows the results of our estimations of signal to noise ratio for NDD with time resolution not worse than 0.1 ms.

Table 2. Diamond detector based SXR diagnostic parameters

Parameter		I	II	III
<Te>	keV	1	2.5	11.8
Detector sensitive area	cm^2	0.2	0.2	0.2
Field of view	ster.	$4*10^{-5}$	$4*10^{-5}$	$4*10^{-5}$
Input SXR power on first wall	W/cm^2	0.05	0.5	15
SXR power on detector	W	$6*10^{-8}$	$6*10^{-7}$	$2*10^{-5}$
Absorbed power	W	$2*10^{-8}$	$2*10^{-7}$	$6*10^{-6}$
Signal current	pA	192	2230	57324
Neutron flux on first wall	n/cm^2/s	10^{14}	10^{11}	10^{14}
Neutron flux on detector	n/cm^2/s	10^{7}	10^{4}	10^{7}
Neutron count rate	n/s	10^{4}	10	10^{4}
Average neutron energy	MeV	2	2	2
Neutron induced current	pA	32	0.04	32
Noise current	pA	2	2	2
Signal/Noise, no neutron		96	1115	28662
Signal/Noise		6	1115	1788

In Table 2 column I presents estimated signal to noise ratio for SXR flux measurement using NDD + capillary bundle system in outer plasma region (electron temperature lower 8 keV) during ITER DT phase, column II - for central chord during first 100 ms of discharge and column III - for central chord during DT phase.

For SXR/UV imaging near X-point and ionization front in diverter the special planar diamond detectors with spectral sensitivity band from several eVs to several keV could be used. It is quite difficult to expect, that electron temperature measurements are possible for that region, because of the predomination of radiation in impurities peaks. The location of ionization front can be estimated due to large changes in concentration and temperature near it. NDD based imaging system can observe bands of high emission near the X-point also. For this measurements glass capillary bundles (about 1 cm in diameter and 4 m in length) can be assembled in focusing configuration along plasma magnetic axis to increase signal. The radius of bundle curvature (above 8 m) is enough to remove NDDs from straight neutron and gamma flux. The results of our testing of some such capillaries SXR transmission properties are shown in Fig.2.

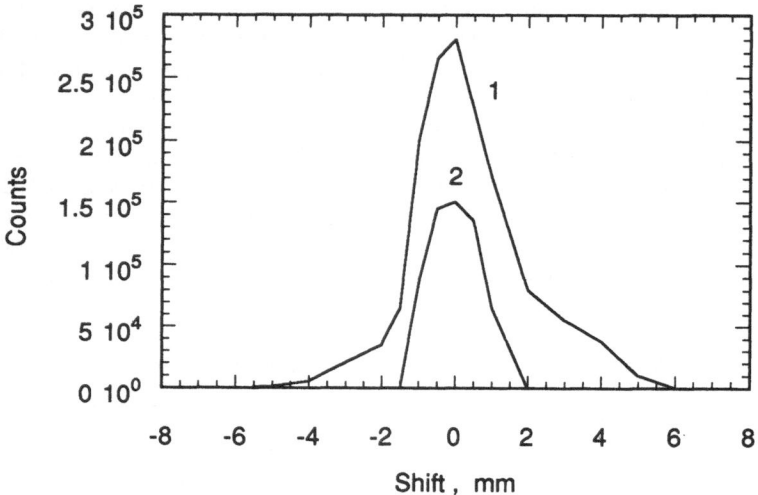

Figure 2. SXR-flux on the detector with (curve 1) and without (curve 2) glass capillary bundle for different detector shifts from collimator optical axis.

The power of SXR flux was increased more than 2 times after insertion of 30 cm long, 0.5 mm inner diameter capillary into optical path that correspond to the increase of solid angle field of view. It has been shown[6], that special capillary concentrators can increase initial SXR flux hundreds times. It is important to stress that capillary bundle can be effectively used for SXR transmission for photons with energy up to 10 keV, that correspond very good to NDD sensitivity range.

So, due to natural diamond detector response function, high radiation resistance and relatively low sensitivity to neutrons and gamma, NDD + glass capillary bundle based imaging system can be used in ITER experiments for: plasma shape measurements and control during current rise (first 100 ms) and down stage of plasma discharge; sawtooth and other MHD oscillations study on periphery of plasma; SXR and UV flux profile measurements in diverter and in the region of X-point.

REFERENCES

1. K.W.Hill, H.Adler, M.Bitter, E.Fredrickson, S. von Goeler, H.Hsuan, A.Janos, D.Johnson, A.T.Ramsey and G.Renda, Analysis of nuclear radiation-induced noise in spectrometric and x-ray diagnostics during high power deuterium-tritium experiments on the tokamak fusion test reactor , *Rev. Sci. Instrum.* 66 (1): 913 (1995)
2. E.A.Konorova, S.F.Kozlov, A diamond detector of nuclear radiation, *Sov.Phys.Semicond.* 4: 1600 (1971).
3. A.E.Luchanskii, S.S.Martynov, V.S.Khrunov, V.A.Cheklaev, Detector Assembly with a diamond detector for recording neutrons, *Sov. Atomic Energy* , 63 : 639 (1987).
4. G.A.Bobrovskiy, Yu.I. Esipchuk, G.E.Notkin, D.V.Portnov and P.V.Savruhine, Internal disruption study in T-10 plasmas, *Proceed. of 12 EPS Conf. on Contr. Fus. and Plasma Phys.*, Budapest-1985, v.1, p.142.
5. A.V. Krasilnikov, Diamond detector based DT neutron spectrometer for ITER, *This Proceedings.*
6. M.A.Kumakhov, I.Yu.Ponomarev, G.N.Popkov, Use of x-ray optical system for x-ray diagnostics, *Fiz. Plazmy* , 18: 496 (1992) [*Sov. J. Plasma Phys.* 18: 258 (1992)].

FUSION PRODUCT MEASUREMENTS IN D-T PLASMAS IN TFTR

L. C. Johnson,[1] Cris W. Barnes,[2] R. E. Bell,[3] M. Bitter,[3] R. V. Budny,[3]
C. E. Bush,[4] D. S. Darrow,[3] H. H. Duong,[5] P. C. Efthimion,[3] R. K. Fisher,[5]
R. J. Fonck,[6] H. W. Herrmann,[3] D. L. Jassby,[3] A. V. Krasilnikov,[7]
G. R. McKee,[3] S. S. Medley,[3] M. Osakabe,[8] M. P. Petrov,[9]
A. L. Roquemore,[3] M. Sasao,[8] S. Sesnic,[3] B. C. Stratton,[3]
E. J. Synakowski,[3] S. von Goeler,[3] and S. J. Zweben[3]

[1]ITER Joint Central Team, San Diego, CA USA
[2]Los Alamos National Laboratory, Los Alamos, NM USA
[3]Princeton Plasma Physics Laboratory, Princeton, NJ USA
[4]Oak Ridge National Laboratory, Oak Ridge, TN USA
[5]General Atomics, San Diego, CA USA
[6]University of Wisconsin, Madison, WI USA
[7]TRINITI, Troitsk, Russia
[8]National Institute for Fusion Science, Nagoya, Japan
[9]Ioffe Physical-Technical Institute, St. Petersburg, Russia

INTRODUCTION

Since the introduction of tritium into the Tokamak Fusion Test Reactor (TFTR) in December 1993, more than 400 D-T plasmas have been produced, with a total fusion energy of over 900 MJ. Fusion power production in D-T supershots is routinely in the 5-10 MW range, within a factor of ~ 200 of that expected in ITER. Neutron flux at the first wall of TFTR has reached about 10% of values projected for ITER, and central fusion power densities are already at ITER levels. Therefore, experience gained in fusion product measurements during the TFTR D-T experiments is directly relevant to the design of diagnostic systems for ITER.

When tritium operation began in TFTR, neutron fluxes at the detectors increased by two orders of magnitude over values typical of D-D plasmas, and the neutron spectrum also changed. The further increase to values expected for an ignited ITER plasma is only about a factor of 10. For neutron measurements in ITER, care must be exercised to maintain system linearity and calibration over many orders of magnitude, as we shall discuss further, but this is a manageable issue. A benefit of the high neutron emission rates on both TFTR and ITER is that they offer new opportunities to observe transient plasma behavior with

Diagnostics for Experimental Thermonuclear Fusion Reactors
Edited by P. E. Stott *et al.*, Plenum Press, New York, 1996

369

remarkable detail. However, neutron fluences arising from the 1000 s pulse duration in ITER will be qualitatively greater than those yet experienced. Radiation damage, material activation, and remote handling will be major concerns for all ITER diagnostic systems, including neutron detectors.

In this paper, we review recent fusion product measurements on TFTR. Some of these results are described in papers listed in the references. Other experiments, particularly some of the neutron profile measurements, have not been previously reported and will be discussed in greater detail.

ESCAPING ALPHA PARTICLES

Of the 1500 MW produced in an ignited ITER plasma, 300 MW goes into alpha particles. Any substantial loss of alphas before their energy is transferred to the bulk plasma will not only reduce the ignition margin but also pose a serious risk of damage to the first wall. In the case of ICRF heating, the potential for local wall damage by fast ions is even greater. In addition to its intrinsic interest for plasma diagnostics, the characterization of alpha particle losses from TFTR D-T plasmas can help establish safe operating conditions for ITER.

A poloidal array of detectors has been operated on TFTR since 1988 to measure escaping D-T alphas, D-D protons and tritons, RF minority tail ions, and NBI ions.[1,2] The detectors consist of planar scintillators, enclosed in protected housings. Fast ions whose gyroradii and pitch angles lie within selected ranges are admitted by an arrangement of entrance slits and impinge onto the scintillators, and coherent fiber optic bundles transmit images of the scintillators to photomultipliers and intensified video cameras in the TFTR basement. The images are then compared with predictions of the distributions of gyroradii and pitch angle from orbit codes.

The measured flux of escaping alpha particles from D-T plasmas in TFTR is generally consistent with first-orbit loss.[3] However, these losses may be significantly enhanced during MHD activity, ICRF heating, or major or minor disruptions, and the particles lost during such events may be localized spatially or in pitch angle. On one occasion, enhanced losses of ripple trapped ions during Toroidal Alfvén Eigenmode experiments caused local melting and leaks at vacuum flange welds.

In preparing for D-T operation on TFTR, the detectors were fitted with more robust scintillators than had been used for D-D experiments. In addition, a radiation shield was installed around the fiber optic bundles, to reduce fluorescence, and the detectors were moved to a shielded location. Radiation-induced damage, heat load, and background light will be much more severe in ITER. It is unlikely that scintillator-based lost-alpha detectors can be developed for use in ITER.

CONFINED ALPHA PARTICLES AND HELIUM ASH

Three diagnostic systems have been used on TFTR to measure concentrations and energy spectra of confined alpha particles. Two of the systems are based on charge exchange recombination spectroscopy (CHERS) of the 468.6 nm spectral line of He^+ (n = 4–3), produced by interactions of He^{2+} with injected deuterium neutral beam atoms. In the case of thermalized alphas, referred to as helium ash, the main component of the spectral line is measured along a number of sight lines intersecting a heating beam.[4] After correcting for HeII emission arising outside the interaction volume and for overlapping CIV and CVI lines, a comparison is made between measurements in high power D-T plasmas and in reference D-D plasmas. The results show that the helium ash confinement time, including recycling effects, is 7–12 times the energy confinement time. Evolution of the local thermal ash density is consistent with modeling based on classical slowing-down of alpha particles and transport coefficients previously measured in helium gas puffing experiments. The ash confinement time is dominated by edge pumping rates rather than core transport and indicates a recycling coefficient of about 0.85 for the cases studied.

The Doppler shifted red wing of the 468.6 nm line is used to measure confined alphas in the 0–0.7 MeV energy range.[5] Since the alpha signal is less than 1% of the bremsstrahlung background and is in the wing of the bright HeII line from thermal helium in the plasma core and edge, a special system (α–CHERS) was developed, combining a high throughput, moderate resolution spectrometer with low-noise, high-quantum-efficiency, high-dynamic-range detectors. For sawtooth-free plasmas, the measured alpha particle densities are consistent with modeling based on neoclassical calculations of the distribution function. After beam turn-off, the alpha density decays with the classical slowing down time of 0.5 s. The measurements also show spatial redistribution of alphas after sawtooth crashes.

Fast confined alphas in the range 0.5–4.1 MeV have been measured in TFTR D-T plasmas using lithium and boron pellets and neutral particle analysis.[6-8] Injected impurity pellets are surrounded by a dense, toroidally elongated ablation cloud. Some of the alpha particles which encounter the cloud are neutralized by charge exchange with impurity ions and escape from the plasma. The neutral particle analyzer views the ablation cloud along a line almost coincident with the pellet trajectory and intercepts some of the escaping fast neutrals. Ions emerging from a stripping foil are mass and energy analyzed and reflect the distribution function of the confined alphas. In these experiments, pellet injection is usually delayed 0.2 to 0.5 s after termination of neutral beam heating in order to allow pellet penetration and to reduce neutron background noise.

Measurements from the pellet charge exchange system in quiescent D-T plasmas are in good agreement with simulations based on classical confinement and slowing down of alpha particles. Special experiments using short 0.1 s neutral beam pulses show a Doppler broadened spectrum at the 3.5 MeV alpha birth energy. In plasmas with sawtooth activity, the measurements show internal redistribution of fast confined alphas after a sawtooth crash. Effects on the fast alphas due to stochastic ripple loss have also been observed. Comparison of experimental data with detailed modeling is underway.

All the measurements of confined alpha particles in TFTR D-T plasmas are difficult and rely on a delicate balance of parameters such as pellet penetration, neutral particle mean-free-path, and background radiation and noise. For ITER plasmas, suitable conditions will be far more difficult to achieve.

NEUTRON MEASUREMENTS

Neutron emission from TFTR plasmas is measured with a wide assortment of detection systems.[9] The primary system for determining global neutron source strength is a set of absolutely calibrated ^{235}U and ^{238}U fission chambers,[10] supplemented by silicon surface barrier diodes[11] and spatially collimated ^{4}He proportional counters[12] and ZnS scintillators.[10,13-15] Neutron yield from each D-T plasma is deduced from a variety of elemental activation foils,[16] and these data are compared with time-integrated source strength measurements to check linearity and consistency.[17] Neutron spectra have recently been obtained from a coincidence scintillator system[18,19] and from natural diamond detectors.[20,21]

In order to make reliable neutron emission measurements on either TFTR or ITER, a number of issues must be addressed. Among these considerations are detector calibration, linearity, stability, dynamic range, spectral sensitivity, and durability. In the case of neutron profile measurements, spatial sampling frequency, plasma shape, and sensitivity to scattered neutrons must also be taken into account. Some of the linearity and stability aspects of experience on TFTR are discussed elsewhere.[9,17] Except for the silicon surface barrier diodes, which must be periodically replaced because of radiation damage, there has been no evidence of deterioration of any of the TFTR neutron detectors. Individual electronic components have failed from time to time, but module replacement has restored normal detector operation.

The counting circuits of the silicon surface barrier diodes and ^{4}He proportional counters on TFTR are configured to discriminate against 2.5 MeV $d(d,n)^3$He neutrons. This is not possible for the fission chambers[10] and ZnS scintillators,[14,15] but they have higher efficiency for 14.1 MeV $t(d,n)\alpha$ neutrons by a factor ranging from 1.3 to ~10, depending upon specific detector and electronic setup. Because of the large cross section of the $t(d,n)\alpha$ reaction, the 2.5 MeV neutrons make a negligible contribution to source strength or neutron spatial profile measurements in plasmas with tritium neutral beam injection. For D-D operation, in which tritium enters the plasma by recycling from the walls or is puffed in during transport studies,[22] data from the fission detectors and scintillators must be corrected.

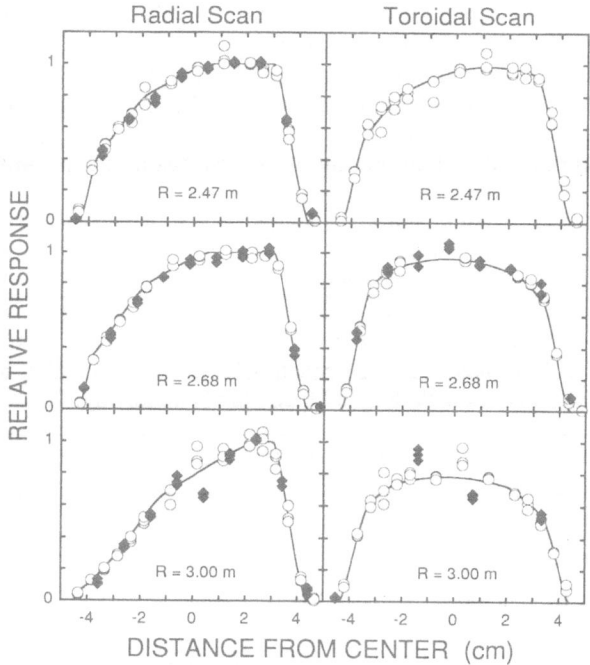

Figure 1. Spatial response functions for three of the central NE 451 (ZnS) detectors of the TFTR multichannel neutron collimator. Radial and toroidal scans using a ^{252}Cf source (open circles) and a D-T neutron generator (solid diamonds) are compared with results of a ray tracing model (solid lines). Distances are measured from the centers of the flight tubes.

Neutron Detector Calibration

Calibration of fission detectors on large tokamaks involves either mapping the detector response to a point neutron source positioned at various locations inside the vacuum vessel[10,23,24] or using activation techniques coupled with neutron transport calculations.[16,25,26] In either case, it is necessary to use multiple detectors with overlapping dynamic ranges (including different electronic operational modes) to span the many orders of magnitude range of neutron source strengths which may be encountered. For ITER, the usual calibration methods will be difficult to apply because of the shielding effects of the blanket and vacuum vessel and the need to protect activation stations from the first-wall heat load.

An alternative procedure is to directly calibrate the neutron cameras, as has been done on TFTR,[10,12,13] and use these data to cross-calibrate the fission chambers. In a direct camera calibration, a neutron point source inside the vacuum vessel is translated in the field of view of each detector to map its response function. Figure 1 shows radial and toroidal scans for three of the central channels of the TFTR system. Measurements for a ^{252}Cf source and for a D-T neutron generator are indistinguishable from one another and are in good agreement with results of a ray tracing model which takes into account geometrical effects, including misalignment of the collimator flight tubes.

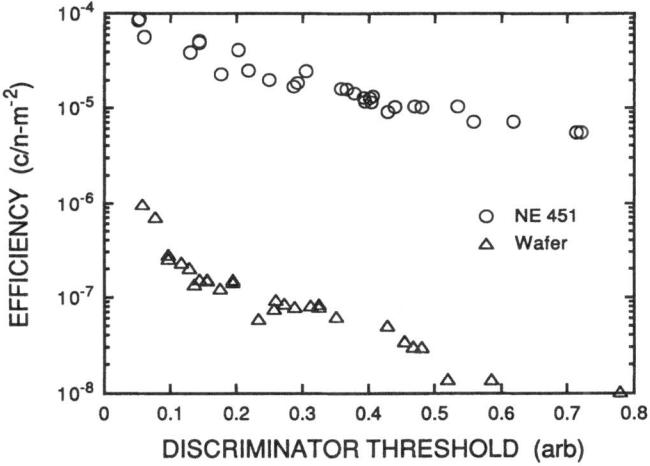

Figure 2. Absolute efficiencies of scintillators in the TFTR multichannel collimator for the detection of 14.1 MeV neutrons. Each point represents a single count-rate mode data channel from one of the detectors.

In order to extend the direct calibration of the TFTR neutron profile monitor to the range of fluxes encountered during high power D-T operation without changing the collimator configuration, three banks of detectors with different efficiencies and operating modes are arranged along the collimated flight tubes: commercially available ^{4}He proportional counters and NE 451 (ZnS) scintillators, and thin, low sensitivity ZnS scintillators (referred to as wafer detectors), which were developed especially for high flux operation. All detectors can be operated in count-rate mode, using multiple discriminator-scaler data channels, and the NE 451 scintillators can also be operated in current mode.[27] Figure 2 shows absolute efficiencies for 14.1 MeV neutrons of the two types of scintillators in count-rate mode. Each point represents one of the three data channels from one of the 10 detectors of each type. During deuterium-tritium operation, all the count-rate mode data channels of the ^{4}He and NE 451 detectors and some of low-threshold channels of the wafer detectors are saturated by pulse pileup, but the high-threshold channels of the wafer detectors and the current mode data from the NE 451 detectors remain linear.

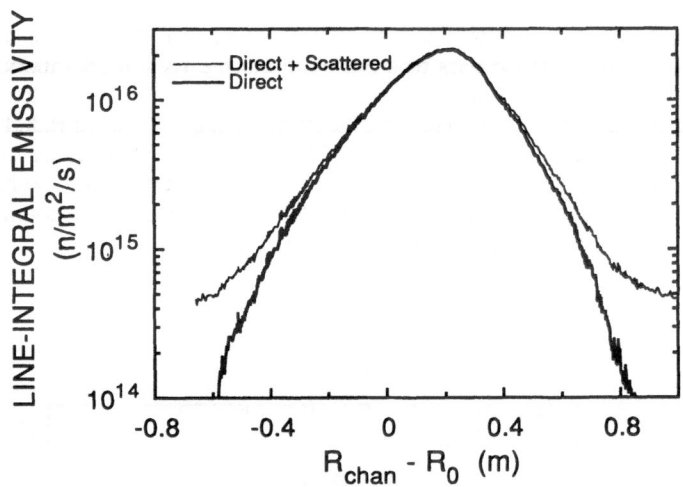

Figure 3. Spatial profiles of line-integral neutron emissivity for TFTR shot 82445 at 3.5 s. The curves are superpositions of data from eight NE 451 detectors during a sweep in plasma major radius.

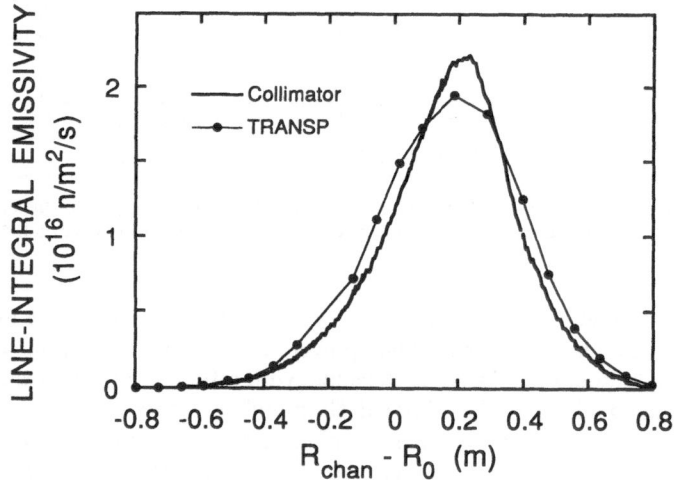

Figure 4. Spatial profiles of line-integral neutron emissivity for TFTR shot 82445 at 3.5 s. The measured profile (thick curve) is more sharply peaked than the profile computed by TRANSP.

Each detector of the TFTR neutron camera is calibrated independently. Because of the collimator configuration (parallel vertical sight lines), the global source strength can be found by spatial integration of line-integral emissivity data. Results are in good agreement with other measurements of source strength and yield, except when the center of a narrow profile lies in the unsampled region above a poloidal field coil. Recent experiments, in which a low power D-T plasma was swept in major radius (so-called jog experiments), have allowed a detailed study of spatial profiles as well as a check on individual detector calibrations and corrections for contributions from scattered neutrons. By taking into account the magnetic flux-conserving change in the profile diameter and the measured change in global neutron source strength, the temporal behavior of each collimator channel during the ~ 50 ms sweep can be mapped to a segment of the spatial profile. If the individual detector calibrations are correct, the segments should overlap, forming a continuous curve. An example is shown in Fig. 3. The thin curve is a superposition of data from the eight NE 451 detectors with current mode capability, before correction for contributions from scattered neutrons, and the thick curve is the result after applying a correction. For the central channels, scattered neutrons constitute 2–3% of the raw detector signal. As a result of these experiments, slight adjustments were required in calibration factors for some of the outer channels, where characterization with the in-vessel neutron generator was incomplete. No changes were needed for the four inner channels.

Jog experiments have not only validated calibration factors but also provided neutron profiles with exceptional spatial resolution. Figure 4 shows the data from Fig. 3, plotted on a linear scale, compared with a TRANSP simulation of the neutron profile. Although there is general agreement between the two, the measured profile is distinctly narrower and more sharply peaked than the simulation. This may be due in part to limited spatial resolution of plasma parameters used in the TRANSP calculations.

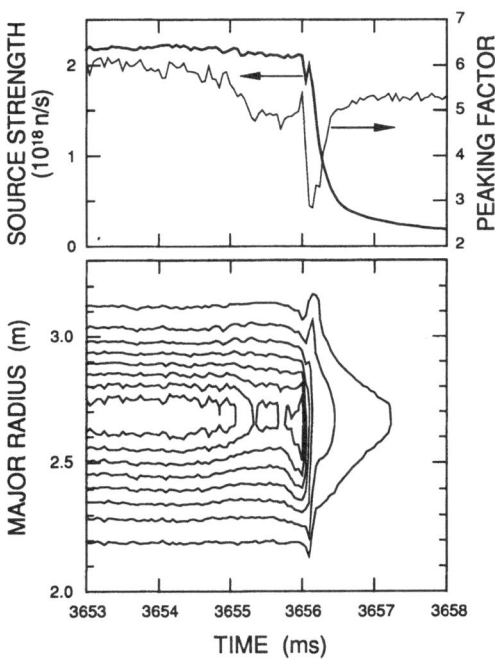

Figure 5. Neutron emission from TFTR shot 85407 during a major disruption. The upper figure shows global source strength (thick curve) and profile peaking factor (thin curve). The lower figure is a contour plot of Abel inverted emissivity.

Transient Behavior of Neutron Emission

As just discussed, by operating some of the fission detectors and scintillators in current mode, it is possible not only to extend the range of linear operation but also to achieve temporal resolution of less than 1 ms for global and collimated D-T neutron emission . This has enabled us to observe transient behavior with remarkable detail. As an example, Fig. 5 shows 5 ms of data from the neutron collimator during a major disruption of a 6 MW D-T plasma. The upper portion of the figure shows the global source strength, obtained by spatial integration of the line-integral emissivity, and the profile peaking factor, i.e., the central emissivity divided by the volume average value. The lower part of the figure is a contour plot of Abel inverted emissivity. The profile begins to flatten about 2 ms before the disruption. At the time of the disruption, the profile broadens, presumably because of the release of deuterium from the limiter.

Two new collimator tubes have very recently been installed on TFTR, separated 180° toroidally from the main neutron camera. The sight lines are slightly inclined from vertical, to allow viewing previously inaccessible major radius positions, and elongated toroidally, to increase neutron throughput. To provide high sensitivity for MHD studies, the flight tubes are fitted with large plastic scintillators, operated in current mode. Although the new channels are not yet characterized, they show the expected increase in sensitivity. Figure 6 shows data for a 5 ms period at a sawtooth crash in a low power (0.4 MW) D-T plasma. The upper curve is the global source strength from a fission detector, operated in current mode with a 1 kHz sampling rate. The middle curve is raw data from the sight line viewing the core of the plasma, and the lower curve is raw data from the channel ~0.26 m further out on the profile. Both channels show a 5 kHz MHD precursor to the sawtooth crash, with a phase difference of 180°. ECE measurements indicate that this is a m=1, n=1 mode.

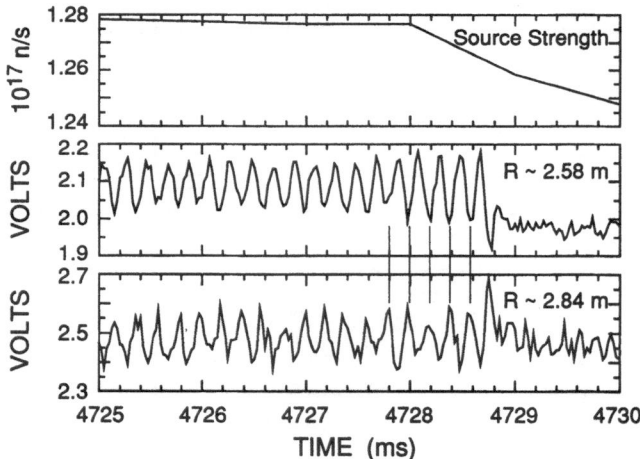

Figure 6. Neutron emission from TFTR shot 87544 during a sawtooth crash. The upper curve is global source strength from a fission detector sampled at 1 kHz. The lower curves are raw data from collimated detectors at the indicated radii and show a 180° phase difference in the 5 kHz precursor to the sawtooth crash.

Measurements of Neutron Spectra

Two systems have been used to observe neutron spectra for TFTR D-T plasmas. The COTETRA[18,19] system (counter telescope with thick radiator) is a proton recoil telescope which uses a plastic scintillator as the radiator and either a silicon surface barrier diode or another plastic scintillator as the detector. Coincidence electronics sum the energy deposited in the radiator and the detector to give the total proton energy. COTETRA has clearly observed the 14.1 MeV neutron line and shown broadening due to ICRF heating, but it has not yet achieved the expected energy resolution.

The second neutron spectrometer system used on TFTR consists of a set of high purity natural diamond detectors.[20,21] Of the many neutron-induced reactions in diamond, the $^{12}C(n,\alpha_0)^9Be$ reaction appears as an isolated line in a pulse height spectrum and reflects the energy of the incident neutron. Figure 7 shows pulse height spectra from one of the detectors for a neutron generator and for a number of D-T plasmas in TFTR. The 14.1 MeV line is at the right of the spectra and clearly shows broadening for the plasma measurements. These and other results using diamond detectors are discussed elsewhere in these proceedings by Krasilnikov.[21]

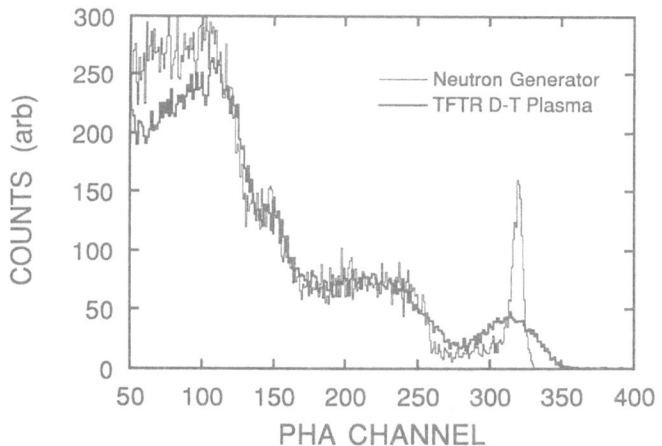

Figure 7. Pulse height spectra from a natural diamond neutron spectrometer. The 14.1 MeV line of the spectrum from TFTR D-T plasmas (thick curve) is significantly broader than that from laboratory measurements using a neutron generator (thin curve).

CONCLUSIONS

A growing number of measurements of fusion products have been made during D-T operation on TFTR. These include observations of escaping alpha particles, energetic confined alphas, helium ash, neutron source strength and emissivity profiles, and neutron spectra. Alpha diagnostics are difficult and will require considerable ingenuity for implementation on ITER. Neutron flux measurements on ITER should be relatively straightforward, although care must be exercised in calibrations, and detector durability needs to be investigated. Natural diamond detectors offer promise for use as compact, rugged neutron spectrometers on ITER. Further development and testing of all these diagnostic systems on TFTR and other large tokamaks would be of great value.

ACKNOWLEDGMENT

This work was supported by U. S. Department of Energy Contract No. DE-AC02-76-CH03073.

REFERENCES

1. S. J. Zweben, Four-channel ZnS scintillator measurements of escaping tritons in TFTR, *Rev. Sci. Instrum.* 60:576 (1989).
2. D. S. Darrow, H. W. Herrmann, D. W. Johnson, *et al.*, Measurement of loss of DT fusion products using scintillator detectors in TFTR, *Rev. Sci. Instrum.* 66:476 (1995).
3. S. J. Zweben, D. S. Darrow, H. W. Herrmann, *et al.*, Alpha particle loss in the TFTR DT experiments, accepted for publication in *Nuclear Fusion* (1995).
4. E. J. Synakowski, R. E. Bell, R. V. Budny, *et al.*, Measurements of the production and transport of helium ash on the TFTR tokamak, submitted to *Phys. Rev. Lett.* (1995).
5. G. McKee, R. Fonck, B. Stratton, *et al.*, Confined alpha distribution measurements in a deuterium-tritium tokamak plasma, *Phys. Rev. Lett.* 75:649 (1995).
6. R. K. Fisher, J. M. McChesney, P. B. Parks, *et al.*, Measurements of fast confined alphas on TFTR, *Phys. Rev. Lett.* 75:846 (1995).
7. M. P. Petrov, *et al.*, Studies of confined alphas using the pellet charge exchange diagnostic on TFTR, submitted to *Nuclear Fusion* (1995).
8. S. S. Medley, R. V. Budny, D. K. Mansfield, *et al.*, Measurements of energetic confined alphas and tritons on TFTR, *Proc. 22nd Euro. Conf. on Contr. Fusion and Plasma Phys.*, Bournemouth, UK (1995).
9. L. C. Johnson, C. W. Barnes, H. H. Duong, *et al.*, Cross calibration of neutron detectors for deuterium-tritium operation in TFTR, *Rev. Sci. Instrum.* 66:894 (1995).
10. D. L. Jassby, C. W. Barnes, L. C. Johnson, *et al.*, Absolute calibration of tokamak fusion test reactor neutron detectors for D-T plasma operation, *Rev. Sci. Instrum.* 66:891 (1995).
11. E. Ruskov, W. W. Heidbrink, H. H. Duong, *et al.*, Measurement of 14 MeV neutrons at TFTR with Si-diode detectors, *Rev. Sci. Instrum.* 66:910 (1995).
12. J. D. Strachan, C. W. Barnes, M. Diesso, *et al.*, Absolute calibration of TFTR helium proportional counters, *Rev. Sci. Instrum.* 66:1247 (1995).
13. A. L. Roquemore, R. C. Chouinard, M. Diesso, *et al.*, TFTR multichannel neutron collimator, *Rev. Sci. Instrum.* 61:3163 (1990).
14. L. C. Johnson, Validation of spatial profile measurements of neutron emission in TFTR plasmas, *Rev. Sci. Instrum.* 63:4517 (1992).
15. A. L. Roquemore, L. C. Johnson, and S. von Goeler, Performance of the upgraded multichannel neutron collimator, *Rev. Sci. Instrum.* 66:916 (1995).
16. C. W. Barnes, A. R. Larson, G. LeMunyan, and M. J. Loughlin, Measurements of DT and DD neutron activation on the Tokamak Fusion Test Reactor, *Rev. Sci. Instrum.* 66:888 (1995).
17. C. W. Barnes, H. H. Duong, D. L. Jassby, *et al.*, DT neutron measurements and experience on TFTR, *These Proceedings* (1995).
18. M. Osakabe, S. Itoh, Y. Gotoh, *et al.*, A compact neutron counter telescope with thick radiator (COTETRA) for fusion science, *Rev. Sci. Instrum.* 65:1636 (1994).
19. M. Osakabe, J. Fujita, J. Kodaira, *et al.*, Development of a neutron spectrometer, counter telescope with thick radiator, for TFTR D-T experiments, *Rev. Sci. Instrum.* 66:920 (1995).
20. A.V.Krasilnikov, Pulse height spectrum of natural diamond detector in DT neutron flux, to be published in *Sov. Voprosi Atomnoy Nauki i Tehniki*, 1 (1995) (in Russian).
21. A.V.Krasilnikov, Diamond detector based DT neutron spectrometer for ITER, *These Proceedings* (1995).
22. P. C. Efthimion, L. C. Johnson, J. D. Strachan, *et al.*, Tritium particle transport experiments on TFTR during DT operation, *Phys. Rev. Lett.* 75:85 (1995).
23. H. W. Hendel, R. W. Palladino, C. W. Barnes, *et al.*, In situ calibration of TFTR neutron detectors, *Rev. Sci. Instrum.* 61:1900(1990).
24. T. Nishitani, H. Takeuchi, T. Kondoh, *et al.*, Absolute calibration of the JT-60U neutron monitors using a ^{252}Cf neutron source, *Rev. Sci. Instrum.* 63:5270 (1992).
25. O. N. Jarvis, E. W. Clipsham, M. A. Hone, *et al.*, Use of activation techniques for the measurement of neutron yields from deuterium plasmas at the Joint European Torus, *Fusion Tech.* 20:265 (1991).
26. M. Hoek, T. Nishitani, Y. Ikeda, and A. Morioka, Neutron yield measurements by use of foil activation at JT-60U, *Rev. Sci. Instrum.* 66:885 (1995).
27. S. von Goeler, A. L. Roquemore, L. C. Johnson, *et al.*, Fast detection of 14 MeV neutrons on the TFTR neutron collimator, to be published.

DT NEUTRON MEASUREMENTS AND EXPERIENCE ON TFTR

Cris W. Barnes[1], Hau H. Duong[2], D. L. Jassby, L. C. Johnson,
A. R. Larson[1], G. LeMunyan, M. J. Loughlin[3], A. L. Roquemore,
S. Sesnic, J. D. Strachan, S. von Goeler, and G. A. Wurden[1]

Princeton Plasma Physics Laboratory
Princeton, NJ USA 08543

[1]Los Alamos National Laboratory, Los Alamos, NM USA 87545
[2]General Atomics ORAU fellow.
[3]JET Joint Undertaking, Abingdon, Oxon OX14 3EA United Kingdom.

INTRODUCTION

Through semi-independent absolute calibrations of multiply redundant neutron detector systems, the Tokamak Fusion Test Reactor (TFTR) has achieved ±7% (one-sigma) accuracy in its fusion power measurements.[1] This has required careful attention to the linearity of detectors up to the present highest fusion power levels achieved on TFTR of over 10 MW. The extended duration of the DT program on TFTR has also tested the stability of the detector systems. These issues of calibration, linearity, and stability will be reviewed for the TFTR experience and how it can be applied to plans for ITER.

ABSOLUTE CALIBRATION

All the absolutely calibrated neutron detectors on TFTR are referenced to the $^{28}Al(n,\alpha)^{24}Na$ cross-section, either directly (via re-entrant activation foil measurements[2]) or through calibration of the fluence of a DT neutron generator[3] by activation foils. The complementary strengths and weaknesses of activation foils, collimated scintillators and proportional counters, and fission chambers provide confidence in the final weighted uncertainty of the DT neutron source strength[4], as well as significant redundancy in the measurements. The issue of absolute accurate calibration of fusion power measurements for ITER is a deserving subject in itself. In general, both the JET[5] and now TFTR experience confirm that activation methods coupled with neutronics calculations in a low-scattering experimental environment can lead to the highest accuracy calibrations. The DT neutron generator used on TFTR may be a source (<10% of the total) of scattered neutrons below

Diagnostics for Experimental Thermonuclear Fusion Reactors
Edited by P. E. Stott *et al.*, Plenum Press, New York, 1996

379

10 MeV and affect detectors without threshold discrimination. With an accurate neutron source, calibration by integrating the spatial profile of collimated detectors is another effective technique for ITER.

STABILITY

Checks of detectors with standard radioactive sources and cross-comparisons of different detectors provides documentation of the stability of the measurements since the time of absolute calibration. Figure 1 shows measurements for an over eight year period of the stability of one of the high purity germanium detectors (HPGe) used in the neutron activation system. A commercial camping lantern mantle inside of a pneumatic system capsule was routinely used in a standard counting location. Three gamma-ray lines at different energies from natural thorium daughters were monitored over this period: 338.4 keV and 911.2 keV from ^{228}Ac, and 2614 keV from ^{212}Po. The gradual increase in the ^{228}Ac activity is consistent with the lantern mantle being manufactured in the late 1970's (~1978), and the natural thorium decay raising the ^{228}Ra level with its 5.75-year half-life. The ^{212}Po level rises a little more slowly from the additional 1.913-year half-life decay of ^{228}Th. Since 1991 a source traceable to the National Bureau of Standards (NBS) has been occasionally used as well; as shown on Fig. 1 the gamma-ray efficiency for ^{60}Co lines near 1.3 MeV (corrected for half-life decay) have shown no variation within the 3% (one-sigma) accuracy.

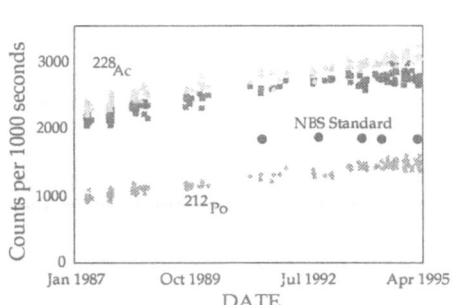

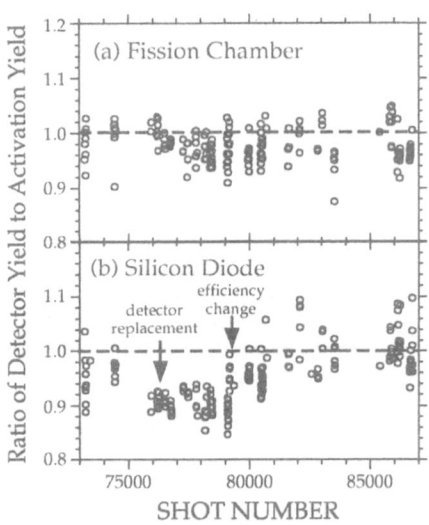

Figure 1. Stability over eight years for a HPGe gamma-ray detector of the TFTR neutron activation system. Shown are counts in 1000 seconds for three lines from thorium daughters in a commercial lantern mantle, as well as more recent measurements from a NBS-traceable source. The apparent increase in efficiency to the lantern mantle represents the natural increase in ^{228}Ra.

Figure 2. Ratio of yields of neutron detectors to the activation system vs. time (shot number on TFTR) since the beginning of the DT program. All DT discharges with aluminum foil data in the re-entrant irradiation end are included. (a) fission chamber yield to activation, showing only variation within the relative precision. (b) silicon diode yield to activation. The silicon diode was replaced at shot 76319 after noticeable degradation of signal, and the efficiency of its calibration adjusted around shot 79100.

The fission chamber detectors[6] have also tracked their sensitivity using a standard radioactive source. The current and Campbell electronic modes appear to behave independently of (and are more stable than) the count-rate mode. Thus, "renormalizations" of the count rate mode by use of low-level radioactive sources does not appear to address the issue of confirming detector stability of fission detector ionization chambers. A cross-comparison from DD discharges of low-sensitivity ^{235}U detectors in count mode with high-sensitivity ^{235}U detectors in current and Campbell mode showed the detectors have been stable for periods over a year. This comparison does not depend on the DT/DD neutron ratio in each shot since both the low- and high-sensitivity detectors are equally responsive to DT or DD neutrons (with a R'$_{235}$ value[7] of 1.30). The final arbiter of stability of the these time-dependent systems is secular comparison of yields to the neutron activation system. Figure 2(a) compares the ratio of yield from the standard high-power fission chamber signal[8] to the neutron activation yield for all DT shots of sufficient yield with aluminum foil measurements in the re-entrant irradiation end location. (The slight non-linearity of the fission chamber signal [see below] has been removed for this comparison.) Within a ±3% (one-sigma) shot-to-shot variation (primarily from foil-positioning irreproducibility in the activation counting station) there is no evidence of changing detector efficiency. Figure 2(b) shows the comparison for a silicon diode detector[9], illustrating effects of changing detectors and efficiencies.

LINEARITY

There has been a continual re-evaluation of the linearity of various detector systems and collection of information on the saturation, pile-up, or dead-time characteristics of detectors operated at high signal levels. Cross-comparisons of detectors have continued as TFTR has increased its peak fusion power.

The neutron activation system has large dynamic range[10] primarily from reducing the mass of the elemental foils while maintaining low deadtimes on the HPGe detectors. Comparing total yield from other detector systems to activation measurements can identify non-linearities in diagnostic response. Figure 3 shows the ratio of the standard high-power fission chamber yield to activation yield, plotted vs. the neutron-source-strength weighted time average source strength, that is the integral (in time) of the square of the source strength divided by the integral of the source strength. Thus discharges of short duration by high average fusion power (with more contribution from possible non-linearities) are plotted further to the right on the ordinate. At low power the ratio of fission-chamber to activation yield is ~0.97 which is the ratio of consensus calibration[4] to activation calibration alone. At the highest average source strengths and highest fusion power, the fission chamber detector response has dropped relative to the assumed linear response of the activation system. The amount is consistent with a 1%—3% decrease per 10^{18} n/sec source strength.

The impact of such a non-linearity is shown in Figure 4(a). Neutron source strength (and hence fusion power) versus time is shown for the highest fusion power shot to date (80539) Signals from the fission chamber, silicon diode, and neutron collimator systems are shown. Also shown is the corrected fission chamber signal assuming a 3% per 10^{18} n/sec source strength non-linearity as from Fig. 3. Figure 4(b) shows a similar comparison but at the lowest (single tritium beam) DT fusion power (shot 79102). The same detectors are shown (but filtered to remove the noise at this low end of their dynamic range), along with a high-sensitivity ^{235}U detector in current mode. In general the scatter between the different calibrated systems confirms the 7% absolute calibration.

IMPACT ON ITER

A neutron activation system would be important for ITER for its accurate absolute calibration, demonstrated detector stability and linearity over a wide dynamic range for comparison to other detector systems. An accurate neutron source will be important for calibrating collimated detector systems. Redundancy in detector systems with time and spactial resolution is important in maintaining accuracy by cross-comparisons.

ACKNOWLEDGMENTS

We thank Ken Young and Dave Johnson for their support and encouragement of the many collaborations in fusion product diagnostics on TFTR. This work was supported by DOE contracts W-7405-ENG-36 and DE-AC02-76-CHO-3073.

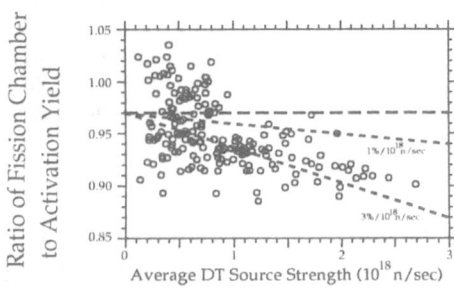

Figure 3. Evidence of non-linearity in fission chamber signals at high average source strength. The ratio of the fission chamber yield to the activation yield is plotted vs. $\int S^2 dt / \int S dt$ where S is the DT source strength.

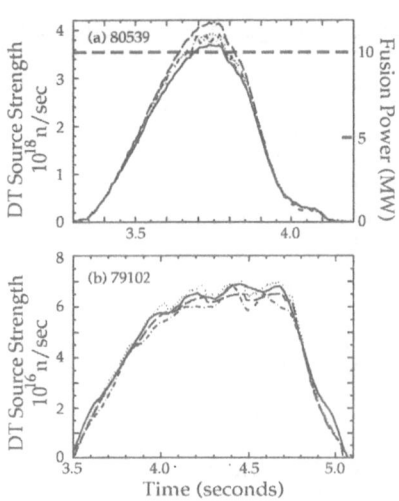

Figure 4. Neutron source strength (and fusion power) vs. time from different detector systems. (a) High fusion power discharge 80539. Solid line is fission chamber, dotted is neutron collimator, dash-dot is silicon diode, and dashed is fission chamber corrected to agree with neutron activation. (b) Low fusion power discharge 79102. First three curves are same as in (a); the dashed curve is a high sensitivity fission chamber detector corrected for an R'_{235} of 1.30.

REFERENCES

[1]J. D. Strachan *et al*. Fusion power production from TFTR plasmas fueled with deuterium and tritium, Phys. Rev. Letts. **72** (1994) 3526.

[2]C. W. Barnes, A. R. Larson, and A. L. Roquemore. Calculations of neutron activation response for the Tokamak Fusion Test Reactor, and absolute calibrations of neutron yield, (submitted to Fusion Tech., 1995)

[3]D. L. Jassby *et al*. Absolute calibration of tokamak fusion test reactor neutron detectors for D-T plasma operation, Rev. Sci. Instrum. **66** (1995) 891.

[4]L. C. Johnson *et al*.,Cross calibration of neutron detectors for deuterium-tritium operation in TFTR, Rev. Sci. Instrum. **66** (1995) 894.

[5]Owen N. Jarvis *et al.* Use of activation techniques for the measurement of neutron yields from deuterium plasmas at the Joint European Torus, Fusion Tech. **20** (1991) 265.

[6]H. W. Hendel, R. W. Palladino, Cris W. Barnes, M. Diesso, J. S. Felt, D. L. Jassby, L. C. Johnson, L.-P. Ku, Q. P. Liu, R. W. Motley, H. B. Murphy, J. Murphy, E. B. Nieschmidt, J. A. Roberts, T. Saito, J. D. Strachan, R. J. Waszazak, and K. M. Young. In situ calibration of TFTR neutron detectors, Rev. Sci. Instrum., **61** (1990) 1900.

[7]D. L. Jassby, H. W. Hendel, and H.-S. Bosch. Relative intensities of 2.5- and 14-MeV source neutrons from comparative responses of U-235 and U-238 detectors. Rev. Sci. Instrum. **59** (1988) 1688.

[8]The so-called NE3 (a ^{235}U detector) current mode signal.

[9]E. Ruskov, W. W. Heidbrink, H. H. Duong, A. L. Roquemore, and J. D. Strachan. Measurement of 14 MeV neutrons at TFTR with Si-diode detectors, Rev. Sci. Instrum., **66** (1995) 910.

[10]C. W. Barnes, A. R. Larson, G. L. LeMunyan, and M. J. Loughlin. Measurements of DT and DD neutron yields by neutron activation on the Tokamak Fusion Test Reactor, Rev. Sci. Instrum., **66** (1995) 888.

A NEUTRON CAMERA FOR ITER: CONCEPTUAL DESIGN

F B Marcus, J M Adams[1], P Batistoni[2], T Elevant[3], O N Jarvis, L Johnson[4], L de Kock[5], G Sadler, P Stott

JET Joint Undertaking, Abingdon, Oxon. OX14 3EA, United Kingdom
[1]AEA Technology, Harwell, Abingdon, UK
[2]ENEA Frascati, Italy
[3]Royal Inst of Tech, Stockholm, Sweden
[4]ITER, San Diego, USA
[5]ITER, Garching, Germany

INTRODUCTION

This paper presents the outline design of an ITER neutron camera, to provide control and plasma diagnostic information based on the deduced neutron emission profile. Information can be obtained on total neutron emission (fuel burn-up rate), alpha-particle birth profile, plasma position and the time dependent effects of plasma instabilities such as sawteeth. The design presented here concerns only a horizontally, radially, viewing (Horizontal) camera. A vertically viewing (Vertical) camera is also necessary for most requirements, and is assumed to have similar design principles and capabilities. This is primarily an engineering design paper, because the physics of neutron production and detection are well known.

There are two existing neutron profile monitor systems which have been used with d-t plasmas, at JET (J.M. Adams et al., 1993) and TFTR (A.L. Roquemore et al., 1990). The JET 2-D fan viewing geometry (with the lack of a viewing dump) is similar to ITER. We specify the number and locations of the profile monitor lines-of-sight, determine the design philosophy for the profile monitor (e.g. variable aperture sizes) and make initial proposals for the detector packages.

Many of the neutron diagnostics (apart from the tangential spectrometer and vertical camera) will be integrated into a single port. In what follows, we consider the main neutron profile monitor; an array of neutron spectrometers (integrated within the profile monitor detector package if possible) and activation systems for absolute neutron yield calibrations.

Diagnostics for Experimental Thermonuclear Fusion Reactors
Edited by P. E. Stott *et al.*, Plenum Press, New York, 1996

385

EXISTING NEUTRON PROFILE MONITORS

The TFTR profile monitor is presently being used to monitor 14 MeV neutron emission from high power d-t fusion experiments on TFTR. It uses a viewing geometry of parallel sight lines, owing to the wide opening ports above and below the circular plasma. Since the opposite port functions as a viewing dump, there is less problem of back scatter, and TFTR detectors do not have Pulse Shape Discrimination (PSD) as at JET. This geometry is not applicable on ITER. Another neutron camera with only vertical sight lines has just been installed on FTU (P Batistoni et al., 1995).

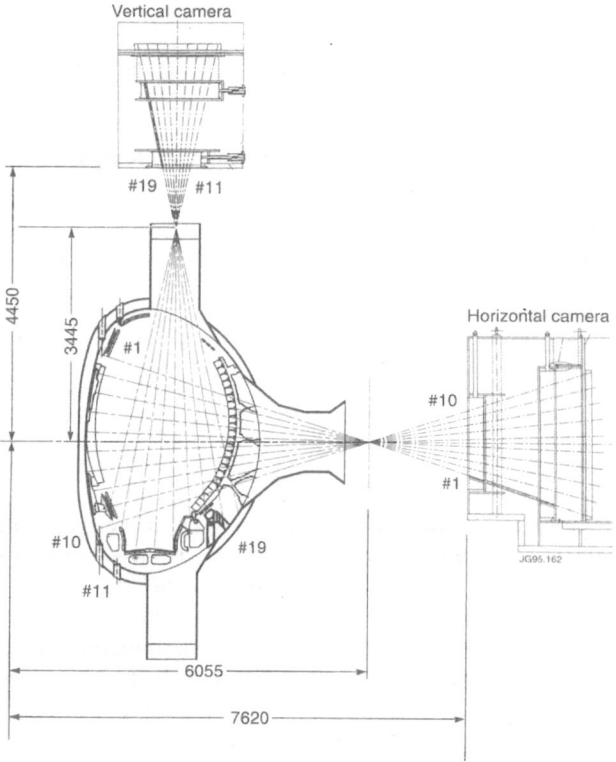

Fig. 1. The JET neutron emission profile monitor - upgrade version.

In the JET system shown in Fig. 1, there are both horizontally and vertically viewing cameras to observe the non-circular plasma, and most of the sight lines intersect the vacuum vessel, where neutron back scatter occurs. This geometry is similar to that proposed for ITER. The original JET system (J.M Adams et al., 1993) was used to observe the neutron emission from the first megawatt level fusion power experiments from a d-t plasma (B. Balet et al., 1993, F.B. Marcus et al., 1993). By using doubled PSD electronics to process the signals from NE213 detectors, it was possible to simultaneously measure 2.5 MeV neutrons from d-d fusion and 14 MeV neutrons at low tritium levels from d-t fusion, and 14 MeV emission alone at higher fusion power levels. The JET profile monitor has been used measure the neutron emissivity profile, which is used to deduce a wide range of physics parameters (F.B. Marcus et al., 1991, 1993, 1994, 1995), many of which are directly relevant to ITER requirements. In particular, 2-D tomography of the time-dependent emissivity profile has allowed a determination of the location of the plasma axis, the α-particle birth profile for 14 MeV d-t neutron emissivity (or fast triton birth profile for 2.5 MeV d-d neutrons), the total neutron emission and the response to MHD instabilities such as sawtooth crashes. An example of such a tomographically deduced emissivity profile is shown in Fig. 2 (from F.B. Marcus et al., 1995). Detailed comparison with TRANSP simulations of JET (B. Balet et al., 1993) and TFTR (S. Scott et al., 1995), and ratio methods (F.B. Marcus et al., 1993) of analysis (in appropriate regimes), also determine the tritium fuel concentration ratios.

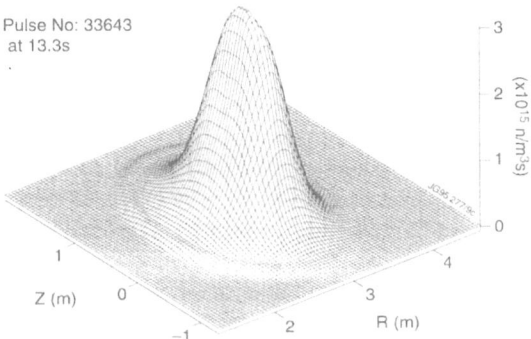

Fig. 2. The 2-D neutron emissivity profile (determined by tomography from the JET neutron profile monitor) from discharge #33643 after 1.3 sec of heating.

The present upgraded version of the JET neutron profile monitor has the same number, (9 vertically viewing and 10 horizontally, radially) of viewing channels as the previous version. Its new features include increased shielding, rotatable front and rear precision drilled cylinders giving adjustable collimation, and three in-line detectors per channel, as shown in Fig. 3 (from P.J.A. Howarth, 1994): a CsI(Tl) photo diode to detect from fast electrons in the 100-400 keV range; a 2.5 cm diameter proton recoil organic scintillator with PSD properties to detect separately 2.5 MeV and 14 MeV neutrons and MeV gamma rays using doubled electronics at low 14 MeV neutron rates, and 1.5 cm diameter fast organic scintillator detectors to detect high 14 MeV neutron rates. Both detectors are 1 cm thick. Each 14 cm diameter detector assembly in JET includes mild steel and other magnetic shielding tubes for the enclosed photo multipliers.

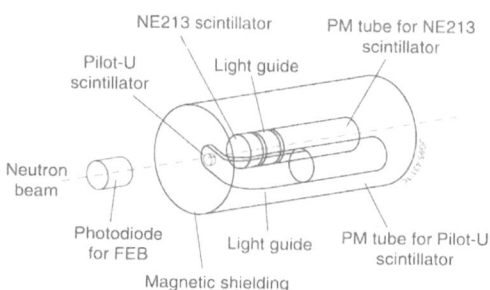

Fig. 3. The in-line detector package for each channel in the JET neutron emission profile monitor-upgrade

Adjustable (by 90° rotation) pre- and post-collimation is fitted to the entrance and exit sides of (oversize) neutron flight tubes for both cameras. The collimators in the horizontal camera consist of 25 cm diameter rotatable cylinders of stainless steel with a fan shaped array of circular holes with two sizes for each channel, 1 cm and 2.2 cm diameter, located at right angles to each other so as to form the defining apertures. The corresponding channels in the two collimation blocks are accurately aligned with each other, with precision end stops. These collimators are rotated by remote control.

The detectors are imbedded in a large shield block to mostly eliminate the gamma ray and neutron flux, apart from those entering through the collimators, from reaching the detectors. The shield comprises 3400 kg/m^3 barytes concrete to attenuate both neutrons and gamma rays and contains fan-like sets of neutron flight tubes. The flight tube channels form cylindrical voids, about 1.0 m long and 3 cm internal diameter. At the ends of the collimators are the neutron detectors, all located inside mild steel detector boxes with water cooling and co-axial electrical cabling. Each detector is enclosed in triple layer magnetic shielding. The concrete shields contain massive internal stainless (to reduce magnetic forces) steel structures to allow accurate alignment of all components. This permitted precision engineering techniques to be used. The concrete shields are enclosed in thin metal to contain cracking. Additional internal strength is provided by re-inforcing bars.

Rear, removable, sections of shielding fit behind and above the detector boxes. These are close fitting, with steps to avoid neutron streaming. A long channel is provided through the removable shields of both horizontal and vertical cameras for locating small neutron sources close to each detector in turn. Polythene beam dumps for each channel are provided.

The approximate total weight of the main horizontal camera shield is 52 tons. The camera is 3.5 m tall, with a cross section in a top view of 2.3 x 1.5 m. The horizontal camera sits upon a single-leg stainless steel support tower 4.2 m tall. The smaller vertical camera sits above the JET vacuum vessel on four support brackets bolted onto the iron transformer limbs on the JET tokamak.

DESIGN FOR ITER HORIZONTAL, RADIALLY VIEWING CAMERA

Geometry of sight lines, pre-shield, and collimators

The problem of sight lines for ITER (as NET) has been considered previously (J.M. Adams et al., 1991 and M Martone et al., 1988). It was concluded that a 2-D fan like geometry similar to JET was the most suitable option, since it allows tomographic reconstruction of the neutron emissivity profile and determination of the axis in R and z of the peak emissivity. A more recent report (L.C. Johnson, 1995) considered the detailed viewing angle requirements of the profile monitor. The report contained preliminary sight line layouts and a study of the fraction viewed of the plasma neutron emission as a function of viewing angle and plasma emissivity profile. A viewing angle of ± 30° about a focal point at $(R,Z) = (13.3,1.1)$ (metres), measured most of the neutron emitting volume.

In the design developed here, shown in Fig. 4, the limiting sight lines are extended to be 0.05 m clear of the points of nearest approach to the port through the vacuum vessel, so as to obtain the maximum possible view. These points are $(R,Z) = (11.67, 2.71)$, $(15.96, 2.72)$, $(11.45,-0.62)$ and $(15.95,-0.20)$, which intersect in a focal point at $(13.90,1.20)$, giving fan angles of 34.23° upwards (on the plasma side) and 36.48° downwards. The nominal plasma magnetic axis is at $(8.14,1.71)$, so the sight lines cover a total vertical height (at the major radius of the plasma axis) of 8.18 m, i.e. over 90% of the plasma height. These extreme sight lines involve intersections with the blanket/shield region facing the plasma, and suitable apertures would need to be designed into these structures.

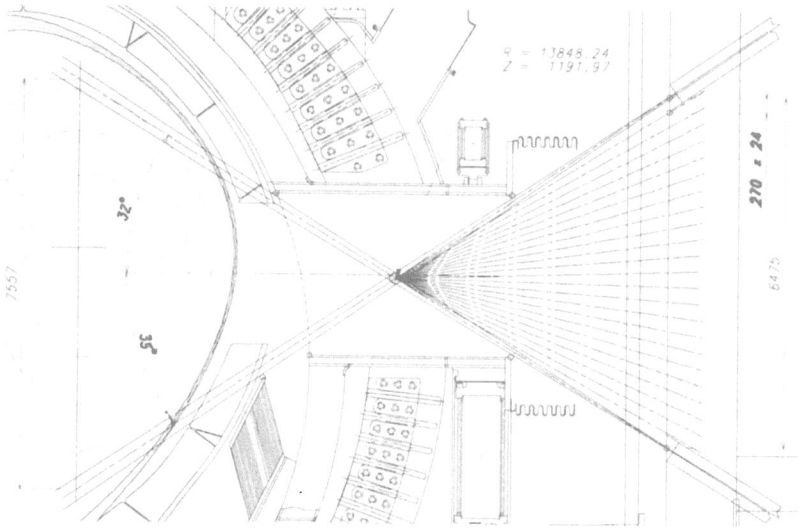

Fig. 4. Sight line limits in the ITER horizontal neutron camera design.

These sight line limits must take account of the finite width of the viewing cones from the individual collimator channels. This calculation requires a choice of the collimator location and geometry. It was decided that the collimators should be outside the cryostat and integrated into the bioshield wall, for reasons of access, maintenance and shielding: If placed at that position, the detectors themselves would be behind the main bioshield wall, hence in a very low radiation region. A pre-shield with a narrow penetration reduces the neutron flux to the bioshield. The finite width of the viewing cones for 2.5 cm diameter collimators (see below) results in a focal point at (13.85,1.19) and fan angles for the viewing cone axis of 32° upwards (on the plasma side) and 35° downwards.

Outside of the cryostat, there is a choice of making the collimators integral with the bioshield or building a separate diagnostic behind the bioshield. The preferred modular system, where each detector system has its own precision machined collimator, allows each system to be maintained or altered as necessary. This design also allows the option of converting all or part of individual selected modules into spectrometers or other instruments.

An initial design is based on JET experience. For ITER, we choose the collimator diameter sizes of 1.0 cm and 2.5 cm, to be able to adjust for the fluxes expected in either d-d or d-t operation, or for low intensity in peripheral channels. It is assumed that ITER will have a d-d phase. The variation in neutron rate (4th power of aperture diameter) is therefore a factor of 39 (compared to a typical ratio of 88 between 50-50 d-t and equivalent 100% d-d plasmas (F.B. Marcus et al., 1993)).

Smaller collimators would present problems. Since the stopping distance of recoil protons from 14 MeV neutrons is of ~0.3 cm, scintillator detectors with any linear dimension much less than 1 cm are subject to "edge effects". It is therefore undesirable to have detectors very much larger than the collimators, since the detector is subject to counts from background gamma and neutron flux. The collimator length is chosen to be 2 m to reduce the flux to the detectors.

The number of channels has been chosen to be 25 in the horizontally viewing camera. This corresponds to slightly better than a 10% resolution in radius. The improved resolution compared to the 10 channels per camera of TFTR or JET is made possible due to the larger system size, and highly desirable, especially if strong gradients in the emissivity are present.

The viewing cone from a collimator of 2.5 cm diameter for both front and rear has its apex half way down the 2 m length, and a full opening angle of 1.43°. With the plasma side of the bioshield at R=18.74 m, the opening of this cone, starting on a horizontal sight line at R=19.74 m, is 0.15 m at the focal point and 0.29 m at the plasma axis, or about 1/28th of the viewing height in the plasma. A 2.5 cm diameter collimator therefore makes the width of the cone nearly equal to the separation between channels at the plasma centre.

The neutron emissivity flux from a 1.5 GW ITER plasma at a detector viewing the plasma axis is 2.5×10^{18} neutrons/m^{2}s, equivalent to 5.6 MW/m^2 of 14.1 MeV neutrons, based on a simple model (J.M. Adams et al., 1991). (The line-integral emissivity deduced by L.C. Johnson, 1995, for the horizontal camera, was about the same for a standard plasma profile, and ranged between 7 to 13 MW/m^2 for different assumptions.) ITER also allows for up to 20% power excursions. For a 1 cm diameter, 2 m long collimator, the maximum flux on a centrally viewing detector is 4×10^{12} neutrons/m^{2}s. An efficiency of 1-2% for typical discriminator bias settings of a 1 cm diameter, 1 cm thick active area would lead to a count rate of about 6 MHz, just within the range of fast organic scintillators. For a full dynamic range, it may be advisable to slightly reduce the collimator diameter below 1 cm. For off-axis channels, the front collimator could be changed to 2.5 cm diameter to increase count rate. For edge channels (especially), background scattering needs to be calculated and subtracted. It limits measurements near the plasma edge below 1% of the peak emission.

Pre-shield

The purpose of pre-shield is to stop all neutron flux to the camera that does not go through the collimators to the detectors and to greatly reduce neutron flux, heating, and radiation damage outside. It is located inside the vacuum vessel port, shown in Fig. 5, with

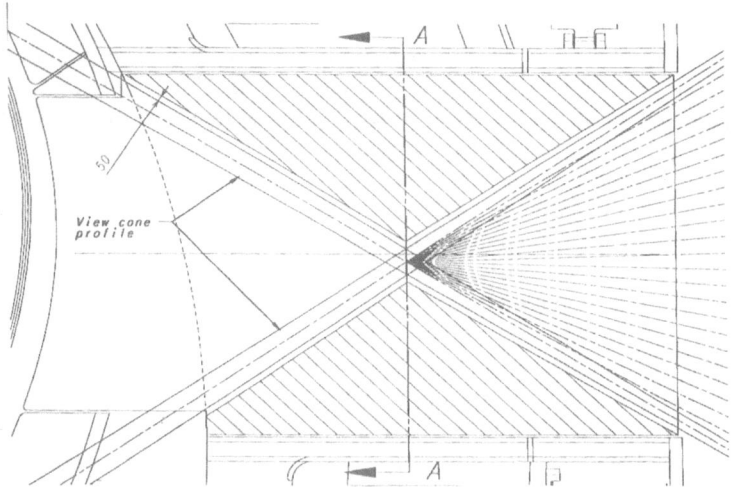

Fig. 5. The pre-shield inside the vacuum vessel port, with a fan shaped opening and a rectangular slit at the focal point.

a fan shaped opening for the sight lines and a rectangular slit at the focal point, with massive shielding elsewhere. The pre-shield (probably) consists of cooled stainless steel and water. The fan is designed for the largest collimators of 2.5 cm diameter. The effectiveness of this pre-shield will be the subject of future calculations. Experimental simulations and neutron transport calculations using the code MCNP are already in progress for the ITER shielding (P. Batistoni et al., 1995) which will be extended to the pre-shield.

The rear of the fan can be used as the high vacuum boundary, with 3-6 mm thick stainless steel, to allow the neutron flux to pass with little distortion, and to have a thin wall only over a small area. Material at the focal point would be undesirable because it would scatter neutrons into other channels.

Neutron detection systems in the pre-shield

The pre-shield serves as the location for a fluid flow neutron activation monitor (J.M. Adams et al., 1991) (or a foil transfer system (O.N. Jarvis, 1991)) incorporated into the plasma facing side of the shield. With MCNP neutron transport calculations, this monitor provides a measurement of integrated neutron flux. When using water as a continuously flowing fluid medium, the activation of the oxygen in the water by 14 MeV neutrons gives a continuous 10 s time resolution. Other fluids should be considered in a special loop since water is only relevant for d-t operation.

With the proposed geometry and position of the pre-shield, detectors could be imbedded in it to measure the global neutron emission, as discussed in the context of slot collimators by Adams et al., 1991.

Individual Collimator and Detector Module Design

A module consists of a cylindrical plug 0.4 m in diameter containing collimators and detectors accurately aligned and inserted into the biological concrete shield along a sight line (with the axis of the module aiming at the focal point). Each module consists of a hollow stainless steel tube (plus stepping to reduce neutron streaming) pointing at the focal point inside the vacuum vessel port. The front of the module does not extend beyond the front of the bioshield, so for angled detectors, the leading edge of the cylinder is level with the plasma side of the bioshield. Since the module diameter is greater than the channel to channel spacing, adjacent modules are slightly offset in the assembly of all the collimators, as shown in Fig. 6.

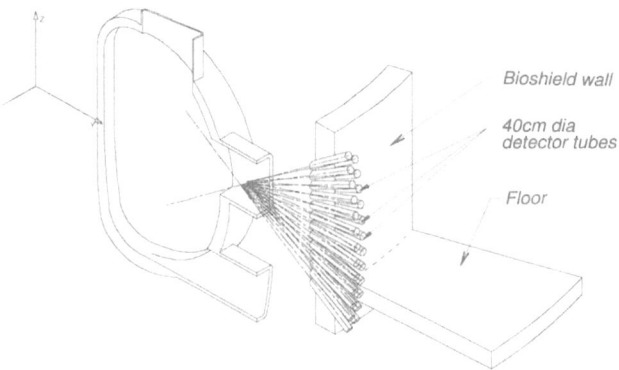

Fig. 6. Overall collimator and detector module layout.

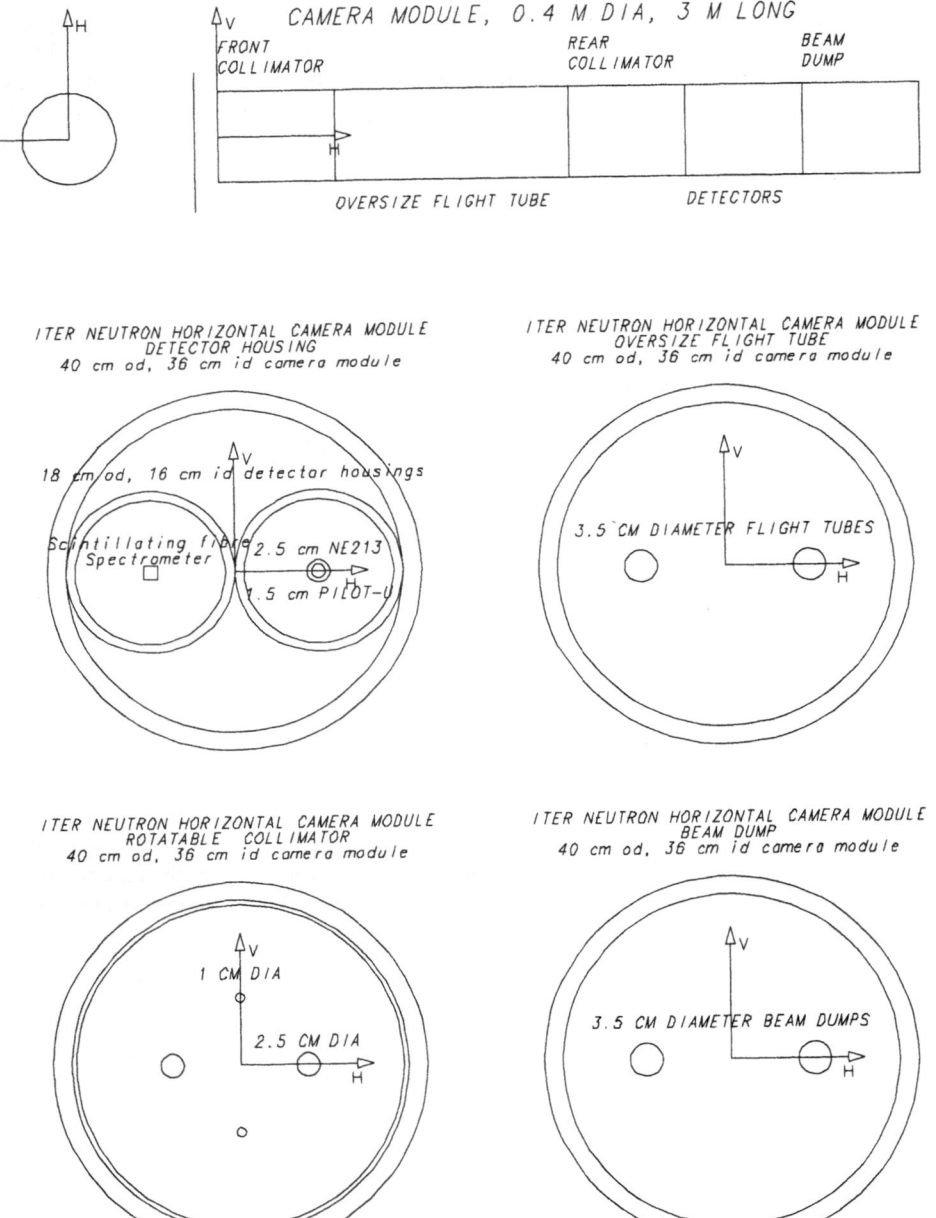

Fig. 7. A collimator and detector module, consisting of a cylindrical plug accurately aligned and inserted into the biological concrete shield along a sight line.

The vertical separation between modules is constant. The overall length of the module is 3 m, including 2 m for the collimator and 1 m for the detectors and neutron beam dump. The module protrudes out the back of the 1.5 m thick bioshield wall. An additional bioshield extension may be required.

In an individual module shown in Fig. 7, the front and rear collimators consist of two cylinders, 36 cm outer diameter (allowing for wall and bearing thickness) and 40 cm in length. Each cylinder has four holes drilled on a circle of diameter 18 cm, two holes of 2.5 cm diameter and two (perpendicular to the first set) of 1 cm diameter. The collimators are individually rotatable to bring the appropriate collimators in line with the detectors, allowing adjustments for d-d or d-t operation, or for peripheral channels with lower flux. (A suitable mechanism needs to be designed.) The intermediate flight tube assembly stops almost all neutrons that do not enter the rear collimator hole, and reduces the radiation outside the bioshield to negligible levels. A collimator rotation of 45° would block neutron flux to the rear of the bioshield and the detectors and allow maintenance of the detector package. The axes of the two collimators are chosen to be parallel, although they could be slightly angled so as to both aim at the focal point, further reducing the slit size in the pre-shield if necessary to reduce neutron flux through the pre-shield.

Several design options are possible. Even smaller diameter holes would be possible with wire erosion machining techniques to obtain precision sizes. The front and rear collimators could be machined from one piece to allow excellent alignment, connected by a machined central shaft allowing the flight tube shielding to be constructed around it. With a precision motor, more than 2 pairs of hole could be included, for example 4 pairs at 45° apart. Careful consideration also needs to be given to aligning the whole module with the vessel and plasma, allowing for both thermal expansion and general movement.

There are two magnetically shielded detector housings inside each module, horizontally side by side, each behind its own collimator path. This system gives great flexibility, allowing two detector packages per sight line, for example one for neutron flux and one for spectroscopy.

Measurement requirements and sources of neutron spectrum anisotropy

The measurements already made with the JET neutron emission profile monitor have been discussed, and the ITER camera design presented here should allow similar measurements, fulfilling the ITER requirements. However, there are regimes of operations where measurement interpretation is difficult. During the ignited phase, without auxiliary heating, the neutron emission is relatively isotropic, and a 2-D profile monitor with many channels is acceptable. Similarly, with a peaked neutron emissivity profile, information about the plasma axial position can be obtained. To unfold the data to obtain a 10% accuracy of the axial neutron rate and alpha birth profile is difficult, since it requires many (accurately calibrated) channels and full 2-D plasma coverage. This design, in conjunction with a similar vertical camera, should allow this accuracy.

Calculations by G.J. Sadler and P. van Belle, 1995, show that if 500 keV neutral beam heating is used, then there will be large distortions in both the 14 MeV neutron energy spectrum and angular distribution. If ICRF heating is used, similar problems may arise. If there is a 3-D distribution of neutron emission, we may need a 3-D viewing system including spectrometer functions.

Neutron Flux Detectors

One of the detector systems could be very similar to those used in JET, with in-line organic scintillator detectors, one with PSD properties and one with fast response. A detailed discussion of detector options is given by Adams et al, 1992. For 2.5 MeV neutron measurements from deuterium plasmas and low level 14 MeV rates, PSD detectors such as NE213 would be appropriate. The development of a solid detector with useful PSD properties and improved electronics should be encouraged. For the 14 MeV neutrons, fast organic scintillators such as PILOT-U allow MHz count rates.

In addition, for emission and control measurements, a highly reliable 14 MeV neutron monitor is desirable, even if it is sensitive to gamma rays or background neutrons. An ionisation chamber working in current mode could be appropriate.

Integrated Neutron and Gamma Ray Spectrometers

The above design for detector modules can be used without modification if compact neutron spectrometers can be developed to fit into one of the detector housings. T. Elevant et al., 1995a, have considered a range of compact spectrometers. The radiation damage sustained by silicon diodes tends to limit their utility, leaving diamond or scintillating fibre detectors as candidates. If successfully developed, scintillating fibre spectrometers (T. Elevant et al., 1995b) could be incorporated into every profile monitor channel. Diamond detectors exist and are being tested. If compact spectrometers are unsuitable, then it would be appropriate to consider mounting high efficiency spectrometers (T. Elevant et al., 1995a) either behind or just slightly to the side of the profile monitor detector stack. The detector modules could contain gamma ray or x-ray flux detectors or spectrometers (G.J. Sadler, 1994), primarily for use in the d-d phase. Gamma ray detectors could be incorporated behind the neutron detectors, with absorbers between them.

CONCLUSIONS

This paper presents a conceptual design for a neutron profile monitor for ITER. Its modular, multiple detector system located in and behind the bioshield makes it flexible and able to accommodate a wide range of detector systems, which could be maintained and modified. The same range of measurements could be provided as in existing profile monitors, and extended by the incorporation of additional instruments.

ACKNOWLEDGEMENTS

We gratefully acknowledge the contributions to configuration control, design and CATIA drawing of Allan Basile, Jim Mann and Mark Mills at the ITER San Diego Site.

REFERENCES

J.M. Adams et al., 1990, The JET neutron emission profile monitor, Nucl. Instrum. Methods A, 329:277.

J.M. Adams et al., 1991, Neutron diagnostics for NET, AEA Technology 411/90.2/FU-UK, Harwell, UK.

J.M. Adams et al., 1992, Investigation of potential fast neutron detectors for use in the JET neutron emission profile monitor in the D-T phase, Report AEA FC 305, AEA Technology, Harwell Lab, UK.

B. Balet et al., 1993, Particle and energy transport during the first JET tritium experiments, Nucl Fus 33:1345

P. Batistoni, B. Esposito, M. Martone, S. Mantovani, 1995, Design of the Neutron Multicollimator for Frascati Tokamak Upgrade, accepted for publ. in Rev. of Sci. Instrum.

P. Batistoni et al., 1995, ITER bulk shield experiment at the Frascati neutron generator, ITER task 28.

T. Elevant et al., 1995a, Ion temperature profile measurements in ITER by means of neutron spectroscopy, Workshop on ITER Diagnostics, Varenna, Italy.

T. Elevant et al., 1995b, Scintillating fibre neutron spectrometer, ibid.

P.J.A. Howarth, 1994, The JET neutron emission profile diagnostic: data exploitation and the development of an upgraded version, Ph.D thesis, University of Birmingham, England.

O.N. Jarvis, 1991, Neutron and gamma-ray diagnostics, Int. School of Plasma Phys., Varenna, Italy.

L.C. Johnson, 1995, Study of generic access for neutron cameras in ITER, Report ITER/US/95/PH-07-13.

FB Marcus et al., 1991, JET Neutron emission profiles and fast ion redistribution from sawteeth, Plasma Phys. and Cont. Fusion 33:277

F.B. Marcus et al., 1993, Neutron emission profile measurements during the first tritium experiments at JET, Nucl. Fusion 33:1325.

F.B. Marcus et al., 1994, Effects of sawtooth crashes on beam ions and fusion product tritons in JET, Nucl. Fusion 34:687.

FB Marcus et al., 1995, Evolution and optimisation of neutron emissivity profiles in JET", Proc. EPS Conference, Bournemouth, U.K.

M. Martone, P. Batistoni, L. Bertalot, M. Pillon, S. Podda, 1988, Analysis and Preliminary Design of Neutron Diagnostics for NET - First Part: Plasma Diagnostics, NET Report EUR-FU/XII-80/88-86.

A.L. Roquemore et al., 1990, TFTR multichannel neutron collimator, Rev. Sci. Instrum. 61:3163.

G.J. Sadler, 1994, Gamma ray and X-ray profile cameras, ITER Working Group, JET, UK.

G.J. Sadler and P van Belle, 1995, The distribution of fusion products at birth, Proc. EPS Conference, U.K.

S. Scott et al., 1995, Isotopic scaling of confinement in deuterium-tritium plasmas, Phys. Plasmas 2:2299.

NEUTRON SPECTROMETRY FOR ITER

J Källne[1] and the NPF Collaboration*

[1]Department of Neutron Research, Uppsala University, Uppsala, Sweden

INTRODUCTION

Neutron diagnostics can be expected on ITER in the form of total yield (Y_n), neutron spectrometry (NS) and camera profile measurements. On present tokamaks, the total yield is done routinely in a way similar to what would apply to ITER[1]. NS diagnostics also exist but these would not be up to the ITER standards with regard to functions, quality and scope of measurements. The same is true for profile measurements. Here we shall be concerned with the new NS diagnostics that ITER will require considering also linked camera aspects.

The planning of NS diagnostics for the present generation of large tokamaks was presented at the 1982 Varenna conference[2]. This summarized proposals for spectrometers and what they could be used for. Since then, NS has been applied mostly at JET where several spectrometer types have been tested in D-plasmas and two of the original program chosen for DT plasmas[3]. NS for ignition plasmas have later been formulated with case studies in relation to the next step tokamaks such as NET[4] and Ignitor[5], besides the work done for ITER as part of the Conceptual Design Activity of 1990[6]. The most concrete manifestation of the NS development work for ITER is the construction of a prototype spectrometer (the MPR) for ignition plasma studies[7]. The present Varenna conference is about ITER and diagnostic functions that must be performed in order to operate the machine for high power, self-heated plasmas.

While present NS is occasionally used for ion temperature (T_i), it is not a routine diagnostic up to ITER information standards and those T_i-diagnostics that are, may not be practical on ITER. Thus, a switch of diagnostic methods may be needed for regular plasma parameters. NS is a potential switch-to candidate on practical grounds but its function standards must be proven. Moreover, some ITER tasks are new such as measurements of dual fuel densities and confined α-particles for which NS is a prime candidate. The experience base is small for the big challenge of developing NS diagnostics for ITER.

The planning of the ITER NS diagnostics involves two main elements namely, the design of the spectrometer and the deduction of the plasma diagnostic functions that can be performed. The linkage comes from describing the observational capabilities, derived from the spectrometer specifications, and expressing them in terms of observed plasma quantities and phenomena with their accuracies and resolutions. Here, this interrelationship approach is used to illustrate the role of NS diagnostics on ITER, the principle capabilities and limitations of instrumentation, and the scope and accuracy/resolution of the observations. The focus of this paper is on 'burning plasma' NS diagnostics operating in the range 1 to 100 % of peak power with expected performance maximum at the peak, i.e., ignition conditions.

Diagnostics for Experimental Thermonuclear Fusion Reactors
Edited by P. E. Stott *et al.*, Plenum Press, New York, 1996

PLASMA PROBING AND MEASUREMENT

NS is a plasma probe of the product parameter of densities (n_d and n_t) and fusion reactivity (ρ), i.e., the neutron emissivity, $y_n = n_d \cdot n_t \cdot \rho$. We assume that the basic equilibrium of the fuel ions is thermal so that ρ is a single valued function of temperature, $\rho = \rho(T_i)$. Non-thermal situations are treated as the sum of thermal (T) and supra-thermal (ST) components, $y_n = y_T + y_{ST}$. This implies an underlying assumption of an unperturbed bulk emissivity y_T with the fast ions (d^* and t^*) contributing an extra emissivity y_{ST}; $y_n = y_T$ unless stated otherwise.

Fast fuel ions can arise from (i) acceleration (RF-heating), (ii) injection (neutral beams), (iii) reactions (e.g., $d+d \rightarrow t^* + p$), and (iv) collisions (e.g., $\alpha^* + d/t \rightarrow d^*/t^* + \alpha$). These produce Doppler <u>shifted</u> ((i and ii)) and Doppler <u>broadened</u> ((iii and iv)) emission peaks in the neutron spectrum. The corresponding y_{ST} emissivities can be probed and distinguished depending on the circumstances. Here we shall only consider the diagnostic use of (iv).

NS measures the neutron energy distribution of the neutron emission (amplitude and shape), i.e., the spectrum S, which for a homogeneous plasma is a Gaussian function

$$S(E_n) = G(E_n) = \frac{1}{\sqrt{2\pi}\sigma} \exp\left[-\frac{(E_n - E_0)^2}{2\sigma^2}\right] \tag{1}$$

Here, $E_0 = 14$ MeV is the peak center and the plasma state is determined by $\sigma = 75.2\sqrt{T_i}$ (keV).

The viewed plasma is not homogenous in a steady state so the emissivity is a function of space and time, $y_n(\underline{r}, t)$; special non-stationary situations can also be considered (e.g., due to rotation) resulting in y_n with vector components. The measurement is a line integral of y_n along the cord λ through the plasma defined by the lateral coordinate (the distance ζ from the plasma center in relative scale of the minor radius). Since $T_i = T_i(r)$, the resulting spectrum S_λ is a sum of Gaussians G_{T_i} with weight from $y_n(\underline{r}, T_i)$ along λ.

Distinction can be made between plasma core measurements with $\lambda = \lambda(\zeta = 0)$ and profile measurements with a sight line array $\lambda(\zeta)$ extending to the peripheral limit, typically, $\zeta^{MAX} = 2/3$. The most detailed measurement can be done for $\lambda(\zeta = 0)$ with one dedicated installation while the profiles would be done using a camera with spectrometer functions.

FUSION NEUTRONS FOR DIAGNOSTICS

The spectrum of fusion neutrons from DT plasma (of fixed T_i, $T_i = 20$ keV) consists of two Gaussian peaks at 14.0 and 2.5 MeV of the reactions $d+t \rightarrow \alpha+n$ and $d+d \rightarrow {}^3\text{He}+n$; the shape of the latter is related to T_i through $\sigma = 35.1\sqrt{T_i}$ (in keV) (see Figure 1a). A third component is due to 3-body reactions $t+t \rightarrow \alpha+2n$ producing a phase space distribution (no peak). This is of no diagnostic use as it is obscured by background (see below). The useful neutron signal of the plasma is the flux of the $d+d$ and $d+t$ reactions which can be measured using a high resolution spectrometer. The resolution requirement scale is set by the thermal Doppler width $\Gamma_{DT}(T_i)$, as the fusion neutron spectrum is always smoothed by this factor.

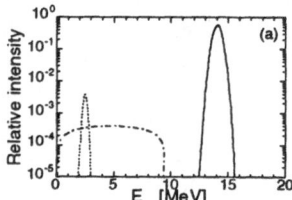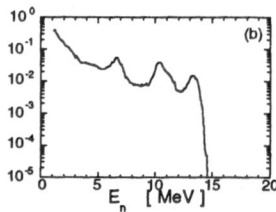

Figure 1. (a) Neutron spectrum due to thermal ion reactions in DT plasma ($T_i = 20$ keV). (b) Spectrum of the neutron wall emission due to primary $d+t \rightarrow \alpha+n$ reactions in the plasma[8]

A spectrometer viewing the plasma will also see the back wall of the plasma chamber. This acts as a wall emission (WE) source due to the incidence of plasma neutrons[8]. The shape of the WE spectrum can vary mainly due to the relative abundance of light (e.g., carbon) and medium weight (e.g., iron) wall elements, and also on the detailed light nuclide composition; an example of a calculated spectrum for C+Fe is shown in Figure 1b. The amplitude for given Y_n varies little with composition and plasma conditions. Studies of the WE spectrum in terms of principle scatterings and reaction mechanisms and their dependence exist. There is a need for a better data bank, especially neutron back scattering cross sections[8].

Similar in nature is the in-scattering (IS) background. It is due to neutron scattering in the collimator wall so that the solid angle of acceptance is larger than that of the physical aperture which defines the flux considered as signal (S). The WE+IS background (B) is practically inescapable. For instance, attempts to decrease the WE amplitude by recessing the wall would be ineffective in lowering the background as IS may be increased with the remedy of WE. B has a weak $\lambda(\zeta)$ dependence[8] contrasting sharply with the ratio B/S.

The amplitude of the neutron spectrum mirrors changes occurring in the emissivity expressed by $y_n=n_i^2 \cdot \rho(T_i)$ if $n_i=n_t=n_d$. The temperature swing (viz., $T_i=4$ to 40 keV) will cause the neutron signal to vary two orders of magnitude; dt and dd track each other but dt is dominating by a factor of ≈ 100 (Figure 2)[9]. To compare the strengths of the T_i and n_i dependencies we introduce a local strength α for the former, i.e., $y_n=n_i^2 \cdot T_i^\alpha$. The variation of α with T_i is shown in Figure 1b for the dt reaction. We thus find that the T_i dependence in terms of α_{dt} varies from 4 to 0.2 for the operating range $T_i=4$ to 40 keV, being equal to the n_i dependence ($\alpha_{dt}=2$) at $T_i=14$ keV.

Excursions in T_i and n_i will have very different impact on the emissivity depending on the T_i working point. The burning plasma of ITER will, at high temperature, be most sensitive to density changes. Moreover, the density profile would have a relatively greater impact on the emissivity profile than at lower T_i. At low temperatures, the emissivity is most sensitive to T_i-changes. Below 1 keV it becomes so strong ($\alpha=6$) that it can overwhelm other dependencies which is the justification for the practice to use Y_n-measurements as a T_i monitor.

The neutron signal intensity can contain transients (amplitude changes $\Delta S/S$ over a short time period τ_p) related to time variations in the level of the total yield $Y_n(t)$ or in the profile ($y_n(r,t)/Y_n$), the latter being mostly felt for peripheral sight lines. These are observed in successive measurements over periods Δt whose length determines the time available to acquire counting statistics ($\sigma=1/\sqrt{N_S}$) and the time resolution to 'see' (resolve in time) plasma transients. Ability to measure fast transients with respect to time ($\Delta t<\tau_P$) and amplitude ($\sigma<\Delta S/S$) rests with the count rate ($C_n=N_S/\Delta t$) which translates into the requirement $C_n>(S/\Delta S)^{1/2}$. To measure the transient would require about 3 observations during τ_P with an accuracy of better than $\sigma>1/3(\Delta S/S)\cdot 1/\Delta t$ which can be performed if $C_n>9\cdot(\Delta S/S)^2\cdot(1/\tau_p)$. For instance, with $C_n=100$ kHz one can measure transients of $\tau_p=100$ ms and $\Delta S/S=3$ %. To provide the required minimum count rate (C_n^{MIN}) at low fusion power and keep observing, as the rate increases orders of magnitude up to C_n^{MAX} is the fundamental problem of dynamic range that NS diagnostics must face. The dynamic range is given by the ratio C_n^{MAX}/C_n^{MIN} which is maximized by matching spectrometer and collimated flux[10].

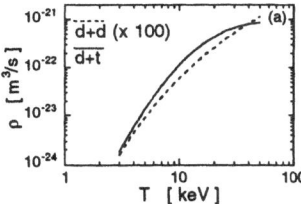
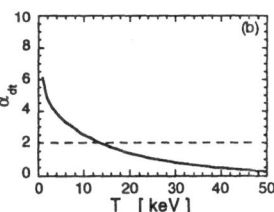

Figure 2. The fusion reactivity of d+t and d+d reactions (a)[9] and the parameter α_{dt} (b) as function of T_i.

SPECTROMETER CHARACTERIZATION FOR PLASMA DIAGNOSTICS

The NS instrumentation can be described by specifying values for the relevant basic parameters: count rate capability limit (C_n^{CAP}), (flux) detection efficiency (ε), energy resolution ($\Delta E/E$), energy bite covered (E_{BITE}). The performance of the diagnostic derives from these. What should then the target values for developing a spectrometer design be? This is hard to define strictly but can be indicted by noting that undesirable performance restrictions would result if $C_n^{CAP}<1$ MHz or $\varepsilon<10^{-5}$ cm^2 at the reference $\Delta E/E =2.5$ % (FWHM-value, matching the Doppler broadening at $T_i=4$ keV), or if E_{BITE} is narrower than ±15 %. The latter requirement derives from the realization that separate measurements of the d+d and d+t components seem most practical. It is interesting to note, in this context, that the quality of the data and the diagnostic scope will generally increase with increasing count rate.

The count rate stands out as the main figure of merit[10] which translates into a dual parameter requirement: the efficiency should be as high as possible to maximize the count rate for available collimated neutron fluxes and the count rate capability should be sufficiently high so that it does not restrict count rate. It is this combination of high efficiency and count rate capability that makes the design of a spectrometer for fusion neutrons special[11] and indeed different from developing one for nuclear laboratory experiments which is yet the place where most of the relevant technical know-how originates.

The characterization should also include quality aspects such as type and accuracy of calibrations, stability in operation, etc. These are very important but must be assessed for each spectrometer proposal and judged with respect to required standards.

SPECTROMETER DESIGNS

One can distinguish between three main categories of fusion neutron spectrometry based on the nuclear reaction that is used to make the neutron manifest itself in a detector and the method of energy measurement: (i) the energy of proton recoil (PR) of neutron scattering, (ii) the time of flight (TOF) of the scattered neutron, and (iii) the total exit channel energy of the neutron reaction products (NRP); hybrids such as (ii) combined with (i), i.e., TOF-PR are also possible. Further distinctions are based on design differences that are important for the spectrometers characteristics in terms of parameter specifications and qualitative aspects. Some of the main designs are listed in Table 1 and the principle target and detector configurations relative to the neutron beam are shown in Figure 3.

Of the recoil spectrometers, there is one case where the proton energy is measured with a magnet spectrometer (#1, MPR) while all other cases (2, 3, 4, 7) use pulse height determination which is also used in all NRP methods (8,9,10). The TOF spectrometers are based on measuring the timing difference of events in two detector sets; this is, essentially, a coincidence measurement as opposed to detector events that are unrelated (accidental).

Some of these spectrometer types have been used on present machine (2,3,4,8,10) or built for use in the near future (2,7)[3], while one (1) is under construction[7]. Three are proposed

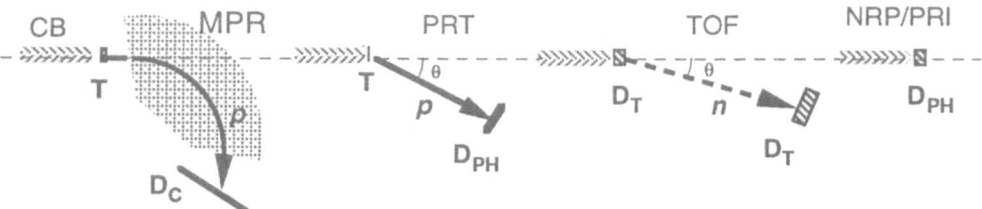

Figure 3. Configurations of targets (T) and detectors (D for counting, D_C, and pulse height measurements, D_{PH}) in fusion neutron spectrometers designs (cf. Table 1).

400

Table 1. List of neutron spectrometers of different categories and configurations.

	METHOD	TARGET	REACTION	DESCRIPTION	ITER USE
1	MPR	Polythene	$n+p \rightarrow p'+n'$ [1]	Magnetic analyzer	A
2	PRT	Polythene	$n+p \rightarrow p'+n'$	Telescope[2]	A
3	PRI	Scint.	$n+p \rightarrow p'+n'$	Inclusive	B
4	TOF	Scint.	$n+p \rightarrow p'+n'$	Coincidence	
5	TOF	Deut. scint.[4]	$n+d \rightarrow d'+n'$	Coincidence	
6	TOF	Polythene	$n+p \rightarrow p'+n'$	Coincidence	A
7	TOF-PRT	Polythene	$n+p \rightarrow p'+n'$	Coinc. + Telescope	
8	NRP	^{3}He+Ar gas	$n+^3$He$\rightarrow p+t$	Ion chamber	
9	NRP	Diamond	$n+^{12}$C$\rightarrow^9$Be$+\alpha$	Diode	B
10	NRP	Silicon	$n+^{28}$Si$\rightarrow^{25}$Mg$+\alpha$	Diode	B

[1] Bold face recoils and reaction products are used for neutron energy determination. [2] Telescope viewing the target. [3] Inclusive neutron energy determination; elemental construction needed for spectrometry. [4] Deuterium instead of hydrogen based plastic scintillator. A = for principal ITER use; B= auxiliary.

as principal spectrometers for ITER (1,2,6) and three for auxiliary use (3,9,10). The proposals are covered in separate papers of these proceedings[7,12,13,14]. Some comments are made below on the principal spectrometer proposals which are in different development stages.

The MPR is being built and its parameter specifications and qualities have been defined by computer simulations[7]. Its performance on ITER has been projected for assumed operating conditions which in part can be tested on present tokamaks. The TOF method has been used at JET for D-plasmas but the design for ITER requires that the first coincidence detector be moved out from the beam. This puts very tight demands on the first detector which must be placed very close to the neutron beam and must include regional angular selection. These are principle problems whose solutions[12] must be assessed with regard to feasibility before parameter specifications can be made for this type of design. The PRT method is used in a 14-MeV spectrometer recently installed at JET[3]. It has a complicated annular configuration which one would attempt to avoid in a design for ITER. A design with an angular selective recoil detector is described in[13].

The count rate capability is the aspect that is most difficult to combine with the others. The proton recoil method can 'theoretically' reach $\varepsilon \approx 10^{-4}$ cm^2 at 2.5 % as compared to a value of $\varepsilon = 0.6 \cdot 10^{-4}$ cm^2 for the MPR instrument; higher values can be obtained by giving up energy resolution, e.g., $\varepsilon = 10^{-3}$ cm^2 at $\Delta E/E = 15$ % is projected for the MPR. The count rate capability is high for the MPR ($C_n^{CAP} >> 10$ MHz) but is otherwise a limitation of the proton recoil spectrometers because of the pulse height based energy determination used in the PRT:s. The TOF-method can offer higher efficiency than the PR-methods by up to a factor of 100 but the measured neutron spectrum consists <u>intrinsically</u> of both coincidence (signal) and accidental events (background). The B/S ratio can be small (B/S<<1) but increases proportionally as the signal rate and escalates into a run-away situation as the detector single rates exceeds a critical level. The C_n^{CAP} is set by the maximum acceptable B/S-value and must always be well below the runaway level where the data become unreliable. The PRT and TOF methods are not suitable for high count rate applications unless used in combination with a spectrometer without count rate capability limitation.

NEUTRON SPECTROMETRY DIAGNOSTIC FUNCTIONS FOR ITER

We will first consider central measurements performed with a spectrometer system looking through the plasma axis. A system with a radial and two tangential (counter to each other) sight lines through a horizontal port is sketched in Figure 4. Such a concept (TRIAD) and its potential ITER diagnostic functions are outlined below[15].

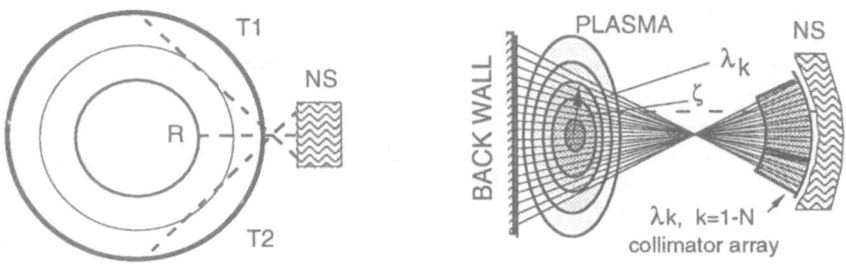

Figure 4. Schematics of sight lines for the TRIAD[15] and RAINES[16] neutron spectrometer systems.

(I) The ion temperature can be determined from the shape of the neutron spectrum peak due to thermal d+t reactions[17]. The central part of the peak is used, i.e., the range -0.5 MeV<$E-E_0$<+1.5 MeV including the high energy tail dropping a factor of 500. Traditionally, the FWHM-value of the peak is used to obtain T_i-value which indeed is necessary in case of limited accuracy (statistics or otherwise). In the new approach, the data of the shape can be analyzed to provide a specific T_i-value, namely, the maximum of the spatial profile, T_i^{MAX}. The result is practically model independent suggesting the possibility of obtaining very high overall accuracy levels (i.e., better than ±5 %). Information would also be obtained on how the emissivity of the plasma within the volume of the line of sight is distributed as a function of T_i. It is assumed that both density and temperature profiles are centrally peaked, i.e., $T_i(0)=T_i^{MAX}$. Hollow profiles may or may not be manageable reflecting the generic limitations of line integrating measurements. In this case, model based deduction of T_i from the data would be needed. Experimentally, the peak shape measurement requires a spectrometer that is very well calibrated (well known response function) and that is able to provide data of reasonably good statistics (i.e., high count rate).

(II) Determination of the fast α-particle component in the plasma is based on measuring the high energy range 15.5 <E_n<20 MeV of the neutron spectrum, i.e., the slope of the dt peak starting at $E-E_0\geq1.5$ MeV (see Fig.5).[18] This region has been predicted to be dominated by the d^*+t and t^*+d reactions where d^* and t^* belong to supra-thermal populations created by collisions between fast α:s and bulk ions in plasmas with significant α-heating. This 'tail' spectrum (amplitude and steepness of slope) is affected by the $α^*$, d^* and t^* populations. Information about the velocity distributions and amplitudes, and the underlying slowing-down and confinement conditions presented by the plasma, is embedded in these data some of which can be extracted by means of model based analysis. Specifically, the α-particle pressure affects the neutron spectrum rather directly which should be possible to determine[19].

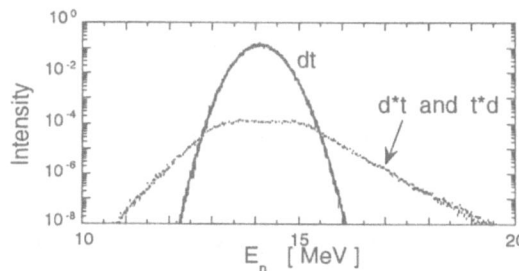

Figure 5. Detail of the neutron spectrum showing the main peak due to thermal reactions d+t→α*+n and a broad distribution due to supra thermal d*+t and t*+d reactions: the relative intensities represent the balance between creation and slowing down of the α*, d* and t* populations at T=20 keV.

Generally, measurement of the neutron emission, due to dt and d*t plus t*d reactions, is a way to monitor continuously the amplitude of the α^*-particle population compared to the α^* production rate and including, especially, its time variation. The use of NS as an α-particle diagnostic (as well as a fast fuel ion diagnostic) requires a spectrometer that can measure the d+t peak in the range E_n=13 to 18 MeV covering an intensity swing of 10^5. High count rate operation in the MHz range is necessary. Moreover, the spectrometer must have high sensitivity which is secured through a high degree of control of extraneous radiation.

(III) Determination of plasma toroidal rotation (V_T) is based on absolute measurement of the neutron emission in two opposing tangential directions (T1 and T2, see Figure 1) and determined from an <u>observed difference</u> quantity[15]. This can be expressed, for the sake of illustration, as a peak position difference $\Delta E_T = E_{T1} - E_{T2} = 2\Delta E_{Ds} + \varepsilon_{T1} - \varepsilon_T$ letting ε_{T1} and ε_{T2} denote off-sets in the energy calibrations of the two spectrometers. The Doppler shift ΔE_{Ds} is proportional to V_T and amounts to 0.27 keV ($\Delta E/E = 4 \cdot 10^{-5}$) at $V_T = 1$ km/s (the ITER lower limit diagnostic requirement), or a measured difference of $2\Delta E_{Ds} = 0.54$ keV. It is not feasible to attempt an energy calibration at the 10^{-5} level. Instead, the acquired data for different plasma conditions should be used to find the limit (ΔE^{MIN}) of lowest values of $|\Delta E_T|$ and use ΔE^{MIN} as an estimate for the calibration difference, i.e., $2\Delta\varepsilon_T \equiv \varepsilon_{T1} - \varepsilon_{T2} \approx \Delta E^{MIN}$ as $2\Delta\varepsilon_T \rightarrow \Delta E^{MIN}$ for $V_T \rightarrow V_T^{MIN} \approx 0$ (assuming mono-directional V_T for small values). One can now use the obtained ΔE^{MIN} value to determine V_T for any plasma with $V_T \gg V_T^{MIN}$ from the measured ΔE_T using $\Delta E_{Ds} = (\Delta E_T - \Delta E^{MIN})/2$. This leads to a systematic underestimation of the true rotation value amounting to the lowest rotation the plasma has had in any observed difference spectrum. Experimentally, a dual spectrometer is required with very special calibration characteristics. It is not unfeasible that this can be attained and it hinges mostly on finding solutions to rather specific stabilization problems. This is a well defined technical R&D task. It can be noted that x-ray and neutron spectrometry are, in principle, similar with respect to plasma rotation measurements. Crystal spectrometry, benefiting from passive photon energy (wave length) determination, is presently used to measure V_T. The use of two cross calibrated and stabilized instruments with passive neutron energy determination (such as the MPR:s) is what could make neutron spectrometry a plasma rotation diagnostic. A switch from x-ray to neutron spectrometry on ITER would be motivated by practicality factors.

(IV) Determination of the <u>fuel ion densities</u> n_d and n_t is based on the absolute measurement of the neutron emission spectrum contained in the 2.5- and 14-MeV peaks (referred to as 1 and 2)[20]. The quantities measured are the peak intensities (I_1, I_2), and the peak widths (Γ_1 and Γ_2) giving an effective T_i. This could be improved with the shape derived T_i described above. Also measured are the WE+IS background around the 2.5-MeV peak, B_1', and the absolute magnitude of WE (B). The information on I_1 is obtained through subtraction of background based on B' and B and supported by model predictions. The I-quantities are products of the density factors ($I_1 \propto n_d^2$ and $I_1 \propto n_d \cdot n_t$), the core reactivities ($\rho_1(T_i)$ and $\rho_2(T_i)$), and the effective plasma volumes intersected, i.e., V_{eff} being essentially the product of chord length (Λ) and the cross section area (defined by the geometry and distance of neutron collimator). The absolute densities n_d and n_t are determined by the measured I-values, the measured T_i-value, inserted into the known functions $\rho(T_i)$ in Figure 1a, and Λ. The Λ-value must be provided for, the best source being the neutron camera. The n_d/n_t ratio can be determined without the Λ input. The errors come from three main sources: (i) The effective T_i to determine the reactivity (less than ±5 %); (ii) the chord length Λ (less than ±10 %); (iii) the background subtraction and absolute efficiency (ε) calibration. Error sources (i) and (ii) are general to the method while (iii) depends on the instrument (can be made small). Thus, the estimated systematic uncertainty of the density diagnostic is less than ±15 % in n_d and n_t and less than ±5 % in n_d/n_t; the error on n_d increases as n_d/n_t falls below 0.1. Instrumentally, a spectrometer is required that can measure the 2.5-MeV peak superimposed on the WE+IS background. The signal is weak (Figure 6) and a factor of 3.5 is gained by using a tangential rather than a radial signal line. This improves the data quality (a factor 2 in statistics) and extends the dynamic range (factor of 2) and the n_d range to 4 times smaller n_d/n_t-ratios.

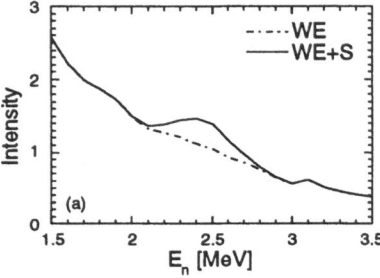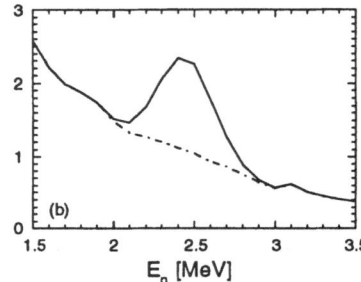

Figure 6. Detail of the neutron spectrum showing the peak due to thermal d+d→^{3}He+n reactions (at T_i=20 keV) superimposed on the plasma wall emission (WE) background as computed for carbon+iron and radial sight line through plasma center (a)[8]; the estimated signal to noise enhancement for a tangential sight line is also shown (b).

We now turn to spectrometry profile measurements which can be done, for instance, with an advanced neutron camera with spectrometer function[16]. Here, we point out some general aspects relating to the discussion above to illustrate how these profile measurements can complement the 'central' ones.

(i) The <u>temperature</u> can be measured as function of the sight line $\lambda(\zeta)$. As the peak temperature would be derived for each sight line one would actually obtain $T_i(\zeta)$, i.e, the radial dependence. The temperature range would be limited because both intensity and Doppler width decrease with decreasing T_i and therefore fall with increasing ζ (assuming peaked profiles). It would be difficult to go below the 3 to 4 keV level as a lower limit for T_i measurements. At the radius where $T_i \approx 3.5$ keV, the intensity should have fallen by two orders of magnitude given a peak T_i of 20 keV besides additional intensity reduction due to the density profile peaking. However, it is not only the extended ζ-range that is of interest but the core and the proximity where one would like also the <u>precise</u> information on the gradient $dT_i(\zeta)/d\zeta$ to be combined with the results of the peak shape analysis mentioned in (I).

(ii) It is possible that the signature of the <u>α*-population</u> could be observed in the spectra for regions a bit outside the core. Here one could probably take an intensity drop by a factor of 10 but not more which, however, could be enough to bring the measurement outside the radius of the q=1 surface. The motivation for this would be to compare the behavior of the α*-populations as function of radius sampleds at a time resolution of seconds.

(iii) Clearly, insight on the toroidal rotation cannot be contributed. The poloidal rotation is here assumed to be small at the level of having no significant effect on the measurement in (i). Should this rotation be significant it would manifest itself as a Doppler shift of the spectra recorded with the spectrometer array. It could then be detected by difference measurements as discussed in (III). This possibility remains to be studied. If the result was positive, it would provide further motivation for a camera with spectrometer function on ITER.

(iv) The spectrometer array would provide information for reducing the systematic errors in the density measurements discussed in (IV). The array would determine the $T_i(\zeta)$ profile which can be used to take out the reactivity factor of the line integrated emissivity profile measured simultaneously. This leaves a quantity proportional to $n_d \cdot n_t \cdot \Lambda$ whose ζ-dependence is measured which can be iterated to determine $\Lambda(\zeta=0)$ used in (I).

Finally, the spectrometer array can resolve signal neutron emission from the WE+SI background as function of ζ which is essential for peripheral channels where the background would dominate conventional camera detectors (see Figure 7). The spectrometer array would provide calibration for camera detectors which could then be used for lower power shots beyond the reaches of spectrometers because of efficiency limitations.

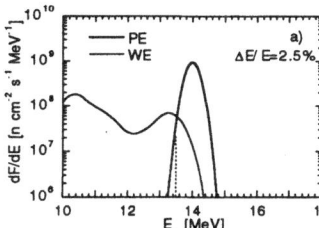 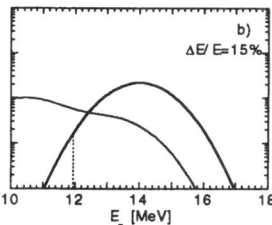 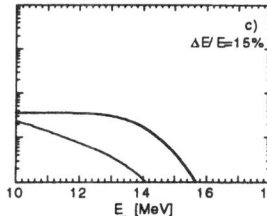

Figure 7. The plasma emission of neutrons from d+t reactions (at T_i=3.5 MeV) and the wall emission (WE for iron+carbon) as seen using a spectrometer with $\Delta E/E$=2.5 % and 15 % energy resolution (a and b) or stopping recoil detector with $\Delta E'/E'$=15 % pulse height resolution (c). The signal amplitude is that of a peripheral sight being a factor 1/100 times that of the central line. The dotted line shows the energy threshold excluding 1 % of the signal intensity.

CONCLUSION

A short survey of the status and progress in neutron spectrometry (NS) diagnostics has been presented. The potential of NS has been discussed from new analyses of principle capabilities and instrumental developments. It is found that NS could be a very powerful method for burning plasmas and could potentially provide several diagnostic functions on ITER exemplified by four essential ones referring to fuel ion densities, ion temperatures, toroidal rotation and the fast alpha particle population. This assessment is based on the new instrumentation and methodological approaches hitherto untested on tokamaks. It assumes that a high performance magnetic proton recoil (MPR) method works as projected and that it is practical on ITER. Moreover, it is assumed that two large spectrometer installations can be accommodated in high neutron flux positions, one (TRIAD) for central plasma measurements and one (RAINES) for profiles (incorporated in the camera). The study of these systems continues to explore yet other diagnostic uses of NS. An important further step of ITER EDA work on NS is to verify that the interface on ITER permits the attainment of projected measurement conditions, especially with regard to sight line directions and proximity to plasma. Finally, technical R&D should be done to specify operation of the NS diagnostics in the tokamak environment.

* The Neutron Physics for Fusion Collaboration of EURATOM-NFR and EURATOM-ENEA-CNR Associations: L Ballabio, S Conroy, J Frenje, G Ericsson, J Källne, Department of Neutron Research, Uppsala University, S-75121 Uppsala, Sweden; E Traneus, Department of Radiation Sciences, Uppsala University, S-75121 Uppsala, Sweden; PU Renberg, The Svedberg Laboratory, Uppsala University, S-75121 Uppsala, Sweden; G Gorini, M Tardocchi, Department of Physics, University of Milan, I-20133 Milan, Italy, and Istituto di Fisica del Plasma, CNR, I-20133 Milan, Italy.

REFERENCES

1. C. Barnes et al, DT neutron measurements and Experience on TFTR, these Proceedings.
2. O. N. Jarvis, Neutron and gamma ray diagnostics, in Diagnostics for Contemporary Fusion Experiments (Proc. Course and Workshop, Varenna, 1991), P.E. Stott, D.K. Akulina and E. Sindoni, eds., Vol. ISPP-9, Società Italiana di Fisica, Bologna (1992), p 541.
3. O. N. Jarvis, Neutron measurement techniques for tokamak plasmas, Plasma *Phys. Contr. Fusion* 36(1994)209-244.
4. J. Källne, The Role of Neutron Measurements in the Study of Fusion Burning Plasmas, *Comments Plasma Phys. Contr. Fusion* 12:235 (1989); J. Källne, Neutron diagnostics for burning DT plasmas of ITER/NET, Rep. UU-NF 91/1, Uppsala University, Uppsala, Sweden (1991).
5. G. Gorini and J. Källne, Neutron spectrometry for compact ignition experiments, *Il Nuovo Cimento* 14D:1115 (1992); G. Gorini, J. Källne and M. Nassi, Neutron emission diagnostics in compact ignition experiments, Rep. UU-NF 93/7 , Uppsala University, Uppsala, Sweden (1993).
6. V. Mukhovatov, ITER operation and diagnostics, *Rev. Sci. Instr.* 61:3241 (1990).
7. G. Ericsson and the NPF Collaboration, The MPR neutron spectrometer project, these Proceedings
8. P. Antozzi et al, Neutron wall emission in tokamaks and diagnostic effects, to appear in *Nucl. Instr. Meth.* (1995).
9. H.S. Bosch and G.M. Hale, *Nucl. Fusion* 32:611 (1992).
10. G. Gorini and J. Källne, Neutron spectrometry diagnosis of DT fusion experiments, *Comments Plasma Phys. Contr. Fusion* 15:193 (1993).
11. J. Källne and G. Gorini, Count rate performance of spectrometers for fusion neutrons, *Rev. Sci. Instr.* 64:2765 (1993).
12. T. Elevant et al, Ion temperature profile measurements in ITER of neutron spectrometer and Scintillating fibber neutron spectrometer, these Proceedings.
13. T. Nishitani et al, Neutron spectrometers for ITER, these Proceedings.
14. A. Krasilnikov et al, Diamond detectors based DT neutron spectrometer for ITER, these Proceedings.
15. J. Källne, L. Ballabio, S. Conroy, G. Ericsson, J. Frenje, P.U. Renberg, E. Traneus, G. Gorini and M. Tardocchi, Multiple neutron spectrometer diagnostic for ITER, in: "Controlled Fusion and Plasma Physics", P.E. Stott, ed., European Physical Society, Petit-Lancy, Switzerland (1995).
16. E. Traneus and the NPF Collaboration, Neutron camera with spectrometer function, these Proceedings; J. Källne and G. Gorini, A Neutron Camera for Burning Plasmas, *Rev. Sci. Instrum.* 63:4545 (1992).
17 G. Gorini and the NPF Collaboration, Detailed ion temperature information from the neutron emission spectrum, these Proceedings.
18. G. Gorini, L. Ballabio and J. Källne, Alpha particle kinetic effects in the neutron emission of burning DT plasmas, *Rev. Sci. Instr.* 66:936 (1995); J. Källne and G. Gorini, Neutron observations and alpha particles in high power deuterium-tritium plasmas, *Fusion Technology* 25:341 (1994).
19. L. Ballabio et al, Alpha particle kinetic effects in the neutron emission of burning DT plasmas, Rep. UU-NF 95/7, Uppsala University, Uppsala, Sweden (1995).
20. J. Källne, P. Batistoni and G. Gorini, On the possibility of neutron spectrometry for determination of fuel ion densities in DT-plasmas, *Rev. Sci. Instr.* 62:2871 (1991).

ADVANCED NEUTRON CAMERA WITH SPECTROMETER FUNCTIONS

E. Traneus[1] and the NPF Collaboration*

[1] Department of Radiation Sciences, Uppsala University, Uppsala, Sweden

INTRODUCTION

The main purpose of the neutron camera is to measure the line emission brightness of the plasma (B) as a function of the lateral distance from the plasma center (ζ)[1]. This is done with an array of collimators (corresponding to sight lines λ_k, k=1 to k^{MAX}) lying in the poloidal plane each equipped with a detector. The resulting data, $B_\lambda = B(\zeta)$ with $\zeta = \zeta(\lambda_k)$, are related to the radial profile of the plasma neutron emission. The collimators define the ζ_λ-values and the spatial (lateral) resolution $(\Delta\zeta)$, while other performance characteristics of the camera reside with the detectors. The detectors measure the signal neutrons in the channels (the flux reflecting B_λ) which are blended with extraneous radiation coming down the collimators (low energy neutrons and γ-radiation). Several detector aspects are crucial for the camera: (i) accuracy of calibration (the flux detection efficiency of signal neutrons, ε); (ii) insensitivity to extraneous radiation or discrimination capability; (iii) stability in calibration and discrimination; (iv) time resolution (Δt); (v) dynamic range (D) as specified by the interval of neutron flux change for which required performance can be maintained.

Ability to determine the energy distribution of the collimated neutron flux is an implicit requirement and, in practice, essential for substantiating the quality of the data[2-5]. It is compelling to investigate the spectrometer function of neutron cameras from the double motivation of (i) securing highest performance in the camera's basic diagnostic mission and (ii) enabling functional mission extension. This would be an advanced camera to view the plasma through a horizontal port and intended mainly for operation with burning plasmas over a dynamic range of up to $D \approx 10^2$; complementary detectors could be fitted to increase the D-value to include low power plasma conditions. In this contribution we shall discus the design of an advanced neutron camera with spectrometer functions and its diagnostic role on ITER.

The entertained hopes that small and simple 'conventional' detectors can meet the perceived basic diagnostic mission should not deter considerations of an advanced neutron camera on ITER. The high interface price of the large installation affords potential benefits in terms of increased quality and extension in scope of the data. This is the system of high cost

Diagnostics for Experimental Thermonuclear Fusion Reactors
Edited by P. E. Stott *et al.*, Plenum Press, New York, 1996

407

and benefit to be compared with the alternative opposite of the conventional detector approach representing low cost and benefit. Within these extremes, the cost/benefit ratio can be adopted as a gauge for system evaluation and subsequent implementation, should the fortunate situation arise that both advanced and basic camera detectors prove realizable for ITER. This work contributes material to the discussion of the role neutron systems in the ITER diagnostic complement on the important cost/benefit basis.

COLLIMATOR ARRAY

The spatial resolution (for a horizontal camera) is determined by channel separation (ΔZ) and width (Δz), which are given by the collimator parameters D_c, $\emptyset_c$ and l_c besides the plasma distance L (Fig. 1). The choice $\Delta z = \Delta Z$ optimizes the flux and this specifies the relationship $L=(\Delta z/\emptyset_c) \cdot l_c$ between resolution, viewing distance and collimator[6]. With a typical collimator diameter of $\emptyset_c$=3.6 cm and a required resolution value of 30 cm for ITER (Fig. 1c), the viewing distance should be $L=8.4 \cdot l_c$. With a minimum value for the collimator length of $l_c \approx 80$ cm, the optimum viewing distance would be about 6 m. The total profile span is limited by the viewing aperture through the port which for ITER has been estimated (Nov. 1994) to allow a total Z-span of 6 m in ITER for an aperture of about ±1 radian and a minimum distance of about 11 m (see Fig. 1c). The total profile span is thus limited to about 2/3 of the minor radius a_Z, $Z/a_Z \equiv \zeta < \zeta_p = 2/3$. It is desirable to have about 20 channels over the range $-\zeta_P \leq \zeta \leq \zeta_P$ and, hence, a resolution of $\Delta \zeta \approx 1/15$ which, together with $\zeta_p=2/3$ and $L=11$ m, represent our reference geometry. Since $B(\zeta)$ falls below the measurable level at some point $\zeta = \zeta^{MAX}$. This can imply a span restriction to $|\zeta| \leq \zeta^{max} < \zeta_P$ for broad profiles.

The measured brightness distribution can be written $B(\zeta)=2a_z y_0 I_\alpha [1-\zeta^2]^{\alpha-1/2}$ where $I_\alpha = 2\alpha!!/(2\alpha+1)!!$; y_0 and the peaking factor α come from $y_n(\zeta)=y_0(1-\zeta^2)^\alpha$ representing the plasma iso-emissivity curves shown in Fig. 1a. The ratio $B(\zeta)/B(0)$ can vary orders of magnitude for $\zeta \leq \zeta_P$ depending on the plasma profile. The aim is to cover the brightness range $B(\zeta^{max})/B(0) \geq 10^{-2}$ which falls within the aperture limit $\zeta^{MAX} < \zeta_P$ for profiles of $\alpha \leq 7.3$.

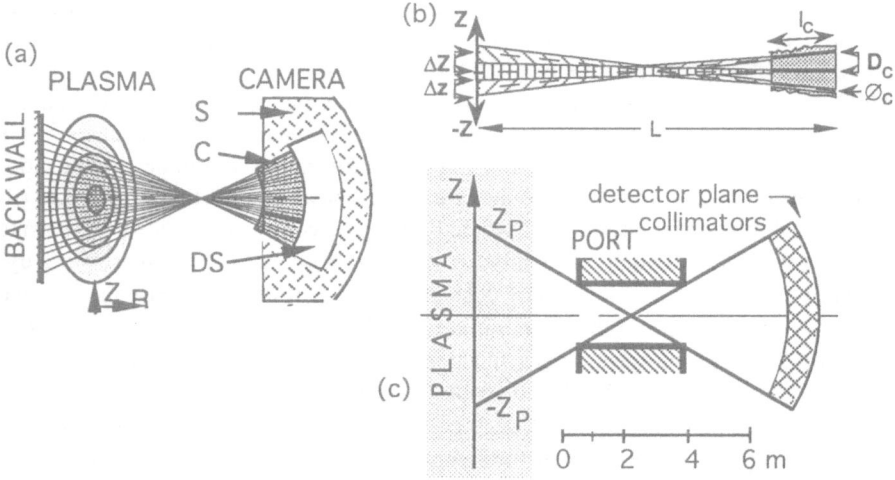

Fig 1. Schematics of a horizontal neutron camera: (a) The plasma and the shielding (S), collimator array (C) and space for detectors (DS) in the camera. (b) The camera channel parameters defining the plasma sight lines. (c) Principle interface limitations with reference to ITER.

The brightness gradient $B'=dB(\zeta)/(d\zeta=-2(\alpha-0.5)\cdot\zeta\cdot[1-\zeta^2]\alpha-3/2$ can be used to illustrate two important aspects of profile measurements. First, determination of the slope of the profile in the center with finite error (say ±10%) requires very high statistical accuracy. For instance, at $\zeta=1/15$ for $\alpha=7.3$, it takes $N_\zeta\approx10^5$ counts in two adjacent channels which corresponds to count rates of $C_n=N_{\zeta=0}/\Delta t$, viz. $C_n>1$ MHz for $\Delta t<0.1$. The detailed shape is the general aim of the measurement and the slope is especially important as input to the detailed peak shape analysis that gives qualified temperature information[7,8]. Second, the count rate requirement can be lowered by increasing the number of channels. However, the detector efficiency must then be increased to maintain accuracy; for instance doubling the number of channels requires doubling the detector efficiency for things to be equal. On the other hand, in order to exploit the nominal benefit of increased spatial precision, this has to be matched by corresponding better accuracy of the measurement; the detector efficiency has to be increased another factor of 4 and hence also the channel count rate. This would give a sensitivity of ±10% in the B'-determination for a lateral resolution of $\zeta=1/30$ but we are talking about orders of magnitude augmentation in detector performance above those required for a camera of the reference geometry with $\zeta\approx1/15$. This suggests that reference is the practical limit for the lateral resolving power of neutron cameras with respect to lateral resolving power that it is very difficult to surpass.

SPECTROMETER PRINCIPLE AND SPECIFICATIONS

A magnetic proton recoil (MPR) instrument[2] is being developed as a prototype for a high resolution neutron spectrometer for ITER[10]. It is primarily meant for measurement of 14-MeV neutrons of $d+t->\alpha+n$ reactions utilizing the high fluxes expected from burning plasmas. It consists of a CH_2-foil at the end of a collimator that emits proton recoils. The forward proton flux is deflected and momentum analyzed in a magnetic spectrograph and detected in focal plane proton detector. The proton spectrum is read out as a position histogram[10] which is closely related to the neutron spectrum covering an energy bite, $\Delta E/E_{BITE}=±20$ %. The neutron flux detection efficiency (ϵ)is variable. The maximum is $\epsilon\approx10^{-3}$ cm^2 for an energy resolution of $\Delta E/E\approx15\%$ and $\epsilon\approx7\cdot10^{-5}$ cm^2 the typical high resolution setting $\Delta E/E\approx2.5$ %. The count rate capability is very high ($C_n^{CAP}>>100$ MHz) which could be exploited given an favorable viewing position close to the plasma; for instance, a neutron flux of $F_n>10^{11}$ n/cm$^2\cdot$s gives a count rate, estimated from $C_n=\epsilon\cdot F_n$, of $C_n>0.7$ to 10 MHz depending on the resolution setting. The MPR can also be set for measuring 2.5-MeV neutrons in DT or D-plasmas and can be used where limited count rate is acceptable. The MPR technique would furnish the ITER camera with a powerful spectrometer function.

DESIGN AND PROPERTIES OF SPECTROMETER ARRAY

The MPR is rather bulky so that special multiple-unit designs must be developed for the camera application. For instance, the minimum separation between the units would be 78 cm if they were to form the array of channels in the camera. An integrated design is possible by observing that adjacent spectrometers can share pole pieces, coils and yokes (see Fig. 2) thus reducing the channel separation to about 40 cm. This can be further reduced by a factor of 2 by forming the camera channel array based on two stacks where the channels of one

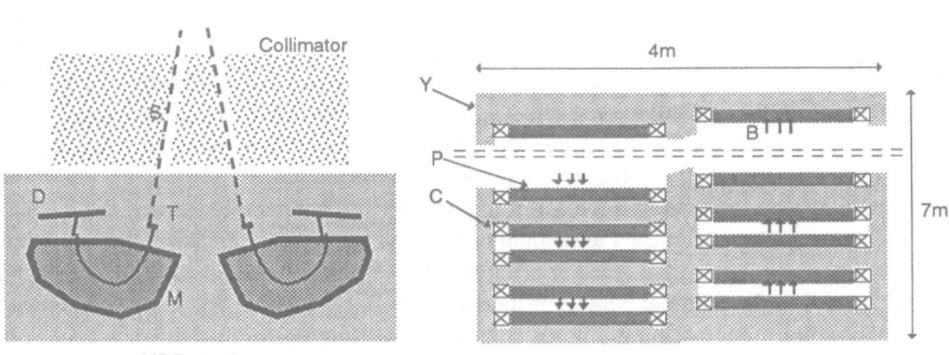

Fig.2 Design and envelope dimensions of a horizontal neutron camera based on MPR spectrometers. The magnet configuration of the left and right hand spectrometer stacks are shown in top (left panel) and back view (right panel). The market components are detector(D), target (T), magnet foot print (M), collimated sight line to plasma (S), yoke (Y), coil (C), pole face (P) and magnetic filed gap (B).

stack are interleaved with those of the other; for instance, all odd numbers in the left hand stack and all even ones in the right hand stack (Fig. 2b). This provides the additional advantage that the magnetic field in one spectrometer array provides the return field path for the other resulting in a significant savings on the amount of return yoke needed. The manufacturing cost of an integral multiple spectrometer should be much lower per channel than the cost for a single channel instruments. The magnetic spectrometers stack is a known concept from nuclear reaction studies where it is usually referred to as 'multi-gap' design. What is needed here is an adaptation of the concept to the fusion neutron application. This requires development work, especially, to incorporate the new dual stack idea.

The MPR type of camera has the following advantages: (1) Any channel can be run in camera mode (sufficient energy resolution for background discrimination), high efficiency mode (for high time resolution studies, $\Delta t < 100$ µs in central channels)* and high energy resolution mode (for spectrometry studies). (2) The camera can be set for measuring different parts of the neutron flux including, in principle, the 2.5-MeV neutron emission from d+d reactions. (3) Each channel has an absolute calibration in energy and intensity independent of operating conditions, event rate and energy setting (4) Prompt information out-put and high reliability/availability with expected suitability to use in on-line feed back control system. These features make the MPR camera unique and conventional in-beam detectors in use today are simply not up to the same performance standard for high power plasmas.

BACKGROUND

The collimator flux consists of direct neutrons from the plasma (signal S) and scattered ones (background, B_x)[1,12]. Since B is rather independent of ζ and α, the signal-to-background ratio S/B_x will vary with the signal strength, i.e., reflecting the profile $B(\zeta)$. The

---

*Can be complemented with a high efficiency detector (up to 10 cm^2) in the MPR beam dump. Operated in current mode, sampling rates could be increased by a factor of 10^3 times compared with the MPR spectrometer by itself could be attained. The supplemented system would be used for studies of fast time variations in the neutron emission including fluctuations. The underlying assumption is that the signal-to-background ratio does not change over the time scale that can be detected with the MPR spectrometer as this is the calibration standard for the fluctuation monitor.

S/B$_x$-ratio is highest in the central channel for a peaked profile and becomes smaller if the profile is flattened (small α) and, especially, the lowest S/B$_x$-values are encountered in the plasma periphery. There is also a γ-background that follows the neutron background. Therefore, the neutron camera counters must be γ-insensitive and must be able to separate the energy degraded neutron background flux. Alternatively, the measurements must be limited to those plasma experimental conditions for which S/B$_x$ is inherently large. However, this requires a detector that monitors that the bounds of the safe operating conditions are not exceeded. This means, in practice, that the camera detector system must include spectrometer a function. The latter is also needed in order to assure reliability in the calibration.

PROJECTED PERFORMANCE AND DIAGNOSTIC FUNCTIONS

The MPR camera could, if placed in a suitable position with high neutron fluxes, provide information on two required essential diagnostic functions: (1) the absolute neutron yield and, hence the fusion power, to an accuracy of 5 % or better at a time resolution of 100 ms or better. (2) the α-particle birth profile as reflected in the brightness distribution, $B(\zeta)$. Beyond this, information can also be provided on fast fluctuations in the neutron emission (down to the 100 μs level)[12]. Moreover, the MPR camera is a radial array instrument for neutron emission spectrometry (RAINES)[1]. This will provide information on the ion temperature radial dependence and constitutes together with the TRIAD system a comprehensive ion diagnostic for ITER. The spectrometry aspect of the camera is not discussed here.

The accuracy of $B(\zeta)$-measurements with MPR cameras relies on the ε-calibration and statistical variance of the data σ_N. The variance is conjugate to the data accumulation time, i.e., Δt. In addition, the B_x-error enters the measurements in regions of low relative brightness, for instance, for peripheral sigh lines where $B(\zeta)/B(0) \to 10^{-2}$. This is the problem that can be handled with the MPR-camera and is the objective quantitative assessment. Here we shall limit ourselves to illustrating the relationship between σ_N and Δt for the envisaged burn phase conditions on ITER for $Y_n^{MAX} = 4 \cdot 10^{20}$ n/s (producing $F_n^{MAX} = 3 \cdot 10^7$ n/s for $\zeta = 0$) considering also that $\psi = Y_n/Y_n^{MAX}$ can vary a factor of 10. As an example, for $\sigma_N = 1$ %, one obtains $\Delta t = 0.3$ ms at $\zeta = 0$ and 30 ms at $\zeta = \zeta^{MAX}$ and for $\sigma_N = 3$ % one gets 10 times better (smaller) Δt. Thus the **full** profile shape can be determined at the sensitivity level of 3 % at $\Delta t < 30$ ms for ITER's ignition plasmas of 10 to 100 % of maximum fusion power, or $\Delta t < 300$ ms for D=100. This is sufficient for observing profile changes taking place on the time scale of about a second of the α-particle slowing down time.

Similarly, saw-tooth oscillations would evolve on a comparable time scale. Of particular interest here is to observe $B(\zeta,t)$ in the central plasma region to study its collapse and subsequent slow recovery. This happens within the radius of the q=1 surface (i.e., $\zeta \leq \zeta_{q=1} << \zeta^{MAX}$) so that plasma transients in this partial profile could be recorded at a sensitivity level of $\sigma_N = 3$ % at the time scale of $\Delta t < 3$ ms. With regard to fluctuations in the neutron flux (spatially undifferentiated), one could reach $\Delta t < 10$ μs at $\sigma_N = 3$ % (or $\Delta t < 1$ μs at $\sigma_N = 10$ %) by combining the data in the central camera channels. This would thus approach the time scale of fast plasma instability oscillations. If it is a valid to assume that S/B$_x$ does not vary on the time scale $\Delta t < 10$ μs, one could observe fluctuations as fast as 1 to 10 MHz with a auxiliary current integrating detector in the MPR beam dump.

Another task for the MPR camera is to determine $Y_n(t)$. This can be done with the same time resolution as for the full profile. We note that the neutron camera channels sample the neutron emission across the plasma[3,12]. The sampling fraction can be determined with high accuracy (cf. Fig. 1) as can the ε-calibration which together provide for an absolute determination of Y_n with an error of between 2 and 5 % (depending on circumstances). The fact that all factors relating Y_n to $B(\zeta)$ have an empirical base ensures that the absolute error value ascribed is relevant. The MPR camera is a unique Y_n-diagnostic and can serve as a calibration standard for the fusion power measurements in DT plasmas. The time resolution is modest for low power levels but would still permit calibration of conventional neutron flux monitors that can follow time variations also at low power. It should be noted that the MPR camera can also be used for measurement of the absolute yield of 2.5-MeV neutrons which would put yield measurement in both D and DT plasmas on the same calibration standards (absolute in both cases). There is no other known technique offering this capability.

It can be noted in this context that the development for the next step neutron diagnostics implies a change in direction away from global flux measurements, performed with flux monitors, to differential (collimated) neutron emission measurements with cameras. The reason for this is that differential neutron measurements have high precision for high power plasmas while the flux monitors will be practically impossible to calibrate because of the effective radiation containment required by burning plasmas such as ITER's.

CONCLUSION

It has been shown that a spectrometer function can be incorporated in the horizontal neutron camera for ITER using the MPR technique. A dual-stack concept is proposed that allows these large spectrometer units to be integrated into a compact design so that the same lateral resolution can be attained as with conventional small detectors. The potential diagnostic capabilities are superior to those of conventional cameras. To realize these functions in full, the instrument should be placed in a high flux viewing position close to the plasma. The interface of the MPR at ITER is therefore a critical issue both with regard to available space and the mutual magnetic comparability of this diagnostic and the machine.

* The Neutron Physics for Fusion Collaboration (see ref. [1])

REFERENCES

1. J Källne and the NPF Collaboration, Neutron spectrometry for ITER, these Proceedings.
2. J. Källne, The Role of Neutron Measurements in the Study of Fusion Burning Plasmas, *Comments Plasma Phys. Contr. Fusion* 12(1989)235-47.
3. J. Källne, Neutron diagnostics for burning DT plasmas of ITER/NET, Rep. UU-NF 91/1, Uppsala University, Uppsala, Sweden (1991).
4. J. Källne, P. Batistoni and G. Gorini, Neutron camera diagnostics for burning plasmas, Rep. UU-NF 90/3, Uppsala University, Uppsala, Sweden (1990) and in Proc. 5th topical conference on high temperature plasmas, Minsk, Belo-Russia (1990).
5. J. Källne and G. Gorini, A Neutron Camera for Burning Plasmas, *Rev. Sci. Instrum.* 63:4545 (1992).
6. G Gorini and J Källne, Neutron emission diagnostics in compact ignition experiments, Rep. UU-NF 93#7, Uppsala University, Uppsala, Sweden (1993).
7. G Gorini and the NPF Collaboration, Detailed ion temperature information from the neutron emission spectrum, these Proceedings.
8. J Källne and G Gorini, Neutron observations and alpha particles in high power DT plasmasa, *Fusion Technology* 25:341 (1994).
9. G. Ericsson and the NPF Collaboration, The MPR neutron spectrometer project, these Proceedings.
10. J Frenje and the NPF Collaboration, The MPR neutron diagnostic at JET- an ITER prototype study, these Proceedings.
11. P Antozzi, G Gorini, J Källne and E Ramström, Neutron wall emission in tokamaks and plasma diagnostics, to appear in *Nucl. Instr. Meth.* (1995).
12. J Källne and G Gorini, Alpha particle information on burning plasmas from neutron measurements, *Fusion Technology* 22:439 (1992).

THE MPR NEUTRON SPECTROMETER PROJECT

G. Ericsson[1] and the NPF Collaboration[*]

[1] Dept. of Neutron Research, Uppsala University, Uppsala, Sweden

INTRODUCTION

A neutron spectrometer of new type for fusion diagnostics is presently under construction at the Dept. of Neutron Research, Uppsala University[1]. The spectrometer is based on the Magnetic Proton Recoil (MPR) method and combines high efficiency with good energy resolution and high count rate capability. The MPR spectrometer will be installed as a diagnostic on JET and operated during the coming DT phase[2]. It is also a prototype for an ITER diagnostics, aimed at providing information on a range of plasma quantities, such as ion temperature (T_i), fuel ion densities (n_D, n_T), α-particle effects, and total neutron yield[3]. Due to its high count rate capability, the MPR can offer high quality data at kHz sampling rates. In this paper we will mainly deal with the status of the MPR construction programme.

THE MPR PRINCIPLE

The MPR spectrometer is shown schematically in Fig. 1. A collimated beam of neutrons from the plasma intersects a CH_2 foil. Some neutrons scatter elastically on the target hydrogen nuclei. The recoil protons knocked out within a narrow cone in the forward direction are selected for analysis in the magnetic field. They are focused, deflected, momentum analyzed and transported to the focal plane where they are registered. The image of the target on the focal plane must be small in order to subtend a small energy range. This requires a fairly long particle orbit to give high dispersion. Moreover, the magnet optics must provide this imaging with a minimum of aberrations. See Fig. 2a for examples of protons trajectories.

The focal plane detector is situated about 150 cm from the collimated neutron beam. It is divided into thin strips in the magnet dispersive plane, providing energy binning of the neutron energy spectrum. As no complicated signal processing or data reduction is necessary, the information could be almost instantly available, e.g., for feedback into the tokamak control system. Using strips of fast plastic scintillator, operated in singles counting mode, allows rates of up to several MHz per scintillator. The scintillators are protected from the background of direct and indirect neutrons and γ's by an extensive radiation shield around the spectrometer[2].

---

[*] See contribution by J.Källne, these proceedings.

Diagnostics for Experimental Thermonuclear Fusion Reactors
Edited by P. E. Stott *et al.*, Plenum Press, New York, 1996

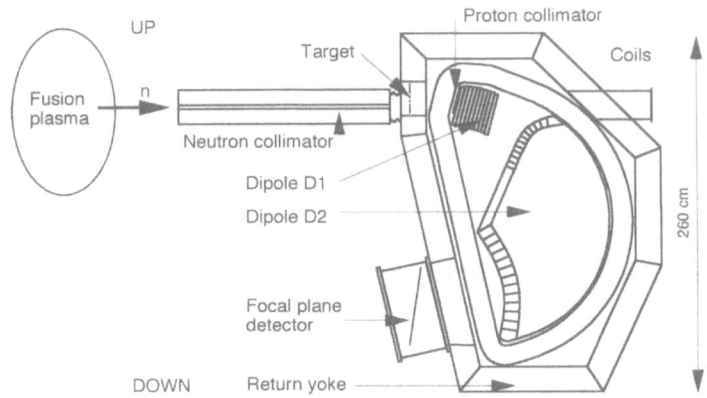

Figure 1. Schematic drawing of the main components of the MPR spectrometer.

MPR DESIGN AND CONSTRUCTION

The spectrometer return yoke is constructed as a large soft iron box, completely enclosing the magnetic coils and poles inside. Four small penetrations are included, for the neutron beam entrance and exit, for the detector hodoscope, and for water and power feed-throughs. The magnet weighs about 20 tonnes with dimensions 1.9 x 2.6 x 0.8 m^3.

The **magnet configuration** of the MPR is based on the clamshell spectrometer concept from nuclear physics. In this application for fusion neutrons, the spectrometer must combine large target area and large recoil proton solid angle with good energy resolution. To this end, the magnet is built in a split-pole configuration (Fig. 1). The small D1 "dipole" produces strong quadrupole and sextupole magnetic components. These provide strong vertical focusing and corrections for higher order imaging aberrations. The large D2 dipole is of the proper clamshell type and performs most of the bending, while also assisting with some focusing and higher order corrections. For 14 MeV neutrons, the reference magnetic field is 1.08 T, but the MPR can be tuned to any energy (magnetic field), e.g., 2.5 MeV. The energy band width is ±20 %.

The **focal plane detector** is divided into 37 scintillator strips (channels), arranged as a straight plane hodoscope, 50 cm long and 10 cm wide. The central 20 channels have a width of 8 mm, giving an energy binning of 88 keV (0.63% of 14 MeV). The non-central channels have a width of 20 mm, giving a binning of 200 to 280 keV. The scintillators come in three different thicknesses (2.0 to 3.5 mm) to minimize the amount of material used, while still fully stopping the incoming protons. To check the background, an additional three scintillators are placed behind the hodoscope, so as not to be seen by the signal protons. Small PM tubes are used to make the hodoscope compact. The need for good signal linearity with count rate and energy requires careful design of the PM bases. The aim is 1% non-linearity for the expected JET experimental conditions. The specifications and design are based on tests with radioactive sources, accelerator beams, and computer modeling[2].

Auxiliary systems, like the neutron collimator, target and recoil-proton collimator, have been designed based on careful modeling and simulation of the MPR performance. A large degree of flexibility has been built into these systems. The length of the neutron collimator can be varied over the range 70 - 200 cm, corresponding to a range in solid angle between 0.5 and 4 msr for the JET installation. Six targets and five different proton collimator apertures are available at any given time by remote control. New sets can easily be installed whenever access to the spectrometer is possible. Table 1 shows some combinations of targets and proton collimators that have been selected for the JET operations.

Table 1. Combinations of target thicknesses and proton collimator apertures for maximum count rate at set energy resolution values. The neutron collimator length is fixed at 200 cm and the target size at 10 cm². The combinations chosen for JET operations are shown.

Combination #	1	2	3	4
Nominal resolution [%]	2.0	2.5	4.0	10
Empirical* resolution [%]	1.86	2.41	4.05	9.89
Target thickness [mg/cm²]	6.0	8.0	18	50
Proton collimator [msr]	20	40	55	55
Empirical* relative count rate	0.41	1	2.8	7.8

* Empirical values calculated from the measured "14-MeV" magnetic field maps.

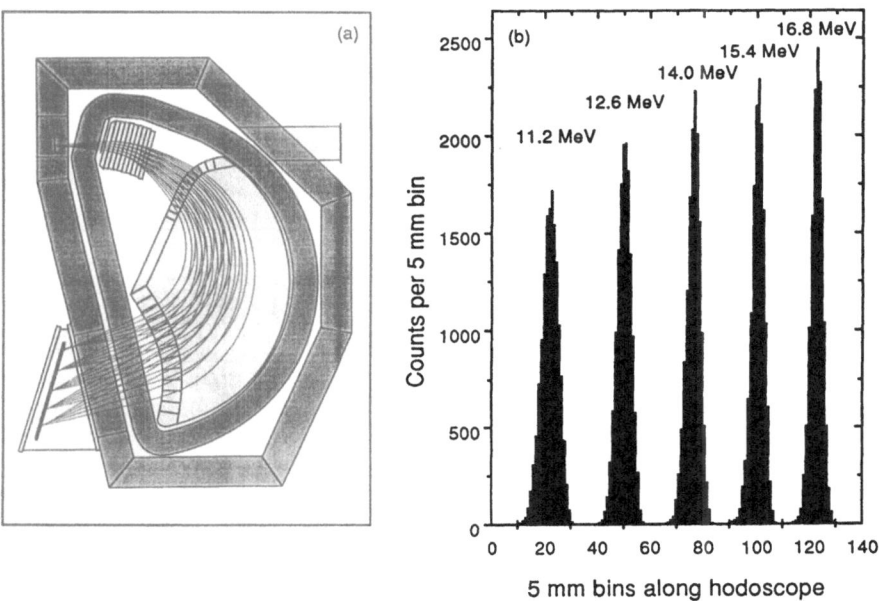

Figure 2. Tracking in the measured field maps. a) Complete mid plane proton trajectories from the target to the hodoscope. A point target was used. b) Spectral line shapes for full geometry.

Detailed **magnetic field maps** of the delivered magnet have been measured. The first set of maps indicated some small deviations from the design configuration and one iteration of shimming of the pole pieces has been undertaken. This has rectified most of the deviations from design and the field configuration is now remeasured.

To **optimize the performance** of the spectrometer careful simulations and raytracing have been performed. Performance was optimized with respect to count rate for various running conditions, e.g., high resolution or high efficiency. All factors affecting the performance have been identified and included into the simulation and raytrace codes. The major effects are the geometry, (n,p)-scattering kinematics, proton energy loss in the target, and magnet optics. The combinations of values of target thickness and recoil proton solid angle shown in Table 1 have been selected in this way.

Fig. 2 shows results from raytrace calculations for protons in the measured field map with the spectrometer set to accept neutrons in the 14-MeV energy band. Five distinct neutron energies were used with the reference spectrometer settings. In Fig. 2a, where for clarity only tracks in the magnetic mid plane from a point source are used, is shown the

415

complete proton trajectories from target to the hodoscope. Fig. 2b depicts the spectral line shapes for the five energies on the hodoscope, using the full reference geometry (combination #2 of Table 1). Table 2 gives the energy resolution and transmission for each energy. The performance of the spectrometer is satisfactory, exceeding the design goal in the high energy part of the range while being slightly worse for the lowest energies. A detailed understanding of the relationship between magnet configuration and optics has now been gained. The present prototype thus forms a basis from which modifications can be made according to the specific performance requirements in subsequent designs. However, the present performance of the MPR is adequate to proceed with the JET installation.

Table 2. Performance of the MPR spectrometer for the reference case using the measured "14-MeV" magnetic field map.

E_n [MeV]	11.2	12.6	14.0	15.4	16.8
Transmission [%]	85.8	83.5	83.6	79.7	76.2
Position on FP detector [cm]	-27.1	-13.1	0.0	12.1	23.2
Dispersion [cm/MeV]	10.33	9.69	9.00	8.28	7.53
Peak FWHM [cm]	3.84	3.20	3.05	2.93	2.64
ΔE [MeV]	0.372	0.330	0.339	0.354	0.350
ΔE [% of E_n]	3.32	2.62	2.41	2.30	2.08

JET INSTALLATION

Interface work is in progress, with the aim of installing the MPR prototype on JET in the beginning of 1996. The MPR has been assigned a position with a sight line in the equatorial plane at 22° to the major radius and with the target at about 6 m from the plasma center. A concrete radiation shield has been designed for the JET installation. The design is based on extensive calculations with the MCNP neutron transport code, where detailed modeling of the MPR and the Torus Hall have been included. The total weight of the shield is some 75 tonnes, and the dimensions 4.5 x 2.5 x 3.5 m^3. Requests for cables, power, water, etc. have been filed at JET. Space for electronics and for the magnet power supply has been allocated. The MPR data acquisition will be handled by CODAS and integrated into the JET data base. A local acquisition system will also be available for tuning, tests, monitoring, and system development between experimental campaigns.

CONCLUSIONS

The construction of the prototype Magnetic Proton Recoil (MPR) spectrometer is reaching its final stages. The MPR performance has been shown to meet or exceed the design goals. Interface work is going on with the aim of installing the MPR on JET for operation during the coming DT phase. The planning, design and construction of this instrument provides prototype experience that is used as input for work on neutron diagnostics for ITER.

REFERENCES

1. J. Källne, H. Enge, *Magnetic Proton Recoil Spectrometer for Fusion Plasma Neutrons*, Nucl. Instr. Meth. A311 (1992) 195.
2. J. Frenje, these proceedings.
3. J. Källne, these proceedings, and references therein.

THE MPR NEUTRON DIAGNOSTIC AT JET -
AN ITER PROTOTYPE STUDY

J Frenje[1] and the NPF Collaboration*

[1]Department of Neutron Research, Uppsala University, Uppsala, Sweden

INTRODUCTION

The DT plasma of ITER will be a very strong neutron source whose core (for volume of minor radius scale, a^3) will shine with a brightness of up to $B=10^{15}$ n/cm^2·sr. If viewed through narrow collimators (solid angle of the order msr) with spatial resolution (1/10 of the core size), neutron fluxes (F_n) of up to $F_n \approx 10^{11}$ n/cm^2 can be attained at ignition ($Q \rightarrow \infty$). Present DT plasmas ($Q \rightarrow 1$) can produce fluxes of $F_n < 10^{10}$ n/cm^2 which is a factor of >100 above that of D plasmas. High fluxes ($F_n > 10^9$ n/s·cm^2) and narrow collimation are the basic prerequisites for bringing neutron spectrometry into the category of advanced diagnostics which now can start to be developed and tested for the first time[1]. This implies exploration of new diagnostic functions and instrumentation with opportunities for use now in pre-ignition plasma studies at the 1 to 10% performance level of that to be reached on ITER.

A high resolution spectrometer is being constructed[2] as a prototype instrument for measurement of fusion neutrons of high intensities and fluences. It is based on the magnetic proton recoil (MPR) technique[3]. This was selected after evaluation of various alternatives to handle the principal <u>conversion</u> of neutron beams to charged particles, and for detection and energy determination of the neutrons. The following points were considered:
(i)   ability to use high neutron fluxes for high count rate operation, and robustness;
(ii)   transparency of characterization and calibration accuracy/stability;
(iii)   simple data acquisition and fast input/output conversion;
(iv)   suitability for on-line diagnostic information and signals for control purposes;
(v)   suitability to explore/exploit the potential of neutron spectrometry diagnostics.

The MPR technique is being developed for use in multiple-spectrometer diagnostic systems[4] whose principles can be investigated with the present diagnostic based on a single-unit MPR instrument designed for installation and operation at JET. Although it is a prototype diagnostic for ITER, significant measurement power is expected to be added to the JET diagnostic complement. This paper describes the MPR diagnostic and its relevance as an R&D contribution to the ITER Experimental Design Activity in this field.

PRINCIPAL DIAGNOSTIC COMPONENTS AND FUNCTIONS

The MPR diagnostic includes, besides the spectrometer magnet (described in ref[2]), a focal plane detector, a radiation shield and neutron collimator (Fig. 1), and signal and control

Diagnostics for Experimental Thermonuclear Fusion Reactors
Edited by P. E. Stott *et al.*, Plenum Press, New York, 1996

417

electronics. The planning and design are based on computer simulations[4] regarding radiation fields, shielding attenuation, and signal and background response of the detector; the functional system characterization, including calibration and stability, is equally substantial.

The MPR diagnostic at JET will be placed in the Torus Hall radiation enclosure of JET and conforms to the rules of restricted access. The sight line is near tangential ($\approx 22^\circ$) relative the plasma axis which, together with port vignetting, limits the brightness at which the plasma can be observed for given Y_n; yet fluxes $F_n \rightarrow 10^{10}$ n/s·cm^2 can be expected. The radiation field consists both of direct and ambient components (Fig. 1). A concrete block house is used to protect the detectors but also keep the neutron background flux low compared to the collimated neutron flux; the collimated flux at the diagnostic is 1/100 of that of the direct radiation field without shielding. The stray magnetic field is ≤ 200 G at the spectrometer location which the MPR can tolerate. The machine can also tolerate the MPR presence. The installation is quite bulky measuring just under 100 tons in weight and 2.4×4.3×4.0 m^3 in envelope. The focal plane proton detector is a 37-channel scintillator (volume 10×0.8×0.3 cm^3) hodoscope with PM-tubes . The target is a CH$_2$ foil of 3.6 cm diameter the same as the collimator. The diagnostic can be adjusted to suit varying measuring objectives and plasma conditions. This allows settings so as to maximize the count rate at the $\Delta E/E$-value of choice in the range $1.5 \leq \Delta E/E \leq 5$ %. The normal setting would be $\Delta E/E = 2.5$ % (FWHM, Gaussian equivalent) giving an estimated maximum count rate of several hundred kHz at $P_f = 30$ MW. Finally, the magnet can be set to observe neutrons of up to $E_n \approx 19$ MeV with the down limit set by the proton energy ($E_p \geq 1.5$ MeV). The normal setting would be to observe the 14-MeV peak (covering ±20 % range) and but also the 2.5-MeV peak[1].

SYSTEM CHARACTERIZATION AND RESPONSE FUNCTION

The data recorded are a proton histogram (see below) closely related to the energy distribution of the incident neutron flux, $F_n(E_n)$, through the response function $R(E_p, E_n)$. This is illustrated in Fig. 2. The response function contains three calibration aspects, namely, intensity (efficiency, ε), spectral shape ($S(E_p/E_{p_1})$) and energy scale ($\kappa_E = E_{p_1}/E_{n_1}$) which indeed factorizes into $R = \varepsilon \cdot S$ for a symmetric shape function so that E_{p_1} is the mean and peak energy of S if E_n is assumed fixed ($E_n = E_{n_1}$). The response function is calculated based on physical input data and beam optics[2,8] indicating an achievable accuracy of 10^{-3} in κ_E, 1 % in ε and 5 % in shape expressed as width error $\Delta\Gamma/\Gamma$. In reality, the response is not perfectly symmetric and the incoming beam has a finite divergence which are incorporated as adjustments to the symmetric case so that the error analysis is still representative.

It is worth noting that this diagnostic does not rely on calibration of any active <u>neutron</u> detector. Instead, the neutron calibration is limited to the neutron-proton conversion factor

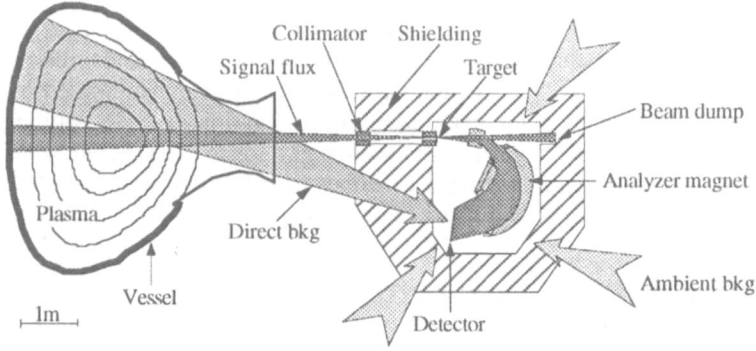

Figure 1 Outline of the MPR diagnostic planned for JET.

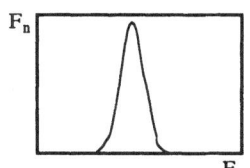 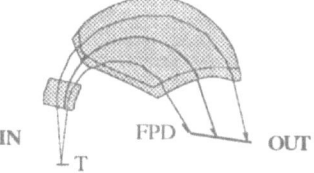 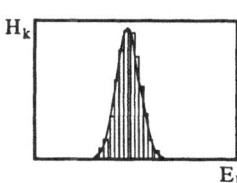

Figure 2. Schematics of the diagnostic response to neutrons falling on the target (T) producing recoil protons that are imaged on the focal plane detector (FPD) whose output is a proton energy histogram.

given by the forward proton scattering cross section and the effective proton solid angle. These are both known to better than 1 %. Moreover, the energy determination is passive. Thus, this diagnostic will be calibrated before installation so that in situ one needs only to monitor that prescribed settings are upheld and checked for drifts, normally, with a calibration stability requirement of ±30 % or better. However, it is envisaged that this diagnostic can be operated with enhanced relative accuracy on ITER for detailed shape and difference measurements. In this case, the stability will be a crucial performance issue.

RADIATION FIELDS, COLLIMATION AND SHIELDING

The signal and background radiation fields vary both with the neutron source strength (Y_n). The collimator controls the brightness of the plasma signal, which at JET, is estimated to be factor of 100 smaller than the direct background. The **diagnostic flux detection efficiency for signal neutrons** is of the order 10^{-4} cm^2 as compared to a **detector flux sensitivity to background neutrons** of the order 10^0 cm^2. The shielding reduces the background a factor of 10^6 and the pulse height discrimination of the detector takes the S/N-ratio up a factor of 10^4 to S/N$\approx 10^4$. It should be noted that the shielding brings down the background to the level where its pulse rate load on the detector is lower than that of the signal. The attainment of S/N>1 can then be obtained solely by pulse discrimination as illustrated in Fig. 3. Neutrons dominate the background as the γ:s events of significant pulse heights are very rare. Although already regular scintillators can give very good S/N values, it is yet possible to improve the ratio by a factor of about 20 using deuterated scintillators. These would work excellently for our application as will be reported[9].

The energy measurement and discrimination functions are decoupled in this diagnostic. Moreover, the signals are represented by protons of specific energy and hence also detector pulse height (within a resolution of about 5 %). The background is represented by neutrons with an energy spread of $0 \leq E_n \leq E_n^{MAX} = 14$ MeV producing pulses unspecific to E_n. This situation has two important consequences: One can set a high discrimination level so that (i) very little of the background spectrum appears above the threshold and yet (ii) relatively few signal events fall below the threshold with little change due to instabilities. Thus, effective discrimination is not incompatible with high calibration accuracy in efficiency.

There is a close relationship between the input neutron spectrum, $F_n(E_n)$, and the output histogram, H_k, of channels k=1-37 (Fig. 2) as defined by the response function which is, in our case is passive, apart from the proton detector. The detector performs event counting and pulse height discrimination which could be affected (and hence the ε-calibration) at high rates due to pulse-pile up corruption of the resolution. However, the critical level is well defined and can be arranged to be well above projected operating conditions; the count rate is automatically monitored to alert of any unforeseen approach the critical rate level. Thus, this diagnostic is robust against rate variations which is otherwise a notorious problem for high count rates (MHz) diagnostics. The MPR diagnostic is made to operate over a large dynamic range (D=10^2 or more) over which the response calibration must not vary.

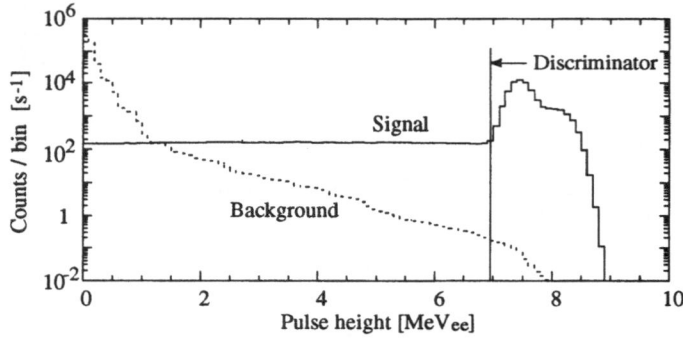

Figure 3. Detector pulse height spectra of signal protons and background neutrons from numerical simulations for the planned diagnostic set up at JET for DT plasmas at T_i=20 keV; a discrimination threshold is indicted.

STATUS AND IMPLICATIONS FOR ITER

The MPR diagnostic project has now entered interface stage. The design specifications have been met or exceeded as assessed from extensive measurements and computer simulations. Specific technical issues affecting the implementation of MPR based diagnostics on ITER have been identified besides areas requiring R&D. These relate to (i) magnetic interference between the machine and the diagnostic (the diagnostic uses magnetic material and precise internal magnetic fields); (ii) finding space and sight line supplying the high neutron fluxes on which the performance depends; (iii) the focal plane detector; (iv) interference free transmission of signals from the diagnostic. These factors will in part be tested at JET except for (i) as the JET stray fields are relatively week (10^2 G) and much higher at ITER. It should be noted that the target accuracies for ITER are set very high so that fusion neutron measurements can be pushed to the fundamental limits as we know them today.

* The Neutron Physics for Fusion Collaboration (see ref.[4])

REFERENCES

1. G. Gorini and J. Källne, Neutron spectrometry for compact ignition experiments, *Il Nuovo Cimento* 14D:1115 (1992).
2. G. Ericsson and the NPF Collaboration, The MPR neutron spectrometer project, these Proceedings.
3. J. Källne, The Role of Neutron Measurements in the Study of Fusion Burning Plasmas, *Comments Plasma Phys. Contr. Fusion* 12:235 (1989).
4. J Källne and the NPF Collaboration, Neutron spectrometry for ITER, these Proceedings.
5. S Conroy et al, Neutron transport calculations for the MPR spectrometer, Rep. UU-NF 95#3, Uppsala University, Uppsala, Sweden (1995).
6. J. Frenje et al, Signal and background response functions of the MPR detector, Rep. UU-NF 94#10, Uppsala University, Uppsala, Sweden (1995).
7. E. Traneus and the NPF Collaboration, Neutron camera with spectrometer function, these Proceedings; J. Källne and G. Gorini, A Neutron Camera for Burning Plasmas, *Rev. Sci. Instrum.* 63:4545 (1992).
8. E Traneus et al, Optimization model for the MPR spectrometer, Rep. UU-NF 94#1, Uppsala University, Uppsala, Sweden (1994).
9. J Frenje et al, Deuterated plastic scintillator for proton detection in neutron background, Rep. UU-NF 95#4, Uppsala University, Uppsala, Sweden (1995).

DETAILED ION TEMPERATURE INFORMATION FROM THE NEUTRON EMISSION SPECTRUM

Giuseppe Gorini[1] and the NPF Collaboration*

[1]Department of Physics, University of Milan, Milan, Italy and
Istituto di Fisica del Plasma, CNR, Milan, Italy

INTRODUCTION

Neutron emission recorded with a spectrometer comes from the plasma within the region defined by a collimator and is a line integral along the chord viewed. The resulting spectrum, $S(E_n)$, is, for a thermal plasma, a superposition of Gaussian distributions $(G(E_n,T))$ each specified by the temperature of the plasma region of emission. The superposition is weighted by the emission brightness distribution, $B(T)$, reflecting the integrated emissivity over the regions with temperature T, $y_n(T)$. This gives a heavy dominance to the high temperature regions (normally the core) but only as long as the peak temperature is not too high, say $T_M < 20$ keV. $S(E_n)$ can be significantly narrower than $G(T_M)$ so that from its width one deduces an apparent temperature $T_A < T_M$ which is not a well defined plasma parameter. Various prescriptions have been presented relating T_A to T_M (see e.g. ref.[1]). It has also been noted[2] that much of the ambiguity in the conventional (**width** based) analysis of neutron spectra can be removed by an elemental decomposition of the spectrum given that data of high statistical accuracy can be recorded. This should be possible in neutron spectrometry diagnostics of DT plasmas on ITER[3]. Here we will therefore illustrate the temperature information that can be obtained from a detailed **shape** analysis of the neutron spectrum using the slope from the peak at $E_0 \simeq 14$ MeV towards the high energy tail where the hot plasma components dominate as one approaches $E_n \approx 15.5$ MeV; the latter is the estimated limit due to the onset of α-particle kinetic effects[2,4,5].

NEUTRON BRIGHTNESS AS A FUNCTION OF TEMPERATURE

We consider a radial, narrowly collimated line of sight in the plasma equatorial plane. The ion temperature and density profiles are assumed to be of the form $T(r)=T_M(1-\rho^2)^{\alpha_T}$ and $n(r)=n_M(1-\rho^2)^{\alpha_n}[1+\beta_n(\rho^2+\rho^4)]$. In these expressions, r is the distance from the plasma centre, $\rho=r/a$, $a=2.15$ m is the ITER plasma horizontal minor radius, T_M and n_M are scale parameters, and α_T, α_n and β_n are shape parameters. The emissivity profile is given by

Diagnostics for Experimental Thermonuclear Fusion Reactors
Edited by P. E. Stott *et al.*, Plenum Press, New York, 1996

421

$y_n(r) = n(r)^2 \langle \sigma v \rangle$ where $\langle \sigma v \rangle$ is the reactivity and $n_t = n_d = n$ has been assumed for the densities.

For the purposes of this paper, n_M is always set to $n_M = 5 \; 10^{13} \; cm^{-3}$, while T_M can take the values $T_M = 20$ keV (reference case), 10 keV, and 30 keV. Four combinations of values for the parameters α_T, α_n and β_n are considered (see Table 1) representing extreme changes in profile shapes. The quantity w in Table 1 is the full width at half maximum of $y_n(r)$ divided by 2a. It is a simple measure of the peakedness of $y_n(r)$ for the case of REF and BT but not for the HD and VHD cases which have flat and hollow $y_n(r)$, respectively.

Table 1. Summary of parameter values defining the Reference (REF), Broad Temperature (BT), Hollow Density (HD) and Very Hollow Density (VHD) profiles used in the calculations.

SHAPE	α_T	α_n	β_n	w
REF	3.9	0.5	0	0.3
BT	0.36	0.5	0	0.6
HD	3.9	0.5	5	0.47
VHD	3.9	0.5	10	0.55

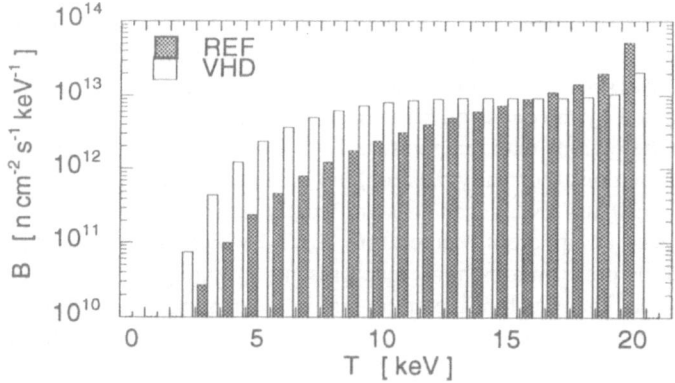

Figure 1. Calculated neutron emission brightness as a function of temperature for a plasma of $T_M = 20$ keV and profile cases of REF and VHD; the two cases are normalized to give the same integrated brightness.

The energy spectrum of neutrons from a volume element of temperature T is

$$G(E_n, T) = \frac{1}{\sqrt{2\pi}\sigma} \exp\left[-\frac{(E_n - E_0)^2}{2\sigma^2} \right] \tag{1}$$

where $\sigma = 75.2 \sqrt{T}$ (in keV) and $E_0 = 14.10 \pm 0.03$ MeV for $T = 20 \pm 10$ keV. The observed spectrum can be written as

$$S(E_n) = \omega \int_0^{T_{MAX}} B(T) \cdot G(E_n, T) \cdot dT \tag{2}$$

where ω is a proportionality factor. The neutron brightness distribution is a product of the emissivity and the inverse of the temperature gradient:

$$B(T) = y_n \cdot \left(\frac{dT}{dr}\right)^{-1} \tag{3}$$

B(T) expresses the ion temperature information that is contained in the plasma neutron energy spectrum. Examples of B(T) distributions, illustrating the variability of this quantity for different plasma profile conditions, are shown in Figure 1 for two of the four plasma profile shapes of Table 1.

DIAGNOSTIC INFORMATION FROM THE NEUTRON ENERGY SPECTRUM

The variability of a measured neutron spectrum with plasma profile conditions is illustrated in Figure 2. This shows $S(E_n)$ plotted vs $s=(E_n-E_0)^2$ on a log scale, so that the slope would be proportional to $-1/T$ for a plasma of uniform T. The data points shown illustrate a measurement with a typical bin width of $\Delta E_n=70$ keV and the statistical accuracy corresponding to a total of 10^7 counts in the spectrum; the effect of finite instrumental energy resolution is neglected here.

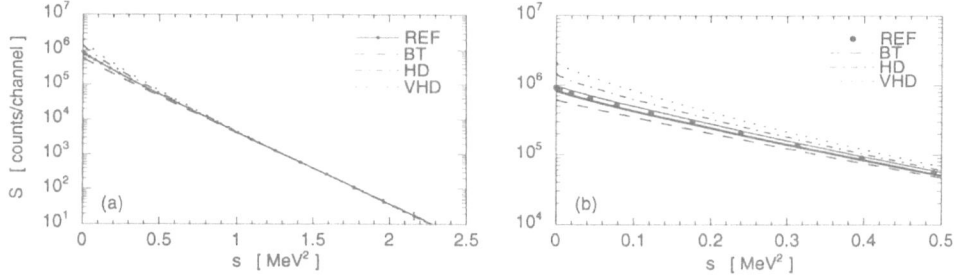

Figure 2. Neutron energy spectra for different profile conditions normalized at $s=1.5$ MeV^2. The data points and error bars illustrate a possible measurement with a spectrometer of the MPR type with statistics as for a total of $N=10^7$ counts. The spectra are shown for a wide (a) and narrow (b) energy range. The shaded area around the reference profile case illustrates a change of $\pm 1\%$ in the T_M value used in the simulation.

Table 2. Values of T_T and T_A for different profile shapes and different T_M values.

SHAPE	T_T [keV]	T_A [keV]	δ_T [%]	δ_A [%]	T_M [keV]	T_{Tl} [keV]	T_A [keV]	δ_T [%]	δ_A [%]
BT	18.8	18.7	5.9	6.4	10	9.5	8.9	5.0	11.4
REF	18.5	17.0	7.6	14.8	20	18.5	17.0	7.6	14.8
HD	18.2	15.1	9.0	24.6	30	27.1	24.7	9.6	17.7
VHD	17.9	13.7	10.3	31.3					

The results show that the four spectra take nearly the same slope in the tail region. Because of its profile independence, the slope is a good estimator of T_M. A straight line fit to the data in the region 1 MeV^2<s<2 MeV^2 provides the temperature values T_T of Table 2. The statistical error in T_T is at the 1% level for the case considered. The relative difference $\delta_T = (T_M - T_T)/T_M$ has a weak profile dependence, so that it is possible to relate T_M and T_T to an estimated accuracy of 1% or better (excluding statistics). By comparison, the apparent temperature T_A (determined from the variance of $S(E_n)$ around E_0) deviates significantly from T_M and is profile dependent.

The low energy part of the spectrum (Figure 2b) can be used for determining additional ion temperature plasma information as contained in $B(T)$. This can be done by comparing a measured spectrum with model simulations using different plasma profile shapes. A basis for this analysis is the knowledge of T_M. An error of 1% in T_M gives the uncertainty shown in Figure 2b. This uncertainty gives an idea of the sensitivity of the measurement to plasma profile changes. Enhanced sensitivity would be possible if the same analysis was made on a set of data from simultaneous spectrometry measurements along different sight lines with different $B(T)$ distributions. This would be on offer, for instance, with the proposed TRIAD neutron spectrometer system[6] observing the ITER plasma along both radial and tangential sight lines. Morover, emissivity spatial profiles of the neutron camera would be useful input to the emissivity temperature distribution analysis proposed here. Finally, ion temperature **spatial** profile measurements, which could be obtained from an advanced camera as discussed in a separate contribution[7], would complement the $B(T)$-analysis described here.

CONCLUSION

High statistics neutron spectrometry should be possible on ITER which can be combined with detailed analysis of the shape of the neutron emission spectrum to give new information on the plasma ion temperature. It is shown that this reflects directly the maximum ion temperature of the plasma viewed. Moreover, plasma profile information can be derived from the central region around the 14 MeV peak. Greater sensitivity to changes in the plasma profile shapes can be achieved by combined use of several neutron diagnostics.

* The Neutron Physics for Fusion Collaboration (see ref.[3])

REFERENCES

1. M.J. Loughlin, Profile effects on ion temperature measurements derived from neutron spectrometry, in: "Diagnostics for Contemporary Fusion Experiments", P.E. Stott, D.K. Akulina, G. Gorini and E. Sindoni, eds., Societa Italiana di Fisica, Bologna, Italy (1992).
2. J. Källne and G.Gorini, Neutron Observations and Alpha Particles in High Power Deuterium-Tritium Plasmas, *Fusion Technology* 25:341 (1994).
3. J. Källne and the NPF Collaboration, Neutron spectrometry for ITER, these Proceedings.
4. G. Gorini, L.Ballabio and J.Källne, Alpha Particle Kinetic Effects in the Neutron Emission of Burning DT Plasmas, *Rev. Sci. Instr.* 66:936 (1995).
5. R.K. Fisher, P.B. Parks, J.M. McChesney and M.N. Rosenbluth, Fast Alpha-Particle Diagnostics Using Knock-on Ion Tails, *Nucl. Fusion* 34:1291 (1994).
6. J. Källne, L. Ballabio, S. Conroy, G. Ericsson, J. Frenje, P.U. Renberg, E. Traneus, G. Gorini and M. Tardocchi, Multiple neutron spectrometer diagnostic for ITER, in: "Controlled Fusion and Plasma Physics", P.E. Stott, ed., European Physical Society, Petit-Lancy, Switzerland (1995).
7. E. Traneus and the NPF Collaboration, Neutron camera with spectrometer function, these Proceedings.

NEUTRON SPECTROMETERS FOR ITER

Takeo Nishitani, Tetsuo Iguchi[1], Eiji Takada[1], Jun-ichi Kaneko[2],
Satoshi Kasai and Tohru. Matoba

Naka Fusion Research Establishment, Japan Atomic Energy Research
Institute, Naka-machi, Naka-gun, Ibaraki-ken 311-01, Japan
[1]Nuclear Engineering Research Laboratory, University of Tokyo,
Tokai-mura, Naka-gun, Ibaraki-ken 319-11, Japan
[2]Tokai Research Establishment, Japan Atomic Energy Research Institute,
Tokai-mura, Naka-gun, Ibaraki-ken 319-11, Japan

INTRODUCTION

Neutron spectroscopy was proposed[1] early on in the fusion research as a direct method of ion temperature measurement in deuterium or deuterium-tritium plasmas. If the ion velocity distribution is Maxwellian, the neutron energy spectrum is centered at 14 MeV for the $d(t,n)\alpha$ reaction with a full width at half-maximum ΔE_n (in keV) represented by

$$\Delta E_n \approx 177 T_i^{1/2} \qquad (1)$$

where T_i is the ion temperature in keV. The ion temperature can be derived from the broadening of the neutrons spectrum using this relation.

In present large tokamaks such as JET, TFTR or JT-60U, ion temperature measurements have been tried using a ^{3}He neutron spectrometer[2-4] and a time-of-flight neutron spectrometer[5,6] for ohmically heated discharges. In those tokamaks, main experiments are carried out with intense auxiliary heating of neutral beam (NB) injections or ICRF heating, where the ion velocity distribution is no longer Maxwellian and consequently it is difficult to derive the ion temperature from the neutron spectrum. The competing diagnostic, Charge Exchange Recombination Spectroscopy (CXRS), is employed as a reliable method of the ion temperature measurement with high temporal and space resolutions. The International Thermonuclear Experimental Reactor (ITER) aimis to demonstrate a 1000 sec ignited plasma, where the ion velocity distribution is expected to be Maxwellian. Although CXRS needs a diagnostics neutral beam during the ignited phase, the this have not been planed for at this time Furthermore more, CXRS has problem with the

Diagnostics for Experimental Thermonuclear Fusion Reactors
Edited by P. E. Stott *et al.*, Plenum Press, New York, 1996

425

radiation damages of optical components such as fiber optics. Therefore we beliieve the neutron spectroscopy is the most promising T_i measurement technique for ITER.

The neutron emission spectrum of a DT plasma consists of the 2.5 MeV and 14 MeV peaks, which are from d(d,n) ^{3}He and d(t,n)α reactions, respectively. The 14 MeV neutron emission is ~100 times larger than 2.5 MeV neutrons. The broader component of neutrons centered at 14 MeV is predicted to be produced by super-thermal ions generated by knock-on collision with fast alpha particles. The neutron spectroscopy is useful not only for the ion temperature measurement but also the ion density from 2.5 MeV and 14 MeV neutron intensities, the fast alpha particle population from the broader spectrum of 14 MeV neutrons and the plasma toroidal rotation measurements from the shift of the 14 MeV peak in the tangential measurement. Several types of the 14 MeV neutron spectrometer have been proposed or developed. The proton recoil counter telescope (COTETRA)[7] and the diamond detector[8] have been demonstrated in the TFTR DT experiments. The tandem-radiator spectrometer[9], the associated-particle time-of-flight spectrometer (TANSY)[10] and the magnetic proton recoil spectrometer(MPR)[11] are or will be installed on JET for the DT experiments to be carried out in 1996. Those neutron spectrometers are too large to install in the cryostat of ITER except the diamond detector. We are developing a compact neutron spectrometer[12] based on a recoil proton counter-telescope technique to consist the radial neutron spectrometer array in the cryostat concentrating on the ion temperature profile measurement.

DESIGN OF COMPACT NEUTRON SPECTROMETER

Requirements for Neutron Spectrometer

The ion temperature profile measurement aiming the burn optimization and transport studies is categorized in the measurements for the performance evaluation and the optimization (Category II) in ITER. The physics requirements for the ion temperature measurement are listed in Table 1.

Table 1. Requirement for ion temperature profile measurement.

Temperature range	Spatial resolution	Time resolution	Accuracy
0.5-50 keV	30 cm	100 ms	10%

In order to measure the ion temperature from the neutron spectrum, a high-energy-resolution spectrometer should be employed. To be useful, the intrinsic energy resolution of the spectrometer, ΔE_{det} needs to be less than the thermal broadening represented by equation (1). Thus, for an ion temperature of 3 keV, ΔE_{det} should be better than 2.2% which is achievable. The accuracy of the ion temperature is related to the total counts of 14 MeV neutron peak, N, by

$$\frac{\Delta T_i}{T_i} = \sqrt{\frac{2}{N}\left[1 + \left(\frac{\Delta E_{det}}{\Delta E_n}\right)^2\right]} \tag{2}$$

When $\Delta E_{det} \leq \Delta E_n$, 400 counts will satisfy the 10% accuracy of the ion temperature. So the detection efficiency should be high enough to obtain a counting rate of 4000 cps to result in 100 ms time resolution. Table 2 shows the design target for radial neutron spectrometers.

Table 2. Design target for radial neutron spectrometers.

Energy resolution	Detection efficiency	Spatial resolution	Time resolution
2% at 14 MeV	1×10^{-5} counts/(n/cm^{-2})	< a/5	100 ms

Concept of Spectrometer

Figure 1 shows the schematics of the compact neutron spectrometer based on a recoil proton counter-telescope technique. The neutron spectrometer will be installed on a well-collimated neutron beam line. A large-area recoil proton emitter(radiator) is placed in parallel to the incident neutron beam and a multi-collimating plates are inserted between the radiator and the recoil proton detectors away from the neutron beam in order to limit the scattering angle of protons to the recoil proton detectors. Here a very thin polyethylene film and a silicon surface barrier detector(SSD) are employed as the radiator and proton detector, respectively. Recoil proton energy is measured with the SSDs using conventional pulse height analysis electronics. Incident neutron energy E_n is related to the recoil proton energy E_p by

$$E_n = E_p / \cos^2\Theta \qquad (3)$$

where Θ is the tilting angle of the channel in the multi-collimating plate against the neutron incident direction.

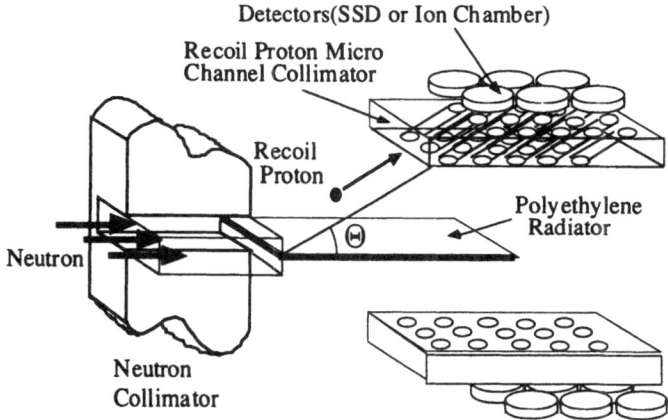

Figure 1. Schematics of the compact neutron spectrometer based on a recoil proton counter-telescope technique.

The high energy resolution characteristic can be realized by the use of a very thin radiator and a recoil proton collimator to subtend the small solid angle of recoil protons incidence to the SSDs. While the detection efficiency can be freely controlled by adjusting the detection area under the high energy resolution geometry.

Numerical Calculation of Detector Response

The design parameters were widely surveyed on the radiator thickness, the diameter of a recoil proton collimator and the incident angle of the recoil protons against the neutron incident direction, using simple Monte Carlo calculations. In this calculation, we considered only the loss of the recoil proton energy in the radiator. The radiator thickness dependence of the detection efficiency and the energy resolution is shown in figure 2, where the detection efficiency is defined as the counts for single collimator channel. We can see that the detection efficiency and the energy resolution are strongly trade-off. Also the collimator radius dependence of the detection efficiency and the energy resolution isshown in figure 3.

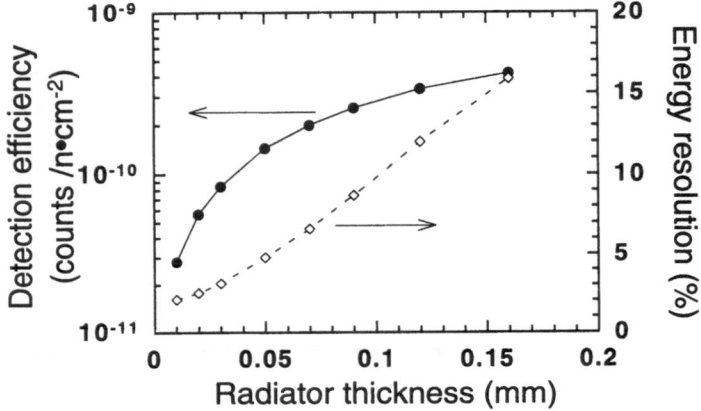

Figure 2. Radiator thickness dependence of the detection efficiency and the energy resolution of the neutron spectrometer evaluated by Monte Carlo calculations.

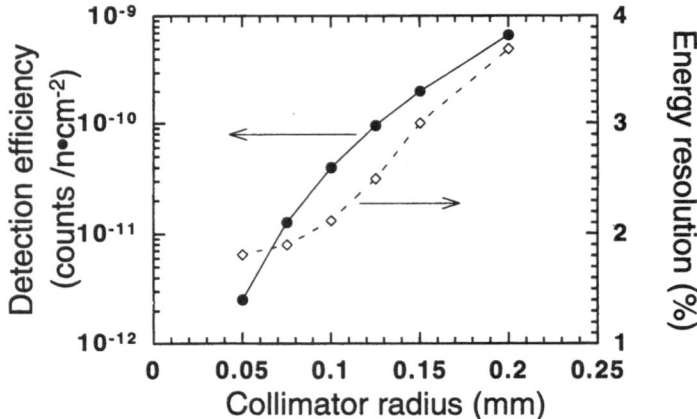

Figure 3. Collimator radius dependence of the detection efficiency and the energy resolution of the neutron spectrometer evaluated by Monte Carlo calculations.

Figure 4 shows the calculated recoil proton spectrum for 14 MeV neutrons with 2% of the FWHM assuming 1% of the SSD's energy resolution. From the error propagation in equation(3), the accuracy of the incident neutron energy is evaluated to be 2.1%. The detection efficiency is also calculated to be 3.64×10^{-7} counts/(n/cm^{-2}) for the single collimator channel. If we can open 0.15 mm diameter channels with 0.2 mm pitch, 550 cm^2 of the active area provides the 1×10^{-5} counts/(n/cm^{-2}) of the required detection efficiency.

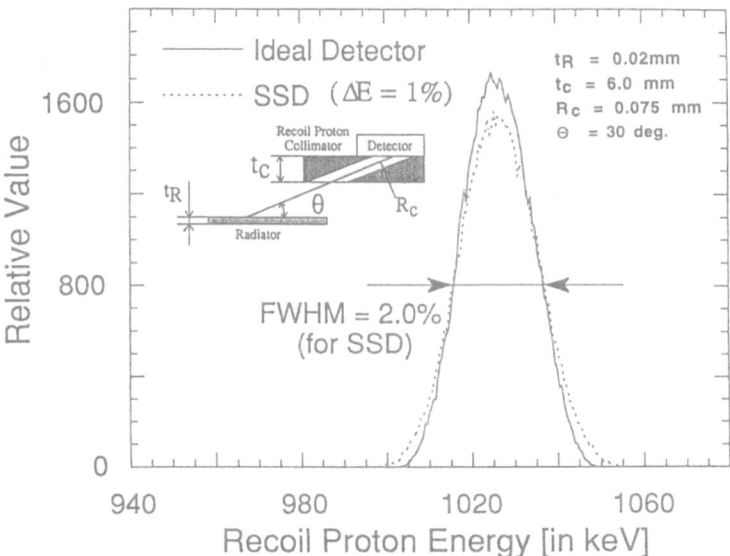

Figure 4. Calculated recoil proton spectrum for 14 MeV neutrons. FWHM of the spectrum is 2% when the SSD's energy resolution of 1% is assumed.

Neutron Calculation

In this spectrometer, radiation damaged of the SSDs is most serious problem. We expect that the radiation resistance will be ensured by placing the detector at sufficient long distance from the incident neutron beam plane. The shielding performance of a neutron collimator with rectangular cross-section was calculated using the Monte Carlo code MCNP[13] with the neutron cross-section library JENDL-3[14] for the model of the single neutron collimator. When the detector behind the collimator is placed at a distance of 15 cm from the neutron beam plane, the scattered neutron flux above 100 keV is estimated to be 1.5×10^4 neutrons cm^{-2}s^{-1} for the 1 MW m^{-2} neutron wall load. In this neutron flux, the life time of the SSD due to the radiation damages reaches to about 18,000 hours, corresponding to a 14 MeV neutron fluence of ~10^{12} cm^{-2}. The SSD can be used without replacement in the ITER life because the total operation time of ITER with high power is expected to be less than $50,000 \times 1000$ sec $\approx$ 14,000 hours. The noise signal induced by the background neutron is also important to keep the spectrometer performance. The signal to noise ratio is also estimated to 47, where not only Si(n,p) and Si(n,α) reactions inside the silicon plate but also many charged particle production reaction in the housing materials of the SSD are taken into account as noise production processes.

CHARACTERISTICS OF PROTOTYPE SPECTROMETER

The prototype of this spectrometer has been tested its performance at the FNS 14 MeV neutron generator[15]. Figure 5 shows the schematics of the experimental arrangement of the prototype spectrometer at FNS. The spectrometer unit consists of a polyethylene radiator film and two sets of the multi-collimating plate and the SSD with an active area of 2000 mm^2 and a depletion depth of 0.5 mm. The vacuum chamber where the spectrometer unit is mounted in is installed behind the neutron collimator. The axis of the system is tilted 100° against the deuteron beam line, where the spread of the source neutron energy is almost minimum. We used two radiator films with 0.05 mm and 1 mm thickness in order to confirm the validity of the numerical calculation. The multi-collimating plates are made of 2.4 mm thick aluminum plate with 1.2 mm radius collimator channels tilted 30 °. The energy resolution of both SSDs are calibrated using ^{241}Am source to be 47 keV for 5.5 MeV alphas.

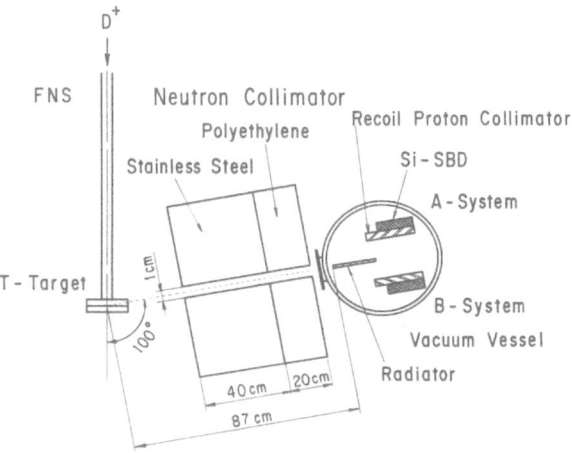

Figure 5. Schematics of the experimental arrangement of the prototype spectrometer at the FNS neutron generator. The vacuum chamber with the spectrometer unit is installed behind the neutron collimator.

Figure 6 shows the recoil proton spectra using 0.05 mm and 1 mm thick radiators comparing with numerical calculations assuming 1% of the SSD energy resolution. The measured spectrum with a thick radiator has good agreement with the calculation. In the case of the thin radiator, the measured spectrum has a low energy tail larger than the calculated one. We have three candidates to explain the difference; the small angle scattering of the recoil proton in the collimator, the energy loss of the recoil proton inside the collimator material at the collimator edge, and the energy loss of the recoil proton in SSD out of the active area. We revised the numerical calculation code to evaluate the former two processes. The calculation results indicate that the energy loss in the collimator edge is dominant process in the low energy tail formation, but the intensity is only 1% of the direct component. So those two process can not explain the measured low energy tail. We will investigate the energy loss in the dead area of SSD by measuring the recoil proton spectrum with SSD of which dead area is covered by a foil.

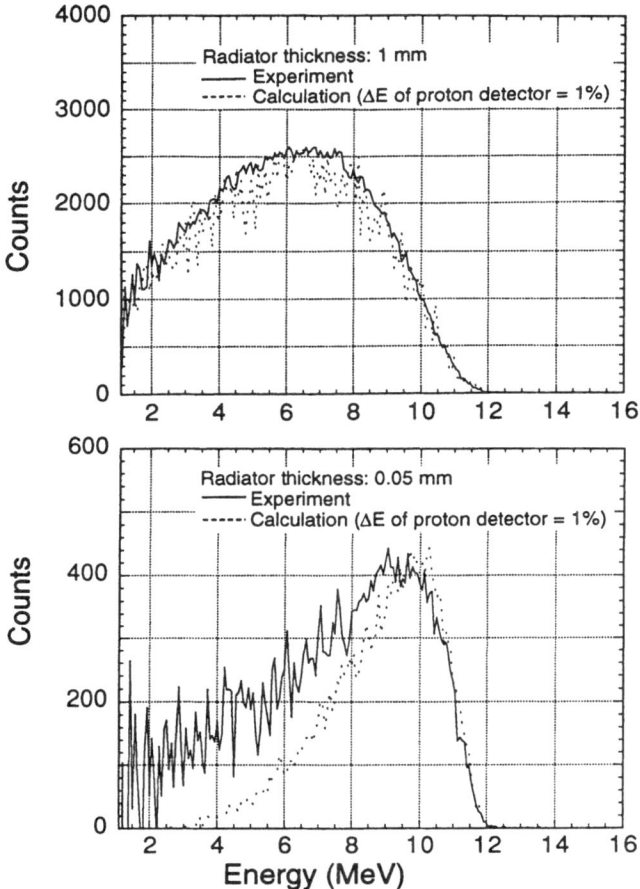

Figure 6. Recoil proton spectra using 0.05 mm and 1 mm thick radiators comparing with numerical calculations assuming 1% of the SSD energy resolution.

DESIGN OF RADIAL NEUTRON SPECTROMETERS

We designed the radial neutron spectrometer array to be installed in a horizontal big port such as a remote handling port, which has a rectangular space with a cross section of 1.6m x 1.6m and a length of 3.4m. Figure 7 shows the schematics of the neutron collimator for the neutron spectrometer array. The stainless steel(SS316) and coolant water occupying 30% of the volume are adopted as a structural and shielding material. If the guideline shielding performance is assumed 10^{-6} as an attenuation rate of 14 MeV neutron flux, the shield thickness of more than 120 cm is needed. So the 14 MeV neutron flux is estimated to be 1×10^{10} neutrons cm^{-2} s^{-1} at the radiator position of the central chord for the 1 MW m^{-2} neutron wall load. A fan-like system is adopted to obtain a field of view as wide as possible. Under the structural constraints on the access port, the view angle is limited vertically between 30 degrees upward and downward orientation covering about 2/3 of the plasma volume in sight.

The neutron collimator consists of five parts. The first part is a pre-collimator of 200 cm in length. The path of neutrons is 25 cm wide and squeezed vertically in the middle of the way for minimizing background radiation. The second part is a multichannel collimator equipped with 120 cm long multichannels. There are three sets of multichannel collimators arranged in parallel. The central channels have a rectangular cross section 20 cm wide and 3 cm high to supply a plane 14 MeV neutron beam mainly for a 14 MeV neutron spectrometer. The other two side channels are prepared for other diagnostics such as hard X-ray detectors in order to share the big collimator space, which have a circular cross section of 5 cm diameter. The maximum number of channels in the central and the side ones will be around 5 and 15, respectively, from the viewpoint of detector spaces necessary to avoid cross talks between the neighboring channels. The spatial resolution of the neutron spectrometer is about 1.5 m on the vertical line at R = 8.14 m. The third part is a detector space for arranging detectors, of which size was determined by the neutron spectrometer. The fourth part is the post-shielding against back-scattered neutrons. Most of the primary neutrons from the collimator exit pass through the detector part and causes the source of background radiation. The post-shielding has penetration holes corresponding to each collimator channel to let the unnecessary primary neutrons escape to another room and prevent a counter flow of back-scattered neutrons. The last part is for additional shielding to confine and absorb neutrons leaked out from the detector part and will be available for other diagnostic systems which need a large set-up space.

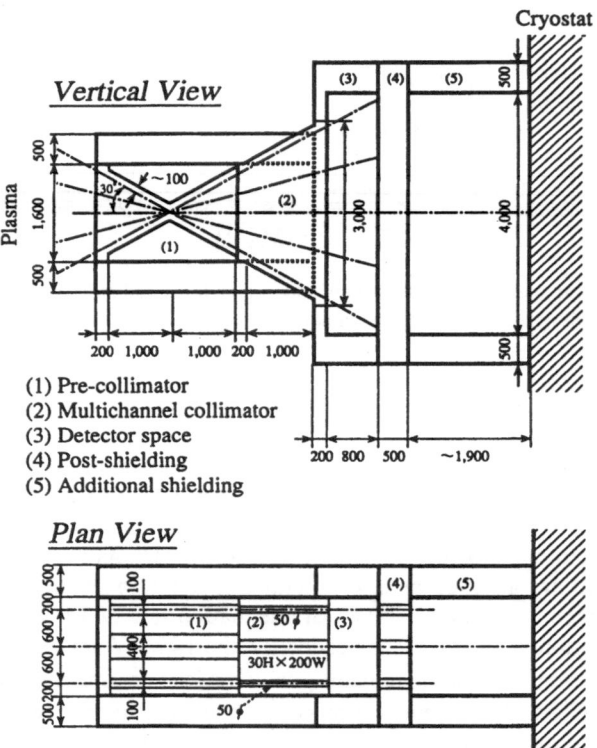

Figure 7. Schematics of the neutron collimator for the neutron spectrometer array made of stainless steel(70%) and water (30%). The collimator will be installed in a horizontal remote handling port in ITER.

The shielding geometry on the present collimator system was designed only by the preliminary calculation results based on one-dimensional radiation transport code ANISN[16], and MCNP with single collimator channel modeling. It is, therefore, necessary to make a detailed evaluation and optimization on the shielding performance by using three-dimensional radiation transport calculation such as MCNP.

SUMMARY

A new type of DT neutron spectrometer based on a recoil proton counter-telescope technique has been developed for the ITER radial neutron spectrometers. The neutron spectrometer will be installed on a well-collimated neutron beam line. A large-area radiator is placed in parallel to the incident neutron beam and a multi-collimating plates are inserted between the radiator and the recoil proton detectors away from the neutron beam in order to limit the scattering angle of protons to the recoil proton detectors. Here a very thin polyethylene film and a silicon surface barrier detector(SSD) are employed as the radiator and proton detector, respectively. The energy resolution and detection efficiency are estimated to be 2.1% and 1×10^{-5} counts neutron^{-1} cm^2, respectively for DT neutron through Monte Carlo calculations. The radiation damages of SSDs is an essential issue for ITER diagnostics. The life of SSDs is also estimated to be about 18,000 hours which is longer than total high power operation time of ITER through the neutron Monte Carlo code MCNP. The prototype of this spectrometer has been tested at the FNS 14 MeV neutron generator. The measured spectrum with a thin radiator has a low energy tail larger than the calculation, which is probably due to the energy loss of the recoil proton in the dead area of SSD. We designed the radial neutron spectrometer array with fan-like geometry be installed in a remote handling port. The stainless steel(SS316) and coolant water occupying 30% of the volume are adopted as a structural and shielding material. The detail evaluation and optimization on the shielding performance should be carried out by three-dimensional radiation transport calculation.

ACKNOWLEDGMENTS

The authors would like to appreciate Dr. Y. Ikeda for his arrangement for the prototype spectrometer experiments in the FNS neutron source. The authors would like to thanks Dr. G.A. Wurden for his useful comment on this work.

REFERENCES

1. G. Lehener and F. Pohl, Reaktionneutron als hifsmittel der plasmadiagnostik, Z. Phys. 207: 83 (1967)
2. O.N. Jarvus, G. Gorini, M. Hone, J. Källne, G. Sadler, V. Merlo and P. van Bell, Neutron spectroscopy at JET, Rev. Sci. Instrum. 57:1717 (1986).
3. J.D. Strachan, T. Nishitani and C.W. Barnes, Neutron spectroscopy on TFTR, Rev. Sci. Instrum. 59:1732 (1988).
4. T. Nishitani and J.D. Strachen, Neutron spectroscopy with a ^{3}He ionization chamber on TFTR, Jpn J. Appl. Phys. 29:591(1990).
5. T. Elevant, D. Aronsson, P. van Bell, G. Grosshoeg, M. Hoek, M. Olsson and G. Sadler, The JET neutron time-of flight spectrometer,Nucl. Instrum. Methods A306:331(1991).

6. T. Elevant, P. van Bell, G. Grosshoeg, M. Hoek, O.N. Jarvis, M. Olsson and G. Sadler, The new JET 2.5-MeV neutron time-of-flight spectrometer, *Rev. Sci. Instrum.* 63:4586 (1992).

7. M. Osakabe, S. Itoh, Y. Gotoh, M. Sasao and J. Fujita, A compact neutron counter telescope with thick radiator (COTETRA) for fusion experiments, *Rev. Sci. Instrum.* 65:1636 (1994).

8. A.V. Krasilnikov, Diamond detector based DT neutron spectrometer for ITER, in this workshop.

9. N.P. Hawkes, P. van Bell, M. Hone, O.N. Jarvis, M.J. Loghlin and M.T. Swinhoe, A 2.5 MeV neutron spectrometry system with a tangential line of sight for the D-D phase at the JET tokamak, *Nucl. Instrum. Methods* A335:533(1993).

10. G. Grosshög, D. Aronsson, K.H. Beimer, R. Rydz, N.G. Sjöstrand, Ö. Skeppstedt and L.O. Pekkari, The use of the neutron-proton scattering reaction fro D-T fusion spectrometry, *Nucl. Instrum. Methods* A249: 468 (1986).

11. J. Källne and H. Enge, Magnetic proton recoil spectrometer for fusion plasma neutrons, *Nucl. Instrum. Methods* A311: 595 (1991).

12. T. Iguchi, J. Kaneko, M. Nakazawa, T. Matoba, T. Nishitani and S. Yamamoto, Conceptional design of neutron diagnostics systems for fusion experimental reactor, *Fusion Eng. Design* 28: 689 (1995)

13. LANL Group X-6, *MCNP-a general Monte Carlo code for neutron and photon transport version 3A, Report LA-7396-M, Rev.2*, Los Alamos National Laboratory, Los Alamos (1986)

14. K. Shibata, T. Nakagawa, T. Asami, et al., *Japanese evaluated nuclear data library version-3, Report JAERI 1319*, Japan Atomic Energy Research Institute, Tokai (1990)

15. T. Nakamura, H. Maekawa, Y. Ikeda and Y. Oyama, Present status of the fusion neutronics source (FNS), in: *Proc. 4th Symp. on Accelerator Sci. Technol.*, p155 RIKEN, Saitama (1982)

16. W.W. Engle, Jr., *A User's Manual for ANISN, A One Dimensional DIscrete Ordinates Transport Code with Anisotropic Scattering, K-1693*, Union Carbide Corporation (1976)

DIAMOND DETECTOR BASED DT NEUTRON SPECTROMETER FOR ITER

Anatoli V. Krasilnikov

Troitsk Institute Innovating and Fusion Researches
Troitsk, Moscow region, Russia

INTRODUCTION

The neutron diagnostic of future deuterium-tritium (DT) burning experiments [such as ITER] will be of fundamental importance in determination and control major parameters of plasma ion component. Two-dimensional neutron emission measurements will give information for burn and fuelling control and optimization, disruption avoidance and instabilities study. Perpendicular and tangential DT-neutron spectrometers will give information about dynamic of tritium and deuterium distribution function (f(E,r)) for different pitch angle ranges and α-particle birth energy distributions on different magnetic surfaces. Such information gives possibility for burn optimization, determination of α-particle heating profile and study of high energetic ions transport.

Here, I focus on the possibility to use[1] Natural Diamond Detectors (NDD)[2,3] in ITER perpendicular neutron cameras[4,5] and tangential neutron collimator for the purpose of DT-neutron spectrometry and flux profile measurements. The performance of the neutron spectrometers and cameras is closely related to the neutron flux level. In this paper I shell make an analysis for diagnostic geometries, which are optimized for ITER maximum fusion power (P=1.5 GW, Y=10^{21} n/s) performance[6]. But natural Diamond detectors based DT-Neutron Spectrometer (DNS) will operate down to the neutron yield of the order of 10^{19} n/s. This will give the possibility to use this diagnostic in a wide range of ITER performances for alpha-particle heating, ignition and burn optimization.

The prototype of DNS (with standard electronics) have been tested on SNEG-13[1], FNG[7], PPPL's[8] and FNS[9] neutron generators. It was successfully used for DT-neutron spectrum and flux measurements in TFTR DT[8] and JT-60 DD[10] experiments.

NATURAL DIAMOND DETECTOR PROPERTIES

Pure natural diamond[11,12] is one of the most simple non metallic crystal insulator. Table 1 summarizes and compares with silicon and best artificial diamond some of the most important properties of natural IIa diamond. High forbidden energy gap, resistivity and break

Diagnostics for Experimental Thermonuclear Fusion Reactors
Edited by P. E. Stott *et al.*, Plenum Press, New York, 1996

435

down resistance, large saturation velocity and mean free drift time (> 10 ns)[13] are very important properties for its uses as material for radiation detectors. NDDs have a very high radiation resistance (maximum acceptable fluence about $5*10^{14}$ n/cm^2)[14], a low noise due to leakage current, can operate at extremely high temperatures and in aggressive chemical media.

Table 1. Properties of natural diamond in comparison with other materials

Property	Natural Diamond	CVD & DC arc jet Diamond[15]	Si
Atomic number	6	6	14
Density (g cm^{-3})	3.5	3.5	2.32
Thermal Conductivity (W m^{-1} K^{-1})	2000		150
Band gap (eV)	5.5	5.5	1.12
Resistivity (W cm)	$>10^{15}$	$>10^{11}$	$2*10^5$
Breakdown Voltage (V cm^{-1})	10^7	10^7	10^3
Dielectric constant	5.6	5.6	11.7
Operation temperature (K)	100-600		
Mobility of: Electrons (cm^2 V^{-1} s^{-1})	2400	~ 500	1500
Holes (cm^2 V^{-1} s^{-1})	2100	~ 500	500
Saturation Velocity for: Electrons (cm s^{-1})	$2.2*10^7$	$5*10^6$	10^7
Holes (cm s^{-1})	$2*10^7$	$5*10^6$	$3*10^6$
Charge collection distance (mcm)	3300	45	10^7
Mean free drift time (ns)	10 - 30	1	10^5
Energy per e-hole pair creation (eV)	13.2	13.2	3.6

First observation of pulses due to discrete radiation using NDDs was made in 1941 by Stetter. Recently efforts have been made in the characterization of the natural diamond and in using the ion implantation in fabrication of the electrical contacts[2] to remove the effect of charge polarization.

Diamond detectors have metal-insulator-metal structure. The insulator is natural very pure (IIa) diamond (area ~15mm[2], thickness L~200mcm), the metals are thin (~50 nm) gold layers ~ 9 mm[2] in area, which form ohmic contact with diamond. Diamonds for detectors are selecting by photoconductivity and absorption coefficient measurements to ensure a nitrogen concentration not exceeding 10^{19} cm^{-3} and bulk carries lifetime not less than 10 ns.

Any radiation, that generates charge carries in the diamond, can be detected. This incorporates UV, X-rays, γ-rays, high energy particles, such as: α-particles, electrons, neutrons and others. The mechanism of detection based on collection of charge carriers, which were created by the interaction of the radiation with diamond, using external applied electric field. An important detector characteristics are the charge (electrons and holes) collection distances $d_{e,h}$, which are the products of the material parameters $\tau_{e,h}$ -electrons and holes mean free drift times, $\mu_{e,h}$ - electrons and holes mobilities and applied electrical field E

$$d_{e,h} = \mu_{e,h} * \tau_{e,h} * E$$

To provide detector energy resolution better then 3%, the ratios of $d_{e,h}$ to detector thickness should be more then 15. So only natural diamond and silicon can be used for creation detectors for high resolution particle spectrometry now. The value of 71±18mm for NDD full charge collection distance d_c ($d_c \approx 0.03*d$ for detector with energy resolution ~3%) was measured in FNG experiments[6]. Best detectors produced from artificial diamonds (chemical vapour deposition (CVD) or DC arc jet)[15] have essentially lower mean free drift time and charge collection distances (see table 1) and so can't be used for spectrometry.

Mean free drifting time of the order of 10-30 ns for holes and 15-50 ns for electrons[16] and Full Width Half Maximum (FWHM) of 82-90 keV[16,7] using Am, Pu and Cm α-sources have been measured for NDD.

The possibility to use NDD for DT neutron spectrum measurements have been investigated experimentally and analytically during last three years[1,7-9,17]. This possibility based on the shape of NDD pulse height spectrum in DT neutron flux[1]. High energetic particles (α, ^{12}C and ^9Be) are creating in diamond due to intrinsic nuclear reactions, neutrons elastic and inelastic scattering, when detector is placed in the DT neutron flux. Possible channels of reactions, as well as the corresponding cross sections[18] are gathered in Table 2. The charged products of these reactions produce electron - hole pairs in the diamond crystal. The energy required to produce one electron - hole pair is 13.2 eV[16], so charge collected on the positive electrode will be:

$$Q = e * E_{kin}/13.2 *1/L * \int exp(-x/d_e)dx$$

for $d_e >> L$

$$Q \approx e * E_{kin} /13.2 * (1-0.5 * L/d_e)$$

where e is electron charge, E_{kin} (in eV) is the sum of charged products kinetic energy. Thus the sum of the kinetic energies of the charged products defines the NDD pulse height spectrum of 14 MeV neutrons. The comparison of calculated, using double differential cross sections[18] for reaction channels from Table 2, and measured in SNEG-13[1], FNG[7] and FNS[9] experiments pulse height spectra for 14.1 and 14.8 MeV neutrons are shown in Fig.1.

Table 2. Reaction channels which occur in diamond with DT neutrons

	Reaction channel	Energy threshold (MeV)	Cross section (mb)		Range of charged products sum kinetic energy (MeV)	
			E_n=14.1MeV	E_n=14.8MeV	E_n=14.1MeV	E_n=14.8MeV
1	Elastic Scattering[1] $^{12}C(n,n')^{12}C$	0	800	800	0.00 + 4.00	0.00 + 4.20
2	Inelastic Scattering $^{12}C(n,n')^{12}C*$					
	(E* = 4.439 MeV)	4.81	200	200	0.04 + 3.29	0.03 + 3.49
3	$^{12}C(n,n')^{12}C*(\alpha)^8Be(\alpha)$					
3.1	E12$_{C*}$ = 7.65 MeV	8.29	10 ± 2	9 ± 2	0.50 + 3.07	0.49 + 3.28
3.2	E12$_{C*}$ = 9.64 MeV	10.4	76 ± 8	69 ± 8	2.60 + 4.64	2.57 + 4.86
3.3	E12$_{C*}$ = 10.8 MeV	11.7	47 ± 4	36 ± 4	3.87 + 5.52	3.83 + 5.76
3.4	E12$_{C*}$ = 11.8 MeV	12.8	39 ± 12	50 ± 5	5.01 + 6.25	4.94 + 6.46
3.5	E12$_{C*}$ = 12.8 MeV	13.8	13 ± 6	48 ± 12	6.29 + 6.80	6.11 + 7.17
3.1	E12$_{C*}$ = 14.0 MeV	15.1	0	5 ± 1		7.06 + 7.52
4	$^{12}C(n,\alpha)^9Be*$					
	(E* = 2.43 MeV)	8.81	32 ± 20	18 ± 14	5.10 + 6.30	5.70 + 7.00
5	$^{12}C(n,\alpha_0)^9Be$	6.19	72 ± 9	72 ± 9	8.40	9.10

[1]Only elastic scattering with cross section 1.6 b and range of carbon recoil kinetic energy 0 + 0.8 MeV takes place for 2.5 MeV neutrons.

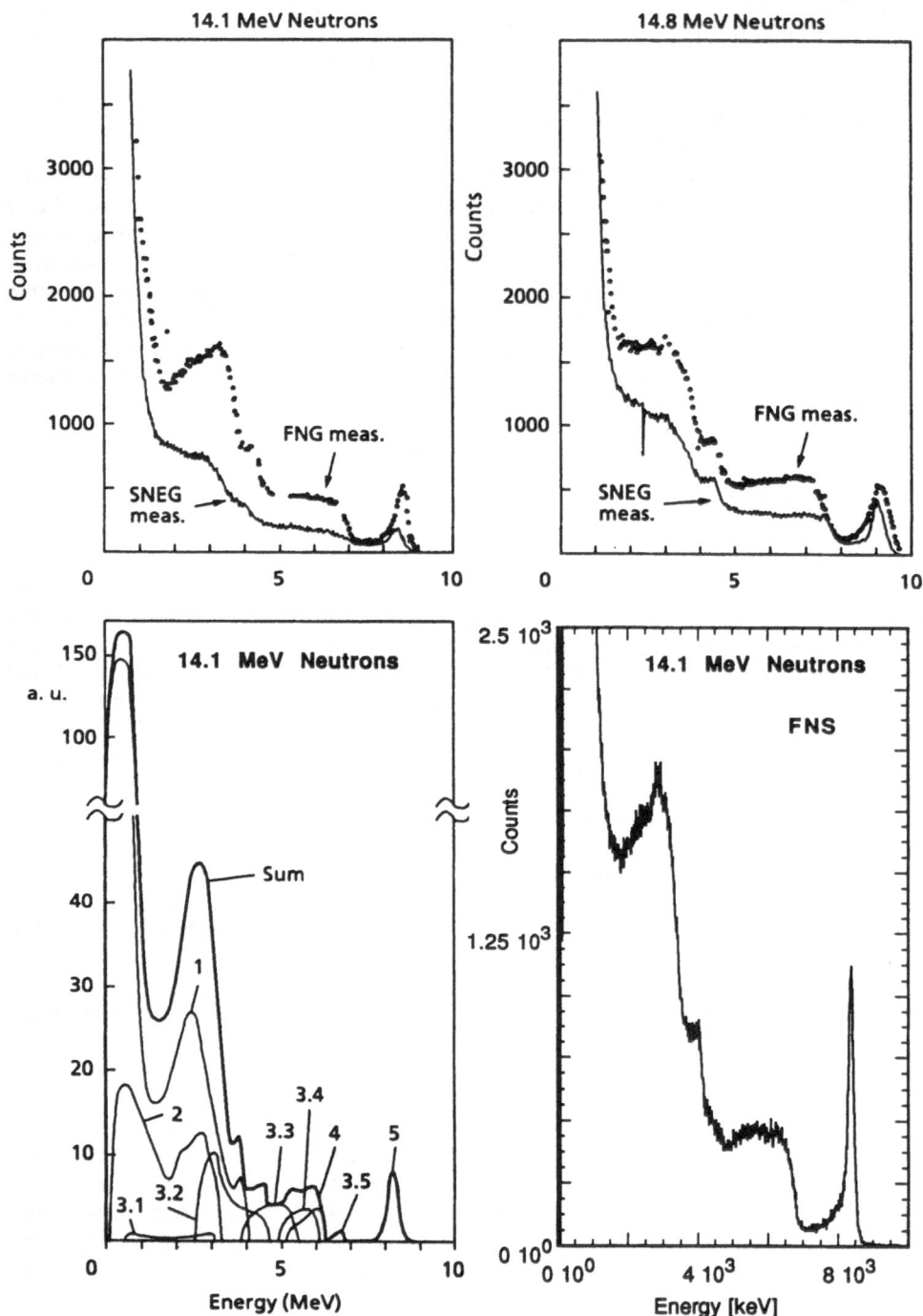

Figure 1. Measured on SNEG-13[1], FNG[7] and FNS[9] and calculated[1] NDDs pulse height spectra for 14.1 and 14.8 MeV neutrons. Numbers in the calculated lines refer to Table 2.

The $^{12}C(n,\alpha_0)^9Be$ reaction (channel 5 in Table 2) is responsible for the useful for spectrometry E_n-5.7 MeV peak in the NDD pulse height spectrum. The absence of contribution from channels 1 - 4 (Table 2) in the pulse height spectrum around the $^{12}C(n,\alpha_0)^9Be$ peak, namely in the range $E_{\alpha 0+9Be} \pm 2$ MeV make NDD very useful DT neutron spectrum measurements of hot (up to $T_i \sim 50$ keV) burning plasma.

Energy resolutions (FWHM) of some NDDs have been measured during DT neutron spectrum measurements on neutron generators[1,7-9] and were in the range 1.95 - 3.9 % for best detectors. Measured[9] DT neutron pulse height spectra, produced in NDD by the FNS neutron generator for 160, 95 and 0 degrees, are shown in Fig.2. The energy scale corresponds to the sum of the energy of the charged products in the NDD. The peaks energy shifts, respect to the 100^0 peak energy was +848, +52 and -588 keV for 0^0, 95^0 and 160^0 respectively, that agree very well with the results (+853, +55 and - 587 keV) of calculations[9]. Due to good enough (1.95 %) energy resolution of using in this measurements NDD, it was possible to measure not only broadening of DT neutron spectra for 0^0 and 160^0 angles but the shapes of distributions also[9].

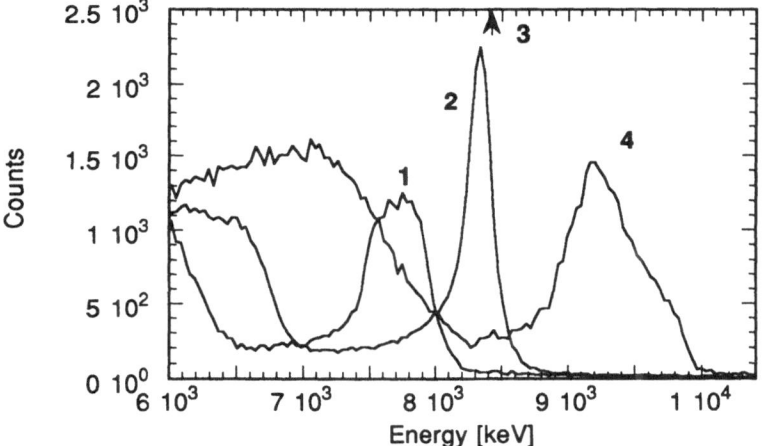

Figure 2. The measured NDD pulse height spectra in FNS experiments for detector positions located 20 cm far from target at 1 - 160^0, 2 - 100^0, 3 - 95^0 and 4 - 0^0 angles with respect to 350 keV deuteron beam.

There always exist some broadening of neutron generator DT neutron spectrum at 95^0 angle. The main reason of this broadening is scattering of deuterons inside target. This broadening can be essentially diminished by diminishing deuteron energy close to peak energy of DT reaction. Corresponding measurement have been made on FNG[7] and FNS[9] neutron generators. In both cases NDDs have detect changes in the DT neutron width from ~2% to ~1%.

The sensitivity of NDD with mentioned above dimensions to neutrons and gammas were measured during FNG experiments[6]. The values: $5*10^{-4}$ counts*cm^2/n - total for DT neutrons, 10^{-5} counts*cm^2/n - due to $^{12}C(n,\alpha_0)^9Be$ reaction for DT neutrons, 10^{-3} counts*cm^2/γ - for ^{137}Cs and ^{60}Co gammas (for compton electrons pulse height spectrum range > 52 keV) and 10^{-3} counts*cm^2/n - for 2.5 MeV neutrons were measured.

It has been shown[19], that NDD operates with high energy resolution in the temperature range 100-600 K with the best performance at 300 K. Upper limit of this temperature range of

NDD spectrometric operation and extremely high thermal conductivity of diamond make it possible to use NDD in ITER temperature conditions without any problems.

High radiation resistance of diamond ($5*10^{14}$ n/cm^2)[14] is a very important property for it's application in ITER radiation conditions. Radiation damage of NDD placing into DT neutron flux are taking place due to creation defects in crystal lattice mostly by high energetic ^{12}C recoil nuclei and atoms. Two kinds of crystal lattice defect are most important: vacancies and ^{12}C interstitials. The ^{12}C displacement threshold energy for diamond lattice is in the range 37.5 - 47.6 eV, depending upon the direction of initial motion of the displaced atom[20]. This value is essentially higher then displacement threshold energy (11-22 eV) of silicon[20]. The high diamond displacement threshold energy is due to a rapid dissipation of kinetic energy from bombarded atom into incoherent vibrational energy of its neighboring atoms before the displaced atom can reach the top of the energy barrier for defect formation. The transfer of kinetic energy is effective, because the speed of the displaced carbon atom is comparable to the speed of sound in diamond. Thus the extremely high radiation hardness of diamond is the consequence not only its large binding energy, but its high sound velocity also.

It is important to notice that NDD should have essentially higher radiation resistance to ITER γ-flux than to DT neutrons. This should be due to the fact, that instead of neutrons γ-rays create inside diamond compton electrons with relatively low energy. Most of these electrons will have energy not more then some hundreds keV. To create defect (displacement of ^{12}C atom) in diamond lattice by electrons, their energy must be higher then 180-220 keV, depending upon direction of displacement[20]. So only some electrons will produce not more than one or two defects. Extremely high NDD radiation resistance in γ-flux is very important for its application as the γ-flux monitor.

Very low NDD leakage current (~ 1 pA/mm^2 at bias 100 V and low radiation background conditions) provide extremely low level of intrinsic current noises and therefore high signal to noise (due to leakage current) ratio in the case of current measurements. This property in the complex with the high radiation resistance and relatively low (with respect to Si detectors) sensitivity to neutrons and γ-rays make it interesting to consider NDD current mode application for UV and SXR flux measurement in energy range 0.006 -10 keV[21].

ELECTRONICS

The useful for DT neutrons spectrometry NDD sensitivity (due to ^{12}C(n,α)^{9}Be reaction) in ITER collimator conditions will be of the order of 2% of total sensitivity to neutrons and gammas. So, to receive high useful count rate and so good time resolution, it is necessary to have as fast as possible pulse height analyzing electronics. This electronic system will include fast charge sensitive preamplifier, ultra fast spectrometric amplifier with pileup rejection, high throughput ADC with discriminator and also single channel analyzer for flux measurements.

The duration of NDD current pulses is of the order of 1 ns, and so diamond detector can successfully operate at count rates up to 10^8 counts/sec.

Fastest standard (ORTEC-142A, CANBERRA-2004) preamplifiers, when using with NDD for DT neutron spectrometry, can operate at count rate up to 10^6 counts/sec. Specially developed pulse-drain feedback charge sensitive preamplifier[22] can operate even at higher count rates.

Fastest standard gated integrator amplifier (ORTEC-973U) with integration time 1.5 ms in the case of input count rate $3.2*10^5$ can provide output count rate $1.2 * 10^5$ counts/s.

The degradation of NDD based spectrometry system energy resolution due to pileup effect in amplifiers have been noticed during TFTR high beam power DT experiments[8]. This effect was investigated in FNS measurements[9] using fastest standard (CANBERRA 2024) unipolar shaping (0.25 ms shaping time) amplifier. Whole system energy resolution changes (without using pileup rejection procedure) from 1.95 % to 3.58 % (see Fig.3), when input amplifier count rate changes from 10^4 to $5.7*10^5$ counts/s. These experiments show, that standard amplifiers with shaping time constant 0.25 ms can be used in diamond detector based spectrometers up to input count rates $4*10^5$ counts/s. Pileup rejection procedure can slightly increase this limit. For diamond detector based spectrometry at count rates of the order of 10^6 counts/s it is necessary to develop and use more fast amplifiers and ADCs. The possibility to create, using present hardware fast pulse signal processor operating at input count rate 10^6 and providing output count rate $5*10^5$, have been shown by Dudin[23].

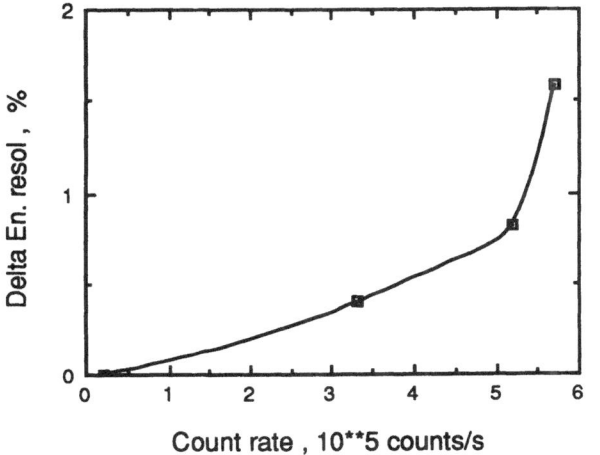

Figure 3. The dependence of increase of NDD based DT neutron spectrometric system energy resolution upon CANBERRA 2024 (shaping time is equals to 0.25 ms) amplifier input count rate, measured in FNS[9].

This analysis shows, that for today's hardware upper level of DNS count rate is limited by amplifier. In the case of fast pulse signal processor[23] usage and NDD placing in the position with DT neutron flux $2*10^9$ n/(cm^2*sec) DNS will have useful for DT neutron spectrometry count rate 10^4 counts/sec. This means, that DNS can provide 0.1 sec time resolution in ITER DT neutrons spectrum measurements.

The creation of the more fast electronics especially for NDD can provide time resolution better than 0.1 sec, and is the subject for future investigations. Some possibilities here exist. First - development more fast pulse signal processor with the width of spectroscopy pulses shorter than 1.25 ms. Second - development fast low noise preamplifier with pulse duration ~50 ns. Use such preamplifier in a scheme NDD - preamplifier - pulse height discriminator - amplifier - ADC can provide useful count rate of the order of 10^5 counts/s. During NDD testing in FNG neutron flux[7] the prototype of such electronic scheme has been investigated. The duration of the output pulse from very low noise logarithmic dc preamplifier[24], connected by 100 m superscreen cable to NDD placed in DT neutron flux, was equal to 80 ns (10ns rise time, 70 down time). Energy resolution of this electronic scheme was 5.5% instead of 3.9% for scheme with 142A ORTEC preamplifier.

Single channel analyzer can be used in discussed above spectrometric system for counting pulses with energy higher than 1.4 MeV (to count only DT neutrons, but not gammas). NDD will have sensitivity to DT neutrons of the order of $2.2*10^{-4}$ counts/(n/cm^2) for counting in such pulse height spectrum range. The single channel analyzer count rate will be equal to $2.2*10^5$ counts/s.

MULTICHANNEL COLLIMATOR

Very small dimensions (some cm^3) of NDDs housings make it possible to place them inside channels of top and midplane ITER neutron cameras[4,5]. As discussed in previous chapter, the optimum NDD position in ITER's cameras will be the positions, where DT neutron flux will be of the order of $2*10^9$ n/(cm^2 *sec). The values of the count rates for NDDs placed inside the channels of midplane and top ITER neutron cameras[4] outside the cryostat for 1.5 GW fusion power ITER plasma with flux surfaces taken from a Garching JWS study have been calculated. The relatively broad neutron emissivity profiles with 2.5 MW/m^3 central fusion power density was used. Neutron cameras with 20-25 8m (midplane) and 15-20 10m (top) long channels with cross section areas from 50cm^2 (for center channels) to 100cm^2 (for periphery channels) will provide DT neutron fluxes $\sim 2*10^9$ n/cm^2 (for center channel) and $\sim 5*10^8$ n/cm^2 (for r=2/3*a surface channels) in the detector positions. So NDDs placed in the channels of cameras, which will see the plasma core (r<2/3*a), will provide in ITER experiments useful for DT neutron spectrometry count rates in the range 10^4 - $5*10^3$ counts/s. This will gives the possibility to measure the ion temperature (effective perpendicular ion temperature for the case of unisotropy plasma) distribution for r<2/3*a and T_i on the plasma axis with time resolution $\sim$100 ms with spatial resolution $\sim$ 30 cm. Very similar numbers will be received in the case of NDDs allocation in the 2.5 cm in diameter collimator channels of ITER neutron camera proposed by F.Marcus[5].

The single channel analyzer can be used for simultaneous, with spectrometry, DT neutron counting with count rates $2.2*10^5$ and 10^5 counts/s for center and r=2/3*a channels correspondingly. Such count rates will give possibility to measure the total DT neutron flux and core emission profile with time resolution 1 ms, spatial resolution $\sim$30 cm and accuracy 10 %. The use of compact NDDs in one ITER neutron camera channel with several other detectors will provide adequate dynamic range and redundancy.

The ion distribution function in ITER can be essentially unisotropic during 100 MW additional heating. Tangential NDD based neutron spectrometer can be used for measurements of the effective temperature of ions with small pitch angles and plasma rotation velocity.

First measurements of DT neutron spectra and consequently effective ion temperature for different pitch angle ranges have been made recently on TFTR using natural diamond detectors[9]. Corresponding NDD pulse height spectrum for 90^0 pitch angle is shown on Fig.4.

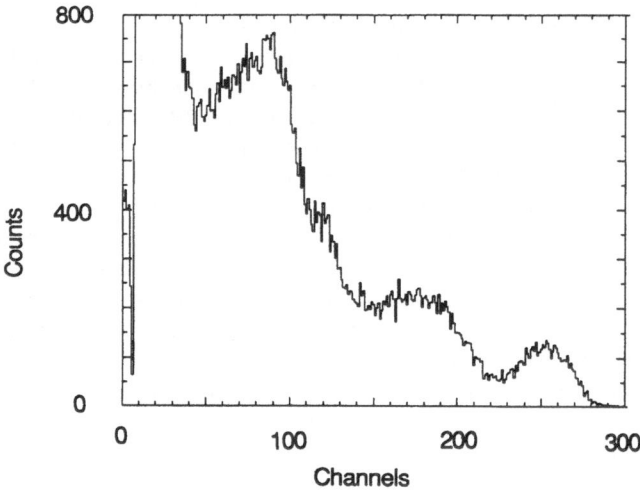

Figure 4. NDD pulse height spectra, accumulated during 15 TFTR DT shots of 22 MW beam power.

Fig.5 represent the values of effective temperatures, derived from measured by NDDs TFTR DT neutron spectra for three ranges of pitch angles. These measurements were done in experiments with heating neutral beam power in the range from 10 to 30 MW, with and without energetic beam's tritons in plasma. Results of these measurements are in a good agreement with mathematical modelling[17] and CHERS ion temperature measurements.

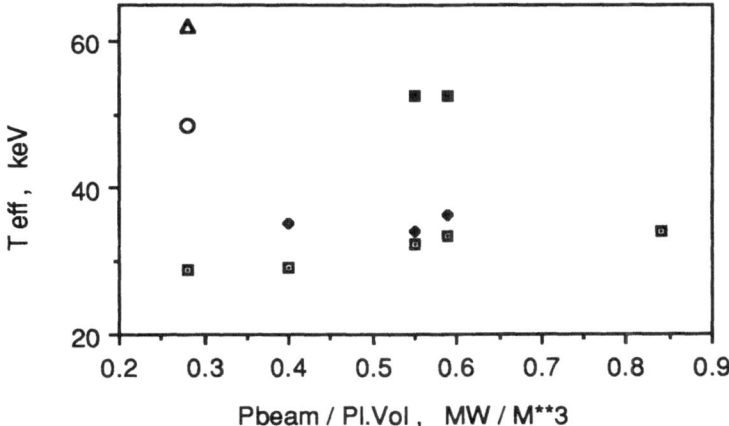

Figure 5. The effective ion temperature derived from DT neutron spectra measured by NDDs during 10, 15, 17, 22 and 30 MW TFTR neutral beam heating experiments. $\boxdot$ - pitch angle is equal to 90^0 (NDD N3 - inside TFTR neutron collimator 8.5 m far from plasma axis), $\bigcirc$ (tritium is present in a beam) and $\blacklozenge$ (no tritium in a beam) - pitch angle range 60^0 - 120^0 (NDD N1 - in Test Cell, 12 m far from plasma axis without any collimation), $\triangle$ (tritium is present in a beam) and $\blacksquare$ (no tritium in a beam) - pitch angle range 0^0 - 180^0 (NDD N2 - in Test Cell, 5 m far from plasma axis without any collimation).

Very small dimensions of NDD will give a possibility to place it in the same tangential collimator channel with magnetic proton recoil[25] or time-of-flight[26] neutron spectrometer. Due to the fact, that DNS count rate is restricted by electronics, NDD should be placed in the position with DT neutron flux $2*10^9$ n/cm². Time resolution will be equal to 100 msec.

CONCLUSION

The scheme of natural diamond detector based multichord DT neutron spectrometer has been described. When applied to ITER plasmas for perpendicular core ion temperature distribution and tangential effective ion temperature measurements, DNS provides (with a relative error of 10%) a time resolution of 100 ms and spatial resolution ~ 30 cm. This measurements will be of grate importance for burn optimization, alpha particle birth energy and spatial distribution, transport study. Tangential DNS channel can be used for evaluation of plasma rotation velocity.

Simultaneous with spectrometry, NDD use for DT neutron counting will give possibility to measure the total DT neutron flux and core (r<2/3*a) emission profile with time resolution 1 ms, spatial resolution ~30 cm and accuracy 10 %. Such measurements will be very useful for burn and fuelling control, disruption avoidance and instabilities study.

The use of compact NDDs in the same ITER neutron cameras channels with several other detectors and in tangential collimator with other neutron spectrometer will provide adequate dynamic range and redundancy.

ACKNOWLEDGMENTS

I am grateful to Drs E.A.Azizov, V.S.Khrunov, V.I.Trotsik, M.Pillon, M.Angelone, O.N.Jarvis, K.M.Young, L.C.Johnson, A.L.Roquemore, R.V.Budny, N.N.Gorelenkov, T.Nishitani, J.Kaneko, F.Maekawa and M.Isobe for collaboration in diamond detectors testing, TFTR and JT-60 DT neutrons spectrum and flux measurements and data analysis.

REFERENCES

1. A.V.Krasilnikov, Pulse height spectrum of natural diamond detector in DT neutron flux, to be published in Sov. Voprosi Atomnoy Nauki i Tehniki, 1 (1995) (in Russian).
2. E.A.Konorova, S.F.Kozlov, A diamond detector of nuclear radiation, *Sov.Phys.Semicond.* 4:1600 (1971).
3. A.E.Luchanskii, S.S.Martynov, V.S.Khrunov, V.A.Cheklaev, Detector Assembly with a diamond detector for recording neutrons, *Sov. Atomic Energy* , 63 : 639 (1987).
4. L.C.Johnson, Study of general access for neutron cameras in ITER, ITER/US/95/PH-07-13 (1995).
5. F.B. Marcus, A neutron camera for ITER, *These Proceedings* (1995).
6. *ITER Interim Design Report*, 12 of July 1995.
7. M.Pillon, M.Angelone, A.V.Krasilnikov, 14 MeV neutron spectra measurements with 4 % energy resolution using type IIa diamond detector, to be published in *Nucl.Instrum.Methods* in 1995.
8. A.V.Krasilnikov, A.L.Roquemore, R.Budny, N.N.Gorelenkov, L.C.Johnson, TFTR DT neutron spectra investigations using natural diamond detectors, to be published.
9. A.V. Krasilnikov, J.Kaneko, M.Isobe et al, FNS DT neutron spectrum measurements using natural diamond detector, to be published in *Rev. Sci. Instrum.*
10. M.Isobe, T.Nishitani, A.V.Krasilnikov et al, First measurements of triton burnup neutrons spectra using natural diamond detector on JT-60U. to be published on *7th Intern.Toki Conf. (ITC-7) on Plasma Phys. and Contr. Fus. "Fusion Plasma Diagnostics"* , Toki-city (1995).
11. V.S. Vavilov, Semiconducting diamond, *Phys. Status. Solidi (a)* 31: 11 (1975).
12. *The Properties of Natural and Synthetic Diamond* Ed. by J.E.Field, Academic Press, London (1993).
13. E.A.Konorova, S.F.Kozlov, V.S.Vavilov, Ionization currents in diamonds during irradiation with 500 - 1000 keV electrons, *Sov. Phys. Solid State* , 8: 1 (1966).
14. S.F.Kozlov, R.Stuck, M.Hage-Ali, P.Siffert, Preparation and characteristics of natural diamond nuclear radiation detectors, *IEEE Trans. Nucl. Sci.* NS 22: 160 (1975).
15. L.S.Pan, S.Han, D.R.Kania, M.A.Plano, M.I.Landstrass, S.Zhao, M.Kagan, Comparison of high electrical quality CVD diamond and natural single crystal IIa diamond, *3rd Int. Symp. on Diamond Materials (Honolulu Hawaii, May 16-21*, p.735 (1993).
16. C.Canali, E.Gatti, S.F.Kozlov, P.F.Manfredi, C.Manfredotti, F.Nava, A.Quirini, Electrical properties and performances of natural diamond nuclear radiation detectors, Nucl. Instrum. Methods, 160: 73 (1979).
17. A.V.Krasilnikov, N.N.Gorelenkov, R.V.Budny, A.L.Roquemore, Simulation and analysis of measured energy spectra of DT neutron emission in TFTR, to be published in November APS meeting (1995).
18. B.Antolkovic, G.Dietze, H.Klein, Reaction cross sections on a carbon for neutron energy from 11.5 to 19 MeV, *Nucl. Sci. Eng.*, 107: 1 (1991).
19. F.Nava, C.Canali, M.Artuso, E.Gatti, P.F.Manfredi and S.F.Kozlov, Transport properties of natural diamond used as nuclear particle detector for a wide temperature range, *IEEE Trans. on Nucl. Sci.*, NS-26: 308 (1979).
20. J.Koike, D.M.Parkin, and T.E.Mitchell, Displacement threshold energy for type IIa diamond, Appl. Phys. *Lett.*, 60: 1450 (1992).
21. A.G.Alekseev, V.N.Amosov, V.S.Khrunov, A.V.Krasilnikov, D.V.Portnov, A.Yu.Tsoutskikh, E.G.Utjugov, "Diamond detector based SXR-array for ITER", *This Proceeding* (1995).
22. O.V.Ignatyev, A.I.Kosse, A.D.Pulin, Yu.A.Shevchenko and N.F.Shkola, An update system of electronic modules for X-ray spectrometers with cooled semiconductor detectors, *Nucl. Instr. Meth.* A 282: 734 (1989).
23. S.V.Dudin, O.V.Ignatyev and A.D.Pulin, A fast signal processor for NaI(Tl) detectors, *Nucl. Instr. Methods* A 352: 610 (1995).
24. Preamplifier has been supplied to JET by "Harwell Instruments, A.E.A. Technology"
25. J.Kallne and H.Enge, Magnetic proton recoil spectrometer for fusion plasma neutrons, *Nucl. Instrum. Methods*, A311: 595 (1992).
26. T.Elevant, N.Garis, R.Chakarova and P.Linden, A neutron spectrometer for ITER, *Rev. Sci. Instrum.* 66: 881 (1995).

ION TEMPERATURE PROFILE MEASUREMENTS BY MEANS OF NEUTRON SPECTROSCOPY

T. Elevant[1,3], H. Brelén[1], P. Lindén[2] and J. Scheffel[3]

[1] JET Joint Undertaking Abingdon, Oxfordshire, OX14 3EA, U.K.

[2] Department of Reactor Physics, Chalmers University of Technology, S-412 96 Göteborg, Sweden, (EURATOM/NFR Fusion Association)

[3] Alfvén Laboratory, Royal Institute of Technology, Stockholm, S-100 44, Sweden, (EURATOM/NFR Fusion Association)

ABSTRACT

Information on ion temperature profiles will be needed for burn optimisation and transport studies in ITER. The feasibility of deriving these profiles for the core plasma $(r < 0.75a)$ directly from the width of measured 14-MeV neutron energy spectra is demonstrated for Maxwellian ion distributions. Neutron energy spectra and fluxes generated under different heating conditions are calculated by means of Monte-Carlo technique. The computation takes the reaction kinematics and the velocity distributions of the reacting ions into account and calculates the resulting neutron energy distribution and flux into a defined collimator. Energy spectra of neutrons emitted along a line-of-sight are superimposed. The associated correction factor, which depends on the measured ion temperature, can be given an analytical form when fitted to code data and is insensitive to large variations in temperature-, density and magnetic flux profile shapes. The accuracy in ion temperature evaluation is expected to be better than $\pm$ 10% and can be improved to $\pm$ 5% provided information on fuel density profiles are made available.

Features of several spectrometer candidates are briefly presented in relation to ITER conditions and measurement requirements. An array of 5-9 Time-of-Flight spectrometers can provide ion temperature profiles satisfying ITER measurement requirements, i.e. $T_i \geq 2.5$ keV, 10% accuracy and spatial and temporal resolutions of 30 cm and 100 ms respectively.

Diagnostics for Experimental Thermonuclear Fusion Reactors
Edited by P. E. Stott *et al.*, Plenum Press, New York, 1996

445

INTRODUCTION

ITER [1] plasmas with central ion temperatures ranging from 4 to 50 keV and central fuel ion densities from $0.2 \cdot 10^{20}$ to $2 \cdot 10^{20}$ m^{-3}, will generate 10^{17} to $5 \cdot 10^{20}$ neutrons per second. Maxwellian ion distributions generate Gaussian shaped neutron spectra with the energy spread directly related to the ion temperature[2]. Evaluation of the width of a spectrum therefore provides a simple and direct technique to determine the ion temperature. Furthermore, the reaction rate depends strongly on the temperature and is also proportional to the square of the fuel density.

Requirements and conditions for evaluation of ion temperature profiles from analysis of neutron energy spectra are here investigated with respect to diagnostic demands at ITER. Ion temperature and density profile data will be used for burn optimisation, particle and energy transport studies and temperature and density gradients are of particular importance in this type of analysis. Diagnostic requirements for ITER have been specified[3]. Predictions[4] yield parabolic density and temperature profiles; $n_e(r) = n_e(0)[1-(r/a)^2]^\alpha$ where $\alpha = 0.5$, and $T_i(r) = T_i(0)[1-(r/a)^2]^\beta$ where $\beta = 1.0$, respectively. The requirements on T_i measurements by means of neutron spectroscopy are[3,4]: T_i ranging from 2.5 to 50 keV with an accuracy of 10%, temporal resolution $\Delta t \leq$ 100 ms, spatial coverage $r \leq 0.75a$ and spatial resolution $\Delta r \leq 30$ cm.

Measurement conditions will be particularly difficult at ITER due to high levels of ionisation- and electromagnetic radiation. Furthermore, severe constraints are applied to the first wall and the blanket by the cooling system. Torus ports will be occupied by equipment that need to be close to the torus such as cooling tubes and lines-of-sight tubes for various diagnostics. The ports will also be used for access for exchange of blanket modules and for inspections. Therefore, instruments that can operate adequately at large distances, such as the neutron diagnostics, will preferably be located outside the main shielding.

CALCULATIONS OF NEUTRON ENERGY SPECTRA AND FLUXES

Neutron energy spectra have been calculated by means of a Monte-Carlo technique using the NSPEC code[5]. The code takes the reaction kinematics, the velocity distributions and densities of the reacting ions into account and calculates the resulting neutron energy distribution and flux into a defined collimator. The fusion cross-sections given by Bosch[6] are used. Energy spectra of neutrons emitted along a line-of-sight are obtained by superimposing contributions from a large number of sub-volumes. To obtain the maximum temperature along a line-of-sight a correction to the observed spectrum has to be applied. To estimate the value of the associated correction factor, simulations using different ion temperature and density profiles have been carried out for ohmically heated and ignited D(50%)-T(50%) plasmas in ITER configuration.

Resulting ion temperature correction factors are shown in fig.1. One encouraging fact is illustrated: even if the density and temperature profiles are completely unknown, the maximum temperature along each line-of-sight can be derived from the measured T_i with an accuracy better than 10%. If the magnetic equilibrium $\psi(r,z)$ is known, $T_i(\psi)$ can be obtained taking the line-of-sight geometry into account. Consequently, the coefficient β in a parabolic temperature profile can be determined. It is obvious from fig. 1 that the correction factor is profile dependent. The top full line, with $(\alpha,\beta)=(0.05,2.0)$, represents an extreme case where the density profile is nearly flat, giving large neutron contributions from regions outside the temperature maximum. The lower full line shows an other extreme where $(\alpha,\beta)=(1.0,0.5)$. The central curve represents standard ITER parameters as above and the resulting correction factor is approximated by:

$$C(T_i^{obs}) = 1.06 + (0.003 \pm 0.003)T_i^{obs}(keV) \tag{1}$$

where $C(T_i^{obs})$ is T_i^{max} / T_i^{obs}. T_i^{max} is the maximum temperature along a line-of-sight and T_i^{obs} denotes the ion temperature evaluated from the accumulated neutron spectrum. The correction factor shown in fig. 1 correspond to four different lines-of-sight which all pass through a pivot point in the median plane at $R=12$ m. The lines intersect a vertical line passing through the magnetic axis at 0, 1, 2 and 3 m above the geometric centre axis respectively and the results shown in fig. 1 apply to any line-of-sight. This is of particular importance for evaluations of the ion temperature gradients. The errors in the line integral corrections are canceled out in this case and the relative accuracy is estimated to better than ±5%.

In order to estimate the sensitivity of the correction factor on deviations from the true $\psi(r,z)$, we have computed $C(T_i^{obs})$ both for the ITER standard equilibrium and for an inverse-D shaped equilibrium, indicated by full and dashed lines respectively in fig.1. The reasons for the small differences in $C(T_i^{obs})$ are that only lines-of-sight within $r=0.75a$ are considered here, making flux shape effects small, and that contributions to the neutron flux are small near the periphery.

Improved accuracy in the evaluation of $T_i(r)$ can be achieved by including information on neutron fluxes Φ_n. Given the ion temperature profile shape β, measurement of flux ratios R_{Φ_n}, between different lines-of-sight, provides information on plasma density profile. Such measurements provide accurate estimates for low ion temperatures, i.e. $T_i \leq 10$ keV and the uncertainty in the evaluation of $T_i(r)$ is brought down to ±5%. At temperatures exceeding 10 kev, $\alpha+\beta$ becomes a sensitive function of R_{Φ_n}, and the uncertainty increases from 5 to 10%.

SPECTROMETERS CONSIDERED

Characteristics and operating ranges of several spectrometers located outside the main shielding and for which the ITER diagnostic requirements are fulfilled are shown in Table 1. Estimates of radiation damages are based on ratios of 10^3 for flux in the neutron beam over flux at a position 2-5 cm outside the neutron beam. At a distance of 15 cm this ratio is equal to $10^{4.7}$.

Table 1. Operating ranges for spectrometers located outside the ITER main shield.

Spectrometer type Large spectrometers	Efficiency (cm^2)	Energy resolution (%, FWHM)	Oper. range		Radiation damage, at Γ_{max} (n/cm^2) [3)]	Dimensions W H L (m^3)
			T_i (keV)	$\Phi_{n,max}$ [1)] $/\Phi_{n,min}$		
Proton recoil [8]	10^{-4}	2.2	≥ 5	$5\cdot10^9$ $/5\cdot10^7$	10^{15} [3a)]	$0.4\cdot0.4\cdot2$
Time-of-Flight [9]	10^{-3}	1.6	≥ 2.5	$5\cdot10^{10}$ $/5\cdot10^6$	10^{16} [3b)]	$1\cdot1\cdot4$
Associated Particle, Time-of-Flight [10]	10^{-5}	1.5	≥ 7	$5\cdot10^{10}$ $/5\cdot10^8$	10^{15} [3a)]	$2\cdot2\cdot1$
Magnetic proton recoil [11]	$5\cdot10^{-5}$	2.5	≥ 7	$5\cdot10^{10}$ $/10^8$	10^{16} [3b)]	$1\cdot2.5\cdot1$
Proton recoil with micro channel [12]	10^{-5}	2	≥ 7	$5\cdot10^{10}$ $/10^8$	$5\cdot10^{15}$ [3a)]	$0.2\cdot0.2\cdot0.4$
Compact spectrometers						
Silicon diode [13]	10^{-3}	0.8	2 - 20	---- [2a)]	10^{12} [3c)]	Small
Diamond diode [14]	10^{-5}	2.5	≥ 7	---- [2b)]	$5\cdot10^{14}$ [3c)]	Small
Scintillating Fibre detector [15]	$3\cdot10^{-3}$	3.3	≥ 7	10^8 $/10^6$	$7\cdot10^{15}$ [16] [3d)]	Small

[1] Fulfilling $\Delta t \leq 100$ ms. [2] Due to intrinsic spurious reactions the best obtainable time resolutions are: [2a] 250 ms and [2b] 200 ms. [3] Numbers refer to fluence in neutron beam and are given by damages in [17]: [3a] silicon diode outside neutron beam, [3b] polyethylene foil in neutron beam, [3c] diodes in neutron beam, [3d] scintillating fibres in beam.

The time-of-flight technique fulfils all ITER plasma measurement requirements and combines a large operating range with high resistance against radiation damages and high degree of flexibility. A version suitable for 2.45 MeV neutrons has provided information on ion distribution functions in the JET tokamak.[18]

RADIAL ARRAY OF TIME-OF-FLIGHT SPECTROMETERS

To meet the ITER requirements on $T_i(r)$ measurements, several lines-of-sight must be utilised. An array of five spectrometers, shown in fig. 2, provides ion temperature measurements for $r \leq 0.75a$, for ion temperatures above 2.5 keV and with a temporal resolution of 100 ms. Radial resolution better than 30 cm with 80 cm spacing is provided.

The number of spectrometers chosen is determined by the desire to have accurate information on the ion temperature profile on one hand and practical considerations such as neutron flux crosstalk between adjacent lines-of-sight and detector space requirements on the other. To allow for a large variety of temperature profiles we assume that a description of the form $T_i(r) = T_i(0)[1-(r/a)^\gamma]^\beta$ is adequate. However, up-down symmetry can not *apriori* be assumed and the number of degrees of freedom must correspond to at least five spectrometers.

SUMMARY

Calculations show the feasibility to derive ion temperature profiles up to r = 0.75a directly from the width of multiple line-of-sight-integrated neutron energy spectra in a fusion experimental reactor like ITER. A correction factor which value is a function of the observed ion temperature, must be applied. It can be given an analytical form when fitted to code data, see eq.(1). Broad variations in density-, temperature- and magnetic flux surface profile shapes have negligible effects on the correction factor.

An array of 5-9 time-of-flight spectrometers, with maximum count-rate of 1 MHz, provides ion temperature profile data in accordance with the ITER diagnostic requirements. The absolute accuracy in ion temperature measurements is estimated to ± 10% and the relative accuracy is better than ±5%. Utilisation of information on fuel ion profiles can improve the absolute accuracy from ±10% to ± 5%.

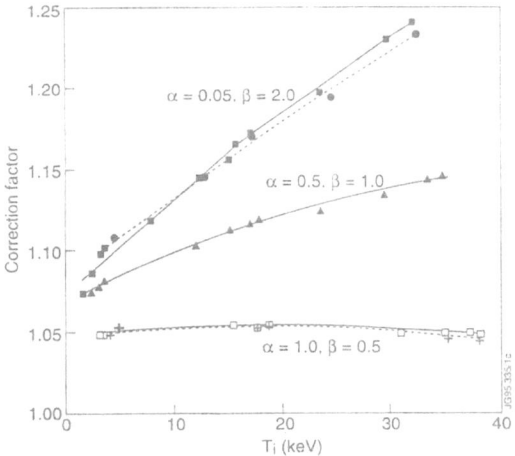

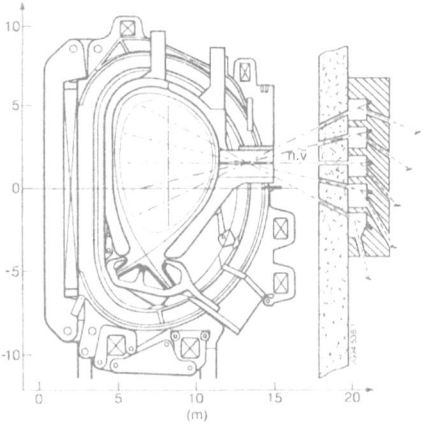

Fig.1 Correction factors $C(T_i^{obs})$ for different parabolic plasma density (α) and temperature (β) profiles are shown. The full lines represent the standard ITER equilibrium; R_0=8.0 m, a=3.0 m, Δ_{Shaf}=0.5 m, κ=1.6 and z_{axis}=4.8 m and the dashed lines an inverse-D shaped ITER equilibrium with Δ_{Shaf}=0.3 m, κ=1.6 and z_{axis}=4.3 m.

Fig.2 Shown is a radial array of five Time-of-Flight Spectrometers[10] for measurements of ion temperature profiles.

REFERENCES

1   K. Tombaechi et.al, ITER conceptual design, *Nuclear Fusion,* Vol. 31.No.6 (1991). ITER Joint Central Team, Parameters of the ITER EDA Design, *Plasma Phys. Control. Fusion* 35 (1993) B23.

2   H. Brysk, Fusion Neutron Energies and Spectra, *Plasma Physics*, Vol. 15, (1973), 611.

3   Minutes of 1st Meeting, Physics Expert Group on Diagnostics. July 18-22, 1994; ITER Joint Work Site, San Diego, USA. S CX M1 94-08-03 F1.

4   Mukhovatov et.al., ITER Diagnostics, *ITER Documentation series,* No. 33, IAEA, Vienna, 1990.

5   J. Scheffel, Neutron spectra from beam-heated fusion plasmas, *Nucl. Instr. and Meth.* 224 (1984).

6   H. S. Bosch, Report IPP I/252, Max-Planck Inst. für Plasmaphysik, Garching bei Munchen 1990.

7   K-H. Beimer, Studies of Neutron Measurement Methods for Fusion Plasma Diagnostics, Dep. of Reactor Physics, Chalmers University of Technology, Dissertation Thesis (1986).

8   N. Hawks, Harwell Laboratories, U.K. Private communication, June 1995.

9   T. Elevant et. al., A neutron spectrometer for ITER. *Rev. Sci. Instrum.* 68 (1), (1994).

10 M. Hoek et.al., Simulation of the neutron and proton transport in the 14 MeV neutron time-of-flight spectrometer TANSY, *Nucl. Instr. and Meth.* A322, 248 (1992).

11 J. Kallne and H. Enge, Magnetic Proton recoil spectrometer for fusion plasma neutrons, *Nucl. Instr. and Meth.* A311 (1992) 595.

12 T. Iguchi et.al., Conceptual design of neutron diagnostic systems for fusion experimental reactor, *Fusion Engineering and Design* 28 (1995) 689.

13 T. Elevant et al., Silicon surface barrier detector for fusion neutron spectroscopy. *Rev. Sci. Instrum.* 57 (8), (1986).

14 F. Borchelt et. al., *Nucl. Instr. and Meth.*, A 354 (1995) 318-327, and A. V. Krasilnikov, Troitsk Institute of Innovating and Fusion Researches, Russia. This conference.

15 T. Elevant, Scintillating Fibre Neutron Spectrometer. This conference.

16 M. Adinolfi et. al., *Nucl. Instr. and Meth.*, A315 (1992).

17 J. F. Kircher and R. E. Bowman. *Effects of Radiation on Materials and Components*. Reinhold Publishing Corporation, New York. See also J.F. Baur et. al., Radiation Hardening of Diagnostics for Fusion Reactors, General Atomic Company, GA-A16614 UC-20, (1981).

18 T. Elevant et.al., The new JET 2.5-MeV neutron time-of-flight spectrometer, *Rev. Sci. Instrum.* Vol. 63, No.10 (Part II), October (1992).

SCINTILLATING FIBRE NEUTRON SPECTROMETER

T. Elevant[1], R. Chakarova and J. Karlsson[2],
[1]Alfvén Laboratory, Royal Institute of Technology, S-100 44 Stockholm,
Sweden, (EURATOM/NFR Fusion Association).
[2]Dep. of Reactor Physics, Chalmers University of Technology, S-412 96
Göteborg, Sweden, (EURATOM/NFR Fusion Association)

ABSTRACT

A neutron spectrometer consisting of a number of square shaped hydrogen based scintillating fibres is described. The considered design will have an efficiency of $3 \cdot 10^{-3}$ cm^2 for 14-MeV neutrons and is suitable for measurements of energy spectra of collimated fluxes ranging from 10^6 to 10^8 n/(cm^{2}s). The energy resolution is $\cong 3.3\%$ (FWHM) which permits ion temperature measurements for $T_i \geq 7$ keV with $\Delta T_i/T_i \leq 10\%$. The detector has high efficiency that makes it particularly useful for measurements of energy spectra of collimated neutron beams in a radial profile monitor in ITER. The objective of these measurements is to evaluate ion temperature radial profiles. A time resolution of 100 - 1 ms and a radial resolution of 1-3 cm is provided.

INTRODUCTION

Collimated neutrons generate recoil protons through elastic (n,p) scattering in the fibres. The energies of the incident neutron E_n and the recoil proton E_p are related through

$$E_n \cong E_p/ \cos^2 (\theta_p) \qquad (1)$$

where θ_p is the recoil angle (in the laboratory system) relative to the direction of the incident neutron. The mass of the proton is here set equal to the neutron mass. The non relativistic approximation used is sufficiently accurate for the present purpose. Furthermore, the proton range l_p, has a well defined dependence on the proton energy,

$$l_p = aE_p{}^b \qquad (2)$$

Diagnostics for Experimental Thermonuclear Fusion Reactors
Edited by P. E. Stott *et al.*, Plenum Press, New York, 1996

451

and the number of fibres fired therefore provides a measure of E_p. For scintillating material like NE102 and l_p given in cm, a is equal to $2.25 \cdot 10^{-3}$ and b is 1.76[1]. Thus, through eqs.(1) and (2) information on θ_p and l_p provide information on the energy of the incident neutron E_n.

DETECTOR ARRANGEMENT

Alternate orthogonal planes of square scintillating fibres form a compact detector with a volume of a few cm^3, see fig.1. All fibre axes are oriented perpendicular to the incident collimated neutron beam direction. Neutrons undergo elastic (n,p) scattering, and the pattern of fibres fired provides a measure of the proton range and recoil angle and therefore of the energy of the incident neutron.

The design to be studied consists of square shaped, 100 μm, scintillating fibres of 10 mm lengths. The scintillating fibres are spliced to clear fibres and connected to a multi-channel PMT. Only the proton track length and its orientation are important but not its position. Therefore, two sides of the detector are divided into 40 modules. Each module consists of 16·12 fibres that are individually connected to 192 channels on a multi-anode PMT. Fibres from all modules are connected in parallel to the same PMT, see fig.1. This is repeated for an adjacent detector side with all fibres connected to a second PMT. Thus, two orthogonal 2D pictures form a 3D picture of the proton track. A readout grid on the PMT's with 16 anode wires orthogonally oriented to 12 anode strips will be used.

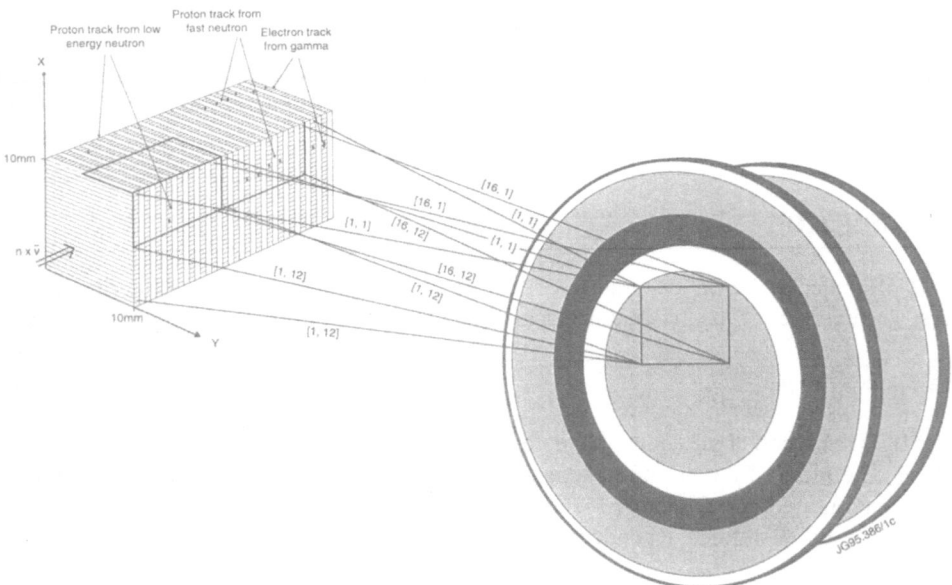

Fig. 1. The scintillating fibre detector is shown with only a few modules indicated. Transmission fibres, indicated with their module position numbers, connect the scintillating fibres with the multichannel photo-multiplier tube.

Signal electronics provides pulse amplitude discrimination for each of the read-out channels and generates logical pulses. For each of the two detector sides a 28 bit word is created and the resulting 56 bit word contains information on the proton range and the recoil angle. A fast digital filter will be applied to select events with neutron energies ranging from 11-17 MeV and a fast digital procedure will thereafter be used for evaluation of the neutron energy. The total light-output signal from the loose fibre ends may be used as a first trigger signal. A collimator with a few cm^2 cross-section area and a length of a few meters will be used. The collimator must be tapered in both ends. Results from neutron and gamma transport calculations for a similar neutron detector are given by Yariv[2].

Efficiency and count-rates

With the detector dimensions considered (i.e. $1 \cdot 1 \cdot 1.7$ cm^3), proton density of $4.8 \cdot 10^{22}$ cm^{-3} and the (n,p) elastic cross-section equal to 0.7 barns, the total efficiency for 14-MeV neutrons is $5 \cdot 10^{-2}$ and the useful efficiency is $\cong 3 \cdot 10^{-3}$ cm^2. The fast response of the scintillating fibres ($\cong 10$ ns) enables useful count-rates of approximately 600 kHz to be obtained.

Energy resolution

For evaluation of the energy resolution accuracy in both proton range and recoil angle matters. Differentiation of eq. (1) yields:

$$dE_n/E_n = [(dE_p/E_p)^2 + (2tg(<\theta_p>) \cdot d\theta_p)^2]^{1/2}, \quad \theta_p \cong 0, \tag{3}$$

where $<\theta_p>$ is the proton flux weighted average over accepted recoil angle range. Assuming independent contributions, differentiation of eq. (2) yields:

$$dE_p/E_p = b^{-1}[dl_p/l_p] = b^{-1}[(dl_p/l_p)^2_{fib.dim.} + (dl_p/l_p)^2_{strag.}]^{1/2}, \quad l_p \cong 2.4 \text{ mm}, \tag{4}$$

where the two contributions originate from the finite fibre cross-section dimension and the proton range straggling respectively. With $(dl_p/l_p)_{fib.dim.} = 0.1/2.4$, $(dl_p/l_p)_{strag.} = 3.2 \cdot 10^{-2}$ as given by Janni[3], $<\theta_p> = 0.4/2.4$ and $d\theta_p = 0.1/2.4$ the energy resolution becomes $dE_n/E_n = 3.3\%$ (FWHM). Thus, $T_i(r) > 7$ keV can be evaluated with an accuracy $\Delta T_i/T_i < 10\%$[4].

Light-output and neutron-gamma separation

The light-output is essential for the performance of this detector. Considering scintillating-, trapping- and light conversion efficiencies, [0.024, 0.04 and 0.17 respectively], a 14-MeV recoil proton, traversing the active part of the fibre ($\cong 75$ μm in square) generates approximately 15 photo-electrons[5,6]. An electron traversing the fibre generates 2-3 photo-electrons independent of its energy. The specific energy loss for protons increases for lower energies, giving an increased photon yield from successive completely penetrated fibres. Thus, different specific light-output from protons and electrons enables neutron and gamma events to be separated by means of simple pulse amplitude discrimination. Tests with 200 μm square fibres have resulted in signal amplitude ratios equal to 10/2.5/1 for neutrons (1-4 MeV), gamma rays (0.8-1.1 MeV) and noise, respectively.

Radiation resistance

Glass capillaries with liquid scintillators can sustain very high radiation doses, i.e. $5 \cdot 10^5$ Gy[7]. This corresponds to fluxes of 10^8 n/(cm^{2}s) during 70 000 full power discharges of 10^3s duration each. Plastic fibres can sustain 200-400 discharges before suffering from radiation damages[7].

FUTURE WORK

To obtain information necessary for a detail design of a prototype a few important issues have to be studied:

- Neutron transport by means of Monte-Carlo calculations of efficiency, energy resolution and effects of reactions with carbon.
- Simulation and tests of the read-out system.
- Design and construction of a prototype (plastic) detector.
- Tests of efficiency and energy resolution of prototype detector.
- Tests of neutron and gamma separation capabilities and energy resolution at high count-rates using a neutron generator and at JET.

REFERENCES

1   J.W. Watson and R.G. Graves, Finite-Size Effects in Neutron Detectors, *Nucl. Instr. and Meth.* 117 (1974) 541.
2   Y. Yariv et. al., Simulations of Neutron Response and Background Rejection for a Scintillating-Fibre Detector, *Nucl. Instr. and Meth.* A292 (1990).
3   J.F. Janni, Proton Range-Energy Tables, *Atomic Data & Nucl. Tables*, v.29 (1982).
4   N. S. Garis, R. Chakarova and T. Elevant, Simulation of particle transport in a novel 14 MeV neutron time-of-flight spectrometer, Lab. rep. CTH-RF-97. Dep. of Reactor Physics. Chalmers University of Technology, S-412 96, Göteborg, Sweden.
5   J. Karlsson, Scintillating Fibre Tracking Neutron Detector, Lab. rep., CTH-RF-114. Dep. of Reactor Physics. Chalmers University of Technology, S-412 96, Göteborg, Sweden.
6   J.B. Birks, *The Theory and Practice of Scintillating Counting*, Pergamon Press 1964.
7   M. Adinolfi et. al., Progress on High-Resolution Tracking with Scintillating Fibres: a new Detector based on Capillaries with liquid Scintillator, Nucl. Instr. and Meth. A315 (1992).

NEUTRON IMAGE SYSTEM TO MEASURE NEUTRON
PROFILE OF ITER PLASMA

S.V. Trusillo, V.A. Agureev

Russian Federal Nuclear Center "Experimental
Physics Institute," Arzamas-16

Utilization of overheated liquid properties has caused in the 50-th years creation of bubble chamber (BC).

Enormous intensification and high spatial resolution that were achievable in overheated liquid has made it possible to view interaction acts of fast neutral and charged particles.

In the 80-th years in our institute we have successfully applied BC with collimator-obscure (CO) for neutron brightness measurements in cameras of plasma focus [1].

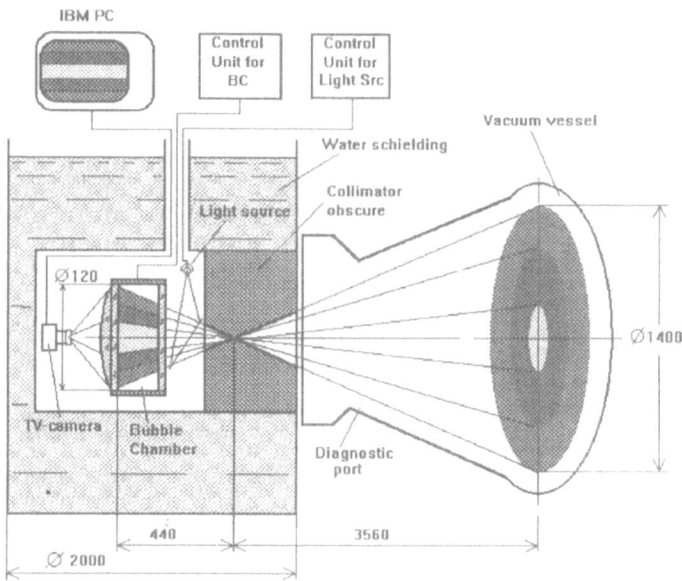

Fig. 1. Neutron image system

Diagnostics for Experimental Thermonuclear Fusion Reactors
Edited by P. E. Stott *et al.*, Plenum Press, New York, 1996

455

At interaction of fast neutrons with work liquid (without hydrogen) heavy recoil nuclei with short range are generated giving rise to one bubble at a point of neutron interaction. By this mean spatial distributions of bubble density are consistent with distribution of neutron flux density.

Through transmitted light the bubbles appear as dots and the formation of optical image is quite similar to the process that has place in photographic materials.

The distinguishing feature of the optical image formation is a large thickness of the registration medium. Difficulties associated with this fact are overcome by a special configuration of photography - see Fig.1. Here the proposed scheme of neutron image system for plasma profile on tokamak T-15 (Moscow) is shown.

As a consequence of the high penetrating ability of fast neutrons it is impossible to make CO that would allowed to achieve the spatial resolution of overheat liquid. Nevertheless achieved spatial resolution exceeds noticeably the multichannel collimator system one. In the last systems the separately distinguished elements quantity is equal just to the channels quantity (usually 10-15).

In the neutron image system based on BC the separately distinguished elements quantity is defined :

$$m = \frac{d}{b_{EF}(1+\beta)} \; ;$$

m- the elements quantity, d - diameter of BC view window, $\beta=d/H$ - magnification of the system; H - vertical size of the diagnostic port, b_{EF} -effective slit width of CO, that is above the geometrical one by value, at that transparency of the slit edge for neutrons decreases by a factor of "e". For Fig.2 the numerical estimate of m equals 43 elements / BC.

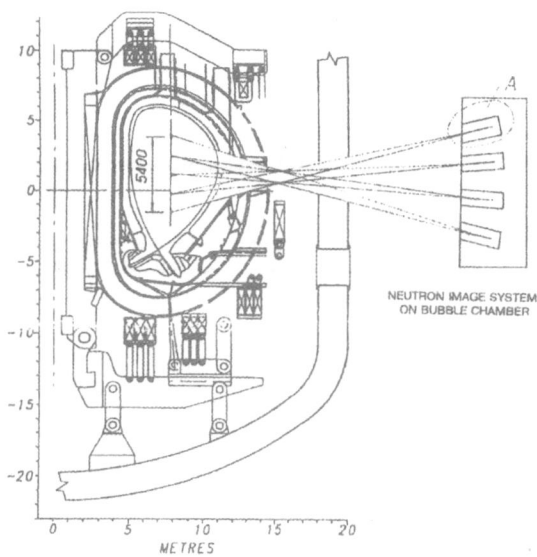

NEUTRON IMAGE SYSTEM
ON BUBBLE CHAMBER

Fig.2. Horizontal neutron system based on BC

Requirements of radiation shielding strongly restrict possibility to see by individual BC the whole plasma profile - there is required too tall and wide window. Because of it assumed scheme will use 3-5 BC worked synchronously. In Fig.2 the horizontal neutron system based on BC is shown.

The image registration cycle involves the neutrons registration phase, the phase of rising bubbles to needed sizes, TV-writing image, the phase of condensation (disappearing image), repeat TV-writing background image. The cycle period defines time resolution of neutron image system.

For realization of such cycle a sine shaped pressure is created in BC work volume (see Fig.3).

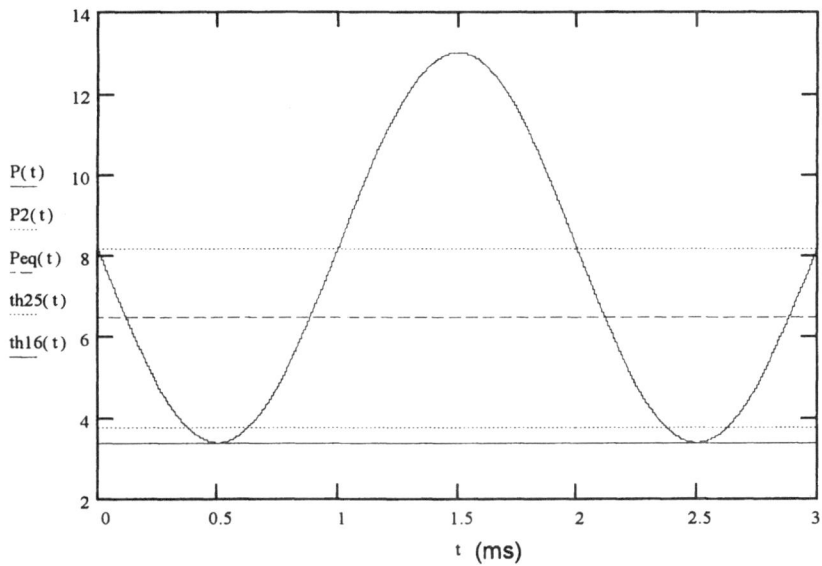

Fig.3. Diagram of pressure in work volume of cyclic BC.
P(t) for FR-12, Peq(at) - equilibrium pressure at work temperature (grad C), P2(at) - average pressure in liquid volume, P1(at) - amplitude of sine part of pressure, th16 - threshold pressure for registration of neutrons with $En=1.6$ Mev, th25 - threshold for DD-neutrons.

Choosing cycle period depends of BC sizes and properties of work liquid - sound speed, time of bubbles growth and condensation [2]. For supposed system the optimal period has been estimated to 2 ms.

Rough estimate of neutron flux value forming the element image in Fig. 2 ($N=10^{20}$ s^{-1} - neutron intensity, $t=250$ μs - neutrons registration time during one cycle) constitutes $N_{EL} \approx 5.2 \cdot 10^5$ neutrons/cycle.

Special feature of neutrons registration process is existence of energy threshold for bubble appearance. This threshold depends of specific energy loss of recoil nucleus. Its value is defined at chosen work temperature by pressure in liquid and can be set for sensitivity suppression to neutrons with energy less then selected one.

Of course, this decreases registration efficiency of neutrons with energy more than selected one. For example, at threshold En=1.5 Mev DD-neutrons registration efficiency decreases by 2.4 times. Naturally, at such threshold BC is insensitive to γ-radiation.

Because of work medium large thickness registration efficiency of BC is enough high. For example, for thickness 120 mm at above indicated the threshold and neutron energy it is 25%. In accordance with this value the bubbles quantity in the image element is $N_B = N_{EL} \cdot 0.25 = 1.3 \cdot 10^5$, that provides statistical accuracy of measurements better 1%.

At periodical changing threshold, for example, with period 20 ms, it is revealed in concept possibility of DT-neutrons spectrometry. Together with spatial resolution this can allow to reconstruct spatial distribution of plasma temperature and its temporal evolution.

Parameter measured in the BC is optical image density that is tightly coupled with bubbles density by relation

$$D = 0.434 \cdot n \cdot \sigma,$$

n - the bubbles quantity per area unit, σ - the bubble cross section.

Because of the bubbles cross sections vary in wide limits, it is conveniently the optical image density to define by the relative one:

$$\tilde{D} = \frac{D}{\overline{D}} \equiv \frac{n}{\overline{n}},$$

$\overline{D}$ and D - average upon the whole image and the local optical densities.
$\overline{n}$ and n - average and local bubbles quantity per area unit.
And it is conveniently to make σ automatically controlled for obtaining the image optimal density by choosing moment of TV-writing.

Naturally, for specific neutron image system it will be required work out of reconstruction algorithm of plasma profile.

Creation of BC with high frequency of registration cycles was confronted up to present by a problem of parasitic boiling of overheat liquid on BC walls [2]. It is caused by existence on surface structure defects that are centers of boiling. This problem has overcome by technology means worked out in our institute, that allows to hope to create cyclic BC with frequency restricted by only processes of growth and condensation of bubbles.

It is supposed to work out BC with cycles high frequency and to decide all technical problems of creation of whole neutron image system in the context of project "Neutron plasma diagnostic for T-15". This project is directed at present to ISTC.

REFERENCES

[1] S.V.Trusillo, B.Ya.Guzhovskii, N.G.Makeev, V.A.Tsukerman "Determination of Neutron Production Region in the Plasma Focus Chambers", JET Letters, v.33 (1981), No.3,p.148-151 (in Russian).

[2] Yu.A.Aleksandrov and others, by edition N.B.Delone "Bubble Chambers", GOSATOMIZDAT, Moscow, 1963.

ITER PLASMA POSITION AND POWER DETERMINATION FROM NEUTRON MEASUREMENTS

A. P. Navarro[1], R. Carrasco[2], I. Labrador[2], M. A. Ochando[2]
and A. L. Romero[1]

[1] Universidad Alfonso X el Sabio, 28691 Villanueva de la Cañada, Spain
[2] Asociación EURATOM / CIEMAT, 28040 Madrid, Spain

INTRODUCTION

Neutron diagnose in deuterium-tritium (D-T) plasmas[1] results increasingly important as these become hotter and more dense because other more conventional techniques will experience difficulties, due to the increasing plasma opacity and to the radiation damage to the instrumentation. In addition, neutron measurements could provide the only direct way to observe fusion reactivity. They have been already used in the D-T phase of the JET[2] and the TFTR[3] tokamaks, deducing from them the total fusion power and the ion temperature. In the ITER conceptual design[4] (CDA) neutron detection systems are candidates for determination of plasma position and shape, fusion power, ion temperature, D/T density and disruption precursors.

Neural networks are algorithms which enable for non-linear transformations between multidimensional spaces. Therefore, processing techniques based on them are a promising tool for analysis of plasmas[5]. Artificial Neural Nets (ANN) are able to learn a set of rules providing fast, more efficient and less complex control structures. In addition, they can process information from incomplete data sets, (v.g.: malfunction of some sensors), signals with a high level of noise, and, due to their capability to generalize, cases for which they have not been previously trained. These properties make ANN specially adequate to manage neutron detector signals in fusion reactors providing a real-time system for power and position control.

We have already explored[6] the possibility to determine plasma power and position in ITER using the CDA neutron profile diagnostic. In this paper, we summarize the results of that technique and extend it by using neural networks.

The assumed detection system[4] consists of two monolithic cameras, each with 20 independently collimated neutron detectors in a fan-like arrangement. Local neutron emissivity and fusion power are deduced using a 1-D transport model[7] to simulate the power balance between α-particle heating, radiation, convective and conductive losses for specific electron

Diagnostics for Experimental Thermonuclear Fusion Reactors
Edited by P. E. Stott *et al.*, Plenum Press, New York, 1996

459

density and temperature profiles. Figure 1 displays the experimental set-up showing the electron density contours and the viewing chords defining the region of interest for the neutron emission for the ITER-CDA parameters, (a = 2.15 m, R_0 = 6 m, B = 5 T). Figure 2 shows the deduced signals at the two arrays and their time evolution for a typical plasma discharge.

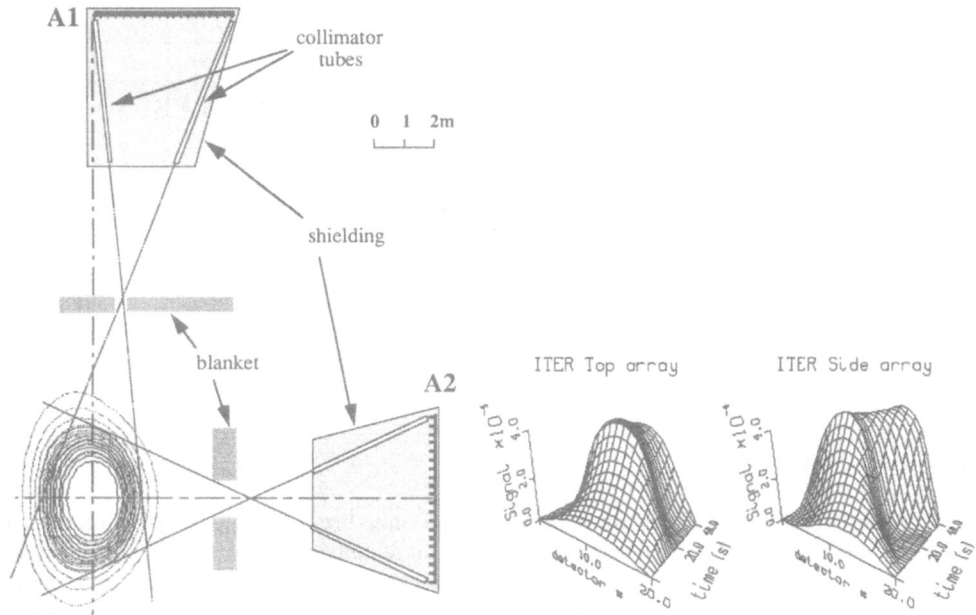

Figure 1 Experimental set-up   **Figure 2** Simulated signals

PROFILE ANALYSIS BASED METHOD (PABM)

This method[6] is based on the analysis of the neutron profile signals. Introducing the array profile asymmetry, defined as:

$$Ars(a) = \frac{\sum_{d=1}^{N/2} S(a,d) - \sum_{d=N/2+1}^{N} S(a,d)}{\sum_{d=1}^{N/2} S(a,d) + \sum_{d=N/2+1}^{N} S(a,d)} \tag{1}$$

being S(a,d) the signal at the detector d in the array a and N the total number of detectors in each array, it can be used to determine plasma displacements.

Simulation results prove that this parameter is independent of plasma power and shape. So, we can relate plasma horizontal and vertical displacements with the two array relative asymmetries by the formula:

$$\begin{vmatrix} \Delta x \\ \Delta y \end{vmatrix} = A \cdot \begin{vmatrix} Ars(1) \\ Ars(2) \end{vmatrix} + B \tag{2}$$

where A is a 2x2 matrix that defines the effect from the displacements in both arrays and the 2x1 matrix B takes into account asymmetries due to the array geometry or internal shifts for a well centered plasma.

From the simulation results, we have obtained for the ITER arrangement:

$$A = \begin{vmatrix} 0.45 & -0.09 \\ -0.01 & -0.46 \end{vmatrix} \qquad B = \begin{vmatrix} 0.046 \\ 0.000 \end{vmatrix} \tag{3}$$

The total number of detected neutrons can be used to determine fusion power. It is defined as:

$$At = \sum_{d=1}^{N} \left[S(1,d)/\eta\,(1,d) + S(2,d)/\eta\,(2,d) \right] \tag{4}$$

with $\eta(a,d)$ being the (a,d) detector sensitivity. We have deduced from the simulations a linear relationship between both magnitudes, independently of plasma ellipticity and displacements. For ITER:

$$P_\alpha = 9.05 \cdot 10^{-6} \cdot At \qquad \left[MW, neut.s^{-1} \right] \tag{5}$$

Plasma turbulence introduces fluctuations in the density and temperature levels and, as a consequence, on the neutron emissivity that could distort the position and power determination. We obtained that plasma displacement evaluation is much less sensitive to turbulence than the power estimation. Power is determined with an uncertainty of around 10% for the usual values of 1 to 2% in central density and temperature fluctuations and both of them have similar importance on the power determination error.

Absolute calibration of the detectors is not required for the position determination where only relative calibrations are needed. Power estimation can be also made under these circumstances if an independent power measurement is available to calibrate this technique results.

NEURAL NETWORK BASED METHOD (NNBM)

A Multilayer Perceptron (MLP) with back-propagation learning algorithm[8] was implemented for this application. MLP is a feed-forward architecture where nodes are associated in different layers. In each layer, every node is connected, with an adequate weight, to all of the nodes in the next layer to perform a non-linear transformation relating output and input. After an optimization process, the selected configuration consist of an input layer with 40 nodes, one for each of the detectors in the two arrays, a single hidden layer with 15 nodes, and an output layer with 3 nodes, for power, horizontal and vertical position, respectively. The neural network was simulated using a software written in C language, running on a VME equipment with a 68030 CPU, main clock at 30 MHz, and OS-9 as operating system.

Training was performed using neutron signals from 90 different snap shots with power ranging 100 to 1000 MW ±5% of the plasma radius for posible displacements. It takes 3 days to complete. Once trained, the network was used to determine plasma position and power for several simulated discharges. Correlation coefficients between the values imposed in each simulation and the deduced from the networks are: 0.934, 0.928 and 0.999 for horizontal, vertical and power estimations, respectively. Figure 3 compares the results from NNBM and PABM.

In order to test the effects from detector malfunctioning, we have crashed each of the inputs, one by one, and deduced the effect on the parameter determination. Similar study was made for two detector malfunctioning. Table I summarizes the obtained correlation coefficients. Neural network was also checked by corrupting every input with a random error for a simulated discharge with no horizontal and vertical displacements. Sensitivity is at least a factor 2 less than in PABM, as shown at table 2.

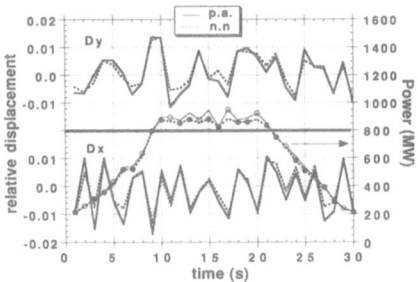

Figure 3 Comparative results from pabm and nnbm techniques

Table 1. Detector malfunctioning effects on NNBM.
(regression coefficient input-output)

	One detector failure		Two detectors failure	
	Position	Power	Position	Power
Worst case	0.85	0.996	0.67	0.992
Average case	0.93	0.998	0.75	0.995

Table 2. Signal noise effect on both methods
(Displacement in % plasma radius)

	NNBM	PABM
Horizontal position	0.32	0.49
Vertical position	0.35	0.53

SUMMARY

A technique to determine plasma position and fusion power in ITER from neutron measurements has been proposed and checked. Simulation results confirm its validity and its realization through neural network, using a perceptron with a backward propagation algorithm, has proved the potential of this technique as a control system. Additional efforts are in progress in several areas to fully assure the capabilities of this technique. Experimental evidence for the applicability of this technique could be possible using, as substitute for neutron measurements, X-ray array signals in well documented tokamak discharges.

REFERENCES

1. O.N. Jarvis, Rep. JET-IR (90)01, JET Joint Undertaking, (1990)
2. JET Team, Nucl. Fusion 32, 187, (1992)
3. J.D. Strachan, et al. Rep. PPPL-2978, (1994)
4. ITER PHYSICS, Iter Documentation series, 21, IAEA, Vienna, (1991)
5. C. Bishop, Rev. Sci. Instrum. 63, 10, (1992)
6. A.P. Navarro, M.A. Ochando and J.B.Blázquez, Rev. Sci. Instrum. 66, 878, (1995)
7. A.P. Navarro, Fusion Technology, 27, 152, (1995)
8. R.P. Lippmann, IEEE ASSP, 4, 4, (1987)

ITER GAMMA DIAGNOSTICS: 2-D NEUTRON AND GAMMA CAMERA

V.G.Kiptily,[1] I.A.Polunovskii,[1] V.O.Naidenov,[1] I.N.Chugunov,[1] V.S.Zaverjaev,[2]
S.V.Popovichev,[2] A.V.Khramenkov,[2] S.N.Abramovich,[3] A.G.Zvenigorodskii[3] and
M.V.Savin[3]

[1]A.F.Ioffe Physical Technical Institute, 194021 St.Petersburg
[2]RNC "Kurchatov Institute" - Nuclear Fusion Institute, 123182 Moscow
[3]All-Russia Research Institute of Experimental Physics, 607200 Arzamas-16
Russia

INTRODUCTION

Gamma-ray spectroscopy of fusion plasma is the technique of fundamental and vital importance for ITER. The method is based on the detection and analysis of high energy gamma rays resulting from fusion reactions (D+H, D+D, D+T etc.) and other ones that occur between fast ions and background impurities (e.g. Be) or target atoms injected in the plasma (Table 1).

It is well known that interaction between fast RF driven ions and low-Z plasma impurity ions leads to nuclear reactions with the following gamma-ray emission. The cross section of the reactions may reach as much as 1 barn, therefore gamma-ray yield is high despite the low impurity density. This fact has been used at JET to study the fast ion confinement[1]. In the measurements performed, the gamma-ray spectra from threshold reactions were recorded. The variation of intensity of the gamma rays during the discharge allowed to derive the degree of confinement of the fast ions.

Table 1. Main diagnostic reactions and associated gamma rays.

Fusion Reactions	Background Reactions	"Artificial" Reactions
$D(p,\gamma)^3He$, $\gamma 5.5$ MeV	$^9Be(p,\alpha\gamma)^6Li$, $\gamma 3.5$ MeV	$^7Li(p,\gamma)^8Be$, $\gamma 17.6$ MeV
$T(p,\gamma)^4He$, $\gamma 20$ MeV	$^9Be(d,n\gamma)^{10}B$, $\gamma 2.8$, 2.2 MeV	$^{11}B(p,\gamma)^{12}C$, $\gamma 11.7$ MeV
$D(d,\gamma)^4He$, $\gamma 24$ MeV	$^9Be(d,p\gamma)^{10}Be$, $\gamma 3.3$ MeV	$^{19}F(p,\alpha\gamma)^{16}O$, $\gamma 6.1$, 6.9, 7.1 MeV
$D(t,\gamma)^5He$ $\gamma 17$ MeV	$^9Be(^3He,p\gamma)^{11}B$, $\gamma 6.7$ MeV	$^{10}B(\alpha,p\gamma)^{13}C$, $\gamma 3.1$, 3.7 3.85 MeV
$D(^3He,\gamma)^5Li$, $\gamma 17$ MeV	$^9Be(^3He,n\gamma)^{11}C$, $\gamma 6.3$ MeV	
	$^9Be(\alpha,n\gamma)^{12}C$, $\gamma 4.44$ MeV	

In DD plasma at auxiliary ICRF and NBI heating, both fast minority protons and fusion protons at 3.0 MeV and 14.7 MeV may be formed. To execute the technique, the LiF and boron pellet should be injected into the plasma. Information about behavior of fusion protons can be derived from 17.64-MeV and 6.13-MeV gamma rays which arose during resonant reactions $^{19}F(p,\alpha\gamma)^{16}O$, $^7Li(p,\gamma)^8Be$ and $^{11}B(p,\gamma)^{12}C$.

One of the prospective method for ITER to measure fast alpha particles is gamma spectroscopy of 4.4-MeV emission from $^9Be(\alpha,n\gamma)^{12}C$ reaction in energy range above 1.7 MeV. As to alpha-particle reaction with artificial impurity that very useful $^{10}B(\alpha,p\gamma)^{13}C$ (above 1.2 MeV) which is still under study by the Ioffe-Kurchtov-Arzamas collaboration. The interaction of fast alphas

Diagnostics for Experimental Thermonuclear Fusion Reactors
Edited by P. E. Stott *et al.*, Plenum Press, New York, 1996

463

with the target leads to the nuclear reaction with the following gamma-ray emission 3.09, 3.68 and 3.85 MeV. At auxiliary heating of the DT plasma when energy of initial alphas may reach 4.0 MeV or more, gamma-rays 6.86 and 7.55 MeV will be radiated during the reaction. Preliminary in-beam study of the $^{10}B(\alpha,p\gamma)^{13}C$ reaction have been carried out in the energy range 1.4-4.0 MeV. Cross sections and Doppler line shapes have been measured with Ge spectrometer. We found out that this reaction may be used for the fusion alpha-particle study.

FUNCTION OF THE SYSTEM

The primary application of the diagnostics is measurement of time-dependent spatial gamma-ray and neutron emissivities to derive (i) fusion reactivity and alpha-particle birth profile by means of reactions $D(t,n)^4He$ and $D(t,\gamma)^5He$; (ii) profile of fast confined alpha particles - $^9Be(\alpha,n\gamma)^{12}C$ and $^{10}B(\alpha,p\gamma)^{13}C$; (iii) D/T fuel ratio - $D(d,\gamma)^4He$ and $D(t,\gamma)^5He$; (iv) degree of the fast ion confinement under ICRF or/and NBI plasma heating; (v) profile of run-away electrons from bremsstrahlung hard x-rays. Main adventures and shortcomings of the diagnostics are listed in Table 2.

Technical approach and methodology are based on 2-D Neutron & Gamma Diagnostic system for the 14-MeV neutron counting and nuclear reaction gamma-ray spectroscopy in the mode of on-line data processing and tomographic reconstruction. In the case of Be background impurity in ITER the system will furnish information about alpha-particle birth profile by means of 14-MeV neutrons and 17-MeV gamma rays detection, and 2-MeV alpha-particle profile by measuring 4.4-MeV gamma rays from $^9Be(\alpha,n\gamma)^{12}C$ reaction.

DESIGN DESCRIPTION AND INTERFACE

Two versions of the Neutron & Gamma Diagnostic system are developed[2]. In first one the cameras are located into the cryostat. According to other version cameras are placed outside cryostat. Cameras would be separate from the tokamak vacuum vessel. There would be steel windows up to a total of 50 mm.

The Neutron & Gamma Diagnostic system comprises two multi-collimator cameras. One camera will be located at a horizontal port to view the vertical profile and to cover about 2/3 of full plasma height. The front shielding would be about 2.5 m thick with 15 collimator penetrations; the cross section of each would be about 3-12 cm depending on detector unit. The collimator geometry will be fan-shaped. The collimator channels would be plugged by low-Z material (^6LiH is essential).

Gammas transmitted through the collimators reach the detectors, which are to be housed in shielding structure appropriate for their size. The horizontal camera would be a massive shielding structure of envelope dimensions - approx. 4mLx4mWx8mH. Own port is required for the horizontal camera.

The next one will be located in a vertical port to view the radial profile. The front shielding would be about 2.5 m thick with 12-15 collimator penetrations. The inner location vertical camera is preferable since the flux channel can be install through the side segment of the top blanket.

Table 2. Comparison with other diagnostics

Advantages	Shortcomings
Multi-task diagnostics	Massive structure of Cameras
Spatial & Time-dependent information	Needs additional information on Be-profile
Can be used as Plasma Control System	Sophisticated calibration
Conventional nuclear technique	Complicated data processing & reconstruction
No necessity to use any injected beams	No operational experience
Simple access to the plasma	
No any vacuum problem	

GAMMA DETECTORS

Different kind of gamma-ray detectors was considered including NaI, BGO, BaF_2, NE226. They vary in size and have differing performance characteristics with regard to energy resolution, efficiency, count rate, and radiation sensitivity and resistivity. Although a final choice of gamma detector has not been made at present time, feasibility of detectors and their relative sensitivity to neutrons will be studied and prototypical detectors suggested.

As first step, a full energy absorption spectrometer (gamma calorimeter) *GAMMACELL*, which we are working out, can be regarded as a detector unit for the ITER gamma cameras.

The *GAMMACELL* array consists of nine detector units based on BaF_2 scintillation crystals and operating in regime in which the energy of the gamma rays is summed, along with one more detector in front of the central unit of the array to serve as a active-passive shielding of the spectrometer.

A computer study and design of *GAMMACELL* for gamma diagnostics has been performed. Figure 1 shows the response functions of the spectrometer according to calculations by the Sigma 1.4 program[3] for the geometry of the proposed camera. Preliminary evaluation of realization of the detector radiation shielding will be made and signal-to-background assessments performed.

The application of the *GAMMACELL* operated in the mode of full gamma-ray energy absorption will allow to accomplish the best peak-to-background ratio, high detection efficiency in the energy range up to 30 MeV and high counting rate.

The *GAMMACELL* must secure:

- energy range: 1 - 30 MeV;
- energy resolution: 15% - 20% @ 0.662 MeV;
- efficiency up to 80% @ 17 MeV;
- total counting rate: > 1 MHz;
- best quality of active shielding;
- maximum peak to background ratio;
- minimum sensitivity to low energy scattered gammas and neutrons.

BaF_2 detector have been chosen for the ITER Neutron & Gamma Camera because of the following features:

- fastest scintillator (fast decay component - 700 ps, slow component - 630 ns);
- sufficient gamma-ray detection efficiency, which is nearly as high as that of BGO detectors;
- low sensitivity to neutrons;
- highest radiation resistance (limit is about 10^8 rad or equal neutron fluency - $3 \cdot 10^{17}$ cm^{-2});
- recovery of the optical quality almost up to the initial state.

We study a feasibility to use a combine neutron and gamma detector unit in the 2-D camera. Figure 2 depicts a scheme of the detector unit. We propose to measure simultaneously the yields both neutrons from the reaction $D(t,n)^4He$ and gamma rays from reactions $D(t,\gamma)^5He$ (17 MeV), $D(d,\gamma)^4He$ (24 MeV), $^9Be(\alpha,n\gamma)^{12}C$ (4.4 MeV). It is expediently to use the neutron spectrometer with the NE213 or plastic scintillator.

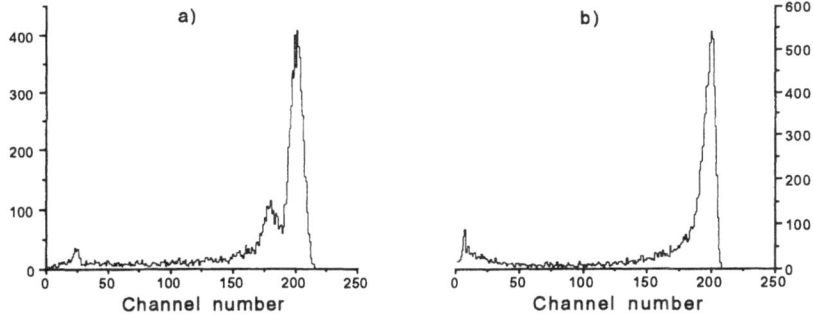

Figure 1. Calculated response function of the *GAMMACELL*: a) 4.44-MeV gamma rays; b) 16.7-MeV gamma rays. The energy resolution of the spectrometer is 15% at 662-keV.

FEASIBILITY OF APPLICATION IN ITER

According to the design of the diagnostic system, spatial resolution of horizontal camera would be equal to 20-30 cm; for vertical camera the resolution is about 15-30 cm. The time resolution of the system can be estimated as about 100 ms at the 30% accuracy for 1.5-GW ignited plasma in ITER at Be impurity content of 1%.

The main drawback of the using reaction $^9Be(\alpha,n\gamma)^{12}C$ to measure alpha-particle profile is background 4.44-MeV gamma rays originating from inelastic scattering $^{12}C(n,n_1\gamma)^{12}C$. The intensity of this undesirable radiation has been estimated. The proposed technique requires the following signal-to-background relation: $I_S/I_B \geq 10$. Taking into account that cross section of the scattering is equal to 0.18 b, and the neutron flux is about 2×10^{14} cm^{-2} s^{-1}, we obtained that the average carbon density in the field of view of the detector must not exceed 10^{17} cm^{-3}. That is, the proposed technique is feasible in the case of non-carbon first wall of ITER.

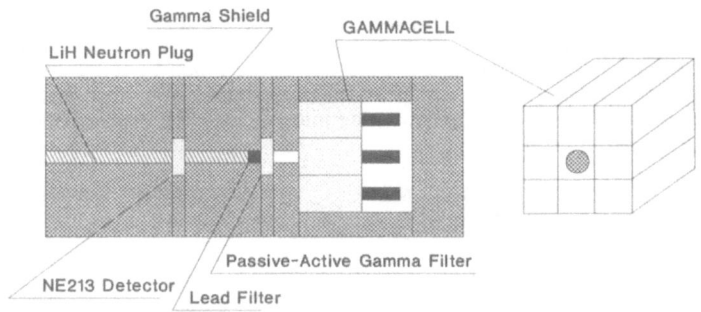

Figure 2. Cross section of the *GAMMACELL* spectrometer and collimator.

ACKNOWLEDGEMENTS

Thanks go to many people from Ioffe Institute, Kurchatov Institute and Arzamas-16 who contributed to this project.

This work was supported by the International Science and Technology Center, under ISTC Contract No. 161-94.

REFERENCES

1. G.J.Sadler, S.W.Conroy, O.N.Jarvis, P.van Belle, J.M.Adams and M.A.Hone, Investigations of fast-particle behavior in JET plasmas using nuclear techniques, *Fusion Technology 18:556 (1990)*.

2. V.G.Kiptily, Two-dimensional neutron&gamma camera, in: *Proposals on Diagnostic Access for ITER, ITER Meeting on Generic Access Routes for Diagnostic, Garching, 1994. GSS IP 58 94-06-16F*.

3. Alexander Bykov, SIGMA version 1.4, Copyright 1993.

ALPHA-PHYSICS AND MEASUREMENT REQUIREMENTS FOR ITER

S.J. Zweben[#], S. Putvinski[*], M. P. Petrov[†],
G. Sadler[+], K. Tobita[¶], and K.M. Young[#]

[#] Princeton Plasma Physics Laboratory, Princeton, NJ 08540 USA
[*] ITER Joint Central Team, La Jolla, Calif. 92037 USA
[†] Ioffe Physical-Technical Institute, St. Petersburg, 194021 Russia
[+] JET Joint Undertaking, Abingdon, Oxfordshire, United Kingdom
[¶] JAERI, Naka-machi, Naka-gun, Ibaraki, 311-01, Japan

INTRODUCTION

Sustained ignition in ITER requires about 300 MW of alpha particle heating power. Since the alpha particle creation rate will depend strongly on the plasma conditions, this plasma heating will be more difficult to predict and control than existing heating systems such as NBI and ICRH. In addition, alpha particle heating and loss will also depend upon the transport of alphas during their relatively long thermalization time (≈ 1 sec), during which they may be affected by MHD activity and other non-axisymmetries in the ignited plasma.

This paper reviews alpha particle physics issues in ITER and their implications for alpha particle measurements. First, a comparison is made between alpha heating in ITER and NBI and ICRH heating systems in present tokamaks. Then the alpha particle issues in ITER will be discussed in three physics areas: "single particle" alpha effects, "collective" alpha effects, and RF interactions with alpha particles. Note that this paper will not cover the important subject of alpha particle ash, which is more related to thermal plasma transport.

ALPHA HEATING IN ITER vs. PRESENT FAST ION HEATING SYSTEMS

The fast ions used for NBI and ICRH minority heating in present experiments have not been diagnosed in great detail, since they usually heat the plasma without problems[1]. It is interesting to compare the parameters for these fast ion heating systems with those expected for alpha particle parameters in ITER to help identify potential problem areas and appropriate alpha particle measurement requirements. This comparison is shown in Table 1.

The heating power density of alphas in ITER will actually be *lower* than that for present tokamak heating systems, since the plasma energy density will be only slightly larger, but the plasma energy loss rate will be lower. Thus the relative alpha particle density

Diagnostics for Experimental Thermonuclear Fusion Reactors
Edited by P. E. Stott *et al.*, Plenum Press, New York, 1996

467

in ITER is typically an order of magnitude lower than the fast ion density in NBI[2] and ICRH[3] heated tokamaks, also in part due to the larger alpha particle energy. The alpha particle beta in ITER is also expected to be *smaller* than the fast ion betas normally obtained with present auxiliary-heated large tokamaks[2,3], but somewhat larger than presently obtained for alphas in the TFTR DT experiment[4].

Table 1: Fast Ion Parameters for Various Heating Systems

Parameter	NBI*	ICRH‡	Alphas* (TFTR)	Alphas (ITER)
$P_f(0)$ [MW/m^3]	3	1-3	0.3	0.3
δ/a (orbit shift)#	0.05	0.3	0.3	0.05
$n_f(0)/n_e(0)$ %	13	1-10	0.3	0.3
$\beta_f(0)$ %	0.9	1-3	0.26	0.7
$<\beta_f>$ %	0.4	0.5	0.03	0.2
$R\nabla\beta_f$	0.04	≈0.1	0.02	0.06
$V_{fo}/V_{Alf}(0)$	0.35	≈1-2	1.6	1.9

* TFTR with 40 MW of 100 keV NBI in a 5 T. DT plasma (#76770)[2]

‡ JET with ≈15 MW ICRH He^3 minority heating with $<E_f> ≈ 1$ MeV[3]

\# δ (orbit shift from magnetic flux surface) ≈ $q(R/r)^{1/2} \rho_{tor}$ ≈ $5 \rho_{tor}$ @ q≈2

The similarity of the fast ion orbit parameter δ/a and the fast ion pressure gradient $R\nabla\beta_f$ between NBI in TFTR and alphas in ITER suggests that the single-particle and collective fast ion physics should also be similar. However, a crucial difference is that the fast ion speed relative to the Alfven speed V_{fo}/V_{Alf} is much larger for alphas in ITER than for NBI in TFTR, leading to the possibility of Alfven instabilities in ITER[5-7]. Although super-Alfvenic fast ions were simulated by ICRH minority heating in JET, these ions were mainly trapped particles and so were potentially different from the nearly isotropic alpha distributions expected in ITER.

If the single-particle and collective alpha physics in ITER were similar to NBI and ICRH ions in existing tokamaks, then the alpha particle *heating* in ITER should be predictable from the DT neutron source profile measurements. However, alpha particle *loss* to the first wall might still be a problem due to ITER's 1000 sec pulse length, which requires that every part of the first wall be actively cooled to avoid impurity influx or wall damage. For example, even a few-percent alpha loss fraction due to TF ripple loss is a concern for ITER[5,8], whereas a similar level of fast ion loss in present experiments is generally undetectable. Hints of this problem were found in the ≤60 sec long pulses in Tore-Supra, in which localized wall heating due to fast ion loss was observed[9].

The conclusion from this comparison is that all the relevant physics parameters are *not* similar between alphas in ITER and fast ions in present experiments, so diagnostics to evaluate the alpha particle heating and loss must be considered for ITER.

"SINGLE PARTICLE" ALPHA INTERACTIONS

Individual alpha particles can be affected by many types of background plasma fluctuations and non-axisymmetries. A list of potential single-particle alpha effects in ITER is shown in Table 2. The timescales of these effects range from steady-state TF ripple loss to sub-millisecond bursts of alpha loss associated with sawtooth crashes or disruptions. *All* of these effects have been observed for fusion products in present tokamaks[1], and all of these have caused some measurable DD or DT fusion product loss in TFTR[10,11].

Table 2 - Potential "Single-Particle" Alpha Interactions in ITER

Interaction	Frequency (kHz)	Toroidal mode #	Loss in TFTR
TF ripple	0	20	10%
locked modes	0	1 (?)	1%
tearing modes	0.1	1-3	1-10%
ELMs	1	1-3	1%
fishbones	10	1	1%
sawtooth crash	10^2	1 (?)	<<1%
disruptions	10	1 (?)	>10%

The best understood of these alpha interactions is TF ripple loss, which was calculated using Monte Carlo codes specifically for ITER[5]. The conclusion from this study was that the peak alpha heat loads there were in the range ≈ 0.1-1 MW/m^2, depending on the plasma current and the direction of the toroidal field (for 24 coils). This is comparable to the average heat flux expected on the ITER first wall from plasma radiation (≤ 0.5 MW/m^2).

The most important measurement of alpha ripple loss in ITER would be a spatially-resolved surface temperature measurement of the first wall, since this information could be used for real-time operational control as well as for physics purposes. A good example of such a measurement is the IRTV image of the JT-60U wall shown in Fig. 1. The localized heating of $\approx 50°C$ between the TF coils is consistent with the expected beam ion ripple loss, as calculated by a Monte Carlo guiding center code[12]. However, the heat deposition pattern is also influenced by a slight misalignment in the tiles, which causes the leading edges to become much hotter than the average wall temperature in the ripple loss region. This type of localized heating is likely to occur in ITER due to the difficulty of aligning the wall tiles and limiters over such a large and complicated vacuum vessel.

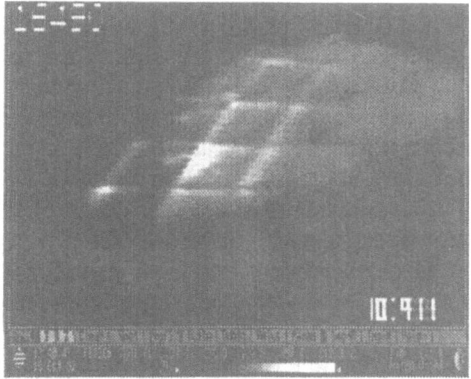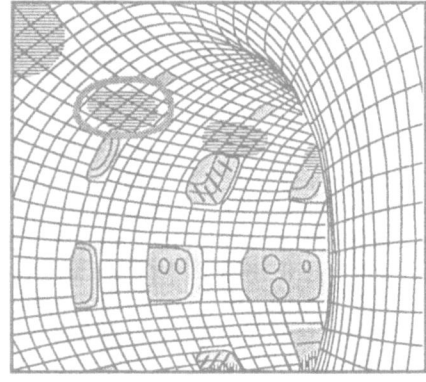

Fig. 1 - At the left is an IRTV image of the surface temperature of the JT-60U wall showing TF ripple loss of beam ions during perpendicular NBI. These hot spots occur just above the outer midplane between the TF coils (right). The heat flux in this case is ≈ 1 MW/m^2, similar to that expected for alpha ripple loss in ITER.

Also potentially important for ITER is the effect of sawtooth crashes on the alpha particle heating profile, and also on the alpha loss to the first wall. Evidence for a weak radial redistribution of alpha-like 1 MeV tritons during a sawtooth crash in JET is illustrated in Fig. 2[13]. Sawtooth-induced radial redistribution of H-minority tail ions has also been measured in JET using impurity charge exchange[14], and with alphas in TFTR using the pellet charge exchange (PCX)[15] and alpha-CHERS[16] diagnostics. Direct measurements of alpha particle loss during sawteeth in TFTR show a very fast ($\approx$0.1 msec) but small (<<1% loss) burst of alpha particle loss coincident with a sawtooth crash.

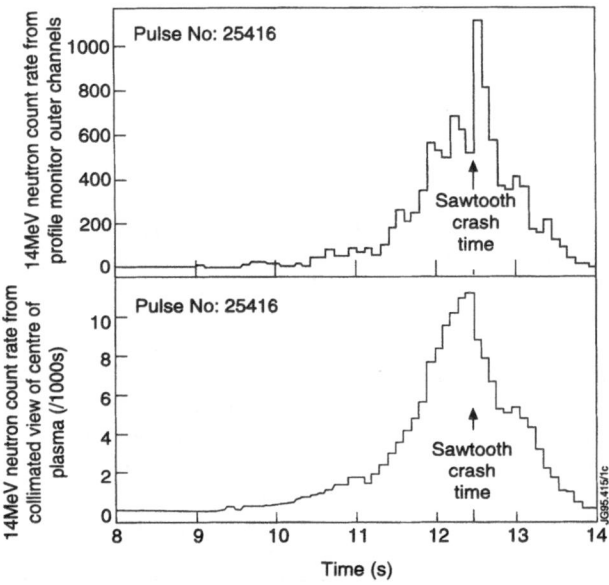

Fig. 2 - Measurements of triton burnup in JET showing a weak redistribution of confined tritons during a sawtooth crash. The top trace shows an increase in tritons near the plasma edge, and the bottom trace shows an decrease in tritons nearer the plasma center.

The effect of sawtooth crashes on the confined alpha population in the core of ITER will be difficult to measure directly, since there is no "burnup" of the alpha particles analogous to that for the 1 MeV tritons used for the measurement of Fig. 2, and neutralizing atoms for charge exchange will not normally be present at r/a<0.5. In principle, the sawtooth effects can be measured using charge exchange from neutral beams or pellets in the plasma periphery at r/a$\geq$0.5[15]. Ideally, confined alpha diagnostics could help clarify the physics of the sawtooth-alpha interaction through measurements of the energy and pitch angle dependence of the alpha particle redistribution during a crash. However, even if these measurements were available, the interpretation of the radial alpha transport will also depend on the physics of the sawtooth crash itself, which is not yet understood. Most likely, the effect of sawtooth crashes on confined alphas will be measured indirectly in ITER through the plasma temperature profiles, which will respond slowly to a redistribution of the alpha particle heating. It is reassuring that ITER will probably maintain ignition even if most of the alpha particles were lost from inside the sawtooth inversion radius[8], since the timescale for alpha particle re-creation ($\approx$ 1 sec) is shorter than the energy confinement time ($\approx$ 4 sec).

The effect of sawtooth crashes on alpha particle loss in ITER should be measurable using a wide-angle IRTV camera. For a clearer identification of the loss mechanism, this should by supplemented by discrete IR channels viewing the first wall or limiters with sub-millisecond time resolution, or by low-mass "foil" bolometers such as used at DIII-D[17]. The sawtooth-induced alpha loss might consist of two components; namely, a prompt loss to the wall on the timescale of the sawtooth crash (<1 msec), and a slower diffusive loss of alphas which were redistributed into the TF ripple loss region ($\approx$10-100 msec).

Most of the other single-particle alpha interactions in Table 2 are some form of coherent MHD activity. These modes are usually present in high-powered tokamaks, but normally do not cause a significant redistribution or loss or fast heating ions. However, large MHD levels in TFTR have caused up to a $\approx$10% fusion product loss, often modulated with this activity over the range 0-20 kHz[10]. Modeling of this type of MHD-induced alpha loss with a Monte Carlo guiding center code predicted that (for a given low-n mode) the alpha loss decreased about an order-of-magnitude as the orbit shift parameter δ/a decreased from the TFTR value to the ITER value, suggesting that the fractional alpha loss in ITER should be much lower in ITER than in TFTR.

Measurements of the influence of MHD activity on confined alphas in ITER will be difficult, particularly since this activity is usually not steady or reproducible. If such measurements were available, it would be interesting to correlate the spatial location and mode number of the fluctuations with the radial redistribution of alphas vs. energy and pitch angle. An indirect analysis of the MHD-induced alpha transport effects from the temperature profile measurements will be more problematic than for sawtooth crashes, since the MHD activity will simultaneously affect the thermal plasma transport. Measurements of alpha loss should be made with enough time resolution to see the bursts of alpha loss correlated with ELMs, such as seen during limiter H-modes in TFTR DT discharges[18].

The most serious single-particle alpha losses in TFTR occurred during major disruptions, as illustrated in Fig. 3, when over 10% of the confined alphas were lost within a few milliseconds[11]. This alpha loss was apparently concentrated 90° below the midplane in the ion ∇B drift direction, and so might cause damage at the entrance to the ITER divertor dome. The mechanism of this loss is probably similar to that for coherent MHD modes, but so far there is no modeling to predict its size or location in ITER. It is interesting that any alphas which are not immediately lost during a major disruption will probably be thermalized before the end of the current quench, since the alpha thermalization time will be reduced proportionally to $T_e^{3/2}$ during the thermal quench, i.e. probably to <1 msec.

Measurements of alpha loss during disruptions in ITER will be extremely difficult, since IRTV measurements will be ambiguous due to simultaneous plasma loss and radiation. Energy resolving alpha loss detectors could discriminate alpha loss from thermal plasma loss, but these detectors may not survive the heat loads during an ITER disruption.

COLLECTIVE ALPHA PARTICLE INTERACTIONS

A large amount of experimental work has been done on collective fast-ion instabilities[1], and some of the associated theory has been applied to predict alpha particle stability in ITER[5-7]. In general, collective alpha instabilities may have a more serious effect on alpha particle heating and loss than the single-particle effects, but they are less likely occur. Measurements of the *fluctuations* due to these instabilities will be as important as the direct measurements of the alpha particles themselves.

471

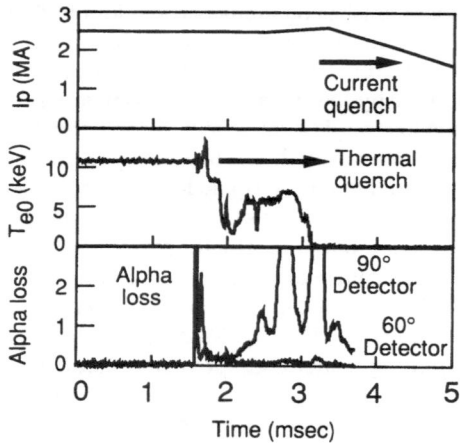

Fig. 3 - Measurement of DT alpha particle loss during a major disruption in TFTR. The disruption at 1.5 msec causes a rapid loss of ≥10% of the confined alphas before the current quench, mostly to the 90° detector. The loss at ≤1.5 msec is the normal first-orbit loss level.

A list of potential collective alpha effects in ITER is shown in Table 3. *All* of these instabilities have been observed on existing tokamaks using fast ions from NBI or ICRH tails. However, so far *none* of these instabilities has been driven by DT alphas, except for the rather benign ICE (ion cyclotron emission) such as seen at JET and TFTR.

Table 3 - Potential "Collective" Alpha Particle Effects in ITER

Instability	Freq.(kHz)	n-mode #	Location (r/a)	Some Observations
sawtooth	0	1	≈0.5	JET, TFTR, JT-60U
fishbones	10	1	≈0.5	PDX, JET
KBM	50 (?)	10-20	≈0.5	TFTR
BAE/EAE	50 (?)	10-50	≈0.5	DIII-D, JET
TAE mode	100	10-50	≈0.5	TFTR, DIII-D, JT-60U
AFM	100	0	≈1	TFTR
ICE	10^5	-	≈1	JET, TFTR, DIII-D

The TAE mode (toroidicity-induced Alfven eigenmode) has caused up to an ≈70% loss of injected NBI ions in TFTR and DIII-D, but only when the toroidal field was lowered to make the NBI ions super-Alfvenic. An examples of TAE mode spectrum is shown in Fig. 4, taken during NBI heating in DIII-D[19]. There are multiple TAE modes at a frequency $f=2\pi V_A/2qR$, which are separated by toroidal rotation. In ITER the most unstable alpha-driven TAE modes are so far expected in the range n≈5-50, with multiple m-modes for each n, so there may be very many such TAE mode peaks in ITER (5-7).

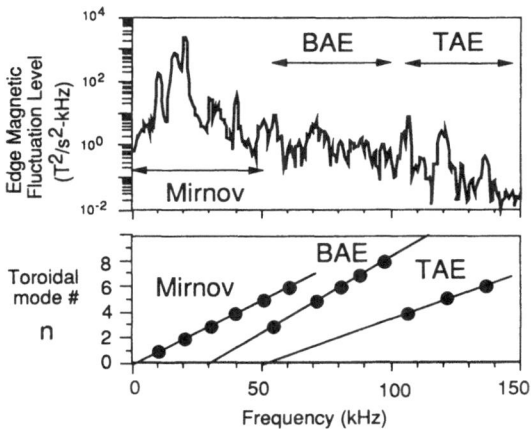

Fig. 4 - Edge magnetic fluctuation spectrum showing TAE modes driven by NBI in DIII-D. The TAE (and BAE) have mode numbers in the range n≈3-8. The TAE modes in ITER are expected to be unstable over the range n=5-50, implying a more complex frequency spectrum in ITER (figure courtesy of W. Heidbrink).

Measurement of TAE mode fluctuations in ITER should probably be done using external magnetic loops and internal density fluctuation measurements, as for existing NBI and ICRH experiments. The TAE spectrum in ITER should be localized in the frequency range f≈100 kHz at the radial location of the highest alpha particle pressure gradient near r/a≈0.5. The level of the internal electron density and external magnetic fluctuations would probably be similar to that in present NBI and ICRH TAE simulation experiments, i.e. ñ/n≈0.1-1% and $\delta B/B_T$≈10^{-6}-10^{-5}[20].

Confined alpha particle measurements for TAE studies will be limited by the difficulty of the diagnostic techniques. Present experiments diagnose the fast ion content mainly by changes in the neutron rate (for NBI) or by magnetic measurements of the fast ion stored energy, neither of which would be useful for this purpose in ITER. Ideally, it would be desirable to have a confined alpha diagnostic which could focus on the specific range of alpha particle pitch angles and energies which are expected to interact with TAEs in ITER, e.g. alphas with a parallel speed equal to the Alfven speed. Ideally, such a diagnostic would be able measure the alpha profile changes on the timescale of the individual "bursts" of TAE mode activity, which are typically ≈10 msec apart in present NBI experiments.

Measurements of TAE-induced alpha loss should attempt to distinguishing three different processes: radial diffusion of passing alphas to the outer or inner midplane regions of the first wall, loss across the passing-trapped boundary over a wider range of poloidal angles in the ion-∇B drift direction, and TF ripple trapping and loss of alphas between the TF coils. The last of these was responsible for making a hole in the TFTR vacuum vessel during ICRH hydrogen minority tail-driven TAE modes[21]. Wall and limiter temperature measurements using a wide-angle IRTV would be useful for operational purposes, and ≈10 msec time response would be useful to correlate these temperature rises with TAE fluctuation measurements.

Measurement requirements for the alpha-driven KBM (kinetic ballooning mode), BAE/EAE (beta- and ellipiticity-induced Alfven eigenmodes) and the alpha-driven fishbone modes are generally similar to those for TAE modes. The fluctuation diagnostics should be designed to cover the whole range up to ≈300 kHz (the toroidal transit frequency of 3.5 MeV alphas). Since alpha particle diagnostics will probably not respond on the timescale of the mode frequency, the specific diagnostic requirements for various modes would differ only in their focus in pitch angle, energy, and r/a, and first-wall location. Theoretical predictions of these characteristics have not yet been made for alphas in ITER.

The collective alpha effect on sawteeth is unique in having a stabilizing influence. However, it may be difficult to identify this effect in ITER, since there will be no alpha-free comparison discharges. The AFM (Alfven frequency mode) was recently identified at TFTR as a magnetic fluctuation at the edge Alfven frequency occurring with or without DT alpha particles[22]. A measurement of the edge electron density would be useful in ITER to help distinguish the AFM mode from a TAE mode. ICE is apparently due an unstable distribution function of fast ions near the outer midplane, but will probably be benign for alphas in ITER. Ion cyclotron absorption (ICA) measurements may be useful as an alpha particle diagnostic in ITER [23].

RF INTERACTIONS WITH ALPHA PARTICLES

There have been several recent proposals for the use of various RF waves to *control* alpha particles in order to improve the performance of a tokamak reactor. If these schemes are to be tested in ITER, then measurements need to be made to monitor this control.

One idea is to "channel" the alpha particle's energy directly into the fuel ions, rather than using the collisional ion heating from the electrons. In principle, this can be done using an externally launched IBW (ion Bernstein wave) resonant with the alphas on a spatial scale of the alpha gyroradius and a timescale of the alpha gyroperiod[24]. This interaction could be studied in ITER with a spatially-resolved IBW wave detector capable of measuring density fluctuations of ñ/n≈1% on these time and space scales, and by confined alpha diagnostics with good spatial and energy resolution. Evidence for such an interaction has come from IBW experiments in TFTR, in which an increase in the fast ion loss and energy was measured by the lost alpha scintillator detectors[25].

Another idea is to launch low frequency waves which can resonate with the transit frequency of the alphas in order to remove epithermal alpha ash or control the burn. Calculations have been done which demonstrate that alphas can be selectively and efficiently transported using swept-frequency waves of a few hundred kilohertz[26,27], but no experimental evidence for this interaction exists at present. If this idea is to be tested in ITER, measurements of the wave fields and energy-dependent radial profiles of the confined alphas would be useful to monitor this interaction.

Even if these novel alpha control schemes are not tested in ITER, there will be many possible interactions between alphas and the normal RF heating and/or current drive systems. For example, it is well known that some fraction of ICRF minority heating power will be absorbed by the fast alphas, and it has been demonstrated in TFTR that alphas are lost during ICRH through RF-induced transport across the passing-trapped boundary[28]. Spatially asymmetrical ICRH can also be used to cause radial transport of fast or epithermal alphas[29], and alphas can also resonate strongly with lower hybrid waves. Therefore measurements of confined and lost alphas should be done during any RF heating of ITER to insure that the alphas stay well confined.

SUMMARY OF ALPHA PARTICLE MEASUREMENT REQUIREMENTS

These alpha physics issues suggest the measurement requirements summarized in Table 4. These are prioritized in terms of their operational value rather than their intrinsic physics interest. For example, the most serious alpha particle problem in ITER is the potential impurity influx or damage due to overheating of the first wall by alpha particle loss. Therefore the highest diagnostic priority is for measurements of the first wall temperature, which could alert the operators to an impending problem. The next highest priority would be measurements of the internal plasma fluctuations which might become a source of this problem. Perhaps of a greater physics interest would be direct measurements of the confined and lost alpha particle populations, e.g. their spatial, energy, and pitch angle distributions.

Table 4 - Summary of Alpha Physics Measurements Needed in ITER

Physics area	Spatial resolution	Time resolution	Possible Techniques
Alpha heat loss	≈3-30 cm	≈10-100 msec	IRTV, thermocouples
Alpha instabilities	$\Delta(r/a) \approx 0.1$	≤300 kHz	B loops, reflectometer
Confined alphas	$\Delta(r/a) \approx 0.1$	≈0.1 Hz	CX, EM wave scattering
Lost alphas	a few points	≈100 kHz	Faraday cups, foils

Fortunately, the highest priorities on this list generally have the lowest diagnostic difficulties. Surface temperature measurement such as shown in Fig. 1 could be made with imaging mirrors and conventional IRTV detectors. High frequency fluctuation diagnostics can be made rugged and radiation resistant, and will be useful for other areas of ITER physics. The confined and escaping alpha particle diagnostics may be quite difficult to implement on ITER, but they are certainly worth pursuing for the sake their unique contributions to the understanding of burning plasmas.

Finally, it is important to recognize that these measurements in themselves will not guarantee satisfactory control over the ≈300 MW of alpha particle heating in ITER. Practical and reliable methods must be developed to use this information in feedback control schemes, for example, to abate an alpha-induced hot spot on the wall by moving it to another place, or to quench the ignition of a plasma which is becoming strongly unstable to TAE modes. In this regard, any potential alpha particle control scheme may have substantial value for ITER.

ACKNOWLEDGEMENTS

We thank R. Budny, J. Candy, Z. Chang, D. Darrow, H. Duong, L.-G. Eriksson, E. Frederickson, G. Fu, W. Heidbrink, D. Majeski, F. Marcus, R. Nazikian, M. Redi, M. Saigusa, R. Wilson, and P. Van Belle for contributions to this paper. This work was supported in part by US DOE Contract #DE-AC02-76-CHO-3073.

REFERENCES

1) W. W. Heidbrink and G. Sadler, Nucl. Fus. **34**, 535 (1994)

2) R. Budny, et al, Princeton Plasma Physics Lab Report PPPL-3112 (1995), submitted to Nuclear Fusion

3) L.-G. Eriksson, et al, Nucl. Fusion **33**, 1037 (1993); also, JET Team, IAEA-CN-50/A-1-3 (1988)

4) R. Hawryluk, et al, Proc. 15th IAEA Meeting Seville, 1994, IAEA-CN-60/A-1-I-1

5) S. Putvinski, et al, Proc. 15th IAEA Meeting, Seville, 1994, IAEA/CN/60 E-P-4

6) C.Z. Cheng, et al, Proc. 15th IAEA Meeting, Seville, 1994, IAEA/CN/60 D-3-III-2

7) H.L. Berk, et al, Proc. 15th IAEA Meeting, Seville, 1994, IAEA/CN/60 D-P-II-1

8) Design Description Document for the ITER Plasma, ITER JCT

9) V. Basiuk, et al, Fusion Technology **26**, 222 (1994)

10) S.J. Zweben, et al, Proc. 13th IAEA Meeting, Wurzburg 1992, IAEA-CN-56/A-6-3

11) S.J. Zweben, et al, PPPL-3045, Jan. 1995, to be published in Nuclear Fusion (1995)

12) K. Tobita, et al, Phys. Rev. Lett. **69**, 3060 (1992), also, to be published in Nuclear Fusion, Dec. 1995

13) F.B. Marcus, et al, Nucl. Fusion **34**, 687 (1994)

14) A. Gondhalekar, IAEA Technical Committee on Alpha Particles, Princeton, Apr. 1995

15) M. Petrov and R. Fisher, this conference; also PPPL-3121 (July 1995), to be published in Nucl. Fusion

16) G. McKee, et al, Phys. Rev. Lett. **75**, 649 (1995), also B. Stratton, private communication (1995)

17) H.H. Duong, W.W. Heidbrink, E.J. Strait et al., Nuclear Fusion **33** 749 (1993)

18) C. E. Bush, et al, Phys. Plasmas **2**, 2366 (1995)

19) W.W. Heidbrink, et al, Phys. Rev. Lett **71**, 855 (1993)

20) E.D Fredrickson, et al, Proc. 15th IAEA Meeting Seville, 1994, IAEA-CN-60/A-II-5

21) R.B, White, et al , Phys. Plasmas **2**, 1 (1995)

22) Z. Chang, et al, Princeton Plasma Physics Laboratory Report PPPL-3115, July 1995

23) G. Cottrell, et al, JET Report JET-R (94)10 1994

24) N.J. Fisch and J.-M. Rax, Phys. Rev. Lett. **69**, 612 (1992)

25) D.S. Darrow, et al, Proc. 11th Topical Conf. on RF Power in Plasmas, Palm Springs, 1995

26) C.T. Hsu, et al, Phys. Rev. Lett. **70**, 13 (1993)

27) H.E. Mynick, et al, Nuclear Fusion **34**, 1277 (1994)

28) D.S. Darrow, et al, PPPL- 2975, to be published in Nuclear Fusion 1995

29) C.S. Chang, et al, Phys. Fluids **B 3**, 3429 (1991)

INITIAL OPERATION OF AN ITER COMPATIBLE
LOST ALPHA DETECTOR AT JET [*]

M.J.Loughlin, F.E.Cecil[1], M.Hone, O.N.Jarvis,
G.J.Sadler, P.van Belle, G.Whitfield.

JET Joint Undertaking, Abingdon, Oxfordshire. OX14 3EA, UK
[1]Colorado School of Mines, Golden, Colorado 80401, USA

INTRODUCTION

Alpha particles produced in a magnetically confined fusion plasma must be well contained to heat the plasma and to avoid damage to the first wall. One technique in the study of alpha particle behaviour is the measurement of the flux of escaping particles. This has been done at TFTR [1] using scintillators. However, scintillators cannot be used at ITER if it is to operate with a high wall temperature ($>250^\circ$C) because of the temperature dependence of light output from scintillators. For a detector to be compatible with the projected ITER operation it must not only be able to operate at a temperature of 350°C but it must also be insensitive to 14MeV neutrons and x-rays, be sufficiently robust to withstand disruptions but be sensitive enough to measure the low particle flux.

In this paper the initial results, obtained at JET during a period of operation in pure deuterium operation, are presented. The detector appeared to operate satisfactorily and showed no sensitivity to neutrons, nor any response to neutral-beam, RF or LHCD heating.

THE DETECTOR

A prototype detector was installed at JET primarily to measure the background response since no significant alpha production was expected. The detector is a thin foil Faraday collector consisting of four Nickel foils each 2.5×10^{-3} mm thick with an area of 20cm^2. The housing is also Nickel. The foils are stacked in parallel, separated by 3mm. Between each foil is a bias ring (Figure 1). It is possible to apply voltages to each of these

---

[*] Supported in part by the U.S. Department of Energy

Diagnostics for Experimental Thermonuclear Fusion Reactors
Edited by P. E. Stott *et al.*, Plenum Press, New York, 1996

477

bias rings although this facility was not used. All components are electrically isolated from each other [2].

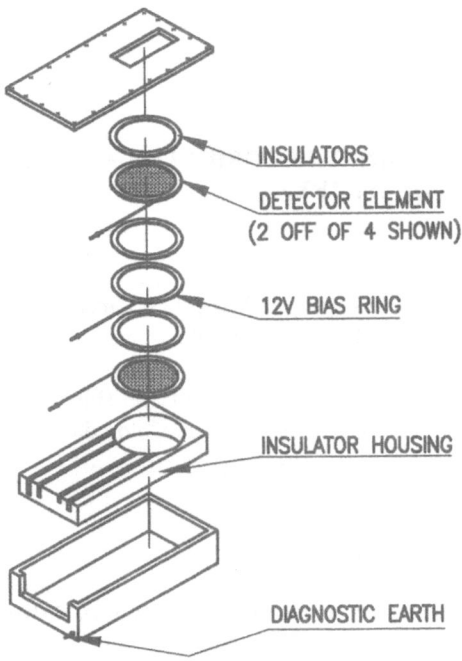

Figure 1. Exploded view of detector

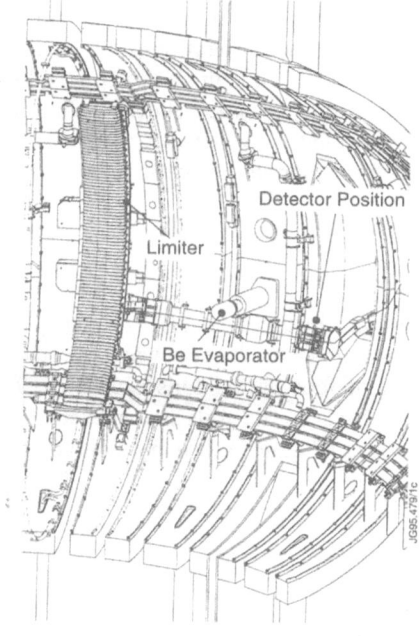

Figure 2. Location of detector inside JET

The detector is mounted on the JET first wall, on the outboard side 37.5cm below the vessel midplane. It is in the shadow of a poloidal limiter, 26cm behind the last closed flux surface (Figure 2). The orbits of fusion products heading for the detector are partially intercepted by the poloidal limiter and the beryllium evaporator head. Thus, in this position and for dd plasmas, the response is expected to be less than 1 nano-amp (compared to 1 micro-amp from a dt plasma) unless the detector proved to be sensitive to neutrons, x-rays, neutral particles or particles originating very close to the detector in the plasma edge. The detector is electrically insulated from the vacuum vessel but is in thermal equilibrium with it and hence is operating at a temperature of 600K.

The current from each foil is measured by digitising the output of a logarithmic current to voltage amplifier. The current from each foil was measured either with respect to the adjacent bias ring, the next foil or the body of the detector. Low pass filters were used to reduce noise and pickup and low current supplies were connected in series to the amplifier input to reduce the sensitivity and, as logarithmic amplifiers can cope with unipolar signals only, to avoid the saturation of the signal. These current supplies could be removed to increase the sensitivity.

OBSERVATIONS

Sensitivity to neutrons

The sensitivity of the detector to 2.5MeV neutrons (from pure deuterium plasmas) was examined. No response was observed up to total neutron emission levels of 2.1×10^{16} n/sec (see figure 3). This corresponds to a flux of approximately 2.5×10^{10} n/sec/cm^2 incident on the detector. The detector response shown in figure 3, and subsequent figures, is the digitized voltage output from a logarithmic amplifier with a calibration of 2volts/decade.

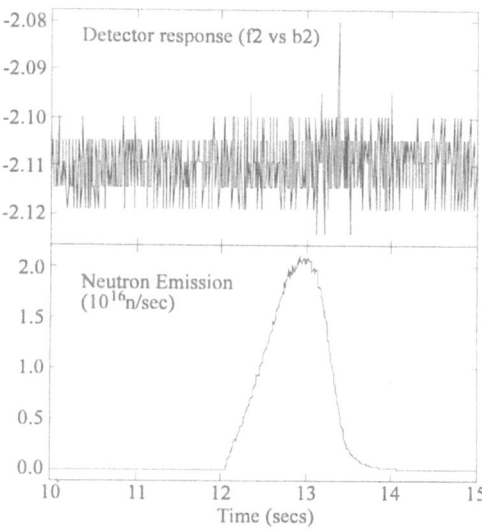

Figure 3. No response to neutrons up 2.1E16 n/sec (35368). f2 vs. b2 means the current was measured in foil 2 with respect to bias ring 2. The same scheme applies in subsequent figures.

Response during Neutral Beam Heating (NBI)

A fraction of the beam particles will be ionised close to the plasma edge. This depends on plasma edge density, the density of the scrape off layer, and the background neutral gas in front of the ducts. If the orbits of these ions (which depend on the toroidal field) took them to the detector this could result in either excessive power loading or a decreased signal to noise ratio at the time of interest, namely during the period of maximum alpha production. No response was observed during pure NBI into deuterium plasmas, up to power levels of 17.5MW (figure 4). However, a signal was seen when neutral beams were injected simultaneously with heavy gas puffing (figure 5). This response was not highly reproducible. An analysis of the beam deposition for a number of discharges is required.

Radio Frequency (RF) heating.

A response during RF heating was observed. RF power is delivered to the plasma via four modules. The response proved to be a pickup from one of them (module D) (figure 6) when the current in each foil was measured with respect to the body of the detector. This signal was not observed when the current was measured with respect to the adjacent bias ring (fig.7). No response during lower hybrid heating was observed up to power levels of 6MW (fig. 8)

Some response was observed during the operation of the saddle coils. It was determined that this response was caused by one set of saddle coils (fig 9). As this set was fed via cables in the same octant as the detector cables it seems likely that this was a pickup problem exterior to the vessel. Re-routing of the cable should eliminate this problem.

Response before disruptions

Observations were made of what appears to be a disruption precursor. Approximately 200-300msec before the disruption an increased current in foil 2 with respect to foil 3 was observed (fig. 10). Although was seen in a number of disruptive discharges it was not seen in all. It appears, from a limited data-base that this "precursor" is only discernible if the disruption occurs either during, or within 250ms of the end of additional heating. i.e. when there is population of fast particles in the plasma.

CONCLUSIONS

Detector current could detected down to the level of <1 namp. In dd operation, currents as large as this were not expected. The initial operation of the prototype detector indicates that the detector can be successfully operated in JET DT plasmas. Four new detectors will be constructed and installed and tested during the JET DTE1 experiment.

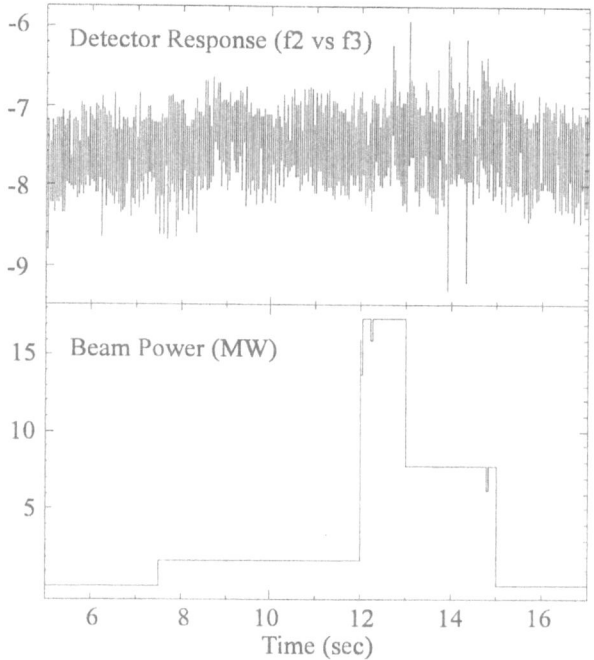

Figure 4. No response up to power levels of 17.4MW of NBI (35702)

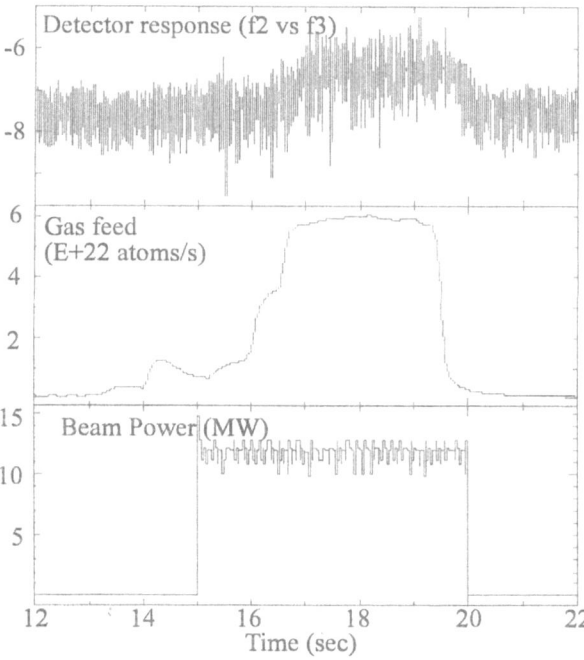

Figure 5. Observed DC shift during neutral beam heating and gas puffing (35764)

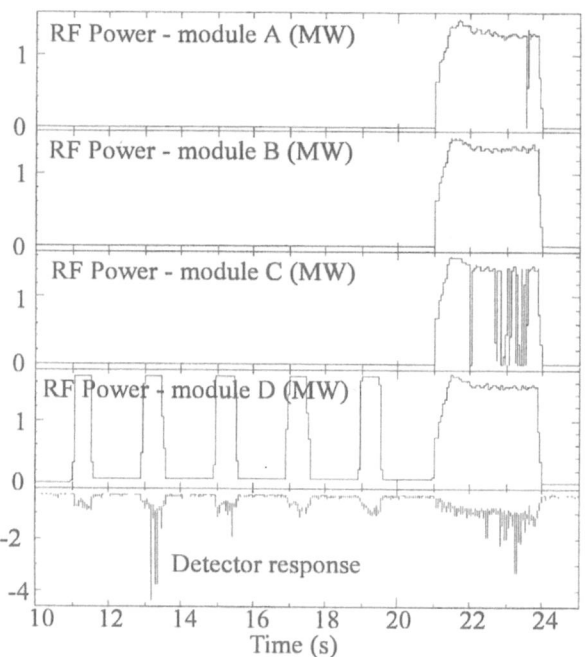

Figure 6. The response (foil vs. detector body) to RF module D (35316)

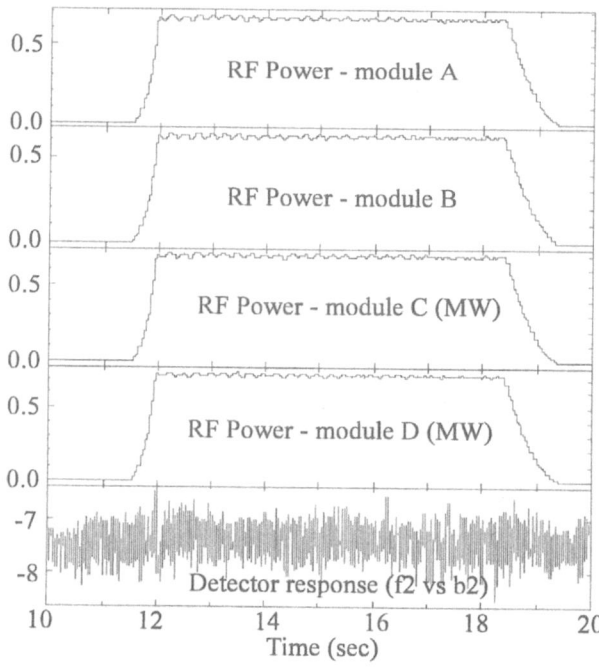

Figure 7. The response (foil vs. bias ring) to RF module D (35491).

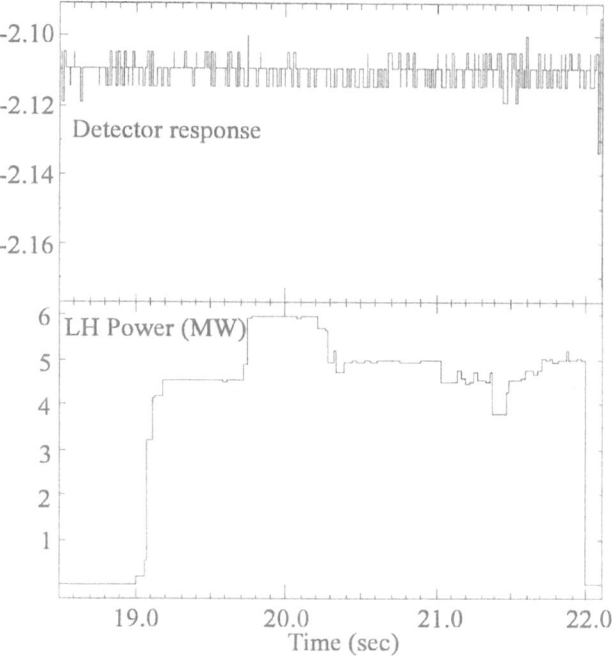

Figure 8. Detector response during lower hybrid heating (35539).

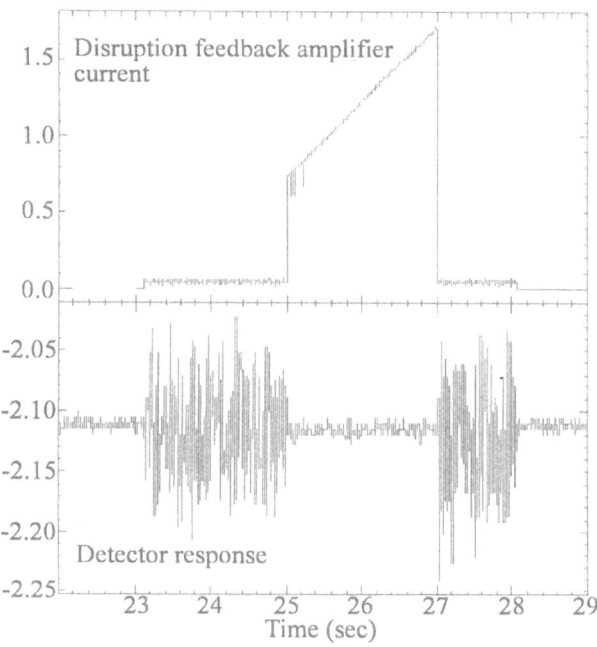

Figure 9. Pickup from disruption feedback amplifier (35521).

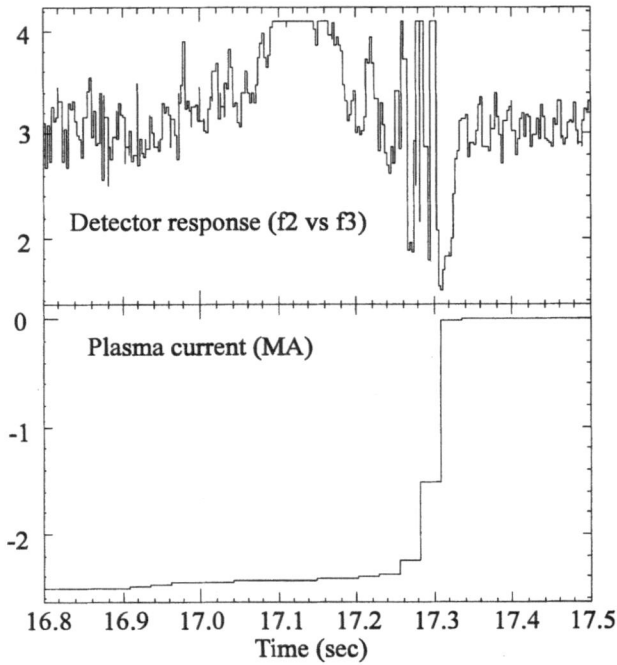

Figure 10. A disruption precursor (35762)

REFERENCES

[1] S.Zweben. Observations of alpha particles during D-T experiments on TFTR, at *Workshop on DT Experiments*, Princeton NJ (March 2-4 1994).

[2] F.E.Cecil, P.van Belle, O.N.Jarvis, G.Sadler. Proc. 21st EPS Conference on Controlled Fusion and Plasma Physics. *Europhysics Conference Abstracts*, Vol 18B (III) 1340.

ITER ALPHA PARTICLE DIAGNOSTICS
USING KNOCK-ON ION TAILS

R.K. Fisher,[1] C.W. Barnes,[2] A. Gondhalekar,[3] P.B. Parks,[1]
J.M. McChesney,[1] A.L. Roquemore,[4] and M.N. Rosenbluth[5]

[1] General Atomics, San Diego, California, U.S.A.
[2] Los Alamos National Laboratory, Los Alamos, New Mexico, U.S.A.
[3] JET Joint Undertaking, Abingdon, Oxfordshire, U.K.
[4] Princeton Plasma Physics Laboratory, Princeton, New Jersey, U.S.A.
[5] Also at University of California, San Diego, California, U.S.A.

INTRODUCTION

Alpha particles will play a critical role in the physics and successful operation of ITER. Achieving fusion ignition requires that the α-particles created by deuterium-tritium (D–T) reactions deposit a large fraction of their energy in the reacting plasma before they are lost. Toroidal field ripple can localize any alpha particle losses and cause first wall damage. Obtaining information on the α-particles in ITER is a high priority diagnostic goal.

We have proposed a new method of measuring the fast confined α-particle distribution in a reacting plasma.[1] The same elastic collisions that transfer the alpha energy to the D–T plasma ions and allow fusion ignition will also create a high energy tail on the deuterium and tritium ion energy distributions.[2,3] Some of these energetic tail ions will undergo fusion reactions with the background plasma producing neutrons whose energy is increased significantly above 14 MeV due to the kinetic energy of the reacting ions. Measurement of this high energy tail on the D–T neutron distribution as a function of plasma minor radius would provide information on the alpha density profile with a time response equal to the ion slowing-down time ($\sim$ few seconds).

Although this technique may provide only limited information on the α-particle energy distribution, experimental studies of fast ions on existing tokamaks have shown that the observed slowing-down is essentially classical. Hence the α-energy distribution is expected to be classical except in situations where the α-confinement is poor. The confinement of α's can be affected by ripple losses and a number of instabilities.[4] Toroidal field ripple can cause both prompt orbit losses and stochastic ripple diffusion losses. Magnetohydrodynamic (MHD) activity, including fishbone instabilities, toroidal Alfvén eigenmodes (TAE), and sawtooth oscillations, may also affect alpha confinement. The diagnostic proposed here, by monitoring the confined

Diagnostics for Experimental Thermonuclear Fusion Reactors
Edited by P. E. Stott *et al.*, Plenum Press, New York, 1996

485

alpha population, can provide valuable information on the confinement of fast alphas in a reacting plasma. Measurement of the confined alpha density profile will be especially important in ITER where it is needed to calculate the alpha heating profile. Studies of alpha heating will be an important objective of the ITER experiment.

SIZE OF THE KNOCK-ON TAILS

The first step is to calculate the size of the energetic plasma ion tail populations due to knock-on collision with alphas. Our results are reported in Ref. 1. Results of similar calculations are reported in Ref. 5. Using the classical slowing-down alpha distribution under ITER Conceptual Design Activity (CDA) conditions $T_e = T_i = 17$ keV, $n_e = 2\,n_D = 2\,n_T = 8 \cdot 10^{13}$ cm^{-3}, we find the ion tail spectra given in Fig. 1. This figure shows the fractional tritium ion density per MeV versus the ion energy. A very similar result is obtained for the deuterium ion tail due to knock-on collisions with alphas.[1] Figure 1 shows two curves, the result with only Coulomb scattering included and the result with nuclear plus interference forces included. Nuclear forces are important for these small impact parameter knock-on collisions. The alpha knock-on tail breaks out of the Maxwellian ion tail at ion energies above approximately 200 keV. At this energy the fractional tritium ion density $n_t^{-1} dn_t/dE \sim 10^{-3}$, falling relatively slowly to 10^{-4} at ~ 800 keV and 10^{-5} at ~ 2 MeV. Also shown in Fig. 1 is the classical slowing-down distribution of 1 MeV tritons from D–D fusion reactions occurring in this D–T plasma. Fortunately, the alpha-induced tail is larger than the D–D triton population by over a factor of five, so that measurements reflecting the ion tail can be used to provide information on the alphas.

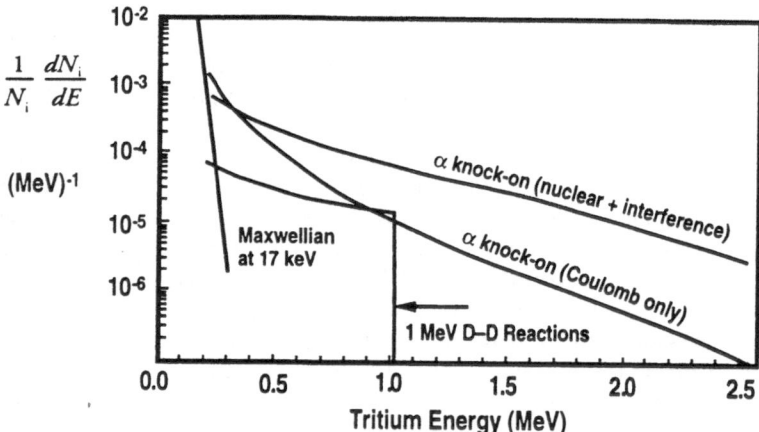

Figure 1. Calculated ion energy distributions due to alpha knock-on collisions under ITER conditions. $T_e = T_i = 17$ keV, $n_e = 2\,n_D = 2\,n_T = 8 \cdot 10^{13}$ cm^{-3}. Figure shows the fractional tritium ion density in units of (MeV)$^{-1}$. The effects of the knock-on collisions with Coulomb forces only and with nuclear plus interference forces are shown. Also shown is the classical 1 MeV triton distribution from D–D reactions.

Next we calculate the D–T neutron energy spectrum resulting from these energetic knock-on tail ions undergoing fusion reactions with their much more populous thermal counterparts. Because of their higher mass, the tritium tail ions contribute significantly more to the energetic neutron tail than do the deuterium tail ions.[1] The neutron energy resulting from a triton of energy E_T colliding with a thermal D ion $(E_D \ll E_T)$ is given by $E_n = \langle E_n \rangle + \Delta E_n \cos\theta$ where θ is the CM-frame angle of emission of the neutron with respect to the incident triton direction and

$$\langle E_n \rangle = \frac{Q}{1 + m_n/m_\alpha} + a\,E_T \quad , \tag{1}$$

$$\Delta E_n = b\left(Q\,E_T + d\,E_T^2\right)^{1/2} \quad , \tag{2}$$

where

$$a = \frac{1}{(1 + m_n/m_\alpha)(1 + m_T/m_D)} + \frac{m_n\,m_T}{(m_T + m_D)^2} \quad , \tag{3}$$

$$b = 2\left[\frac{m_T\,m_n}{(m_D + m_T)^2(1 + m_n/m_\alpha)}\right]^{1/2} \quad ,$$

$$d = (1 + m_T/m_D)^{-1} \quad .$$

This result comes from conservation of energy and momentum. The Q for the D–T reaction is 17.6 MeV so that

$$E_n = 14.1 \text{ MeV} + 0.44\,E_T + 0.62\left[E_T\left(17.6 \text{ MeV} + 0.4\,E_T\right)\right]^{1/2} \cos\theta \quad . \tag{4}$$

Since a tritium ion can receive up to 3.4 MeV in a single collision with a 3.5 MeV alpha, Eq. (4) indicates neutrons with energies up to 20.6 MeV will result from α-particle collision-induced tail ions.

Figure 2 shows the D–T neutron energy spectra calculated using Eq. (4) and the tritium ion distribution of Fig. 1 reacting with the bulk deuterium ions.[1] The ion distributions were assumed to be isotropic in θ. The vertical axis is the fractional neutron distribution $N_{DT}^{-1}\,dN_{DT}/dE_n$ in units of $(\text{MeV})^{-1}$, while the horizontal axis is neutron energy. The neutron tail is smaller than the ion tail due to the drop in the D–T fusion cross-section with tritium ion energy above $E_T \sim 180$ keV.

Other possible sources of high energy neutrons in a fusion plasma include thermal broadening of the D–T neutron emission and reactions involving fast ions from neutral beam or ICRF heating. Alpha knock-on effects begin to dominate the thermal broadening of the D–T neutron spectrum above neutron energies of ~ 15.5 MeV. Beam-beam reactions will not produce neutrons above 16.0 MeV at beam energies of up to 140 keV so will not be a problem on TFTR or JET. Beam-target neutrons will also not be a problem at these beam energies according to our calculations. If ITER uses neutral beams for heating or current drive at energies ~ 0.5 to 1 MeV, the knock-on neutron technique would remain useful during the latter stages of the heating phase of ITER and during the ignited burn phase, after the beams have been turned off. ICRF heating can easily produce energetic tails on minority ion distributions but will not produce a significant energetic tail on the majority deuterium or tritium ion distributions unless the plasma density is low ($\sim$ a few 10^{13} cm^{-3}). The question of

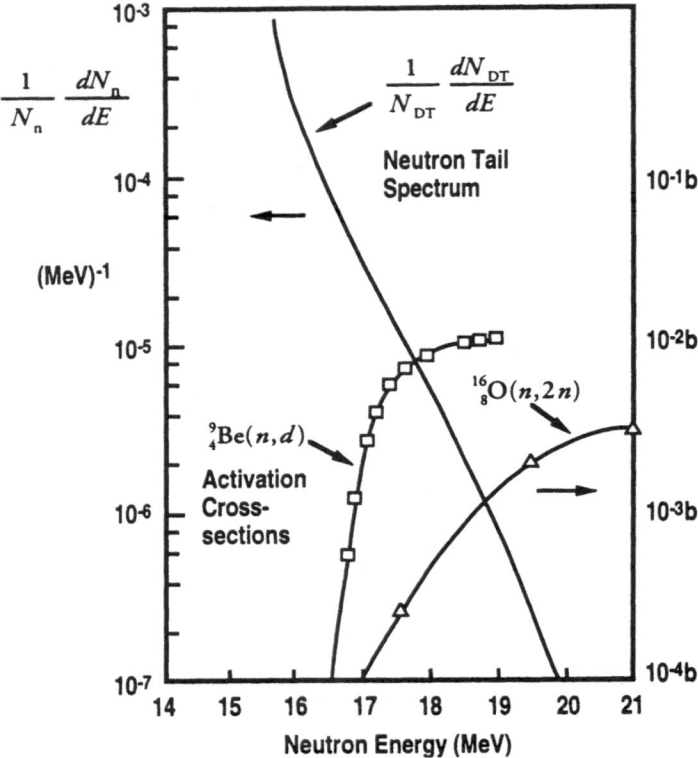

Figure 2. Calculated D–T neutron energy spectrum for the tritium ion distribution of Fig. 1 reacting with thermal deuterium ions. Also shown are the neutron activation cross-sections for the $^9\mathrm{Be}(n,d)$ and $^{16}\mathrm{O}(n,2n)$ reactions.

whether ICRF heating will enhance the tail created by the alpha knock-on collisions needs to be examined. We expect the alpha-knock-on effects to dominate the neutron tail production above 16 MeV under most plasma conditions on TFTR, JET, and ITER.

NEUTRON THRESHOLD ACTIVATION DETECTORS

One of the most straightforward and attractive methods of measuring the D–T neutron tail is activation detectors. This method utilizes nuclear reactions which require that the incident neutron energy be above a threshold energy to produce the reaction. A very attractive candidate reaction is:

$$n + {}^{16}_{8}\mathrm{O} \rightarrow {}^{15}_{8}\mathrm{O} + 2n \quad ,$$

with a threshold energy of 16.7 MeV. Figure 2 also shows the $^{16}\mathrm{O}(n,2n)$ cross-section[6] as a function of neutron energy. The $^{15}\mathrm{O}$ end product subsequently decays to $^{15}\mathrm{N}$ via positron emission with a half-life of 2.1 min.

By exposing an activation target containing oxygen to the D–T neutron flux from a reacting plasma, only those neutrons with energies above 16.7 MeV can excite the $^{16}O(n, 2n)$ reaction. By only exposing the activation target for a few seconds, we can monitor the alpha-induced tail with a time response limited only by the energetic deuterium and tritium ion slowing-down times ($\gtrsim 1$ sec in these plasmas). The target is then transported to a radiation-quiet area where the ^{15}O decay gammas are measured. The positron emitted by ^{15}O will annihilate with an electron inside the activation target producing two coincident 511 keV gamma-rays emitted in opposite directions. Hence the activation target would be placed inside a split well detector which would require both gamma-rays to be detected in coincidence. Measuring the 511 keV coincident gamma-rays that were observed to decay in a 2.1 min half-life would ensure that we were monitoring the desired $^{16}O(n, 2n)$ activations, and allow discrimination against other background sources of gamma-rays. Background gammas from $^{16}O(n, p)$ reactions excited by the much larger flux of 14 MeV neutrons would be a problem if not for the 7.4 sec half-life of the resulting ^{16}N. The $^{16}O(n, 2n)$ reactions resulting from the neutron tail will dominate the decay gamma spectrum 4 min after the exposure to the tokamak neutron flux. It will also be necessary to load the transport capsule used to transfer the oxygen activation target in a helium or other non-air atmosphere. Otherwise the 14 MeV neutrons will produce $^{14}N(n, 2n)$ reactions with the air trapped inside the capsule. The resulting β^+ decays of ^{13}N, which has a 10 minute half-life, would dominate over the $^{16}O(n, 2n)$ signal reactions from the knock-on neutron tail. Preliminary tests of this approach are underway on TFTR.

Using the calculated neutron spectrum and the activation cross-section in Fig. 2, we have calculated the expected signal levels for various tokamaks. The existing activation target transfer systems on TFTR and JET restrict the target volume to approximately 20 cm. This size target exposed for 1 sec during D–T operation of TFTR or JET at $4 \cdot 10^{18}$ n/sec would produce about 9,000 total $^{16}O(n, 2n)$ activations. Hence the detected coincident gamma-ray signal rate would be approximately 50/sec $E^{-t/182\,sec}$ for a high-efficiency well detector. ITER at a neutron output of $4 \cdot 10^{20}$ n/sec would produce $2 \cdot 10^4$ total activations after a 1 sec exposure yielding a detected gamma signal rate of $1.3 \cdot 10^5$/sec $e^{-t/182\,sec}$ for a 1000 cm^3 target.

Using other target elements with different neutron energy thresholds could provide a cross-check on the results and also possibly provide some information on the α-particle energy distribution. A classical slowing-down alpha distribution was assumed in the results of Figs. 1 and 2. For example, the reaction with beryllium

$$^9_4\text{Be} + n \rightarrow {}^8_3\text{Li} + d \ ,$$

has a threshold energy of 16.3 MeV and a cross-section[6] about ten times larger than the $^{16}O(n, 2n)$ cross-section as shown in Fig. 2. The resultant ^8Li decays via β^- decay with a half-life of 0.84 sec. The short half-life means that the target transfer must be done as quickly as possible. The existing JET transfer system requires about 8 sec but it may be possible to reduce this.[7] A Cerenkov detector is attractive because it only produces a signal if the decaying β^- has a velocity greater than the velocity of light in the Cerenkov medium. Even high purity Be contains impurities such as boron, carbon, and oxygen which can be activated by the much larger flux of 14 MeV neutrons and also result in β^- decays. Fortunately, we have calculated that the purest forms of Be should be useable if the Cerenkov detector is designed to detect only β^-'s with energies above 10.4 MeV. The Cerenkov β^- energy threshold is controlled by the index of refraction in the chamber, which is determined by the gas pressure. Under ITER conditions, the calculated $^9Be(n, d)$ activation using Fig. 2 results in $5 \cdot 10^5$/sec decay β^-'s for a 20 cm^3 Be target exposed for 1 sec. Since the half-life is only 0.8 sec the activation rate quickly saturates at this level and then tracks the knock-on tail population with ~ 1 sec time resolution. Under JET and TFTR conditions, the corresponding signal size is $6 \cdot 10^4$/sec. The signal drops to ~ 70/sec after the 8 sec transfer time required by the existing JET transfer system.

By installing a number of threshold activation detectors in a collimated neutron detector array, it may be possible to monitor the neutron tail size as a function of minor radius in the plasma. The neutron tail is linearly proportional to the local alpha density in the region viewed by the detector. Using the profiles of electron temperature and density measured by other diagnostics, we can then calculate the local alpha density producing the knock-on tail. Since the neutron emission profile is highly peaked, the neutron tail measured by each detector in the array will reflect the alpha density at the point of highest neutron emission in its view, i.e., the point where the detector sightline is tangent to the plasma flux surface.

NEUTRON SPECTROSCOPY TECHNIQUES

We have examined a number of other methods to observe the energetic D–T neutron tail due to alpha knock-on collisions. A magnetic proton recoil (MPR) spectrometer similar to that discussed by J. Källne et al.[8] has promise for measurement of the D–T neutron tail. Ideally it would be designed to maximize the small neutron tail signal by increasing the detection efficiency at the expense of reducing the energy resolution. Hence it might differ somewhat from an MPR system optimized for ITER ion temperature and D/T fuel density ratio measurements. This technique requires that the detector background due to neutrons and gamma rays from ITER be minimized using a combination of shielding and detectors with very low neutron and gamma ray sensitivity.

Another technique we are studying is neutron time-of-flight spectroscopy. Time-of-flight spectroscopy has been successfully used on JET to measure ion temperature. Figure 3 shows a high count rate time-of-fight spectrometer proposed for ITER by T. Elevant et al.[9]. The collimated incident DT neutron flux interacts with a thin (0.3 mm) polyethylene scattering foil oriented along the neutron path. Neutrons scattered at 45 deg from probes in the foil are detected in the lower set of fast scintillator detectors D_2 which are approximately 5 m from the foil. The start signal is obtained by detecting the foil protons which are scattered upward at 45 deg into a linear array of fast scintillation detectors. Both the scattered neutrons and protons carry half the incident neutron energy, because of the 45 deg scattering angle. The projected overall neutron signal rate is $\dot{N}_{DT} \sim 10^6$/sec at an incident neutron flux of 10^9/cm^2-sec. Using the results in Fig. 2, we can expect

$$\frac{dN_{\text{tail}}}{dE} \sim 10^{-4}/\text{MeV} \, (10^6/\text{sec}) \sim 10^2/\text{MeV-sec} \quad ,$$

counts in the neutron tail at $E_n \sim 16.5$ MeV.

The expected time-of-flight spectrum is schematically shown in Fig. 4. The ion temperature broadened DT neutron peak appears above the flat random coincidence spectrum. On the high energy (short flight time) side of the spectrum, the knock-on tail neutrons appear above the flat random coincidence background when

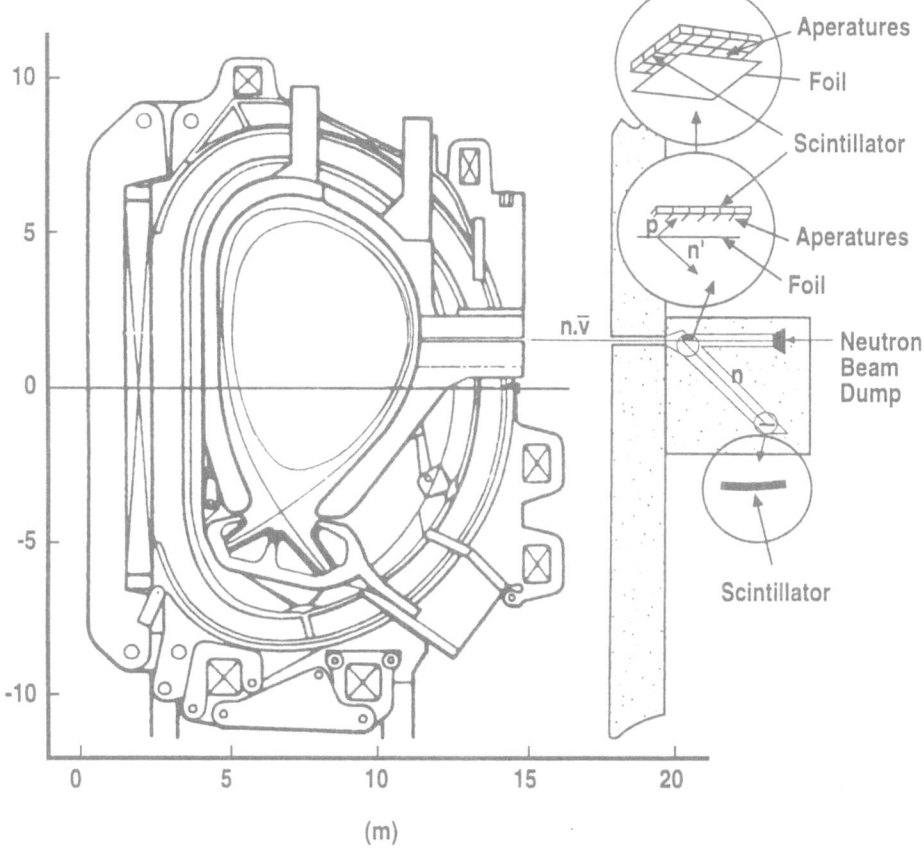

Figure 3. High count rate time-of-flight neutron spectrometer for ITER proposed by T. Elevant, *Rev. Sci. Instrum.* 66:881 (1995).

$$N_{\text{tail}} = \dot{N}_{\text{tail}} \, \Delta t \gg \left(\dot{N}_1 \, \dot{N}_2 \, \tau \, \Delta t \right)^{1/2} \quad , \tag{5}$$

where $\dot{N}_{\text{tail}}$ is the knock-on signal rate, Δt is the data collection time, $\dot{N}_1$ and $\dot{N}_2$ are the random (background) counting rates in the start and stop detectors, and τ is the time window ~ 1 nsec. Hence it is very important to minimize the background counting rates, $\dot{N}_1$ and $\dot{N}_2$. The intrinsic backgrounds due to neutron scattering from the scattering foil will require integrating the data for ~ 30 sec to observe the knock-on tails.

The time-of-flight spectrometer described in Ref. 9 is designed for ion temperature measurements. It should be possible to sacrifice some energy resolution and reduce the required data collection time for knock-on tail measurements. For example, if the neutron detectors are moved closer to the scattering foil, both the signal rate $\dot{N}_{\text{tail}}$ and the intrinsic background rate $\dot{N}_2$ should increase. This should result in a reduction in the data collection time necessary for measuring the knock-on tail. Trade-offs are

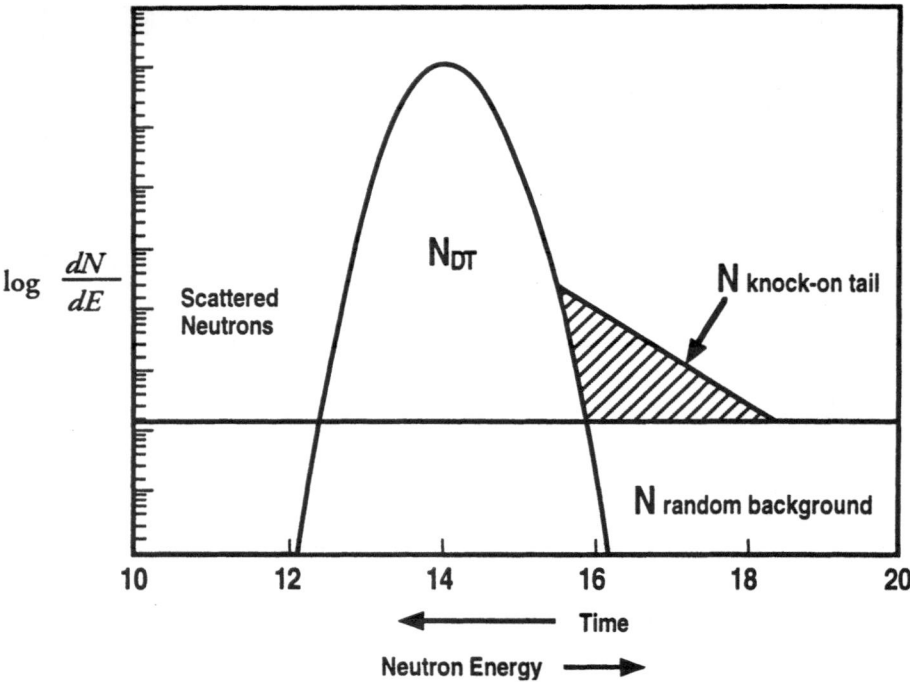

Figure 4. Expected time-of-flight neutron spectrum showing knock-on tail and "flat" background due to random coincidences.

also possible in the neutron scattering angle, proton detector distance and angular collimation, and foil thickness and composition.

Other neutron spectroscopy techniques under consideration as a result of this workshop include scintillating fiber neutron spectrometers[10] and bubble chamber threshold detectors.[11]

OBSERVING ION TAILS DIRECTLY

In addition to measuring the D–T neutron tail due to alpha knock-on collisions, it may be possible to directly measure the knock-on plasma ion tails. Experiments on JET have shown that energetic hydrogen neutrals are created when ICRF heating generates H–ion tails and these MeV tail ions subsequently charge exchange with single electron impurity ions Be^{3+} and C^{5+} in the center of JET.[12] We have observed energetic H neutrals during similar experiments on TFTR, and have also observed tritium neutrals from 1 MeV tritons produced in D–D reactions in TFTR. JET experiments have also resulted in observations of high energy tritium neutrals from 1 MeV tritons interacting with single electron impurity ions during D–D experiments. Since the predicted D and T energetic ion tails due to alpha knock-on collisions are somewhat larger than these 1 MeV triton populations, we can expect to observe measurable high energy neutral signals due to alpha knock-on collisions during D–T operation of JET. This assumes we add neutron shielding to the JET neutral particle analyzer to keep the D–T neutron background near present D–D levels. This should be possible since we have achieved a reduction of more than a factor of 100 in D–T neutron background by adding neutron shielding to a nearly identical Ioffe analyzer on TFTR.

Hence it may also be possible to observe the D– and T–ion tails due to alpha collisions as they also can pick up a single electron from impurity ions. Calculations show this technique should produce neutral signals on JET comparable to those produced for alpha observations while not requiring a helium diagnostic beam. Reionization of the newly created tritium neutrals as they exit the plasma will require correction of the measured spectra. For example, approximately 50% of the 1 MeV tritons created at $r/a \sim 0.5$ will escape to the edge of ITER. Tests of the direct ion tail measurements using a high energy neutral particle analyzer are planned on JET in collaboration with JET scientists.

CONCLUSIONS

Measurement of knock-on tails should provide valuable information on alpha particles in ITER. Direct measurements of the energetic ion tails would provide information on both the alpha population and energy spectrum, and could be done using the same high energy neutral particle analyzers proposed for alpha charge exchange measurements using neutral beams or pellets.

Measurements of the energetic neutron tail on the DT neutron emission from ITER offers a very attractive approach to alpha particle diagnostics. It is non-perturbing to the plasma, relatively low cost, and requires no vacuum penetrations. A collimated array of neutron threshold activation detectors can be used to provide information on the alpha density profile. Time-of-flight and/or magnetic proton recoil neutron spectroscopy should provide information on the alpha energy spectrum as well as the alpha population in ITER.

It would be highly desirable to test these approaches during D–T operation of TFTR or JET. Preliminary tests of the oxygen neutron threshold activation system are planned for TFTR, while tests of the direct ion tail measurements using charge exchange neutrals are proposed on JET.

ACKNOWLEDGMENTS

We would like to thank S.S. Medley, M. Petrov, and H.H. Duong for their efforts on the observations of 1 MeV tritons and MeV hydrogen ICRF tail ions in TFTR.

This is a report of research and work sponsored by the U.S. Department of Energy under Grant No. DE-FG03-92ER54150 and Contract No. DE-AC03-76CH03073.

REFERENCES

1. R.K. Fisher et al., Fast alpha particle diagnostics using knock-on ion tails, *Nucl. Fusion* 34:1291 (1994).
2. P.B. Parks et al., private communication (General Atomics, 1991).

3. D. Ryutov, Energetic ion propulsion formed in close collision with fusion alpha-particles, *Physica Scripta* 45:153 (1992); P. Helander *et al.*, Formation of hot ion population in fusion plasmas by close collisions with fast particles, *Plasma Phys. and Contr. Fusion* 35:363 (1993). J. Källne *et al.*, Alpha particle information from neutron observations, *Fusion Technol.* 25:341 (1994).

4. W.W. Heidbrink and G.J. Sadler, The behavior of fast ions in tokamak experiments, *Nucl. Fusion* 34:535 (1994).

5. G. Gorini *et al.*, Alpha particle kinetic effects in the neutral emission of burning DT plasmas, *Rev. Sci. Instrum.* 66:936 (1995).

6. Victoria McLane *et al.*, *Neutron Cross Sections*, Academic Press, San Diego (1988), Vol. 2, BNL-325.

7. O.N. Jarvis, private communication (JET, 1994).

8. J. Källne, Neutron spectroscopy for TFTR, talk I14 of this meeting; J. Källne *et al.*, On the possibility of neutron spectrometry for determination of fuel in densities in DT plasmas, *Rev. Sci. Instrum.* 62:2871 (1991); see also Ref. 5.

9. T. Elevant, A neutron spectrometer for ITER, *Rev. Sci. Instrum.* 66:881 (1995); also Ion temperature profile measurements in ITER by means of neutron spectroscopy, this meeting.

10. T.. Elevant, Scintillating fiber neutron spectrometer, this meeting.

11. S.V. Trusillo, Neutron image system for neutron diagnostic of ITER plasma, this meeting.

12. A. Gondhalekar *et al.*, Measurements of MeV energy ICRF driven minority ions and D–D fusion protons in JET using neutral particle analysis, in *Proc. IAEA Technical Committee Meeting on Alpha Particles in Fusion Research, Trieste* 1993; Ya.I. Kolesnichenko and G.J. Sadler, Summary of IAEA technical committee meeting on alpha particles in fusion research, Trieste, *Nucl. Fusion* 33:1912 (1993).

CHARGE-EXCHANGE (C-X) DIAGNOSTICS OF FAST CONFINED ALPHAS: PRESENT SITUATION AND PROSPECTS FOR ITER

M.P.Petrov[1], R.K.Fisher[2]

[1]A.F.Ioffe Physical-Technical Institute, St.Petersburg , Russia.
[2]General Atomics, San-Diego, CA, USA

INTRODUCTION

The effective operation of ITER requires that the alpha particles generated in DT fusion reactions be well confined to allow deposition of most alpha energy in the plasma and not to cause first wall damage. So the radially resolved measurements of density and energy spectra of fast confined alphas are very important.

There are very few diagnostics which can be used to study fast confined alphas in a tokamak plasma. These include: a) Charge-Exchange (C-X) alpha particle diagnostic based on the neutralization of fast confined alphas in the plasma and subsequent analysis of escaping helium atoms using high energy Neutral Particle Analyzers (NPA), b) Charge-Exchange Recombination Spectroscopy (CHERS) based on measurements of profiles of He^+ spectral lines excited by single C-X reactions of alphas with injected hydrogen or helium atoms and c) Collective scattering (CS) of microwave radiation from confined alphas. These methods are in preparation (CS) or in use (C-X and CHERS) on large tokamaks for the studies of DT alphas, other fusion products or RF driven MeV minority ions. In this paper we describe the status and prospects for the use of C-X alpha diagnostics on ITER.

At the moment two methods of active neutralization of alphas are in use. First is the neutralization of fast alphas by double C-X reactions with helium atoms of a 50 - 100 keV/amu diagnostic beam injected into the plasma. This method is now in successful use at JET for the studies of RF driven minority hydrogen and helium MeV ions [1,2]. The second method of alpha neutralization is based on the use of pellet ablation clouds. A fraction of alphas incident on the cloud is neutralized by sequential single or by double electron capture [3]. The Pellet Charge-Exchange (PCX) alpha diagnostic is successfully being used now in DT experiments on TFTR. PCX using lithium and boron pellets gives radially resolved DT alpha energy spectra in the range of 0.5 - 3.5 MeV and alpha density radial profiles.

Diagnostics for Experimental Thermonuclear Fusion Reactors
Edited by P. E. Stott *et al.*, Plenum Press, New York, 1996

495

Recently the possibility of passive neutralization of fast confined alphas in DT plasma due to double electron capture from helium-like beryllium and carbon ions and also atoms of helium ash has been estimated [4]. Calculations show that this possibility may be useful for ITER.

Below we will describe the existing MeV range Neutral Particle Analyzers and estimate their compatibility for ITER. We will also briefly discuss the methods for alpha neutralization mentioned above, will present the existing experimental results of such diagnostics on JET and TFTR and discuss the prospects of these alpha diagnostic methods for ITER.

NEUTRAL PARTICLE ANALYZERS FOR MEV RANGE

The Neutral Particle Analyzers of GEMMA-2 type for 0.5 - 4 MeV range [5] relevant for C-X alpha diagnostic were developed at the Ioffe Institute, St.Petersburg, Russia and have been delivered in 1991-1994 to JET, TFTR and JT-60U for studies of RF driven minority ions and fusion alpha particles. The NPA GEMMA-2 is based on the stripping of MeV atoms emitted by plasma in thin diamond (carbon) foil (200 A). The secondary ions are deflected in the analyzing magnetic field by 90°. The maximum magnetic field necessary to deflect 4.0 MeV alphas to the most energetic NPA channel is equal 1.1T. After passing the magnet a parallel ion beam with momentum dispersion is formed. This ion beam passes through the electrostatic condenser. The trapezoidal condenser deflects ions with the same Z/E value by a distance h=5 cm from the middle plane of the instrument where eight detectors are located. So the NPA can measure the energy spectrum of mass-selected H,D,T and He atoms simultaneously at eight different energies. The maximum electric field applied to the analyzing condenser is equal +/- 20 kV. The detectors are phototubes attached to 3 cm^2 very thin CsI(Tl) scintillators (8 microns), covered by Al (500A) which gives nearly 100% detection efficiency for the high energy ions and very small sensitivity for background neutrons and gammas. For reducing the neutron sensitivity of the detection system special materials having very low cross-sections of n-p and n-alpha reactions are used in the vicinity of the detectors. Tests made with neutron radioactive sources and performed on JET in the presence of DD and DT neutrons showed that the contrast in detection efficiency between secondary ions and neutrons is less than 10^{-7} for ion energies E>0.5 MeV. The NPA's were absolutely calibrated at the Ioffe Institute with the use of cyclotron accelerator. Overall detection efficiency of He and H atoms versus energy was measured during the calibration. For He it varies from 0.07 to 0.2 in the range 0.5-4 MeV. For H atoms it is constant in the same range and equal to 0.17. Energy widths of the NPA channels were also measured in the calibration and vary from 9% for the low energy channel to 6% for the high energy channel.

Use of NPA GEMMA-2 on ITER is defined by very low sensitivity of the detectors to background neutrons and gammas. According to [6] we can expect the total neutron flux $F_n=10^8 cm^{-2}s^{-1}$ at the possible NPA detector position on ITER. It means that without any additional shielding we can expect a background neutron counting rate <30 counts per s which will allow us to measure reliably the alpha counting rates of 100 counts per s and less.

ALPHA-BEAM C-X DIAGNOSTIC ON JET

Recently the cross-section of double electron capture by alpha particle from helium atoms was measured in Ioffe Institute with the use of cyclotron accelerator in the energy range of alphas 0.5 - 4 MeV [7]. Calculations performed at the Ioffe Institute on the basis of this cross-section showed that the detection of neutralized DT alphas and RF driven hydrogen atoms in the MeV range is possible on JET using JET heating beams operated in helium and deuterium.

In 1991, NPA GEMMA-2 was delivered from the Ioffe Institute to JET and was installed at the top of the torus with its vertical line of sight intersecting the injected heating neutral beams at the plasma center. After installation and commissioning, the NPA was successfully used on JET for studies of RF driven H-minority ions in MeV range. As a result of the very first experiments, the presence of a rather strong passive flux of MeV H atoms during ICRF heating was discovered [1]. The origin of this passive H^0 flux was explained as a result of electron capture by H^+ minority ions from hydrogen-like plasma impurity ions (C^{+5} and Be^{+3}). Another objective of the MeV neutral particle analysis experiments on JET was checking the capability of the alpha-beam C-X diagnostic by observing $^3He^{+2}$ RF driven minority ions neutralized on a $^4He^0$ beam. These experiments were made and the energy spectrum of $^3He^{+2}$ minority ions in a ^4He plasma with $^3He^{+2}$ RF driven minority and $^4He^0$ injected beam was measured [1]. The main parameters of the plasma were $n_e=3.10^{19}$ m^{-3}, $T_e(0)=4.6$ keV, $P_{ICRF}=1.8$ MW, $P_{BEAM}=2.5$ MW, $E_{BEAM}=120$ keV. The measured energy spectrum was derived from the detected He^0 flux using the cross-section for double electron capture by the reaction $He^{+2} + He^0 = He^0 + He^{+2}$ mentioned above [7]. This was the preliminary demonstration of the capability for measurements of double C-X fast helium ions using a helium atomic diagnostic beam. Final tests of this diagnostic can be expected during future DT experiments on JET.

ALPHA-PELLET C-X DIAGNOSTIC ON TFTR

The PCX alpha diagnostic was developed in a collaboration between General Atomics, the Ioffe Institute and the Princeton Plasma Physics Laboratory in 1992-94 and is now operated routinely on TFTR during DT experiments [8-10]. The PCX diagnostic on TFTR uses Lithium and Boron pellets of 1.7 mm diameter by 3 mm length injected at velocities 500-700 m/s (Li) and 300-400 m/s (B). The NPA views the pellet from behind with a sight line at a toroidal angle of 2.75^0 to the radially injected pellet trajectory. Consequently, only near perpendicular alphas are detected in these experiments. Upon entering the plasma, the pellet forms a toroidally elongated ablation cloud which travels with the pellet into the plasma. A small fraction of the alphas incident on the cloud is neutralized either by sequential single electron capture from [H]-like ions or by double electron capture from [He]-like ions of the cloud. The escaping helium neutrals are analyzed by the NPA GEMMA-2. By measuring the energy distribution of helium neutrals dn_0/dE the energy distribution of the incident alphas, dn_a/dE, can be determined as $dn_a/dE = dn_0/dE$ $[F_0(E)]^{-1}$ where $F_0(E)$ is the equilibrium fraction of incident alphas neutralized in the cloud as a function of alpha energy. The values of $F_0(E)$ are obtained from modeling calculations [8].

The radial position of the pellet as a function of time is measured using a linear photodiode array situated on the top of tokamak vacuum vessel. By combining this measurement with the time dependence of observed helium neutral signal, radially resolved alpha energy spectra and alpha density radial profiles can be derived with a radial resolution of ~ 5 cm.

Pellets were injected 0.2 to 0.5 s after the neutral beam heating was turned off. This timing delay leads to deeper penetration of the pellet as a result of post-beam decay of the electron temperature and plasma density.

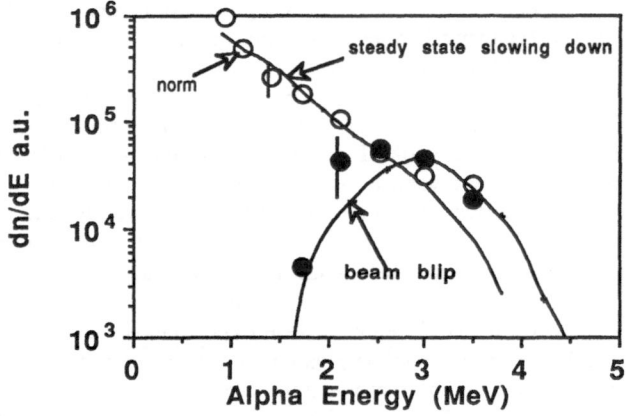

Figure 1. PCX alpha energy spectra for steady state slowing down and "beam blip" case on TFTR. Solid lines are TRANSP-FPP modelings.

Fig 1 presents PCX alpha energy spectra in the plasma core obtained with the use of Boron pellets during the birth ("beam blip") and steady-state slowing down cases [10]. For the slowing down case (I=1.5 MA, P_{NB}=15 MW) alpha energy spectrum was obtained using a pellet 0.2 s after termination of a 1.0 s beam pulse. For "beam blip" case (I=1.5MA, P_{NB}=20 MW) the pellet was injected 20 ms after a 0.1s beam pulse.

In this figure the PCX data are compared with the TRANSP Monte Carlo Code modeling[11] and Fokker-Planck Post Processor Code (FPP) [9]. TRANSP-FPP follows the orbits of alphas as they slow down and pitch angle scatter by Coulomb collisions, includes the influence of toroidal field ripple on the alpha orbits and takes into account the spatial and temporal

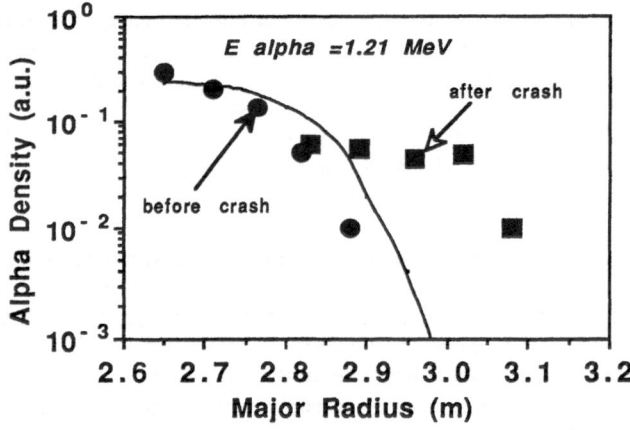

Figure 2. PCX alpha density radial profiles before and after the sawtooth crash. Solid line - TRANSP-FPP modeling result for the case before crash.

distributions of background plasma parameters for each particular shot. The normalization of PCX data with TRANSP-FPP modeling was made only once as noted in the Fig.1. It is seen that the shape of PCX steady state alpha slowing down spectrum is in good agreement with the TRANSP-FPP simulation. This result confirms the expectation that fusion generated alphas are well confined and slow down classically in the core of the TFTR DT plasma. Reasonable agreement is seen also between PCX data and TRANSP-FPP Code in the case of short "beam-blip" experiment. Note that code results include Doppler broadening of the alpha birth energy distribution, given by $dE(keV)=182(T_{DT\ eff})^{0.5}$ where $T_{DT\ eff}=33$ keV is effective temperature of deuterium and tritium ions.

The data presented in Fig.1 were obtained in the plasma core of quiescent TFTR plasmas. PCX data also show the effects of magnetic field ripple and of large sawtooth oscillations which usually occur after the termination of neutral beam injection on TFTR.

For the studies of sawteeth influence on the alphas, Li pellets were injected before and after the first sawtooth crash which occurs after NB termination. Fig.2 presents PCX alpha density radial profiles before and after the sawtooth crash for alpha energy $E_a= 1.21$ MeV in two sequential shots (I=2 MA, P_{NB}=20MW)[9].

TRANSP-FPP Code results with ripple stochastic diffusion for sawtooth free case are also shown in this figure. It is clearly seen that sawtooth crashes transport alphas outward which can lead to enhanced alpha losses. The control of radial transport of trapped alphas by sawtooth oscillations can be very important for ITER where the attendant alpha energy loss may cause first wall damage.

THE PROSPECTS OF ALPHA-BEAM DIAGNOSTIC ON ITER

Fig.3 presents the TRANSP simulations of alpha density profiles in ITER with and without sawteeth oscillations [12]. For modeling flat density radial profile with $n_e(0)=1.25.10^{20}$ m^{-3}, and ion and electron temperature profile with flat top within r/a= +/- 0.4, $T_e(0)$=22 keV,

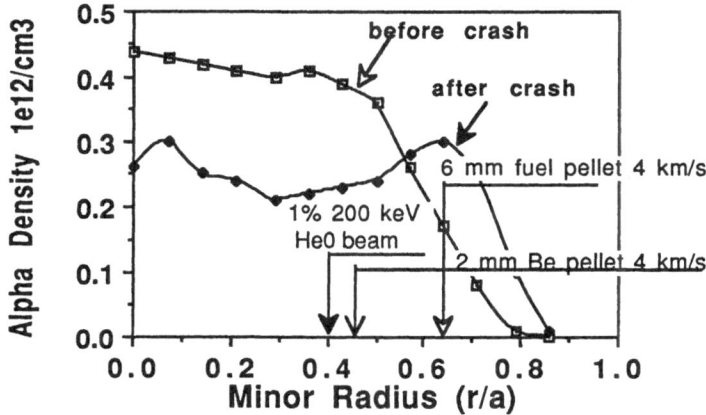

Figure 3. TRANSP simulation of ITER alpha density profiles together with helium diagnostic beam and pellets penetrations into ITER plasma

$T_i(0)=18$ keV, a=2.8 m were used. For the estimations of the relevance of alpha beam diagnostic on ITER the diagnostic He^0 beam injector was supposed to be located in the middle plane of the torus and from three to five NPA's to be installed at the top. The beam parameters are: energy 200, 400 keV, injected power 2.8 MW, diameter of the beam in the plasma 40 cm. Each NPA has geometrical factor $3.7.10^{-4}$ cm^{2}str and sees a 40 x 40 cm area in the plasma center. The penetration of the beams into ITER plasma (the most critical issue for this diagnostic) for r =0.5 a is 0.03 and 0.15. The transparency of the plasma for escaping neutralized alphas varies from 0.1 to 0.6 in the energy range 0.5 - 3.5 MeV. Fig.4 shows the predicted active count rates in the NPA channels versus alpha energy for sawtooth free case at r=0.5 a . Calculations were performed by V.I.Afanasiev, A.V.Khudoleev, M.P.Petrov et al. at the Ioffe Institute. It appears that fast alpha signal is rather insensitive to the beam energy in presented range. The counting rate of the alphas is equal to 10^3 - 10^1 counts per sec in the energy range 0.5 - 3 MeV. For the region r=0.4 a the counting rate of the alphas decreases by approximately 2 times. On the same figure the counting rate of expected neutron background signal is also presented.

It is seen from Fig.3 that sawtoothing in ITER can increase C-X alpha-beam diagnostic signal by 4-5 times in the range r/a =0.6 - 0.8. We can conclude that proposed alpha-beam charge-exchange diagnostic allows to measure radially resolved alpha density and energy spectra in ITER plasma within the range r/a= 0.5-1.0, with radial resolution ~40 cm, time resolution 0.1 s, statistical accuracy 10-15%.

THE PROSPECTS OF PASSIVE ALPHA C-X DIAGNOSTIC ON ITER

As we mentioned above the possibility of passive neutralization of fast confined alphas in ITER plasma due to double electron capture from helium-like beryllium and carbon ions and also atoms of helium ash was recently estimated [4]. In [4] the cross-sections for double charge-exchange of alphas with [He]-ions Be^{+2}, C^{+4} and He^0 atoms calculated using the model of the independent electrons (IEM) were presented. It was shown that the cross-sections for Be^{+2}

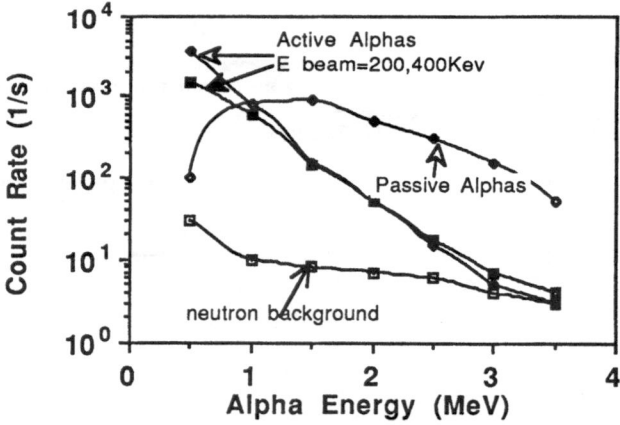

Figure 4. Active and passive alpha count rates in the NPA channels on ITER at r=0.5a.

and C^{+4} are comparatively high and attain 10^{-18} - 10^{-20} cm^2 at E alpha=1- 3.5 MeV. Then the densities of Be^{+2}, C^{+4} and He^0 in ITER plasma were calculated with the use of a model of the impurity ionization equilibrium for different impurity compositions. Radial ion transport in this model was described by empirical confinement time T_z taken to be equal for all species. For calculations the value of T_z= 5 s was used. Calculations show that the density of [He]-ions of Be and C and the density of He atoms appear to be in the range of $(10^{-5} - 10^{-6})$ n_e. This yields a neutralization probability for alphas equal to ~10^{-2} 1/s in the energy range 1 - 2 MeV. The counting rates of the NPA GEMMA-2 channels for the flux in vertical direction corresponding to the alphas neutralized by passive double electron capture are presented on Fig.4 for the following impurity composition: 15% He, 6% Be, Zeff=2.05 together with active count rates. It is seen that passive count rate exceed active signal at the energy range above 1 MeV.

The possibility of measuring the passive neutralized alpha flux seems to be very attractive. The predicted alphas count rate far exceeds the neutron background at Ealpha>0.5 MeV. The most promising diagnostic approach is a multichord measurement of the radial alpha flux. It is very important to check this diagnostic possibility on pre-ITER DT experiments (TFTR and JET).

THE PROSPECTS OF ALPHA-PELLET C-X DIAGNOSTIC ON ITER

Both impurity (beryllium, lithium, boron and carbon) and fueling (tritium and deuterium) pellets are of interest for the PCX alpha diagnostic. Below we present some estimations of penetration of impurity pellets into ITER. The pellet velocity of 4 km/s achieved using two

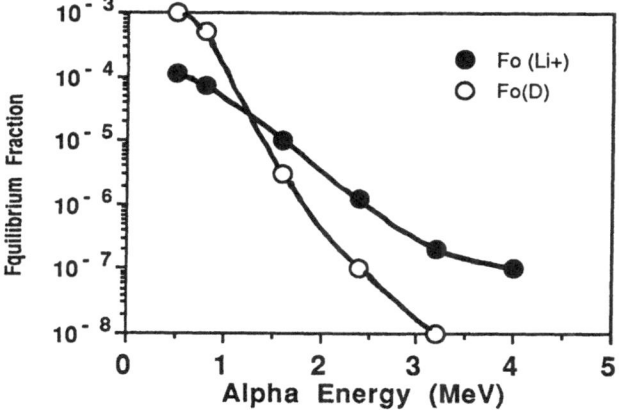

Figure 5. Alpha neutral equilibrium fractions for D and Li pellets.

stage gas gun technology was used for estimations. Even higher pellet velocities may be possible using other injector technologies. The results are based on the pellet ablation model by Parks, et al [13]. The model calculations were made by J.M.McChesney. A 1.2 mm radius beryllium pellet injected at 4 km/sec, which would produce a 1% rise in volume averaged electron density would penetrate about 113 cm into ITER (r/a=0.6). A 2 mm radius beryllium pellet (5% density perturbation) would penetrate about 149 cm (r/a= 0.46, see Fig.3). In both cases this should produce measurable PCX signals and allow significant alpha physics studies.

Although all PCX results to date have been obtained using lithium and boron pellets [8-10], fueling pellets are also of interest. Fig.5 compares the calculated alpha particle neutral equilibrium fractions for a neutral deuteruim or tritium cloud target and singly ionized lithium pellet cloud target. While the D^O and T^O results fall below the Li^+ $F_O(E)$ above 1.2 MeV, the fueling pellet $F^*(E)$ is still large enough to predict measurable alpha signal over much of the energy range shown. The presented estimations of the fuel pellet penetration are based on the ORNL pellet ablation model and the Parks neutral gas shielding model [14]. To penetrate to r/a=0.5 requires a 9 mm radius pellet at 4 km/sec. But this would increase the average density by approximately two times. A 6 mm pellet at 4 km/sec would penetrate ~100 cm (r/a=0.64) with a density perturbation ~25%. This is enough to be interesting for alpha diagnostic.

CONCLUSIONS

1. Active charge-exchange alpha particle diagnostics are in preparation (alpha-He^O beam on JET) or in successful use (alpha- impurity pellets on TFTR).

2. Alpha-He^O beam diagnostic can be used on ITER for r/a>~0.5. It requires the use of 200-400 keV, 2.8 MW He^O diagnostic beam. It can provide the spatial resolution ~40 cm and time resolution around 0.1 s.

3. Alpha-pellet diagnostics can also be used on ITER for r/a>~0.6 -0.7. It requires the use of beryllium or boron pellets at 4 km/sec velocities. Fueling pellets (6 mm, 4 km/sec) can also be used. Further tests of the PCX using both the impurity and fueling pellets   should be looked into on existing large tokamaks (JET and JT-60U). PCX diagnostic can provide radial resolution ~10 cm and time resolution 10^{-3} sec.

4. Passive alpha CX diagnostic can also be relevant for ITER. With the use of the array of several NPA's it can provide the same radial and time resolution as active CX-He^O beam diagnostic. It also needs the tests on existing tokamaks with DT plasmas (TFTR, JET).

5. The existing NPA's for MeV range which are required for all modes of alpha CX diagnostics are basically suitable for ITER conditions.

REFERENCES

1. M.P.Petrov, V.I.Afanasyev, S.Corti, et.al ., Neutral particle analysis in the MeV range in JET, *International Conference on Plasma Physics and 19-th EPS Conference, Innsbruck, 1992, Report 6-2, I-1031*.
2. A.Gondhalekar, S.Dalla and A.Stuart , Interaction of high energy ICRF driven ions with sawtooths and ELM events, *4-th IAEA Meeting on Alpha-particles in Fusion Research, Princeton, April 25-28, 1995*.
3. R.K.Fisher, J.M.McChesney, A.W.Howald, et al.., Alpha particle diagnostics using impurity pellet injection, *Rev. Sci. Instrum. 63, 4499, (1992)*
4. A.A.Korotkov, A.M.Ermolaev, Impurity induced neutralization of alpha particles and application to ITER diagnostics, *22-nd EPS Conference on Controlled Fusion and Plasma Physics, Bournemouth, 1995, report R-109*.
5. A.B.Izvozchikov, A.V.Khudoleev, M.P.Petrov et al., Charge-exchange diagnostic of fusion alpha particles and ICFR driven minority ions in MeV range in JET plasma, *JET Report JET-R(91) 12, (1991)*.
6. ITER Diagnostics, IAEA, Vienna, 1991.
7. V.V.Afrosimov, D.F.Barash, A.A.Basalaev et al., Single and double-electron capture from many-electron atoms by alpha particles in the MeV energy range, *Zh. Ekper. Teor. Fiz. 104, 3297-3310 (October 1993)*

8. R.K.Fisher, J.M.McChesney, P.B.Parks *et al.*, Measurements of fast confined alphas on TFTR, *Phys. Rev. Lett. 75, 846 (1995)*

9. M.P.Petrov, R.V.Budny, H.H.Duong *et. al.*,"Studies of energetic confined alphas using the pellet charge-exchange diagnostic on TFTR", *Report PPPL-3121 (July 1995), submitted to Nucl. Fusion, (1995)*

10. S.S.Medley, R.V.Budny, D.K.Mansfield *et al.*, Measurements of energetic confined alphas and tritons on TFTR, *22-nd EPS Conference on Controlled Fusion and Plasma Physics, Bournemouth, 1995, report P-111.*

11. R.V.Budny, A standard DT supershot simulation, *Nucl. Fusion 34, 1247 (1994)*

12. R.V.Budny, D.C.McCune, M.Redi and R.M.Wieland, TRANSP simulations of alpha parameters in ITER, *submitted for publication in Nucl. Fusion (1995)*

13. P.B.Parks, J.S.Lefler, R.K.Fisher, Analysis of low Z impurity pellet ablation for fusion diagnostic studies, *Nuclear Fusion 28, 477 (1988)*

14. P.B.Parks and R.J.Turnbull, Effect of transonic flow in the ablation cloud on the lifetime of a solid hydrogen pellet in a plasma, *Phys. Fluids 21, 1735 (1978)*.

A DIAGNOSTIC He⁰ BEAM FOR ITER AND LHD

M. Sasao,[1] A. Taniike,[1] and M.Wada[2]

[1]National Institute for Fusion Science, Chikusa Nagoya-464-10, Japan
[2]Department of Electronics, Doshisha Univ. , Tanabe, Kyoto-610-03, Japan

INTRODUCTION

The local measurement of alpha-particle density and its velocity distribution will play an important role in understanding physics of a burning plasma and also in controlling a high temperature plasma of a D-T or a D-^{3}He fusion device. Especially the fusion reaction rate of a D-^{3}He plasma can be monitored only by detecting direct capture gamma rays or by measuring alpha particles confined in a plasma.

Alpha-particle measurement using a high-energy neutral beam seems to be feasible on the D-^{3}He experiment of LHD or on the D-T experiment of ITER.[1] In this scheme, alpha particles are neutralized by beam particles through two-electron transfer processes,

$$^4He^{++} + A^0 ------> \, ^4He^0 + A^{+2},$$

escaping from a plasma, and are detected in a high energy neutral particle analyzer. Here A^0 represents a beam particle, which should be a helium or a heavier atom. The most serious problem of a diagnostic neutral beam is penetration into the plasma center of a large device, such as ITER, where the plasma density will be over the range of (1-4) $\times 10^{20}$ m^{-3}. Among various kinds of atomic beam, the highest penetration will be obtained with a He^0 beam. It is known, however, that a fraction of helium atoms are in the long-life metastable state. The metastable component in a helium atomic beam will be easily ionized and lost in the plasma periphery.[2]

In this paper several methods to produce a helium atomic beam are compared in view points of both the neutralization efficiency and the metastable fraction. The beam energy of our interest lies in the region from several tens keV to 2 MeV.

Diagnostics for Experimental Thermonuclear Fusion Reactors
Edited by P. E. Stott *et al.*, Plenum Press, New York, 1996

PRODUCTION OF A He⁰ BEAM

There are several methods to produce such a high-energy He^0 beam; (1) the gas neutralization of a He^+ beam of the MeV region, (2) the gas neutralization of a high energy He^- beam, (3) the time of flight neutralization of a He^- beam, and (4) the gas neutralization of a high energy HeH^+ beam. In general, a helium atomic beam is contaminated with a certain fraction of atoms in the long-life metastable state He(2S), which can be easily ionized in a gas cell or in a plasma.

When a beam of He^+, He^-, or HeH^+ enters in an helium gas cell, the beam fractions of $He^-(I^-)$, $He^0(I^0)$, $He(2S)(I^*)$, $He^+(I^+)$, $HeH^+(I^{HeH+})$, and $He^{++}(I^{++})$ can be calculated as a function of the target gas thickness x. Taking the most of possible charge transfer and excitation/deexcitation processes into account, the calculation has been performed by solving the following rate equations,

$$dI^-/dx = -(\sigma_{-,0}+\sigma_{-,*}+\sigma_{-,+}+\sigma_{-,++})I^-$$

$$dI^0/dx = -(\sigma_{0,*}+\sigma_{0,+}+\sigma_{0,++})I^0 +\sigma_{-,0}I^- +\sigma_{*,0}I^* +\sigma_{+,0}I^+ +\sigma_{++,0}I^{++}$$

$$dI^*/dx = -\sigma_{*,+}I^* +\sigma_{-,*}I^- +\sigma_{0,*}I^0 +\sigma_{+,*}I^+ +\sigma_{++,*}I^{++}$$

$$dI^+/dx = -(\sigma_{+,0}+\sigma_{+,*}+\sigma_{+,++})I^+ +\sigma_{-,+}I^- +\sigma_{0,+}I^0 +\sigma_{*,+}I^* +\sigma_{++,+}I^{++}$$

$$dI^{++}/dx = -(\sigma_{++,0}+\sigma_{++,*}+\sigma_{++,+})I^{++} +\sigma_{-,++}I^- +\sigma_{0,++}I^0 +\sigma_{*,++}I^* +\sigma_{+,++}I^+$$

$$dI^{HeH+}/dx = -(\sigma_{3,0}+\sigma_{3,*}+\sigma_{3,+}+\sigma_{3,++})I^{HeH+}$$

Here $\sigma_{3,j}$ denote the dissociation cross sections for HeH^+ ions in He into a helium neutral or an ion of the charge state j, and measured values obtained by Sterarns et al. are used.[2] The ionization cross section for He(2S) in collision with a helium atom $\sigma_{*,+}$ has not yet be measured. Here it is assumed that $\sigma_{*,+}$ shows the same velocity dependence as the electron impact ionization cross section for He(2S), and the absolute value of the cross section is multiplied by a ratio of the ionization cross section for ground state helium in a He gas to the electron impact ionization cross section for ground state helium.[3] Other cross sections are refereed from the review paper of K.Okuno[4].

Charge fractions when a He^- beam of 600 keV is injected into a helium gas are shown in Fig. 1 as a function of the target gas thickness. There exists an optimum thickness for the production of ground state helium. The metastable fraction in the beam is not negligible when the target thickness is less than 2×10^{16} cm^{-2}, but can be diminished at the greater thickness. The maximum neutralization efficiencies starting from various beams calculated with the present equations are shown in Fig. 2.

Negative helium ions are known to auto-ionize into ground-state neutral atoms. The life-times of two components of negative helium ions have been reported to be about 10 μs (t_1) and 300 μs (t_2).[4] The initial fraction of each component is about 50%. When negative ions fly over a distance L, the neutralized fraction η can be expressed as,

$$\eta = I^0/I^- = 1 - 0.5 \{ \exp(-L/v_B/t_1) +\exp(-L/v_B/t_2) \}$$

where v_B represents the beam velocity. The neutralization efficiencies for a 30m flight path is also shown in Fig. 2.

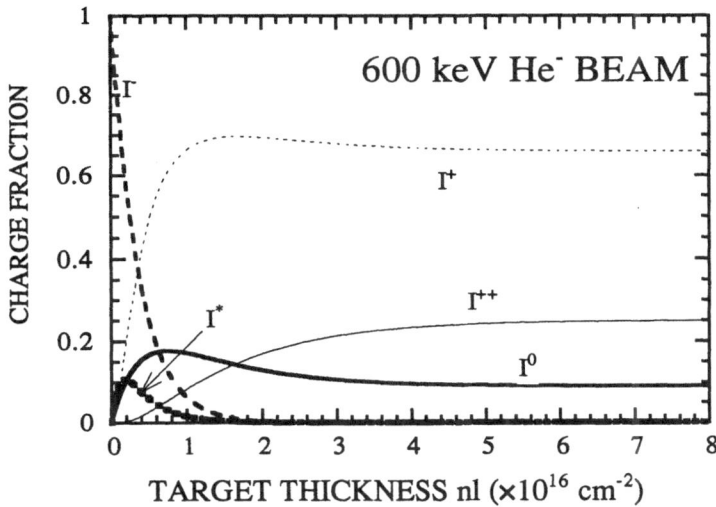

Figure 1. Charge fractions when a He⁻ beam of 600 keV is injected into a helium gas

DISCUSSIONS AND SUMMARY

Alpha particles confined in an ITER and LHD plasma can be neutralized by a high energy neutral beams of ^{3}He with beam velocities of 0.6-0.8 V_α. The overall optimization of the diagnostic beam are given in the Ref. 1 The counting rate would be sufficient to measure the neutral particle spectra with good statistics for both cases if the port-through current density of diagnostic beam (10 cm in diameter) is more than 1 mA/cm^2

In order to produce a ^{3}He^0 beam in the energy region above 1 MeV, the conventional method of electron attachment to positive ions becomes very inefficient. From detailed calculations of charge fractions and the fraction of long-life metastable states in a gas cell, starting with various positive ions and negative ions of helium, the production of a ground state beam of He^0 from He⁻ is most efficient for energies greater than 0.4 MeV. The typical efficiency is about 15 % at E = 1-2 MeV.

A He⁻ beam can be produced using a two step process in an alkali metal gas cell, such as Li, Na, Mg, K, Rb, or Cs.[5] The maximum value of the He⁻ fraction of 1.7% is obtained through collisions with a Rb target, at an He$^+$ ion incident energy of 6-9 keV. Using a sodium gas cell, production of a He⁻ beam more than 10 mA are reported.[6] Development of an He⁻ source using a Rb gas cell in a DC operation has been recently initiated for the purpose of application to the alpha-particle measurement.[7]

ACKNOWLEDGMENT

The authors are also grateful for useful discussions with Dr. K. Tobita and Professors T.Kato. This work was carried out under the Collaboration Research Program at the National Institute for Fusion Science, and the Collaboration Research Program of the Graduate Univ. of Advanced Studies.

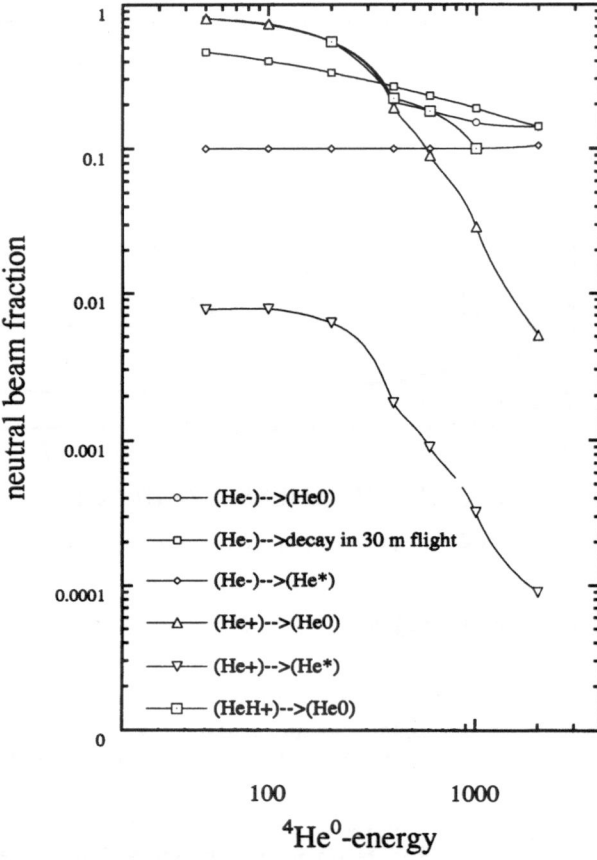

Figure 2. Neutralization efficiencies of various methods.

REFERENCES

1. M. Sasao, *et al.*, Development of Diagnostic Beams for Alpha Particle Measurement on ITER, submitted to *the Proceedings of the IAEA technical workshop on Alpha Particle Physics 95 Nuclear Fusion*, Suppl.)
   M. Sasao, *et al.*, Diagnostic Plan of Fusion Product Measurement on LHD, will be submitted to *the Proceedings of the 6th Toki Conference on Fusion Plasma Diagnostics,(1995)* (Nuclear Instrum. Methods, Suppl.)

2. J. W. Steraarns, K.H.Berkner, and R.V.Pyle, Dissociation Cross Sections for 0.5-10 1-MeV HeH+ Ions in H2, He, N2,and Ne gases. Phys. Rev. A, 4 (1971) 1960

3. T.Kato, private communication, and T.Kato and R.K.Janev, Parametric Representation of Electron impact Excitation and Ionization Cross sections for Helium atoms, *Nuclear Fusion Suppl.* 3 (1992), 33

4. K. Okuno, Charge Changing Cross Sections for Heavy-Particle Collisions in the Energy Range from 0.1 eV to 10 MeV. *IPPJ-Am-9* , (1978)

5. A.S. Schlachter et at, Phy. Rev. 174 (1968) 201
   H.B.Gilbody et at, J. Phys. B 2, (1969) 465
   R.A.Baragiola et at, Nucl. Instrum. Methods, 110 (1973) 507

6. G.I. Dimov and G.V.Roslyakov, Pribory i Tikhnika Eksperimenta, No. 3, (1974), 31
   E.B.Hooper et al., Rev. Sci. Instrum. 51-8 (1980) 1067

7. M.Sasao et al. Development of a double charge exchange He⁻ source. *Annual Report of National Institute for Fusion Science 1994.*

ALPHA SLOWDOWN MEASUREMENT FOR ITER

Alessandro Fubini,[1] Ezio Bittoni,[2] Marcel Haegi,[3] Sofia Rollet[4]

[1]ENEA, Dip. Innovazione, Settore MACO,
  Centro Ricerche Bologna, 40138 Bologna, Italy
[2]ENEA, Dip. Innovazione, Settore Fisica Applicata,
  Centro Ricerche Frascati, C.P. 65 - 0044 Frascati, Rome, Italy
[3]Associazione EURATOM-ENEA sulla Fusione,
  Centro Ricerche Frascati, C.P. 65 - 0044 Frascati, Rome, Italy
[4]Present address: CERN, Geneva (Switzerland)

THE NUCLEAR REACTIONS

The cross section of a nuclear reaction producing a photon is generally several orders of magnitude smaller than that of a reaction producing a nucleon.[1,2]

Let us consider the nuclear reaction:

$$^4He_{fast} + T_{thermal} \rightarrow ^7Li + \gamma_{2.5+0.478}\ MeV$$

The fusion energy Q is 2.5 MeV; the energy of the secondary γ-ray is 0.478 MeV. Since there are no compound states in ^7Li near the T+^{4}He threshold, the capture reaction will be dominated by the non-resonant, direct capture (DC) mechanism into the ground state and the 478 keV first exited state of ^7Li, as both the final states have a significant T $\otimes$ ^{4}He cluster configuration.[3] We would therefore expect to observe two primary γ-ray transitions, DC $\rightarrow$ 0 and DC $\rightarrow$ 478 keV, and a 478 keV (isotropic) secondary transition (Fig. 1). Due to the non-resonant nature of the T(α,γ)^7Li reaction, the above primary γ-ray transitions can be observed at all the initial kinetic energies $E_{c.m.}$(energy in the centre of mass); their energies vary with the initial kinetic energy according to the relation:

$$E\gamma\ (DC \rightarrow E_x) = Q + E_{c.m.} - E_x$$

where E_x is the lower level of transition.

The difference in kinetic energy in the laboratory framework before and after the collision is transformed into internal energy of the produced nucleus. Thus, the DC level comes from the sum of the fusion energy Q and the kinetic energy $E_{c.m.}$. As the collision

Diagnostics for Experimental Thermonuclear Fusion Reactors
Edited by P. E. Stott *et al.*, Plenum Press, New York, 1996

509

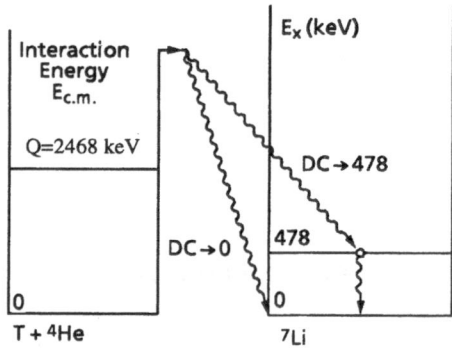

Figure 1. Level diagram of ^{7}Li near T+^{4}He threshold. The T(α,γ)^{7}Li capture reaction proceeds predomi-nantly via the direct capture (DC) mechanism into the ground state and the 478 keV excited state. [4]

takes place between the decelerating fast fusion products that have energies up to several MeV and the nuclei of the thermal plasma, $E_{c.m.}$ varies in an energy range of about 1 MeV. Therefore, the two primary γ-ray transitions have each a very large energy spectrum. This, associated with the small reaction cross-section, makes them practically impossible to observe in thermonuclear plasma conditions. Therefore we consider the secondary transition between the 478 keV excited state and the ground state. This line emitted by a nucleus at rest is intrinsically thin. In our case the γ-rays are emitted by nuclei with kinetic energy up to several MeV and are therefore affected by Doppler broadening. The shape of the line gives us information both on the emitting nucleus and the interacting α particle.

Negliging in this presentation the momentum of the thermal T and that of the photon, the conservation of the momentum leads to:

$$W_{cin\ Li} = 4/7\ W_{cin\ \alpha}$$

The primary γ-ray lines are so wide that they are practically impossible to observe. For the maximum speed of the resulting lithium nucleus, the maximum Doppler shift of the secondary γ-ray is ~12 keV.

FEASIBILITY EVALUATION

The considered feasibility conditions are:
1) that gamma-rays produce a sufficient number of counts in the detector (for example > 1000);
2) that the background counting in the region of experiments (from 475 to 485 keV) is not higher than the useful counting rate in the detector.
3) that the total counting rate in the detector is < 10^5 counts per seconds, which is the maximum for a solid state detector.
4) that the maximum integrated neutron flux at the detector is < 10^{10} neutrons/cm^2, to avoid performance degradation.

For this high background radiation scenario it is absolutely necessary that the line of observation of our diagnostics does not hit the opposite wall of the vacuum chamber. Without this condition this diagnostic would not be possible because of the high intensity γ-rays pro-duced by neutron capture in the structure of the device. Moreover the high level background radiation prohibits the use of a Germanium spectrometer to detect the 478 keV photon.

We have therefore developed a radiation-resistant detector based on the photoelectric effect. In this method, the photon crosses a very thin bismuth foil. The emitted photoelectron has exactly the same energy as the incoming electron, less the binding energy of the 91 keV k-level electron. It is thus possible to expose the foil of bismuth to the high flux of neutrons and gamma produced by the reactor and to bend the trajectory of this relativistic electron out of the high flux stream, with a magnetic field. The magnetic field which focus the electrons from the bismuth foil into an image on the detector acts also as a first electron spectrometer. Thus the detector receives only electron from the relevant energy band. The electron are then successively detected by a shielded high resolution Si(Li) cooled electron spectrometer. A schematic section of the diagnostic is shown in Fig.2.

For the 478 keV photon, the bismuth mass attenuation coefficients for Compton scattering and the photoelectric effect are nearly the same at about 0.1 $cm^2 g^{-1}$. For Compton scattering, the ejected electron has a continuous spectrum, but for the photoelectric effect the electron is mono-energetic. Thus, the effects are well separated. The mass stopping power coefficient, for the 478 - 91 = 387 keV electron in bismuth, is 1.14 $MeV cm^2 g^{-1}$. Thus, if we want a maximum of 1 keV attenuation, the thickness has to be nearly 1/1000 mm. for which a support grid is necessary. For this thickness, only one photoelectron is produced for every 10^4 photons.

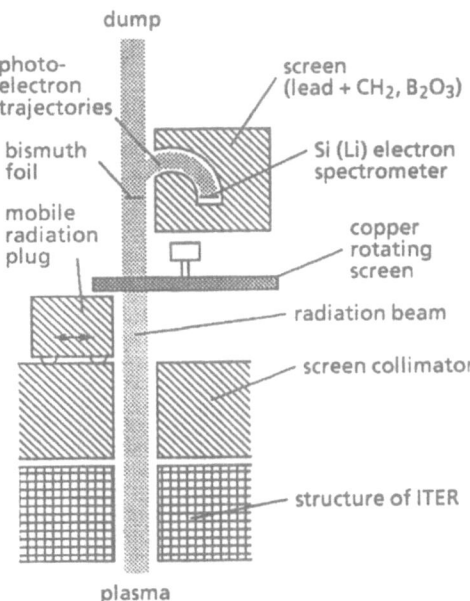

Figure 2. Schematic section (not to scale) of the radiation-resistive gamma diagnostic spectrometer. A magnetic field, perpendicular to this section, is present in the region of the electron trajectories. The relative magnetic coils (not shown) are magnetically screened from the vertical magnetic field of the tokamak.

For the ITER parameters,[4] producing 6×10^{20} 14-MeV neutrons per second (1360 MW), the flux of 478 keV photons at the bismuth foil is 4.10^4 s^{-1}, assuming a distance of 7 m from the magnetic axis, and a 1% visibility of the plasma volume. Thus, the flux of photoelectrons out of the foil is 4 cm^{-2} s^{-1}. If we want a counting rate of 10^3 s^{-1} and have a 100% electron detection efficiency on the Si(Li), we need a bismuth surface of 250 cm^2. The resolution of the Si(Li) spectrometer for 450 keV electrons, is better than 1,5 keV.[5]

The capacity for survival of the thin bismuth foil under the intense radiation coming from the thermonuclear plasma has been carefully investigated. The attenuation of gamma rays and neutrons in thin foils is proportional to their thickness. In the foil, the absorbed energy is transferred to particles, which if the foil is sufficiently thin, can escape from the foil. Thus, the energy that remains in thin foils can be much smaller than the energy lost in the foil from the incoming radiation. In particular, the gamma rays interact by means of the photoelectric effect, Compton effect, or pair production. In all these effects the energy is transferred from the photons to the electrons, which escape, each one leaving, in our case, only about 1 keV in the foil. The same is true for the scattered gamma in the Compton effect. Bismuth has very small cross sections for neutrons. The (n,n'), (n,α), (n,γ) and (n,2n) reactions concerned also produce particles which escape from the foil at nearly full energy. For elastic scattering the energy transfer between the neutron and the bismuth is inefficient because of the large atomic mass of the latter. However, for electromagnetic radiation of energy smaller than about 80 keV, the energy may be dissipated within the foil and cause its meltdown. Therefore, this radiation must be screened out, e.g. with a copper filter 2 mm thick. The electrons produced within this copper screen don't reach the foil because of the deflection due to magnetic field of the device. We have calculated that the rate of temperature increase in the copper-protected bismuth foil, exposed to the radiation of the thermonuclear plasma, is less than 10°C/min. For the copper filter, instead, the temperature grows so rapidly that it is necessary to mount this filter on a rotating wheel in order to distribute the heat along a circumference, as shown in Fig. 2.

To conclude, in spite of the condition that the line of sight of the detector must not hit the wall, which may put some restriction on the luminosity of the system, and thus on the resolution time, this γ-ray diagnostic system based on the photoelectric effect of a thin bismuth foil appears a promising tool for the investigation of future thermonuclear plasma.

REFERENCES

1. F.E. Cecil et al. Nucl. Instrum. & Meth. **22** 1 (1984)
2. G.Sadler et al. Proc. 15th Europ. Conf. Control. Fusion Plasma Heat. Dubrovnik (1988) Vol. I p. 131 Published by European Physical Society.
3. H. Kräwinkel et al. Zeitschrift Physik, **A304**, 307 (1982)
4. G. Simbolotti et al. Fusion Eng. & Des. **22** 323 (1993)
5. I. Ahmad and F.Wagner Nucl. Inst. and Methods **116** (1975) p.465

COLLECTIVE THOMSON SCATTERING DIAGNOSTIC
OF CONFINED ALPHA PARTICLE DISTRIBUTIONS IN ITER

Umberto Tartari

Istituto di Fisica del Plasma
CNR-ENEA-Euratom Association
via Bassini, 15 - 20133 Milano, Italy

INTRODUCTION

The presence of diffused, strongly non-Maxwellian and possibly anisotropically-distributed heating sources, the fusion-produced alpha particles, makes the real difference between the heating phase of an ignited plasma and present-day fusion plasmas. This stresses the importance of alpha particle diagnostics for ITER. The information required will be manifold. It is unanimously recognized, however, that knowing the space and velocity distribution of the confined alphas during slow down, up to their birth velocity $v_{\alpha 0} = 1.3 \times 10^7$ m/s ($E_{\alpha 0} = 3.5$ MeV), will be essential. Collective Thomson scattering (CTS) has the potential to provide this piece of information. Radiation from a high-power coherent source is launched into the plasma and the Doppler-shifted radiation scattered at a given angle is collected and analyzed. Under suitable conditions, the scattering process is primarily due to the shielding electrons moving along with the ions and therefore the spectrum is strictly related to the ion velocity distributions, including the alpha's. The schematic of a CTS diagnostic, shown in Fig. 1 with reference to a proposed experiment based on the free electron laser (FEL)[1], is conceptually similar at all frequencies.

Proof-of-principle demonstrations and preliminary results of thermal (i.e. non turbulent) CTS have been achieved first in TCA[2] in 1989 using a FIR laser, then in ATF[3] using a CO_2 laser and, more recently, in W7-AS[4], TFTR[5] and JET[6] using mm-wave gyrotrons. However, the wide spread in frequency of the experiments mentioned above provides in itself indirect evidence of the fact that thermal CTS is still far from being a consolidated diagnostic in fusion plasmas.

In recent years, the high-power mm-wave gyrotron seemed to open the way to a definite solution of the CTS problems. While it will be most probably so for the present generation of fusion-relevant devices, however, it is now becoming increasingly clear that the extrapolation of this option to ITER meets some inherent difficulties. Accordingly, which source frequency

Diagnostics for Experimental Thermonuclear Fusion Reactors
Edited by P. E. Stott *et al.*, Plenum Press, New York, 1996

513

is optimum for CTS in ITER is still a matter of debate. Besides the mm-wave gyrotron, the CO_2 laser, emitting in the infrared (IR) range, and the future FEL operated in the far infrared (FIR) range are being positively considered as candidate sources. The technology and the impact on the machine design will be very different in the three cases.

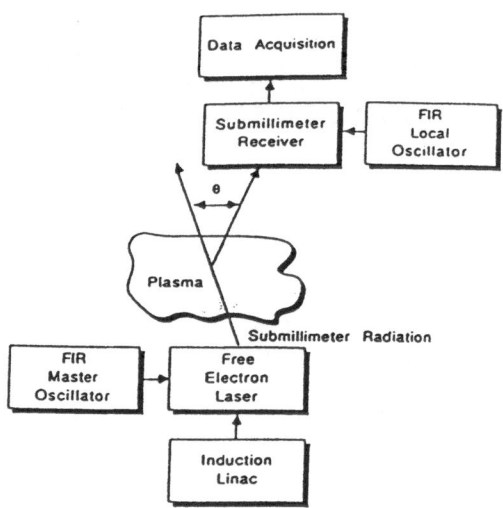

Fig. 1. Block diagram of a FEL-based CTS diagnostic experiment[1] illustrating the main features of a CTS set-up.

Accounting for the situation depicted above, the main aim of this work is to compare the CTS performances in the different options. More emphasis will be nevertheless given to the gyrotron option, of which a possible configuration will be discussed as an example, since this represents the most natural extrapolation to ITER of the thermal CTS diagnostic experiments now in course, or being implemented, in existing fusion devices.

BASIC CTS PRINCIPLES

A few basic principles apply in CTS[7]. First, given an incident wave (ω_o, $\mathbf{k}_o$), a scattered wave (ω_s, $\mathbf{k}_s$), and an electron density fluctuation (ω, $\mathbf{k}$), the energy and momentum conservation in the scattering process are expressed by $\omega = \omega_s - \omega_o$ and $\mathbf{k} = \mathbf{k}_s - \mathbf{k}_o$. A Bragg approximation $k_i \cong k_s$, giving $k \cong 4\pi \sin(\theta/2)/\lambda_o$, where $\lambda_o = 2\pi c/\omega_o$ and θ is the scattering angle, is usually adequate. Second, for the scattering process to take a collective character the Salpeter condition $\alpha \equiv 1/(k \lambda_{De}) \geq 1$, λ_{De} being the Debye length, must be satisfied. Third, plasma particles with velocity $\mathbf{v}$ will be mapped onto the Doppler-shifted spectrum as $\omega = \mathbf{k} \cdot \mathbf{v}$. Particles of different species but with the same velocity will therefore provide overlapped contributions. Finally, a requirement of plasma transparency (or quasi-transparency) is typical of any form of scattering process.

Apart from the constraints related to the limited number and size of the ports, a scattering geometry is basically determined by two angles: the scattering angle θ and the 'magnetic' angle φ, with $k_{//} = k \cos\varphi$, $k_{//}$ being the component of the fluctuation wavevector parallel to the magnetic field.

Let us now draw some consequences of the simple principles above as regards CTS on ITER at the different frequencies starting with the problem of accessing the plasma center.

PLASMA ACCESSIBILITY

By the antenna reciprocity principle, in CTS the incident beam and the scattered radiation can be described as two beams crossing each other at the measuring point (scattering volume). These beams are required to propagate in the plasma avoiding cut-offs and, as far as possible, plasmas resonances, i.e. suitable 'propagation windows' must be found for CTS.

As it is easily understood, high frequency CTS is widely favoured in this respect thank to the possibility of unperturbed line-of-sight propagation. The only problem there is avoiding the cyclotron resonances, $n\omega_{ce} < \omega_o < (n+1)\omega_{ce}$, or making them crossed at the plasma edge, and even in this last case the plasma will be optically thin, if not fully transparent.

On the contrary, ensuring suitable propagation conditions is the most difficult problem of CTS at mm-wavelengths in ITER. Dielectric and kinetic plasma effects become appreciable and may even result prohibitively strong. Due to refraction, optical ray-tracing codes must be used to properly track the beams and describe their absorption. To jointly account for diffraction, quasi-optical (Gaussian) ray-tracing codes should be better used[8]. In general, the 'hot' (i.e. accounting for absorption) propagation windows will be specific of a single plasma mode and significant differences also exist between horizontal (e.g. equatorial) and vertical windows since, thank to the near-constancy of the magnetic field along the beam path, only slightly affected by beam bending due to refraction, the latter are wider and independent of the aspect ratio.

If electron cyclotron absorption is neglected, two 'cold' propagation windows exist at mm-waves for the ITER magnetic field, $B_{T0} = 5.7T$ ($f_{ceo} = 156GHz$), and density range. The lower window is for $\omega_o < \omega_{ce}$, the higher for $\omega_{ce} < \omega_o < 2\omega_{ce}$. Actually, it is readily verified that OM propagation has significant limitations in the former window due to too low cut-off densities and that XM propagation has to be excluded in the latter due to the upper cut-off and the need of injection/collection from the l. f. s..

No further mention of the propagation problem will be made in the following as regards the IR and FIR cases. Instead, examples will be given in a later section of possible hot propagation windows for mm-wave CTS in ITER.

SOURCE FREQUENCY AND SCATTERING ANGLE

An important relationship directly follows from the Salpeter condition $\alpha > 1$. Normalizing to ITER parameters one in fact finds

$$f_o[\text{THz}]\, sin(\theta/2) < 0.18 \left(\frac{n_e/10^{20}\,\text{m}^{-3}}{T_e/30\text{keV}} \right)^{1/2}. \tag{1}$$

The complementary role of source frequency and scattering angle in CTS is emphasized. Decreasing the source frequency will allow larger θ values and therefore improved spatial resolution. At any given θ, the α value will be higher the lower the source frequency. It is also easily verified that in ITER conditions scattering angles as large as $\theta \leq 90°$ are allowed only in mm-wave CTS ($f_0 \leq 250\text{GHz}$). For FEL radiation at $\lambda_0 = 200\mu\text{m}$ ($f_0 = 1.5\text{THz}$) one has $\theta \leq 12°$; with a CO_2 laser at $\lambda_0 = 10.6\mu\text{m}$ ($f_0 = 28\text{THz}$) $\theta \leq 1°$. Large θ values are the main advantage of the gyrotron option. Conversely, too small θ values are by large the main drawback of the CO_2 laser option. As a matter of fact, beam alignment becomes difficult, stray radiation hardly manageable, and spatial resolution practically null. It should be nevertheless mentioned that the capability of managing most of these difficulties has been experimentally demonstrated in one case[3].

The complementarity between frequency and scattering angle is reflected in the complementarity of advantages and drawbacks of the mm-wave and IR options. As underlined yet in a early ITER diagnostic plan[9], operation in the FIR range would represent a very good compromise. However, achieving the required power and energy (pulse duration) capabilities would imply going well beyond the performances of present-day optically-pumped FIR lasers, i.e. an FEL would be necessary. As it is known, no such a source is available so far.

A parameter which qualifies the source performance in CTS is its acceptable bandwidth, which is also readily estimated from the basic principles. Accounting that $f = \mathbf{k} \cdot \mathbf{v}/2\pi$, the total (single side) CTS bandwidth of the bulk particles can be conveniently defined as $\Delta f_{\text{tot},j} \cong k v_{\text{th},j}/\pi$, with $v_{\text{th},j}^2 = T_j/m_j$, and that of the alphas as $\Delta f_{\text{tot},\alpha} = k v_{\alpha 0}/2\pi$. Taking for simplicity $T_e = T_i$ this gives

$$\Delta f_{\text{tot},j}[\text{GHz}] \cong 22\, \mu_j^{-1/2}\, (T_i/30\text{keV})^{1/2}(f_0/\text{THz})\, sin(\theta/2) \quad (j=e,D,T); \tag{2}$$

$$\Delta f_{\text{tot},\alpha}[\text{GHz}] \cong 112\, (\mu_\alpha/4)^{-1/2} (E_{\alpha 0}/3.5\text{MeV})^{1/2}(f_0/\text{THz})\, sin(\theta/2), \tag{3}$$

with $\mu_\gamma = m_\gamma/m_p$ ($\gamma = j, \alpha$). From eqs. (1) - (3) then it is seen that for any given α and plasma parameters, hence $f_0\, sin(\theta/2)$ factor, the total bandwidths become independent of source frequency. Actually, different α values will be achievable at the different frequencies, but nevertheless the CTS bandwidths do not vary much with frequency. What will significantly change is accordingly the *fractional* CTS bandwidth, $\Delta f_{\text{tot},\gamma}/f_0$, and the same will be true for the fractional source bandwidth, $\Delta f_{\text{source}}/f_0$, this being bound to be much smaller than the total CTS bandwidth.

With reference to the alpha feature, taking $\Delta f_{\text{source}}/\Delta f_{\text{tot},\alpha} \leq 1/10$ one finds that percentually $\Delta f_{\text{source}}/f_0 \leq sin(\theta/2)$. For instance, at $f_0 = 200\text{GHz}$ and with $\theta = 90°$ one will need $\Delta f_{\text{gyr}}/f_0 \leq 0.7\%$, which requirement is largely satisfied by existing mm-wave gyrotrons ($\Delta f_{\text{gyr}}/f_0 \cong 0.1\%$). In contrast, at $f_0 = 1.5\text{THz}$ and with $\theta = 10°$ one would need $\Delta f_{\text{FEL}}/f_0 \cong 0.09\%$ ($\Delta f_{\text{FEL}} \leq 1.3\text{GHz}$), which condition is far from being satisfied by the laboratory (Linac-driven) FELs of more promising power performance.

No problem of source bandwidth seems to exist as regards the CO_2 laser. However, a relevant limitation here is that the available detector bandwidths ($\leq 2\text{GHz}$) do not cover the total CTS bandwidth. Hence, a whole set of receivers, with several isotopic CO_2 or N_2O lasers used as local oscillators, will be required. The measurement of the k (or θ) spectrum instead of the frequency spectrum has been proposed[10] as an alternative to overcome both this difficulty and the strong limitation in θ.

From eqs. (2) and (3) it also follows that the bandwidth ratios are independent of source frequency and scattering angle. One has

$$\Delta f_{\text{tot},j}/\Delta f_{\text{tot},T} = (\mu_j/\mu_T)^{-1/2}; \quad \Delta f_{\text{tot},\alpha}/\Delta f_{\text{tot},T} = (E_{\alpha 0}/T_i)^{1/2}(\mu_\alpha/\mu_T)^{-1/2}/\sqrt{2}; \quad \text{etc..} \tag{4}$$

This gives $\Delta f_{tot,D}/\Delta f_{tot,T} = 1.2$, $\Delta f_{tot,e}/\Delta f_{tot,T} = 74$, $\Delta f_{tot,\alpha}/\Delta f_{tot,T} = 6.6$ (at $T_{io} = 30 KeV$) and $\Delta f_{tot,e}/\Delta f_{tot,\alpha} = 11$. The last of these figures is of special interest since it shows that the alpha feature has necessarily to be detected against an electron background. Consideration of magnetic effects would not substantially modify the picture. For CTS to be feasible, therefore, conditions must be found which make the alpha feature dominant compared to the electron feature.

CONTRIBUTIONS OF SINGLE PLASMA SPECIES TO CTS SPECTRA

According to standard theory[7], the power scattered into a solid angle $\Delta\Omega$ and a (sufficiently narrow) resolution channel of bandwidth $\Delta\omega_{ch}$ writes

$$P_s(\omega) = \frac{1}{2\pi} r_e^2 P_o L_s n_e \Gamma S(\mathbf{k},\omega) \Delta\omega_{ch}\Delta\Omega, \tag{5}$$

where r_e is the classical electron radius, P_o the incident power, spread over an area A, and $L_s = V_s/A$ the scattering length, V_s being the scattering volume. Apart from the constant factors, $P_s(\omega)$ is seen to be the product of a 'geometrical form factor', Γ, related to the differential scattering cross-section and dependent on polarization and scattering geometry, and a 'spectral density function', $S(\mathbf{k},\omega)$, which determines the spectral shape. Both these quantities are provided by theory in a number of approximations. Significant spectral enhancements may be produced due to $\Gamma > 1$ in particular cases[11], but the simplest approximation $\Gamma \approx 1$ is adequate for order-of-magnitude estimates.

Under sufficiently wide conditions, $S(\mathbf{k},\omega)$ takes the form of a sum of single particle contributions (see Fig. 2). Neglecting impurities, in an ignited DT plasma one has

$$S(\mathbf{k},\omega) = S_e(\mathbf{k},\omega) + \sum_i S_i(\mathbf{k},\omega) + S_\alpha(\mathbf{k},\omega) \qquad (i = D, T). \tag{6}$$

The relationship with the velocity distributions is made clear when the (unmagnetized) $S(\mathbf{k},\omega)$ is written in the more explicit form:

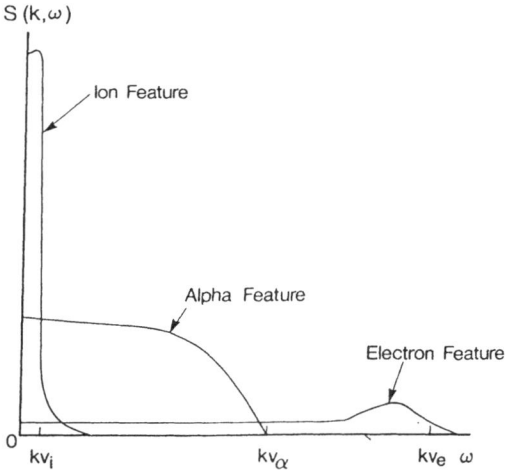

Fig. 2. Illustrating the different particle contributions to the CTS spectrum in an ignited plasma.

$$S(\mathbf{k},\omega) = \frac{2\pi}{k}\left\{\left|1 - \frac{G_e}{\epsilon_L}\right|^2 f_{ke}\left(\frac{\omega}{k}\right) + \left|\frac{G_e}{\epsilon_L}\right|^2 \sum_{j,ions} Z_j^2\, f_{kj}\left(\frac{\omega}{k}\right)\right\}. \tag{7}$$

Here the f_{ke} and f_{kj} are the one-dimensional distribution functions along $\mathbf{k}$, G_e is the electron susceptibility and

$$\epsilon_L = 1 + G_e + \sum G_j \quad (j = D, T, \alpha) \tag{8}$$

the longitudinal dielectric function, the G_j now being ion susceptibilities. If $\alpha \ll 1$ (the so-called 'incoherent' scattering) the dominant term in eq. (7) is just the first half of the electron contribution. If $\alpha > 1$ (CTS), instead, the electron term approximately cancels and it is mostly the ion terms which remain. The presence of ϵ_L in the 'screening functions' of eq. (7) explains the sensitivity of the scattered spectrum to the plasma resonances[12].

The separation of the alpha from the bulk ion features is readily achieved since $v_{th,j} \ll v_{\alpha 0} \ll v_{th,e}$. Whenever the alpha feature is made dominant with respect to the electron feature then we have

$$S_\alpha(\mathbf{k},\omega) = \frac{2\pi}{k} Z_\alpha^2 \left|\frac{G_e}{\epsilon_L}\right|^2 f_{k\alpha}\left(\frac{\omega}{k}\right). \tag{9}$$

If a parameter range is found where the screening function $\left|G_e/\epsilon_L\right|^2$ becomes almost independent of ω then the full frequency dependence in $S(\mathbf{k},\omega)$ will be due to the one-dimensional distribution function. Numerical simulations[13] with magnetized spectra confirm that conditions with $S_{tot} \cong S_\alpha \propto f_{k\alpha}$ are achievable for sufficiently high α and $\varphi \le 60°$.

When magnetic effects are included, in $S(\mathbf{k},\omega)$ the one-dimensional distribution functions f_{ke} and f_{kj} are replaced by a sum over Bessel harmonics. This accounts for the unperturbed particles travelling on helical orbits. The magnetized spectra then take more complicated forms and modulations at ω_{cj} appear in their lower frequency part for $k_{//} \to 0$.

ALPHA-TO-ELECTRON FEATURES RATIO

To separate the alpha from the electron feature one basically relies on the correlation effects, contained in the particle susceptibilities, which concentrate most of the power in the electron feature about $\omega \cong \omega_{pe}$, hence at larger frequencies than covered by the alpha feature, while its area remains constant. In the unmagnetized approximation, the alpha-to-electron features ratio can be approximately written as[14]

$$R_{\alpha e} = Z_\alpha^2\, \alpha^4\, \frac{f_{k\alpha}(\omega/k)}{f_{ke}(\omega/k)}. \tag{10}$$

The strong dependence on α will be noticed. Assuming for simplicity a uniform alpha distribution and Maxwellian electrons eq. (10) reduces to

$$R_{\alpha e} = (2\pi\mu)^{1/2} Z_\alpha^2 \, \alpha^4 \left(\frac{T_e}{E_{o\alpha}}\right)^{1/2} \frac{n_\alpha}{n_e} exp\left(\frac{\omega^2}{2k^2 v_{th,e}^2}\right), \tag{11}$$

with $\mu = m_p/m_e$. Since in the range of interest $\omega \ll kv_{th,e}$, in eq. (11) the exponential term is actually ≈ 1. Thus it is readily found that in ITER conditions achieving $R_{\alpha e} \geq 10$ will require

$$\alpha \geq \frac{2.2}{(n_\alpha/\%)^{1/4}(T_e/30keV)^{1/8}}, \tag{12}$$

i.e. typically $\alpha > 2$. This makes the limitations in scattering angle stricter. In particular, for CTS at $f_o = 28THz$ now one has $\theta \leq 0.6°$ and the criticality of the CO_2 laser option is further emphasized.

When magnetic effects are included, due to their heavy mass the alphas still behave as unmagnetized but strongly magnetized electrons must be considered. This modifies the alpha-to-electron features ratio both indirectly, through the electron pedestal, and directly, through the magnetization of the electron clouds screening the alphas. The main change with respect to eq. (10) is that now the one-dimensional electron distribution along **k** is replaced by the distribution parallel to the magnetic field and one finally finds that

$$\frac{R_{\alpha e}^{(B)}}{R_{\alpha e}^{(0)}} \cong |w(x)|^2 \, exp \, x^2 \, cos^2 \, \varphi, \quad \text{with:} \quad x = (\omega/kv_{th,e})/(\sqrt{2} \, cos\varphi), \tag{13}$$

where $w(x)$ is a function related to the Fried-Comte function. Making magnetic effects negligible requires $x \ll 1$ in eq. (13), which in turn implies a sufficiently small magnetic angle ($\varphi \leq 60°$). When this is so, $R_{\alpha e}^{(B)} \cong R_{\alpha e}^{(0)} cos^2 \, \varphi$. CTS at relatively small φ values has the advantage that the spectra become sensitive to the electron-drift and toroidal bulk plasma velocities, so making these quantities potentially measurable. A price is to be paid, however, in the sense that most natural scattering geometries as those in a poloidal plane now must be excluded since there $\varphi = 90°$.

Once the alpha feature has been isolated, the problem becomes its detectability in terms of signal-to-noise ratio (SNR). This in turn requires separate consideration of the alpha signal and the noise contributions relevant in CTS.

DETECTABILITY OF THE ALPHA SIGNAL

Still assuming for simplicity $|G_e/\varepsilon_L|^2 \cong 1$ and $\Gamma = 1$, and for uniformly distributed alphas, the scattered signal power of eq. (5) can be put in the form

$$P_{s\alpha}(\omega) = \frac{r_e^2 Z_\alpha^2}{2} P_o n_\alpha \frac{\Delta\omega_{ch}}{kv_{\alpha o}} L_s \Delta\Omega \, . \tag{14}$$

Due to the requirement of spatial coherence at the detector, in heterodyne detection the product $L_s \Delta\Omega$ is actually limited to $L_s \Delta\Omega = \lambda_o^2/(2w_o|sin\theta|)$, w_o being the beam radius (normally the waist) at the scattering volume. Thus one readily finds that, in terms of alpha signal temperature $T_{s\alpha} \equiv P_{s\alpha}/\Delta f_{ch}$:

$$T_{s\alpha}[eV] = 0.005 \frac{(P_o/MW)(n_\alpha/\%)(n_e/10^{20}\,m^{-3})}{(w_o/cm)(f_o/THz)^3 |\sin\theta \, \sin(\theta/2)|}. \tag{15}$$

The alpha concentration n_α is referred to the electron density as usual. Typically, assuming $n_\alpha = 2\%$ and $n_e = 10^{20}\,m^{-3}$, and with realistic (or expected) source power levels, for gyrotron-based CTS ($f_o = 200GHz$, $\theta = 90°$, $w_o = 3cm$, $P_o = 1MW$) one finds $T_{s\alpha} \cong 0.6eV$; for FEL-based CTS ($f_o = 1.5THz$, $\theta = 10°$, $w_o = 1cm$, $P_o = 100MW$), $T_{s\alpha} = 20eV$; for CO_2 laser-based CTS ($\theta = 0.5°$, $w_o = 0.5cm$, $P_o = 2MW$), $T_{s\alpha} = 0.05eV$. The importance of the very high power capability predicted for the future FEL is emphasized by these estimates.

The detectability of the alpha signal is eventually determined by its capability of overcoming the background noise, hence by the achievable SNR, defined as the ratio of the mean signal power in a given channel to the total r. m. s. fluctuations. In heterodyne detection this quantity is usually expressed by the well-known formula

$$SNR = \frac{GT_s}{T_s + T_{n,\,tot}}, \tag{16}$$

where $T_{n,\,tot}$ is the total noise temperature and $G \equiv \sqrt{\Delta f_{ch} \, \tau_{int}}$ the radiometric gain, τ_{int} being the integration time. Comparison with the general definition of SNR puts in evidence that in eq. (16) the r.m.s. fluctuations have been expressed by the ratio $(T_s + T_{n,\,tot})/G$, hence through the associated average (signal+noise) powers (temperatures). It is worth stressing here that this result only applies to input processes characterized by a fully Gaussian statistics, since no more general theory is available at our knowledge. While this is often the case in CTS, relevant consequences may follow whenever a non-negligible noise contribution of doubtful statistics is produced by the source of radiation itself (source noise).

Experimentally the signal temperature is estimated as $T_s \cong T_{s+n} - T_n$, T_{s+n} being the total signal in the presence of noise and $T_n = T_{n,\,tot}$. For unmodulated gyrotron output and in the CO_2 laser case the integration time will be coincident with the measurement time, $\tau_{int} = \tau_{meas}$. For multi-pulse operation instead $\tau_{int} = F_{pulse}\tau_{pulse}\tau_{meas}$, F_{pulse} being the repetition rate and τ_{pulse} the pulse length.

Eqs. (15) and (16) can be conveniently used to estimate the maximum total noise temperature which still allows to achieve $SNR_\alpha \geq 10$ in a suitably defined average spectral channel. Accounting that normally $T_{s\alpha}/T_{n,\,tot} \ll 1$, and taking for simplicity $\Delta f_{ch} = kv_{o\alpha}/(2\pi N_{ch})$, N_{ch} being the total (single side) number of equal channels, for single-pulse operation it is required that

$$T_{n,\,tot}[eV] \leq 5 \frac{(P_o/MW)\,(n_\alpha/\%)(n_e/10^{20}\,m^{-3})(\tau_{meas}/ms)^{1/2}}{N_{ch}^{1/2}(w_o/cm)(f_o/THz)^{5/2}\,|\sin\theta \, \sin^{1/2}(\theta/2)|}. \tag{17}$$

For multi-pulse operation the substitution $(\tau_{meas}/ms) \to 10^{-6}(F_{pulse}/kHz)(\tau_{pulse}/ns)(\tau_{meas}/ms)$ applies. With the same power levels, scattering angles, beam and plasma parameters as above and taking $N_{ch} = 10$, for single-pulse gyrotron-based CTS with $\tau_{meas} = 100ms$ one then typically finds $T_{n,\,tot} \leq 700eV$. Achieving the same noise performance in the same τ_{meas} using an FEL supplying $P_o = 100MW$ will require $F_{pulse} = 100kHz$ and $\tau_{pulse} = 10ns$. Finally, for

CO_2 laser-based CTS with $\tau_{meas} \cong 10\mu s$ one finds $T_{n,tot} \leq 0.5eV$. These estimates must be compared with the actual noise temperatures met in the different CTS configurations.

NOISE CONTRIBUTIONS AND STRAY RADIATION

The main noise contributions in CTS are: i) receiver noise, $T_{n,rec}$; ii) ECE noise, $T_{n,ece}$; and, possibly: iii) source noise, $T_{n,source}$. Therefore, $T_{n,tot} = T_{n,ece} + T_{n,rec} + T_{n,source}$.

Receiver noise temperatures at the mixer input of $T_{rec} \leq 1eV$ are achievable with present-day receiver technologies in each of the frequency ranges of interest. Low-noise Schottky diode mixers are used in the mm and FIR ranges and HgCdTe mixers in the IR range. Comparison with previous estimates thus seem to indicate that receiver noise will be of some concern only in CO_2 laser CTS, where nevertheless it will also presumably be the dominant contribution. It is worth stressing, however, that this noise may play a much larger role than indicated by the performance mentioned above. Being 'equivalent' in character, in fact, receiver noise is conventionally referred to the mixer input while as regards the SNR formula all other noises (and the signal) may be intended as referred to the receiving antenna input. Therefore, any attenuation factor $A_{in,tot}$ inserted in the antenna-to-mixer path will increase the relative weight of receiver noise up to $T_{n,rec} = A_{in,tot}T_{eq}$. At high frequencies this is actually no problem since propagation is as in vacuum. In contrast, low-loss propagation either of quasi-optical type or in suitable oversized waveguides will be necessary in mm-wave CTS on ITER to cover the long distance from the plasma to the receiver site.

As regards ECE noise, simulations show[15] that even in a plasma with $T_{e0} = 30keV$ radiation temperatures $T_{rad} \leq 10eV$ for the XM and $T_{rad} < 1eV$ for the OM are easily achieved if $f_o \geq 1.5THz$. Full transparency of the ITER plasma is easily predicted in the IR range. In contrast, blackbody ECE constitutes by large the main, and most dangerous, noise contribution at mm-wavelengths. The margins available for overcoming this noise will be shortly discussed in a later section.

Even if hardly quantifiable, the stray radiation problem in CTS is worth of special mention also in view of its connection with the noise problem. Due to more than ten orders of magnitude unbalance between the power level of the incident radiation and the received signal, some stray radiation from the source will always enter the receiver. At mm-waves this is so due to both direct (cross-talk) and indirect (through multiple reflections) antenna coupling. At higher frequencies diffractive effects are more limited, nevertheless stray-radiation remains a primary problem since the scattering angles involved are small. Actually, the problem is twofold. Most of the stray radiation will obviously be at f_o, the central source frequency, but some spurious radiation at very low levels is emitted by any real radiation source also at off-set frequencies, hence possibly within the CTS bandwidth, and the associated r. m. s. fluctuations will contribute what we have termed source noise. While very different in nature, neutralizing the negative effects of both types of stray radiation largely depends on the same factor, the decoupling between the transmitting and the receiving 'antenna' systems.

If incident on the mixer at power levels comparable with the local oscillator power, the stray radiation at f_o will give rise to spurious signals and may even produce irreversible damage. A high-attenuation notch filter is expressly inserted to reject it in all CTS schemes. As it has been shown with reference to the gyrotron case[16], high total 'antenna' decoupling is normally still required to fully solve the problem. This explains the importance of beam dumps in CTS. High frequencies are favoured in this respect since line-of-sight propagation makes the use of external dumps easier. At mm-wavelengths, on the contrary, this is a big problem since the use of rather large internal dumps, as required, made of highly absorbing materials, is easily recognized to be hardly compatible with the ITER environment.

Paradoxically enough, one could say that the main problem about source noise is that, apart from its probable existence, quite nothing is known about it. Measurements of relative spectral noise levels of order $10^{-19}\,\text{Hz}^{-1}$ and decreasing with frequency have been reported only recently on a mm-wave gyrotron[4]. The total antenna decoupling required to make this noise negligible in an alpha CTS diagnostic would still be of about 60dB. At higher frequencies the situation might be more favourable, but account must be taken of the fact that, as already mentioned, there the CTS bandwidths are fractionally much smaller. The treacherous nature of source noise is well evidenced by the fact that it is generated only as long as the source is switched on. A dedicated noise measurement is therefore required whenever this noise is non negligible to allow recovering the true scattered signal. Moreover, the Gaussian character of this noise being easily understood to be far from granted, in a source-noise dominated CTS measurement the actual SNR would be not reliably predictable using eq. (16).

In conclusion, assuming that stray radiation is adequately rejected, it can be stated that a high-frequency alpha CTS diagnostic will be feasible in ITER conditions: i) with good SNR if the expected high power and energy performances of the future FEL will be fully exploitable; ii) with only marginally acceptable SNR if a CO_2 laser, with only slightly extrapolated performances, is considered. The viability of the gyrotron option is discussed in some more detail in the following section.

A POSSIBLE CONFIGURATION FOR MM-WAVE CTS IN ITER

Accounting for previous results we can also state that for an alpha CTS diagnostic at mm-waves to be feasible: i) suitable hot propagation windows must be found for both beams; ii) a scattering geometry must be defined, exploiting the available windows, with sufficiently large scattering angle ($\theta \leq 90°$) and relatively small magnetic angle ($\varphi \leq 60°$); iii) beam refraction must be limited or otherwise controlled; iv) a sufficiently high stray-radiation rejection factor must be achieved; v) ECE must be prevented to cause unacceptably low SNR. No general agreement exists so far on the existence of a reliable solution capable of accounting for all of these problems in ITER conditions. Accordingly, to illustrate a possible approach, we shall limit ourselves to discuss a preliminary investigation for OM propagation in the range $\omega_{ce} < \omega_o < 2\omega_{ce}$.

The plasma is taken to have circular magnetic configuration, Maxwellian distributions, $B = 5.7\text{T}$, $n_{eo} = 1 \div 2 \times 10^{20}\,\text{m}^{-3}$, $T_{eo} = 10 \div 30\text{keV}$, flat density profile, $n_e = n_{eo}(1-x^5)$, and parabolic temperature profile. The plasma center is displaced $d = 50\text{cm}$ with respect to the geometrical axis of the machine to roughly account for the Shafranov shift in the actual ITER configuration. Quasi-optical (Gaussian) ray-tracing is used in all cases.

To start with, the existence of a vertical transmission window (A) is put in evidence. As shown in Fig. 3, where the total transmission factor, t_o, for propagation across the whole plasma is plotted vs. frequency, in the range $f_o \cong 160 \div 180\text{GHz}$ transmission remains high up to $T_{eo} = 30\text{keV}$. This preliminary result suggests checking if a second window (B) exists in the same frequency range but now for horizontal propagation from the l. f. s.. A window at $f_o = 180\text{GHz}$ is put in evidence in Fig. 4, where t_o is plotted vs. minor plasma radius. Transmission now is total up to well beyond the plasma center on the h. f. s., but thereafter absorption quickly becomes total. Finally, the existence of a third propagation window (C), still at $f_o = 180\text{GHz}$, is put in evidence for beams propagating horizontally but now parallel to the magnetic field at their starting point, set on a plasma radius, where the beam waist is located. As shown in Fig. 5, total transmission in the whole parameter range, up to $n_{eo} = 2 \times 10^{20}\,\mathrm{m}^{-3}$ and $T_{eo} = 30\mathrm{keV}$, now is achieved within the outer half plasma radius on the l. f. s..

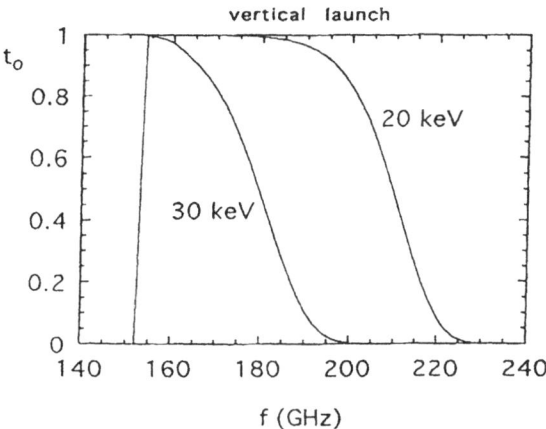

Fig. 3. Transmission factor t_o vs. frequency for vertical OM propagation with $n_{eo} = 2 \times 10^{20}\,\mathrm{m}^{-3}$, $T_{eo} = 20$ and 30keV. A hot propagation window where $t_o \geq 0.8$ is evidenced in the range $f_o \cong 160 \div 180\mathrm{GHz}$.

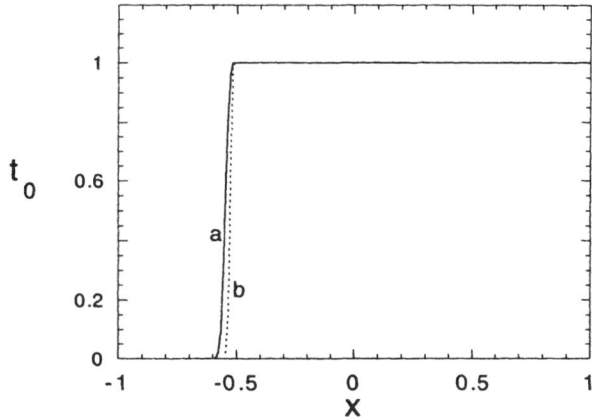

Fig. 4. Transmission factor t_o vs. minor plasma radius for horizontal OM propagation in radial direction at $f_o = 180\mathrm{GHz}$. a) $T_{eo} = 30\mathrm{keV}$, $n_{eo} = 2 \times 10^{20}\,\mathrm{m}^{-3}$; b) $T_{eo} = 10\mathrm{keV}$, $n_{eo} = 2 \times 10^{20}\,\mathrm{m}^{-3}$.

Window B is suitable for radial injection of the gyrotron beam. The use of window A for radiation collection is prevented by the requirement of sufficiently low magnetic angle. On the contrary, and in spite of the limitation in the accessible region, window C is basically suited. By the reciprocity principle, what is actually meant here is that a receiving antenna system must be used such that its radiation pattern fully reproduces the 'receiving' beam considered above. A convenient scattering geometry is defined by windows A and B since at the measuring point the basic angles are $\theta = 90°$ and $\varphi = 45°$. It is also worth noting that a frequency range including $f_o = 180GHz$ is proposed also for ECRH on ITER[17].

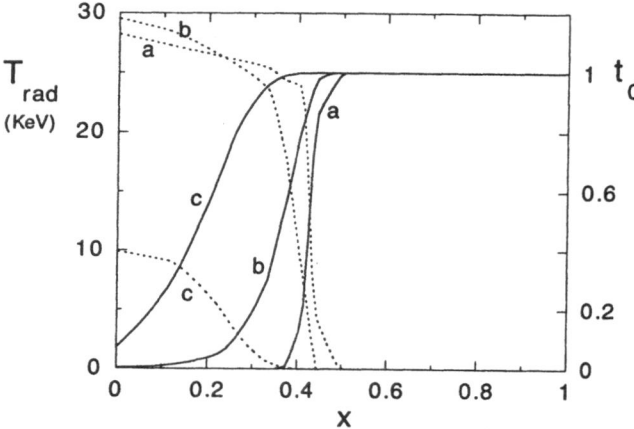

Fig. 5. Total transmission factor, t_o, (*solid lines*) and radiative temperature, T_{rad}, (*dotted lines*) vs. outward displacement from the plasma center for beams propagating horizontally and parallel to the magnetic field at their initial point, where the waist is located. Three cases are considered: a) $n_e = 2 \times 10^{20} m^{-3}$, $T_e = 30keV$; b) $n_e = 1 \times 10^{20} m^{-3}$, $T_e = 30keV$; c) $n_e = 1 \times 10^{20} m^{-3}$, $T_e = 10keV$.

As clearly shown in Fig. 5, in the geometry considered the plasma region accessible to the CTS measurements will be limited to the outer half plasma radius on the l. f. s.. It can be stated, however, that as regards the alpha velocity distribution this region is sufficiently representative of the whole ITER plasma. Accounting for the ITER profiles, at x=1/2 the electron density will still approach the central density, while the electron temperature will never be less than 2/3 of the central value. On the other hand, no significant differences are expected to be introduced in the shape of the alpha distribution in going from x = 0 (center) to x = 1/2 since, according to theory, most of the phenomena leading to possibly anisotropic and/or localized alpha losses are expected to occur in the outermost plasma. A further expansion of the absorption zone toward the l. f. s., with respect to the one considered here, might be produced in the presence of strong suprathermal tails in the electron distribution induced by additional heating. Sufficiently wide margins should be nevertheless available in most of the plasma parameter range to avoid this danger.

The 'receiving' beam propagating in window C is affected by refraction in the form of a progressive bending toward the vessel wall. The bending will of course increase with plasma density. However, as shown in Fig. 6 where the refracted beams are overlapped in the whole density range of interest ($n_e = 0.5 + 2.5 \times 10^{20} m^{-3}$), all the 'receiving' beams will cross the vessel wall in a relatively limited region, of no more than 1m size. Signal collection should therefore be possible, in principle, by using a single 'quasi-elliptical' mirror. This mirror should be suitably shaped to provide in all cases, i. e. independent of density, beam focusing in a small second focal region set outside the vessel. Suitable corrections to the propagation direction of the 'receiving' beam with respect to $\theta = 90°$ will be needed in a more detailed design to properly match an equatorial port. A vertical displacement due to poloidal field effects must be also accounted for.

The stray-radiation problem is effectively solved in the geometry in question since total beam absorption on the h. f. s. provides a very efficient beam dump. Unwanted rises in ECE level due to such power absorption can be avoided by modulation of the gyrotron beam. So, only the most difficult problem of ECE noise remains. As a matter of fact, should blackbody ECE produce noise temperatures as high as $T_{n,ece} = T_{rad} = T_{e0} = 30keV$, in single-pulse gyrotron operation with $\tau_{meas} = 100ms$ and $P_o = 1MW$, for an off-axis measurement at x = 1/2 this would give SNR ≈ 0.1. The failure then would be complete since no suitable and ITER-compatible viewing dump is expected to be available.

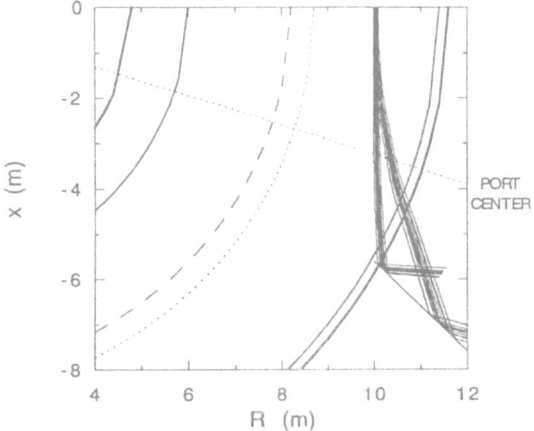

Fig. 6. Overlap of refracted 'receiving' beams in the density range $n_e = 0.5 + 2.5 \times 10^{20} m^{-3}$ to put in evidence the limited region of wall crossing. The bending of the beams increases with density.

However, two kinds of remarks are worth to be made in this respect. First, the predicted long burn times of the ITER plasma will provide significant margins to increase the SNR. In the limiting case with $\tau_{meas} = \tau_{burn} = 10^3 s$ a gain of 10^2, leading to SNR ≈ 10, would be achieved without any dump. Second, in the geometry considered an ECE-noise temperature $T_{n,ece} = 30keV$ will be attained only provided that the blackbody ECE is fully isotropized as a consequence of multiple reflections at the vessel wall. Now, this is very questionable since any radiation emitted from a point of the optically-thick plasma region will be actually totally reabsorbed at a point set symmetrically in the same region after a single wall reflection. Multiple reflections are therefore expected to be present at most in connection with: i) ECE coming from the optically thin outermost plasma region; ii) the usually small diffusive component always present even in a specular reflection process. If this is so, the overall ECE radiation entering the 'receiving' antenna will be made of: i) a *direct* contribution constituted by radiation which propagates within the radiation pattern of the antenna itself; ii) an *indirect* contribution due to multiple reflections. As suggested by the fact itself that total transmission is achieved, and as shown in Fig. 5, however, in the plasma region accessible to the CTS measurement the radiative temperatures (obtained by integration from wall to wall following the 'receiving' beam) associated to the direct contribution are actually negligible. So, only the indirect contribution should be left, leading to $T_{n,ece} \ll 30keV$. The degree of uncertainty always associated to conclusions that widely depend on the validity of a wall-reflection model, as ours, could be eliminated by suitable simulation tests in a high temperature plasma.

FINAL REMARKS

In spite of many progresses, the CTS community is not yet in the condition of providing a fully reliable design of an alpha CTS diagnostic for ITER. While difficulties of inherent character are present, the lack of experience remains a primary cause of the present delay. It is therefore important that, following the availability of mm-wave gyrotrons, in recent years thermal CTS diagnostic esperiments at mm-wavelengths have been started in a number of devices (W7-AS[4], JET[5], FTU[18], etc.). So, new experimental results will be available in the near future in this frequency range and a number of ITER-relevant technical problems will find a more precise definition. It is nevertheless also easily predicted that any definite answer as regards mm-wave CTS on ITER will unavoidably still rely at a large extent on simulation work and thoroughful modeling of the ITER plasma. The preliminary investigations of possible CTS scenarios carried out so far quite erratically should therefore evolve toward a more coordinated and systematic work up to the achievement of clear cut and unanimously agreed conclusions on the viability of this option.

Since rapid convergence of the work mentioned above cannot be *a priori* granted, one could alternatively rely since now on either the CO_2 laser option or on the FEL option. The intrinsic limitations of the former solution have been already stressed. It is also worth stressing here that, even apart from the source problem, the FEL solution is by no means devoided of relevant difficulties. As an example, the implementation of an otherwise convenient radial geometry would require a mirror recessed in the inner blanket, which is easily recognized to be rather hazardous. More generally, even in this case the necessity of a database and of preparatory experimental work is not eludible. If pure academy is to be avoided, therefore, the strongest near term implication of choosing the FEL option for ITER would be that the parallel necessity will rise of implementing new and more advanced FIR-laser based thermal CTS experiments (of the TCA type) in some of the existing devices. Should meanwhile the gyrotron be demonstrated to provide a convenient solution for thermal CTS at a non-ITER scale, making this need a real fact will represent in itself a very significant test of the credibility of the ITER project.

ACKNOWLEDGEMENTS

Precious collaboration of S. Nowak in preparing this work and stimulating discussions on several CTS-related topics with E. Suvorov, I. Fidone, G. Giruzzi, S Cirant, and M. Lontano are acknowledged.

1. R.E. Shefer et al., *Rev. Sci. Instrum.*, 61:3214 (1990).
2. R. Behn et al., *Phys. Rev. Lett.*, 62:2833 (1989).
3. R.K. Richards et al., *Appl. Phys. Lett.*, 62:28 (1993).
4. E.V. Suvorov et al., *IX Joint Workshop on ECE and ECRH,* Borrego Springs (1994).
5. J.S. Machuzak et al., *IV IAEA Meeting on Alpha Particle in Fusion Research,* Princeton (1995).
6. J.A. Hoekzema et al., *XXII EPS Conf. on Contr. Fusion and Plasma Phys.,* Bournemouth (1995).
7. J. Sheffield, *Plasma Scattering of e.m. Radiation,* Academic Press, New York (1975).
8. S. Cirant, S. Nowak and A. Orefice, *IFP Rep.* FP 94/3 (1994).
9. A.E. Costley, J.A. Hoekzema and T.P. Hughes, *ITER Rep.* IL-PH-7-0-10 (1990).
10. D.P. Hutchinson et al., *Rev. Sci. Instrum.*, 56:1075 (1985).
11. H. Bindslev, *Plasma Phys. and Contr. Fusion*, 33:1775 (1991).
12. S. Galbiati, M. Lontano and U. Tartari, *Plasma Phys. and Contr. Fusion*, 33:1049 (1991).
13. L. Vahala, G. Vahala, and D. J. Sigmar, *Nucl. Fus.*, 28:1595 (1988).
14. I.H. Hutchinson, *JET Rep.* R(87)07 (1987).
15. I. Fidone and G. Granata, *Phys. Plasmas*, 1:1231 (1994).
16. F. Orsitto and U. Tartari, *Rev. Sci. Instrum.*, 66: 2 (1995).
17. M. Makowski, *Meeting on ITER Tasks on ECRH,* Cadarache (1995).
18. M. Lontano et al., *V Joint Russian-German Meeting on ECRH and Gyrotrons,* Garching (1993).

REVIEW OF POSSIBILITIES FOR A COLLECTIVE THOMSON
SCATTERING SYSTEM ON ITER

F. Orsitto

Associazione EURATOM-ENEA sulla FUSIONE
C.R.Frascati-C.P.65 - 00044Frascati(ITALY)

INTRODUCTION

The radiation source wavelength useful for a Collective Thomson Scattering (CTS) on ITER is analyzed in this paper, and the related scattering geometry , signal to noise ratio and technology needed for the measurement of the alpha particles are specified. The ITER plasma parameters taken for reference[1] in the calculations are: n_e= 1.2 $10^{20}m^{-3}$, T_e = 22keV , T_i = 20KeV , n_α/n_e = 0.0017 , B_T=5.7T , R =8.1m,a=3.0m,k=1.55.The plasma, machine and alpha parameters are quite different from those included in the first proposal[2] for a Collective Scattering System, where a FEL(Free Electron Laser) source at 1.5THz(λ=200μm) was proposed.In the present work three sources are considered with wavelength λ=10.6μm(CO2 laser P(20)line), 200μm(FEL),4.61mm(65GHz gyrotron). For the evaluation of source figures the calculation of ECE background radiation level is needed, after some geometry constraints are discussed,and then the evaluation of scattered power and signal to noise ratio is performed, adding observations on the technology needed.

ECE BACKGROUND RADIATION EVALUATION

The evaluation of the plasma emissivity $\eta^{(X,O)}$ (expressed in $W/(m^2$ sr (rad/sec) m) for the X and O modes in the direction perpendicular to the magnetic field, where it is maximum, is performed using the formulas given in ref.3, in this way the power entering the detection system can be calculated.

ECE background radiation evaluation for laser based CTS.

For laser based systems(λ=10.6μm,200μm) the condidtions $\omega*=\omega/\omega_c$>>1 and $\omega*>>(v_{th}^2/2c^2)^{2/3}$ are met ($f_c=\omega_c/2\pi$ is the EC frequency, vth the electron

Diagnostics for Experimental Thermonuclear Fusion Reactors
Edited by P. E. Stott *et al.*, Plenum Press, New York, 1996

527

thermal velocity), so the evaluation of $\eta^{(X,O)}$ in the perpendicular (to the magnetic field) direction can be performed using the high temperature tenuous plasma approximation for the continuum part of the spectrum[3].So the plasma emissivity is:

$$\eta^{(X,O)} = \omega^{*2} \, T_{kev} \, (f_p^2 \, f_c \, / c^3) \, 1.6 \, 10^{-16} \, G(\omega^*, \mu)$$

where $\mu = (c/v_{th})^2$, f_p the plasma frequency, c the speed of light. The equivalent temperature of the radiation entering detection system is evaluated using the following formula:

$$T_{N,eV}^{(X,O)} = 2\pi \, \eta^{(X,O)} \, \lambda^2 \, a \, / \, 1.6 \, 10^{-19}$$

where a is the ITER minor radius. It results that $T_N \sim 0$ for 10.6µm while $T_N \sim 1 keV$ for 200µm.

ECE evaluation for 65GHz gyrotron CTS.

The evaluation of the plasma emission at 65GHz ca be performed computing the optical depth for both modes. Approximate calculations show that the X-mode optical depth at 65GHz is of the order of $\tau_X = 10^{-5}$ with a maximum noise temperature of $T_N \sim 10 eV$.

CONSTRAINTS ON VALUES OF α AND SCATTERING GEOMETRY

The collective scattering regime[4] is identified by a value of the Salpeter parameter $\alpha_s > 1$, and this give a constraint on the scattering angle θ at a given incident wavelength.For the previous plasma parameters, we find $\alpha_s = 790$ $\lambda(m)/\sin(\theta/2)$, where θ is the scattering angle, and λ the incident wavelength; so $\theta(10.6µm) < .9^0$ while $\theta(4.61mm)$ is not limited and $\theta(200µm) < 18^0$. The density of alpha particles is so low that the scattered signal due to alphas(S_α) can be of the order of that due to the plasma electrons(S_e), even at angles where the collective scattering signal is maximized.The ratio of alpha to electron signal is given by [4]

$$R_{\alpha e} = S_\alpha / S_e = 2.7 \, \alpha_s^4 \, (n_\alpha / n_e)(v_{th} / V_\alpha) \cos \phi$$

where maxwellian alpha scattered spectrum was used($\sim e^{-(\omega/\omega_\alpha)^2}$, $\omega_\alpha \sim KV_\alpha$) V_α the alpha velocity , ϕ the angle between the magnetic field direction and the scattering wavevector K. The condition of detectability of the alpha feature can be $R_{\alpha e} > 2$. This implies a constraint : $\alpha_s > 3 / \cos^{1/4}\phi$,($n_\alpha = 210^{17} m^{-3}$, $V_\alpha = 1.3$ $10^7 m/s$, $v_{the} = 6 \, 10^7 m/s$, $n_e = 1.2 \, 10^{20} m^{-3}$).If an angle $\phi < 50^0$ is chosen, we have $\alpha_s > 3$; and this value implies that

$$\theta(200µm) < 3^0 \, , \quad \theta(10.6µm) < 0.3^0, \quad \theta(4.61mm) \text{not limited.}$$

Using this constraint, the tangential access must be chosen for the 10.6µm laser for easier separation between the principal and scattered beam, while the backscattering for the 65GHz gyrotron is possible.

EVALUATION OF THE SCATTERED POWER AND SIGNAL TO NOISE RATIO

The temperature of the scattered radiation $(Ts)^4$ is given by:

$$Ts(eV)= C(\lambda) \, (n\alpha(m^{-3}) \, / \, 10^{18})(P0/10^6 W)$$

P0 is the incident power and $C(\lambda) = 48.6 \cdot 10^4 \, \lambda(m)^2 \, / \sin \theta$. Table I reports the values of the parameters entering the scattered signal calculations, corresponding to the three wavelength chosen, the temperature of the scattered radiation is calculated for 1MW of incident power.

TABLE I
Scattering Parameters

λ	200μm	10μm	4.61mm
K(m-1)(deg.)	2905(θ=5)	3300(θ=.3)	1927(θ=90)
LΩ(sr*m)	2510-6	1.3 10-7	803 10-6
$\omega_\alpha/2\pi$(GHz)	6	7	9
C(λ)	.21	9 10-4	10
Ts(eV)	.042	1.8 10-7	2
w0(m)	0.01	0.96 10-2	16 10-3
L(m)	.23	3.6	0.03

where $\omega_\alpha/2\pi$ is the alpha bandwidth, w_0 the beam waist radius into the plasma, L the spatial resolution.

SIGNAL TO NOISE(SNR) CALCULATION FOR 200 μm AND 10.6μm

The well known formula of the SNR for a heterodyne detection system equipped with a filter bank, can be used for determining the power needed to obtain a SNR=2 on a channel 1GHz wide:

$$SNR=Ts/(Ts+NEP/\varepsilon + T_N) \, (B\tau)^{1/2}$$

where τ is the integration time ($\tau \sim 2\mu s$), NEP the noise equivalent power of the detector (NEP~1eV), ε the heterodyne efficiency(which is a measure of the mismatch between the wavefront of the scattered radiation and that of the local oscillator, η<0.5),B the bandwidth of the considered channel of the filter bank (B=1GHz),T_N is the ECE radiation noise temperature. At τ=2μs, for obtaining a SNR=2 an incident power of 45MW(90J) is needed for 10.6μm laser while 1.5GW for 200μm.

FEASIBILITY OF KILOJOULES CO_2 LASER SOURCES

For the measurement of the alpha particles a source at 10.6μm is needed with the following figures: E~1kJ, $\tau \sim 2\mu s$,$2w_0 > .02m$. This laser (at least the components) is available commercially: it can be built by an hybrid oscillator, one preamplifier(active volume length~2m) in double pass and a final high pressure electron beam pumped amplifier (active volume length ~2m) . This is a compact and (relatively)low cost source[5].

The detection system can be built using a photoconductive HgCdTe with a $NEP{\sim}10^{-19}W/Hz$ @10μm.So no development is needed for the laser source and the detector.The problem here is the spectral purity of the laser(this problem is met also with the other sources considered in this paper): it must be guaranteed that eventual spurious longitudinal modes do not fill the scattering band. The laser system must be designed for the best contrast between the main amplified mode and spurious modes, so notch filters made with hot CO_2 or N_2O cells must be used along the laser chain, and the hybrid oscillator must very well stabilized on the CO_2-P(20) line(λ=10.6μm).

FAR INFRARED LASER SOURCE AT 200μm

For this source a FEL of 1.5GW(2μs) is needed, so for a pulse length of 100ns a repetition rate of 1kHz is enough.The bandwidth required for the laser must be less than 100MHz and the noise level of the source in the scattering band must be at least 80dB down the peak power: it can be speculated that in principle the FEL at high power(i.e.using LINAC) have an intrinsic limit in these quantities.

CONCLUSIONS ABOUT THE CTS LASER BASED SYSTEMS

The analysis shows that laser sources and detection systems can be built for the measurement of the alpha spectrum achieving enough signal-to-noise ratio for 10.6μm.

MICROWAVE SOURCE (65GHz,λ=4.61mm)BASED CTS SYSTEM.

The 65GHz gyrotron source has been chosen because this frequency is well below the fundamental EC frequency for ITER, where low background radiation emission from plasma can be hoped, and the propagation of extraordinary mode can be allowed for the ITER plasma parameters.
The equivalent temperature of the ECE emission limits the SNR of the measurement in the microwave region.Calculations of ITER ECE spectra show complete overlap of the ECE first and second harmonic, so only below the first harmonic (f_0 = 159.6 GHz) a region with low ECE emission can be found.It turns out that (see Tab.I) also at ECE noise level of 1keV, a SNR=2 can be obtained with ms integration time on a channel 1GHz wide.

REFERENCES

1. 1st Meeting ITER Exp. Group - S CX MI 1 94-08-03 F1.
2. Iter Information S92 RE2 21-oct-94 F1.
3. M.Bornatici et al.Nucl.Fus.23(1989)1153.
4. F.Orsitto- Rev.Sci.Instr.61(1990)3093 and Ref.therein.
5. R.Behn et al.-Phys.Rev.Lett.62(1989)2833.

DIVERTOR DIAGNOSTICS FOR JET

Paul R. Thomas for Experimental Divisions I and II
and the Diagnostic Engineering Group

JET Joint Undertaking
Abingdon, OX14 3EA
United Kingdom

INTRODUCTION

The installation of the Pumped Divertor in JET has brought about a change in emphasis of the diagnostic capability. Twelve new systems were installed to diagnose the divertor plasma and many others were enhanced to improve their contribution to this effort. This paper will describe the thinking behind the divertor diagnostics and the experience of their use which has been obtained in the 1994/5 experimental campaign.

Some of the diagnostic designs have involved novel responses to problems which are characteristic of JET and give a foretaste of issues in ITER; operation at high temperature, remote handling, neutron irradiation, tritium compatibility and large plasma fluences. Examples of systems in which most of these are concerns are the target Langmuir probes and the divertor bolometers. Instrumentation within the vacuum vessel has required electrical feedthroughs and wiring which are compatible with this working environment. Some of the difficulties encountered are indicative of the attention which will have to be paid in ITER to such detail.

Putting data together from all of the divertor diagnostics is essential to the formation of a coherent picture of the divertor plasma. This requires both sophisticated analysis programmes and simulation codes to fill in the gaps in the diagnostic capability. The development of these programmes is proceeding apace; particularly now that operations have finished for the rest of 1995.

REQUIREMENTS OF THE JET DIVERTOR DIAGNOSTICS

It has become apparent that the thickness of the scrape-off layer does not increase with the size of the tokamak. This simple fact means that it will not be possible to exhaust by conduction the hundreds of MW of power from a burning ITER plasma to the divertor target because the power density would be too high. The JET Pumped Divertor programme was motivated by the need to find a solution to this difficulty which is, at the same time, compatible with the removal of helium ash from the plasma core.

The only viable approach which has been proposed is to radiate and charge exchange the plasma loss power from the divertor legs and to allow it to fall on an extended target region. The details of how this may be accomplished are unimportant for this discussion and may be regarded as the essential output of the divertor programme. However, it can be said in general that the domination of atomic physics in the divertor region requires a radical decrease of plasma temperature along the divertor channel from the 100-200eV typical of the SOL at the outboard seperatrix to a few eV at the target. This is accompanied by a commensurate

Diagnostics for Experimental Thermonuclear Fusion Reactors
Edited by P. E. Stott *et al.*, Plenum Press, New York, 1996

531

increase in plasma density. When such conditions are pushed to their extreme limit, the plasma can detach from the target and to which the flows then become negligible. The plasma pressure need not be a constant along the divertor field lines because of the momentum loss by charge exchange.

As well as an experimental investigation of this divertor solution, the JET divertor programme called for an intense modelling effort, coupled to experiment, so that properly validated predictions of the performance of future machines, such as ITER, could be made. The divertor models are required to simulate the diagnostic signals so that direct comparisons can be made with experiment.

The essential features of the desired divertor regime defined the main requirements of the diagnostics. In so far as it was possible, measurements should allow the determination of plasma parameters along the divertor legs. Not only is this required for model validation but, more importantly, to steer the operation of the tokamak and divertor towards the required plasma conditions and spreading of the heat load. At the point of detachment, the divertor plasma is either unstable or close to instability. Thus diagnostics are required as sensors for feedback loops to control gas flow, impurity injection or input power.

Clearly, since it is of primary concern, the power balance in the divertor region must be monitored. Target Langmuir probe arrays, infra-red cameras, thermocouples and bolometer cameras are required to determine where the power is going and through which channel. Particle flows must also be monitored; in particular, the species arriving at the pumps must be measured so that the ash removal capability can be determined.

Unfortunately, tokamak plasmas are not quiescent and steady! Of particular concern to divertor diagnosis are ELMS and disruptions. These can have an important effect on the balance of power and its distribution in the divertor region. Thus, it is an essential requirement that the divertor diagnostics have the time resolution to be able to resolve these events; something of order $10\mu S$ being useful.

THE JET DIVERTOR DIAGNOSTICS

JET has given diagnostic designers the first taste of the problems which will be encountered in a device such as ITER. Divertor diagnostics are particularly difficult because, in many cases, they must be built into the divertor structure and so are exposed to the high operating temperature (300°C), tritium and irradiation by 14MeV neutrons. Clearly, neutron irradiation is only a concern in JET so far as the correct operation of the sensors is concerned. Also, all equipment in the vessel must be compatible with remote handling. A later section will single out electrical feedthroughs and in-vessel cabling as an example which encapsulates most of these environmental concerns and is indicative of the further strides which will have to be made in the instrumentation of ITER.

Since JET was not designed for a divertor it had some characteristics which made access for divertor diagnostics rather difficult; there are no divertor ports, as in ASDEX-U, and the large number of TF coils make for a rather restricted view from the midplane. Thus, in order to make visible spectroscopy measurements, six periscopes were installed which peer over the lips of the divertor side-walls and bolometer cameras had to be integrated into the divertor structure. This greatly complicated what might otherwise have been relatively simple jobs. Nonetheless, the resulting systems are not unlike those which will be mounted in ITER's divertor cassettes and much useful experience has been gained.

Installation of the Pumped Divertor resulted in the loss to existing diagnostics of nearly all the bottom ports. All those systems had to be moved to other locations. In a number of cases this had a knock-on effect because in meant rearranging other diagnostics. On top of this, other systems had to be modified to cope with the new plasma configuration or were upgraded for other reasons. In association with the divertor diagnostics, a lithium beam probe and an improved reflectometer system were added to improve coverage of the SOL. The net result was that almost every existing diagnostic was being worked simultaneously with the new ones. The Experimental Divisions and the Diagnostic Engineering Group were rather stretched and it is a tribute to their efforts that the diagnostic coverage obtained in the 1994/5 experimental campaign was as good as it was. This must surely indicate that in order to adhere to the advertised programme, ITER must devote adequate resources to the preparation of diagnostic systems.

The scope of this paper is not sufficient to give detailed coverage of all the divertor diagnostics. Table I lists them all with some indication of their purpose and status. A brief description of each and the results obtained will be given in the following.

Table 1. The JET divertor diagnostic systems.

System	Purpose	Status
Target viewing (KLn)	Strike point positioning, target temperature and spatial distribution of line emission using filters	operational
Divertor bolometry (KB3)	Distribution of radiation in the divertor and total power radiated from the divertor	operational
Target calorimetry (KD1D)	Power to the divertor targets	operational
Divertor LIDAR (KE9)	Absolute n_e and T_e	technically operational but providing no usable data (see text)
Interferometer (KG6)	Line density on three chords	operational
Comb reflectometer (KG7)	Peak density on three chords	operational
Electron cyclotron absorption (KK4)	T_e profile on three chords (in conjunction with density data)	operational
Scanning VUV monochrometer (KT1)	Spatial distribution of brightest VUV lines	operational in previous campaigns but suffered vacuum failure at start of 94/5 (see text)
VUV/XUV Survey (KT7)	Spectral survey 7-1000eV with slow spatial scan (4Hz)	operational
Target Langmuir probes (KY4)	V_f, T_e and I_{SAT} along target	operational
Divertor pressure gauges (KY5)	neutral pressure around target and pump	operational
Thermal helium beams (KY7)	n_e and T_e near target using plasma electron excitation of visible helium lines	operational
Magnetic pick-up coils (KC1)	magnetic equilibrium	operational
Periscopes (KT6)	Multichord visible spectroscopy and divertor plasma viewing	operational
Toroidal viewing (KT5)	Flow velocities in divertor	installed but not brought into operation

Target Viewing

Several systems are involved here: CCD cameras equipped with filters (usually CII and D_α); a CCD viewing the divertor plasma at a tangent to the flux surfaces using one of the periscopes; some IR diodes to provide temperature measurements for target protection; and an IR array, sensitive in the range 800-1650nm and giving a 3mm resolution across the target. The IR measurements have been of importance for the protection of the target tiles, particularly during the beryllium experiment. Still, colour images of the target have been available between pulses; partly to monitor target damage and partly to investigate possibilities of measuring erosion and deposition processes.

Divertor Bolometry

The bolometer cameras are a development of the design used previously on JET to enable them to operate at the ambient temperature of the divertor without cooling. The resistance meanders and absorbers are deposited upon mica instead of kapton. The deposition process required considerable development, particularly in light of the irreversible phase transition in gold at 350°C (which changes the resistivity of the deposited film), as did the wiring and electrical connections. Particular care had to be taken to eliminate microphonicity in the wiring. Each of the 7 cameras has 4 channels. Measurements of very high quality have been taken with this system, in conjunction with its main plasma counterpart (KB4). Inversion for the radiation distribution has been accomplished using a method which constrains the short wavelength variations to be perpendicular to the flux surfaces [1].

Target Calorimetry

This consisted of conventional thermocouples set into the target tiles. Unfortunately, bad thermal and electrical contacts greatly diminished the usefulness of this diagnostic.

Divertor LIDAR

A mirror system mounted in one of the main radial ports is used to view the divertor region. The divertor LIDAR was technically very ambitious and, whilst technically successful, has not provided useful data primarily because the electron density was much lower than had been anticipated and because of detector overload from the flash of reflected light from the divertor target.

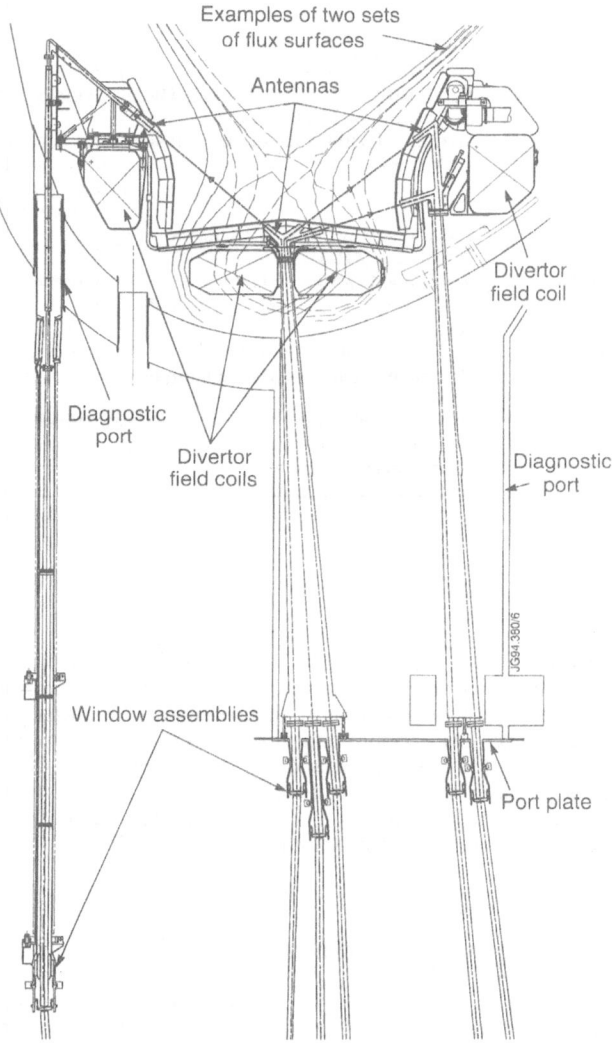

Figure 1. The waveguide assemblies and antennae for the divertor microwave systems.

Microwave Interferometer

The interferometer shares waveguide access with the reflectometer and the electron cyclotron absorption (figure 1). There are two lines of sight across the outer divertor leg and one across the inner. Quasi-optical components have been used extensively to keep transmission losses to a minimum. In order to eliminate phase shifts due to mechanical displacements, a two frequency (130 and 200GHz) system was used. The system works well up to densities of 10^{20}m^{-3} where refraction causes unacceptably large signal losses and consequent, unrecoverable fringe jumps. This occurs in a sufficient fraction of useful plasmas that the low frequency interferometer is being replaced with another at >200GHz. Excellent agreement is obtained between the interferometer line density and the peak density inferred from the cut-off seen by the ECA diagnostic, as shown in figure 2, if it is assumed that the plasma has a transverse scale length of 5cm.

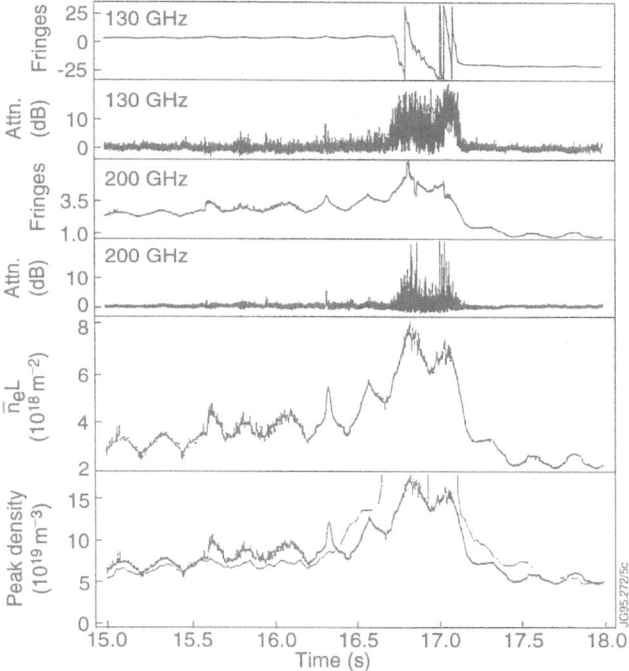

Figure 2. Traces comparing the line integral density from the interferometer with the peak density inferred from the cut-off seen by the ECA system. The fringe loss and attenuation suffered by the 130GHz channel at high density can be seen clearly in the top two traces. The next two traces show that the 200GHz channel fares much better. The next trace shows the line density obtained using the 200GHz channel alone. The 4Hz movement due to the plasma sweeping can be seen. Finally the interferometer is compared with the ECA cut-off frequency by assuming a density scale length of 5cm..

Comb Reflectometer

The comb reflectometer is intended to give a crude but reliable measure of the peak density. It has 8 sources between 50GHz and 100GHz whose reflections from the plasma are picked up with heterodyne detectors. The peak density is determined from the frequency at which plasma reflection is lost; ie. a direct measurement of the cut-off frequency. Good

535

agreement is obtained with the peak density obtained from the cut-off seen with the ECA system.

Electron Cyclotron Absorbtion

The brightness of the electron cyclotron emission from the main plasma is such that it could not be used for diagnosis of the divertor plasma. However, a measurement of electron cylotron absorbtion is possible and should be able to provide a profile of electron temperature, when used together with the electron density. The diagnostic uses a BWO which is swept from 120GHz to 180GHz and a heterodyne detector. Since the BWO is swept at a constant rate, the length of the waveguide run ensures that the ECA signal is detected at a fixed frequency That the system works is attested to by the good agreement between the measured cut-off density and those determined by the reflectometer and interferometer sytems, as described above. However, thus far, no electron cyclotron absorbtions has actually been seen. This is a puzzle which has, so far, not been solved. Apparently the electron temperature is much lower than anticipated.

Scanning VUV Monochromator

This diagnostic is not new; it was the so-called "day-1" spectroscopic system. However, it came into its own when used to provide spatial scans of the emission of the brightest VUV lines in the divertor region. It is a Roland circle spectrometer, equipped with a microchannel plate and multi-anode detector. The spatial scan is achieved by rotating a multi-facetted mirror at about 200rpm. Unfortunately, the rotary feedthrough, which had previously proven reliable, failed catastrophically at the start of the '94/5 campaign and, in so doing, admitted a small amount of TMP oil into the torus. The instrument was banished until satisfactory, tritium compatible, rotary vacuum seals were produced to replace the previous design. The development process turned out to be rather tough; a double sealed feedthrough with an adequately small amount of mechanical play and a lifetime of >10^{6}cycles proving to be surprisingly difficult. This has now been accomplished and the diagnostic will be back in operation in the '96 campaign.

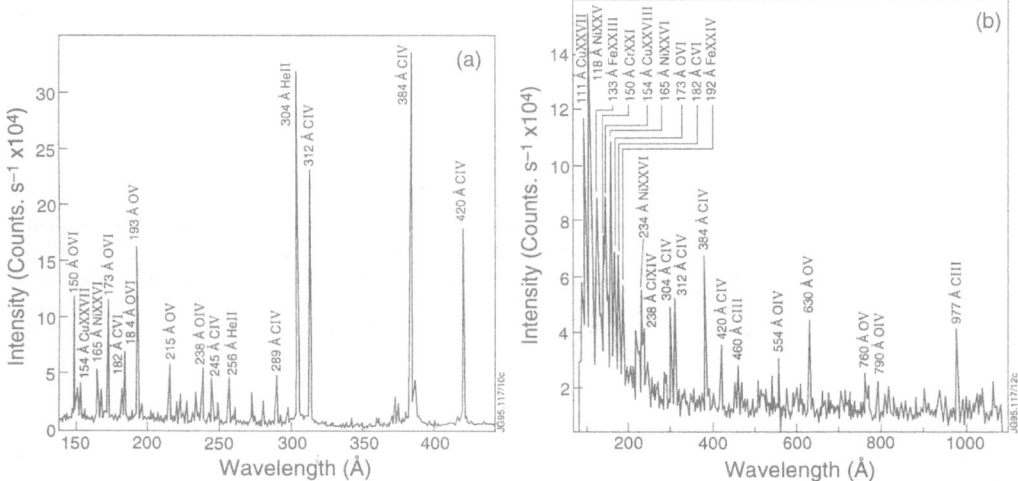

Figure 3. (a) The VUV spectrum obtained with the divertor XUV/VUV spectrometer and (b) the core plasma spectrum obtained with a similar instrument, looking across the midplane. Note that the metallic lines and the shortwavelength enchancement are negligible in the divertor spectrum.

VUV/XUV Spectrometry

This diagnostic is intended to provide spectral coverage in the range of photon energies which are characteristic of the electron temperatures encountered in the divertor. It consists of a double SPRED instrument, which has been specially designed to incorporate a significant amount of neutron shielding, and a SOXMOS spectrometer, looking between the SPRED channels. It includes a visible spectrometer to provide calibration by the branching ratio method. It has proven to be an extremely reliable instrument although some of its more advanced features, such as mechanical sweeping, have not yet been tested. A comparison of spectra from the main plasma and the divertor is shown in figure 3. It can be seen that the low wavelength "quasi-continuum" is relatively unimportant in the divertor view and that the intensity of the CVI line at 182 Angstrom is surprisingly small. This latter, it was thought, should have been excited by charge exchange between incoming, fully stripped carbon and neutral hydrogen in the divertor. The reason why this is not a bright line is, at present, unknown; it does however underline the need for very complete analysis and simulation packages to fully use diagnostics such as this.

Target Langmuir Probes

Experience in previous campaigns had shown that Langmuir probes set into the target tiles tended to have a very limited lifetime during high performance operation. This time, a careful design, paying attention to the thermal characteristics of the probe tips, has proven to be highly successful. All the probes survived the entire campaign, apparently unscathed. A drawing of a probe is shown in figure 4. There are 39 triple and 32 single Langmuir probes in the divertor. They were set in the gaps between the divertor target tiles. The triple probes normally make measurements at 5kHz but a few channels have been operated at 250kHz to resolve ELM structure.

A curious feature, which has been observed at plasma high density, is that the positive and negative saturation currents become more or less equal. It is thought that this results from the extra resistance caused by cross field transport and, if this is the case, the electron temperature needs considerable downward correction. Also, it is clear that the diagreement in hot-ion H-modes between power to the target inferred from probe measurements and the difference between input power and radiation implies substantial ion temperatures in the divertor plasma or secondary electron emission from the probe tip. This quantitative difficulty notwithstanding, the probes have proven to be a highly reliable and useful diagnostic. The ion saturation current has been used as a feedback sensor during detached plasma operation to control the feed of impurity doped deuterium.

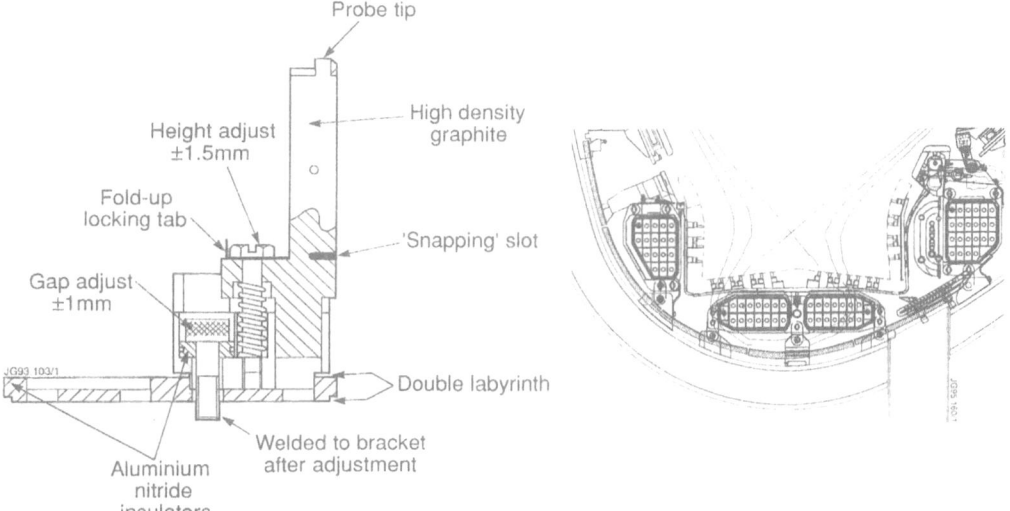

Figure 4. On the left, a drawing of a Langmuir probe and the layout of the probes in a poloidal section on the right.

Divertor Pressure Gauges

Fifteen ionisation gauges are placed in three groups at different toroidal locations. In each group, two are placed at the centre of the target, one each are placed at the inboard and outboard corners and one is underneath the cryopump. Apart from the latter, the time response is limited to 5ms by sampling tubes between the gauges and gaps in the target. A good correlation is seen between the ion current from the main plasma and the neutral pressure, particularly during ELMs, as shown in figure 5.

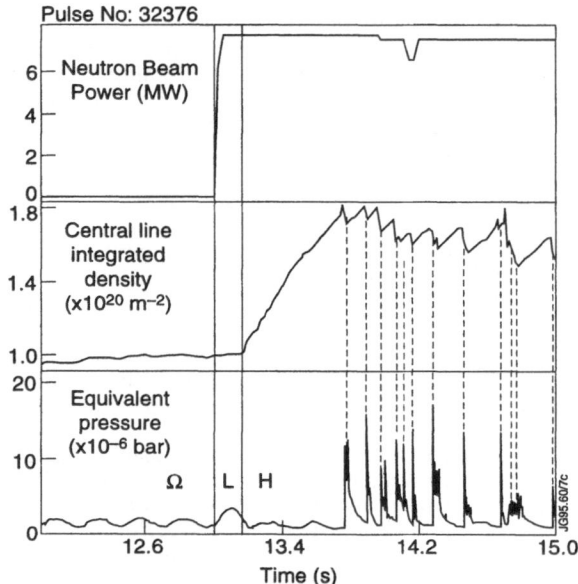

Figure 5. The time traces for an ELMy H-mode, showing the neutral beam power, the core plasma line integral density and the pressure in front of the divertor cryopump. There is a good correspondance between the loss of density from the core with the flux to the divertor.

Thermal Helium Beams

Two nozzles have between placed at nominal inner and outer strike zones, within the field of view of two periscopes, for the injection of helium gas. By measuring the ratios of intensities of the 668, 707 and 728nm lines of HeI, the electron density and temperature can be determined. This method has been established at TEXTOR [2] and, from early results is showing signs of promise at JET as well.

Magnetic Pick-up Coils

Since the plasma is now further away from the vessel walls than in previous configurations, new magnetic measurements were required in order to be able to recover the accuracy of equlibrium. A total of 93 new sensors have been added, including pick-up coils in the poloidal limiters and both pick-up coils and flux loops under the divertor target

ELECTRICAL FEEDTHROUGHS, WIRING AND CONNECTORS

The advent of large amounts of in-vessel instrumentation meant that the problem of providing feedthroughs, wiring and connectors, compatible with the high operating temperature (300°C), tritium and remote handling, had to be tackled. The requirements tend to produce conflicting demands on design, as will be apparent in the following.

JET was originally specified to be bakeable to 500°C. Thus, mineral insulated cable was chosen as the standard for use in the vacuum vessel. Although the vessel has only ever been baked to 350°C it is still very much the best choice because of its good insulation properties and mechanical toughness. Most of the cable in use has an OD of 1.5mm. It does have the drawback that it is quite stiff when grouped in twisted pairs. Consequently, cable looms are very unwieldy. To relieve this, glass insulated braided cable has been used for the majority of the cabling in conduits within the vacuum vessel. Whilst more compliant, the glass insulation is not particularly robust and this cable has a rather large OD of 3.5mm. In some parts of the installation, most notably the bolometers, slight inaccuracies in the length of the terminated cables tended to accumulate in junctions between conduits; so called "spaghetti junctions" (named after a notorious highway junction near Birmingham!). This was so bad in one case that the installation nearly had to be abandoned because of time pressure.

An early design of vacuum feedthrough had mineral insulated cables brazed into vacuum flanges with the space in between forming a pumpable interspace. Within the interspace, the outer sheath and insulation was removed so that tritium diffusing down the cable could be pumped. Vacuum breakdown between the leads rendered this design impractical. The design in use has 64 lengths of >0.5m of mineral insulated cable, brazed into a flange with "Quartex" terminations at each end forming tritium barriers and the slow diffusion rate through the cable providing an additional inhibition to tritium leakage. However, the terminations are fragile and the conductance to nitrogen through 2m of the cable is approximately 10^{-10} ℓs^{-1} at 300°C. It seems probable that this will have to be significantly improved for ITER.

Remote handling connectors proved to be quite a headache. Crimping of the pins on the cable ends proved to be unreliable; apparently good connections going resistive after thermal cycling. Some could be recovered by "zapping" them with short, high voltage pulses. However, some were not recovered and better crimping procedures will be followed in future. Difficulties during the manufacture of the feedthroughs lead to a change in design of the pins. Whilst reasonably accurate, their tips were insufficiently rounded and a number snagged and bent in the socket, in spite of checking the alignment in a jig before engagement. Other plugs and sockets, to the original specification, were completely successful.

The reason for dwelling on this seemingly trivial detail is that the most perfect sensor can be completely ruined if the wiring is inadequate! Conditions in ITER will be still more difficult than in JET and the requirements for remote handling more stringent. Much development and testing will be needed before commiting designs to active operation. Such considerations are important in many other areas as well but this one has been prominent in the JET experience.

CONCLUSIONS

In spite of the amount of work which had to be performed simultaneously, the majority of JET's divertor diagnostics have been brought into operation successfully. In some cases, the installation work continued well into the experimental campaign so the amount of data which was obtained is quite limited. Nonetheless, some interesting results have been obtained from all the systems and analysis is proceeding apace.

As with the core plasma diagnostics, a policy of duplication of measurements of the main plasma parameters has been adopted for the divertor diagnostics as well. This is particularly the case for the electron density and temperature diagnostics because of their importance for the analysis of other diagnostics and for model validation. It will be apparent from the discussion above that this was useful for the microwave systems. The ageement between the ECA cut-off and the interferometer density was such that it is clear that the ECA system works and that the lack of any absorbtion is, of itself, an interesting physics result.

Some of the diagnostics will need to be adapted to some of the non-ideal features of divertor plasmas which are encountered in practice. An example of this are the effects of refraction of the interferometer beams by the density gradients encountered at high density and the losses associated with ELMs.

Comparison of the more complex diagnostic signals with simulation packages and model codes has barely begun. Even at the level of comparison between D_α profiles and

Monte Carlo neutral simulations, some puzzling inconsistencies have been found. Some of them can be put down to missing details in the modelling of the divertor and its target. Others might be due to incomplete rendition of the atomic physics. Much of this will need to be sorted out if VUV/XUV spectra are to be understood and spectroscopic ion temperatures obtained, although considerable benfit is gained from developing both in parallel.

In principle, the diagnostics used in the JET divertor can be transferred to ITER. However, even the limited attempts in JET to deal with high temperature, tritium containment and remote handling have shown how important it will be to pay attention to even what seem to be the most trivial details. The effects of activation and neutron damage will make the engineering still more challenging! Experience on JET has shown that satisfactory diagnosis of the ITER plasmas will require a full and timely commitment of resources. This is especially true of the divertor diagnostics because of their close proximity to a reacting plasma.

ACKNOWLEDGEMENT

As well as the parts of the JET Team cited under the title, the diagnostic effort on JET has benefitted enormously from the assistance of our colleagues in the institutions of the Euratom Associations. This has ranged from responsiblity being taken for the preparation of entire systems to advice and sharing of experience being freely given. Too many individuals are involved to sensibly cite any but a heartfelt thanks are due to all.

REFERENCES

1. J.C.Fuchs, K.F.Mast, A.Herrmann and K.Lackner, Two Dimensional Reconstruction of the Radiation Power Density in the ASDEX Upgrade, 21st EPS Conference on Controlled Fusion and Plasma Physics, Montpellier, 1994. Volume 18B, Part III, page 1308.
2. B.Brosda, Modellierung von Helium-Atomstrahlen und ihr Einsatz zur Plasmadiagnostik der Tokamakrandschicht, PhD. Thesis presented to the Faculty of Physics and Astronomy, Ruhr-Universitat Bochum, 1993.

SPECTROSCOPY OF DIVERTOR PLASMAS

R. C. Işler

Fusion Energy Division
Oak Ridge National Laboratory
Oak Ridge, TN 37831, USA

ABSTRACT

The requirements for divertor spectroscopy are treated with respect to instrumentation and observations on present machines. Emphasis is placed on quantitative measurements.of impurity concentrations from the interpretation of spectral line intensities. The possible influence of non-Maxwellian electron distributions on spectral line excitation in the divertor is discussed. Finally the use of spectroscopy for determining plasma temperature, density, and flows is examined.

INTRODUCTION

Studies of impurities in plasma confinement devices have always constituted an integral part of fusion research efforts since it will be necessary to develop methods of controlling the core concentrations of these ions in a reactor to minimize the dilution of the D-T fuel and to suppress the bremsstrahlung radiation. As the fusion program moves toward power producing devices, such as ITER, it is becoming even more imperative to understand the means by which pruduction of impurities can be ameliorated and the techniques by which their transport into the core plasma may be inhibited. At present, it is envisioned that such impurity control, as well as power and particle handling, will be achieved primarily by poloidal divertors. However, the development of divertors that can deal with extreme heat loads presents a serious technological challenge because the power deposition tends to concentrate in narrow regions where the field lines strike the target plates, a situation that leads to severe demands on materials as well as to potentially large impurity production rates as a result of erosion. One solution proposed for successful operation is the creation of high-density, highly-radiating plasmas from which power can be distrbuted through charge exchange or through spectral line emission more uniformly over the divertor region. Such situations might be realized by the deliberate injection of impurities. The current ITER designs call for 400 MW of heating power with 300 MW delivered to the divertor. But the power incident on the target plates must be less than 50 MW, so the strategy is to transfer up 250 MW to the first wall and divertor chamber by atomic processes.[1] In presently operating tokamaks, where the efficacy of this approach must be verified, it is anticipated that nitrogen, neon, argon, and krypton will be tested as radiating targets and that beryllium, boron, carbon, molybdenum, and tungsten will serve as plasma facing materials on target plates.

Diagnostics for Experimental Thermonuclear Fusion Reactors
Edited by P. E. Stott *et al.*, Plenum Press, New York, 1996

Therefore, both from the necessity to understand intrinsic impurity production and the need to evaluate the feasibility of radiating divertors for dissipating power benignly, spectroscopy of divertor plasmas is becoming an intensive research area.

The goal of this paper is not to present specific designs for spectroscopic instrumentation in the ITER divertor but to outline the requirements that will be necessary for understanding impurity erosion and transport in light of the results that are rapidly accumulating from presently operating machines.

INSTRUMENTATION

A major goal of divertor spectroscopy in present machines is the determination of absolute concentrations and radiative losses of impurities as a function of location in order to benchmark predictive codes that can be used to model ITER operation. Unlike the main plasma where this problem can often be reduced to a one-dimensional characterization in terms of the minor radius, it is necessary to obtain rather complicated two-dimensional (or even three-dimensional) spatial distributions for divertors. An ideal experimental arrangement would incorporate multiple views over the entire vacuum ultraviolet region of the spectrum to observe the resonance lines of impurity ions (optically allowed excitations from the ground or metastable states) and to determine their distributions. The calculated excitation rates for the resonance lines are generally more accurate than for other transitions, and they are often only weakly dependent on electron temperature in the regions of interest, in contrast to lines originating from highly excited states. Morover, it is frequently impossible to detect suitable spectral lines in the visible region as a means of obyaining a comprehensive analysis of a particular impurity, i.e., to find lines which are both intense enough to be measured and for which accurate excitation rates are available. Visible spectroscopy is most useful for impurities with $Z \leq 7$ (nitrogen); it is difficult even to detect neon ions other than Ne II at long wavelengths. Nevertheless, the ease of obtaining spatial resolution from visible transitions, including ones that cannot be easily interpreted in terms of densities or radiated power, still renders them valuable for understanding the functioning of divertors. Also, since the viewing access for the vacuum ultraviolet region in the ITER divertor may be quite restricted, one of the prime objectives of current research should be to establish a firm basis for analysis of visible lines if it turns out that they are the main source of divertor impurity data from future machines.

Several types of spectrometers provide various capabilities in the vacuum ultraviolet region below 2000 Å.[2] Czerny-Turner mounts can operate from the infrared region down to almost 1150 Å and are particulary useful for detecting strong transitions in Be-like and Li-like ions in carbon and lighter elements. This type of spectrometer provides nearly stigmatic images with excellent spectral resolution. Normal incidence devices can generally be employed from 6000 Å to 350 Å, which is adequate to detect the resonance lines of all the low-Z impurities except for the helium-like and hydrogen-like species. The requirements for optics and detectors varies considerably over these wide spectral ranges, of course. Grazing incidence spectrometers can cover the region from 10 Å to about 1200 Å, and the multichannel SPRED[3] spectrometers are most useful from about 100 Å to 1100 Å. Recent developments with multi-layer mirror[4] spectrometers have also proved promising for monitoring specific spectral regions, and it is possible they could serve as compact, versatile instruments for ITER. An excellent example of employing the combination of a grazing incidence and a normal incidence spectrometer to assess divertor impurities in JT-60U can be found in Ref. 5, where comprehensive measurements of the resonance line intensities allowed the determination of the fractions of radiation attributed to deuterium, carbon, and oxygen.

The ideal array of spectrometers is difficult to implement on most machines because of the extensive spectroscopic hardware required and because views into the divertor are often quite restricted. However, it is possible to design systems that provide most of the information necessary to determine production mechanisms, erosion rates, and general features of the transport. Figure 1 illustrates, in a highly schematic fashion, the sightlines for the divertor spectrometers on ASDEX-U. A dual system consisting of a Czerny-Turner spectrometer to detect visible emissions and a normal incidence spectrometer for the vacuum ultraviolet are installed so that they view the divertor area through a movable mirror (boundary-layer spectrometer- BLS)[6]. This setup permits scans across the outer target plate and also operates to produce a limited scan in the toroidal direction. A second spectrometer array consisits of a 16 channel visible system viewing almost parallel to the face of the outer target plate and having vertical chordal resolution of 1 mm or alternatively, 16 channels that cover the region from 5 mm to 100 mm above the target (divertor spectrometer - DIV)[7]. These two systems combine to provide data with excellent spatial resolution in nearly orthogonal directions.

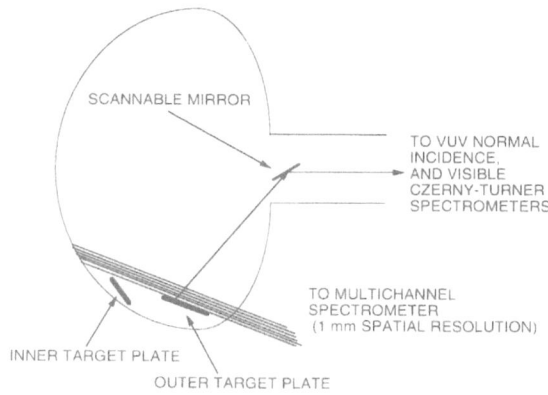

Fig. 1. Schematic of ASEDX-U showing spectroscopic views into the divertor

In addition to high resolution spectrometers, it is possible to employ tangentially oriented cameras equipped with narrow band filters for determining the 2-dimensional emission patterns of individual lines in the visible region of the spectrum. A diagnostic camera of this type has proven to be very useful on DIII-D[8] both for comparing carbon emissions to bolometric data and for helping to benchmark modelling codes.

MODELLING AND ATOMIC PHYSICS

Because of the complex magnetic field structure outside the last closed flux surface, quantitative analysis of spectral emissions to obtain fluxes and densities in a divertor and scrape-off layer is more difficult than the analysis of the core plasma. Neutral impurity atoms produced at the inner and outer strike points become ionized in the divertor region and are then transported along field lines. Frictional forces tend to drive them back toward the plate, but thermal-gradient forces push them toward the midplane. In general, Monte-Carlo codes, such as DIVIMP[9] or MCI[10] are really needed to model the impurity behavior, nevertheless, it is useful to use a one-dimensional code like NEWT-1D[11] to gain some insight into the impurity ion distributions expected in the divertor and SOL.

Figure 2 illustrates predicted plasma parameters and radiated power deduced from the NEWT-1D code for different ionization stages of carbon as a function of distance Z measured from the midplane of the DIII-D tokamak.The target plate lies at -136.2 cm and the X-point is approximately at -123 cm. The simulation is based on 8 MW of input power with sputtering at the plate as the carbon production mechanism. The details of such results depend on a variety of inputs to the code, but the general characteristics of the impurity radiation are typical of what can be expected. The electron temperature is 40 eV at the target plate and rises to 80 eV at the midplane. The electron density is a maximum of $9 \times 10^{13}\,\mathrm{cm}^{-3}$ at 6 cm above the target and falls to about 1/2 this value at the midplane.The radiation from the easily ionized stages, C II - C IV, is constrained to lie near the divertor floor, whereas C V and C VI spread upward into the scrape-off layer. The total radiation from carbon is calculated to be 285 kW, or about 3% of the input power.

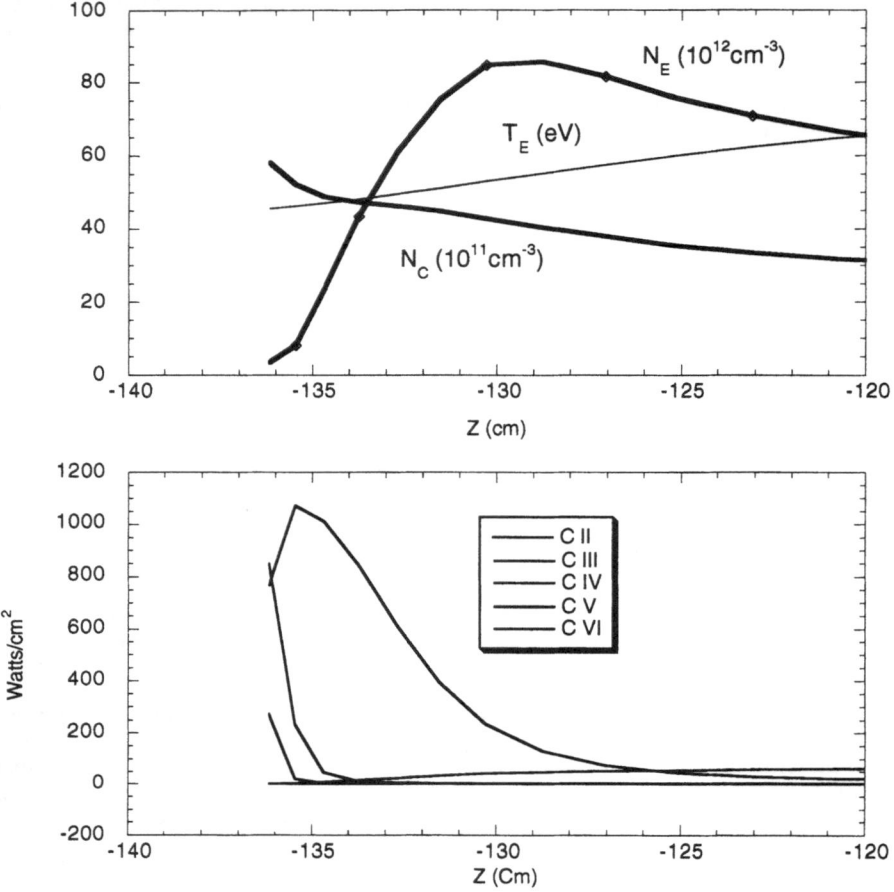

Fig. 2. NEWT-1D simulation of carbon radiation for 8 MW of power in DIII-D.
The target is a-136.2 cm and the X-point at -123 cm.

As already noted, it is preferable to have access to the vacuum ultraviolet region of the spectrum for determining the radiated power from a particular ionization stage since theresonance lines are dominant. It is not always possible to achieve this goal, however, and many interpretations must be made from visible emissions. This problem is likely to become even more severe in the future with experiments utilizing closed divertors employing relatively deep and narrow slots. The atomic physics for calculating ionization and recombination rates for low-Z ions appears to be on a relatively firm footing, but some questions still exist concerning the computation of radiation rates. This is a particularly sensitive question if one relies on the weak visible lines to infer the total radiation emitted from a particular ion. Figure 3 shows a Grotrian diagram for C III (not to scale in energy). The strongest feature of carbon in the visible region is the 4650 Å triplet, but only about 0.25 % of the radiated C III power comes from these lines. The radiated power from the strong lines must be modelled using a detailed-balance (collisional-radiative) calculation if only the visible signals are measured. For many transitions the data to perform such calculations is not available. Also, at present there seem to be discrepancies in different atomic data bases. The results shown in Fig. 2 were obtained using the ADAS[12] data base; the same calculations using the ADPACK[13] data base give results that are roughly a factor of 2 greater for the low ionization stages of carbon. The recommended cross sections or rate coefficients for various isoelectronic sequences together with estimated uncertainties are discussed in a recent overview of the available computations.[14]

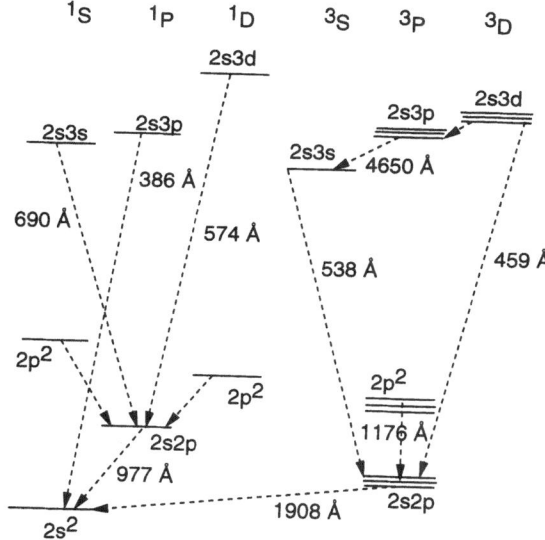

Fig. 3. Grotrian diagram for C III.

EXPERIMENTAL RESULTS OF IMPURITY STUDIES

In typical single null operation in DIII-D, bolometer measurements show that the total radiated power is peaked at the inner, high-field leg of the divertor near the X-point,[15] and data from filtered tangentially viewing cameras indicate that the C III emission exhibits roughly the same structure.[8] Preliminary results of modelling with the MCI code imply that the radiation peak near the X-point results from the influx of neutral carbon into this area even if the source is located at the outer strike point. This radiating region apparently mitigates the power deposited on the inner target plate whereas the outer plate is observed to have a narrow hot spot where it is intersected by the field lines. With the addition of strong D_2 puffing a partial detachment occurs, during which time the power delivered to the outer target plate is reduced by a factor of almost 5, and a strongly radiating marfe-like feature forms near the X-point. Such features are commonly observed, and in ASDEX-U a completely detached high-confinement mode (CDH-mode) can be achieved with the simultaneous injection of neon and deuterium where pumping of the deuterium in the divertor is important for entrainment of the neon.[16]

Measurements of both carbon and boron using the DIV system on ASDEX-U (Fig. 1) indicate that the Be-like and Li-like stages are narrowly peaked only a few mm above the divertor target plate[17,18] as predicted by the DIVIMP code or by NEWT-1D simulations. But scans of the BLS mirror across the face of the target plate show that the distributions of neither the boron nor the carbon line intensities can be explained completely by sputtering as a source since the emission from lines of sight passing outside of the strike point appear too great. A good fit to the carbon data has been obtained by postulating an additional flux of carbon from the pumping duct, which is presumably produced as a result of chemical erosion. Similarly, a source of boron from the walls, perhaps in the form of boron hydrides, has been suggested as a possible factor contributing to the boron observations.On the other hand, analysis of C II and C III radiation from JET shows that the number and distribution of carbon ions in the divertor can be explained almost exclusively by physical sputtering[19]; chemical process appear to be responsible for only 10% of the C II ions near the target plate supposedly as a result of prompt , localised redeposition of CD_4. Therefore, it appears that the details of carbon ion production may depend on characteristics specific to individual machines. A wall source is believed to be responsible for at least 50% of the carbon in the

JET core plasma;[20] a similar conclusion is reached for Alcator C-Mod where it is found that the stronger divertor source is screened very efficiently.[21]

The 2^3P - 2^3S transition of B IV (2823 Å) was also observed in the ASDEX-U experiments, and somewhat curiously, it was also localized close to the target plate rather than having the expected extended distribution similar to the radiation shown for the isoelectronic C V ion in Fig.2. Since the typical electron temperatures in the region of the target plate are 20-40 eV, whereas the energy required for excitation of this ion from the ground state is more than 200 eV, it has been proposed that charge exchange from neutral hydrogen is resonsible for the observations. This explanation seems unlikely, however. At the low temperatures near the target plate charge exchange of B V and electron ionization of B IV primarily establish the relative populations. The characteristic times for charge exchange are much shorter than for electron ionization so the density of B V ions is low; the concentration of B IV ions is maintained by ionization from B III which has an ionization potential of only 37.9 eV. If every ionization of B IV leads rapidly to production of the 2823 Å line through charge transfer, the ratio of electron-excited to charge-exchange-excited emission is given by

$$I_{cx} / I_{elect} = S / R,$$

where S is the ionization rate coefficient for B IV and R is the excitation rate coefficient computed from the collisional radiative model. If S is taken as the ionization rate from the ground state, this ratio is less than 0.01 for all temperatures below 50 eV. A more detailed calculation accounting for charge exchange increasing the population of the ^3S metastable state, from which only 4.4 eV is required to excite the 2383 Å line, indicates no more than a 15% influence on the emission. These conclusions are substantiated by the full numeric modelling with NEWT-1D, but they.do not explain why this transition seems so peaked near the target plate. The rapid drop in temperature along the field lines should lead to a monotonically decreasing intensity despite the fact that the boron concentration may be rising sharply at the plate. The explaination for the observed distribution has not been firmly established, but recent theoretical work has pointed out that deviations from Maxwellian distributions can occur as electrons drift from the midplane toward the target.[22] High-energy electrons, which have low collisionality, do not have time to thermalize as they do in the core plasma where they make many toroidal revolutions on the time scale for cross field drifts. At present, this seems to be a promising avenue to explore for explaining the excitation of these lines originating on highly excited levels in He-like ions.

Although most diverted tokamak experiments have concentrated on low-Z materials at the target plates, Alcator C-Mod uses molybdenum and some tests have been performed with a tungsten panel in ASDEX-U.[23] The heavy metals have much higher sputtering thresholds than low-Z materials and should be eroded very little if ion temperatures can be maintained around 10 eV. In fact, the temperatures in the divertor of Alcator C-Mod usually do not reach the molybdenum sputtering threshold in pure deuterium plasmas. Molybdenum is produced more readily when argon, having a lower sputtering threshold, is added.[20] The heavy metals present problems for spectroscopic characterization; normally the neutral species can be observed with direct viewing onto the strike points, but lines from the low ionization stages likely to be found in the divertor cannot be distinguished except perhaps for Mo II. Tungsten has been detected in the core plasma so far only by pseudocontinuua[24] around 50 Å and 28 Å, but distinct lines from charge states between Mo XXIV and Mo XXXIV are readily observable.[25]

As already noted, gas puffing of deuterium and impurities, either separately or in combination, is employed in all machines to test the radiative divertor concept. Nitrogen, neon, and argon have all been used for such experiments.This procedure can lead to a so-called detached state where the plasma near the target plates drops to temperatures of only a few eV and very little heat is transferred, while a marfe forms in the vicinity of the X-point. As expected, very little radiation is observed near the target plates while ionization stages up to the lithium isoelectronic sequence are of the low-Z impurites emit strongly in the marfe region. Puffing with deuterium alone and relying upon intrinsic impurities to generate the radiative losses does not produce the desired results. In JET it has been possible to achieve detachment in this fashion only in ohmic and L-mode discharges where it is possible to radiate up to 70% of the input power for carbon tragets and 85% for beryllium targets, but further puffing leads to a density limit disruption.[26] Similar attempts during H-modes showed that the ion flow to the plates was reduced between ELM's, however, the plasma reverted to L-mode when P_{rad}/P_{in} reached approximately 50%. In ADSDEX-U, feedback controlled puffing of both deuterium and neon permitted the achievement of stationary, completely-detached, high-confinement operation in which 90% of the power was radiated without degrading the H-mode parameters.[16] Only small amplitude, high-frequency ELM's that transferred little power to the targets were observed. Experiments using nitrogen as the impurity showed it to have a more favorable ratio of power radiated outside the separatrix to that radiated inside than did neon, but large-amplitude compound ELM's could not be eliminated. When using argon, the fraction of radiation inside the separatrix was larger than for the other two gases employed.

DIAGNOSTIC APPLICATIONS

Aside from impurity studies *per se*, spectroscopy may also prove useful for determining certain properties of divertor plasmas. The most obvious application is the measurement of ion temperatures from Doppler broadening. This method is used quite successfully for the analysis of core plasma temperatures from charge-exchange excitation of hydrogen-like ions. The analysis of data from relatively cold regions of the plasma, such as near the target plates in a divertor, may not be quite so straightforward. The Zeeman splitting of the several components of a transition can be reponsible for a large fraction of broadening and, it may be necessary in some cases to account accurately for the Paschen-Back effect in order to obtain a satisfactory fit to the lineshape. Moreover, the birth distributions of velocities from sputtering or from molecular dissociation are not Maxwellian, and the low ionization stages of an impurity may not thermalize before being ionized. In such cases, the linewidth represents some measure of the average energy rather than a true temperature.

Electron temperatures are also measurable, in principle, from the ratios of $\Delta n=0$ and $\Delta n \neq 0$ line intensities because the excitation energies are quite different. The ratio of the 2s - 2p and 2s - 3p transitions have been employed in JT-60U as a qualitative means of following the evolution of the electron temperature.

Stark broadening of hydrogen spectral lines in plasmas denser than those generated in tokamaks is frequently used to determine the electron density. In tokamak plasmas this technique is usually not feasible, but recent investigations in the divertor of Alcator C-Mod[28] have shown that the n=8 through n=11 transitions in the Balmer series of deuterium can be useful for obtaining electron densities from $5 \times 10^{13} - 1 \times 10^{15} \, cm^{-3}$. Further investigations of this spectroscopic application should be pursued.

Finally, it would be highly desirable to measure impurity flows in the divertor in order to gain insight into problems of erosion and redeposition and to benchmark the codes that are being applied to predict the performance of the ITER divertor. Laser-induced fluoresence could serve as the means for such measurements. Although this diagnostic has been applied with limited success in tokamaks, continuing improvements in lasers seem to make the application promising and some preliminary experiments for developing such a diagnostic have been started on the ENCORE tokamak at the California Institute of Technology.[29]

ACKNOWLEDGEMENTS

I should like to thank the following people for contributing to this presentation: M. Fenstermacher, A. Field, J. Fuchs, R. Harvey, S. Hirshman, L. Horton A. Kallenbach, C. Klepper, K. Krieger, G. McCracken, B. Napiontek, J. Terry

REFERENCES

1. D. E. Post, presented at the Workshop on Plasma Edge and Divertor Physics, Garching, Germany, May 16-17, 1995.
2. J. A. R. Samson, *Techniques of Vacuum Ultraviolet Spectroscopy,* John Wiley and Sons, New York, 1967.

3. R. J. Fonck, A. T. Ramsey, and R. V. Yelle, Appl. Optics **21**, 2115 (1982).
4. A. P. Zwicker, S. P. Regan, M. Finkenthal, and H. W. Moos, Rev. Sci. Instrum. **61**, 2786 (1990).
5. H. Kubo, T. Sugie, M. Shimada, N. Hosogane, A. Sakasai, S. Tsuji, K. Itami, N. Asakura, and K. Shimizu, Nucl. Fusion **33**, 1427 (1993).
6. A. R. Field, R. Dux, G. Fussmann, C. Rempel, U. Schumacher, and U. Wenzel, Rev. Sci. Instrum. (in press)
7. B. Napiontek, private communication
8. M. E. Fenstermacher, private communication
9. P. C. Stangeby *et al.*, Nucl. Fusion **28**, 1945 (1988).
10. T. Evans, private communication.
11. R. B. Cambell, T. W. Petrie, and D. N. Hill, J. Nucl. Mat. **196-198**, 426 (1992).
12. H. P. Summers, *Atomic Data and Analysis Structure,* User Manual, JET Joint Undertaking.
13. R. A. Hulse, Nucl. Technol./Fusion **3**, 259 (1981).
14. Atomic Data and Nuclear Data Tables **57,** 1-332 (1994), Edited by J. Lang.
15. A. Leonard, private communication.
16. O. Gruber *et al.*, Phys. Rev. Lett. **74**, 4217 (1995).
17 K. Krieger, H. S. Bosch., W. Eckstein, J. D. Elder, A. R. Field, G. Lieder, C. S. Pitcher, J. Roth, R. Schneider, and P. C. Stangeby, J. Nucl. Mat. **220-222**, 548 (1995).
18. K. Krieger, D. Elder, A. R. Field, A. Herrmann, D. Hildebrandt, G. Lieder, B. Napiontek, C. S. Pitcher, D. Reiter, P. C. Stangeby, and W. West, *Proceedings of the 22nd EPS Conference on Controlled Fusion and Plasma Physics,* Bournemouth, 1995 (in press).
19. H. Y. Guo *et al.*, *Proceedings of the 22nd EPS Conference on Controlled Fusion and Plasma Physics,* Bournemouth, 1995 (in press).
20. G. F. Matthews *et al.*, J. Nucl. Mat. **196-198**, 374 (1992).
21. C. Kurz, B. Lipschultz, G. M. McCracken, M. Graf, J. Snipes, J. L. Terry, and B. Welch, MIT Plasma Fusion Center Report PFC/JA-94-013.
22. K. Kupfer, R. W. Harvey, and O. Sauter, Abstarcts of the 1995 International Sherwood Fusion Theory Conference, Incline Village, Nevada, 3-5 April, 1995.
23. W. Engelhardt, private communication
24. R. C. Isler, R. V. Neideigh, and R. D. Cowan, Physics Letters **63A**, 295 (1977).
25. J. E. Rice, J. L. Terry, K. B. Fournier, M. A. Graf, M. Finkenthal, M. May, E. S. Marmar, W. H. Goldstein, and F. Bombarda, (to be published).
26. R. D. Monk *et al.*, *Proceedings of the 22nd EPS Conference on Controlled Fusion and Plasma Physics,* Bournemouth, 1995 (in press).
27. C. C. Klepper, R. C. Isler, S. J. Tobin, J. T Hogan and W. R. Hess,*Proceedings of the 21st EPS Conference on Controlled Fusion and Plasma Physics,* Montpelier, 27 June - 1 July, pp. 1300-1303.
28. B. L. Welch, H. R. Griem, J. Terry, C. Kurz, B. LaBombard, B. Lipschultz, E. Marmar, and G. McCracken, submitted for publication.
29. J. McChesney, private communication.

BOLOMETRY FOR DIVERTOR CHARACTERIZATION AND CONTROL

A.W. Leonard,[1] J. Goetz,[2] C. Fuchs,[3] M. Marashek,[3] F. Mast,[3]
R. Reichle[4]

[1]General Atomics, P.O. Box 85608, San Diego, CA 92186-9784
[2]Massachusetts Institute of Technology
[3]Max Planck Insitute
[4]Joint European Torus

INTRODUCTION

Operation of the divertor will provide one of the greatest challenges for ITER.[1] Up to 400 MW of power is expected to be produced in the core plasma which must then be handled by plasma facing components. Power flowing across the separatrix and into the scrape-off-layer (SOL) can lead to a heat flux in the divertor of ≈ 30 MW/m^2 if nothing is done to dissipate the power. This peak heat flux must be reduced to ~5 MW/m^2 for an acceptable engineering design. The current plan is to use impurity radiation and other atomic processes from intrinsic or injected impurities to spread out the power onto the first wall and divertor chamber walls. It is estimated that 300 MW of radiation in the divertor and SOL will be necessary to achieve this solution.

Measurement of the magnitude and distribution of this radiated power with bolometry will be important for understanding and controlling the ITER divertor. Shown in Fig. 1 is a sketch of the ITER divertor with possible regions of strong radiation. Present experiments have shown intense regions of radiation both in the divertor near the separatrix and in the X–point region. The task of a divertor bolometer system will be to measure the distribution and magnitude of this radiation.

The bolometric measurements will have a number of uses. First, radiation measurements can be used for machine protection. Intense divertor radiation will heat plasma facing surfaces that are not in direct view of temperature monitors. Measurement of the radiation distribution will provide information about the power flux to these components. Secondly, a bolometer diagnostic is a basic tool for divertor characterization and understanding. Radiation measurements are important for power accounting, as a cross check for other power diagnostics, and gross characterisation of the plasma behavior. A divertor bolometer system can provide a 2-D measurement of the radiation profile for comparison with theory and modeling. Finally a bolometer system can provide real-time signals for control of the divertor operation. Important characteristics for control might include the magnitude of X–point radiation, the balance of radiation in the inboard and outboard divertors, or the location above the divertor floor of the most intense radiation.

Specifications for measurement of divertor radiation to meet the above goals has been put together by the ITER diagnostic group.[2] They are listed in Table 1.

Table 1. Specifications for measurement of divertor radiation.

Total Radiated Power	Max. Amplitude	Time Response	Accuracy	
X–point region	≤0.6 GW	10 ms	±10%	
Divertor	≤0.6 GW	10 ms	±10%	
Radiation Profile	Max. Amplitude	Spatial Resolution	Time Response	Accuracy
X–point	≤300 MWm^{-3}	20 cm	10 ms	±20 %
Divertor	≤ 00 MWm^{-3}	5 cm	10 ms	±30 %

Diagnostics for Experimental Thermonuclear Fusion Reactors
Edited by P. E. Stott *et al.*, Plenum Press, New York, 1996

549

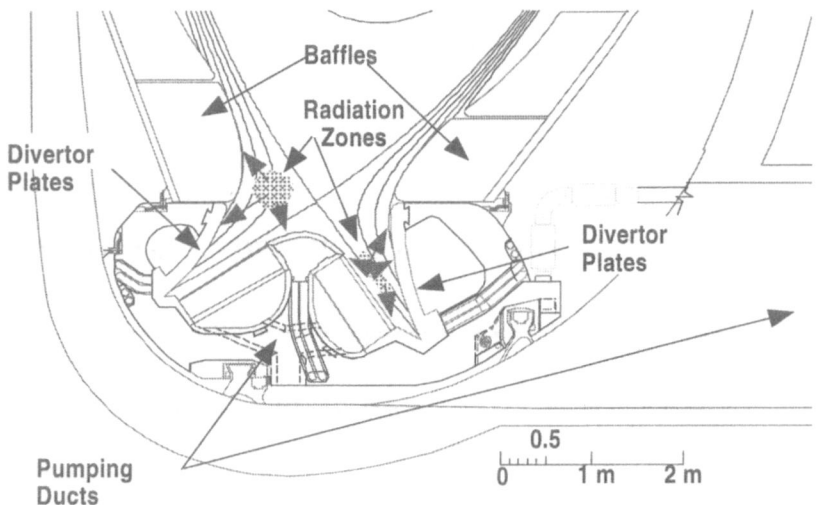

Fig. 1. A sketch of the ITER divertor. Shown are the vertical target plates, the private flux structure and possible regions of intense radiation.

The desired resolution of emission intensity has not been specified, but given experience with current divertor experiments a line-averaged emissivity of 0.1 MWm^{-3} should be resolved. The specification for 5 cm resolution in the divertor is driven by the desire for radial profiles in the divertor leg. For gradients parallel to the poloidal field, the 20 cm specification may be adequate.

For this discussion of divertor bolometry for ITER, a description of bolometer systems currently in use on the world's major tokamaks will first be presented in Section II. This description will serve to highlight techniques and limitations of extracting 2-D information from a limited number of views. A discussion of applying these techniques to ITER will be presented in Section III with a summary and conclusions in Section V.

BOLOMETER SYSTEMS CURRENTLY IN USE

Bolometer systems for measuring divertor radiation are currently in use on the tokamaks ASDEX–U and JET in Europe, ALCATOR–CMOD and DIII–D in the U.S. and JT–60U in Japan. All these systems are based on detectors consisting of small-sized platinum or gold resistors that change resistance as they are heated by plasma radiation.[3] The detector power is then calculated using the temperature measurement and the detector thermal characteristics. A 2-D emissivity profile can then be calculated by careful analysis of all the detector signal levels.

The common difficulty with inverting the measured signals into a 2-D poloidal profile is the limited number of views of these systems. Consider a divertor region that is partitioned into the desired resolution of $m \times n$ cells. Typically only two or three cameras, each with n chords covering the region, have been installed for this measurement. However, m cameras with distinct views are needed to unambiguously specify the emissivity in each cell. The inversion of the measurements into a radiation profile is given by the following equation:

$$\vec{S}_i = \bar{\bar{T}} \vec{I}_{m \times n} \tag{1}$$

where S is a vector of i signals, I is the vector of $m \times n$ cells and T is a transformation matrix containing the i th sensor's response to a unit of radiation in the m,n th cell. The elements of S and T are known and the matrix T must be inverted to solve for I, the radiation distribution. When the number of radiating cells, $m \times n$, greatly outnumbers the number of measurements, i, the matrix T is very singular and cannot, in general, be inverted without additional constraint equations. Each divertor program has developed individual methods to handle this inversion problem. Three divertor bolometer systems and their associated analysis will be briefly described below to illustrate what information can be extracted from a given set of chords.

An example of a divertor bolometer system is that of ALCATOR-CMOD[4] shown in Fig. 2. This system consists of three cameras with four viewing chords each. The cameras are well

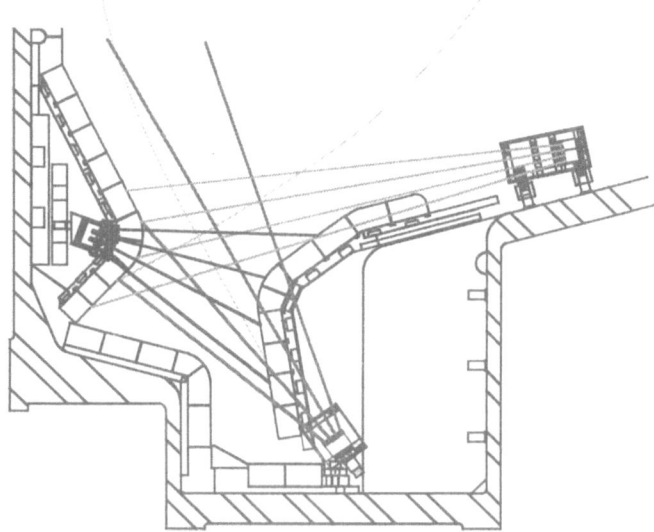

Fig. 2. The geometry of the ALCATOR-CMOD divertor bolometry system. Three bolometer cameras are each installed with four detectors.

separated poloidally to acquire unique information. Before inverting the divertor radiation, the core plasma radiation is measured by a midplane system and subtracted from those divertor chords which also view the main plasma. The divertor region is divided into a 9x12 grid, 2 cm a side for each cell and additional smoothing equations are added to the transformation equation, Eq. (1), to make the matrix T less singular. The smoothing equations effectively limit the second derivative of the radiation profile. Finally the matrix is inverted using singular value decomposition (SVD) techniques.[5] Matrix inversion with SVD is useful for inverting near singular matrixes and extracting the principle values.

An example of a radiation profile produced by the ALCATOR–CMOD bolometer system and subsequent analysis is shown in Fig. 3. The inversion sets the emissivity level in the center of the cell, but for this figure the contour levels are interpolated between cell centers. This example of a highly radiating divertor illustrates the detail that can be extracted from the data set. The greatest emissivity is seen in both divertors near the strike point locations. In the outboard divertor strike-point emissivity can be distinguished from emission near the X–point, but not details of the radial profile from the separatrix.

Another example of a bolometer system is that of DIII–D,[6] shown in Fig. 4. Here there are two cameras each of which spans the main plasma and the divertor region. A total of 10–12 of the 48 channels view the divertor plasma, similar to that for ALCATOR–CMOD. To obtain the divertor radiation profile the contributions from the main plasma radiation must first be subtracted from the divertor chords of the upper camera. The main plasma radiation is assumed constant on a flux surface and is fit to a spline function with bolometer channels that do not view the divertor. Additional constraint equations are then added to the divertor radiation signals in vector S of Eq. (1). For DIII–D the divertor radiation is described by a 2-D spline function of the magnetic flux surface and the height z above the divertor plate. With this constraint, plasma radiation can be limited to the SOL region where significant power is flowing, excluding radiation from the private flux region and far SOL. Such constraints are necessary for the DIII–D geometry with only two poloidal views. The spline function adds a series of nonlinear equations that have not yet been put into the form of matrix Eq. (1). To obtain the radiation profile, the spline parameters are adjusted through iteration to minimize the error given by:

$$\chi^2 = \sum_i^l \frac{(S_i - M_i)^2}{(l-h)(\eta_i S_i)^2} \tag{2}$$

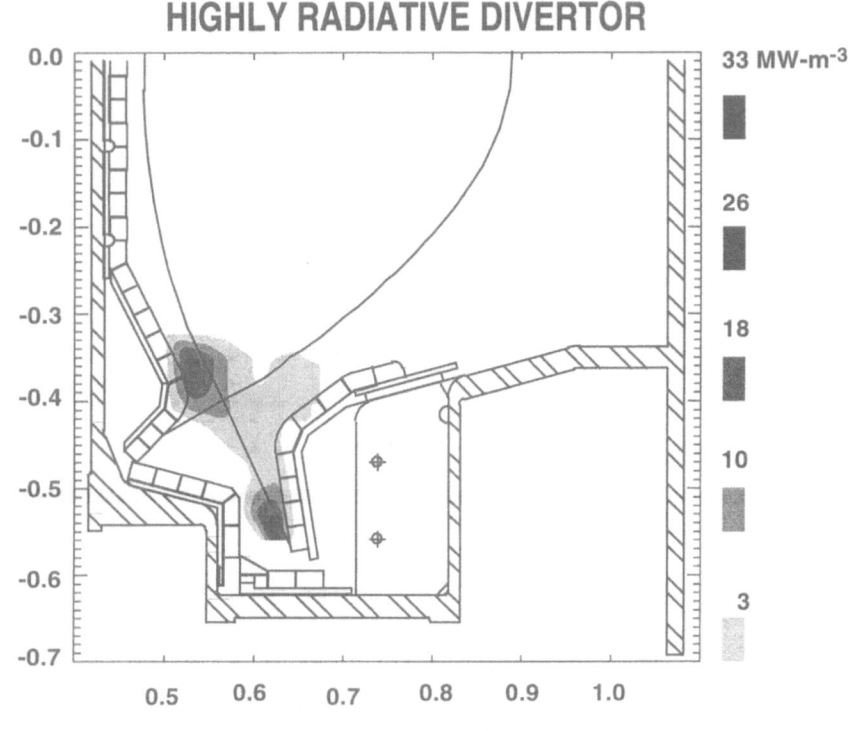

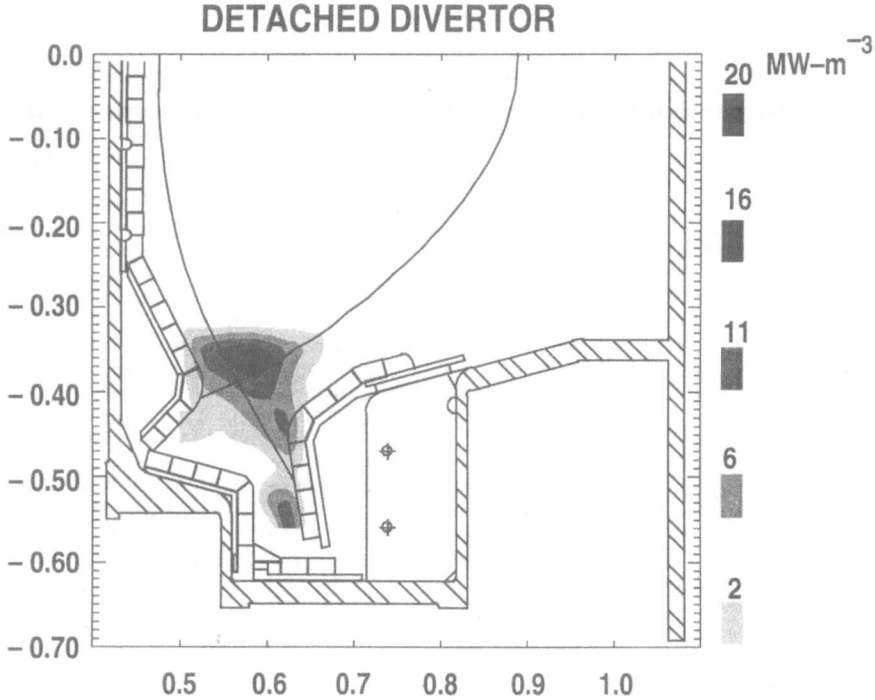

Fig. 3. A radiation profile obtained from inversion of ALCATOR–CMOD bolometry data.

where S_i is the measured signal and M_i is signal that would be obtained with the fitted profile and η is the uncertainty in the measurement. The number of spline parameters is given by the variable h.

An example of a radiation profile obtained by this procedure is shown in Fig. 5. The scale of the resolved features are similar to that for the ALCATOR–CMOD example. In this case, the strong radiation can be seen extending from the X–point region to the strike-point in the outboard divertor. Once again radiation from the X–point can be distinguished from the strike-point, but the radial resolution from the separatrix is limited. This would be expected from the chord geometry of Fig. 4. From the side camera there are three chords which intersect the outboard divertor, but the SOL is only as wide as one of the upper camera chords.

A final example is the bolometer system of ASDEX–U[7] shown in Fig. 6. A total of 72 channels are now operational in five cameras. The 40 channels of horizontal cameras span the main chamber and divertor region as do the 24 channels of the upper vertical camera. An additional eight channels viewing the lower divertor with a finer spatial resolution have been installed but are not yet operational. For the inversion of the ASDEX–U data additional constraint equations are also added to the transformation Eq. (1). In this case the second derivative of the radiative solution is limited, but by different amounts in the direction perpendicular and parallel to the poloidal field lines. This smoothing constraint is modelled as a diffusion equation. By adjusting the "diffusion" constants one can obtain a solution free of numerical artifacts yet with an acceptable fit to the data. The solution is then obtained by minimizing the following function:

$$\Im_{min} = \int (\nabla \bullet (D\nabla I)) + A \sum_{i}^{l} \frac{(S_i - M_i)^2}{(\eta S_i)^2} \tag{3}$$

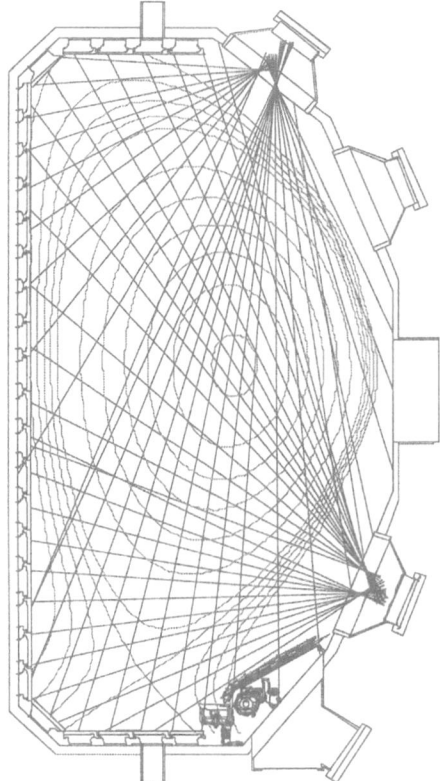

Fig. 4. The bolometer system on DIII–D contains two cameras each with 24 channels. The view of each camera spans the entire plasma with finer resolution of the divertor regions.

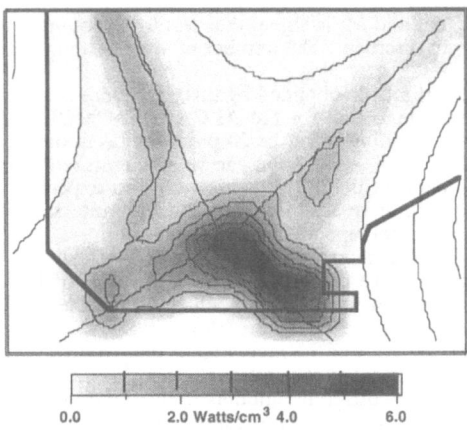

Fig. 5. An inversion of DIII–D bolometry data shows radiation extending from the X–point to the divertor floor during radiative divertor experiments.

where D is the anisotropic "diffusion" tensor, and again S_i is the actual detector signals and M_i is the fitted signal due to radiation profile I. The variable A determines the trade-off between the smoothness of the solutions and the agreement of the fit to the data. By using finite differences for the gradient in the radiation profile a linear system of equations is obtained that can be solved numerically. For ASDEX–U the entire radiation profile, the main plasma and divertor is solved in this manner simultaneously.

An example of a radiation profile obtained for ASDEX–U is shown in Fig. 7. This example is a highly radiating plasma produced by neon injection. Finer scale resolution in the divertor is more difficult because the views of the midplane and upper cameras are becoming more parallel in the divertor region. The lower horizontal camera will be of significant help when it becomes operational.

The ASDEX–U program has also done additional work on using the bolometer signals for real-time control of impurity gas puffing for divertor experiments.[8] For these experiments a real-time signal of the main plasma radiation was desired. Because hardware was available for only 10 channels of real-time processing an SVD analysis was performed on the bolometer data from a range of relevant plasmas to determine the most appropriate channels. Finally the main plasma radiation was reconstructed with all the available channels from which a linear regression analysis was performed to determine the best linear combination of the chosen channels to represent the total main chamber radiation. This method has produced a sufficiently accurate real-time signal of main chamber radiation for control of impurity gas injection for enhanced radiation experiments. It is easy to conceive of using a similar method for real-time signals of the divertor radiation profile.

DIVERTOR BOLOMETRY FOR ITER

Using the systems described above as a guide one can produce a conceptual design for an ITER bolometer system. The basic goal will be to cover all plasma of the divertor and X–point region with at least three views from well separated cameras. The chord spacing within each camera should be such as to achieve the desired resolution. A sketch of such a system is shown in Fig. 8. This concept includes six cameras for each divertor, three viewing from the private flux region and three from the outside the vertical divertor plate. To achieve the desired spatial resolution each camera would need approximately 20 detecting channels.

From the simple sketch of Fig. 8 several design issues become apparent. The bolometer cameras are mounted into the divertor cassette structure itself. The divertor design should include apertures for the bolometer camera views as well as mounting structure and signal cable routing. This is likely the case for other divertor diagnostics as well. The divertor cameras are mounted close to the divertor plasma for two reasons. The first, and most important, is to obtain complete divertor plasma coverage. By bringing the defining aperture of the camera closer to the plasma a much greater field of view of the plasma can be obtained with a smaller opening in the divertor structure. The second reason is that close to the plasma a greater solid angle, and larger signal, is possible for a given spatial resolution. For these reasons it is important to integrate the bolometer diagnostic into the divertor design.

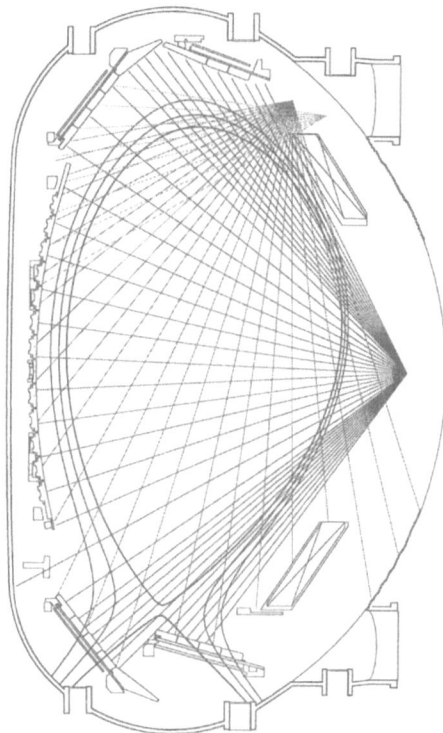

Fig. 6. The operational bolometer cameras on ASDEX–U contain a total of 72 channels in three cameras.

Design of the bolometer cameras into the vertical target plate appears difficult. The vertical plate is a high heat flux component taking direct heat flux from the divertor plasma. The edges of any holes in the surface of this plate would be preferentially heated to intolerable levels. A small gap between divertor cassettes may be sufficient for the bolometer view. The views through the vertical plate are most important for measuring the radial radiation profile. Probably only the profile parallel to the poloidal field can be measured if cameras are installed in the private flux region only. Another possible difficulty is providing apertures in the divertor dome. This structure is designed to take significant heat flux and provide neutron shielding. A camera in this region is important for measuring radiation in the upper divertor and X–point region. A study of the feasibility of such placement should be incorporated into the divertor design.

From the bolometer concept outlined above, one can estimate the signal levels that would be detected. The power onto a detector is given approximately by:

$$P_d \approx \frac{\varepsilon A_p}{4 \pi L^2} \tag{4}$$

where P_d is the power per unit area on the detector, A_p is the area of the viewing aperture, L is the distance of the detector to the aperture and ε is the line integral of the plasma radiation in power per unit area. If the spacing between detectors arranged behind the aperture is ~2 cm, the detector to aperture radius, L, will be about 25 cm if the spatial resolution of 5 cm is kept over the bulk of the divertor plasma. The poloidal width of the aperture is set to keep the chords of neighboring channels just overlapping. That would be an aperture of 2 cm in this example. The toroidal extent of the aperture is only limited by the solid angle of view available to the camera. Due to spatial limitations this will probably also be about 2 cm. Finally the desired resolution in radiative intensity is estimated to be 0.1 W/cm^3 through the line of sight. This example would produce a heat flux of 0.5 mW/cm^2 on the detector. This is very much in the range of current detectors which can measure fluxes as low as 1.0 µW/cm^2.

The greatest concern for noise on the bolometer detectors comes from neutrons produced by the main plasma and the associated γ-rays. Estimates, made during the EDA phase of ITER, of neutron flux to bolometer detectors indicated neutron and γ-ray noise could reach the 0.5 mW/cm^2

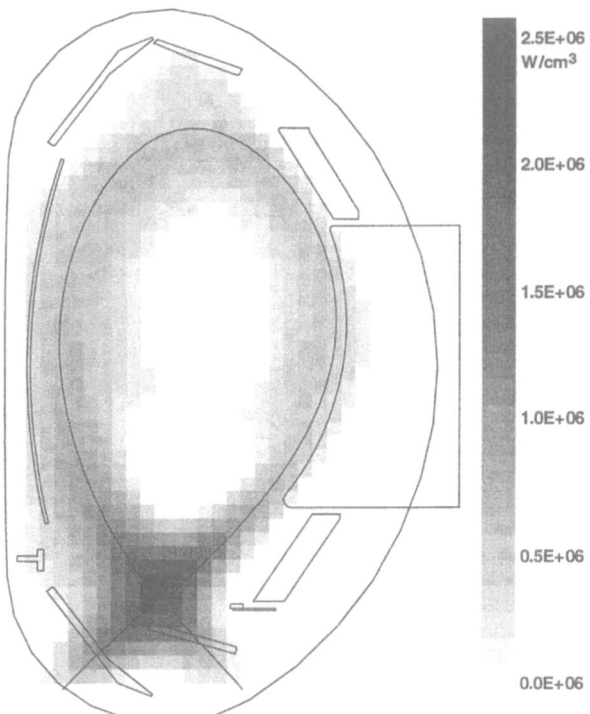

Fig. 7. An inversion of ASDEX–U bolometry data during neon injection experiments.

level. With the placement of detectors close to the divertor plasma and the additional shielding of the divertor from the main plasma neutrons this problem should be less in the current design. However, this issue does deserve further study.

Another concern is the neutral particle environment of the divertor. Neutral pressures of 5 mtorr to 100 mtorr can be expected in the divertor region[1] where the bolometer cameras will be mounted. Neutral pressures this high affect the thermal characteristics of the bolometer sensors. If the gas is at wall temperature it will cool the active sensor imitating a lower incident radiation. For neutral gas at a higher temperature the opposite effect will take place. Since the sensors are enclosed in a camera structure with only a small viewing aperture for gas inlet, the gas should be at the temperature of the surrounding walls. This pressure effect might be taken into account, if it is known. It may be necessary to have a neutral pressure measurement in each of the cameras in order to keep an accurate calibration.

Higher energy neutrals, produced by charge-exchange or other ion-neutral collisions, may also be of some concern. A high energy neutral will deposit energy on the sensor just as well as a photon. However the neutrals represent a non-isotropic source, the hot neutrals directed toward the plasma are reionized where their energy remains in the plasma. This non-isotropic radiation still carries energy out of the plasma, but the inversion algorithms described above are no longer valid. If sufficient neutral gas is present between the source and the detector these particles may become thermalized before they strike the bolometer sensor. These processes will have to be modeled with an accurate divertor plasma before the magnitude of the problem can be identified.

CONCLUSION

A conceptual design for a divertor bolometer system for ITER has been sketched out. This design places the bolometer cameras close to the divertor plasma in the divertor structure. The proposed twelve cameras, six for each divertor leg, provide good coverage of the divertor and X–point region. The specified spatial resolution of 5 cm is obtained by placing 20 channels in each camera. The signal levels for this geometry are adequate for current detector technology. However, the neutron and associated γ-ray noise will have to be studied to determine its magnitude.

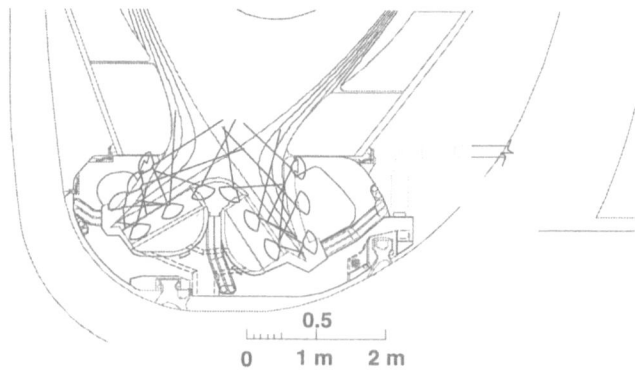

Fig. 8. A conceptual sketch of an ITER bolometer system. A total of six cameras are installed in each divertor leg with twenty detector channels in each camera.

If the bolometer cameras are moved back from the plasma and out of the divertor structure then the design will become much more difficult. From outside the divertor structure the signal levels will be much smaller if 5 cm spatial resolution is maintained. Also complete coverage of the divertor and X–point plasma would require even greater apertures in the divertor structure. Finally, a bolometer system moved out of the divertor structure would not likely provide the number of views necessary for tomographic analysis of the spatial distribution of the radiation.

A successful bolometer system will be necessary not only for understanding the divertor, but for plasma control as well. For plasma control , eal-time processing of the bolometer sensor data will be needed. A linear combination of channels could produce any number of the divertor radiation characteristics. One combined signal could represent X–point radiation, the distance from the divertor target of peak radiation, or signal that some surface is receiving too much radiative heating. These signals could then be used to control gas puffing from different regions, or change the pumping speed, as an example. The combination of signals could be produced by the statistical methods used on ASDEX–U, or possibly by neural network methods. However, a better understanding of divertor operation in general, and the ITER divertor in particular, is necessary before such control schemes can be envisioned.

ACKNOWLEDGEMENT

This is a report of work supported by the U.S. Department of Energy under Contract No. DE-AC03-89ER51114.

REFERENCES

[1] K.J. Dietz, *et. al.*, Proc. 15th Intl. Conf on Plasma Phys. and Contr. Nucl. Fusion Research, 1994, Seville Spain IAEA-CN-60/E-1-I-4.

[2] A. Costley ITER Diagnostic Expert Group.

[3] K. F. Mast, *et. al*, Rev. Sci. Instrum. **63**: 744 (1991).

[4] J. Goetz, *et al.*, J. Nucl. Mater **220–222**: 971–975 (1995).

[5] W.H. Press, B.P. Flannery, S.A. Teukolsky, W.T. Vetterling, "Numerical Recipes, the Art of Scientific Computing," Cambridge University Press 1986.

[6] A. W. Leonard, *et. al.*, Rev. Sci.. Instrum. **66**:1201 (1995).

[7] J.C. Fuchs, *et. al.*, Proc of 21st EPS Conf. on Contr. Fusion and Plasma Physics, Montpellier (France).

[8] M. Marashek, to be submitted for publication.

BOLOMETER FOR ITER

R. Reichle[1], J.C. Fuchs[2], R.M. Giannella[1], N.A.C. Gottardi[3], H.J. Jäckel[1], K.F. Mast[2], P.R. Thomas[1], P. Van Belle[1]

[1] JET Joint Undertaking
   Abingdon, Oxfordshire, OX14 3EA, UK
[2] Max-Planck Institut für Plasmaphysik
   Boltzmannstraße 2, D-85748 Garching, Germany
[3] Commission European Community
   DGXVII, E2 Bâtiment Cube, L 2920 Luxembourg

INTRODUCTION

Bolometers are indispensable diagnostic instruments for ITER. They allow to control the radiated power fraction and the position of the radiation maximum in the divertor, both key issues for the minimisation of erosion and achievement of high performance. To fulfil the ITER requirement of a reasonably good tomographic reproduction of the divertor radiation distribution one needs not only a dense coverage with lines of sight in the divertor but also a substantial coverage of the main plasma. To achieve this, cameras will have to be placed into a harsh environment. This is, where JET's experience with the same approach is very useful. The JET detector based on a proprietary [1, 2], compact 4-channel bolometer which has been modified so that active cooling is not required [3, 4, 5, 6] leads the way towards the solution of these problems. Drawing from JET's experience with this new system a concept for ITER is outlined with the intention to take all relevant aspects into account.

ITER REQUIREMENTS

The source for the ITER requirements is the presently used Design Description Document (section 5.5 on Diagnostics) based on previous ITER diagnostic meetings [7, 8]. Table 1 gives an overview. A system with a spatial resolution of 10 to 15 cm in the divertor and the plasma edge and 50 cm in the centre of the plasma at a time resolution of 1 ms with a tomographic accuracy of 30% and an integral accuracy of 10%, which is the used as target system for this proposal, fulfils the requirements apart from the tomographic resolution requirement of 5 cm in the divertor. The requirement for 0.1 ms time resolution at the current quench can be met due to the better signal to noise ratio in these instances.

Diagnostics for Experimental Thermonuclear Fusion Reactors
Edited by P. E. Stott *et al.*, Plenum Press, New York, 1996

559

The requirements for the construction are: compact size, resistance to radiation, high temperature and high vacuum compatibility, low sensitivity to temperature changes in the mounting structure, compensation of neutron and γ radiation, negligible electrical interference and magnetic pick up and simple in situ calibration. Most of these requirements are obvious. The requirement for high temperature compatibility stems from the difficulty to estimate the temperature at the locations where the bolometers will be mounted and in order to have a safety margin, it is best to have detectors which can work under the hottest possible conditions. Detectors working without cooling can be positioned more flexibly. The low sensitivity to temperature changes is needed because during a 1000 sec pulse the structure surrounding the bolometer which forms its heat-sink will warm up. This has to be taken care of. The calibration requirement favours resistor based bolometers because they can be simply calibrated by sending a heating current through the resistors. Any other temperature measuring technique would require extra cables and resistors to produce a local heat source in the detector. The conclusion is that metal resistor bolometers are presently the best candidates for this task [9, 10]. Apart from the proof that the insulating materials are sufficiently resistant to radiation damage, JET's new bolometer system [5] is probably the only system that fulfils these requirements and has proven to work in a large tokamak.

Table 1. ITER requirements for bolometer measurements

A) For machine protection and plasma control: Disruption avoidance, Marfes, detached plasmas, ELM's, feedback control

	Power range	Resolution in space	in time	Accuracy
P_{rad}(main plasma):	≤ 0.6 GW	integral	10 ms	10%
	< 100 GW[†]	integral	3 ms	20%
P_{rad}(X-point/Marfe):	≤ 0.6 GW	integral	10 ms	10%
P_{rad}(divertor):	≤ 0.6 GW	integral	10 ms	10%
	≤ 0.4 GW	10 cm	1 ms	10%
	≤ 1000 GW[†]	10 cm	0.1 ms	30%

[†] At current quench at a disruption

B) For performance evaluation and optimisation

	Power range	Resolution in space	in time	Accuracy
P_{rad}(main plasma):	0.01-1 MWm^{-3}	20 cm	10 ms	20%
P_{rad}(X-point/Marfe):	≤ 300 MWm^{-3}	20 cm	10 ms	20%
P_{rad}(divertor):	≤ 100 MWm^3	5 cm	10 ms	30%

EXPERIENCE AT JET

JET Bolometer overview

JET has two different bolometer systems, KB1 which is traditionally built with fan shaped lines of sight arrangements [11 and 4], shown in fig. 1 and KB3D/4 with a more modern layout to get detailed divertor information [3, 4, 5, 6, 12], shown in fig. 2. Lines of sight from both systems were used for the tomographic reconstruction of the radiation distribution with the ADMT method [13]. Table 2 gives technical details and performance data of the two systems. Mica was chosen as carrier foil for the new detectors because it has a flat surface and high bending strength and works up to 550°C. The lower resistance in the new system is the reason for its high temperature compatibility. At higher resistance and lower thickness of the film the gold undergoes a phase transition at 350°C which leads to the formation of islands in the meander which destroys the electrical continuity. The smaller size of the new detectors allowed more compact pinhole camera constructions. According to the well known bolometer relation (equation 1) between incident power P and temperature rise δ the new detectors have a larger signal due to their smaller heat capacity C. The sensitivity to a given input powerdensity is however about the same as in the KB1.

$$P = C \,(d\delta/dt + \delta/\tau_c) \qquad\qquad \text{(equation 1)}$$

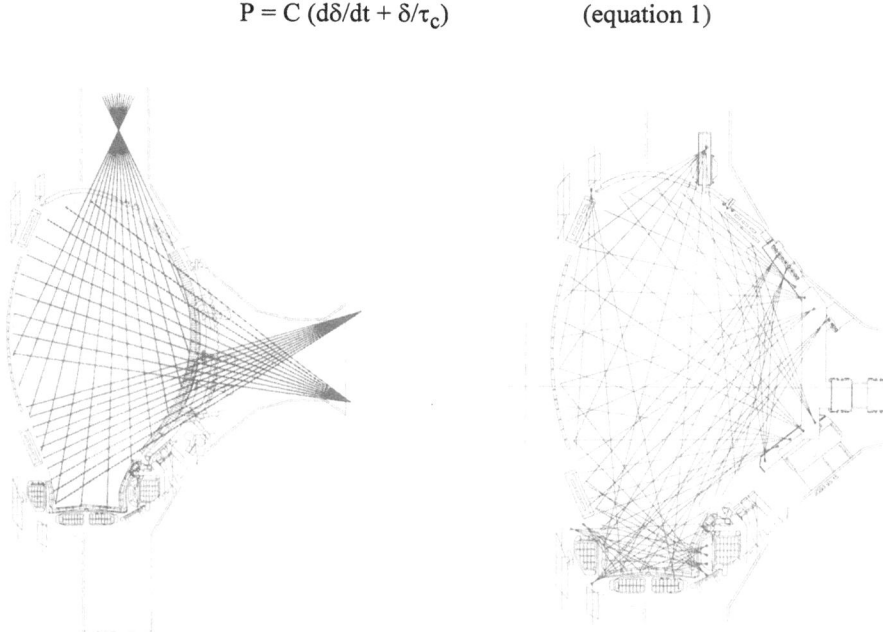

Figure 1. Lines of sight of the old JET bolometer system KB1.

Figure 2. Lines of sight of the new JET bolometer system KB3D and KB4.

Table 2. Comparison of the technical data of KB1 and KB3D/KB4

	KB1	KB3D & KB4
Channels:	34 channels	84 (20 used)
Location:	outside toroidal field	inside toroidal field
Electronics:	DC system	AC system (20 kHz)
Detector operating temperature:	30°C	250°C
Substrate:	Kapton	Mica
Electrical connection:	Springs	Ultrasonic ballbonding & crimps
Resistor:	4.8 kΩ (Gold)	200 Ω (Gold)
Detector area:	121 mm^2	5 mm^2
Heat capacity C:	2.2 mJ/K	0.19 mJ/K
Cooling time constant τ_C:	0.2 sec	0.2 sec
Response time:	0.1ms	0.1 ms
Bridge Voltage:	10 V	3.5 V
Sensitivity:	0.061 W^{-1}cm^2 (20°C)	0.050 W^{-1}cm^2 (20°C)
(bridge output voltage per bridge input voltage/powerdensity)		0.026 W^{-1}cm^2 (250°C)
Equivalent input noise level (25 Hz):	20μW/cm^2	20μW/cm^2
Magnetic pickup:	20μW/cm^2	0.4 mW/cm^2

The equivalent input noise level of the new detectors at 25 Hz however is the same in both systems. The reasons are the smaller detector area in the new detectors and the lower operation voltage imposed by the high operating temperature. The magnetic pickup of the KB3D/KB4 system is also noticeably larger due to its mounting inside the toroidal field coils and next to the divertor coils and the difficulties of minimising pick-up loops with high temperature compatible cables. Without the synchronous detection system the signals could probably not have been used. But with sufficiently strong signals the ultimate time resolution can be as fast as the response time of 0.1 ms.

High temperature detectors

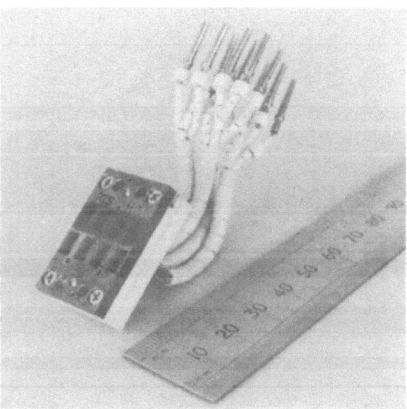

Figure 3. Assembly of 4-channel high temperature detector.

Of the 21 high temperature detector foils used in JET (fig. 3) 20 withstood 18 months of operation in JET mainly at 250°C without any significant change of resistance. The bonded connections inside the detectors survived equally well. The one failing specimen is still under investigation. Problems were caused by the crimp connectors at the back of the detectors. Many failed during the thermal cycling and needed re-crimping. Particularly evident became this problem when the vessel temperature was lowered to room temperature for maintenance work.

Mineral insulated cables, connectors and high frequency interference

The mineral insulated cables used for the new bolometer system have a better cross talk immunity and lower microphonicity than braided cables. They are robust and tritium compatible feed-through cables - at the level foreseen for JET - due to dense packing of the insulating material and the additional sealing provided by metal ceramic connections (quartex) at the ends. Unfortunately no true coaxial construction could be used since coaxial terminations did not seem possible to develop to the same standard in the given time. The welding of the cables' sheaths into the feed-through bulkhead made this point the common ground point for the system. This required complicated screening and shielding methods inside the vacuum vessel in particular since the distance between feed-through and detectors is in some places more than 5 m. The length of the prefabricated cable assemblies was difficult to control. Two twisted pairs of mineral insulated cables twisted around each other are used inside the vessel to serve one channel of the detectors.

This was done because it was not possible to achieve nicely twisted quads of these cables which would have been better for immunity from magnetic interference, given that 4 cables serve one detector channel.

In vessel connectors of cables with crimps failed in a similar way to the crimped connections at the detectors head. With the cameras removed, these connections could be repaired by injecting large currents through the cables. The removable connectors with sliding contacts did show fewer problems than the crimped connections probably due to self cleaning of the contact caused by the machine vibration.

Interference from JET's 3 GHz LHCD system caused initially an apparently negative radiation signal presumably by heating the reference bolometer via eddy currents in the thin copper shield in front of it. With suitable shielding this can be avoided.

Neutral particle fluxes

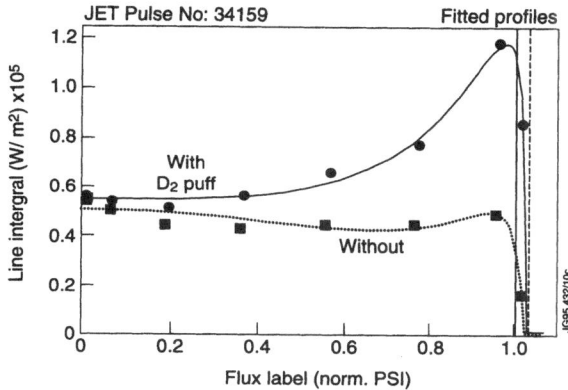

Figure 4. Bolometer signal with and without gas puff.

Gas puffing in the same toroidal plane as measuring with bolometers has a large effect on the bolometer signal (see figure. 4). Since the signal is not noticeable by another set of bolometers in a poloidal plane 2 m away toroidally, the observed effect is believed to stem from local charge exchange particles hitting the bolometer.

Tomographic reconstructions from density limit discharges without additional impurity seeding are compatible with the assumption of 30% more neutral contribution to the signals inside than outside the divertor. In discharges with strong impurity puffing such assumptions are not necessary.

A cooled forerunner to the new bolometer system with a significantly smaller solid angle measured after disruptions the gas release from the vessel walls. The bolometer signal was then roughly proportional to the pressure.

PROPOSALS FOR ITER BASED ON JET EXPERIENCE

In order to achieve a good tomographic reconstruction in the divertor one also needs a similarly good resolution in the main plasma since the lines of sight traverse generally not only the divertor but also the main plasma and the contribution from the rest of the plasma volume has to be subtracted in order to determine the divertor radiation distribution.

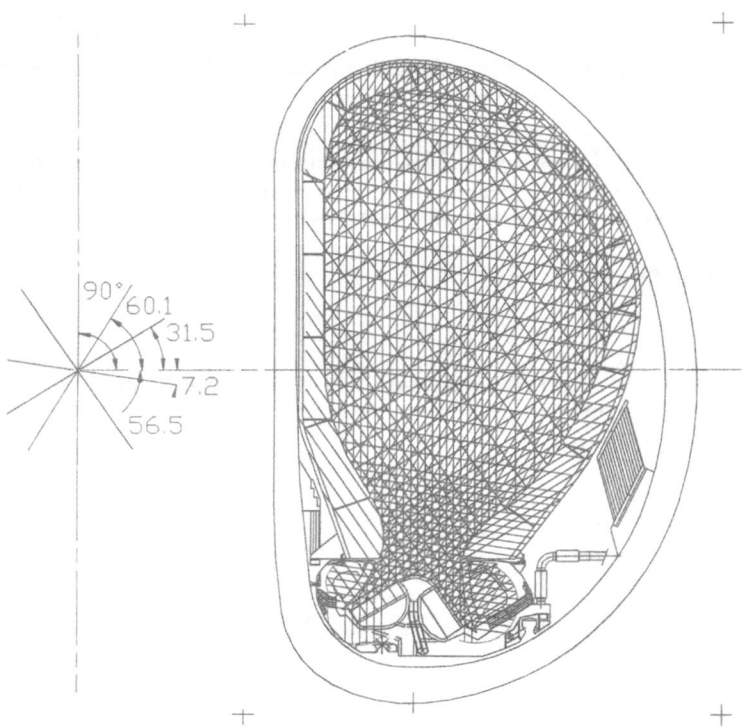

Figure 5. Line of sight proposal for bolometers in ITER - total 208 lines of sight.

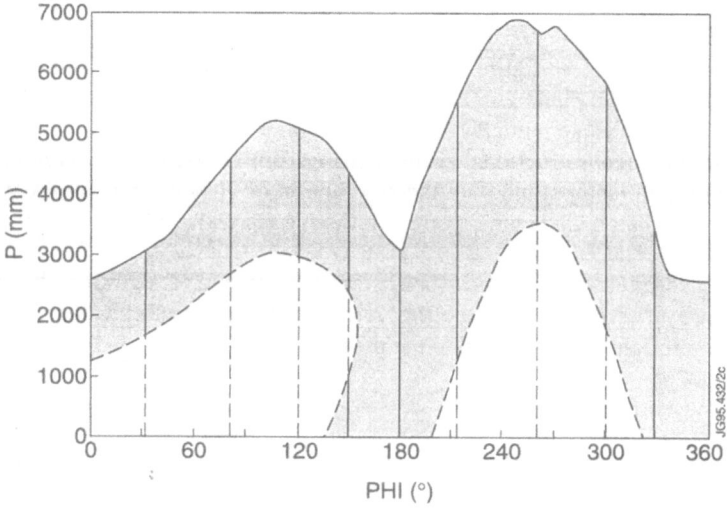

Figure 6. P-ϕ diagram of the lines of sight.

The targeted spatial resolution for this proposal is derived from the ITER requirements (table 1) and is about 10...15 cm in the divertor and the plasma edge and 50 cm in the plasma centre. Assuming parallel lines of sights these requirements lead to 30..50 lines of sight (LOS) for each direction the plasma is viewed from. From the tomographic sampling theorem [14] follows that in these circumstances 50...75 viewing directions are needed to produce a tomographic reconstruction of the radiation without any other information. This would lead to 1500...3750 LOS, which is too much to realise. Therefore sparse data tomography is needed e.g. magnetic flux line oriented tomography like the ADMT method [13]. The compromise that is proposed here, is to keep the spatial resolution given by the distance of the lines of sight and reduce the number of viewing directions to the 5 directions which are immediately useful (see figure 5) namely: parallel (perpendicular) to the 2 inner wings of the divertor, nearly horizontally connecting the lowest points of the divertor, vertical (for simple calculation of the total power and also because this direction is nearly parallel to the outermost divertor wing) and parallel to the innermost divertor wing.

When there is a choice whether detectors look up or down a particular line of sight, looking down is preferable because the neutral particle problem will be smaller and also dirt accumulation in the collimators will be less likely. Despite this, it is advisable to have at least a few lines of sight facing each other for cross checks. The sketch of the P-Φ diagram for this arrangement shown in figure 6 has the plasma centre as reference point. The LOS are points on the vertical lines in this diagram. The shaded area represents the area where the LOS are situated which are either relatively tangent to the edge of the plasma or see the divertor. A denser placement of LOS is needed there. From the coverage of the P-Φ - space one could imagine that by fitting polynomials to isobars in this space checks on the overall magnetic reconstruction could be made.

Signal levels

The worst cases regarding the viewing of the divertor will be over a distance of 10 m into the divertor with a footprint of 10 cm poloidally. For this proposal it is assumed, that the detectors are positioned in the gaps between the blankets with a front face gap of 2 cm in a distance of 55 cm. This gives a solid angle of about $0.35...0.5 \cdot 10^{-3}$ sterad. The situation inside the divertor is more accommodating since the distance is much smaller. Assuming again gap sizes of 2 cm one finds $1.5...3.8 \cdot 10^{-3}$ sterad. Table 3 gives an extrapolation of the bolometer measurements from two steady state ELMy H-mode JET discharges, one with a radiated power fraction of 40% and another one with 75% to ITER. Equation 2 is used, the well known relation between the line of sight integrated measurement of the radiation $\int Edl$ and the power on the detector area P/F.

$$P/F = (1/4\pi) \cdot \Omega \cdot \int Edl \qquad \text{(equation 2)}$$

Table 3. JET - ITER signal level extrapolation

	JET		ITER		units
Apparent surface area of plasma	1		8.5		
Total radiated power (radiated fraction)	5.6 (0.4)	14 (0.75)	350 (0.35);	750 (0.75)	MW
Mid-plane lowest $\int Edl$	0.05	0.08	0.37	0.50	MW/m^2
Divertor highest $\int Edl$	0.6	3.0	4.4	19	MW/m^2
Mid-plane lowest P/F ($\Omega = 0.35 \cdot 10^{-3}$)			1		mW/cm^2
Divertor highest P/F ($\Omega = 3.8 \cdot 10^{-3}$)				574	mW/cm^2

Neutron and γ heating

An estimate of the nuclear heating of a mica detector foil is given in table 4. The assumptions are a neutron rate of 10^{20} n/s (300 MW fusion power), a positioning of the foil at the front of the first wall such that it receives flux from the half sphere in front. The foil (20μm mica) receives 1.8 mW/cm^2 and another 0.3 mW/cm^2 are deposited in the gold foil (7μm) absorber. Heating by γ-radiation is less of a problem than neutron heating and in the worst case is about 30% of the neutron heating.

Table 4. Heating of the detector foil by nuclear radiation

Element	presence (Mica)	weight fraction	heat/mass/neutron J/kg/neutron	power/mass Watt/kg(mica)
H	4	0.01	$1.6 \cdot 10^{-16}$	160
O	12	0.48	$2.0 \cdot 10^{-18}$	96
K	1	0.10	$1.0 \cdot 10^{-18}$	10
Al	3	0.20	$1.0 \cdot 10^{-18}$	20
Si	3	0.21	$0.89 \cdot 10^{-18}$	19
			Sum:	305

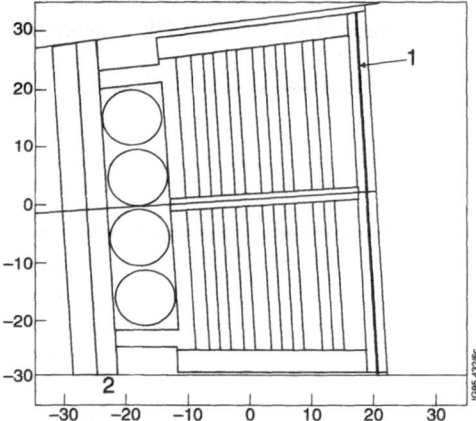

Figure 7. Two positions to compare radiation damage.

Table 5. Damage reduction by shielding [from 15]

Position	dpa/Y	HePRD appm/y	PWD W/cm^3
1 inner wall	7	60	12
1 top	6	52	12
2 inner wall	0.25	2.4	0.5
2 top	0.16	1.7	0.5
2 outer wall	0.14	1.2	0.15

Monte-carlo neutron photon code calculations exist [15] which show the effect of the damage reduction when moving from the front face of the blanket to the remotest part between blanket gaps (figure 7). Table 5 gives values for the displacement per atom and year, the He production and the power density in different poloidal locations at the front of the blanket (position 1 in figure 7) and back in the gap (position 2). On average it appears that the radiation damage is reduced by a factor 24...32. This should be sufficient attenuation.

Detectors

JET's high temperature detectors, which are based on the AUG design [1], which is also used in Tore Supra [16], Textor, Alcator-C-mod and the reversed field pinch in Padua [17], are a good starting point for further developments. The obvious improvement areas are:

Radiation resistance of foils. Mica is known to have poor radiation resistance compared to alumina, because the silica in it causes volumetric distortions. Electrically it may however still be acceptable [18]. It is proposed to test one of JET's high temperature bolometer heads in a high neutron flux test facility to see in particular how the Mica performs. In order not to waste time, at the same time an alternative more promising detector-foil material should be used. From the literature much is known about degradation of the isolating and other properties of Al_2O_3 and AlN [19] another candidate to consider is silicon nitride [18]. Purely on the grounds of excellent heat conductivity thin film diamonds are interesting. The metal parts of the bolometer are prone to much less problems. If primary mirrors for other diagnostics can be used without blistering away it can be assumed that the metallic parts of the bolometers survive also. The time resolution and sensitivity do not seem to pose problems if the new type detectors stands the radiation test. Another concept which does not rely on the detector-foil to be also the mechanical carrier is to use the absorber as carrier foil and to deposit an insulating layer and the resistors on the back [20].

Neutral particle fluxes. Filtered and un-filtered bolometer measurements at both ends of the lines of sight of VUV and SXR survey spectrometers should allow to unravel the various contributions. With time of flight measurements the energy spectrum of neutrals could be determined in the relevant range, but this would require a large diagnostic set-up.

Thermal drift during 1000 sec pulses. During a 1000s pulse the heat-sink of the detectors will drift up in temperature. This would affect the measurement since the detector sensitivity is reduced by a factor 2 from R.T. to 250°C. By measuring not only the resistance difference between measurement and reference bolometer as usual, but also the absolute resistance of the bolometers one can determine the temperature of the bolometer and use the appropriate calibration factor.

Feedback possibility. By externally heating an extra resistor on the detector foil one could keep the measurement always at zero, which generally promises more accuracy [21].

Subassemblies

The proposed scheme for mounting the detectors is that single channel versions of the miniaturised detectors are individually housed in camera units complete with collimator and a single mineral insulated cable. These units are pre-assembled into larger assemblies either of the length of the blanket modules to be located between the blanket modules or of a triangular shape to be mounted inside the triangular filler shield positions (fig. 8) for the main vessel. The first option requires an enlargement of the gap-width next to the inner wall. In the divertor they have to be incorporated into the cassettes. The maintenance and replacement can take place simultaneously with blanket or divertor module exchanges. From figure 5 one can estimate that the densest packing of detectors along the circumference of the vessel wall may be around 25 detectors per m. With a maximum of

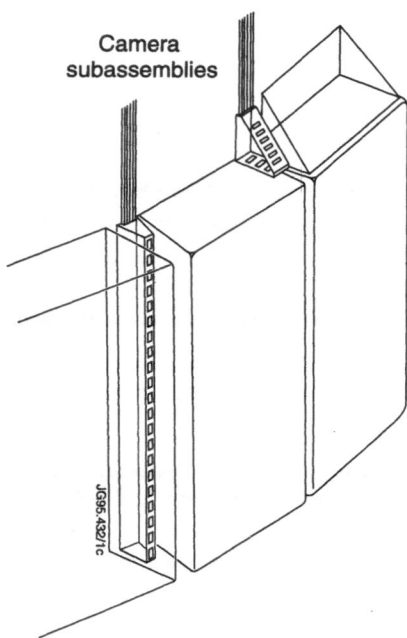

Camera
subassemblies

Figure 8. Sketch of camera subassemblies.

90° between adjacent lines of sight, this leads to a collimator design of up to 5.6 cm length with a collimator foil spacing of 0.6 mm in order to achieve the desired solid angles.

Grounding, cabling and connectors

The local detector potential should be ground reference and the detector potential should be used as screen potential up to the electronics. Around this screen should be another isolated metal sheath, which is in contact with the vessel everywhere. The best cables presently available are mineral insulated cables. Twisted pairs inside a screen are presently quite possible manufacture. The next logical step is to develop cables with more (4...7) twisted wires or coaxial cables (2..3) inside and an additional metal shield around the screen. Redundant cabling can reduce substantially the susceptibility to failure. Connectors should be avoided where possible. But due to the need to handle the subassemblies remotely they may be necessary. Clearly a lot of work is needed in this area still but the fact that in ITER the change of the toroidal field is slow in time alleviates this problem somewhat.

Electronics

JET has started a development on a electronics system based on the concept of one digital signal processor (DSP) per channel coupled to direct digital synthesisers. This intelligent front end points the way forward to a more flexible operation than hitherto usual. If developed further the device could be switched from measurement to calibration in the middle of discharges to allow compensation of thermal drifts. The DSP can process the signals immediately into power traces useful for feedback onto the plasma position in general or the position of radiation zone in the divertor in particular.

SUMMARY AND OUTLOOK

A complete bolometer system for ITER is proposed. The approach is, that in order to reconstruct the radiation distribution in the divertor a tomographic reconstruction in the whole plasma is needed. This lead to a distributed camera layout with the cameras in the gaps between the blanket and divertor modules. It offers some cross checks on the magnetic diagnostic. For the detectors further developments of JET's high temperature detectors are proposed especially towards radiation resistance. Mineral insulated cables with as few as possible connectors should be used. The heating of the detector foil by neutron and γ radiation can be reduced to acceptable levels by positioning the detectors far back in the gap between the blankets. The thermal drift during the 1000 sec pulses can be taken care of by more advanced electronics. The distinction between neutral particles and heat radiation should be investigated at on existing machines. The phase after 1996, when JET becomes dedicated to answer questions related to ITER is ideally suited to pursue further detector and concept development. The envisaged DTE2 phase may be the first test of such a new system under 14 MeV neutron bombardment.

ACKNOWLEDGEMENT

Fruitful discussions with J.K. Ehrenberg (JET), (EC), A. Hofmann (PTS), L. De Kock (ITER), S.F. Dillon (JET), M. Di Maio (JET), E.J. McCarron (JET), D.G. Thompson, R. Webb (JET), R. Wirth (PTS) and S.J. Zinkle (ORNL) are gratefully acknowledged.

REFERENCES

[1] IPP 1/224 E.R. Mueller, G. Weber, K.F. Mast, G. Schramm, E. Buchelt, C. Andelfinger, Max-Planck Institut für Plasmaphysik, Garching, Oct 1985

[2] K.F. Mast et al., Rev. Sci. Instr. 62(3), 1991, p. 774

[3] EUR 15081-EN-C EUR-JET-PR10, JET Joint Undertaking Progress Report 1992, April 1993, p 83

[4] EUR 15722-EN-C EUR-JET-PR11, JET Joint Undertaking Progress Report 1993, April 1994, p 80

[5] 'Divertor Radiation in JET', R. Reichle, N.A.C. Gottardi, R.M. Giannella, H.J. Jaeckel, A.C. Maas, Contrib. to the 36th ann meet. of the DPP/APS 1994, 7-11 Nov. 1994, Minnesota, USA

[6] EUR 16474-EN-C EUR-JET-PR12, JET Joint Undertaking Progress Report 1994, April 1995, p 97

[7] S 56 TD 01 94-09-27 FE , Task agreement 'Review of the initial definition and specification of the ITER diagnostic System' Target measurement resolutions and accuracies for ITER

[8] Proceedings of the EC-Home team meeting/Workshop on Radiation and Spectroscopic Diagnostics for ITER, Culham Laboratory, 18 Oct 1994

[9] 'Bolometric measurements', S. Yamamoto, ITER documentation series, No 33,IAEA, 1990, p62-p66

[10] 'Bolometric diagnostic for ITER', K.F. Mast, N.A.C. Gottardi, P. Martin, R. Reichle, R. Wirth, J.C. Fuchs, Proc. of Workshop on ITER Diagnostics in Culham, U.K., November 1992

[11] 'Bolometric diagnostics in JET' K.F. Mast et al., Rev. Sci. Instrum. 56 (5), May 1985, p 969

[12] 'Radiation in JET's Mark I Divertor', R. Reichle et al., Contr. 22nd EPS on Contr. Fusion and Plasma Phys., Bournemouth, 3-7 July 1995, Poster R022

[13] 'Two dimensional reconstruction of the rafiation power density in ASDEX Upgrade, J.C. Fuchs et al., Contr. 21st EPS Con. Contr. Fus. Plasma Phys., Montpellier, France, June 27 - July 1, 1994, Poster C107

[14] 'The Mathematics of Computerized Tomography' F. Natterer, John Wiley & Sons, B.G. Teubner, Stuttgart, 1986, chapter III 'Sampling and resolution'

[15] '3D neutron transport analysis of the shielding blanket for ITER', L Petrizzi, V. Rado, M. Rapisarda, ENEA, Italy, 1994

[16] 'A low noise highly integrated bolometer array for absolute measurement of VUV and soft x radiation' K.F. Mast et al., Rev. Sci. Instrum. 62(30 March 1991, p 744

[17]   'Fast bolometric diagnostic in the RFX reversed field pinch experiment', K.F. Mast et al., Rev. Sci, Instrum. 63 (10), October 1992, p 4714

[18]   priv. comm. S.J. Zinkle (ORNL)

[19]   'Fusion reactor materials', Proceedings of the Fifth Int. Conf. on Fusion Reactor Materials, Clearwater, FL, USA, Nov 17-22, 1991, Editors: R.L. Klueh, R.E. Stoller, D.S. Gelles, North Holland, 1992:
'Radiation induced changes in the physical properties of ceramics materials', S.J. Zinkle and E.R. Hodgson, p58-p66;
'Radiation induced conductivity in alumina from 100Hz to 10MHZ ', E.D. Farnum et.al., p 548-p551;
'Electrical breakdown in fusion insulators', E.R. Hodgson, p552-p554;
'Reduction of the mechanical strength of Al_2O_3, AlN, and SiC under neutron irradiation' , W. Dienst, p554-p556;
'Dielectric properties in cereamics', R.E. Stoller, p 602-p606.

[20]   'A fast bolometer for the WVII-AS stellarator', H.J. Jäckel, G. Kühner, J. Perchermeier, IPP 2/291 April 1988

[21]   N.A.C. Gottardi, K.F. Mast, Euratom Patent

NEUTRAL GAS DIAGNOSTIC FOR ITER

G. Haas[1], H.-S. Bosch[1], L. de Kock[2]

[1] Max-Planck-Institut für Plasmaphysik Garching
Euratom Association
[2] ITER JCT, JWS Garching
Boltzmannstraße 2, D-85748 Garching/Munich, Germany

ABSTRACT

A crucial point for ITER will be to keep the power load on the walls in the divertor at an acceptable level. Key roles in recent concepts play strong radiation from the plasma edge, high neutral gas density and a low plasma temperature in the divertor region, i.e. ITER has to be operated close to the density limit. Under these conditions a reliable and fast neutral gas diagnostic is mandatory to achieve high performance. For total pressure measurements so called ASDEX gauges are proposed to be installed at about 100 different poloidal and toroidal positions in the divertor. The large number also guarantees a sufficient redundancy. These gauges can cope with the typical conditions inside fusion experiments, if they are operated by a special electronic control unit. No severe problems are expected with the operation under the additional constraints of ITER (high temperature and radiation level). Long term operation at high temperature and nuclear radiation levels can degrade the insulators temporarily or permanently. The gauges work, however, still satisfactorily with a resistance between the electrodes as low as 100 kΩ corresponding to a conductivity of the insulators of 10^{-3} $\Omega^{-1}m^{-1}$. Nevertheless some neutron shielding would be beneficial. Test runs under comparable radiation level are planned.
Besides the total pressure measurement, which is important for a safe operation of the divertor, information on the gas composition in the divertor, especially on the He content is desirable for the exhaust of the He ash and other impurities. Possible candidates for this task will be discussed. (Quadrupol RGA's, Penning and ASDEX gauge based spectroscopic detection systems for He and other gases)

INTRODUCTION

An important task to be solved for a safe and successful operation of ITER is to keep the power load on the target plates in the divertor at an acceptable level. Additionally the power should be transported to the walls by radiation or low energy particles to avoid excessive sputtering. In recent concepts like the CDH mode[1] developed on ASDEX-Upgrade a certain amount of light impurities in the plasma and a high neutral gas density in the divertor play a key role. With such a scenario, i.e. with radiated power nearly approaching the heating power, a disruption free tokamak operation is possible only if the impurity content and the neutral gas density in the divertor are carefully controlled. Whereas the impurity control calls for a good bolometer system which has been dealt with by

Diagnostics for Experimental Thermonuclear Fusion Reactors
Edited by P. E. Stott *et al.*, Plenum Press, New York, 1996

571

Leonard[2] and by Reichle[3], for the neutral gas control a reliable neutral gas diagnostic with good temporal and spatial resolution is mandatory.

On the other hand the flux of neutrals onto the mid plane plasma has to be low for good confinement and easy L => H transition. The efficient retention of neutrals in the divertor necessary for that can not be achieved by structural means alone but calls also for appropriate operational scenarios. One of the objectives of ITER will certainly be to identify them. That task will hardly be possible without a measurement of the neutral flux density in the mid plane.

For the purposes discussed until now a measurement of the total neutral flux density without distinguishing between different gas species will do it. The majority of the gas will be hydrogen isotopes anyway. For other issues, however, like He ash and gaseous impurity removal as well as fuel composition measurements a separate detection of the gas species is desirable.

A CONCEPT FOR THE NEUTRAL GAS DIAGNOSTIC FOR ITER

Total Neutral Flux Density

The ASDEX pressure gauge. For measurements of the total neutral particle flux density inside the vessel so called ASDEX gauges[4] are proposed. These gauges originally developed for ASDEX have been used successfully in several other tokamaks. They are hot cathode ionization gauges. If they are operated by a special electronic control unit, they can cope with the typical conditions inside the vessels of contemporary fusion experiments (magnetic field strength ranging up to 6 Tesla and varying in direction within ± 25°, high e.m. background noise level, pressure range up to 0.1 mbar). They differ from conventional hot cathode ionization gauges (e.g. Bayard Alpert type) in so far as the emission current is chopped by an additional electrode (control grid), the filament is rather thick to withstand the Lorentz force and the electrode arrangement is linear. Fig. 1 shows a sketch of the gauge system.

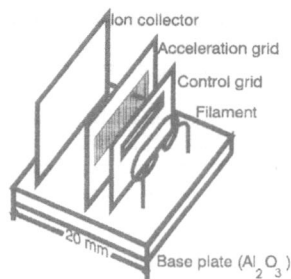

Ion collector
Acceleration grid
Control grid
Filament
20 mm
Base plate (Al$_2$O$_3$)

Figure 1. ASDEX pressure gauge

For measurements of the flux density of neutral particles the gauge system has to be encased by a box with a small opening through which the particles enter the box. The opening may be baffled or connected to a tube. The vacuum time constant, defined by the conductance of the opening and the volume of the box, has to be so large, that neutrals become thermalized before they leave the box again. Then the neutral gas density inside the box is proportional to the flux out of the box. At stationary conditions the flux into the box is identical with the flux out of it. Flux equilibrium is established within the vacuum time constant. The existing gauge system needs a volume of $2.2 \times 1.9 \times 1.8$ cm^3. With such a box volume and an appropriate entrance aperture a response time of 1 msec can be achieved

(50:50 mixture of D_2 and T_2 at 500 K). It will be longer if a tube has to be attached. For low pressures (below 10^{-5} mbar) one has usually to integrate electronically for some time to improve the signal to noise ratio.

The large ion collector is necessary to allow for changes in the field direction. It causes a X-ray limit (in the 10^{-8} mbar range) considerably higher than with a Bayard Alpert gauge. That is, however, not really a problem during tokamak operation when the pressure will be anyway much higher than this limit. The gauges have to be roughly aligned with the toroidal field. Although the gauges work well both with and without magnetic field, on a tokamak they have to be installed inside the toroidal field. Outside of the main field, where the poloidal field dominates, the field direction changes too much. The emission current is stabilized by feedback control of the heating current of the filament. For stationary operation an heating current of 15 to 20 A is necessary. Conditioning the filament and restoring the emission current by the feedback requires occasionally excursions up to 30 A for some seconds. Due to the thermal capacity of the thick filament it is not possible within the required time resolution to avoid deviation of the emission current from the preprogrammed value. To compensate for these deviations besides the ion collector current I^+ the emission current I^e has to be measured, too. The neutral particle density in the box is derived from these quantities by

$$n_o = \frac{I^+}{(I^e - I^+) \cdot s(B, I^e)}$$

The sensitivity s depends on the magnetic field strength B and slightly on the emission current. At weak magnetic fields it increases steeply by about an order of magnitude. Above 1 Tesla it is nearly constant. (Fig. 2) This sensitivity increase due to the magnetic field will additionally mitigate any problems with the X-ray limit.

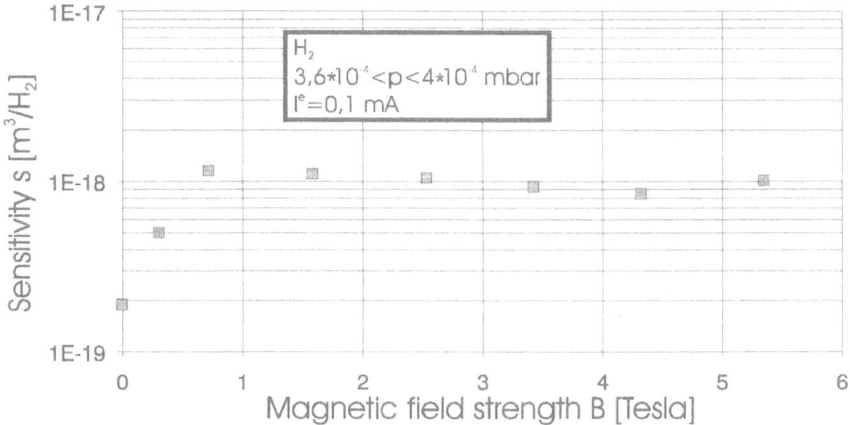

Figure 2. Dependence of the gauge sensitivity on magnetic field strength

The minimum space needed for installation of a gauge is given by the box size (external dimensions typically 2.5^3 cm^3) and about the same volume is required for cable connections on one side perpendicular to the field.

Compared with the environments, in which the gauges are already routinely used and thoroughly tested, the gauges have to work in ITER in a still stronger magnetic field (up to 9.25 Tesla on the high field end of the divertor cassette) and under neutron and γ radiation.

Preliminary tests have been performed up to 12 Tesla. They show basically, that the gauges can be operated in such fields. But during some of the runs the emission current feedback oscillated and the sensitivity showed jumps also at field strength below 6 Tesla. The reasons for that could not be identified yet. Such tests have to be repeated.

The gauge heads contain insulators usually made from alumina. Due to the high heating power of the filament the heads reach temperatures of the order of several hundred °C by which other insulators than ceramics are excluded. Degradation of these insulators has been observed already on the first gauges used on ASDEX. It seems to be caused mainly by reduction of the oxide ceramic by atomic hydrogen produced on the incandescent filament. It is reversible if the gauges are brought to air. The gauges work, however, still satisfactorily with resistance between the electrodes as low as 100 kΩ. For the existing gauge geometry 100 kΩ corresponds to a conductivity of nearly 10^{-3} $\Omega^{-1}m^{-1}$. This value can be slightly improved by further optimization. The maximum electric field strength in the insulators (200V/mm) can be reduced in the course of the same optimization.

In an ionizing radiation field ceramic insulators like Al_2O_3 become not only transiently conducting (Radiation Induced Conductivity), but some times also permanently (Radiation Induced Electrical Degradation), if an electric field is applied in addition[5,6,7]. The dependence of the RIED effect on material, temperature, applied electric field and the kind, dose and dose rate of the ionizing and the displacement damage producing radiation is confusing. In some recent investigations even no RIED effect at all has been found up to a radiation damage of 1.4 dpa in a fission reactor.[8] In this publication also a critical review of the existing RIED measurements is given. As a precaution we will nevertheless assume less favorable results from neutron irradiation approaching 10^{-5} $\Omega^{-1}m^{-1}$ for about 0.3 dpa. In any case no RIED measurements using neutron irradiation have been performed up to conductivities above 10^{-4} $\Omega^{-1}m^{-1}$. But also taking 10^{-4} $\Omega^{-1}m^{-1}$ as tolerable limit for the degradation and leaving the difference up to 10^{-3} $\Omega^{-1}m^{-1}$ as safety margin the gauges would survive the predicted radiation damage for 1000 full power discharges at the inner or outer dump plates or vertical targets[9]. None of the other possible positions for gauges in the divertor show higher radiation levels. Nevertheless some neutron shielding would be beneficial, e.g. by installation inside the divertor cassette instead of on the surface, even if this impairs to a certain degree the time resolution. In particular it is advisable to shield the gauges from 14 MeV neutrons, since the effects of such high energy neutrons can not be tested with the available neutron sources.

So far no severe problems are expected with the operation of the gauges under the additional constraints of ITER. Some modifications of the existing design have to be done which are desirable also for contemporary experiments. Nevertheless the conclusions drawn above have to be confirmed in extended test runs at a radiation level comparable to that expected in ITER and in higher magnetic fields.

Divertor. In the ITER Design Description Document "Pressure Gauges" (WBS 5.5.G.03)[10] a proposal for the distribution and installation of gauges in the divertor is given. Up to 12 poloidal positions are foreseen as shown in Fig. 3. Such sets (except the gauge #11 in the pumping channel and #12 in the divertor port) shall be installed in groves on the flank of the two standard cassettes next to the 4 diagnostic cassettes. The cabling can be routed there, too. This allows good access for installation and maintenance, if the cassette is removed from ITER. The gauges are to be installed 15 cm behind the surface to reduce the radiation level and have to be connected to the divertor volume by tubes 1 cm wide to get a response time of 5 msec. This number of 8 sets of pressure gauges guarantees sufficient redundancy and poloidal and toroidal resolution.

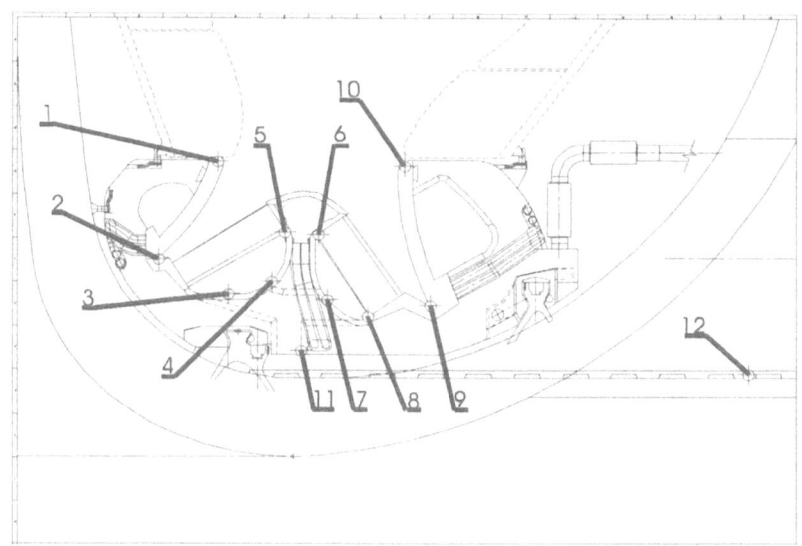

Figure 3. Cross section of the divertor (vertical target type) with 12 poloidal positions for neutral flux density measurements. The gauges shall be installed 15 cm behind the surface and be connected by tubes 1 cm wide with the measuring points except #11 and #12 which can be installed on the surface.

Mid plane. In the mid plane much lower neutral particle flux densities are expected. In the mid plane of ASDEX Upgrade usually less than 10^{20} D_2/m^2sec ($< 10^{-5}$ mbar) have been found. In ITER the flux may be still lower due to the larger size. A correspondingly worse signal to noise ratio will be unavoidable if the time resolution is not reduced by integration. On the other hand the neutron radiation is higher in the mid plane than in the divertor. Therefore it is advisable to embed the gauges deeper in the shielding plug in the horizontal port to shield them from the neutron radiation. It is foreseen to install gauges at the end of a 1 m long channel with one 90° bend as labyrinth on half way. (Fig. 4) With a diameter of 10 cm a vacuum time constant of 10 to 20 msec is expected.

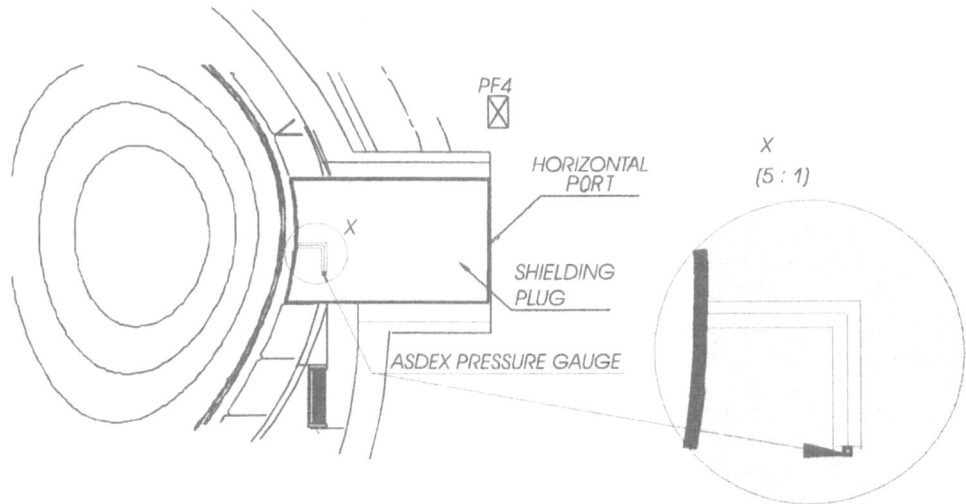

Figure 4. ASDEX pressure gauge embedded in the shielding plug of a horizontal port

Partial Flux Density Measurement

Besides the total neutral flux density the composition of the gas in the divertor, especially the content of He and other light gaseous impurities are important quantities. Their measurement is essential for exhaust studies of He ash and other gases, which may be intrinsic or intentionally added to increase the radiated power from the outer plasma zones. Also the ratio of deuterium to tritium should be monitored routinely.

The standard devices for such tasks are Quadrupol RGA's. They have, however, two essential drawbacks. They are very sensitive to magnetic stray fields and the mass spectrum in the presence of all three hydrogen isotopes will be very complicated.

The RGA's have to be installed outside the toroidal field, but with a high conductance to the point of interest, in the first place the divertor. The divertor ports seem to be most suitable. Four of them contain diagnostic blocks, the others cryo pumps. All are closed at about R = 16.5 m by flanges which are a part of the safety barrier. Due to the close vicinity of two poloidal field coils the poloidal field strength to be shielded will be as high as 1 Tesla for normal operation. If, however, only one of these coils is energized, the field strength can reach several Tesla. One may consider to use a combination of an iron shield with a coil around it as described by W.Herrmann and W.Szyszko[11]. It needs less iron, the field seen by the iron can be reduced by a factor of 2 or 3 and it does not disturb the external field. The RGA's have to be connected by a tube of about 20 cm$^\varnothing$ with the region of interest just outside the divertor cassette, which penetrates the pump or the diagnostic block, respectively. The feasibility of both the shielding and the penetrating tube is rather questionable.

Already with a deuterium background it is hardly possible to detect He with a RGA. Also Ne, frequently used to increase the radiation from the plasma boundary interferes with CD_4 a frequently seen intrinsic impurity. If tritium is added, it becomes still worse. Using known cracking patterns also complicated mass spectra can be decoded. But, as an example, there are more parent molecules from molecular hydrogen, methane and water in the range $1 \le m/e \le 24$ to consider than the spectrum contains mass peaks. One can simplify the problem by reasonable assumptions, e.g. about isotope ratios, but the accuracy will be poor anyway.

The situation improves, if additional input from an other diagnostic is available. An option would be a Penning gauge[12,13] or a modified ASDEX gauge[14] as light source for spectroscopic detection of He or other gases. The Penning gauge based system has been developed on TEXTOR and is used on several tokamaks. The ASDEX gauge based system has been developed for ASDEX Upgrade where such a system is in operation.

Both are not free of interference between different gases, too. They can, however, provide valuable additional information. The main reason for interference is the spectrum of molecular hydrogen, which covers the whole visible range and causes a background for the lines of the gases under study. To compensate for this the densities of hydrogen isotopes have to be known, e.g. from measurement of Balmer lines H_α or H_β.

The intensity delivered by a ASDEX gauge is so weak, that it can be measured only with high throughput interference filters and PMT's. With the Penning gauge which delivers much higher intensities one can use either filters and PMT's or a spectrometer and a MCP camera. It allows also to measure other gases besides He like Ne[13] and to separate the isotopes of hydrogen[15]. An alternative which avoids the problems with the background is to measure the intensity of He resonance lines in the VUV range (e.g. at 58.4 nm) with a spectrometer and a channeltron[13]. For intensity reasons this will be possible only with a Penning gauge.

Conventional Penning gauges contain permanent magnets which can be damaged in a strong magnetic field. Therefore their installation produces similar problems as the installation of RGA's. An alternative may be to install modified Penning gauges without permanent magnets in the vessel using instead the toroidal field. F.Dylla has used in the early 80's such gauges on PDX[16]. There were, however, obviously problems arising from the short pulse length of the toroidal field. This would not be the case in ITER. Tasks still to be solved are to design a system working in magnetic field strength as high as 4 Tesla or more and to test its dependence on changes of strength and direction of this field.

The ASDEX gauge can be operated in the toroidal field. A good position would be in front of a diagnostic block close to the low field end of the divertor cassette. There the flux density of He can be measured with fast time response as far as the signal to noise ratio allows. The flux density of He at this position dominates the He exhaust by the cryo pumps.

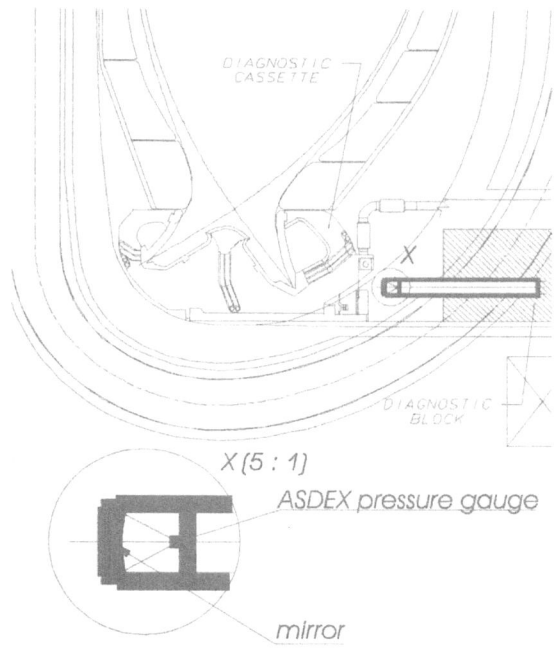

Figure 5. Example for the installation of an ASDEX pressure gauge based spectroscopic He detection system

The signal height depends greatly on the light transport. At the moment in both systems fiber optics are used. This may work for installation in the divertor port of ITER, too, but hardly for an ASDEX gauge (or a Penning gauge) installed inside the vessel. An alternative which may also improve the available signal power is a mirror relay optic as shown in Fig. 5. A concave f # = 1 mirror with a diameter of 25 cm is mounted on the front end of a tube (inner diameter 25 cm) which extends from one of the diagnostic blocks radially inwards to the divertor cassette. The gauge is installed near the focus of the mirror which projects the light produced in the ionization zone of the gauge as a nearly parallel beam through the tube to the detection system in the diagnostic block. The detection system consists of two interference filters, one for a Balmer line (e.g. H_β) the other for the He line at 501.6nm, two high aperture lenses and two PMT's. One of the filters can be used as beam splitter. This allows a very compact design. Radiation damage on the filters should be a minor problem. The PMT's, however, have to be magnetically shielded. If that is not possible one has to try other light detectors like MCP's or CCD chips which are less sensitive to magnetic fields. The focal length of the mirror has to be large compared with the extension of the volume where photons are produced in the gauge. Otherwise the divergence of the light at the filters will be too large. This volume is about 2 cm^3 including the possible movement due to the changing poloidal field. The mirror is well shielded from adverse effects by nuclear radiation (there is the lowest radiation level in the vessel) and plasma which does not see it at all.

SUMMARY

A neutral gas diagnostic for ITER has been proposed, which allows total and partial pressure measurements. For total pressure measurements about 100 ASDEX gauges on different poloidal and toroidal positions are foreseen. For the partial pressure measurement a combination of Quadrupol RGA's and modified Penning and ASDEX gauges with spectroscopic observation should be applied.

For the total pressure measurement with ASDEX gauges only minor R&D effort is needed to get a fully optimized system. The gauges can still be improved to reduce the susceptibility to degradation of insulators. Better choice of some materials may help, too. The ability to work under magnetic fields between 6 and 9 Tesla has to be finally demonstrated. A gauge has to be tested in a neutron irradiation environment.

For a successful partial pressure measurement much more development is necessary. RGA's can only be applied, if one finds a way to shield them in the large poloidal field of ITER. Conventional Penning gauges and the light detectors for both Penning and ASDEX gauges will suffer from the magnetic field, too. But the volume to be shielded is smaller. One could imagine to use Penning gauges without permanent magnets in the toroidal field. They have to be developed and their reproducibility with respect to changing field strength and direction has to be proved. For the ASDEX gauge the severe light intensity problem has to be solved and a mirror optic has to be designed which will work in the vessel.

ACKNOWLEDGMENTS

We like to thank our colleagues G.Janeschitz, S.Yamamoto and T.Ando from ITER, A.Kallenbach, H.Salzmann and O.Gruber from ASDEX Upgrade, K.H.Finken and T.Denner from KFA Jülich and S.Zinkle from ORNL for helpful discussions and providing information.

REFERENCES

[1] A.Kallenbach et al., Radiating boundary in ASDEX Upgrade discharges, 15th Int.Conf.on Plasma Physics and Controlled Nuclear Fusion Research, Seville 1994
A.Kallenbach et al., H-mode discharges with feedback-controlled radiative boundary in the ASDEX Upgrade tokamak, IPP Report IPP I/284 (1995), Nucl.Fusion in press
O.Gruber et al., Observation of continuos divertor detachment in H-mode discharges in ASDEX Upgrade, Phys.Rev.Lett., 74: 4217 (1995)

[2] A.W.Leonard, Bolometry for divertor characterization and control, paper I42 on this workshop

[3] R.Reichle, Bolometer for ITER, paper I43 on this workshop

[4] G.Haas et al, Measurements on the particle balance in diverted ASDEX discharges, J.Nucl.Mat., 121: 151 (1984)
C.C. Klepper et al., Neutral pressure studies with a fast ionization gauge in the divertor region of the DIII-D tokamak, J.Vac.Sci.Technol.A 11: 446 (1993)
G.Haas, Hot-cathode ionization pressure gauge ...,United States Patent 5,300,890, Apr 5, 1994
G.Haas, Hot cathode ionization manometer, UK Patent GB 2 255 442 B, 8th March 1995

[5] S.J.Zinkle et al., Radiation-induced changes in the physical properties of ceramic materials, J.Nucl.Mat., 191-194: 58 (1992)

[6] E.R.Hodgson, Radiation enhanced electrical breakdown in fusion insulators from dc to 126 MHz, J.Nucl.Mat., 191-194: 552 (1992)

[7] T.Shikama et al., Radiation induced conductivity of ceramic insulators measured in a fission reactor J.Nucl.Mat., 191-194: 575 (1992)

[8] L.L.Snead, D.P.White, S.J.Zinkle, Investigation on radiation induced electrical degradation in alumina under ITER-relevant Conditions, to be published in J.Nucl.Mat., (1995)

[9] M.E.Sawan, Neutronics analysis for the ITER divertor cassette, 3/15/1995 Fusion Technology Institute, The University of Wisconsin, Madison, Wisconsin, USA

[10] L.de Kock, ITER Document DDD WBS 5.5.G.03

[11] W.Herrmann et al., Screening of static magnetic fields without external field disturbances, IEEE Transact. on Magnetics, 25: 3278 (1989)

[12] K.H.Finken et al., Measurement of He gas in a deuterium environment, Rev.Sci.Instrum. 63:1 (1992)

[13] T.Denner, Entwicklung eines Verfahrens zur He-Partialdruckmessung in einer Deuteriumumgebung, Bericht des Forschungszentrums Jülich, Jül-3052 (1995) To be published

[14] H.-S.Bosch et al., Helium and hydrogen atom detection in the recycling gas using optical measurements on an ASDEX pressure gauge, J.Nucl.Mat.196-198: 1074 (1992)

[15] D.Hillis, private communication

[16] H.F.Dylla et al., Pressure measurements in magnetic fusion devices, J.Vac.Sci.Technol. 20:119 (1982)

OPTICAL SURFACE TEMPERATURE MEASUREMENT

A. Herrmann

Max-Planck-Institute of Plasma Physics, Berlin Branch,
EURATOM Association
Mohrenstr. 40/41, 10117 Berlin, Germany

INTRODUCTION

A crucial problem of realizing the international thermonuclear experimental reactor — ITER - is to handle the energy flux into the scrape off layer (SOL) and to control the energy flux onto the divertor plates. A main topic in research programs of present day tokamaks is to find operation regimes realizing simultaneously a high energy flux into the SOL and a tolerable divertor load. Such regimes as the recently found CDH-Mode in ASDEX Upgrade [1] or radiating divertor scenarios [2] are candidates and will be further investigated with modified ITER relevant Divertor configurations [3].

Monitoring the surface temperature of the divertor with optical systems (IR-thermography, CCD-systems) is widely used. From the time evolution of the measured surface temperature the power load to the divertor can be calculated by solving the heat conduction equation as it is routinely done on ASDEX Upgrade [4].

Apart from the interests of the physicists in optimizing the working regime, there is the necessity to protect the machine and in particular the divertor against hazardous energy impact.

This paper is divided into three parts. In the first part essential physical basics and problems of optical temperature measurements are discussed. After that, the requirements of ITER and design criteria for a thermography diagnostic are pointed out. A third topic deals with ITER conditions and possible arrangements. At last a summary is given.

PRINCIPLES OF MEASUREMENT AND PROBLEMS

Optical methods open the possibility for a touchless measurement of surface temperatures and are widely used in industry, medicine and science for different purposes. The physical effect used for such thermographic measurements is the variation of the number of photons emitted per surface unit at the wavelength, λ, with the temperature of a blackbody, T, as it is given by Planck's law. In this paper the photon formulation is used because the detectors for optical temperature measurement are photon and not power sensitive. The corresponding formulation of Planck's law is:

Diagnostics for Experimental Thermonuclear Fusion Reactors
Edited by P. E. Stott *et al.*, Plenum Press, New York, 1996

581

$$M_p^b(T, \lambda) = \frac{2\pi c}{\lambda^4} \frac{1}{exp\left(\frac{hc}{k\lambda T}\right) - 1} \tag{1}$$

Where h is Planck's constant, c the speed of light and k the Boltzmann constant.

The emittance, M_p, of real materials is always less the blackbody emittance, M_p^b. The emissivity, ϵ, is a measure of how a real source of radiation compares with a blackbody.

$$M_p(T, \lambda) = \varepsilon(T, \lambda) M_p^b(T, \lambda) \tag{2}$$

The emissivity can be temperature and wavelength dependent.

The relation between the wavelength of maximum emittance and the blackbody temperature is the Wien displacement law (Wien's law) :

$$\lambda_p^m T = 3622\ \mu m\ K \tag{3}$$

Figure 1 shows a contour plot of equation 1. Wien's law is indicated.

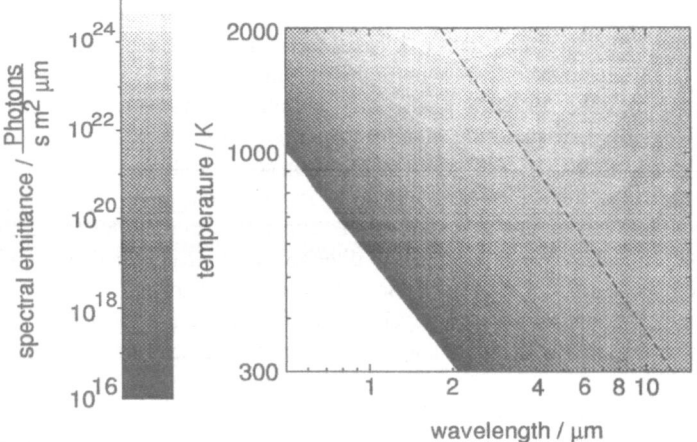

Figure 1. Contour plot of the spectral emittance ($M_p^b(T, \lambda)$, Planck's law). The dashed line indicates Wien's law

From equation 3 and figure 1 follows, that the wavelength of maximum emittance of a body at room temperature is at about 10 μm. The human eye starts to detect heat radiation if the body is at a temperature of about 900 K corresponding to a maximum wavelength of 4 μm. It is also obvious from figure 1, that the flux of photons emitted at a given wavelength varies strongly with temperature. Increasing the temperature from room temperature to 900 K enhance the emittance at a wavelength of 4.7 μm by more than a factor of 100.

The strong nonlinear dependence of the emittance from the temperature is a problem for measurements, which require a large dynamic range and a good sensitivity, as it is normally necessary in divertor measurements. A measure for the sensitivity is the change of the emittance with the temperature as it is shown in figure 2. A linear relation between emittance and temperature would result in a constant sensitivity which is only found at high wavelengths and temperatures above 500 K. Generally, the sensitivity increases for a given wavelength with temperature.

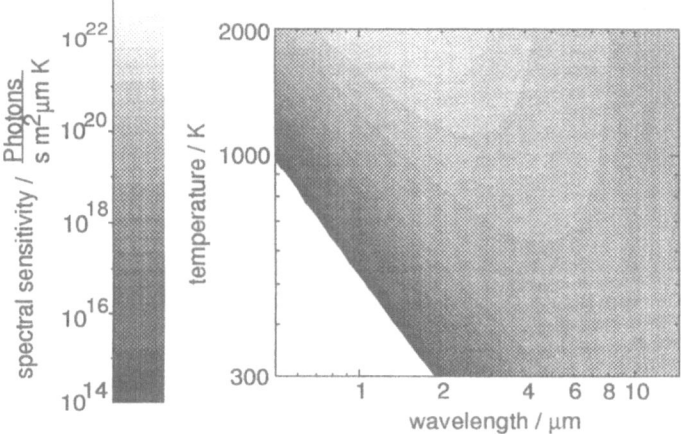

Figure 2. Temperature derivative of the spectral emittance ($\frac{\partial M_\nu^b(T,\lambda)}{\partial T}$).

If only a small dynamic range is requested, this effect is negligible, but in the case of a divertor diagnostic which generally demands a large dynamic range it is troublesome. To illustrate this, figure 3 shows the change in temperature which is necessary to switch one bit of a 10 bit ADC. The parameter is the maximum of the temperature range. It is assumed that the 10 bit range of the ADC is fitted to the dynamic range.

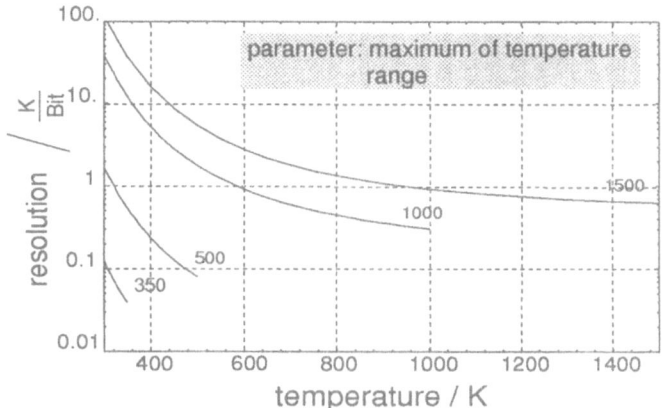

Figure 3. Necessary temperature change to set the lowest significant bit of a 10 bit ADC for different dynamic ranges.

The detection wavelength selected is 4.7 μm, corresponding to the upper end of the spectral response of a InSb detector which is often used for thermography systems. For example, if a dynamic range of 1500 K is requested, the temperature change per bit is 100 K at room temperature in contrast to 0.7 K at the upper limit.

To maximize the sensitivity at low temperatures a detection wavelength above 10 μm should be preferred and the dynamic range has to be adapted to the actual temperature by controlling the integration time of the detector.

Contribution of Bremsstrahlung

More or less bremsstrahlung is produced inside the plasma column and in dense radiating zones, like Marfes, in the edge plasma. Its intensity is essentially proportional to the square of the atomic number, Z, the electron and ion density, n_e, n_i, and inverse proportional to the square of the wavelength, λ:

$$\Phi_{Brems} \sim \frac{Z^2 \, n_i \, n_e}{\lambda^2} \tag{4}$$

In the following the contribution of bremsstrahlung to the temperature measurement is estimated. For this, the emittance of a blackbody at a given temperature and wavelength is calculated (equ. 1) and an additional contribution of photons due to bremsstrahlung is added. This sum is assumed to be the measured temperature information which is used to calculate the temperature (equ. 1) as it will be done in the thermography measurements.

The order of the flux of photons due to bremsstrahlung is taken from the ASDEX Upgrade Marfe measurements at 800 nm [5] and then scaled to the detection wavelength using equation 4. Figure 4 shows the contribution of bremsstrahlung for a thermography diagnostic measuring at 4.7 μm.

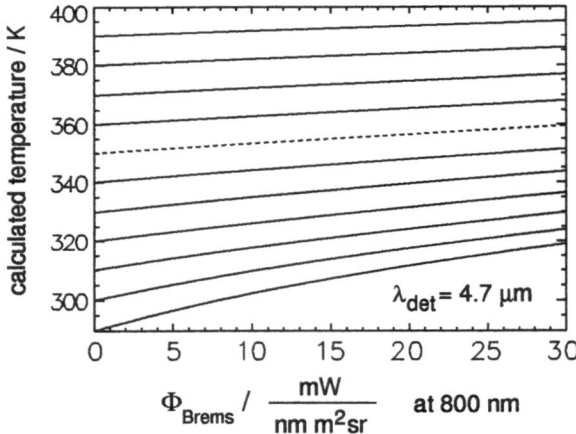

Figure 4. Contribution of Bremsstrahlung to the calculated temperature. The parameter is the blackbody temperature, $T(\Phi_{Brems} = 0)$

The parameter of the curves is the temperature of the blackbody ($\Phi_{Brems} = 0 \frac{mW}{nm \, m^2 sr}$). The error in calculating the temperature is only significant at low temperatures. Because of the strong increase of emittance with the blackbody temperature, a temperature rise of a few ten degrees makes the contribution of bremsstrahlung negligible.

The temperature, where the bremsstrahlung may be neglected depends on the detection wavelength. Figure 5 shows the blackbody temperature, T_{true}, where the contribution of bremsstrahlung to the calculated temperature, T_{calc}, is lower than 10 K as a function of the wavelength.

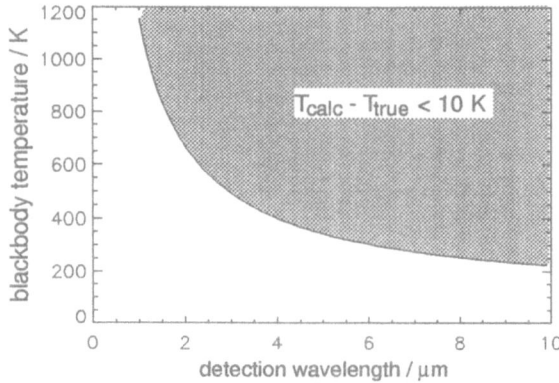

Figure 5. Parameter region, where the contribution of bremsstrahlung is lower than 10 K.

Principles of measurement

There are two principles for optical measurement of temperatures.
(i) measuring the emittance
(ii) measuring the ratio of the emittance for two different wavelengths

(i) Using equation 2 and taking into account the response of the detector and the electronics, R, the optical geometry of the system, F, and a contribution of background emittance, U_{bck}, which is assumed to be constant, then the output signal of the system is:

$$U_{out} = F\, R\, \varepsilon\, M_p^b(T, \lambda) + U_{bck} \tag{5}$$

Calibrating the detection system against the temperature of a blackbody, the contribution of the background and the product of the arrangement dependent parameters, FR, can be determined. If not a blackbody, but a divertor like arrangement is used, the emissivity, ϵ, is also included in the calibration.

(ii) Measuring the photon flux at two different wavelengths and taking the ratio of both measurements eliminates the arrangement parameters and the emissivity, provided they are wavelength independent:

$$I(T) = \frac{(U_{out}(T, \lambda_1) - U_{bck})}{(U_{out}(T, \lambda_2) - U_{bck})} = \frac{M_p^b(T, \lambda_1)}{M_p^b(T, \lambda_2)} \tag{6}$$

A typical change of the intensity ratio, I, with temperature is shown in figure 6. The wavelengths used are conform with the two region of spectral response for a Silicon/Germanium sandwich diode. In equation 6 the radiating area is also eliminated and so the two wavelength method may be used in principle to measure the temperature of small hot areas in a large field of view.

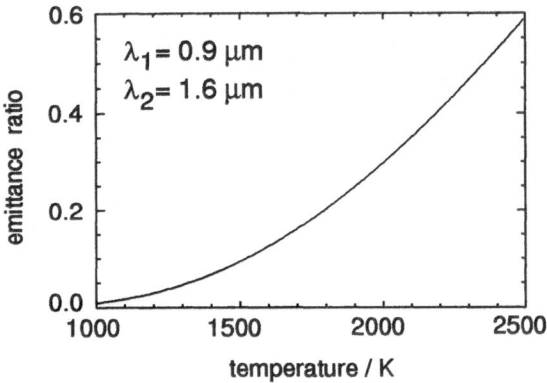

Figure 6. Emittance ratio, I(T), as a function of the blackbody temperature

ITER TASK DEFINITION AND DESIGN CONCEPT

In the ITER task definition two topics are assigned to thermography:
(i) monitoring the plate temperature to avoid divertor overload and destruction
(ii) measuring heat load profiles in the divertor.

It is needed to measure the surface temperature of the divertor for both. From this point of view they are comparable. Essential differences occur in the temperature range to be monitored, as well as time and spatial resolutions. These parameters will be discussed separately for each task in the following.

Monitoring the temperature

A stationary temperature of the divertor of the order of about 1000 K is acceptable for ITER operation. Hazardous temperatures are well above this value. Consequently the measuring range of the monitor system can be restricted to temperatures between 1000 K and 2000 K. As it was shown in the first part of this paper, the emittance in this temperature range is high enough to be detected by detectors with a spectral response in the visible range. The contribution of Bremsstrahlung can also be neglected (figure 5).

The spatial resolution is given by the maximum area of hot spots and hot edges allowed. Based on data of ASDEX Upgrade, a typical extension of hot patterns is 5 to 10 mm. If CCD cameras with a 1024x1024 pixel array are used for the monitor system a minimum of 10 cameras is requested to observe the whole divertor.

If the monitor system is designed for the two wavelength method and a hot spot area of 10% of the field of view is acceptable, then the observed area per detector can be increased by a factor 10 and the amount of detectors can be reduced. Unfortunately until now, no array cameras with sandwich detectors are available.

The standard time resolution of a CCD-system is about 20 ms/frame. Wether or not this fits the demand depends on the heating scenario. To give an estimation it is assumed that an additional load of 20 MW/m^2 is deposited to the plates. Then the surface temperature of CFC-graphite is increased by 350 K in the first 20 ms. After two frames (40 ms) the temperature increase is about 500 K. The sum of the normal divertor temperature of 1000 K and the increase of temperature due to the additional load is still lower after 40 ms than the assumed critical temperature of 2000 K. It seems, that a time resolution of 20 ms fits the monitoring needs. Otherwise it is no technical problem to increase the time resolution.

To achieve a high reliability and availability, the monitor system should be designed as simple as possible. The amount of changeable parameters should be minimized and the parameters once adapted are to be kept constant.

Heat load profiles

The heat load to the divertor has to be derived by searching the inverse solution of the heat conduction equation. To do this, the boundary and initial conditions have to be known. The boundary condition at the backside - the heat transport into the cooling structure - may be determined in the laboratory. The boundary condition at the front side - the time evolution of the surface temperature - is measured by the thermography diagnostic.

In pulsed tokamaks, where the delay between the shots is long enough to cool down the divertor, the initial condition is assumed to be a constant temperature which can be measured also by the thermography diagnostic.

In a steady state machine it is also possible to start the thermography measurement, when the first power is deposited to the target plates. A further possibility to find an initial condition is to use a stationary phase of the discharge where the power deposition is constant and the temperature profile inside the divertor becomes stationary. Stationary discharge phases are also useful to validate or to synchronize the temperature profile, calculated from the previous surface temperature evolution.

It follows, that the lower detection limit of the thermography diagnostic is the initial surface temperature of the divertor, which is normally at room temperature or at a few hundred centigrade Celsius. The upper limit is given by the maximum divertor temperature (segregation or melting temperature of the divertor).

The time resolution of the heat flux diagnostic is determined by the operation scenario for ITER.

Because H-Mode is a possible confinement regime for ITER, ELMs are to be expected and should be resolved. Typical ELM durations [6] are in the order of some hundred μs, the rise of the heat flux at the beginning of an ELM is in the order of 100 μs. Hence, a time resolution better than 100 μs should be aimed at. This time resolution is also sufficient to investigate disruptions [4].

The spatial resolution should be 2 to 3 times better than the expected radial decay length. Furthermore, the structure of the divertor, e.g. gaps between divertor tiles, should be resolved.

The divertor structure of ITER is toroidally symmetric and in principle it is sufficient to measure the heat load at one toroidal location and to verify the symmetry of energy deposition with integral methods such as cooling water calorimetry and divertor thermometry.

Because of the strong sensitivity change, the thermography system should be designed as flexible as possible, enabling the change of integration time and dynamic range during the measurement.

ARRANGEMENT AND SPECIAL ITER PROBLEMS

Installing a thermography diagnostic for the ITER divertor is mainly restricted by the possible geometry of view. The problem to find a proper view to the divertor is engraved by the high level of neutron flux which excludes the use of lenses and optical fibres for image transfer in the unshielded region.

A possibility to monitor the surface temperature of the divertor seems to be transfer optics using metallic mirrors in the high neutron flux region and further away neutron resistant materials to transfer an image behind the neutron shield.

Experiences from the use of IR detectors on ASDEX Upgrade and CCD cameras on JET show that a neutron flux of $10^{13}\frac{1}{m^2s}$ may be tolerated without damaging the detectors, so that the detection system may be placed directly behind the neutron shield.

Possible view ports and a principle arrangement are sketched in figure 7.

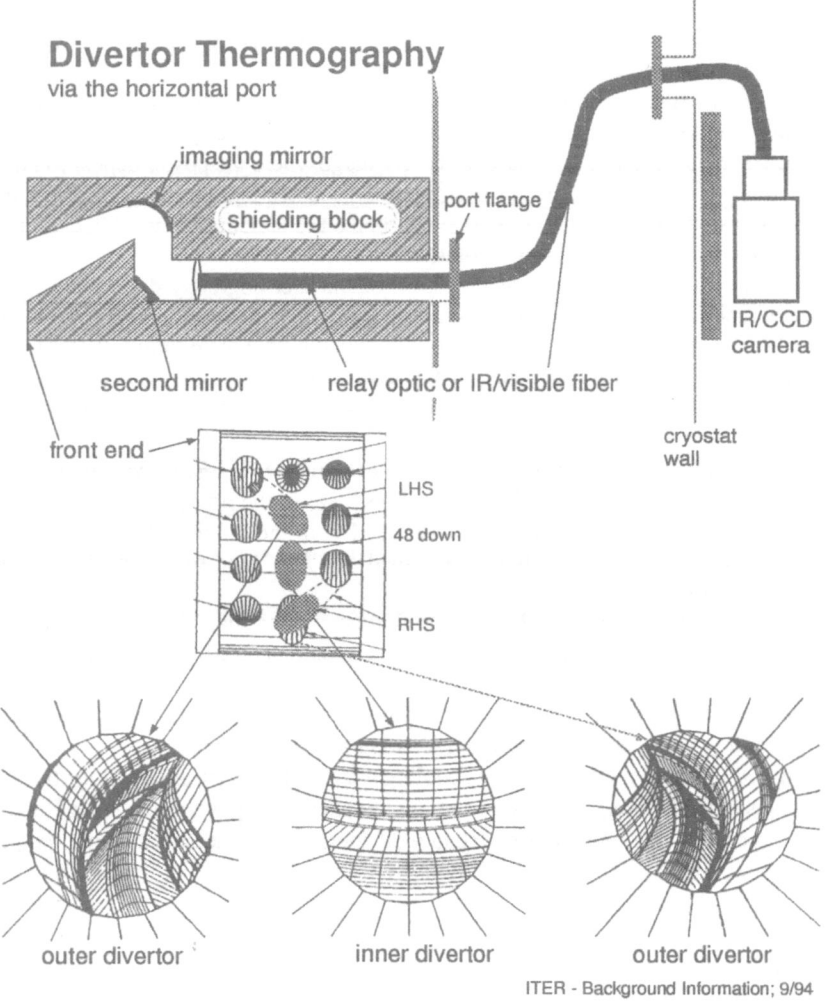

Figure 7. Principle arrangement of a thermography diagnostic at ITER.

ITER should be equipped with at least one heat load measuring system.

Monitoring the surface temperature of the whole divertor should be aimed at. This would require 3 ports per module i.e. nearly a third of the available ports. Such an equipment doesn't seem realistic and a compromise between the endeavour of complete divertor observation and reliability has to be found.

SUMMARY AND CONCLUSIONS

The dependence of the photon emittance and the spectral sensitivity on the detection wavelength and the temperature was discussed. At wavelength above 10 μm the sensitivity tends to become constant in contrast to detection wavelengths lower than 5 μm where it is highly nonlinear. The emittance at 10 μm is high enough to be detected with today's cameras. Unfortunately, IR-cameras working at 10 μm are normally based on mechanical scanned systems and can not be used in magnetic fields. IR-cameras with focal plane arrays are available for the 5 μm region. Focal plane arrays are usable in magnetic fields. A second advantage of the 10 μm region is the vanishing contribution of bremsstrahlung. But bremsstrahlung is no problem at 5 μm if the divertor temperature is above 400 K, as expected for ITER. In spite of bremsstrahlung, the use of visible range detectors for the protection task seems to be possible.

The main problem in realizing thermography diagnostic and protection systems originates from the high neutron flux and the limited fields of view. A 'metal based' relay optic for image transmission should be designed and tested as a next step in diagnostic development.

REFERENCES

1. K. Lackner et al, Recent results from divertor operation in ASDEX Upgrade, *Plasma Phys. Contr. Fusion* 36 (1994) B79
2. V. Mertens, H.-S. Bosch, K. Büchl et al, Divertor performance of high density H-Mode discharges in ASDEX Upgrade, *Europhysics Conference Abstracts, 21st EPS Conference on Controlled Fusion and Plasma Physics, Contributed Papers*, 18B (1994) 326
3. H.S.Bosch, D.Coster, S.Deschka et al, Extension of the ASDEX Upgrade programme: divertor II and tungsten target plate experiment application for preferential support, phase I and II, *IPP Report* 1/281 (1994)
4. A. Herrmann, W. Junker, K. Günther et al, Energy flux to the ASDEX Upgrade divertor plates determined by thermography and calorimetry, *Plasma Phys. Control. Fusion* 37 (1995) 17
5. W. Junker, H.-S. Bosch, K. Buechl et al, Comparison of Experimental MARFE Observations in ASDEX Upgrade with B2 Calculations, *Europhysics Conference Abstracts, 21st EPS Conference on Controlled Fusion and Plasma Physics, Contributed Papers*, 18B (1994) 680
6. A. Herrmann, M.Laux, D. Coster et al, Energy transport to the divertor plates of ASDEX Upgrade during elmy H-mode phases, *J. Nucl. Mater.* 220–222 (1995) 543

LANGMUIR PROBES AND OPTICAL DIAGNOSTICS FOR THE ITER DIVERTOR

L de Kock[1], T.Ando[1], A. Antipenkov[1], G. Janeschitz[1], E. Martin[1], G. Matthews[2], R. Monk[2]

[1]ITER JCT, Garching, Germany
[2]JET Joint Undertaking, UK

INTRODUCTION

Langmuir probes provide a simple measurement of local plasma parameters at the divertor target, information which is difficult to obtain otherwise. In ITER this information is critical for the operation and the control of the divertor in the desired mode.

The probe system will be designed as a triple probe, providing good temporal resolution. The probes need to cover the poloidal cross-section of the target area.

The heat flux to the probe during normal operation and events such as disruptions, vertical displacement events and ELM's is the most important consideration in the design. The unavoidable ceramic insulator is subjected to radiation reducing its life time. The choice of the material is critical.

The design is based on the probes used in JET. The tip can be made of different materials and is attached to a larger body having the role of heat sink. The steady state cooling is to the water-cooled target tiles through the insulating layer. The details of the design will be shown with the thermal analysis.

Spatially resolved optical and IR measurements are important for the recycling and impurity behaviour and the thermography of the high heat flux components. Both require the transmission of images from the divertor cassette to the outside world. The radiation levels preclude refractive components and image transmission will be based on mirror relay systems.

To mitigate problems with the windows,a transition to optical fibres or even detection is planned in the port area where a large shielding block is located in which radiation levels can be low enough.

Diagnostics for Experimental Thermonuclear Fusion Reactors
Edited by P. E. Stott *et al.*, Plenum Press, New York, 1996

591

REQUIREMENTS

Langmuir probes and optical access require quite different solutions for the integration into the diagnostic cassette.

The probe and its wiring must be exchangeable under remote conditions in a hot cell. This requires extremely reliable automatic connections between the cassette and the vessel wiring and possibly between the probe and the wiring harness.

Optical transmission lines must preferably be assembled outside and inserted as one unit with minimum alignment. These assemblies are located under the cassette and use rugged metallic mirrors to reach the area of interest.

The concept of the wiring, the connectors and the use of optical image transmission components together with the necessary remote handling operations will be discussed as a more general approach for all other diagnostics.

PROPOSED TECHNICAL SOLUTION

Langmuir Probes

The outline design of the Langmuir probes with main dimensions is shown in Figure 1. The probe tip is chosen to have a smaller width than the probe body which serves as a heat sink. The body is actively cooled by thermal conduction to the side of the plasma facing component through a ceramic coating which also serves as electrical insulation.

The probes are located in the 10 mm gap between divertor cassettes. The poloidal width of the body is 10 mm, allowing a probe spacing of 10 mm, and the tip is 1 mm wide. Three neighbouring gaps are available to form triple probes from probes at the same poloidal location. A total of 80 probes can be placed in one poloidal cross-section covering inner and outer divertor leg for the two options of gas box and vertical target.

The effective area of the probe tip is determined by the maximum saturation current that one wants to observe and the heat loads that need to be sustained by the probe. High current density is expected for a low temperature, high density plasma and low current density for a fully detached plasma.

The thermo-mechanical and electrical performance are critical for the Langmuir probe. For a first impression of the technical aspects a preliminary one dimensional thermal analysis has been performed on a probe with a 1mm wide short tungsten tip (Figure 2) for 3 different conditions:

1. Normal detached steady state $\quad\quad\quad$ 5 MW/m^2 for 1000s (uniform)
2. Off-normal slow transient attached case $\quad$ 15 MW/m^2 for 10s (along field lines)
3. Disruption $\quad\quad\quad$ 10~140 MJ/m^2(incident heat load) for 0.1~3 ms

$\quad\quad$ in state 1 : detached case + disruption with vapor shield effect
$\quad\quad\quad\quad\quad\quad$ (net/incident=1%)
$\quad\quad$ in state 2 : attached case + disruption with vapor shield effect
$\quad\quad\quad\quad\quad\quad$ (net/incident=1%)

The model is shown in Figure 3 with the materials. The heat flux is assumed to be reduced in proportion to increase in the width of the section. For the disruption case a simple one dimensional time dependent heat penetration model has been used. Table 1 shows the results. For case 1 the temperature of the probe tip (tungsten) remains under the recrystallisation temperature; for case 2 this

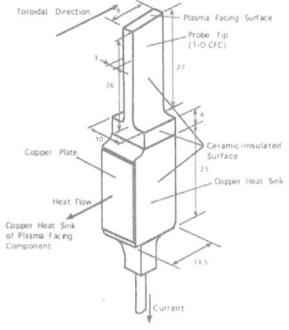

Figure 1: Outline design of the Langmuir probe with main dimensions (long wide tip)

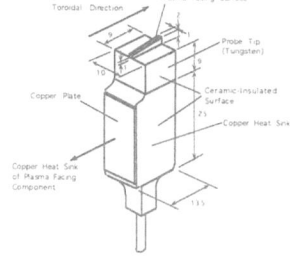

Figure 2: Outline design of the Langmuir probe as used for the analysis (short narrow tungsten tip)

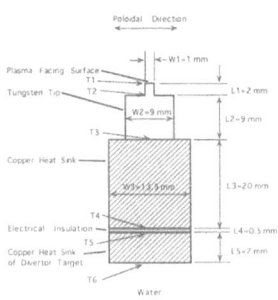

Figure 3: Model for the one dimensional thermal temperature calculation with typical cross-sections and materials

temperature is exceeded but remains under melting temperature. In case 3.1 the tip will remain just under melting temperature, but for case 3.2 strong vaporisation will occur with large erosion. These critical cases need a more detailed analysis.

Table 1: Temperatures at the interface planes of the probe from a one dimensional model

Condition	T_1	T_2	T_3	T_4	T_5	T_6
1. Normal detached	1243	1132	632	353	294	231
2. Slow attached	2131	1132	632	353	294	231
3. Disr. state 1	<3000					
state2	>4000					

From the table one can see that considerable stresses can build up across the insulating layer. A more detailed analysis is required to insure the integrity of the layer.

The probe shown in Figure 1 has a long 3mm wide tip of 1D CFC to take account of the expected erosion and the mechanical stress from the jxB forces. It is a further development which remains to be analysed.

Since the exchange of the probes is planned to be in parallel with the replacement of the target tiles, damage can be repaired. Langmuir probes can fulfil an important function for the initial phases of ITER and help to establish the optimum divertor operating regime.

Optical Imaging in visible and IR

The concept of the optical imaging system is to have a first mirror in the specialised diagnostic cassette. This mirror is either directed towards the the target area for thermography in the range of $\lambda = 3\text{-}5\mu m$ or towards the divertor plasma for the spatial observation of the flame in visible light. The spatial resolution is required to be sub cm.

The intense CX flux and possible impurity deposition may make a shutter mandatory.

The visible light or IR is directed to a removable tray under the cassette where a channel of 15 cm high and 60 cm wide is available for transmitting the images. The radiation levels under the cassette range from 10^{-2} to 10^{-6} of the first wall flux for areas near and far away from a penetration respectively. The images from the tray are relayed to a large shielding block in the divertor remote maintenance port where radiation levels can be low enough for a transition to optical fibre or even detection by suitable instruments. The toroidal and poloidal fields, however, are typically 5 T restricting the range of available options.

A typical arrangement is shown in Figure 4 for a pencil beam system that is proposed as back up for the imaging system and for a dual channel tray to relay images from the inner and outer divertor separately.

Detailed studies need to be undertaken to determine radiation levels which will decide on the lens/mirror option for the transmission system. Once this is known ray tracing calculations are necessary to optimise the optical elements and their locations in the structure.

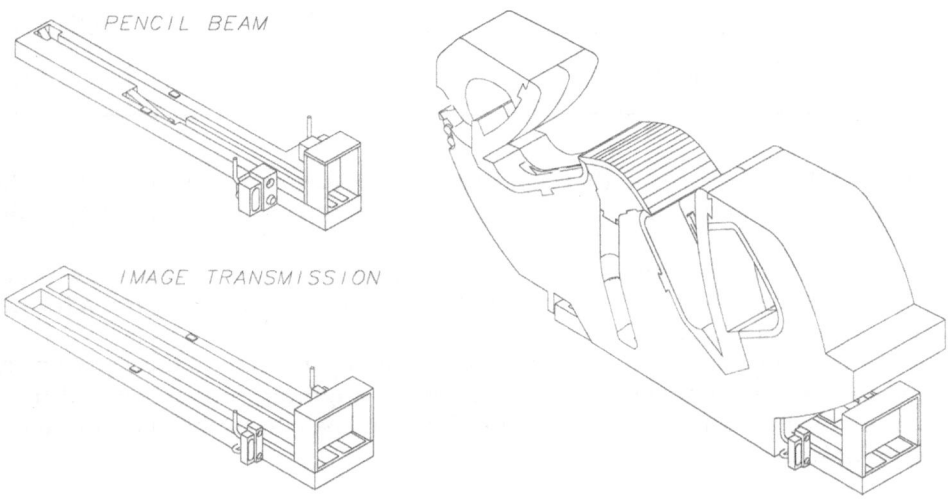

Figure 4 : The diagnostic cassette and the removable tray for the image transmission system

LASER DIAGNOSTICS OF HYDROGEN AND HELIUM ATOMS IN A DIVERTOR REGION

K. Uchino,[1] S. Kuroda,[1] T. Kajiwara,[1] K. Muraoka,[1]
T. Okada[2] and M. Maeda[2]

[1]Kyushu University, Kasuga Fukuoka 816, Japan
[2]Kyushu University, Hakozaki Fukuoka 812, Japan

INTRODUCTION

The divertor operation for density control and impurity suppression is one of the key issues of ITER. In order to evaluate the divertor operation, local measurements of atomic hydrogen densities are very important. Also, atomic helium densities must be measured, because the divertor should exhaust the helium ash produced by D-T burning. In this paper, we propose laser diagnostic techniques for local measurements of hydrogen and helium atoms in a divertor region.

For measurements of hydrogen atoms, the laser induced fluorescence technique based on two-photon excitation (two-photon LIF) from the ground level to the n=3 level is most appropriate. Fluorescence observation is at Hα and the ground level population is directly probed without using VUV light. For helium atom measurements, only excited levels can be probed by laser radiation. In this case, the laser induced ionization (LII) method[1] is suitable because it can yield a local atomic density without elaborate techniques such as frequency tuning and suppression of stray light. The applicability of these techniques to the diagnostics at the ITER divertor were examined through estimations of expected signal-to-noise (SN) ratios.

PRINCIPLE OF MEASUREMENTS AND PRELIMINARY EXPERIMENTS

Two-Photon LIF

The population density at the ground level is to be probed by observing the Hα fluorescence caused by two-photon excitation to the n=3 level from the ground level. Excitation to n=3 level is possible by using the first Stokes shift (205 nm) in D_2 gas of an ArF excimer laser, and/or a combination of the ArF output (193 nm) and the second Stokes radiation (219 nm). The method has clear advantages compared with the conventional LIF technique for hydrogen detection in plasmas, namely: (i) stray laser light can be completely eliminated by a filter, (ii) because the ground level is directly probed by the technique, no plasma model is necessary to yield hydrogen densities from the signal, and (iii) the input wavelength is not in the VUV region.

In our previous publications, we have examined various two-photon excitation schemes for LIF measurements of hydrogen atoms, and shown that the combination of (205nm×2) and (193+219 nm) is most advantageous.[2,3] As for laser light sources, we have so far obtained energies of 63 mJ at 193 nm (fundamental ArF excimer laser beam), 20 mJ at 205 nm (first Stokes component) and 20 mJ at 219 nm (second Stokes component). In

Diagnostics for Experimental Thermonuclear Fusion Reactors
Edited by P. E. Stott *et al.*, Plenum Press, New York, 1996

order to calibrate the detection system, a flow-tube reactor in which atomic hydrogen was generated by dissociation in a microwave cavity. The atomic hydrogen density inside the reactor was calibrated by titration of nitrogen-dioxide with an accuracy of ±20%.[3]

Laser Induced Ionization

There are six sub-levels for the principal quantum number n=3 of the helium atom, and all of them have the possibility of being ionized by a laser beam whose wavelength is shorter than 660 nm because their ionization energy is below 1.9 eV. When the laser beam is injected into the plasma, the emission from the n=3 to n= 2 transition is reduced due to the ionization of the n=3 level. This can be detected quantitatively. Thus, the helium atomic density at the excited state can be obtained and the atomic density at the ground state is calculated using a collisional-radiative model.[4,5] For such an analysis, information of electron density and temperature is necessary.

A frequency doubled YAG laser (wavelength 532 nm) can be used as a light source. The laser power should be strong enough to saturate the photo-ionization transition. The saturation powers are estimated to be around 3 MW/cm^2 except for the level 3^1P. Actual measurements are performed at a power density of several tens of MW/cm^2, which is a power level at which Thomson scattering is possible if the electron density is larger than 10^{18} m^{-3}. This means that electron temperature and density can be determined simultaneously with the LII signal, and so the atomic helium density at the ground state can be obtained without depending on other diagnostic systems.

In order to examine the possibility of detecting helium atoms using the LII method, preliminary experiments were performed on ECR discharge plasmas. The plasmas were produced typically with argon gas at 1 mTorr and microwave input power of 500 W, and the typical electron density and temperature measured using Thomson scattering are $n_e=7\times10^{17}$ m^{-3} and T$_e$=3 eV, respectively. The first measurements were performed by adding helium gas at a partial pressure of 0.2 mTorr. A flashlamp pumped Nd:YAG laser with a second harmonic generator (SHG) was used as the laser source, whose output was 500 mJ, pulse width was 10 ns and repetition was 10 Hz. The output light was injected into the plasma as a parallel beam with a diameter of about 10 mm.

The detection system was the same as that used in Thomson scattering measurements. The observation wavelength was selected by a monochromator and adjusted to HeI 588 nm (transition 3^3D-2^3P). Figure 1 shows an example of the detected signal. In this case, signals were averaged for 2560 laser shots in order to improve the SN ratio. The signal appeared as a decrease of about 5 % of the DC background level of the line emission of an atomic helium (HeI 588 nm).

In order to examine the saturation characteristics of the signal, the laser output energy was changed from 500 mJ to 75 mJ. The signal at an output energy of 500 mJ, which gives 20 times higher power density than the saturation power density, is clearly saturated.

Atomic helium densities at five radial positions in the ECR plasma were evaluated from LII signals through an analysis using the collisional-radiative model. The results agreed with the value determined by the helium pressure (0.2 mTorr) within a factor of two, showing the usefulness of the method.

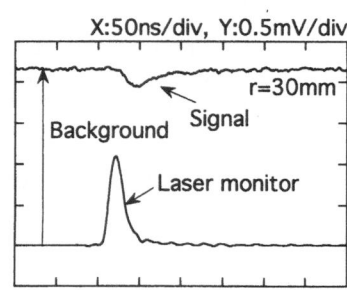

Figure 1. LII signal obtained at the wavelength of 588 nm. The signal is after averaging for 2560 laser shots.

EVALUATION OF SN RATIOS AT ITER DIVERTOR

Possible arrangement of the laser diagnostics proposed here is shown in Fig. 2. Laser beam path is taken to be vertical to the sight line of detection system. Also, laser beam path coaxial to the detection axis is taken into consideration for the two-photon LIF. We assumed that the detection solid angle is 5×10^{-3} sr, plasma width is 300 mm in the direction of sight line, optical transmission of detection system is 0.2, and quantum efficiency of a photomultiplier is 0.1.

Measurement of Atomic Hydrogen Density

In the following estimation of SN ratios, we assumed laser energies at the observation point to be 60 mJ (193 nm), 20 mJ (205 nm) and 20 mJ (219 nm). These are the maximum energies so far we obtained. The rate of two-photon excitation is proportional to square of laser power. However, when the excitation power is increased more and more, photo-ionization from the excited state becomes remarkable and then the fluorescence signal saturates. At a certain excitation energy, the most efficient fluorescence signal can be produced at the power density level where the signal is proportional to the laser power. For the present case, the optimum power density is about 10^9 W/cm^2 and the rate of excited level population to that of ground level n_3/n_1 is 6×10^{-3} at this power density. Because the total laser energy is 100 mJ, the beam diameter should be 1.1 mm. The length of the observation volume was taken to be 20 mm along the laser beam. We took the integration time of the observation to be equal to the fluorescence decay time (20 ns). Using these conditions and assuming the atomic hydrogen density at the observation point to be $n_H = 10^{17}$ m^{-3}, signal intensity was estimated and the expected photo-electron number at the photomultiplier was found to be $S_{LIF} = 84$ for a single laser shot.

The noise component is mainly determined by the photon statistical noise and can be estimated as the square root of the sum of S_{LIF} and background intensity S_B. First, we consider the case of vertical observation against the laser beam. In order to evaluate the background intensity, we examined several cases of different distributions of electron density n_e and electron temperature T_e. Neutral particle distributions were calculated using a Monte-Carlo simulation code, and then, the distributions of the population density at the excited state n=3 were obtained using the collisional-radiative model calculations along the sight line. Estimated values of S_B were around 600, and the resultant SN ratios were found to be SNR=3-4, insensitive to profiles of n_e and T_e. The ratio of the fluorescence signal to the background intensity (enhancement factor) was 0.14 for this vertical observation.

Similar estimations were performed for the case of coaxial observation. In this case, the length of the fluorescence volume of 20 mm is along the sight line and so the cross-

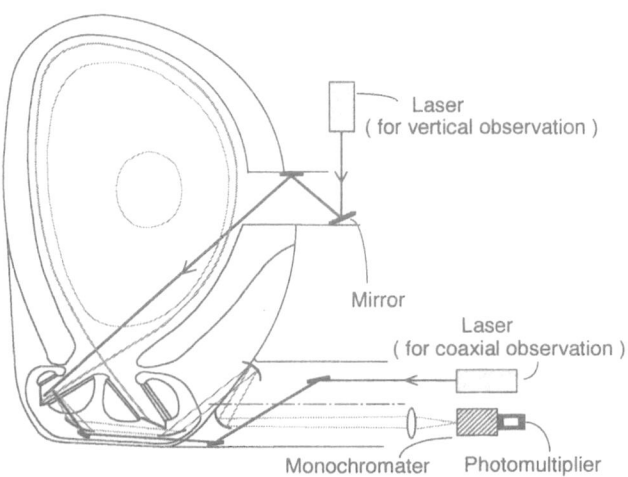

Figure 2. Possible arrangement of two-photon LIF and LII on the ITER divertor.

sectional area of sight field of the observation system can be reduced to the diameter of the laser beam. Here, we took a diameter of 1.5 mm. Then, the background intensity was reduced ($S_B \approx 35$) and the SNR was 2.4 times higher than that for the case of vertical observation. Furthermore, we can expect clear fluorescence signal over the background (enhancement factor ≈ 2.4).

In Fig.3, expected SNRs are shown against n_H for both cases of vertical and coaxial observations. If we require SNR=3 to be the lowest value for the quantitative measurement, the detection limit is $n_H = 1 \times 10^{17}$ m^{-3} for vertical observation, and is $n_H = 2 \times 10^{16}$ m^{-3} for coaxial observation. For both cases, we can expect that the technique will be used for measurements of atomic hydrogen densities in the wide range.

Measurement of Atomic Helium Density

Here, we assumed the simplest situation, that is the uniform distributions of atomic helium density (10^{17} m^{-3}), electron density (10^{19} m^{-3}) and electron temperature (50 eV) along the sight line over the plasma depth of 300 mm.

By using the vertical observation arrangement of Fig. 2, a YAG laser beam with a rectangular cross section of 3×20 mm^2 can be injected. If the laser pulse duration is 10 ns, the output of 100 mJ per a laser shot (power density > 10 MW/cm^2) is sufficient to saturate the LII signal.

We assumed the observation volume to be 2 mm in width, 20 mm in length and 20 mm in depth. In this situation, the LII signal at HeI 588 nm will appear as a transient decrease of 6.7 % of the background radiation. These conditions were fed in the collisional-radiative model calculations of excited state populations of atomic helium, and then, the LII signal intensity was estimated. Estimated photo-electron number for the background radiation was $S_B \approx 140$ and the number corresponding to the LII signal was $S_{LII} \approx 10$. The SNR for a single laser shot is about 1 which is not sufficient for quantitative evaluation of atomic helium density. Therefore, the signals should be accumulated over 100 laser shots in order to obtain SNR=10. If we have a light source with an output of ~100 mJ and a repetition of >1 kHz, we can construct a helium detection system with a response time of <0.1 s. A diode laser pumped YAG laser, which is now rapidly being developed, is expected to fulfill this requirement soon.

REFERENCES

1. V.I. Gladuschak et al., Proc. 19th EPS Conf. Contr. Fusion Plasma Phys., Innsbruck, II-1219 (1992).
2. U. Czarnetzki, K. Miyazaki, T. Kajiwara, K. Muraoka, M. Maeda, and H.F. Doebele, J. Opt. Soc. Am. B 11: 2155 (1994).
3. K. Muraoka, K. Miyazaki, U. Czarnetzki, T. Kajiwara, K. Uchino, H. Takenaga, T. Okada, M. Maeda, T. Yamauchi and T. Shoji, J. Nucl. Mater. 220-222: 563 (1995).
4. H.W. Drawin and F. Emard, EUR-CEA-FC-534, Association Euratom-C. E. A. (1970).
5. T. Fujimoto, J. Quant. Spectrosc. Radiat. Transfer 21: 439 (1979).

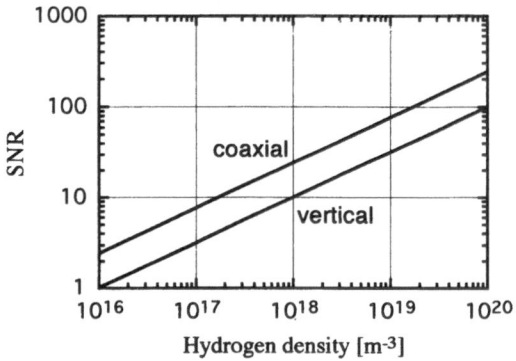

Figure 3. SN ratios plotted against hydrogen density for vertical and coaxial observations.

THERMAL HELIUM BEAM FOR DIAGNOSTIC OF ELECTRON DENSITY AND TEMPERATURE

P. Kornejew, W. Bohmeyer, G. Fussmann
Max-Planck-Institute of Plasma Physics, Berlin Branch
EURATOM Association
Mohrenstr. 40/41, 10117 Berlin, Germany

Plasma diagnostic based on the injection of a thermal helium or lithium beam is a well known diagnostic tool. The advantage of the He-beam diagnostic is a greater penetration depth because of higher ionization energy of the helium atoms and the possibility to built long term stable beam sources.

The helium beam diagnostic has been studied experimentally and theoretically in various places (Schweer et al.,1992, Brosda, 1993). A detailed analysis has been performed using the stationary plasma of the PSI-1 plasma generator device. In this study the results have been compared with Thomson-scattering and Langmuir-probe measurements (Behrendt et al., 1994).

The temperature and density dependence of the population density of the HeI states has been modelled by solving a system of 21 rate equations, including HeI states with the principal quantum numbers up to $n = 6$. The following processes have been taken into account:

- electron impact excitation from the ground level
- electron impact transitions between excited states
- electron impact ionization
- spontaneous emission

Recombination processes can be neglected if the beam propagates perpendicular to the magnetic field.

Because of different temperature dependences of the cross sections for excitation of singlett and triplett levels from the ground state the corresponding line intensity ratios are well suited for determination of the electron temperature. In addition, the electron density can be obtained from the intensity ratio of two singlett lines whose upper levels are preferentially depopulated by different mechanisms (elctron impact or spontaneous emission).

A large number of HeI-lines has been determined by passive spectroscopy and compared with such model calculation. Basing on these studies the following line pairs are recommended for tokamak-diagnostics:
Electron temperature determination :   471 nm / 505 nm   or 706 nm / 728 nm
Electron density determination :   501 nm / 505 nm   or 668 nm / 728 nm

Diagnostics for Experimental Thermonuclear Fusion Reactors
Edited by P. E. Stott *et al.*, Plenum Press, New York, 1996

599

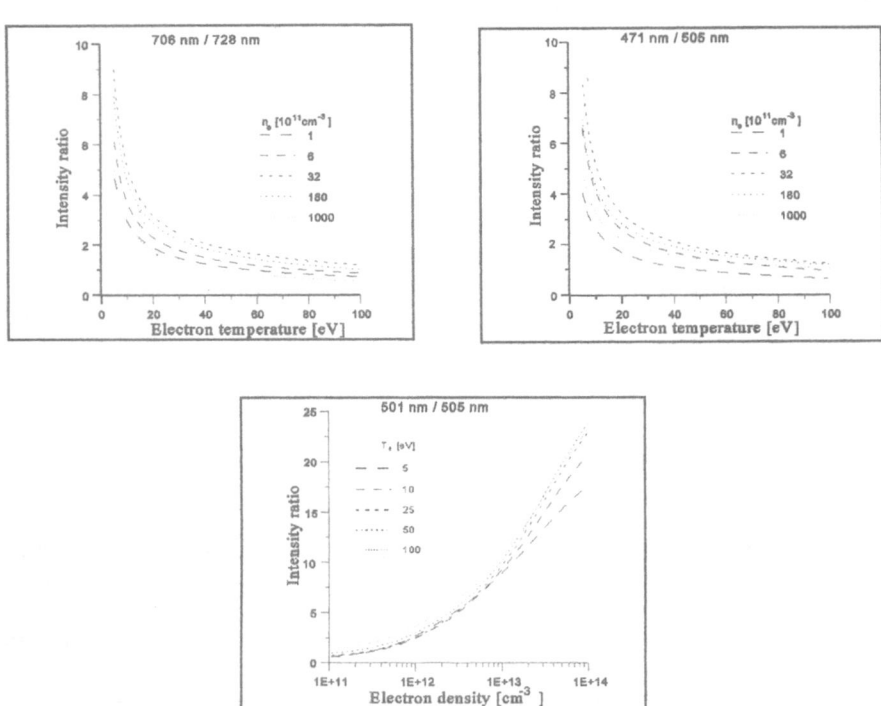

Figure 1. Recommended line ratios for electron temperature and density determination as functions of T_e and n_e

The spatial resolution in beam diagnostic is limited by the divergency of the beam. For this reason different beam sources have been compared in experiments. Single nozzles (diameter 20...500 μm) were used up to a pressure of 1.5 bar and a nozzle scimmer system (Camparque-source) was included in this comparison. In Fig.2 the principal arrangement for both systems is shown.

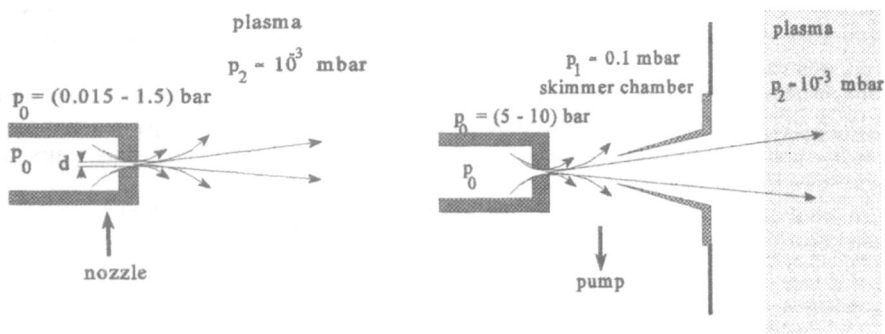

Figure 2. Principal arrangement for a single nozzle source and a nozzle scimmer design

In the case of single nozzles the quality of the beam improves with decreasing the nozzle diameter and using higher helium pressures.

A distinct improvement of the beam quality can only be reached by using a nozzle scimmer system. In such a system the divergency of the beam is determined by the distance between nozzle and scimmer and the inner diameter of the scimmer. This advantage, however, must be paid for by higher pumping speeds, which as a consequence require larger flanges.

The characteristic parameter of different beam sources - the angular width of the intensity distribution - is given in Fig.3. Only with a nozzle scimmer system a small divergence can be reached.

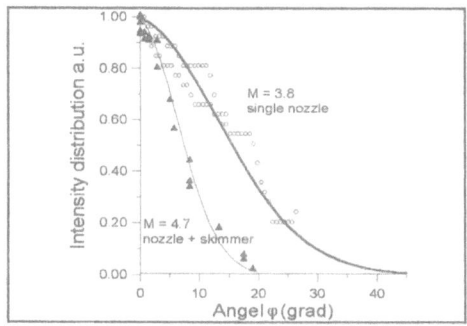

Figure 3. Angular distribution of the beam intensity for a single nozzle and a nozzle scimmer system

A further advantage of this arrangement is a reduction of the width of the velocity distribution (see Fig.4). A nearly monoenergetic He-beam is a prerequesite for successful plasma diagnostics in case of large gradients of density and temperature.

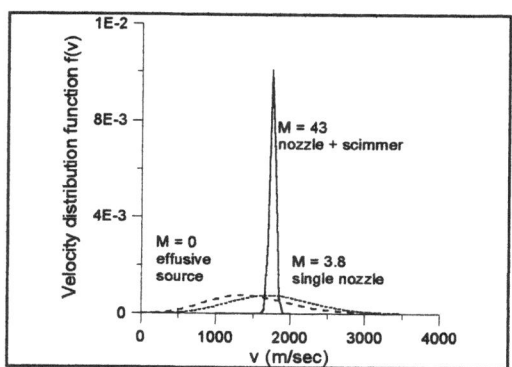

Figure 4. Velocity distribution function of a single nozzle and a nozzle scimmer system

Summarizing the results the He-beam can be applied as a T_e/n_e-diagnostic in plasmas with electron temperatures in the range of 5 eV $\leq T_e \leq$ 30 eV, in addition the electron densities should be below $n_e \leq 1 \cdot 10^{14}$ cm^{-3}. The beam profile depends strongly on the type of the source. On the other hand the penetration depth of the helium beam is determined by the ionization length (e.g.8 mm for $T_e = 10$ eV and $n_e = 1 \cdot 10^{14}$ cm^{-3}) but can be further reduced by a neutral gas background due to elastic scattering).

For tokamak applications the adequate type of beam source depends on the aim of measurement and the available access to the plasma. The greatest advantage of this diagnostic is certainly its simplicity. Moreover, a direct combination with a more comprehensive passive spectroscopy system seems possible. A proposal for such a system using a single nozzle is shown in Fig.5

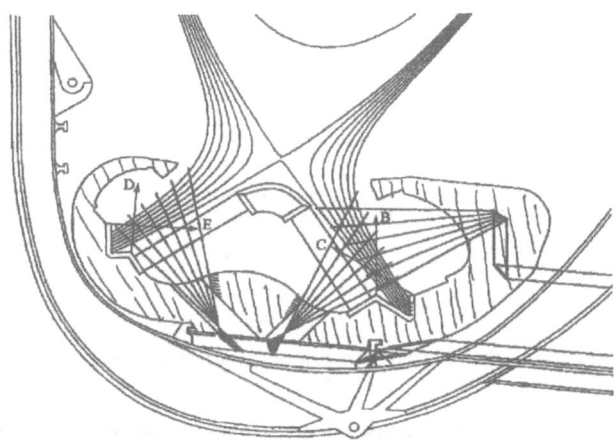

Figure 5. Proposal for a combination of passive spectroscopy and helium beam diagnostic in the divertor chamber; B, C, D, E indicate possible helium beam directions

REFERENCES

Schweer, B., Mank, G., Pospieszcsyk, A., Brosda, B., Pohlmeyer, B. ,J.Nucl.Mat. 196-198 (1992) 174

Brosda, B. Dissertation, Ruhr-Universität Bochum, 1993

Behrendt, .H., Bohmeyer, W., Dietrich, L., Fussmann, G., Greuner, H., Grote, H. Kammeyer, M., Kornejew, P., Laux, M., Pasch, E. "Development and Test of Edge Plasma Diagnostic at the PSI-1 Plasma Generator "21st EPS Conference on Controlled Fusion and Plasma Phys., Montpellier/France 1994, ECA *18B*, Pt.3, 1328-1331.

A RAD-HARD, STEADY-STATE, DIGITAL IMAGING

BOLOMETER SYSTEM FOR ITER

G. A. Wurden

Los Alamos National Laboratory
Los Alamos, New Mexico 87545, USA

ABSTRACT

The concept and design of a new type of bolometer system which can function with excellent spatial resolution and good time resolution in the next generation of long-pulse (or steady-state), harsh-neutron environment, fusion plasmas, is outlined. It uses a cooled pinhole camera design, employing a robust, passive, segmented radiation absorber, cooled from the back-side. Infrared emission from the absorber's front surface is relayed by metal mirror optics to a shielded, high-resolution IR video camera with 0.01 °C temperature resolution. It can make thousands of simultaneous "pixel" measurements at up to 50-60 Hz, without any signal wires through the vacuum interface.

INTRODUCTION

Classical bolometry[1] in a fusion plasma employs discrete thin-foils which are heated by the plasma radiation, and the temperature rise is detected by a metal resistor or thermistor bonded to the back side of the foil, often separated by an insulating film[2-5]. Sometimes a single-channel IR detector has been used to monitor the rise in temperature of the foil instead, so as to provide better electrical noise-immunity[6]. In cases of a short-pulse of plasma radiation, or when studying the energy deposited by by ion, neutral or electron beam, researchers have imaged (in the IR) the backside of a foil or plate target, in order to determine the "instantaneous" distribution of energy in the beam[7]. The problem with using this technique to observe a long-pulse plasma, is that the lateral heat flow in the foil or target plate would spoil the subsequent images, and confuse the measurement. In addition, a foil which is thin enough to have reasonable time response must be cooled over longer time intervals to prevent melting or radiative cooling damage or nonlinear effects from spoiling the measurement. Finally, the detectors must be radiation hardened, to survive the neutron and gamma fluences from a long-pulse DD or DT machine. So-called "silicon bolometers"[8] and pyroelectric detectors have an advantage in that they respond directly to the incident power, but they will not work in this environment. The plasma may have a complex geometry, and multi-channel views (even tomography)[9] are desired for modeling analysis.

Diagnostics for Experimental Thermonuclear Fusion Reactors
Edited by P. E. Stott *et al.*, Plenum Press, New York, 1996

603

DESIGN CONCEPT

The solution to these issues is remarkably simple, and involves a pinhole imaging design, with the plasma radiation striking a compact, segmented, back-cooled "foil". Each segment of the "foil" corresponds to one imaging resolution element or "pixel". Each pixel absorber is raised up from the back cooling block (which can be held at a constant reference temperature), and the height of the pixel (and the material thermal conductivity) is designed for rapid axial heat flow compared to the relatively longer lateral heat flow path over to the next adjacent pixel. The passive absorber matrix is imaged in the infrared, via two metal mirrors, which allow the (neutron-sensitive) state of the art, 12-bit digital video IR video camera to be positioned out of the line of sight of the neutron flux from the plasma. Some of the pixels may be positioned outside of the plasma field of view, to act as "background" pixels if necessary. The front surface of the absorbing matrix may be initially be a "blackened coating" (like any other bolometer design), but in the long term, this may change somewhat due to plasma contamination of the surfaces. No wires exit the vacuum interface, which is a distinct advantage for thousands of channels compared to discrete bolometer arrays.

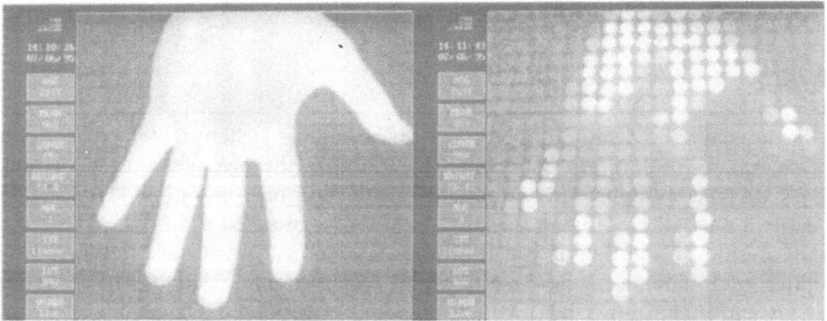

Figure 1: A test matrix (of blackened nails) holds the warmth from a hand, without lateral heat flow degradation.

Tests of a segmented matrix (20x20 array of blackened, galvanized roofing nails) are shown in Figure 1, where the warmth from a human hand held on the matrix for a few seconds is easily detected 30 seconds later. The matrix of nails was heat-sinked into a pool of water, although the heat decay time is much too long in this prototype to be used for plasma imaging. The images in the 3-5 micron band were taken with a 256x256 element InSb focal plane array (Amber Radiance 1) commercially available 12-bit digital IR video camera.

The overall system will tradeoff sensitivity, speed, and spatial resolution; with the ultimate speed being limited by cooling rates (to erase the previous image) or camera framing rates (which can range from 1 Hz to 1000 Hz with today's technology). If the luxury of a chopper is available to block the pinhole aperature, then the heating and cooling timescales of the matrix can be different. For use without a chopper, then some amount of integration of the heat from one frame to the next will occur, and it will be necessary to differentiate subsequent frames (in analogy with normal single-channel bolometers), while taking into account pixel cooling, in order to have time resolution limited by the video system . Metal mirrors form an IR relay telescope. The IR vacuum window can be positioned behind neutron shielding.

Depending on the radiated power density, the desired temporal and spatial resolution, and the plasma access available for diagnostics, one can design a conceptual layout of an imaging bolometer package. A sketch (not to scale) is shown in Fig. 2 below.

CIC-1/95-1857 (8/95)

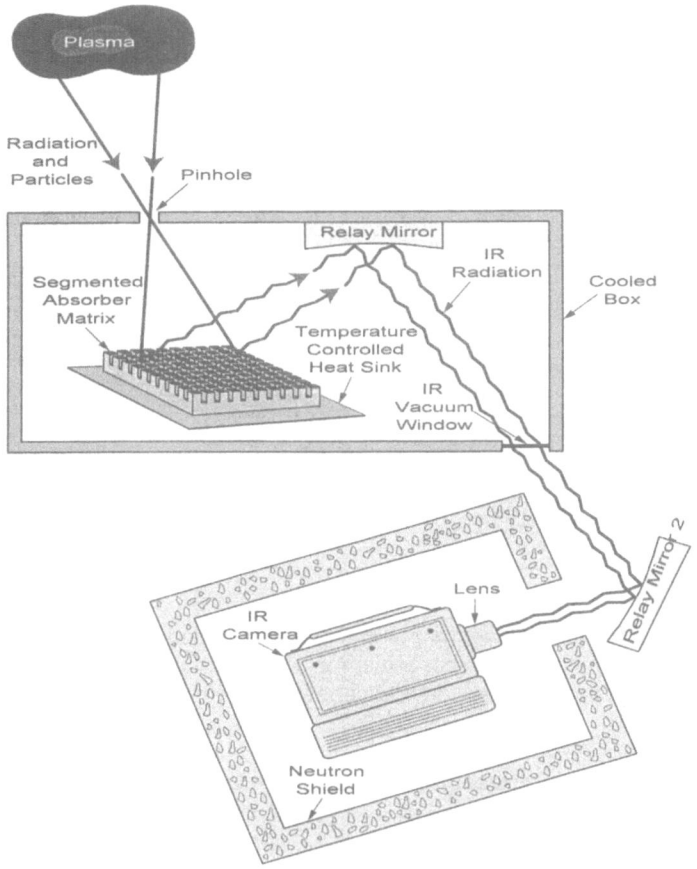

Figure 2: Components of the rad-hard, steady-state, digital imaging bolometer system.

Design choices for pinhole size, demagnification (radial or tangential viewing), target matrix size and number of pixels, location of vacuum interface, distance to IR imager and amount of neutron shielding, must all be made. A sample design point with a 100° viewing angle is shown in Table 1 , assuming an aluminum absorbing matrix, and a pixel spacing 1.1 times the pixel diameter. The possibility also exists to bond a different absorbing material to the "end" of each pixel, thereby allowing various choices for the cooling substrate. There is a potential problem with infrared radiation from hot objects (ie, the inner wall or divertor plates) directly in the field of view, that could be 500-1000 degrees hotter than the segmented matrix in the bolometer itself. These hot objects will radiate ~1000x more strongly in the 3-5 micron band, due to blackbody emission. This light could contaminate the IR coming up off of the segmented matrix. If this is a problem, then either the matrix itself could be run hot, or in the worst case, a design whereby the matrix is viewed from the backside, while cooling is applied from the sides of each pixel could be employed. This would have the advantage that plasma light would never be a problem.

Table 1: One possible design point for an imaging bolometer on ITER.

Plasma Device: ITER	Parameters	Comments
Plasma Volume	2500 m^3	
Radiated Power	600 MW	0.01-1.0 MW/ m^3 core 100-300 MW/ m^3 X point and Divertor
Plasma distance	4 m	average for core; smaller in divertor
Pinhole aperature (diameter)	0.5 cm	
Demagnification	50x	for core system, less in divertor
Distance to back plane (matrix)	0.08 m	larger, if willing to sacrifice compactness
Pixel size (diameter)	0.5 cm	array is 22 cm x 22 cm in size
Number of resolution elements	40x40	25 cm spatial resolution across diameter (could be easily improved 2-4x)
Radiated power at each pixel	10-40 mW/cm^2	1x10^{-7} solid angle at pinhole
Temperature decay time desired	20 ms	
Height of each pixel	~ 5 mm	Distance above cold block, depends on material and desired time constant.
Max temperature rise (target)	10 °C	minimum resolvable 0.01 °C

SUMMARY

A new type of imaging bolometer is presented, which is both robust and flexible, and can operate in the long-pulse and harsh radiation conditions expected to be encountered in ITER. This work is supported by US DOE contract W-7405-ENG-36.

REFERENCES

[1] D. V. Orlinskij and G. Magyar, Plasma Diagnostics on Large Tokamaks, Nuc. Fus., 28 (4), 664 (1988).

[2] J. Shivell, G. Renda, J. Lowrance, H. Hsuan, Rev. Sci. Instrum., 53, 1527 (1982).

[3] G. Miller, J. C. Ingraham, and L. S. Schrank, Bolometer for use in a noisy electromagnetic environment, Rev. Sci. Instrum. 53 (9), 1410 (1982).

[4] K. F. Mast, H. Krause, et. al., ASDEX Bolometers, Rev. Sci. Instrum., 56 , 969 (1985).

[5] Y. Maejima, Y. Hirano, T. Shimada, and K. Ogawa, Bolometric Measurement of Reversed Field Pinch Plasma in TPE-1RM, J. Phys. Soc. Japan, 55 (7), 2203 (1986).

[6] J. C. Ingraham and G. Miller, Infrared calorimeter for time-resolved plasma energy flux measurements, Rev. Sci. Instrum. 54 (6), 673 (1983).

[7] H. A. Davis, R. R. Bartsch, D. J. Rej, and W. J. Waganaar, Intense Ion Beam Optimization and Characterization with Thermal Imaging, in *Proc. of the 10th Intl. Conference on High-Power Particle Beams* (NTIS[PB95-144317), edited by W. Rix and R. White (Springfield, VA, 1994), pp. 668--673.

[8] R. J. Maqueda, G. A. Wurden, and E. A. Crawford, Wide-band silicon bolometers on the LSX field reversed configuration experiment, Rev. Sci. Instrum., 63 (10), 4717 (1992).

[9] A. W. Leonard, W. H. Meyer, B. Geer, D. M. Behne, and D. N. Hill, 2D tomography with bolometry in DIII-D, Rev. Sci. Instrum. 66 (2), 1201 (1995).

OVERVIEW OF W7-X DIAGNOSTICS

J. V. Hofmann, H. J. Hartfuß, H. Ringler, P.Grigull, W7-AS / W7-X Teams

Max-Planck-Institut für Plasmaphysik,
EURATOM Ass.
85748 Garching, Germany

INTRODUCTION

Generally, diagnostic methods and applications in a stellarator are not much different to those in tokamaks. However, requirements for W7-X diagnostics are different in the sense that they must be able to address the specific issues of optimization and to verify the

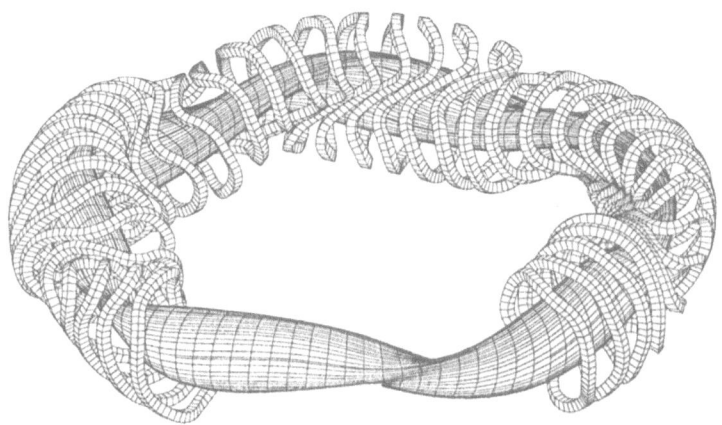

Figure 1. 3D plasma surface and part of the modular coil set. The rotational transform is $\iota = 1$. The plasma has a helical axis and cross sections varying from kidney shaped ($\phi = 0°$) via drop like ($\phi = 0°$) to triangular ($\phi = 36°$).

configuration. This may be comparable to the demands for optimization and diagnostics of

Diagnostics for Experimental Thermonuclear Fusion Reactors
Edited by P. E. Stott *et al.*, Plenum Press, New York, 1996

607

tokamak divertors. We will address optimisation criteria for stellarators and discuss related physics and diagnostic issues.

THE W7-X STELLARATOR

The Wendelstein 7-X experiment is an integrated concept test for properties of reactor relevant plasmas in Advanced Stellarators. The configuration is a toroidal plasma equilibrium of HELIAS type (HELIcal Advanced Stellarator).[1] Prominent geometrical features are a modular superconducting coil assembly (50 non-planar modular field coils, 20 planar auxiliary field coils) a five-fold toroidal symmetry (5 modules of 72°), a helical magnetic axis and, as a result, a truly 3D plasma topology with kidney-shaped plasma cross sections at the positions of strongest curvature ($\phi = 0°$) changing via drop-like cross sections ($\phi = 18°$) into triangular ones in the regions of smallest curvature ($\phi = 36°$), - see Figs. 1,2. The rotational transform on axis $\iota(0)$ and at the boundary $\iota(a)$ can be varied from $0.75 \leq \iota(0) \leq 1.01$ and $0.83 \leq \iota(a) \leq 1.25$, respectively.

Essential characteristics due to optimisation are

1. A high quality vacuum magnetic field with smooth magnetic surfaces and a regular boundary which can be envisioned as a resonant island divertor.

HELIAS FÜR W7-X

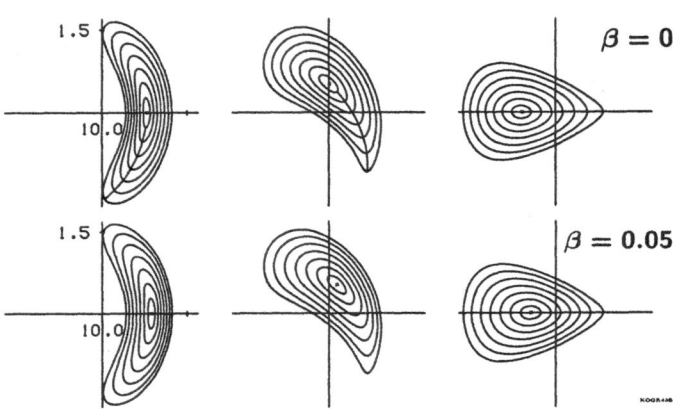

Figure 2. Poloidal cross sections at $\phi = 0°$, 18° and 36°, respectively, for $\beta = 0$ and 5 %. The finite β effect on the flux surfaces is, as can be seen, very small.

2. Improved finite-β equilibrium properties resulting from a small Shafranov shift (strongly reduced Pfirsch-Schlüter current) and a small change of rotational transform and shear with β due to a vanishing bootstrap current which yields a high equilibrium β-limit - see Figs. 2, 3.

3. Good MHD stability properties due to magnetic well stabilization and suitable flux surface shaping (superposition of matched helicities, indentation, triangularity).

4. Reduced neo-classical transport losses in the reactor relevant long mean free path regime (the lower limit analogous to neo-classical transport in axisymmetric systems is close to that of quasi-helically symmetric configurations where the field strength is 2D

and given by B(s,Θ - φ) so that the so called 1/ν-regime does not appear. However, quasi-helical symmetry is only approximated in order to yield low bootstrap current).

5. Negligible bootstrap current in the long-mean-free-path regime. The bootstrap current which increases the rotational transform in axisymmetric configurations but decreases it quasi-helical symmetry can, thus, vanish in a not fully symmetric configuration.

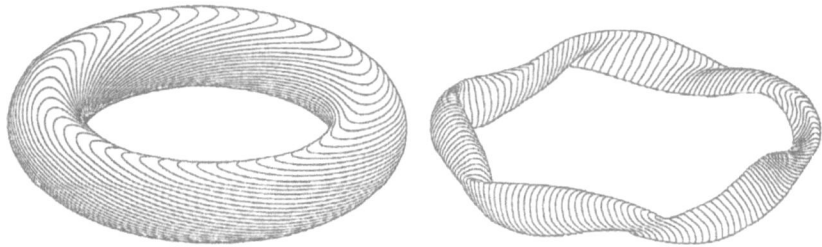

Figure 3. Comparison of magnetic surfaces and current lines (diamagnetic plus Pfirsch-Schlüter currents) of a tokamak at q=3 and the W7-X stellarator. While the parallel currents in the tokamak case are roughly $j_{\parallel} \approx 6 j_{\perp}$, the current lines in the case of W7-X are nearly perpendicular to the field lines, in some regions, indicating very small Pfirsch-Schlüter currents with $<j_{\parallel}^2 / j_{\perp}^2> \approx 0.5$.

6. Good collisionless α-particle confinement (in equiv. reactor) towards higher β - see Fig. 9.

7. Good modular coil feasibility.

With a major and minor radius of 5.5 m and 0.53 m, respectively, and a magnetic field strength ≤ 3 T the envisaged plasma parameters in W7-X are T_e and T_i in the 10 keV range, densities up to $3*10^{20}$ m^{-3} and <β> values of up to 5%. With a heating power of 10 MW ECRH cw (and 20 MW NBI for 10 s) the power load on the divertor tiles will remain below 10 MW/m^2. The cw heating by ECRH will allow studies on neo-classical transport in the long- mean-free-path regime while NBI at low magnetic field will be used for β-limit investigations.

Important aims of the W7-X experiment are the study of energy and particle exhaust in a HELIAS configuration under quasi-stationary conditions and the development of a reactor-relevant divertor system. Scenarios to be investigated will include operation with radiative boundaries. The divertor concept for W7-X exploits the inherent field line diversion property of optimised HELIAS configurations, comprising the variants of ergodic and island field line diversion, also utilising additional loops for island and divertor control.

W7-X CODES

In order to optimize a stellarator fusion experiment a priori an interfaced set of highly optimized three-dimensional codes was developed and has led to an optimized stellarator configuration which can be properly considered as Computational Stellarator.[2]

Stellarators need optimization because classical physics issues seriously limit their viability as fusion devices. A striking example of their optimization potential are quasi-helically symmetric toroidal magnetic fields, which show that the magnetic geometry of a stellarator can be decoupled from its real space geometry. Thus, stellarator optimization does not primarily consist in quantitatively improving some given basic concept but in selecting basic physics properties.

The plasma behaviour in the confinement region can be optimized by noting that the geometry of the confinement boundary within the last closed flux surface completely determines the properties of the confinement region. Thus, boundary value problems must be solved during optimization, the parameters of the boundary being the optimization variables. In a second step optimization of the coils necessary to produce this confinement region can be done, again by solving boundary value problems.

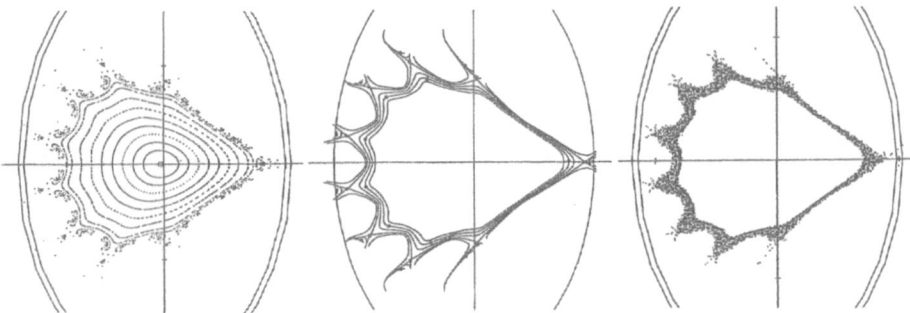

Figure 4. Poloidal cross section of flux conserving surfaces outside the last closed magnetic surface (LCMS) including the "private region" beyond the X-point of an open edge island chain. Shown is a comparison of two Poincaré plots without and with diffusion compared to a new coordinate system for open field lines, which are in good agreement.

In order to provide finite-β equilibria in line with the experiment a data base using function parametrization of equilibria based on VMEC/NEMEC code calculations is set up (presently for W7-AS). However, each stellarator equilibrium to be calculated needs about half an hour Cray time compared to several 10 seconds for a tokamak equilibrium!

Field line mapping and mapping of diagnostics at different cross sections is provided by the TRANS-code, operating on a data base of equilibria from Gourdon/KW- and VMEC/NEMEC- codes. Especially for the ergodized edge region outside the last closed flux surface and in the region where islands are no longer closed, a new coordinate system for open field lines is under development, which just became available at W7-AS - see Fig. 4.

W7-X DIAGNOSTICS

In W7-X standard diagnostics, as known from other tokamaks and stellarators / torsatrons / heliotrons, will be used to measure basic plasma quantities like Thomson scattering (T_e, n_e), ECE (T_e), reflectometry ($n_{e,edge}$), interferometry / polarimetry (n_e), CX-neutral analysis (T_i), CXR-spectroscopy (T_i, n_z, $v_{tor+pol}$, E_r), impurity spectroscopy (n_z, Γ_z), Z_{eff}, bolometry (P_{rad}), H_α arrays (recycling), soft-x and Mirnov loops (MHD, T_e), diamagnetic and flux loops (energy, flux to assess the reduction of the PS-currents), edge probes (n_e, T_e, Φ), calorimetry (heat load), fast Li-beam ($n_{e, edge}$), plasma and IR video (plasma configuration and monitor, target loads) and others.[3]

Most important for ECR-heating and ECE diagnostic is the 1/R magnetic field dependence (equivalent to tokamaks) in the kidney shaped cross sections.

Experience from W7-AS, furthermore shows that mapping of points or lines of sight of one diagnostic in a specific poloidal plane along field lines to other diagnostics and plasma cross sections works well due to the known structure of the magnetic field. Generally, this mapping provides the toroidal and poloidal correspondence of measured quantities which are constant on magnetic surfaces, profiles of different diagnostics as a function of an effective (flux conserving) radius, tracing of information along field lines and radial location of line-integrated measurements via the effective radius of the tangential magnetic surface, Abel inversion and tomographic reconstruction as a function of the effective radius.

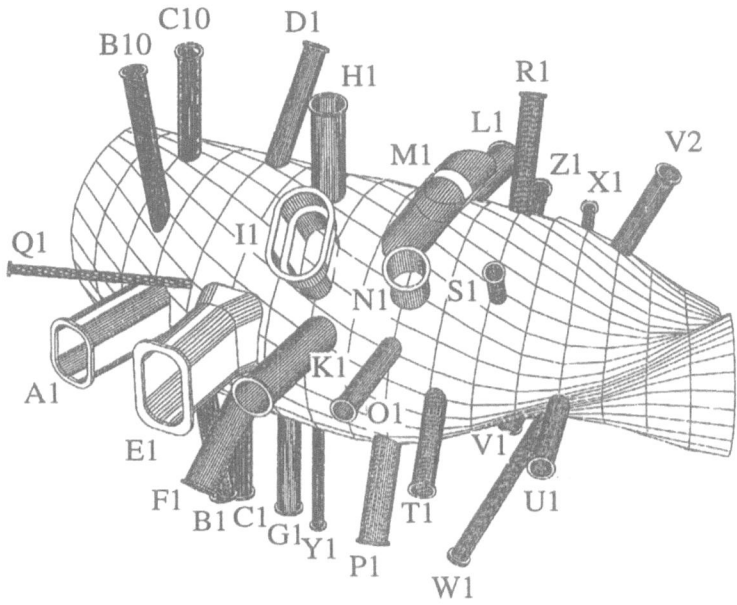

Figure 5. 3D plot of half a torus module (1/10) and ports.

In order to provide a set of basic diagnostics at startup of the machine and to prepare for a later stage of diagnostic needs and developments possible and desirable diagnostics have been categorized in three levels:

level-1 comprizes the basic diagnostics for machine and divertor operation and plasma characterization

level-2 extends the diagnostic potential of the level 1 set with respect to additional parameters and / or higher spatial and / or temporal resolution

level-3 finally includes very special advanced diagnostics which may be of importance for special investigations in a well diagnosed machine and which may include very recent trends and developments towards new diagnostic possibilities

In the following we will concentrate mainly on specific diagnostic issues and possibilities for W7-X.

Ports at W7-X

Despite the very complex coil set and 3D-geometry a huge number of quite large ports is available for heating and diagnostic access to the plasma - see Fig. 5.

On the one hand the loss of toroidal symmetry and the 3D plasma structure, with changing shape in toroidal direction, complicate the diagnostics of the plasma. On the other hand, a five-fold symmetry still exists and the realization as linked mirrors provides nearly straight parts without curvature. Furthermore, due to the large aspect ratio of 10 and the absence of an ohmic transformer, inboard side access to the plasma is possible, which will especially be utilized for beam and divertor diagnostics. Additionally, the φ=18° plane, with its drop like plasma cross section, allows for diagnostic access very far radially into the vessel, which is, again, especially suited for divertor diagnostics - see Fig. 10.

Flux Topology and Optimization (Stability + Equilibrium):

Extended magnetic and soft x-ray diagnostics - see Fig. 6 - are needed to address the optimisation of the configuration with respect to equilibrium, stability and pressure driven currents. On W7-AS a very sophisticated soft x-ray diagnostic has been used extensively to investigate and document the stability and equilibrium of this partially optimized configuration. This plasma core diagnostic is complemented by probe arrays and video diagnostics which image the edge topology of the plasma including islands, as already established for W7-AS. In detail,

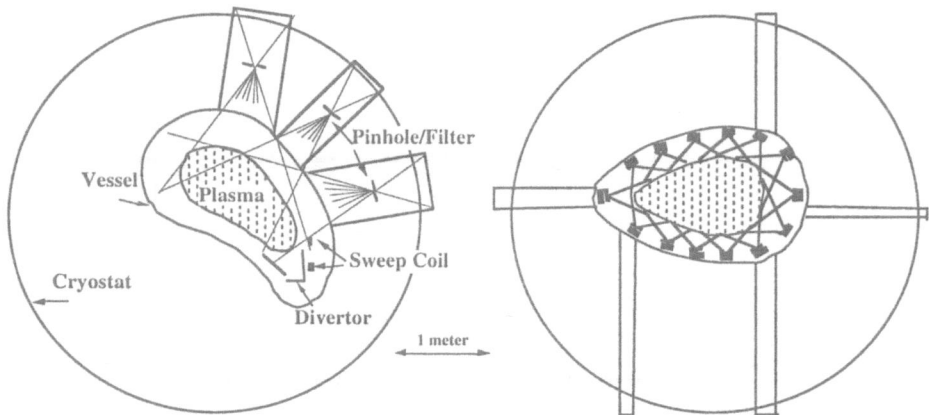

Figures 6. Soft x-ray diagnostic. Left: flexible general purpose camera system at φ=23° with 3 cameras outside the vessel and ~100 channels. Right: tomographic system at φ=36° with 12-15 compact cameras inside the vessel and 300-400 channels.

Verification of the predicted magnetic topology is based on:
♦ tomographic reconstruction of the x-ray emissivity
♦ characterization of different configurations in the whole flexibility range
♦ investigation of high-β operation at reduced magnetic fields
♦ deduction of axis position, plasma radius and Shafranov shift
♦ measurement of the dipole field emerging from the PS-currents (the reduction of which is a major optimization criterium)
♦ verification of low bootstrap current and its flow direction, as it may change sign
♦ characterization of the edge topology
♦ effects of pressure induced magnetic islands (including healing effects)
♦ relation to vacuum field line mapping and transformation codes

Equilibrium and stability identification require:
♦ detection of rational surfaces (MHD activity, stationary islands)
♦ identification of radial and poloidal mode structures
♦ operational limits or enhanced transport due to MHD instabilities
♦ verification of predicted instability thresholds and identification of instability mechanisms
♦ establishment of unfavourable (more unstable) configurations to investigate thresholds and to show effects of stellarator optimization

The low shear allows to address stability and turbulence issues like fluctuation and correlation measurements. These studies will profit from cw operation. Rotational transform ≈ 1 in the plasma core will be of specific interest along with the stability issues around the critical 9/10 and 11/10 resonance.

Classical confinement and neo-classical transport:

Neo-classical transport is a threat to stellarator reactors since particles trapped in local magnetic mirrors can lead to rapid losses. This is especially true in the collisionless long-mean-free-path regime where this neo-classical effect leads to the so-called 1/v-regime. However, approximating quasi-helical symmetry this regime does not appear.

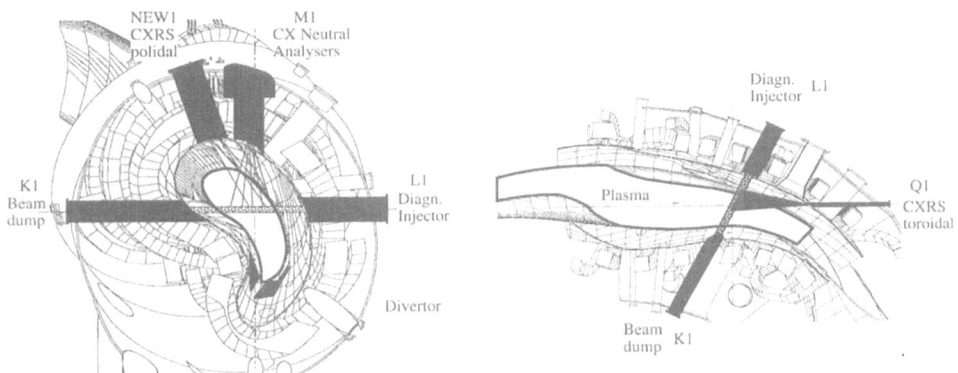

Figure 7. Diagnostic beam, neutral particle analyzer (NPA) and charge exchange recombination spectroscopy (CXRS) in poloidal (left) and toroidal (right) cross sections, respectively. The diagnostic beam penetrates the plasma horizontally from outborad (I1) to inboard (K1) side. In the poloidal plane a 5 channel NPA (M1) and two (core + edge) multichannel CXRS lines-of-sight (NEW1) are viewing the beam. Additionally, a multichannel CXRS system with toroidal lines-of-sight will be available. The equivalent geometry will be used for Thomson scattering, laser blow off and pellet injection.

The key element in these losses is the radial drift of the trapped particles away from magnetic surfaces. This drift is minimized if the particles are localized in regions of small poloidal field variation. In a standard stellarator the helical ripple leads to the enhanced losses. In the long-mean-free-path regime trapped particles dominate neo-classical transport. It is mainly the synergetic effect of the leading Fourier harmonics of a HELIAS configuration which leads to the reduction of the effective helical ripple (by a factor of 10-20 in the 1/v-regime) and radial transport.

For the ions in the 1/v-regime, however, the radial electric field is more effective in reducing the radial diffusion than the magnetic drift. The measurement (and control ?) of the radial electric field is therefore of special importance.

Investigations on neo-classical transport properties will also benefit from the high configurational flexibility of W7-X (mirror ratios) and its influence on trapped particle fractions and neo-classical damping processes.

In order to measure the properties of improved confinement and neo-classical transport very good and precise T_e-, T_i- and E_r- profile diagnostics are required accompanied by accurate measurements of the power balance. This will be provided by Thomson scattering (also in the divertor - see Fig. 10), ECE measurements, interferometry / polarimetry, CX neutral particle analysis, CXRS spectroscopy - see Fig. 7, reflectometry and a fast Lithium beam. In particular, the interferometry will be complemented by polarimetry, since in a stellarator the magnetic field is well known. Polarimetry, however, would overcome the disadvantages of fringe jumps during very dynamic density behaviour!

The development of a heavy ion beam probe (HIBP) is considered; at present suitable ports for this diagnostic and drift orbits are investigated. A decision will be made later.

Confinement issues are addressed via:
- relative changes of pressure profiles
- confinement transitions
- effects of neo-classical transport due to changes in trapped particles (mirrors)
- effects due to magnetic field perturbations at rational iota values
- changes in the radial electric field
- local transport parameters from temporary evolution of injected test particles
- impurity accumulation and ECRH pump out
- radiation losses from plasma bulk, photosphere, islands and X-points

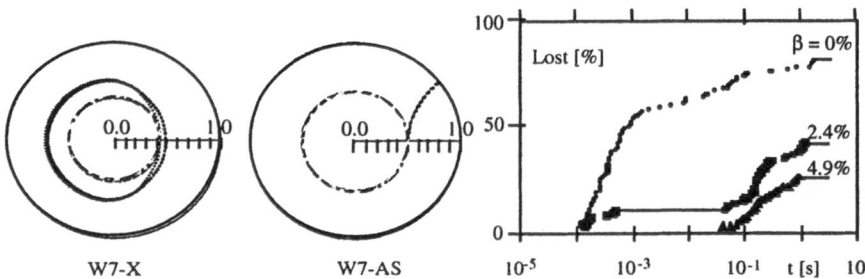

Figure 8. Drift orbits and losses of α-particles. Left: drift orbits in a $s^{1/2}$, θ-plane with flux label s and poloidal magnetic coordinate θ. W7-X: poloidally closed drift orbit of a reflected α-particle starting on the magnetic surface indicated by the inner dashed circle in the direction of the magnetic field (+ positive $v_{\parallel}$; o negative $v_{\parallel}$). W7-AS: Typical drift of a localized α-particle in W7-AS magnetic geometry. Right: α-particle losses as a function of collisionless time of flight. Shown is the fraction of reflected particles which is lost; β = 0% (•), 2.4% ([]), and 4.9% (Δ). Each symbol indicates the loss of a particle.

Collisionless α-particle confinement:

The diagnostic of high energetic particles in W7-X has to be supported by calculations of deposition of fast particles either from a heating or diagnostic beam or from ICRH minority heating. The slowing down of 60 keV deuterons in W7-X is equivalent to 3.5 MeV α-particles on reactor scale because of the same ratio a/ρ of about 20-30 (a = minor plasma radius, ρ = Larmor radius). Comparing the theoretical predictions from particle trajectory codes, on the behaviour of injected 60 keV deuterons, with measurements will allow to assess the predicted improvement of fast particle confinement in HELIAS configurations and thus to extrapolate to reactor scale.

Experimentally the number of drifting particles hitting the wall for small β can be compared to the case for high β. A significant reduction of this particle flux towards higher β is expected for fast particles - see Fig. 8.

Experimental tools can be the measurement of the slowing down spectra of fast injected particles using neutral particle analysers - see Fig. 7, together with suitable measurements of sputtered particles from the wall.

Generally, in strong non-symmetric stellarators all α-particles which undergo reflections are lost very quickly. Quasi-helically symmetric stellarators, however, completely confine collisionless α-particle orbits - see Fig. 8. The equivalent ripple of a configuration, which governs its electron heat conductivity, and the quality of α-particle containment are only weakly correlated quantities. Usually, a large equivalent ripple is accompanied by large α-particle loss. Various β-values of the HELIAS configuration were studied with respect to collisionless α-particle containment. For this purpose a set of particles was started at 1/4 of the plasma radius with random distribution in angular variables and pitch angle. The results, shown in Fig. 8 indicate good α-particle confinement at finite β.

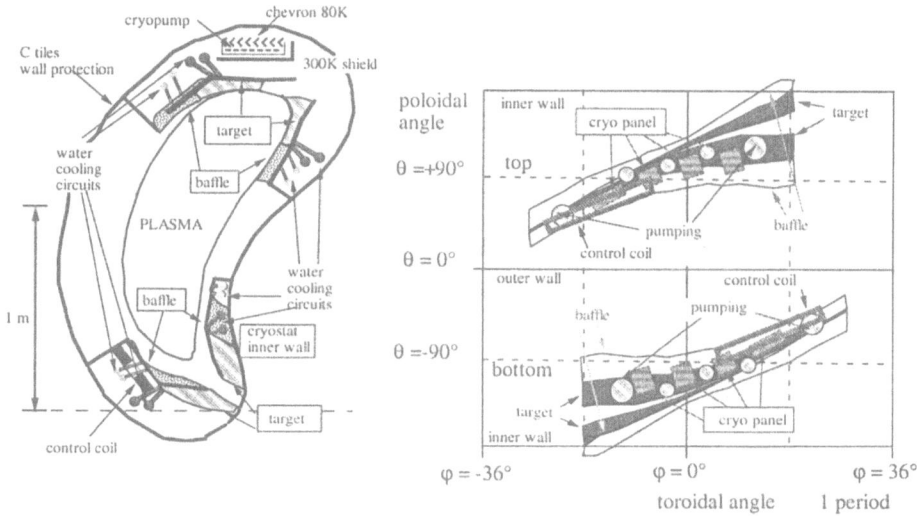

Figure 9. W7-X divertor shown in a 9° poloidal plane (left) and development of divertor structure in a φ-θ-plane of one module.

W7-X DIVERTOR

The properties of the magnetic configuration of W7-X with the formation of the "helical edge" and the associated diversion of magnetic field lines can be used as an inherent divertor - see Fig. 9. Two options for a divertor configuration, ergodic and island divertor, exist.

In case of the ergodic divertor the last closed magnetic surface is surrounded by an ergodic layer without large magnetic islands. Plasma wall interaction occurs close to the helical edges where field line diversion is a maximum. The target plates follow the helical edges and are appropriately shaped in order to minimize the local wall load (helical trough).

The island divertor concept uses the existence of large islands at the boundary established for a magnetic configuration with moderate shear. Streaming along field lines the plasma passes the X-point region of the islands and arrives at the backside of the island after 4 to 5 transits. Whereas for the standard configuration with rotational transform close to 1 the 5/5-islands dominate the boundary, for varied values of rotational transform this role is played by the 5/6- or 5/4-islands.

The high complexity of the surface topology and its utilisation for island divertor operation requires sophisticated local measurement techniques to investigate edge island structures, particle densities and fluxes in the divertor chamber and onto the target plates, and heat load of the plates - see Fig. 10. Fairly good access to the divertor region for a whole set of diagnostics is possible.

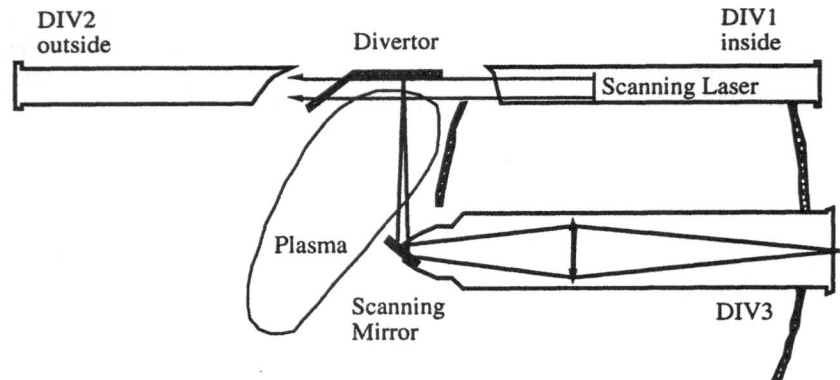

Figure 10. Divertor Thomson scattering. The laser can be scanned from the surface of the divertor plate across the X-point towards the plasma allowing for n_e / T_e measurements in this whole range. Additionally, the observation cord can be scanned across the target plate enabling two-dimensional profile information. The equivalent geometry will be used for laser induced fluorescence, a Lithium beam, video and spectroscopy investigation of the divertor target plates and fast reprocicating Langmuir probes.

STEADY STATE OPERATION

The inherent properties of a stellarator for steady state operation also have implications on the diagnostics. Although stationary operation is the ultimate goal, most of the experiments will be done with limited pulse duration of order 1 minute or 10 seconds in case of NBI. This may not require a principally new diagnostic approach, however intelligent event recognition will be required and especially correlation studies will benefit from long integration times. Good spatial resolution will be possible with diagnostics which can be slowly scanned. Furthermore, temporally separated complex measurements by different diagnostics will be possible in the same volume. A possible scenario is also a cw ECR-heated discharge with periodic NBI pulses of 10 s. Experience from tokamaks, especially Tore Supra, will be incorporated and developments for the LHD heliotron will also strongly contribute.

Development of diagnostics for W7-X: It is planned to develop the diagnostics for W7-X jointly with the Associations. Several meetings have already been held. Fields of interest have been defined so far and diagnostic approach is considered in the phase II proposal for W7-X. It is also planned to jointly realize the data acquisition system for the W7-X diagnostics.

REFERENCES

1. G. Grieger et al., Physics and engineering studies of Wendelstein 7-X, in *Proc. 13th Int. Conf. on Plasma Physics and Controlled Nuclear Fusion Research*, Washington (1990), IAEA, Vienna, 525 (1991); G. Grieger et al., Physics optimization of stellarators, *Physics of Fluids*, B 4: 2081 (1992); H. Wobig, The theoretical basis of a drift-optimized stellarator reactor, *Plasma Physi.Controll.Fusion*, 35: 903 (1993).
2. U. Schwenn, The Computational Stellarator, *Proc. 6th Joint EPS-APS Int. Conf. on Physics Computing*, Lugano 1994, Eds. R. Gruber, M. Tomassini, Europ. Phys. Soc., Geneva, 471 (1994)
3. H. J. Hartfuß and H. Ringler, Eds., *Proc. of the Int. Workshop on W7-X Diagnostics*, Ringberg Castle, Germany, 1995, to be published.

DIAGNOSTICS FOR LHD

J. Fujita and the LHD Diagnostics Group

National Institute for Fusion Science, Nagoya 464-01, Japan

INTRODUCTION

The main research activity in the National Institute for Fusion Science (NIFS), Japan, is directed towards the study on plasma confinement and heating in a helical system, as a complementary approach of tokamak research to the comprehensive understanding of magnetically confined toroidal plasmas.

For this purpose, a helical device called Large Helical Device(LHD) is being constructed at a new site in Toki City located about 30 km North-East of the present site in Nagoya.[1,2]

Although the device is not a tokamak but a helical system, it has a significant meaning related to the ITER program from the following aspects. The role of a satellite machine is known quite important for conducting a big project such as ITER. The LHD machine will be able to serve a plasma for testing new concepts on heating, diagnostics, the divertor scenarios and others. It will be necessary to test newly constructed diagnostics on the operating machine before applying to the ITER device. It is planned to operate LHD in steady state. Then, one can utilize LHD to find possible technical problems with such application in advance. Besides, several years later, LHD might be one of a few large-sized devices under operation in the world which can be utilized for this purpose.

In turn, the achievements obtained in the research and development on ITER must be very valuable and useful for the LHD project, too.

LARGE HELICAL DEVICE(LHD)

The machine parameters of LHD is summarized in Table 1. Superconducting coils of pole number $l = 2$, pitch number $m = 10$, major radius $R = 3.9$ m, produce a steady state helical magnetic field for confinement, together with inner and outer vertical coils. The machine is cooled to liquid helium temperature in a large cryostat.

Because the helical coils are wound continuously around the machine, the windings are carried out on site instead of constructing the device in the firm and transport it to the site. The construction will take two more years, and will be completed in 1997.

Diagnostics for Experimental Thermonuclear Fusion Reactors
Edited by P. E. Stott *et al.*, Plenum Press, New York, 1996

617

An artist's view of LHD is shown in Fig. 1, and the plasma cross-section together with divertor configuration in Fig. 2.

Table 1. Machine parameters of the Large Helical Device (LHD) under construction at NIFS.

Parameter	Phase I	Phase II
Major radius (m)	3.9	3.9
Coil minor radius (m)	0.975	0.975
Averaged plasma radius (m)	0.5 - 0.65	0.5 - 0.65
Plasma aspect ratio	6 - 7	6 - 7
l (pole number)	2	2
m (pitch number)	10	10
Plasma volume (m^3)	20 - 30	20 - 30
Magnetic field		
Center (T)	3	4
Coil surface (T)	6.9	9.2
Helical coil current (MA)	5.85	7.8
Coil current density (A/mm^2)	40	53
Liquid helium temperature (K)	4.4	1.8
Plasma duration (s)	10	10
Repetition time (m)	5	5
Heating power		
ECRH (MW)	10	10
NBI (MW)	15	20
ICRF (MW)	3	9
for steady state operation (MW)	–	3
Neutron yield (n per shot)	–	2.4×10^{17}

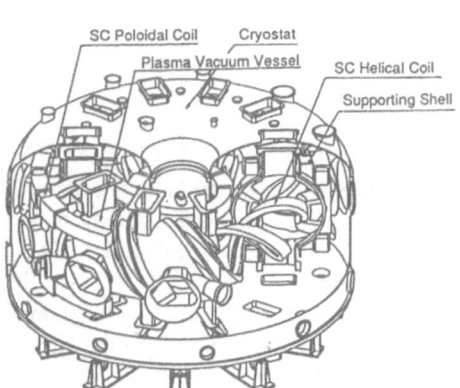

Figure 1. An artist's view of LHD.

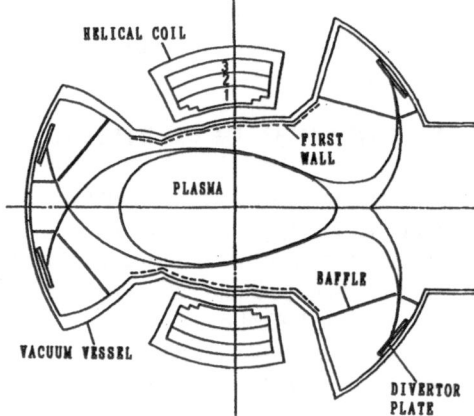

Figure 2. Cross-section of the divertor structure.

DESIGN PRINCIPLE OF DIAGNOSTICS FOR LHD

In general, the diagnostics for LHD are not much different from those for tokamaks. The essential difference is that the plasma shape is not axially symmetric, but 3-dimensional. An elliptic plasma cross-section rotates poloidally along the magnetic axis. Because of the lack of symmetry in the helical configuration, 3-dimensional diagnostics should be provided for LHD. Moreover, the plasma should be diagnosed through a long and narrow observation port drilled across the cryostat. Another restriction is that no observation port is available to look at the plasma from 90 degrees in a plane at one toroidal position, because of the helical coil structure. Therefore, the diagnostics for LHD have various features different from those in conventional tokamaks.

It is also planned in future to produce a steady state plasma in LHD. A proper function of divertor is vitally important in a long pulse machine. Special caution is taken in the design of diverter diagnostics. It is a difficult technical problem to remove the heat from the plasma for the protection of diagnostic components and windows.

Although no D-T operation is planned on LHD, the D-D operation will produce a significant amount of neutrons. Major parts of diagnostics are placed in adjacent rooms of the main experimental hall. A cross-sectional view of the LHD experimental hall including major diagnostics is shown in Fig. 3.

The diagnostics which are in preparation for LHD are summarized in Table 2.

Among various diagnostic developments so far carried out, a multichannel Thomson scattering system which has high temporal and spatial resolutions, a multichannel FIR laser interferometer and a heavy ion beam probe will be explained in detail.

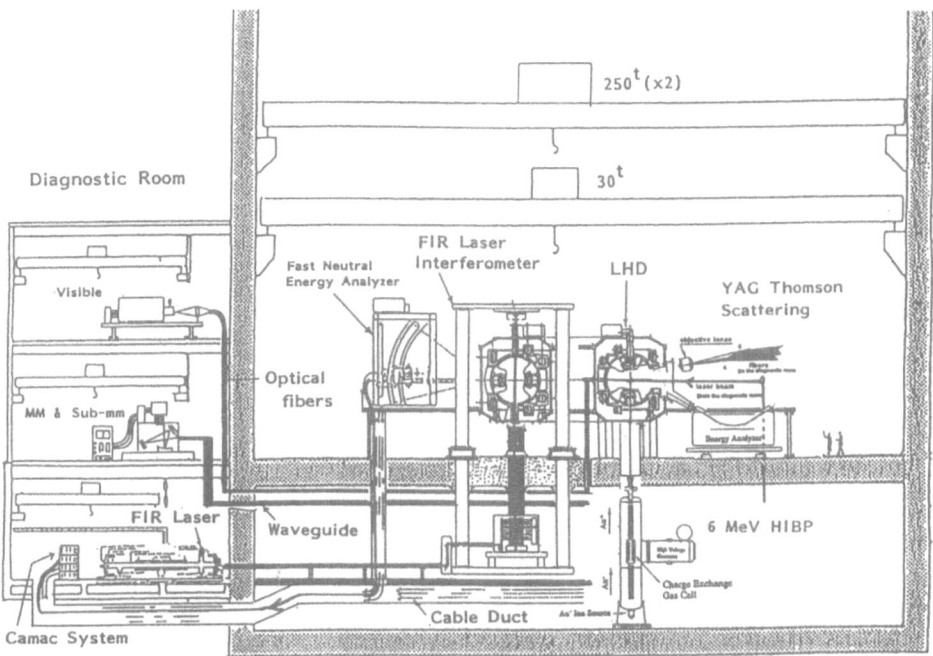

Figure 3.   Cross-sectional view of the LHD experimental hall.

Table 2. List of Diagnostics for LHD

Diagnostics	Purpose	Descriptions
Magnetic Probes	I_p, plasma pressure, position and shape of plasma	Rogowski, Mirnov, Flux loops
Microwave Interferometer	$n_e l$	2mm/1mm wave, single channel
FIR Laser Interferometer	$n_e l$ (r)	119 μm-CH_3OH laser, 10-channel
Microwave Reflectometer	n_e, n_e fluctuation	under development
Thomson Scattering	T_e (r). n_e (r)	200 spatial points
ECE	T_e (r, z)	2-D imaging
X-ray Pulse Height Analysis	T_e, impurities	20-ch Si(Li), 4-ch Ge detectors
Neutral Particle Analyzer	T_i, $f(E)$	radial scan
Charge Exchange Spectroscopy	T_i (r), V_p (r)	use of diagnostic neutral beam
X-ray Crystal Spectroscopy	T_i (r)	0.1-4nm, $\lambda/\Delta\lambda$:10^4
Neutron Diagnostics	neutron flux, T_i	NE-213 detectors, ^{3}He counters, activation foil
Bolometers	P_{rad} (r)	metal film, silicon diode, pyroelectric detector
VUV Spectroscopy	impurities, T_i	1 - 200 nm, $\lambda/\Delta\lambda$: 10^4
Visible Spectroscopy	n_0 (H), Z_{eff}	200 - 700 nm, $\lambda/\Delta\lambda$: 5 x 10^4
Langmuir Probes	T_e, n_e	Fast scanning and fixed probes
Visible/Infrared TV	plasma position, PWI wall/limiter temperature	TV systems
Soft X-ray Diode Array	MHD Oscillations	silicon surface-barrier diodes
MW/FIR Laser Scattering	microinstabilities	1 mm/195 μm multichannel
Heavy Ion Beam Probe	plasma potential, fluctuation	Au^+ or Tl^+, 6 MeV, 100 μA
Diagnostic Pellet	particle transport	Hydrogen/Double layer ice pellet, C, Li
High-energy Particle Diagnostics	high-energy particles	Li/He beam (2 MeV, 10 mA) probe, particle detector probes

MULTIPOINT YAG THOMSON SCATTERING SYSTEM

For physics understanding of a plasma behavior, it is of basic importance to measure time evolution of electron temperature and density profiles with high temporal and spatial resolutions across a full cross-section of the plasma to be studied.

After investigating Thomson scattering systems under operation, and taking a limited access to the plasma into account, a YAG laser Thomson scattering system is chosen.[3] The incident laser beams consist of those from 10 YAG lasers. Each YAG laser operates with a repetition rate of 50 Hz maximum. It is designed to achieve the shortest time interval between consecutive laser beams of 12 μs, and 2 ms for an equal time separation.

The nearly backward-scattered lights are collected through the same port for injection as is shown in Fig. 4. Spatial points as many as 200 from the outer edge to the inner edge along the longer axis of the elliptic plasma cross-section lying along the plasma major radius are imaged through a focussing mirror and multichannel optical fibers.

The spatial resolutions are estimated to be about 15 and 35 mm at the outer and inner edges, respectively.

A testing device has been constructed and applied to the CHS device. The predicted performance is obtained on this application.

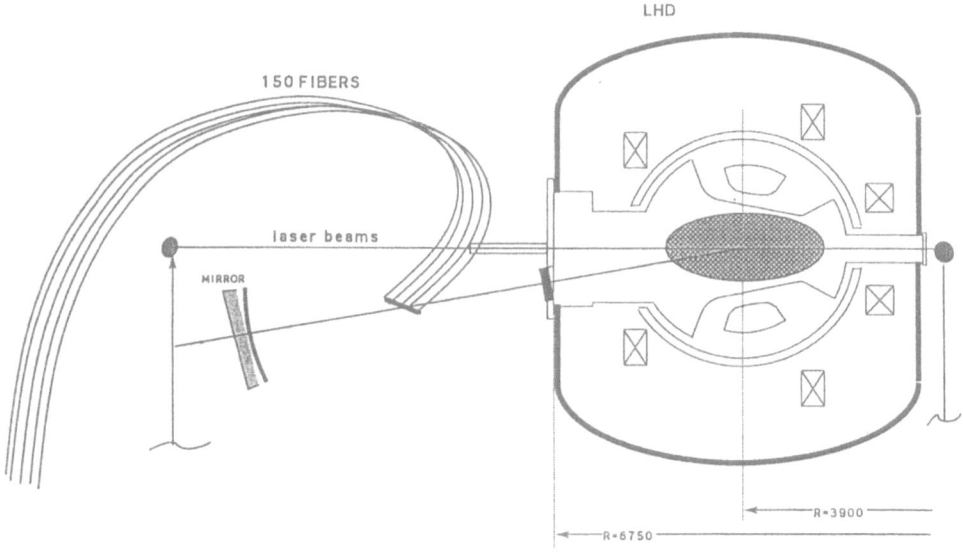

Figure 4. A conceptual drawing of YAG Thomson laser scattering system on LHD.

MULTICHANNEL FIR LASER INTERFEROMETER

Based on a plentiful experience on the interferometer for JIPP T-IIU and CHS, a 10-channel FIR laser interferometer of Michelson type has been designed and constructed for obtaining further information on electron density profiles, especially on time evolution of those along shorter radius of the elliptic plasma cross-section.[4] Because the light pass length from the light source to the optical bench of the interferometer is about 40 meters, and the laser power should be divided into 10 channels, a powerful laser is necessary for the light source. Both 195-μm DCN laser and 119-μm CH_3OH laser have been developed. A CO_2 laser-pumped twin CH_3OH laser is chosen. for the light source of beat-modulated laser interferometer. A laser power of about 0.46 W is obtained. The long term drift of the laser output is less than 2% for 1 hour, and the short term fluctuation is about 1%.

The laser light is transmitted through a wave-guide to the interferometer. The optical system is mounted on a massive frame of mechanical vibration-free structure as is seen in Fig. 3. Ten retro-reflectors are mounted on the upper shelf of the frame which is 17 m high, while the optical bench for the interferometer is installed on the lower shelf. Crystal quartz etalons are used for the beam splitters and windows. The transmissivities are precisely controlled by adjusting their thickness. The signals are detected with GaAs Schottky barrier diode mixers of corner cube mount at the room temperature.

HEAVY ION BEAM PROBE

The plasma potential profile, or radial electric field distribution, is an important quantity to be measured in a helical system, because it is strongly related to the plasma confinement characteristics. Although the measurement of poloidal rotation speed is utilized to find the radial electric field, a direct method to obtain the potential distribution is necessary.

A heavy ion beam probe (HIBP) has been designed for the potential profile measurement on LHD.[5-7] Because of a large size of the plasma and a high magnetic field intensity on LHD, a heavy ion beam such as gold should be accelerated as high as 6 MeV for the beam to penetrate the plasma, even at the magnetic field intensity of 3 T in the first phase. Due to non axi-symmetric configuration of helical magnetic field, the injected beam does not remain in a toroidal plane, but traverses both in toroidal and poloidal directions. The same is for the secondary beam. As a consequence, the secondary beam comes out from the observation port located toroidally next to the beam injection port. The complex geometry of HIBP on LHD is illustrated in Fig. 5.

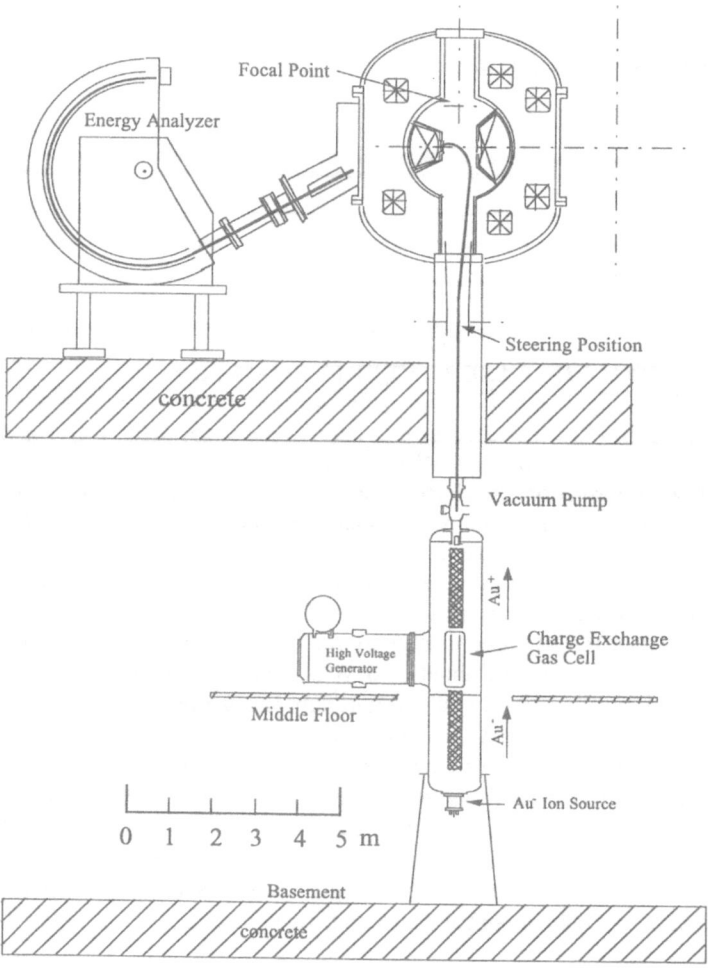

Figure 5. A schematic drawing of the 6 MeV heavy ion beam probe.

For accurate measurement of plasma potential and its fluctuation, the beam should have an energy spread as small as possible. From this viewpoint, a gold negative ion source of plasma-sputter-type has been developed. Negative ions thus produced are accelerated in a tandem manner, and converted into positive ions through a stripping gas cell which is at the high voltage end. A test stand has been constructed for studying negative ion production, extraction, energy dispersion and charge stripping efficiency. An energy analyzer of a high resolution and of high voltage characteristics is also being designed.

OTHER TOPICS

From the work of plasma rotations in tokamak and helical system plasmas carried out on JIPP T-IIU and CHS, it is found quite important to provide diagnostic means for the plasma rotations on LHD. Even in a helical system, it is also important to find out a shift of magnetic field distribution due to high beta effect or plasma current induced. A diagnostic beam will be installed on LHD for charge exchange spectroscopy for ion temperature and rotation velocity profiles, and for motional Stark effect for the measurement of magnetic field distributions.

CONCLUSION

The diagnostics for LHD are in a steady progress. There are many problems to be solved before applying these diagnostics to the actual device. The protection of diagnostic equipments as well as the windows against the heat from the plasma is a severe problem on the long pulse machine. Construction of reliable remote handling systems is another task.

We hope that our effort to solve these problems could contribute to the design and testing of the diagnostics for ITER.

REFERENCES

1. A. Iiyoshi, M. Fujiwara, O. Motojima, et al., Design study for the Large Helical Device, *Fusion Technology*, 17:169(1990).
2. M. Fujiwara and the Large Helical Device Project Team, Status of large helical device project, *Fusion Engineering and Design*, 23:547(1995).
3. K. Narihara, I. Yamada and T. Minami, Thomson scattering system for Large Helical Device, *Proc. 5th Int. Symp. on Laser-aided Plasma Diagnostics (Bad-Honnef)*, 108(1991).
4. K. Kawahata, Y. Hamada, J. Fujita and S. Okajima, FIR laser diagnostics for JIPP T-IIU and Large Helical Device (LHD), *Proc. 5th Int. Symp. on Laser-Aided Plasma Diagnostics (Bad-Honnef)*, 92(1991).
5. M. Sasao, Y. Okabe, A. Fujisawa, H. Yamaoka, M. Wada and J. Fujita, Development of negative heavy ion sources for plasma potential measurement, *Rev. Sci. Instrum.*, 63:2726(1992).
6. A. Fujisawa, H. Iguchi, Y. Hamada, M. Sasao and J. Fujita, Active control of beam trajectories for heavy ion beam probe on helical magnetic configurations, *Rev. Sci. Instrum.*, 63:3694(1992).
7. T. Taniike, M. Sasao, A. Fujisawa, H. Iguchi, Y. Hamada, J. Fujita, M. Wada and Y. Mori, IEEE Trans. Plasma Science, 22:430(1994).

THE T-15 FUSION PRODUCTS STUDY: PLANS AND DIAGNOSTICS

V.S. Zaveriaev[1], V.D. Maisyukov[2], A.V. Khramenkov[1],
V.G. Merezhkin[1], S.V. Popovichev[1], S.V. Trusillo[3]

[1]Russian Research Center "Kurchatov Institute"- Nuclear Fusion Institute,
    Moscow, Russia
[2]Scientific Engineering Center "SNIIP", Moscow, Russia
[3]Russian Federal Nuclear Center "VNIIEF", Arzamas-16

INTRODUCTION

The Tokamak T-15 was designed and constructed in the end of 80's in the Kurchatov Institute. Tokamak T-15 is one of the comparatively large machines in the world and it is suitable for a number of important studies concerning alpha particle problem. It is a classical tokamak with a circular cross section and without a diverter. The main machine parameters and the expected plasma parameters are following: R=243 cm, a=70 cm, - B_{tor}=1.5-3 T, I_p=0.5-1.5 MA, $T_i(0)$=1-2 keV (Ohmic regime), shot duration is 5 s and more. The T-15 has 24 superconducting toroidal magnetic coils, the ripple at the plasma edge is about 0.7%. The main feature of machine is a planned additional Electron Cyclotron Resonance Heating (ECRH), 6-8 MW of absorbed power. The Neutral Beam Injection (NBI) Heating will be also available with a beam energy 40 keV (in future up to 80 keV) and the total beam power about 6-8 MW.

There were several experimental runs of T-15 during last years. In the campaign of 1995 the toroidal magnetic field was 3 T, plasma current about 0.5 MA. The nonohmic additional heating systems were used: 3 gyrotrons provided total ECRH power 0.6-0.7 MW during 0.1 s and the NBI Heating (35 keV, 0.5-0.6 MW). It was shown that the machine may operate successfully, its main components are in order. It is supposed to carry out full-scale plasma experiments in 1996 using ECRH and NBI heating with higher power.

Diagnostics for Experimental Thermonuclear Fusion Reactors
Edited by P. E. Stott *et al.*, Plenum Press, New York, 1996

This paper describes a planned direct fusion study in the T-15. So firstly it is important to define the place that T-15 occupies in the world efforts in the field of controlled fusion.

Because of the additional precautions required for handling tritium, the T-15 will not operate with tritium as a main plasma component. Therefore, a direct confinement study of fusion products from D-T reactions is impossible. Moreover, there is no effective biological shielding at T-15, so the total neutron yield, as a rule, should not exceed 10^{13} neutrons/shot for radiation safety reason. The number of the most powerful shots with the neutron yield about 10^{14} 1/shot is strongly limited during one experimental run. It means that the NBI Heating may be used a hydrogen injection into the deuterium plasma only and there will be no the beam-plasma and beam-beam DD reactions. In spite of an expected good confinement of charged fusion products (CFP) their density in plasma will be far too small to investigate collective instabilities induced by fusion products. The T-15 has no diverter and there is no plan for ICR Heating yet.

In spite of the limited possibilities of the machine a set of important experiments concerning a fusion products behavior may be realized. The authors intend to investigate the confinement, loss mechanisms, thermalisation and diffusion processes of fusion products in the T-15 under various conditions, particularly in the presence of the powerful ECRH.

There is no direct interaction between EC waves and fusion products, but ECRH may influence indirectly on fusion product behavior in two ways.

The first effect of ECRH is the opportunity to vary the plasma current profile. This parameter is of a great importance not only for plasma behavior as a whole but for a CFP also. It is well-known that the prompt losses of fusion particles decrease at high current[1]. The same effect should be expected for more peaked plasma current distribution. Actually, the higher central current density ensures more favorable conditions for the confinement of those fusion products that were born in plasma core. It leads to decrease of the total losses due to the fusion source is very peaked in modern machines. The Monte-Carlo numerical calculations show that this reduction of losses may reach 20%.

The idea of plasma current profile variation in the T-15 is based on the simultaneous application of the current rise and ECRH. The different scenarios of plasma performance result in either flat or sharp current profiles.

The estimated current diffusion time in the T-15 is more than 1 s. Hence the plasma current density would stay almost unchanged and rather flat during the initial phase of the shot when total current increases due to outer voltage. If there is no additional electron heating in the current flattop stage then the current will gradually diffuse into the plasma core, thus a more sharp current profile is formed. The ECRH being applied at any moment during this stage do increase the current diffusion time significantly due to electron heating. So the ECRH may "freeze" a radial distribution of the plasma current. The Fig.1 shows the calculated radial distributions of plasma parameters at the different stage of the shot. The following scenario was considered: the duration of current rise stage is 1.28 s, the ECRH power starts at 2.56 s and finishes at 3.2 s. It should be noted that the evolution of current profile was completed before start of the ECRH. The numerical simulation takes into account the dependence of the electron transport coefficients on electron temperature according to $D_e \sim \chi_e \sim (T_e)^{1/2}$.

Current ramps up or down should result in the change of the peripheral current density rather fast. During the current ramp down experiments the anomalous FP losses in Tore-Supra[2] and T-10[3] tokamaks have been observed.

Hence, the T-15 provides a principal opportunity to study the fusion products confinement and loss under various conditions. We suppose to carry out detailed research of the following phenomena: first orbit losses, ripple-induced and delayed losses, the influence of MHD activities on FP confinement. Again the ECRH together with the current density variations is a useful "tool" to change the position and amplitude of the large scale MHD instabilities including sawteeth oscillations[4]. All these phenomena are of importance for a reactor design.

The confined DD fusion particles - tritons and ^{3}He nuclei - have a finite probability to produce the secondary $T(d,a_1)n$ and $^3He(d,a_2)p$ reactions. The source strength of the secondary reactions depends on the diffusion and slowing down time of the tritons and ^{3}He ions. The latter increases with the electron temperature as $T_e^{3/2}$. Fig.1 indicates that a strong ECRH in the T-15 may increase the slowing down time about 30 times that will cause the essential change of the DT and D^{3}He reactions rates. Taking into account the classical thermalisation, a few percent of tritons burn up in DT reaction. The probability of D^{3}He reaction is approximately two orders less. The Monte-Carlo numerical modelling shows that about 90% of the secondary 14.7 Mev protons and about half of the secondary alphas will escape the T-15 plasma. The detection of these charged particles and DT neutrons gives the information about the diffusion and thermalisation processes of the confined DD fusion products (so-called burnup method[5,6]). The Fig.2 presents the examples of the FP orbits in the T-15.

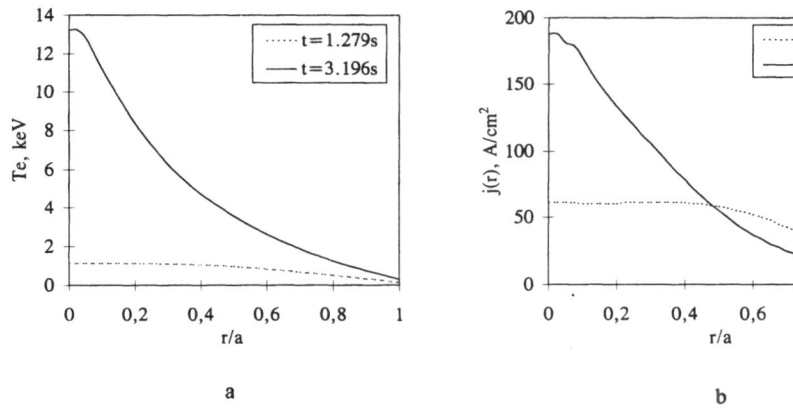

Figure 1. Calculated electron temperature (a) and plasma current (b) profiles for the T-15 regime (3 T, 0.7 MA, 1.5*10^{13} cm^{-3}, 1.5 MW central ECRH). Dashed curves - at the beginning of current flattop, solid curves- at the end of ENRH.

It should be noted that the confined fusion products may produce another kind of the secondary reactions that emit high energy gamma rays - $D(p,g)^3He$, $T(d,g)^5He$ and $^3He(d,g)^5Li$ - and the reactions between primary fusion products and the light impurity nuclei (B, N, C, Be)[7] that has been observed experimentally[8]. The cross section of these reactions is rather small and the usage of the ECRH in the T-15 will increase their yield significantly due to its influence on the slowing down time. It is planned to detect the fusion gamma-rays in the T-15 by means of gamma-spectroscopy methods[9].

Actually fusion program in the T-15 may start not earlier than in a two or three years. For that moment an experimental study and a detailed numerical simulation of the T-15 plasma should be carried out. Perhaps a new results from large machines will force to correct the program presented. An adequate fusion diagnostic complex has to be designed and constructed for the realization of this program. This complex is based on detection of DD neutrons and charged fusion products.

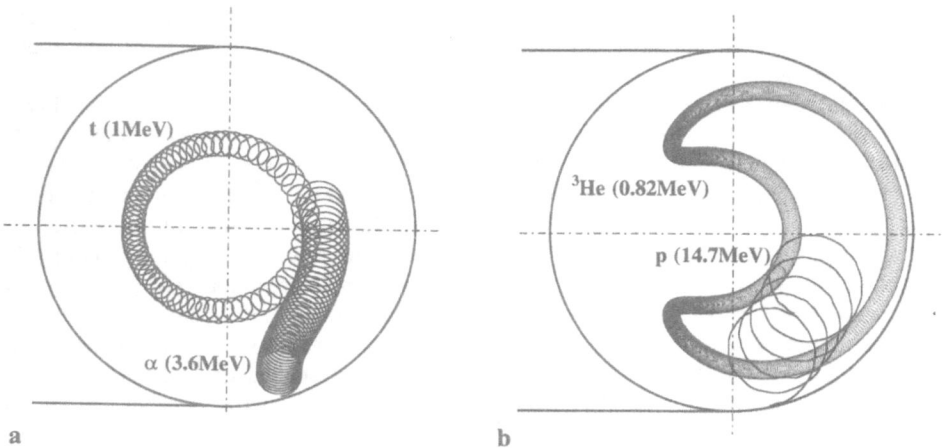

Figure 2. Poloidal projections of typical CFP orbits for the T-15 regime (3 T, 0.7 MA): a - confined 1 MeV triton and escaping DT alpha particle; b - confined 0.8 MeV ^{3}He ion and escaping D^{3}He proton.

THE NEUTRON DIAGNOSTICS

The detection of the thermonuclear neutrons is a direct and standard method to obtain the total fusion source strength. The neutron emission in the T-15·will be measured using 12 absolutely calibrated neutron monitors based on proportional helium counters working in "corona" regimes with different sensitivities to neutrons and with low sensitivity to gamma-rays.

One of the most important task for a neutron diagnostic is the determination of the fusion birth profile. Because of low neutron yield in the T-15 it is impossible to use neutron collimators with cylindrical holes. An alternative approach is therefore sought. We propose two independent original methods based on wide-aperture neutron brightness measurements to determine the fusion birth profile and its temporal evolution in T-15.

A rapid-cycle bubble chamber will be applied as the first method. It will view the plasma volume through a slit-shaped collimator-obscure oriented parallel to the magnetic axis of the torus. Every 20 ms a visible trace of bubbles will be formed in the camera volume along neutron "rays" emitted from plasma. So the neutron source profile forms a contrast TV image through transmitted light and will be recorded by means of video camera. Such a bubble chamber was used for neutronography of small pulsed plasma objects as a "plasma focus"[10], although not in a time-resolved fashion. The advantages of this Neutron Image System (NIS) are described in details by Trusillo[11]. The preliminary estimates show that the effective usage of this method in the T-15 is possible in the most powerful regimes only when the fusion yield exceeds 10^{13} 1/s. So just a demonstration of the NIS operation in the T-15 is planned.

The second method of the neutron profile determination is completely new and may be developed for the machines with a moderate fusion yield. It is based on the application an array of gas discharge neutron counters located one after another along the minor tokamak radius between two parallel neutron moderator plates placed as close as possible to plasma (Fig.3). The basic idea of this method is the using of the partially collimating of the neutron flux that is provided by this "sandwich"-like geometric configuration.

The possibility of an application of such a configuration was analyzed numerically. It was assumed that the neutron source is poloidal symmetric and the plasma volume was divided into ten cylindrical zones. The number of the detectors is chosen to be 12 and the sensitivities of the detectors to an every cylindrical layer have been calculated. Thus the response matrix $M_{k,n}$ was defined (k=1-10, n=1-12). To obtain the count rates for every detector J_n a number of different source profiles S(r) were used as a parabolic to some power $S(r)=S_0*(1-x^2)^b$ and as exponential $S(r)=S_0*[-(x/x_0)^b]$ where x=r/a. The central source density S_0 was assumed to be the same for all profiles, so the total neutron yield varies from $3.6*10^{11}$ to $2.2*10^{12}$ 1/s. The "quasi-experimental" count rates for every detectror were calculated according to $J_n=M_{k,n}*S_k$. Then the statistical dispersion was entered to J_n and the inverse problem was solved to seach the source profile parameters S_0, b and x_0. This procedure has been repeated five times in order to estimate the accuracy of the method. The results of the these simulations are shown in Fig.4. One can see that theoretically the method works well and the accuracy better than 10% and a time resolution about 10 ms even in weak regimes may be achieved. It should be expected that the more detectors may essentially increase the radial resolution.

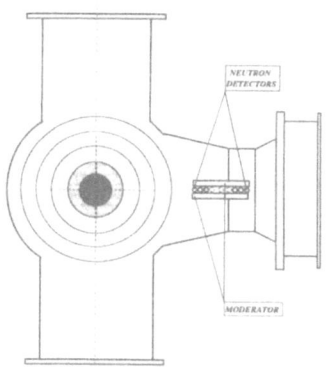

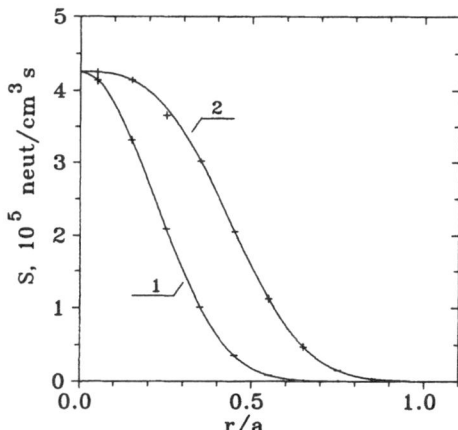

Figure 3. "Sandwich"- radiometer schematic view.

Figure 4. Numerical modelling of the fusion source profile reconstruction. Curves are the original source profiles (x=r/a): 1) $S(r)=S_0(1-x^2)^{11}$. total yield $8.3*10^{11}$ 1/s. 2) $S(r)=S_0*exp[-(x/0.5)^3]$. total yield $2.2*10^{12}$ 1/s. Circles and squares are the calculated values. Interval of "measurements" - 10 ms.

In order to suppress the magnetic field effect the counters would be filled with He-Ar mixture at 4 atm pressure and would operate in the ionization chamber regime with the measurement of the charge intergrated during 10 ms. In any case the counter array would be placed in a special electro-magnetic screen. The calibration of the detection system using movable compact DD neutron generator *in situ* will help to estimate the scattered neutrons effects and to correct the calculated response matrix.

THE ESCAPING FUSION PRODUCTS DIAGNOSTIC

The modest fusion yield in the T-15, and accordingly, the favorable radiation background will allow use of a set of silicon diode detectors with excellent parameters for precise spectrometry of the escaping CFPs. The conceptual design of diagnostic is based on the experience of Si detectors application in the T-10[12,13] and is similar to the FP diagnostic that successfully realized in Tore-Supra[14,15]

The distributions and the parameters of unconfined charged DD fusion products - protons (3.03 MeV), tritons (1.01 MeV) and ^{3}He-nuclei (0.82 MeV) may be obtained using three movable detector assemblies, placed in a diagnostic port near the plasma boundary (Fig.5). Each assembly consists of a detector unit and a preamplifier. The total number of preamplifiers within a single assembly is 16. Thus the maximum number of spectrometric channel circuits is 48.

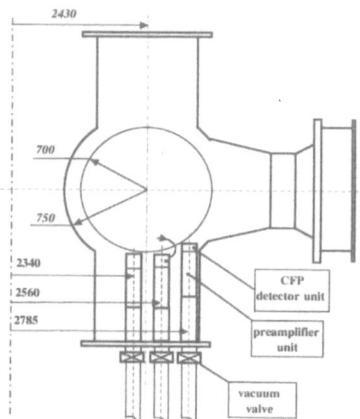

Figure 5. Layout of the T-15 CFP diagnostics.

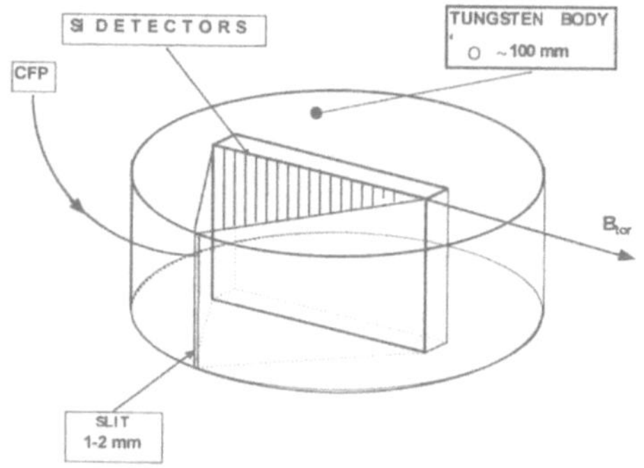

Figure 6. Preliminary version of the detector unit for CFP diagnostics in T-15.

The detector unit (Fig.6) contains multistrip silicon detectors produced by "state-of-the art" planar ion implantation technology. Separate sensitive elements of these detectors are discrete rectangular strips formed with high accuracy on a single silicon plate. The number of strips varies from 16 to 64, and they may arranged according to the best scheme for the particular measurements. All detector units will include weak spectrometric alpha-sources to control normal operation and to carry out energy calibration of the channel circuits *in situ*. The detectors will be water-cooled to prevent high temperature.

The detector unit is constructed as an obscure with a vertical entrance slit. The fusion products will strike the silicon plate at different pitch angles. The plasma visible light and low energy corpuscular fluxes are blocked by a 0.1-0.5 micron thin opaque homogenous carbon and aluminum film evaporated on to the detector surface or placed in front of the entrance slit. One of the strips will be coated with a thick Al foil and will work as a background monitor. The body of detector unit will be made of an alloy containing 80% tungsten that is a good shield against gamma-rays.

The geometric configuration and the multistrip detectors enable, in principle, pitch-angle resolution to about 1-2° or better according to statistics. The own energy resolution of the detectors is expected to be 30-50 keV.

This diagnostic will be used for measurements of both DD fusion products and secondary DT and D^{3}He alphas. The advantage of a thin Si detectors consists in low parasitic gamma ray signal that is originated in the whole detector volume. They could not be applied for a spectrometry of 14.7 MeV D^{3}He protons because they loose only a few hundreds keV in the detector depletion depth. Especially for D^{3}He protons measurements a thick Si diodes should be used. In this case it is possible to perform a simultaneous measurements both D³He protons and DT neutrons that firstly was successfully demonstrated in the DIII-D experiments[6]. The development of this kind of FP diagnostics for the T-15 may be an important addition to the complex described above.

DISCUSSION AND SUMMARY

The diagnostic complex that based on neutrons and fusion products detection may be applied both for plasma parameter measurements and for the direct study of fusion products behavior.

The extensive opportunities of the neutron and CFP diagnostics have been demonstrated in many tokamaks. For instance by means of the JET neutron profile monitor[16.17] the fusion birth profile has been measured in details and the ion temperature radial distribution was obtained in Ohmic regimes. Moreover the detailed analysis of the neutron brightness evolution allowed to determine the ion thermal diffusivity[18]. The influence of sawtooth crashes on neutron source has been observed[19]. Unfortunately this profile monitor may be used in the large machines only. The most advanced system for escaping CFP measurements is that on the TFTR[20] and it is now successfully operating in DT-experiments there. But it suffers from low energy and pitch angle resolutions that does not allow to perform detailed investigations of some phenomena like delayed losses. The FP diagnostics in the DIII-D[6] and Tore-Supra[14] are based on using of the spectrometric semiconductor detectors and operate well. Unfortunately the fusion source profile is unknown, through if the plasma current profile is measured independently then a pitch angle resolved distribution of unconfined CFP gives the source profile. At present such kind of experiments have not been performed yet. In principle the measurement of the poloidal distribution of escaping fusion products in itself allows to restore both the source and plasma current profiles[21], but practically too many detectors are required.

The methods of measurements that are suggested to develop for T-15 jointly form a modern complex for plasma diagnostics and for fusion study. It should be noticed that the simultaneous definitions of the source radial distribution and of the unconfined fusion products parameters (fluxes versus energy, pitch angle and poloidal coordinate) with a time resolution about a few ms are unique.

The most important task in the field of plasma diagnostics is the current profile determination $j(r)$. The independent measurements of the fusion source profile by means of neutrons gives the possibility to solve this problem. Actually the FP probe signal is the fusion source just integrated along particle orbit. The latter is completely defined by the magnetic configuration. So the task of obtaining $j(r)$ from CFP flux versus pitch angle is similar to the inverse Abel transformation. In our case the integration of the source must be performed along complex helical trajectories. The calculated CFP fluxes versus pitch angle for different T-15 current profiles (Fig.7) indicate the pitch angle resolution required. The usage of a few detection systems that are located at different poloidal positions will do this task single valued. Thus the particle orbits become defined and the ion temperature in the central plasma region may deduced from the FP energy spectra due to Doppler broadening. The latter, together with the measured source profile, will allow to estimate the deutron density. This parameter is of importance for T-15 because of hydrogen NBI.

This technique will be developed to observe the current profile evolution during ECRH, ECCD and current ramps. The direct electron heating by means of ECRH increases significantly the MeV ion slowing down time that gives the opportunity to control the burn up of the DD fusion products[22] and to search the optimal conditions for

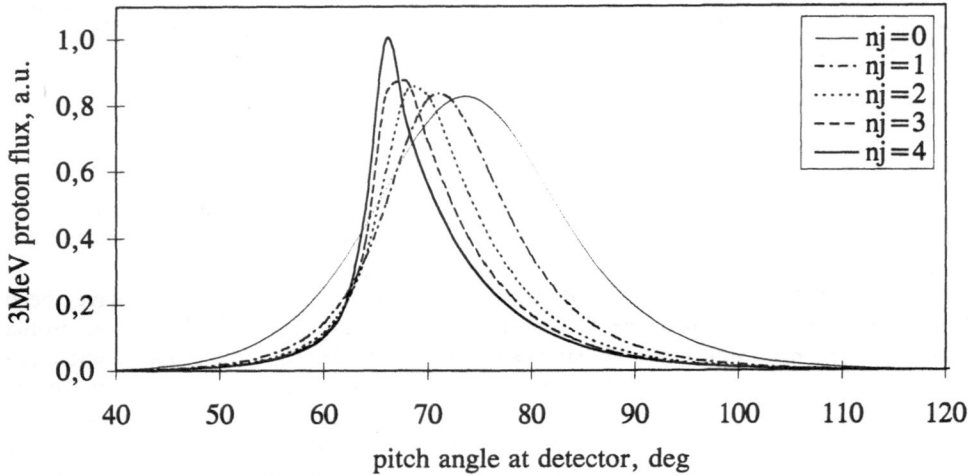

Figure 7. Calculated fluxes of the detected 3 MeV protons versus pith angle for different current profiles $j(r) \sim (1-(r/a)^2)^{nj}$. The T-15 regime: 3 T, 0.7 MA, $T_i(r)$ and $n_d(r)$ are parabolic.

this process. The additional information about confined DD fusion products will be available from the measurements of high energy gamma-rays.

We suppose that the diagnostic complex described is adequate to the T-15 fusion research program and the realization of this program may bring new important results.

ACKNOWLEDGMENTS

This work was partially supported by the Russian Foundation of Fundamental Researches: Project code # 93-02-16910.

REFERENCES

1. S.J. Zweben, D.S.Darrow, E.D.Fredrickson, H.B.Mynick, Anomalous delayed loss of trapped DD Fusion products in TFTR, *Nuclear Fusion*. 33:705 (1993)
2. C.Doloc, private communication
3. V.S.Zaveriaev, V.D.Maisyukov, S.V.Popovichev, et al., Charged fusion products study in the T-10 Tokamak, *Kurchatov Institute Preprint*, be published
4. V.V.Alikaev, A.A.Bagdasarov, A.A.Borshegovskij, et al., Second harmonic electron cyclotron current drive experiments on T-10, *Nucl. Fusion*. 35:369 (1995).
5. O.N.Jarvis, J.M.Adams, S.W.Conroy, et al., Triton burnup in JET - profile effects, *Proc. 18th EPS Conf., Berlin*. 15C: I-21 (1991).
6. H.H.Duong, W.W.Heidbrink, Confinement of fusion produced MeV ions in the DIII-D Tokamak, *Nucl. Fusion*. 33:211 (1993).
7. V.G.Kiptilyj, Nuclear reactions as fusion plasmas diagnostics, *Ioffe Institute Preprint*. No.1176 (1987) (in Russian).
8. G.Sadler, J.P.Christiansen, G.A.Cottrell, et al., ^{3}He-D fusion yield studies in JET, *Proc. 18th EPS Conf., Berlin*. 15C: I-29 (1991).
9. V.G.Kiptilyj, et al, ITER gamma diagnostics: 2D neutron and gamma camera, *this edition*.
10. S.V.Trusillo, B.Ya.Guzhovskii, N.G.Makeev, et al., Determination of neutron production region in the plasma focus chambers, *JETF Letters*. 33:148 (1981) (in Russian).

11. S.V.Trusillo, V.A.Agureev, Neutron image system to measure neutron profile of ITER plasma, *this edition.*

12. V.S. Zaveriaev, V.D.Maisyukov, S.V.Popovichev, S.V.Putvinskii, et al., Plasma diagnostics for the T-10 tokamak based on charged (d,d)-fusion products: preliminary results and problems, *Sov.J.Plasma Phys.* 16:754 (1990).

13. V.S. Zaveriaev, V.D.Maisyukov, S.V.Popovichev, A.P.Shevchenko, Charged fusion product behaviour in T-10 tokamak plasma, *Proc. 20th EPS Conf., Lisboa.* 17C:75 (1993).

14. G.Martin, O.P.Gilles, P.Joyer, Fusion profile measurement on Tore-Supra, *Proc. 17th EPS Conf., Amsterdam.* 14B:1584 (1990).

15. C.Doloc, G.Martin, Interaction between L.H. waves and thermonuclear protons, *Proc. 20th EPS Conf., Lisboa.* 17C:893 (1993).

16. F.B.Marcus, J.M.Adams, D.V.Bartlett, et al., Ti (r) profiles from the JET neutron profile monitor for ohmic discharges, *Proc. 18th EPS Conf., Berlin.* 15C:277 (1991).

17. The JET Team, Fusion energy production from a deuterium-tritium plasma in the JET tokamak, *Nucl. Fusion.* 32:187 (1992).

18. M.Sasao, J.M.Adams, S.Conroy, et al., Determination of the ion thermal diffusivity from neutron emission profiles in decay, *JET Preprint JET-P(92)75 (1992).*

19. O.N.Jarvis, Some aspects of fast ion behaviour in JET plasmas, *Workshop on DT Experiments, PPPL.* Vol. 1 (1994).

20. S.J.Zweben, Pitch angle resolved measurements of escaping charged fusion products in TFTR, *Nucl. Fusion.* 29:825 (1989).

21. N.E.Karulin, S.V.Putvinskij, Some theoretical aspects of charged fusion product plasma diagnostics, *Nucl. Fusion.* 25:961 (1985).

22. D.Anderson, P.Batistoni, M.Lisak, Influence of radial diffusion on triton burnup, *Nucl. Fusion.* 31:2147 (1991).

INDEX